W0050859

Urolithiasis 2

Urolithiasis 2

Edited by

Rosemary Ryall
Flinders Medical Centre
South Australia, Australia

Renze Bais
Institute of Medical and Veterinary Science
Adelaide, Australia

Villis R. Marshall
Flinders Medical Centre
South Australia, Australia

Allan M. Rofe
Institute of Medical and Veterinary Science
Adelaide, Australia

Lynwood H. Smith
Mayo Clinic
Rochester, Minnesota

and

Valerie R. Walker
University of British Columbia
Vancouver, British Columbia, Canada

SPRINGER SCIENCE+BUSINESS MEDIA, LLC

Library of Congress Cataloging-in-Publication Data

Urolithiasis 2 / edited by Rosemary Ryall ... [et al.].
p. cm.
"Proceedings of the Seventh International Symposium on Urolithiasis, held August 24-27, 1992 in Cairns, Australia"--T.p. verso.
Includes bibliographical references and index.
ISBN 978-1-4613-6091-9 ISBN 978-1-4615-2556-1 (eBook)
DOI 10.1007/978-1-4615-2556-1
1. Urinary organs--Calculi--Congresses. I. Ryall, Rosemary L. II. International Symposium on Urolithiasis (7th : 1992 : Cairns, Qld.) III. Title: Urolithiasis two.
[DNLM: 1. Urinary Calculi--congresses. WJ 100 U7695 1994]
RC916.U755 1994
616.6'22--dc20
DNLM/DLC
for Library of Congress 94-20577
CIP

CAIRNS · AUSTRALIA · AUGUST 24th - 27th · 1992

Proceedings of the Seventh International Symposium on Urolithiasis,
held August 24–27, 1992, in Cairns, Australia

ORGANIZING COMMITTEE

Chair	Rosemary L. Ryall
Co-Chair	Villis R. Marshall
Secretary	Pauline Archer
Treasurer	Peter McCabe
Audio Visual Coordinator	Ruth M. Kerr
Trade Coordinators and Consultants	Renze Bais
	Allan M. Rofe
Consultants	Glenn M. Preminger
	John D. Sallis

INTERNATIONAL CONSULTING SCIENTIFIC PROGRAMME COMMITTEE

B Danielson, Sweden
J Dirks, Canada
YM Fazil Marickar, India
J Lemann, USA
F Linari, Italy
M Marberger, Austria
V Marshall, Australia
G Nancollas, USA
CYC Pak, USA
W Robertson, Saudi Arabia
R Ryall, Australia
J Sallis, Australia
P Schwille, Germany
L Smith, USA
R Sutton, Canada

FOREWORD

The first International Symposium on Urolithiasis Research was held in Leeds, England, in 1968. The meeting was the first in what was to become a series of symposia intended to gather together a diverse group of biochemists and physicians, urologists and engineers, with a common interest in factors affecting the formation of human urinary stones. Since its inception the series has threaded a peripatetic course back and forth across the Atlantic Ocean, from Madrid in Spain, to Davos in Switzerland, to Williamsburg in the USA, to Garmisch-Partenkirchen in Germany and Vancouver in Canada, under the guardianship of Drs Nordin, Cifuentes Delatte, Fleisch, Smith, Schwille, Dirks and Sutton, and their colleagues.

In 1992, for the first time, the meeting moved to the southern hemisphere, to Cairns in Northeastern Australia. Unlike most previous symposia, there were no invited papers. Instead, the submitted abstracts were allowed to dictate the content of the meeting so that the conference programme would reflect the flavour of current research in the field. To achieve this, all abstracts were graded anonymously by three referees to determine their categorization as oral, theme poster, or general poster presentations. The 300 or so accepted absracts were then allocated to seven plenary sessions, nine theme poster discussion groups and three large general poster sessions.

In constructing the programme, we adopted some features of past symposia. The theme poster sessions, which had been such a successful venture at the Vancouver meeting, once again proved to be valuable forum for presenting work by combining posters with short, sharp, oral presentations. And Williamsburg's demonstration that a refreshing drink encourages uninhibited, stimulating discussion around the poster boards, was well remembered in Cairns, where the general poster sessions were considerably enhanced by the provision of some excellent Australian wines.

This volume embodies the scientific substance of the Cairns symposium. Because we felt it desirable that these proceedings represent an accurate record of the content of the meeting, the papers have been largely left in the broad categories in which they were grouped in the conference programme, and as far as possible, in the order in which they were presented. Thus, there are only nine main groups comprising:

Physiology and Metabolism
Oxalate metabolism and Transport
Physicochemistry, Promoters and Inhibitors
Crystallization and Proteins
Stone Composition, Matrix and Membranes
Risk Factors and Diet
Epidemiology and Infection
ESWL and Lithotripsy
Investigation, Medical and Surgical Management

Although this list includes some stalwarts in the field - there because they are indispensable, basic, and valuable areas of study, others, such as Oxalate Transport, are relative newcomers. Yet others, Matrix and Urinary Proteins among them, have lain somnolent for some years and have awakened in response to renewed interest or the availability of new technology. This book therefore truly reflects the areas of interest and current status of urolithiasis research, investigation and management, and it is hoped that, like preceding volumes in the series, it will become a valuable resource tool for present and future urolithiasists.

Finally, the members of the Organizing Committee would like to express their sincere gratitude to the various academic and commercial organizations whose participation and generous financial support enabled the meeting to take place, as well as many individuals, too numerous to mention, whose willing assistance behind the scenes eased the burden of organizing the meeting and producing this volume.

Rosemary L. Ryall
Chair

ACKNOWLEDGEMENTS

Australian International Development Assistance Bureau (AIDAB)
Australian Medical Imaging
Beach Pharmaceuticals
Dornier Medizin Technik GmbH
Electro Medical Systems
Flinders Medical Centre
Institute of Medical and Veterinary Science, South Australia
Madaus AG
Merck Sharp & Dohme (Australia) Pty Limited
PSI-Neomedix
Southern Vales Wine Cooperative
The Australian Kidney Foundation
The Flinders University of South Australia
The Institution of Engineers, Australia (International Conference Support Scheme)
The Urological Society of Australasia
Willen Drug Company
Zimmer Australia Pty Ltd

CONTENTS

SECTION I: PHYSIOLOGY AND METABOLISM

SECTION II: OXALATE METABOLISM AND TRANSPORT

SECTION III: PHYSICOCHEMISTRY, PROMOTERS AND INHIBITORS

SECTION IV: CRYSTALLIZATION AND PROTEINS

SECTION VI: RISK FACTORS AND DIET

SECTION VII: EPIDEMIOLOGY AND INFECTION

SECTION VIII: ESWL AND LITHOTRIPSY

SECTION IX: INVESTIGATION, MEDICAL AND SURGICAL MANAGEMENT

SECTION I: PHYSIOLOGY AND METABOLISM

GENETIC INFLUENCES ON URINARY CALCIUM EXCRETION

R.P. Holmes, H.O. Goodman and D.G. Assimos

Department of Urology
Bowman Gray School of Medicine
Wake Forest University
Winston-Salem, NC USA

INTRODUCTION

Urinary Ca is influenced by several factors including dietary intake of Ca, Na, and protein, and vitamin D, parathyroid hormone and thyroid hormone[1,2]. Genetic factors are also involved, as two studies have attributed hypercalciuria to the presence of an autosomal dominant gene[3,4].

Our interest in Ca excretion stems from the well-known over-presentation of hypercalciurics among CaOx stone formers relative to the general population. If the genetic component in CaOx lithiasis is a polygenic system as suggested by Resnick et al[5], one would anticipate identifying two or more loci contributing to the diathesis. Further, the alleles at each locus would be expected to exhibit small, additive effects. The term, "additive", implies lack of dominance. The high frequency of CaOx lithiasis in the general population relative to single-locus Mendelian disorders implies that the number of loci involved are few (3-5 per patient) and that alleles for susceptibility at these loci must be very common ($q>0.2$). If these conditions are not met, there would be fewer patients with nephrolithiasis since they represent the chance coincidence of several susceptible alleles and environmental factors, all of which must be much more frequent than stone formers (who represent their product).

With the aforementioned concepts in mind, we examined the hypothesis that hypercalciuria might represent one of the loci involved in renal stone diathesis. If it were true, we would expect to find three classes of excretors; low, intermediate and high, if a single locus with only two alleles were involved. To minimize the sometimes-wide fluctuations in Ca excretion among both stone patients and normal subjects, we collected three 24 h urine specimens from participants. We elected to study normal persons rather than stone formers because the normals would include larger numbers of low and intermediate excretors compared to stone formers if our hypothesis were correct. We have also studied a small number of families to date to test segregation ratios based on the two-allele, single-locus model for Ca excretion.

MATERIALS AND METHODS

Subjects

Subjects consisted of 101 unrelated Caucasian individuals (54 males and 47 females). The mean age for the total group was 30.5±9.2 yrs (±SD) with a range of 19-64

Urolithiasis 2, Edited by R. Ryall *et al.*,
Plenum Press, New York, 1994

years. The mean age for the males was 29.2 years and for the females, 32.1 years. Exclusion criteria included a history of nephrolithiasis or the presence of renal, hepatic, or intestinal disease. Individuals maintained dietary records during 24 h urine collections which commenced before breakfast. Urine samples were collected in containers with 10 g boric acid and were kept refrigerated. Aliquots of these collections were stored at -70°C until analysis. The study was limited to a three-month period to avoid any seasonal influence on urinary excretion[5].

Assays

Calcium was assayed by atomic absorption spectrophotometry. Creatinine was measured by a kinetic picric-acid procedure using a centrifugal analyzer and a kit supplied by the manufacturer (IL).

RESULTS

The distribution of the mean Ca excretion is shown in Figure 1. A group with high Ca excretion was evident, and those with an excretion above 0.18 g/g creatinine were considered hypercalciuric. A baseline resolution in the distribution was evident at this point. Eight individuals (8%) had hypercalciuria which is consistent with reports of other normal populations[7]. Six of these hypercalciuric individuals underwent Ca-load tests[8] which indicated renal hypercalciuria in two, and type II absorptive hypercalciuria (Pak classification) in four. The distribution of Ca excretion in the remaining individuals was broad and did not appear to fit a normal distribution. At the mean excretion of this group (0.1 g/g creatinine) where the highest frequency should occur, a dip was seen in the distribution. This suggested that the non-hypercalciuric group could more appropriately be divided into two groups, one with low Ca excretion (<0.1 g/g creatinine) and the other with an intermediate Ca excretion (0.1<> 0.18 g/g creatinine). The differences between low and intermediate excretors did not arise from dietary differences since an analysis of the dietary records and urinary excretions showed no significant difference in the intake of protein (dietary analysis and urinary urea-N), Na (urinary excretion) or Ca (dietary analysis) (Table 1). Clearly, our collection of three 24 h urine samples did not produce a complete separation of sub-groups. That is, the antimode at 0.1 did not reach the abscissa.

If the basis for these perceived groups is genetic, they must satisfy several criteria. The first is that inter-individual must significantly exceed intra-individual differences.

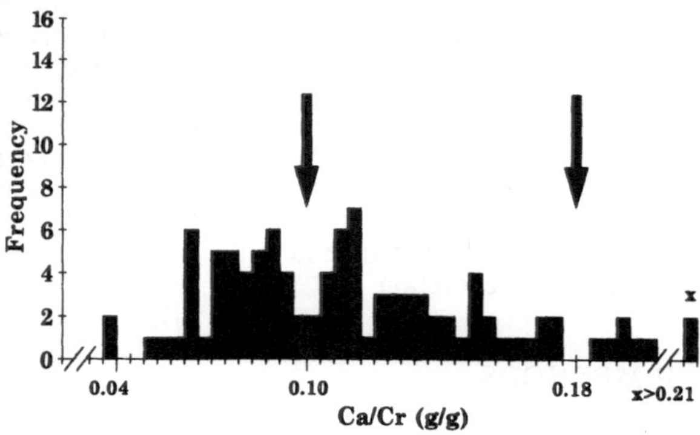

Figure 1. Distribution of Ca excretion in a group of normal individuals. The arrows delineate the proposed antimodes for separating phenotypic classes.

Analysis of variance indicated this to be the case ($p < 0.001$). Secondly, the numbers of individuals in the three sub-groups (44, 49 and 8 in the low, intermediate, and high groups, respectively) should satisfy Hardy-Weinberg equilibrium frequencies[9], and they do ($\chi 2 = 1.21$, $p > 0.2$). The estimate of the frequency of the allele for higher Ca excretion (the susceptibility allele) is 0.32, attesting to the high frequency in the general population of alleles for common disorders as mentioned above.

Table 1. Calcium and protein intake and urinary excretion of Na and urea-N among low- and intermediate-Ca excretors.

Excretor Class	N	Calcium (mg/Kg Bw)	Protein (g/Kg Bw)	Sodium (mmol/g Cr)	Urea-N (g/g Cr)
Low	44	11.8±7.2	1.21±.48	98.3±28.1	6.46±1.96
Intermediate	49	13.6±7.8	1.26±.40	104.5±28.3	6.21±1.43
t		1.10	0.58	1.07	0.69
p		0.27	0.57	0.29	0.49

Values shown are Means±SD. Cr = creatinine; t = test; p = probability

A third criterion is that segregation ratios must be compatible with Mendelian expectancies. Thus far, the spouses and offspring of seven probands from the large study and one family from our clinic files have collected three 24 h urine samples. These relatives were classified as low, intermediate, and high excretors using the same cut-offs described above. Even though data are too few for a rigorous test of segregation ratios, it can be seen in Fig. 1 that four matings between two intermediates (B, C, and two in E) produced three low, five intermediate, and three high excretors, fitting the expected 1:2:1 ratio very well. Similarly, the two matings of two low excretors (D and F) produced eight offspring, all low as expected. The sole exception to the present hypothesis, thus far, occurred in family H. In this family, only intermediate offspring should be found with a mating between a low and high excretor. One offspring was the expected intermediate type, but the exceptional daughter was hypercalciuric. However, she had renal hypercalciuria whereas her mother had absorptive hypercalciuria. Though we believe the more-common absorptive hypercalciuria is regulated by a co-dominant pair of alleles, our estimates of gene frequency may be in error to the extent that data are "contaminated" by renal hypercalciurics. Renal hypercalciuria may require a double dose of an allele that does not manifest itself in heterozygotes, that is, is a recessive condition. Clearly, further genetic studies of both types of hypercalciuria are needed to resolve these and other questions.

To determine whether intermediate Ca excretion was associated with an increased risk of stone disease, the distribution of Ca excretions was examined in a group of 106 individuals attending our Clinic for treatment of CaOx nephrolithiasis. These were selected on the basis of having had Ca and creatinine excretion determined in a 24 h urine specimen. This group consisted of 33% high, 46% intermediate, and 21% low excretors. This classification should be regarded as only preliminary since, for some individuals, it was based on only one collection. From this and the corresponding data of the present study, relative risks were calculated[10]. The risk for high excretors was found to be 5.7. That is, those having a double dose of the allele for high Ca excretion (hypercalciuric) appeared to increase their risk of oxalate stone disease almost six-fold. Omitting the high excretors from both groups, a relative risk of 2.00 was found for intermediates (heterozygotes). As pointed out by Wright[11], though the alleles in polygenic systems generally have additive effects, they need not be additive at the phenotypic level. These relative risks were calculated primarily to determine whether intermediates are, in fact, at increased risk for stone disease since the proportion of intermediates is about the same in both populations (48% of the normal population and 46% of the stone formers were intermediates). However, when one excludes the "hypercalciurics" from both groups, one would expect the proportion of intermediates and lows in the two groups to be comparable if the intermediates are not at increased risk. One sees 53% intermediates among normals and 69% among stone formers, the difference being significant at the 5% level, indicating that those with a single dose of the susceptible allele are at an increased, but lower, risk of stone disease than the classical hypercalciuric.

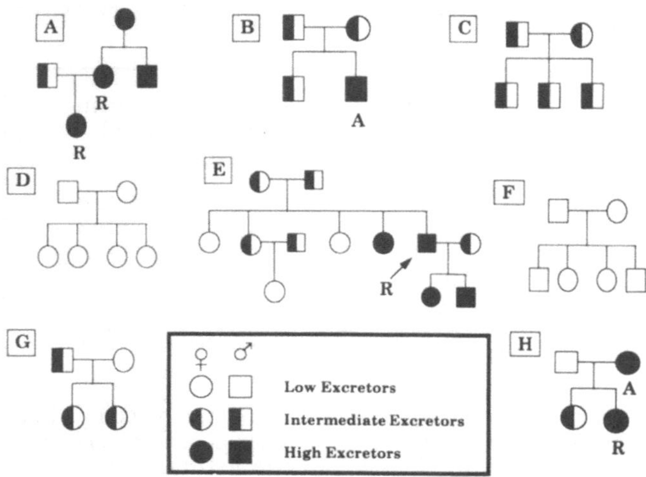

Figure 2. Pedigrees of families showing the inheritance of Ca excretion. The letter A below an individual represents a positive test for absorptive hypercalciuria and R for renal hypercalciuria.

DISCUSSION

Our results indicate that a pair of co-dominant alleles exert a major influence on Ca excretion. Our identification of two sub-populations within the group previously categorized as normocalciuric depended on: a) the collection of three 24 h urine specimens, thereby diminishing the effects of dietary variation, and b) the subdivision of Ca excretion into smaller sub-groups. The size of intervals is important for identifying patterns in frequency distributions. Obviously, if Ca were divided into only two groups, those above and below a given value, one would not perceive sub-classes in either group. In contrast, excessive sub-classification produces too many modes and antimodes to be interpretable.

An important question to consider is whether the Ca excretory groups are associated with absorptive hypercalciuria, renal hypercalciuria, or both. Three findings support the inference that absorptive hypercalciuria is the related entity. They relate to studies on patterns of fasting Ca excretion observed in populations, the pattern of inheritance we observed, and studies on the synthesis of 1,25-dihydroxyvitamin D_3 in individuals. Studies on fasting Ca excretion after equilibration to diets containing 400 mg Ca and 100 mmol Na per day have not shown large differences in excretion in normal (i.e. non-hypercalciuric) individuals. The failure to observe two distinct groups on controlled diets suggests that if renal hypercalciuria is an inherited trait, as family D in Fig. 2 suggests, it is phenotypically silent in heterozygotes under these conditions. Thus, one functional allele may be sufficient to effectively reabsorb the bulk of the Ca reaching the distal nephron. The pattern of inheritance we observed in family H is more compatible with absorptive rather than renal hypercalciuria being associated with calcium excretory classes.

Due to observations of its disordered control in absorptive hypercalciurias, the locus most likely responsible for the major genetic influence on calcium excretion is one associated with the synthesis of 1,25-dihydroxyvitamin D_3[12]. Insogna et al[13] have reported individual production rates of 1,25-dihydroxyvitamin D_3 per day. These results suggested that the 13 normal individuals in their study could be divided into two groups: one (n=7) with an excretion rate above 2.2 µg/day and the other (n=6) with an excretion rate below 2.0 µg/day. The group with absorptive hypercalciuria, however, overlapped with the normal group, indicating that other dietary factors or calculations of production rates may be required to better discriminate between groups. An analysis accounting for differences in body size, may for instance, be more informative. Identification of the gene involved would not only lead to a better understanding of factors involved in absorptive hypercalciuria but would also lead to better diagnostic tests and more directed therapies. The gene could be the 1α-hydroxylase itself (including its promoter region) or it could be one involved in transduction of the parathyroid hormone signal within the cell. It is reasonable to propose that the differences in plasma 1,25-dihydroxyvitamin D_3 levels resulting from different levels of 1α-hydroxylase activity in the Ca-excretor classes will lead to differences in Ca absorption and, hence, in urinary Ca-excretion. Studies of individuals on controlled diets should confirm whether differences in their plasma 1,25-dihydroxyvitamin D_3 levels and fasting urinary Ca are associated with their classification into different Ca-excretory classes.

From the mean urinary Ca/creatinine ratios for low, intermediate, and high excretors (0.072±0.014, 0.128±0.022, and 0.208±0.031 g/g, respectively), one can estimate whether additivity of the two alleles exists. One half of the mean for lows (0.036) is the contribution of one allele for lower excretion. Subtracting this from the mean for intermediates (0.128) yields 0.092 as the contribution of one allele for higher excretion. Multiplying this latter figure by two yields an estimate of the mean for those with a double dose of this allele of 0.184 g/g creatinine. This estimate is about 11% less than the observed 0.207 for higher excretors and is a very good fit with additivity considering that the observed mean is based on only 8 high excretors.

An analysis of the previous reports of an autosomal dominant pattern of inheritance of high Ca excretion[3,4] reveals that each had several families not compatible with this form of inheritance. The pattern of inheritance, however, was compatible with the inheritance of co-dominant alleles in all but one family in the study by Coe et al[3]. The deviation in this family may have been related to an incorrect classification of the Ca excretor type due to abnormal diets, a single urine collection, or an incorrect collection. Alternatively, the family may have had members with renal hypercalciuria for which we have proposed a different form of inheritance. As discussed above, two groups distinguished by their plasma 1,25-dihydroxyvitamin D_3 levels are apparent in a group of non-hypercalciuric individuals.

A complete understanding of the genetic factors associated with urinary Ca excretion is important for several reasons. Firstly, it will assist in deciphering the polygenic nature of urolithiasis. As stated earlier, three to five susceptibility genes may be involved, each with different risks for stone disease. The absence of detectable stone disease in the hypercalciuric individuals in our study clearly indicates that hypercalciuria alone is not a sufficient condition for urolithiasis. If the incidence of forming at least one kidney stone in a lifetime is 6%, if it is assumed that 40% of these individuals are hypercalciuric, and if absorptive hypercalciuria occurs in 75% of these individuals, only 1.8% of the entire population will have absorptive hypercalciuria and form a stone. This is approximately one-third of the individuals with absorptive hypercalciuria. The other two-thirds will remain stone-free. We believe that genetic factors may influence both oxalate and citrate excretion, but the contribution of genes awaits a greater understanding of environmental factors and physiological determinants that modulate excretions. We did not see a tri-modal distribution in either of these latter excretions.

The recognition that three groups of calcium excretors exist, raises questions as to whether Ca balances in the groups are different. Subtle differences in balance may exist and this may have implications for osteoporosis and other disease states. It would be interesting to determine whether the distribution of Ca excretion in an osteoporotic population is different from what we have observed in normal individuals.

Identification of the genetic locus affecting Ca excretion and determining the DNA sequence of the different alleles may lead to blood tests that could unequivocally distinguish the Ca excretory class of an individual. This would eliminate the need for a 24 h urine collection with its inherent problems of dietary disturbances, compliance, and inaccurate collections.

Clinically, with a better understanding of the cellular events involved in the synthesis of the gene product, more directed therapies may surface. With the recognition that not only high excretors but also intermediate Ca excretors have a greater risk of stone disease than low excretors, treatment of individuals in this class to make their Ca excretion similar to that of low excretors may decrease their risk of further stone formation.

REFERENCES

1. J Lemann and RW Gray, Idiopathic hypercalciuria, *J Urol* 141:715 (1989).
2. CYC Pak, M Ohata, EC Lawrence and W Snyder, The hypercalciurias: causes, parathyroid functions and diagnostic criteria, *J Clin Invest.* 43:387 (1974).
3. FL Coe, JH Parks and ES Moore, Familial idiopathic hypercalciuria, *New Engl J Med* 300:337 (1979).
4. K Mehes and Z Szelid, Autosomal dominant inheritance of hypercalciuria, *Eur J Pediatr* 133:239 (1980).
5. MI Resnick, DB Pridgen and HO Goodman, Genetic predisposition to formation of calcium oxalate renal lithiasis. *N Eng J Med* 278:1313 (1968).
6. WG Robertson, A Hodgkinson and DH Marshall, Seasonal variations in the composition of urine from normal subjects: a longitudinal study, *Clin Chim Acta* 80:347 (1977).
7. A Hodgkinson and LN Pyrah, The urinary excretion of calcium and inorganic phosphate in 344 patients with calcium stone of renal origin, *Br J Surg.* 46:10 (1958).
8. CYC Pak, R Kaplan, H Bone, J Townsend and O Waters, A simple test for the diagnosis of absorptive, resorptive and renal hypercalciurias. *New Engl J Med* 292:497 (1975).
9. CC Li, "Human Genetics, Principles and Methods", McGraw-Hill Book Co, New York, (1961). p19.
10. CC Li, "Human Genetics, Principles and Methods", McGraw-Hill Book Co, New York, (1961). p79.
11. SE Wright, "Evaluation and the Genetics of Populations, Vol. 1", Univ. of Chicago Press, Chicago, (1968). p.424.
12. AE Broadus, KL Insogna, R Lang, AF Ellison and BE Dreyer, Evidence for disordered control of 1,25-dihydroxyvitamin D production in absorptive hypercalciuria, *N Engl J Med* 311:73 (1984).
13. KL Insogna, AE Broadus, BE Dreyer, AF Ellison and JM Gertner, Elevated production rate of 1,25-dihydroxyvitamin D in patients with absorptive hypercalciuria, *J Clin Endocrinol Metab* 61:490 (1985).

ROLE OF HYDROXYPYRUVATE IN THE MANIFESTATION OF PRIMARY HYPEROXALURIA L-GLYCERIC ACIDURIA TYPE-II

K.G. Raghavan[1] and K.V. Inamdar[2]

[1]Radiation Biology and
Biochemistry Division Bhabha Atomic Research Centre and
[2]Department of Biochemistry
LTMM College & Hospital
Bombay, India

INTRODUCTION

L-glyceric aciduria primary hyperoxaluria type II, a genetic disorder of oxalate metabolism in man, manifests due to the absence of the enzyme D-glycerate dehydrogenase that acts on hydroxypyruvate[1]. As a result, hydroxypyruvate (OHP) gets reduced to L-glyceric acid by the nonspecific action of lactate dehydrogenase (LDH) with concomitant oxidation of NADH to NAD. This increased formation of NAD is postulated to stimulate LDH to act favourably on glyoxylate and oxidize it to excess oxalate thus leading to secondary manifestation of hyperoxaluria[2]. While this hypothesis could explain both glyceric aciduria and hyperoxaluria, several questions remain unanswered. Whether OHP is reduced to D-glyceric acid by the enzyme D-glycerate dehydrogenase, as in normal individuals or reduced to L-glyceric acid by the enzyme LDH, as occurs in L-glyceric aciduria patients, the amount of NAD produced would be the same in both cases. Since NAD and NADH are common cofactors for other enzymic reactions also, it is difficult to visualize a selective role for NAD in inducing LDH to act on glyoxylate and convert it into oxalate. Unless glyoxylate, the main precursor of oxalate, is made available in excess amounts to LDH by some metabolic pathways, the excess formation of oxalate cannot be explained under such conditions. In view of these considerations we re-examined some of the experimental evidence that led to this hypothesis.

MATERIALS AND METHODS

Reagents

Hydroxypyruvate (lithium salt, 98% purity), sodium glyoxylate, rabbit muscle lactate dehydrogenase and oxalate decarboxylase were obtained from Sigma Chemical Company. [1-^{14}C]glyoxylate was purchased from Amersham, Searle.

Lactate Dehydrogenase Assay

The activity of LDH was followed by determining the amount of radioactive oxalate and glycolate formed from [1-^{14}C]glyoxylate in the presence of NAD or NADH.

Oxalate Assay

[^{14}C]oxalate was estimated using oxalate decarboxylase, an enzyme that degrades oxalate to $^{14}CO_2$ and [^{14}C]formate[2]. Urinary oxalate and oxalate present in other samples were determined by employing oxalate oxidase[3].

Surgical Implantation of Osmotic Pumps containing Hydroxypyruvate

Alzet osmotic pumps (Model 2001, capacity: 200 µL, release rate:1 µL/h) releasing 2 µmole of OHP/h were used for implantation in animals. Laparotomy was performed on male Wistar rats under mild ether anaesthesia by making a 1 cm dorsal incision, an OHP-osmotic pump inserted into the peritoneal cavity and the incision sutured. Following this, the animals were individually housed in metabolic cages for urine collection.

RESULTS

Effect of OHP on LDH

Lactate dehydrogenase catalyses the coupled oxidation of glyoxylate to oxalate and its reduction to glycolate. It also brings about the reversible reduction of OHP to L-glycerate. These two reactions were combined and the effect of OHP on the oxidation of glyoxylate by LDH at pH 7.4 was examined. OHP at concentrations lower than that of optimal concentration of glyoxylate effectively inhibited the LDH catalysed oxidation and reduction of [1-^{14}C]glyoxylate in the presence of added NADH or NAD (Table 1).

Table 1. Effect of OHP on LDH catalysing the oxidation of [1-^{14}C]glyoxylate to oxalate.

	Cofactor added µmole	OHP added µmole	[^{14}C] glycolate formed nmole	[^{14}C] oxalate formed nmole
A				
NADH	0.5	-	1250	78
	0.5	2.5	590	27
	0.5	5.0	230	13
B				
NAD	5.0	-	980	320
	5.0	2.5	445	122
	5.0	5.0	270	41

A: Consisted of 5 µg rabbit muscle LDH and 5 µmole [11-^{14}C] glyoxylate.
B: Consisted of 10 µg rabbit muscle LDH and 10 µmole [1-^{14}C]glyoxylate.
The reaction was carried out at 37°C for 30 min in sodium phosphate buffer pH 7.4.

Glycolic acid oxidase (GAO) is yet another enzyme known to be involved in the synthesis of oxalate in human liver[4]. Addition of OHP to a purified preparation of GAO from human liver (3000 fold) led to a similar reduction in the formation of oxalate, that is being made from [l-^{14}C]glyoxylate (Table 2). A similar reduction in the amount of oxalate formation was observed even when the concentrations of GAO, glyoxylate and OHP were varied in the above reaction mixture.

Aliquots of OHP solutions were allowed to age at pH 7.4 for 1-3 days. Addition of such aged OHP solution to a standard oxalate assay mixture, consisting of a known amount of radioactive oxalate and the enzyme oxalate decarboxylase, was found to reduce the rate of release of radioactive CO_2 from the labelled oxalate (Fig. 1). With the addition of increasing concentrations of aged OHP to the above reaction mixture, a progressive decrease in the rate of release of labelled CO_2 from radioactive oxalate was noted. For a

Table 2. Effect of OHP on GAO.

	GAO added μg	[14C]glyoxalate added μmole	OHP added	Assay pH μmole	[14C]glyoxalate formed nmole
A	16	5	-	7.4	198
	16	5	10	7.4	164
	16	5	20	7.4	121
B					
	32	5	-	8.8	764
	32	5	10	8.8	352

The assay was carried out at 37°C for 1 h.

given concentration of OHP, the magnitude of its inhibitory effect was found to depend on the pH and the duration for which OHP was allowed to age.

The observed inhibitory effect of OHP could be explained on the basis of the dilution of radioactive oxalate by the newly formed cold oxalate, or by the inhibitory activity of the auto-oxidation products on oxalate decarboxylase. HPLC analysis of aged OHP established the new formation of oxalate as one of the major products of auto-oxidation of OHP. No such new formation of oxalate could be observed with a freshly prepared solution of OHP in 0.1 M phosphate buffer pH 7.4 (figure not presented). Similarly, OHP on being allowed to age in rat blood samples or in liver homogenates for one to three days, gave rise to oxalate.

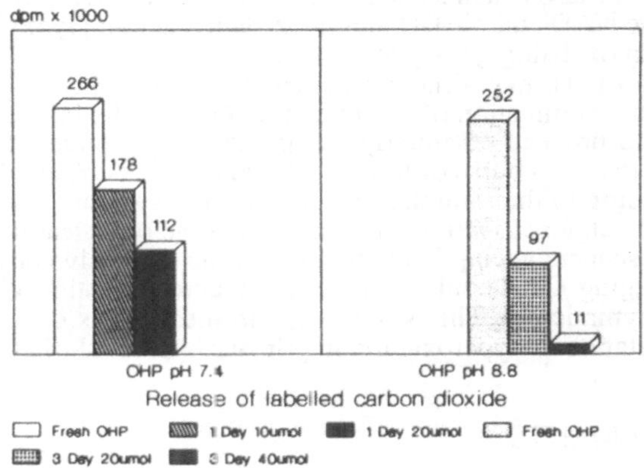

Fig. 1. Release of labelled carbon dioxide from radioactive oxalate.

The possibility that the aging OHP can also give rise to oxalate formation under *in vivo* conditions was examined by employing animal model systems mimicking the symptoms of L-glyceric aciduria. Urine samples obtained from animals implanted with Alzet osmotic pumps releasing OHP at a steady rate of 2 μLh were analysed for oxalate content (Fig. 2). A gradual increase in the daily excretion of urinary oxalate was noted in these animals. The increase in oxalate excretion was sharp from the fourth day onwards following implantation of OHP-pumps.

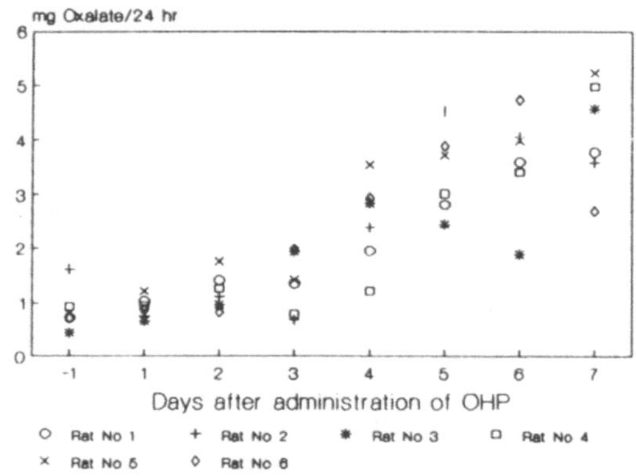

Fig. 2. Daily excretion of urinary oxalate by rats administered hydroxypyruvate.

DISCUSSION

It has been suggested that in L-glyceric aciduria primary hyperoxaluria type II, hyperoxaluria is induced by a favourable shift in NAD:NADH ratio, caused by the reduction of OHP to L-glycerate by LDH coupled with concurrent oxidation of glyoxylate to oxalate[2]. The present study shows that the OHP inhibits LDH as well as GAO. This observed inhibition of LDH activity by OHP is in contrast to an earlier report which implicates a role for OHP in stimulating LDH in oxidising glyoxylate to oxalate[2]. The synthesis of oxalate from glyoxylate *in vivo* by LDH is well established. The present study however points out that the rate of oxalate synthesis by LDH is determined by the affinity and the concentration of other available substrates rather than merely by the magnitude of the ratio of NADH:NAD. The number, affinity and concentration of endogenously available substrates of LDH such as OHP, glycolic acid, pyruvic acid, lactic acid etc, and enzymes that utilise NAD and NADH and those which act on glyoxylate, all can limit the role of liver LDH in oxidising glyoxylate to oxalate.

On the basis of the experimental evidence obtained in the present study, in L-glyceric aciduria, the elimination of accumulating OHP at the initial phase of the disease can be presumed to proceed essentially by its metabolic conversion to L-glycerate by LDH. However in the due course of time, there exists a possibility of gradual build up of the OHP level, leading to the saturation of the enzymic systems acting on it. Under such conditions the unmetabolised OHP would build up in the body leading to its recirculation, deposition and consequent ageing in the tissues. At the conducive physiological pH of the body, ie 7.4, the aging OHP could undergo slow auto-oxidation giving rise to oxalate formation non-enzymatically. This would explain the genesis of L-glyceric aciduria as well as the manifestation of hyperoxaluria in primary hyperoxaluria type II cases.

ACKNOWLEDGEMENTS

The authors express their gratitude to Dr BB Singh, Head, Radiation Biology and Biochemistry Division, for keen interest, encouragement and involvement in this project.

REFERENCES

1. HE Williams and LH Smith Jr, Primary hyperoxaluria, *in*: "The Metabolic basis of inherited disease", JB Stanbury, JB Wyngaarden and DS Frederickson, ed., McGraw-Hill, New York (1972).
2. HE Williams and LH Smith Jr, Possible mechanism of hyperoxaluria in L-glyceric aciduria, *Science* 171: 390 (1971).
3. KV Inamdar, KG Raghavan and DS Pradhan, Development of a procedure for the elimination of ascorbate interference in the enzymatic assaying of urinary oxalate, *Clin Chem* 37:864 (1991).
4. DW Fry and KE Richardson, Isolation and characterization of glycolic acid oxidase from human liver, *Biochim Biophys Acta* 568:135 (1979).

MECHANISMS FOR BI-DIRECTIONAL OXALATE TRANSPORT ACROSS THE LARGE INTESTINE

M. Hatch, R.W. Freel and N.D. Vaziri

Department of Medicine/Nephrology
University of California at Irvine
Irvine, California

INTRODUCTION

The proportion of oxalate in urine that is derived from dietary sources is highly variable and ranges from 2% to 50%[1]. Perhaps the most important factor influencing the amount of oxalate absorbed is the physical chemistry of intraluminal contents which determines both the solubility and hence the availability of oxalate. While the absorption of dietary oxalate in healthy individuals has been shown to occur primarily in the small intestine[2,3], in patients with enteric hyperoxaluria, the large intestine has been identified as the major site where oxalate is hyperabsorbed[4]. In enteric hyperoxaluria, increased formation of soluble oxalate salts is a consequence of malabsorbed fatty acids binding calcium within the colonic lumen resulting in a decrease of the insoluble calcium oxalate product[5]. It is also suggested that malabsorbed fatty acids and bile salts increase colonic permeability to oxalate[6]. These explanations are not mutually exclusive and they underscore the importance of understanding the mechanisms for basal handling of oxalate by the whole intestine. In this report we discuss the absorptive and secretory mechanisms for oxalate that have been identified in the large intestine[7,8]. It is apparent that the proximal and distal colonic segments handle oxalate differently under basal conditions.

MATERIALS AND METHODS

Proximal and distal colonic tissues were obtained from New Zealand White rabbits which were killed via an ear vein injection of pentobarbital. Sheets of intestinal mucosa were stripped of serosal muscularis and mounted in modified Ussing chambers having an exposed surface area of 0.64 cm^2. The solution bathing the tissues contained the following solutes (mmoles/litre): 139.4 Na^+; 123.2 Cl^-; 5.4 K^+; 1.2 Ca^{2+}; 1.2 Mg^{2+}; 21.0 HCO_3^-; 0.6 HPO_4^{2-}; 2.4 $H_2PO_4^-$; and 10.0 glucose. The concentration of oxalate in this solution was 1.5 μM. This solution was maintained at 37°C and circulated by means of a gas-lift system employing 95% O_2/5% CO_2. Flux measurements were made under voltage clamp conditions (Physiologic Instruments VCC600). Tissue conductance (G_T, $mS.cm^{-2}$) was calculated as the ratio of open-circuit voltage (mV) to short-circuit current (I_{sc}, $\mu A.cm^{-2}$), according to Ohm's Law. Unidirectional tracer fluxes of ^{14}C-oxalate (999 MBq/mmole, Amersham) were determined by adding the tracer to one bathing solution (mucosal or serosal side) and measuring its appearance in the opposite saline. Tracer activity (dpm) was measured with quench correction using standard liquid scintillation spectrometry (Beckman LS 9000). Net fluxes were taken as the difference between the two

Urolithiasis 2, Edited by R. Ryall *et al.*,
Plenum Press, New York, 1994

unidirectional fluxes in conductance-matched (±20%) tissue pairs. Net oxalate absorption (from the mucosal to the serosal side) is defined as a positive net flux, whereas net oxalate secretion (from the serosal to the mucosal side) is presented as a negative net flux. After an initial 45 min equilibration period, net fluxes and electrical parameters were measured over a 60 min control period which was frequently followed by a 60 min experimental period. All chemicals were reagent grade and obtained from Sigma Chemical (St. Louis, MO).

RESULTS

In standard buffer, the proximal colon secreted oxalate(-12.7±1.6 pmoles.cm^{-2}.hr^{-1}, n=33), whereas oxalate was absorbed by the distal colon (7.1±0.7 pmoles.cm^{-2}.h^{-1}, n=33). The transport of oxalate across each segment was also shown to involve distinct, coupled, mediated transport systems on the basis of (1), sensitivity to various transport inhibitors and (2), specific dependencies on the major ions, Na^+, Cl^-, HCO_3^-.

In the distal colon, net movements and sensitivities of oxalate parallel those of chloride[9]. Mucosal addition of SITS, an inhibitor of Cl^-:HCO_3^- exchangers (4-acetamido-4-isothiocyano-2, stilbene-2, 2-disulfonic acid, 10^{-4}M) abolished the net flux of both oxalate and chloride. Additionally, bicarbonate-free solutions containing the carbonic anhydrase inhibitor, acetazolamide, reduced the mucosal to serosal flux of both anions.

The magnitude and direction of net oxalate flux could be manipulated in both segments. In the proximal colon, serosal epinephrine (5 X 10^{-5} M) stimulated a significant net absorptive flux of oxalate (from -17.99±3.41 to 21.3±6.34 pmoles.cm-Z.h^{-1}, n=8). Basal oxalate secretion across the proximal colon was enhanced to -43.4±6.9 pmoles.cm^{-2}.hr^{-1}, by adding dB-cAMP (dibutyryl cyclic adenosine monophosphate, 5 x 10^{-3} M, n=5), however, neither basal nor stimulated oxalate secretion were blocked by furosemide (10^{-3} M). In contrast to the proximal segment, dB-cAMP reversed basal oxalate absorption to secretion across the distal colon (from 5.8±2.9 to -19.9±8.8 pmoles.cm^{-2}.hr^{-1}, n=5) and this was fully blocked by furosemide.

DISCUSSION

The early studies of intestinal oxalate transport indicated that transepithelial oxalate fluxes were passive[10]. It is now well documented, by *in vitro* experiments employing shortcircuited sheets of intestine[6-11] and membrane vesicle preparations[12,13], that net oxalate movement across both small and large intestine is a mediated process. A recent study which compared oxalate handling by the various segments of rabbit intestine showed that short-circuit preparations of jejunum, ileum, and proximal colon secrete oxalate, whereas the distal colon was unique in supporting a net absorptive flux.

The data presented here and in a previous report[7,8] identify both absorptive and secretory components in oxalate transport along the intestinal tract. Due to the dietary contribution to urinary oxalate, attention has been focused on absorptive mechanisms for oxalate across intestine and the relevance of oxalate movements in the opposite direction have been largely ignored. The reports[7,8] of existing secretory pathways for oxalate imply that the intestine has a role in the regulation of mass balance of oxalate; that is, the intestine may not operate in an exclusively oxalate-absorptive mode. Indeed, the intestinal tract may provide for a potentially important extra-renal route for oxalate excretion, particularly in the absence of renal function. Perhaps the oxalate secretory capacity of the intestinal route could be enhanced if favourable gradients for oxalate movement from blood to lumen were established. The transport pathways are present in intestinal epithelia to accommodate net flux into the lumen; furthermore, it has been demonstrated that these pathways can be manipulated.

Intestinal secretory pathways for oxalate may also be functionally important in the context of oxalate stone formers. Finlayson[1] estimated that the fraction of dietary oxalate absorbed by calcium oxalate stone formers was more than three times normal and he postulated that some of these individuals must be hyperabsorbing oxalate. It is now conceivable that "dietary" or absorptive hyperoxaluria may be a consequence of a decrease in the secretory component of intestinal oxalate transport, rather than, or possibly in addition to, an enhancement of the absorptive component.

Finally, the oxalate secretory pathways described underscore the importance of the substrate-specific, oxalate-degrading bacteria, which have been isolated from human large intestine[14]. Calculations based upon *in vitro* studies have indicated these microorganisms can make a significant contribution to intraluminal degradation of oxalate[15]. It seems reasonable to speculate that basal oxalate secretory pathways are involved in the dynamics of sustaining the resident microbial population.

REFERENCES

1. B Finlayson."Calcium Metabolism in Renal Failure and Nephrolithiasis", Wiley & Sons New York (1977).
2. DE Barilla, C Notz, D Kennedy, CYC Pak, Renal oxalate excretion following oral oxalate loads in patients with ileal disease and with renal absorptive hypercalciurias, *Amer J Med* 64:579 (1978).
3. M Hatch, PhD Thesis, Trinity College Dublin, Ireland. (1978).
4. JW Dobbins, HJ Binder, Importance of the colon in enteric hyperoxaluria, *N Eng J Med* 296:298 (1977).
5. DL Earnest, Enteric hyperoxaluria, *Adv Int Med* 24:407 (1979).
6. M Hatch, RW Freel, AM Goldner, DL Earnest, Effect of bile salt on active oxalate transport in the colon, *in:* "Colon and Nutrition", H Kaspar and H Goebell, eds, MTP Press (1980).
7. M Hatch, ND Vaziri, Segmental differences in intestinal oxalate transport, *FASEB J* 5:A1138 (1991).
8. M Hatch, RW Freel, ND Vaziri, Comparison of oxalate transport and related mechanisms in the rabbit proximal and distal colon, *FASEB J* 6:A1767 (1992).
9. M Hatch, RW Freel, AM Goldner, DL Earnest, Oxalate and chloride absorption by the rabbit colon: sensitivity to metabolic and transport inhibitors, *Gut* 25:232 (1984).
10. HJ Binder, Intestinal oxalate absorption, *Gastroenterology* 67:441 (1974).
11. RW Freel, M Hatch, DL Earnest, AM Goldner, Oxalate transport across the isolated rat colon :A reexamination, *Biochim Biophys Acta* 600:838 (1980).
12. RG Knicklebein, PS Aronson, JW Dobbins, Oxalate transport by anion exchange across rabbit ileal brush border, *J Clin Invest* 77:170 (1986).
13. RG Knicklebein, JW Dobbins, Sulfate and oxalate exchange for bicarbonate across the basolateral membrane of rabbit ileum, *Am J Physiol* 259:G807 (1990).
14. MJ Allison, HM Cook, DB Milne, H Gallagher, RV Clayman, Oxalate degradation by gastrointestinal bacteria from humans, *J Nutr* 116:455 (1986).
15. RA Argenzio, JA Liacos, MJ Allison, Intestinal oxalate-degrading bacteria reduce oxalate absorption and toxicity in guinea pigs, *J Nutr* 118:787 (1988).

SODIUM-CHLORIDE-DEPENDENT OXALATE ABSORPTION IN THE HUMAN INTESTINE

K. Yamakawa, T. Kato and J. Kawamura

Department of Urology
Mie University School of Medicine
Tsu, Mie, Japan

INTRODUCTION

High urinary oxalate excretion is one of the important determinants of stone formation in idiopathic calcium oxalate (CaOx) stone formers. High oxalate excretion is often observed in CaOx stone formers. In such patients, foods containing high oxalate levels (such as rhubarb, spinach, chocolate) should be restricted.

Urinary oxalate excretion is increased by two major factors in the diet[1]. One is a high animal-protein intake, and the other is a low oxalate/Ca ratio in the diet[1]. High animal-protein diets contain many amino acids such as tyrosine, tryptophan, phenylalanine, and hydroxyproline which may be metabolized to endogenous oxalate[1]. However, it is uncertain that intake of these amino acids increases endogenous oxalate production since high intake of hydroxyproline[2] and glycine[3] could not increase urinary excretion of oxalate. Since oxalate in the diet combines with Ca by chelation in the small intestine, a high Ca/oxalate ratio in the diet can prevent oxalate absorption, thereby decreasing urinary oxalate excretion[4]. Magnesium has been reported to be effective in decreasing oxalate absorption in the small intestine. Magnesium administration markedly decreased urinary oxalate excretion and also inhibited the crystal growth of Ca/Ox[5].

The role of sodium intake in renal stone formation has been well investigated. It has been demonstrated that urinary sodium excretion has a significant correlation with urinary calcium excretion both in normal subjects and stone formers[6]. Rao et al[7] suggested that salt intake should be restricted in order to decrease the risk of hypercalciuria. Singh et al[8] have reported that changes in urinary sodium do not influence the rate of crystallization of calcium, oxalate, and phosphate, both in healthy subjects and stone formers. However, it has not been investigated whether or not high sodium intake in the diet itself increases oxalate absorption in the small intestine.

The purpose of this study was to confirm the effect of sodium chloride on oxalate absorption in the human small intestine. We found that sodium chloride-dependent absorption of oxalate occurs in the human small intestine and that the intake of sodium chloride increases urinary excretion of oxalate.

SUBJECTS AND METHODS

The protocol for this study was approved by the Human Subject Committee of the Mie University School of Medicine. All subjects were fully informed of the nature of this study and volunteered to participate. Nine subjects were obtained from the medical staff of Mie University School of Medicine. None had any history of renal urolithiasis and

Urolithiasis 2, Edited by R. Ryall *et al.*,
Plenum Press, New York, 1994

renal dysfunction. In this study, all subjects fasted for 12 hours after their last meal the previous day, without any dietary restrictions. Urine samples were collected from all subjects prior to the study, and at two and four hours after intake of sodium oxalate. During a 4 h period, all subjects rested and drank 150 mL of water after each urine sample. Eight subjects received sodium oxalate (4 µMol/kg body weight) together with sodium chloride (1.2 Mol/kg body weight). After at least a 1-week interval, the same subjects received a similar dose of sodium oxalate. To elucidate the effect of sodium chloride on urinary oxalate excretion, sodium chloride (1.2 Mol/kg body weight) was administered to three normal subjects. For control purposes at a separate time these same subjects took water alone. Sodium chloride and sodium oxalate were administered to four patients with urolithiasis. Urine samples were collected from three normal subjects and stone formers according to the time course mentioned above. Concentrations of urine calcium were determined using a Ca electrode[9], creatinine by the method of Jaffe[10], and oxalate by the method of Ichiyama[11].

Excretions of oxalate, sodium, potassium, and chloride were calculated as ratio with urinary creatinine. Changes in these individual ratios were determined by ratios obtained at two and four hours after intake of sodium oxalate. The following formula was used:

Ox/Creat (at time +2 h or +4 h) ÷ Ox/Creat (at time 0 h).

All data were expressed as Means ±SD. Student's *t* test for paired data was used in the statistical analyses.

RESULTS

As seen in Table 1, time-dependent increases in urinary oxalate/creatinine ratios were found when NaCl was given with Na oxalate, with the greatest response occurring at four hours post load (p, 0.01) (Fig. 1A). In contrast, no significant changes in urinary oxalate/creatinine ratios were found in response to Na oxalate alone (Table 1 and Fig. 1B). Increases in urinary Na/creatinine, K/creatinine, and Cl/creatinine greater than unity were found at 2 and 4 hours after ingestion of Na oxalate, whether taken with or without NaCl.

Table 1. The changes in oxalate/creatinine ratios in healthy subjects (n=8) in response to NaCl+Na oxalate or with Na oxalate alone.

Time post load, h	NaCl+Na oxalate	Na oxalate
2	2.04±0.74	1.12±0.22**
4	3.25±1.29	0.96±0.19**

Values shown are Means±SD *p, 0.05, ** p, 0.01

The effect of NaCl alone on urinary oxalate/creatinine ratios is shown in Table 2. Here it is seen that ingestion of 1.2 mMol/kg body weight of NaCl caused no change in the ratios at either two or four hours post load. This suggested that administration of NaCl may not affect oxalate handling to any great extent in the kidney.

Table 2. The changes in urinary oxalate-creatinine ratios in healthy subjects (n=3) in response to ingestion of NaCl.

Time post load, h	NaCl+water	Water
2	0.90±0.09	1.19±0.15
4	1.17±0.11	1.07±0.18

Values shown are Means±SD.

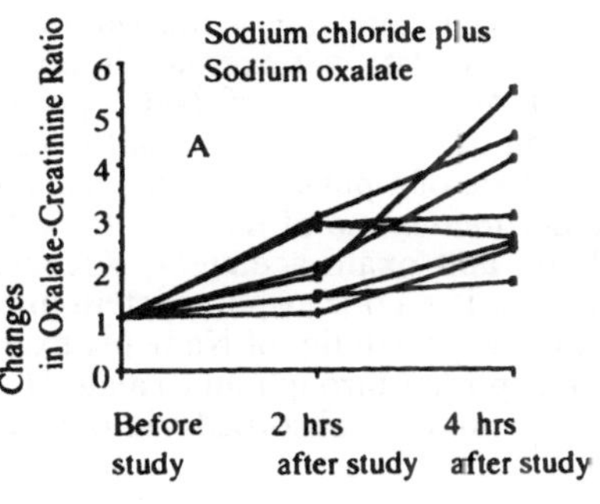

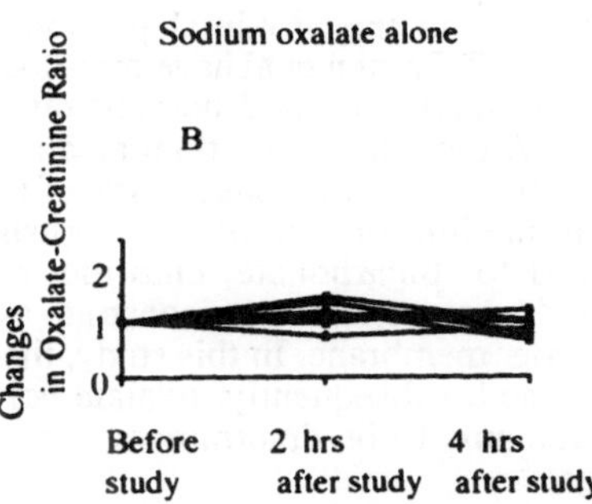

Figure 1. The changes in urinary oxalate/creatinine ratios in healthy subjects (n=8) when consuming NaCl plus Na oxalate or when consuming Na oxalate alone.

Table 3 and Fig 2 show changes in urinary Ca/creatinine ratios in response to NaCl+Na oxalate, Na oxalate alone, or NaCl alone. Salt intake markedly increased the changes in urinary Ca/creatinine ratios at 2 h post NaCl load. In those individuals taking Na oxalate, the changes in urinary Ca/creatinine ratios were less than those taking NaCl alone.

Table 3. Changes in urinary Ca/creatinine ratios in healthy subjects (n=8) in response to Na given in various forms.

Time post load, h	NaCl+Na oxalate	Na oxalate	NaCl
2 h	0.97±1.17	2.11±3.02	49.1±30.0
4 h	2.11±2.55	4.11±9.56	0.36±0.31

Values shown are Means±SD.

These results suggest that oxalate intake might interfere with Ca absorption in the intestine. In Fig. 2A, changes of greater than unity were found in the urinary Ca/creatinine ratio in four subjects. In Fig. 2B, urinary Ca/creatinine ratios did not change in six subjects. This suggests that oxalate might compete with Ca in absorption in the small intestine and that NaCl intake stimulates Ca absorption in the proximal portion of the small intestine.

DISCUSSION

This study showed that oxalate absorption in the small intestine of the human may be dependent upon the presence of NaCl. This conclusion is supported by the following observations: a) the simultaneous intake of Na oxalate and NaCl increased urinary Ox/creatinine ratios, whereas b) intake of Na oxalate alone or NaCl did not. In previous studies, we found that Na bicarbonate given with Na oxalate increased urinary oxalate/creatinine ratios. This suggested to us that oxalate absorption might depend on the presence of Na in the small intestine. However, this does not exclude the possibility that oxalate absorption may also depend on the presence of chloride. Since the urinary oxalate/creatinine ratios increased throughout the 4 h period following the load of Na oxalate and NaCl, this suggests that the NaCl-dependent absorption of oxalate might occur at least in the proximal portion of the small intestine. This supports previous observations[12]. Prenen et al have reported that ^{14}C-oxalate was excreted in urine, peaking between 2-6 h after the oral administration to normal subjects of NaCl given with Na oxalate[12]. Although NaCl stimulated absorption of oxalate in the small intestine of the human, there are no previous reports of the presence of a putative Na-oxalate co-transport system in the intestine. In the ileal brush-border membrane of the rabbit, oxalate can be exchanged for bicarbonate, chloride, sulphate, and oxalo-acetate[13]. Knickelbein and Aronson[14] were unable to demonstrate any Na-oxalate co-transport system in rabbit ileal brush-border membrane. In this study, therefore, the possibility of Na being exchanged for protons[15] and, subsequently, oxalate being transported through an oxalate-H symporter system remains to be determined. Also, oxalate may be absorbed via oxalate-chloride exchange.

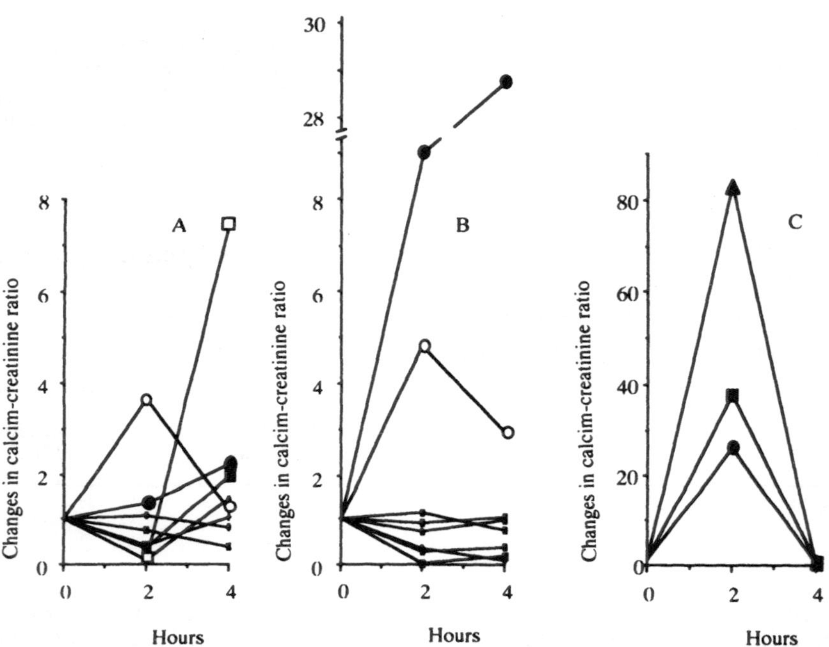

Figure 2. Changes in urinary Ca/creatinine ratios in response to NaCl+Na oxalate (A); Na oxalate (B); and NaCl (C).

The presence of Ca in both the diet and intestine is related to oxalate absorption since oxalate chelates Ca in the intestine. Changes in the urinary oxalate-creatinine ratio at four hours after the simultaneous intake of Na oxalate and NaCl ranged from 1.69 to 5.44 in normal subjects (Fig. 1A). This variation may have been due to an interference of intestinal Ca on the NaCl-dependent absorption of oxalate since we did not impose any dietary restrictions prior to the study. Butz et al.[16] have reported that oral administration of 130 mg oxalate failed to increase urinary excretion of oxalate. The dose of Na oxalate administered in this study was 4 μM/kg body weight which was less than that administered by these authors.

Kleeman et al[17] reported that Na intake correlates with Ca excretion in normal human beings. Walser showed that renal clearance of Ca correlates with that of Na[18]. Calcium transport parallels Na transport throughout the nephron[19]. Silver et al[20] demonstrated Na-dependent idiopathic hypercalciuria in renal stone formers, and their interpretation of Na-induced hypercalciuria was based on the renal handling of both Ca and Na. In this study, only NaCl intake increased urinary Ca/creatinine ratios. The simultaneous intake of NaCl and Na oxalate or Na oxalate alone did not cause any appreciable change in urinary Ca/creatinine ratios, suggesting that oxalate may have interfered with intestinal Ca absorption (Table 3 and Fig. 2) . These results suggest that NaCl might stimulate Ca absorption in the proximal portion of the small intestine via a Na:Ca exchanger[21].

In a previous study, we examined the changes in urinary oxalate/creatinine ratios in response to simultaneous intake of NaCl and Na oxalate in four patients with a history of recurrent idiopathic CaOx nephrolithiasis. Two of these patients showed no increase in urinary oxalate/creatinine in the four hours following ingestion. These results suggest that dietary restrictions should be imposed on stone formers before such studies are performed in order to understand how the NaCl-dependent absorption of oxalate may be involved in the pathogenesis of CaOx nephrolithiasis.

In conclusion, these results indicate that a NaCl-dependent absorption of oxalate may occur in the human small intestine since the intake of NaCl has been shown to amplify both Ca and oxalate absorption. In patients with CaOx nephrolithiasis, excess intake of NaCl may increase the risk of stone formation.

REFERENCES

1. WG Robertson, Epidemiology of urinary stone disease, *Urol Res* 18:S3 (1990).
2. A Hodgkinson, Oxalate metabolism and hyperoxaluria, *in*: "Scientic Foundations of Urology", DI Williams and GD Chisholm, eds, William Heinemann Medical Books, London (1976).
3. PM Zarembski and A Hodgkinson, Some factors influencing the urinary excretion of oxalic acid in man, *Clin Chim Acta* 25:1 (1969).
4. WG Robertson and M Peacock, The cause of idiopathic calcium stone disease: hypercalciuria or hyperoxaluria? *Nephron* 26:26 (1980).
5. W Berg, C Bothor, W Pirlich, V Janitzky, Influence of magnesium on the absorption and excretion of calcium and oxalate ions, *European Urol* 12:274 (1986).
6. RAL Sutton, VR Walker, Relationship of urinary calcium to sodium excretion in calcareous renal stone formers, *in* "Urolithiasis", LH Smith, WG Robertson, and B Finlayson, eds, Plenum Press, New York, (1981).
7. PN Rao, EB Faraghar, A Buxton, V Prendiville, NJ Blacklock, *in*: "Urolithiasis and Related Clinical Research", PO Schwille, LH Smith, WG Robertson, and W Vahlensieck, eds, Plenum Press, New York (1985).
8. PP Singh, AK Pendse, P Hada, PC Nagori, V Rathore, and PK Dashora, Sodium is neither a risk nor a protective factor in urolithiasis? *Urol Res* 15:105 (1987).
9. G Vezzoli, A Elli, P Palazzi, High plasma ionized calcium with normal PTH and total calcium levels in normal-function kidney transplant recipients, *Nephron* 42:290 (1986).
10. K Larsen, Creatinine assay by a reaction-kinetic principle, *Clin Chim Acta* 41:209 (1972).
11. A Ichiyama, E Nakai, T Funai, T Oda, and R Katafuchi, Specrophotometric determination of oxalate in urine and plasma with oxalate oxidase, *J Biochem* 98:1375 (1985).
12. JAC Prenen, P Boer, and EJD Mees, Absorption kinetics of oxalate from oxalate-rich food in man, *Am J Clin Nut* 40:1007 (1984).
13. RG Knikelbein, PS Aronson, and JW Dobbins, Substrate and inhibitor specificity of anion exchangers on the brush-border membrane of rabbit ileum, *J Membr Biol* 88:199 (1985).
14. LP Karniski and PS Aronson, Anion exchange pathways for Cl- transport in rabbit renal microvillus membranes, *Am J Physiol* 253:F513 (1987).
15. R Knickelbein, PS Aronson, W Atherton, and JW Dobbins, Sodium and chloride transport across rabbit ileal brush-border, I. Evidence for Na-H exchange, *Am J Physiol* 245:G505 (1983).
16. M Butz, H Hoffmann, and G Kohlbecker, Dietary influence on serum and urinary oxalate in healthy subjects and oxalate stone formers, *Urol Int* 35:309 (1980).
17. CR Kleeman, J Bohannan, D Bernstein, S Ling, and MH Maxwell, Effect of variations in sodium intake on calcium excretion in normal humans, *Proc Soc Biol Med* 115:26 (1964).
18. M Wasler, Calcium clearance as a function of sodium clearance in the dog, *Am J Physiol* 200:1099 (1961).
19. M Goldberg, ZS Agus, S Goldfarb, Renal handling of phosphate, calcium and magnesium, *in* "The Kidney", BM Brenner, FC Rector, Jr, ed, WB Saunders, Philadelphia (1976).
20. J Silver, D Rubinger, MM Friedlaender, MM Popovtzer, Sodium-dependent idiopathic hypercalciuria in renal stone formers, *Lancet* 27:484 (1983).
21. P Gmaj, H Murer and R Kinne, Calcium ion transport across plasma membranes isolated from rat kidney cortex, *Biochem J* 178:549 (1979).

ERYTHROCYTE URATE SELF-EXCHANGE IN IDIOPATHIC CALCIUM NEPHROLITHIASIS

G. Gambaro, F. Marchini, M. Vincenti, M.A. Nassuato, L. Bacelle, A. D Angelo, A. Tasca and B. Baggio

Institutes of Internal Medicine
Clinical Biochemistry and Urology
University of Padova
Italy

INTRODUCTION

Hyperuricosuria is very common in calcium-oxalate (CaOx) nephrolithiasis[1], and the natural history of this disease in hyperuricosuric patients is much more severe[2]. These considerations led to the thesis that uric acid might play a pathogenetic role in calcium-oxalate stone formation through the epitaxial crystallogenesis of CaOx on sodium urate [3,4], or uric acid crystals in the presence of glutamic acid[5], or through the reduction of urinary inhibiting activity by the binding of colloidal urate to urinary glycosaminoglycans[6,7]. In any case, the role of uric acid in calcium-oxalate stone formation remains a matter of controversy [8-10]. Concerning the etiology of hyperuricosuria in CaOx nephrolithiasis, both an alimentary origin[11,12], and an altered tubular renal handling of urate[13,14] have been advanced. An inheritable anomaly of red cell oxalate transport was found in "primary" calcium nephrolithiasis, and suggested the hypothesis that this disease might be a metabolic disorder characterized by a defect in cellular anion transport[15,16]; we explored the alternative hypothesis that hyperuricosuria might constitute another expression of this generalized anion transport disorder.

METHODS

Erythrocyte urate self-exchange

Sixty seven idiopathic CaOx stone formers, consisting of 41 males and 26 females (age range 18-50 years) were studied; 39% were hyperoxaluric, 31% hypercalciuric, and 10% hyperuricosuric. The control group consisted of 20 medical staff members (12 males and eight females; age range 21-46 years) with no family or personal history of nephrolithiasis.

Erythrocyte urate-exchange was assayed as previously described for oxalate flux[15]. Briefly, washed red blood cells (RBC) were incubated in an isotonic phosphate buffer containing 300 µmol/L sodium urate. After 2 h, the disappearance of ^{14}C uric acid (Amersham, U.K. specific activity 60) mCi/mmol from the same incubation medium was followed in time. The RBC urate flux was expressed as K, an index of urate exchange at equilibrium, and calculated from the slope of the linear relation

$$\ln (A_t - A_\infty) / (A_0 - A_\infty) = - Kt$$

where A is ^{14}C urate in the incubation medium at time 0, t and at isotope equilibrium (∞).

RESULTS

Transmembrane urate flux in controls was $0.25 \pm 0.14 \times 10^{-2}$ min^{-1}, and 0.45 ± 0.40 in patients. Thirty percent of patients had abnormal urate flux, but only two thirds had both abnormalities. Erythrocyte urate flux was correlated with 24 h uricosuria; the red cell anomaly was also associated with hyperuricosuria and more intense disease activity. Transmembrane urate flux was inhibited *in vitro* by stilbenes, pyrazinamide, and heparan sulphate. A family study carried out in two young stone-free sons of two male stone forming probands with abnormal urate RBC flux suggested that the uric acid transport abnormality might be genetically determined.

CONCLUSIONS

Our findings suggest that hyperuricosuria during calcium oxalate nephrolithiasis might be due to a cell defect in transmembrane urate transport.

In view of the fact that pyrazinamide shares common inhibiting activity on urate transport in both the red cell and the kidney, and previous data suggesting the existence of a common tubular secretory site for both urate and oxalate[17], the erythrocyte urate anomaly here described might also be present at the renal level, and thus be the cause of hyperuricosuria in calcium stone formers. Further study with the pyrazinamide test is necessary to confirm this hypothesis.

REFERENCES

1. FL Coe, Uric acid and calcium oxalate nephrolithiasis, *Kidney Int*, 24: 392 (1983).
2. FL Coe, J Keck, E Norton. The natural history of calcium urolithiasis, *JAMA* 238: 1519 (1977).
3. FL Coe, RL Lawton, RB Goldstein, V Tembe. Sodium urate accelerates precipitation of calcium oxalate *in vitro*, *Proc Soc Exp Biol Med* 149: 926 (1975).
4. CYC Pak, LH Arnold. Heterogeneous nucleation of calcium oxalate by seeds of monosodium urate, *Proc Soc Exp Biol Med.* 149: 930 (1975).
5. S Sarig. The hyperuricosuric calcium oxalate stone former, *Mineral Electrol Metab* 13: 251 (1987).
6. WG Robertson, F Knowles, M Peacock. Urinary mucopolysaccharide inhibitors of calcium oxalate crystallization, *in*: "Urolithiasis Research" Plenum Press, New York p. 331 (1976).
7. B Baggio, G Gambaro, A Marchi, E Cicerello, S Favaro, A Borsatti. The role of glycosaminoglycans and uric acid in idiopathic calcium nephrolithiasis, *Contr Nephrol* 37: 5 (1984).
8. B Baggio, G Gambaro, E Cicerello, F Marchini, A Borsatti. Further studies on the possible lithogenetic role of uric acid in calcium oxalate stone disease, *in* "Urolithiasis and Related Clinical Research" Plenum Press, New York p. 785 (1985).
9. B Ettinger. Does hyperuricosuria play a role in calcium oxalate lithiasis? *J Urol* 141: 738 (1989).
10. RL Ryall, PK Grover, VR Marshall, Urate and calcium stones- Picking up a drop of mercury with one's fingers? *Am J Kidney Dis* 18: 426 (1991).
11. FL Coe, E Moran, AG Kavalach. The contribution of dietary purine overconsumption to hyperuricosuria in calcium oxalate stone formers, *J Chron Dis* 29: 793 (1976).
12. WG Robertson, PJ Heyburn, M Peacock, FA Hanes, R Swaminathan. The effects of high animal protein intake on the risk of calcium stone formation in the urinary tract, *Clin Sci* 57: 285 (1979).
13. B Fellstrom, U Backman, BG Danielson, G Johansson, S Ljunghall, B Wikstrom. Renal handling of urate in patients with calcium stone disease, *Nephron* 31: 31 (1982).
14. F Mateos Anton, J Garcia Puig, G Gaspar, et al, Renal handling of uric acid in patients with recurrent calcium nephrolithiasis and hyperuricosuria, *Nephron* 37: 123 (1984).
15. Baggio B, Gambaro G, Marchini F, Cicerello E, Tenconi R, Clementi M, Borsatti A. An inheritable anomaly of red-cell oxalate transport in "primary" calcium nephrolithiasis correctable with diuretics, *New Engl J Med* 314: 599 (1986.)
16. G Gambaro, B Baggio. Idiopathic calcium oxalate nephrolithiasis: a cellular disease, *Scann Microsc* 6: 247 (1992).
17. B Baggio, G Gambaro, E Cicerello, A Borsatti. Oxalate and urate: a possible common secretory tubular site? *Urol Res* 12: 62 (1984).

A NEW CHORDATE MODEL FOR URIC ACID-OXALATE LITHIASIS

M.B. Saffo

Institute of Marine Sciences
University of California
Santa Cruz, CA USA

Because ethical considerations generally preclude research on humans, surrogates are essential to every aspect of biomedical research. Understandably, the similarities ... among mammals [have] made them obvious candidates. ...Less well appreciated has been the ... role of lower vertebrates, invertebrates and microorganisms in biomedical research.

Committee on Models for Biomedical Research (US)
National Academy of Sciences, 1985[1]

Marine biology clearly has much to offer medical research. Yet discourse between these two fields has been all too infrequent.

AC Smith, 1988[2]

INTRODUCTION

Laboratory mammals have served as key experimental models for a number of human diseases, but they have been less useful in the study of urolithiasis. Because they do not produce uric acid as an end product of purine catabolism, most mammals are especially limited models for uric acid urolithiasis and of mixed urate-oxalate stones.

With uric acid as an end-product of purine catabolism, humans and other hominoid primates are unique among vertebrates. Birds and reptiles produce uric acid, but as an end product of purine and protein catabolism. Some nonhominoid mammals resemble humans in patterns of renal secretion or reabsorption of urate[3] but not in urate oxidase activity. Even purine metabolism deficiencies in mice (HGPRT deficiency) which resemble human metabolic diseases (Lesch-Nyhan Syndrome) do not have the same pathological effects as in humans, a difference plausibly correlated with the fact that mice, unlike humans, possess substantial urate oxidase activity[4]. Diseases involving urate overproduction or precipitation are thus difficult to mimic in laboratory mammals and other vertebrates.

Along with clinical studies, physiological investigations of laboratory rodents, including studies of experimentally induced oxalate stones in rats[5] have provided useful data on several aspects of oxalate stone formation. Nevertheless, many basic physiological and mechanical aspects of oxalate precipitation, including the significance of normal oxalate metabolism in humans, and the complex interactive effects of dietary oxalate, dissolved ion concentrations, soluble and crystalline urate, crystallization inhibitors, nucleators, and pH are far from resolved[6,7].

Urolithiasis 2, Edited by R. Ryall *et al.*,
Plenum Press, New York, 1994

Molgulid ascidians as models for urate-oxalate stone disease

In the face of such limitations, ascidian tunicates are especially promising animal analogues to urate and calcium oxalate urolithiasis in humans. Ascidians ("sea squirts") are a globally distributed class of marine invertebrate chordates, whose homologies with other chordates have been of special interest to questions of vertebrate evolution. In addition to the most well known homologies between ascidians and vertebrates - the presence of a notocord, gill slits, dorsal nerve cord, and a thyroxin-producing thyroid gland homologue[8] - urate metabolism and "stone" formation in ascidians and humans show a number of experimentally exploitable (and plausibly homologous) similarities. These similarities are most conspicuous in the *Molgulidae*, a diverse ascidian family widely distributed in the marine benthos, especially in temperate and high-latitude habitats.

Urate is a metabolic end product in molgulids[9] and arises solely from purine catabolism[10]. Urate and calcium oxalate precipitate as solid concretions. Weddellite (calcium oxalate dihydrate; COD) crystals are found free in fluid as well as mixed with urate concretions[11]. Like human kidney stones, molgulid concretions are situated in extracellular fluid compartments[11]. Though ductless, the renal sac shows some resemblance to vertebrate kidneys. Like a kidney, the renal sac is closely associated with the circulatory system. In ultrastructure (infolded basal membranes, apical microvilli and numerous mitochondria) renal sac cells resemble excretory and osmoregulatory epithelia, including several cell types of the mammalian nephron[12]. Oxalate crystal formation is a normal occurrence in both human urine[13] and renal sac fluid[11].

Urate synthesis may be of adaptive value in both molgulids and humans[14]. Allopurinol inhibits urate synthesis in both humans and molgulids. After addition of allopurinol to ambient seawater, HPLC separations of renal sac contents show uptake of allopurinol/oxypurinol by *Molgula*, and buildup of hypoxanthine, xanthine and allopurinol derivatives[15].

Molgulid concretions also show several differences from human kidney stones. Rather than being mere limitations to the system, these differences allow the use of unique experimental manipulations which are difficult or impossible in mammalian models or human subjects. For example: all molgulids make concretions as normal products of metabolism, from early development through sexual maturity. Thus, it is not necessary to screen populations for a small percentage of natural "stone formers" or to resort to special treatments to induce stone formation for study.

Along with concretions, the renal sac of all molgulids contains a heterotrophic protist, *Nephromyces*, in its lumen. This symbiosis is apparently a mutualism, rather than a pathological association. In turn, *Nephromyces* is chronically and hereditarily, infected with Gram-negative, intracellular bacteria, which are probably obligate, non-pathological symbionts of *Nephromyces* [16,17].

Central to my experimental system is the fact that the symbiosis between *Molgula* and *Nephromyces* is a chronic, but non-hereditary symbiosis established anew each generation by infection of young *Molgula* by spores of *Nephromyces*. Thus, in the laboratory, the abundant, estuarine species *Molgula manhattensis* can be raised routinely through sexual maturity, free of *Nephromyces* and its bacteria,along with parallel populations of symbiontinfected counterparts[19]. The symbiont-free groups differ profoundly from the symbiont-infected counterparts in patterns of urate and oxalate metabolism (Table 1), providing useful experimental variants. For instance: although microradioassays of renal sac tissue confirm that *Molgula* itself cannot degrade urate, *Nephromyces* possesses striking urate oxidase activity, with specific activities among the highest values for any organism, including peroxisomal isolates from plant and animal tissue[9,17,19]. Our initial allopurinol trials have shown a greater buildup of urate precursors in infected than in symbiont-free animals, suggesting that rates of purine synthesis may be higher in infected than uninfected animals. Thus, infected and uninfected *Molgula* may differ in rates of urate synthesis as well as urate degradation.

Our data thus far also suggest that the symbionts affect calcium oxalate precipitation. I have not detected mineralized (sodium hypochlorite resistant) crystals in symbiont-free *M. manhattensis*. In contrast, large quantities of mineralized crystals, with morphologies strongly resembling calcium oxalate monohydrate "raphides" of plants[20] are present in concretions from lab-raised, Nephromyces-infected counterparts. Preliminary data are consistent with the hypothesis that "raphides" in laboratory-raised *M. manhattensis* indeed may be calcium oxalate: energy dispersive X-ray spectrophotometry of these crystals show that calcium is the major, and perhaps only, cation in the crystals, and there is no trace of phosphate or any other elements with an atomic weight greater than 19.

In nature, the subtropical *M. occidentalis* contains both weddellite crystals and calcareous "raphides". Laboratory-raised *M. manhattensis* differ in several ways from fieldgrown cohorts: lab-raised animals have slower growth rates, and *Nephromyces*-infected lab animals show denser Nephromyces infections, higher urate oxidase activities, and different calcareous crystal morphology (see above) than their field counterparts. Similarly, different *Molgula* species vary in rates of urate degradation, *Nephromyces* density and crystal morphology. These differences can also be exploited as another set of physiological variants (Table 1).

Table 1. Renal sac variants.

Group	Symbionts	Uricase Activity	Relative Urate Synthesis (hypothesized)	Calcareous Crystals
M.manhattensis				
lab, symb (-)	(-)	(-)	+?	?
lab, symb (+)	+++	+++	+++?	+
field, symb (+)	++	++	++?	+ (COD)
M. occidentalis				
field, symb (+)	+++	+++	+++?	+ (COD and other)

Other possible differences between the composition and environment of molgulid and human stones, the possible absence of phosphate from molgulid concretions, the alkaline [8.1] pH of renal sac fluid[11,21] and unpublished observations simplify some aspects of oxalate analysis and serve as additional natural physiological variants.

In addition to the similarities and differences between molgulids and humans noted above, several other aspects of molgulid biology are experimentally useful: because molgulids are sessile and can be raised in the laboratory, animals can be tracked individually through time. Thus, molgulids are particularly amenable to long-term experiments testing effects of diet, ion composition of ambient sea water and other factors on oxalate and urate precipitation.

Further, the transparency of the renal sac and body wall in *M manhattensis,* coupled with the voluminous renal sac lumen, allow non-invasive observations on concretion formation in individual animals and *in vivo* micro-injections into the blood sinuses and the renal sac[19,21], providing a means for experimental manipulation of the ionic environment of the blood and renal sac fluid of *Molgula*.

Molgulids are thus well-suited as a provocative, experimentally tractable model for study of urate and oxalate urolithiasis. As chordate homologues, they lend a new perspective to the evolutionary significance of urate synthesis in humans. As experimentally useful analogues, they provide a fresh physiological context for uncoupling some of the variables affecting oxalate and urate lithiasis, and especially for probing the links between both oxalate and urate precipitation, and oxalate and urate metabolism, in humans[7,22] and other organisms.

REFERENCES

1. Committee on Models for Biomedical Research, National Research Council, *in* "Models for Biomedical Research.: A New Perspective," National Academy Press, Washington, DC (1985).
2. AC Smith, Marine animals: clear models for medical science, *The Scientist* 2 (10): 18-19 (1988).
3. IM Weiner, Urate transport in the nephron, *Amer J Physiol* 237: F85F92.(1979).
4. TB Friedman, GE Polanco, JC Appold, and JE Mayle, On the loss of uricolytic activity during primate evolution -I. Silencing of urate oxidase in a hominoid ancestor, *Comp BiochemPhysiol* 81B: 653-659 (1985).
5. SR Khan and RL Hackett, Crystal-matrix relationships in experimentally induced urinary calcium oxalate monohydrate crystals, an ultrastructural study, *Calcif Tissue Int* 41: 157-163 (1987).
6. FL Coe and JH Parks, *in* "Nephrolithiasis: Pathogenesis and Treatment," (second edition). Year Book Medical Publishers, Inc. Chicago, (1988).
7. RL Ryall, PK Grover and VR Marshall, Urate and calcium stones -picking up a drop of mercury with one's fingers? *Amer J Kid Dis.* 17 (4): 426-430 (1991).
8. I Goodbody, The physiology of ascidians, *Adv Mar Biol* 12:1-149 (1974).
9. MB Saffo, Nitrogen waste or nitrogen source? Urate degradation in the renal sac of molgulid tunicates, *Biol Bull* 175: 403-409 (1988).
10. JR Nolfi, Biosynthesis of uric acid in the tunicate *Molgu la manhattensis*, with a general scheme for the function of stored purines in animals, *Comp Biochem Physiol.*, 35: 827-842 (1970).
11. MB Saffo and HA Lowenstam, Calcareous deposits in the renal sac of a molgulid tunicate, *Science* 200: 1166 1168 (1978).
12. BD Ross and WG Guder, Heterogeneity and compartmentation in the kidney, *in*: "Metabolic Compartmentation," H. Sies, ed., Academic Press, London, 363-409 (1982)
13. RL Ryall, The formation and investigation of urinary calculi, *Clin Biochem Rev* 10:149-157 (1989).
14. BN Ames et al, Uric acid provides an antioxidant defense in humans against oxidant- and radical-caused aging and cancer: a hypothesis. *Proc Natl Acad Sci* 78: 6858-6862 (1981).
15. MB Saffo, Renal sac function and the *Nephromyces-molgulid* symbiosis: does uric acid play the central metabolic role?, *Amer Zool* 26: 22A (1986).
16. MB Saffo, Symbiosis within a symbiosis: intracellular bacteria in the endosymbiotic protist *Nephromyces*, *Mar Biol* 107: 291-296 (1990).
17. MB Saffo, Symbiogenesis and the evolution of mutualism: lessons from the Nephromyces-bacterial-molgulid symbiosis. *in:*: "Symbiosis as a Source of Evolutionary Innovation: Speciation and Morphogenesis," L Margulis and R Fester, eds. MIT Press, pp. 410-429 (1991).
18. MB Saffo and W Davis, Modes of infection of the ascidian *Molgula manhattensis* by its endosymbiont *Nephromyces Giard Biol Bull* 162: 105- 112 (1982).
19. MB Saffo, Are uricolytic products of *Nephromyces* incorporated by *Molgula? Amer Zool* 28: 168A (1988).
20. HJ Arnott, Calcification in higher plants, *in:* "The Mechanisms of Mineralization in the Invertebrates and Plants," N Watabe and KM Wilbur, eds, University of South Carolina Press, pp. 55-78 (1976).
21. MB Saffo, Studies on the renal sac of *Molgula manhattensis* De Kay (Ascidiacea, Tunicata, phylum Chordata). Ph.D. thesis, Stanford University. pp. 1-128 (1977).
22. PK Grover, RL Ryall, and VR Marshall, Effect of urate on calcium oxalate crystallization in human urine: evidence for a promotory role of hyperuricosuria in urolithiasis, *Clin Sci* 79: 9-15 (1990).

RENAL EPITHELIAL INJURY: A RISK FACTOR IN UROLITHIASIS

S.R. Khan and R.L. Hackett

Department of Pathology
College of Medicine University of Florida
Gainesville
Florida, USA

INTRODUCTION

In humans, crystalluria is a relatively common event, and is an indicator of a urine's supersaturation with regard to a given salt[1]. As long as the movement of these crystals through the urinary system remains free, they do not develop into symptomatic urinary stones. Obviously additional factors, responsible for crystal retention within the renal tubules, are necessary for the initiation of stone formation. What are these factors that can trigger stone formation? Vermeulen et. al[2] suggested that massive crystalluria as a result of high supersaturation could succeed in blocking the papillary ducts and thus initiate the process. Finlayson and Reid calculated[3] that it was not possible for single crystals of calcium oxalate monohydrate to grow large enough to block the urinary tubules within the time it takes the urine to flow from the glomerulus to the pelvis. Thus crystals must be "fixed" at some point within the renal tubules so that they have enough time to grow and initiate stone formation.

What are the factors then, that can promote crystal retention and development of stone disease? Since idiopathic urinary stone formation is an episodic event, the causes for its inception should also be episodic. Based on our observations of cellular damage in experimental models of nephrolithiasis[4] we concluded that one of the major factors responsible for the development of this disease is an insult to the kidney that results in temporary tubular damage. We postulated[5] that injury is preliminary to urinary stone disease and urinary stones form in the presence of renal epithelial injury and adequate supersaturation. Later we showed[6] that renal epithelial injury induced by gentamicin sulphate administration facilitates calcium oxalate crystalluria in hyperoxaluric rats. These findings were confirmed by Kumar et al[7] who showed that a strategy combining moderate hyperoxaluria and mild renal proximal tubular injury caused a marked increase in the incidence of calcium oxalate crystal deposition and papillary stone formation in rats over that seen in rats with hyperoxaluria or renal injury alone.

An association between clinical nephrolithiasis and renal injury is also well documented[8-10]. It is however, unclear whether the renal injury is primary, representing a predisposing factor, or secondary, a result of crystal deposition and stone formation. It is also unclear how renal damage affects the process of crystallization and stone formation. This project was undertaken to further our understanding of the relationship between renal tubular injury and calcium oxalate nephrolithiasis.

Urolithiasis 2, Edited by R. Ryall *et al.*,
Plenum Press, New York, 1994

MATERIALS AND METHODS

Hyperoxaluria was induced in male Sprague-Dawley rats by administration of 0.25% or 0.5% (v/v) ethylene glycol in drinking water. Renal tubular injury was produced by daily subcutaneous injections of gentamicin sulphate at a dose of 100 mg/kg rat body weight[6] and was determined by histological and electron microscopic examination of kidneys[11]. Three hour urine samples were collected for the examination of crystalluria. Rats were sacrificed on day 7 or 10. Crystals were isolated from the bladder aspirate by filtering the urine through a 0.22 μm nucleopore filter and were then examined by scanning and transmission electron microscopy.

RESULTS

Administration of gentamicin sulphate alone resulted in injury to the proximal tubular epithelial cells and an increased appearance of cellular degradation products in the urine. Ethylene glycol treatment alone did not cause significant morphological changes in the kidneys and no crystals were detected in the urine or anywhere in the renal tubules. But all rats receiving the combined treatment of ethylene glycol and gentamicin had calcium oxalate crystalluria by day 3. The urinary crystals were associated with cellular degradation products which appeared as amorphous sheets when examined by scanning microscopy (Fig. 1). All the kidneys removed from these rats at the time of sacrifice, demonstrated some degree of tubular necrosis. There was localized loss of proximal tubular epithelial cells exposing their basement membrane. Renal tubules were focally dilated and contained calcium oxalate crystals intermingled with cellular degradation products. Most of the crystals were located in the renal medulla.

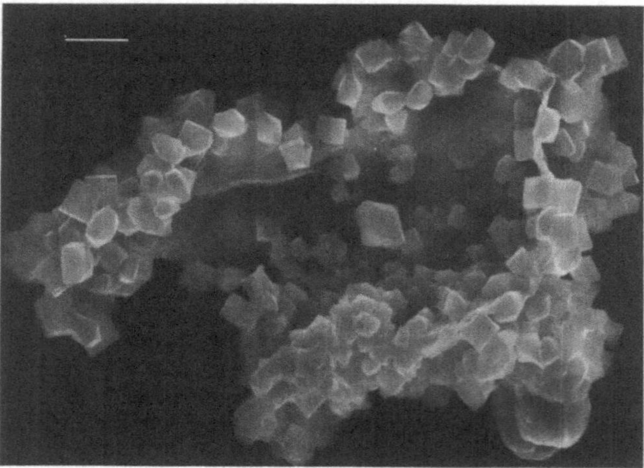

Fig. 1. Calcium oxalate dihydrate crystals isolated from the bladder aspirate are associated with an amorphous sheet-like entity. Bar = 5 μm

DISCUSSION

Administration of 0.25% or 0.5% ethylene glycol to male Sprague-Dawley rats in drinking water results in mild hyperoxaluria without crystal deposition in their kidneys[12]. Gentamicin sulfate at the dose we used here, is injurious to the kidneys and causes focal necrosis of renal proximal tubules resulting in the appearance of membranous profiles and myeloid bodies in the urinary sediment[13]. In our study, treatment of rats with either compound did not result in nephrolithiasis, but when the two were given together, all rats became crystalluric and had calcium oxalate nephrolithiasis. Kumar et al have earlier shown that rats given gentamicin at a smaller dose of 40mg/kg for four days and then in addition given ammonium oxalate for the next ten days developed nephrolithiasis with abundant papillary stones[7]. These data support our earlier hypothesis that a combination of tubular injury and a mild hyperoxaluria can produce nephrolithiasis.

Apparently injury predisposes the kidney to nephrolithiasis under suitable supersaturation. What are the reasons for this predilection? Development of stone disease involves a cascade of events including crystal nucleation, growth, aggregation, and retention[12,14]. Each of these events is controlled by a combination of factors, some of which promote while the others inhibit the process. For renal injury to work as a catalyst for urolithiasis, injury should raise the urine's crystallization promotory potential and lower its inhibitory potential.

Structurally, a nephrotoxic challenge, depending on agent and dosage, can damage the renal cell membrane, cause cellular necrosis, shed membranes in the urine, and produce focal loss of the epithelial lining of the renal tubules resulting in exposure of the underlying basement membrane. Damage to the plasma membrane can cause an influx of calcium into the epithelial cells resulting in increased cytosolic calcium which in turn will result in the formation of calcific crystals. This is what generally happens in dystrophic calcification[15]. In the hyperoxaluric state, crystals formed will most probably be calcium oxalate. In rats made hyperoxaluric by chronic administration of ethylene glycol we have often seen crystals inside the cells lining the collecting ducts of the renal papillae (unpublished observations). Such a formation of intracellular crystals may assist in their retention inside the nephron.

Nephrotoxic challenges including acute hyperoxaluria[4] or gentamicin treatment[13] cause renal tubular cellular necrosis, shedding of the epithelial brush border membrane or the entire cells into the urine. These membranes may act as heterogeneous nucleators of both calcium phosphate and calcium oxalate crystals. In *in vitro* experiments we have demonstrated[16] that brush border membrane vesicles isolated from rat kidney proximal tubules can induce crystallization of calcium oxalate from a metastable solution of low supersaturation which on its own will not support such a crystal formation. In similar *in vitro* experiments, calcium phosphate crystallization has also been shown to be promoted by cellular membranes. According to current concepts[15,17] initial calcium phosphate crystal deposition in a number of calcific diseases occurs on cellular membranes which are present at the calcification site either as limiting membrane of the so-called matrix vesicles or as cellular degradation products. In rat models of calcium oxalate as well as calcium phosphate nephrolithiasis, crystal deposition appears to be mediated by the cellular membrane. Intranephronic calculosis in the laboratory rats, caused by feeding purified diet, is membrane-dependent. Vesicles derived from the brush border of the proximal tubules provide a nidus for intratubular deposition of calcium phosphate[18]. Experimental studies of calcium oxalate nephrolithiasis in rats have shown that irrespective of the method utilized, nature of the hyperoxaluria inducing agent used and the location of crystal formation within the urinary system, calcium oxalate crystals are always found intermingled with cellular degradation products[4]. In this study too, calcium oxalate crystals present in the renal tubules or isolated from the bladder aspirate, were found associated with cellular degradation products. Our studies of decalcified human calcium oxalate urinary stones have shown cellular degradation products as a constituent of stone matrix[7]. A number of renal stones start at the tip of renal papillae on plaques called Randall's plaques[8]. These plaques contain cellular degradation products and crystals of calcium phosphate. Stones formed on such plaques were once thought to be rare. But a recent study by Cifuentes-Delatte[19], of five hundred spontaneously passed small stones collected in Spain, has revealed that 142 (28.4%) of them contained nidi of necrotic material identifiable as calcified tips of the renal papillae indicating that at least in these cases the stones could have originated by heterogeneous nucleation of crystals on membranous cellular degradation products.

As mentioned earlier, an important step in the development of stone disease is retention of crystals inside the kidneys. Single crystals cannot grow large enough to be retained within the renal tubules. Thus for crystal retention to occur, they must either aggregate and become large deposits or attach to the tubular epithelium. Renal injury can promote both these mechanisms. We have earlier shown attachment of calcium oxalate crystals to luminal membrane of the damaged renal epithelial cells[20]. We have also shown retention of crystals within the nephron because of aggregation[20]. Even in the bladder, crystals are surrounded by membranous cellular degradation products[21]. These membranous entities may initially trap and hold the crystals together and bring them closer to each other to facilitate the formation of crystal bridges.

In a rat model of calcium oxalate nephrolithiasis where crystal deposition was induced by an acute hyperoxaluric challenge[11], papillary tubules were denuded of their epithelium and the crystals were seen anchored to the tubular basement membrane. A similar phenomenon has been observed in a chronic model of calcium oxalate nephrolithiasis (unpublished observations). The mechanisms involved in crystal attachment to either the damaged epithelium or the denuded basement membrane are not well understood. In studies of calcium oxalate crystal/papillary epithelium interaction, crystals were found to preferentially adhere to cells which lost their attachment to the basement membrane[22]. It was suggested that crystal adherence was accomplished by binding molecules of basement membrane origin which were exposed because of the loss of cellular attachment.

Gentamicin administration also results in a decline in the urine's capacity to inhibit the growth of calcium oxalate crystals[23]. The decline appears to be connected with a decrease in urinary excretion of low molecular weight crystallization inhibitors and citrate. This is yet another way that renal injury can influence the process of crystal deposition in the kidneys and the development of urinary stone disease.

It is obvious from the above discussion that renal injury can produce a number of changes in the kidneys. These changes can facilitate crystal formation, growth, aggregation and retention within the urinary system. Thus renal injury in the presence of adequate supersaturation can be a risk factor in urolithiasis.

ACKNOWLEDGEMENTS

This work was supported by NIH grants #ROIDK41434 and #POIDK20586

REFERENCES

1. PG Werness, JH Bergert, and LH Smith, Crystalluria, *J Crystal Growth* 53:166 (1981).
2. CW Vermeulen, ES Lyon, JE Ellis, and TA Borden, The renal papilla and calculogenesis, *J Urol* 97:573 (1967).
3. B Finlayson and F Reid, The expectation of free and fixed particles in urinary stone disease, *Invest Urol* 15:442 (1978).
4. SR Khan and RL Hackett, Calcium oxalate urolithiasis in the rat: is it a model for human stone disease? A review of recent literature, *Scan Electr Microsc* 2:759 (1985).
5. B Finlayson, SR Khan, and RL Hackett, Mechanisms of stone formation - an overview, *Scan Electr Micros.* 3:1419 (1984).
6. RL Hackett, PN Shevock, and SR Khan, Cell injury associated calcium oxalate crystalluria, *J Urol* 144:1535 (1990).
7. S Kumar, D Simon, T Miller, B Carpenter, S Khan, R Malhotra, C Scheid, and M Menon, A new model of nephrolithiasis involving tubular dysfunction/injury, *J Urol* 146:1384 (1991).
8. A Randall, The etiology of primary renal calculosis, *Intl Abst Surg* 71:209 (1940).
9. B Baggio, G Gambaro, E Ossi, S Fararo, and A Borsatti, Increased urinary excretion of renal enzymes in idiopathic calcium oxalate nephrolithiasis, *J Urol* 129:1161 (1983).
10. P Jaeger, L Pertmann, J Ginalski, A Jocqout, E Temler, and P Burckhardt, Tubulopathy in nephrolithiasis: consequence rather than cause, *Kid Intl* 29:563 (1986).
11. SR Khan, B Finlayson, and RL Hackett, Experimental calcium oxalate nephrolithiasis in the rat, role of renal papilla, *Am J Path* 107:59 (1982).
12. SR Khan, Pathogenesis of oxalate urolithiasis: lessons from experimental studies with rats, *Am J Kid Dis* 17:398 (1991).
13. C Josepovitz, R Levine, B Lane, and GJ Kaloyanides, Contrasting effects of gentamicin and mercuric chloride on urinary excretion of enzymes and phospholipids in the rat, *Lab Invest* 52:375 (1985).
14. B Finlayson, Physico-chemical aspects of urolithiasis, *Kid Intl* 13:344 (1978).

15. HC Anderson, Calcification processes, *Pathol Annu* 15:45 (1980).
16. SR Khan, PN Shevock, and RL Hackett, Membrane associated crystallization of calcium oxalate *in vitro*. *Calcif Tissue Intl* 46:116 (1990).
17. AL Boskey, Current concepts of physiology and biochemistry of calcification, *Clin Orthop* 157:225 (1981).
18. HT Nguyen and JC Woodard, Intranephronic calculosis in rats, *Am J Pathol* 39:56 (1980).
19. L Cifuentes-Dellate, JLR Minon-Cifuentes, and J Medina, Papillary stones, calcified renal tubules in Randall's plaques, *J Urol* 133:194 (1985).
20. SR Khan and RL Hackett, Retention of calcium oxalate crystals in renal tubules, *Scan Microsc* 5 :707 (1991).
21. SR Khan and RL Hackett, Crystal-matrix relationship in experimentally induced urinary calcium oxalate monohydrate crystals, an ultrastructural study, *Calcif Tissue Intl* 41:157 (1987).
22. NS Mandel and R Riese, Crystal-cell interactions: crystal binding to rat renal papillary tip collecting duct cells in culture, *Am J Kid Dis* 17:402 (1991).
23. RL Hackett, PN Shevock and SR Khan, Inhibition of calcium oxalate monohydrate seed crystal growth is decreased in renal injury, *These proceedings*.

OXALATE TRANSPORT ACROSS THE RENAL PAPILLA: IMPLICATIONS IN SITES AND MECHANISM OF INITIAL KIDNEY STONE FORMATION

P.S. Chandhoke[1] and K.A. Hruska[2]

[1]Division of Urologic Surgery
[2]Renal Division
Jewish Hospital of St. Louis
Washington University Medical Center
St. Louis
Missouri, USA

INTRODUCTION

Although we have a good understanding of the physico-chemical processes of stone nucleation and growth, disagreement exists regarding the site of nucleation[1]. The only systematic evidence for the site of initial stone formation has been provided by Carr who used microradiographic techniques to demonstrate that the initial stone formed within the fornix and in time grew large enough to be detected by conventional radiographic techniques[2].

The mammalian renal pelvis is a urinary space surrounded by a transitional epithelium which lines its outer wall, and a simple cuboidal epithelium which covers the surface of the papilla and lines the pelvic recesses (fornices and secondary pouches)[3]. The part of the simple epithelium that covers the surface of the papilla has previously been referred to as the "papillary surface epithelium (PSE)"[4]. In the rabbit, a similar simple epithelium also lines the pelvic recesses and the inner part of the pelvic septum[5]. The human PSE is also a simple cuboidal epithelium[6].

The pelvic recesses are cul-de-sacs that are removed from the direct path of urinary flow in the collecting system and presumably are relatively poorly mixed[5]. Studies of pelvi-calyceal dynamics in patients with a history of nephrolithiasis have demonstrated that in these patients, areas of stagnation exist that are more pronounced than in normal subjects[7]. Physico-chemical factors contributing to initial stone formation include the state of saturation with respect to crystal components, local pH, availability of matrix, modifiers of crystal formation such as promoters and inhibitors[8], and possibly the existence of specific surfaces for crystal attachment[9]. If the epithelium lining the pelvic recesses is involved in the modification of these physico-chemical factors, it may have an important influence on initial stone formation within poorly mixed regions of the pelvis.

METHODS

The papillary surface epithelium was isolated from the surface of the rabbit renal papilla and then mounted in a specially-designed Ussing chamber which allowed the composition of fluid bathing both the apical and basolateral surfaces to be controlled and/or measured. The PSE dissection procedure and the mounting of the tissue in the modified Ussing Chamber for transepithelial transport studies have been described previously[10].

The PSE was mounted in the chamber with the apical surface upward and the epithelial area available for transport was 7 mm^2. The basolateral side of the tissue was continuously perfused via ports drilled into the lower half of the chamber. The effluent could be sampled for chemical or radioactivity analysis. The apical aspect of the epithelium was bathed with a 100 microliter standing droplet which was covered by water-equilibrated, CO_2-equilibrated mineral oil. Unless otherwise stated, all studies were done at 37°C. Transepithelial voltage was measured throughout each study using agarose bridges coupled to calomel half cells. Current pulses were injected via a separate pair of electrodes for determination of transepithelial resistance. According to previously established criteria to eliminate epithelia with nonspecific leaks[10], epithelia were used only when they exhibited a resistance >70 $ohm.cm^2$.

The basic dissection and perfusion solution contained the following (in mM): 110 NaCl, 25 $NaHCO_3$, 2.5 K_2HPO_4, 1.2 $MgSO_4$, 2 $CaCl_2$, 5 glucose, 100 urea, and 5 creatinine (osmolality 360 mosmol/kg). The solutions were bubbled with 95% O_2 and 5% CO_2 and had a pH of 7.45. ^{14}C-oxalate (New England Nuclear, Boston, MA) at a concentration of 1.7×10^{-5} M was added either to the apical or basolateral surfaces, or both, to study the oxalate transport. This concentration of oxalate was used such that the solubility product of calcium oxalate was not exceeded as determined by the Finlayson's EQUIL computer program[11].

Appearance or disappearance of oxalate in the apical droplet was used as an indicator of secretion or absorption, respectively, of oxalate across the PSE. The change in oxalate concentration in the apical droplet over a 30 minute time period was determined by measuring the radioactivity of ^{14}C-oxalate (disintegration per minute, DPM) with a liquid scintillation counter (TriCarb, Packard Instrument Company, Meriden, CT). Oxalate flux, J_{ox}, calculated in units of nanomolesJhr/cm^2, was obtained from the following equation:

$$J_{ox} = \frac{\text{change in DPM of apical droplet}}{S \times t \times A}$$

where, S = specific activity of ^{14}C-oxalate (DPM/nanomoles), t = experimental time in hours (0.5 hours), and A = available PSE surface area for transport (0.07 cm^2).

In some experiments, 1 mM 4,4'-diisothiocyanostilbene-2,2'disulphonic acid (DIDS, Sigma Chemical, St. Louis, MO) was added to the perfusion solutions. The DIDS was dissolved in dimethylsulfoxide (DMSO); a final concentration of 0.2% of DMSO was necessary to keep 1 mM of DIDS in solution. The same amount of DMSO was also added to the control solution in DIDS related experiments. Ion substitution experiments were also conducted to characterize the mechanism of transport as described in the results section. In sodium free experiments, NaCI was replaced with N-methyl-D-glucamine (NMDG) chloride. In chloride free experiments, NaCI and $CaCl_2$ were replaced with sodium gluconate and calcium gluconate, respectively. In these studies, the experiment was first conducted in the absence of the ion being studied, followed by a 30 minute equilibration period with the presence of the ion, and another 30 minute period over which J_{ox} was again measured. Thus, all ion substitution experiments yielded paired experimental data.

RESULTS

Initial experiments were done to determine if there was measurable transport of oxalate across the PSE. Transepithelial oxalate transport was first studied in the apical to basolateral direction by placing 1.7×10^{-5} M radioactively labeled ^{14}C-oxalate in the apical droplet. This concentration of oxalate was chosen to ensure that the relative calcium oxalate (CaOx) supersaturation was less than 1.0[11]. In a similar fashion, ^{14}C-oxalate was added only to the basolateral perfusion solution to determine basolateral to apical oxalate transport. These two sets of experiments demonstrated disproportionate rates of transport:

the basolateral to apical flux was 7.1±1.5 (mean±SE) nanomoles/cm²/hr (n=5), whereas the apical to basolateral flux was 1.6±0.7 nanomoles/cm²/hr (n=5). When the PSE was exposed to the same concentration of oxalate on both the apical and basolateral sides, experiments demonstrated a net apical to basolateral flux of oxalate of 3.0±0.7 nanomoles/cm²/hr (n=4). The differences in oxalate fluxes in all three groups were statistically significant from each other ($p < 0.05$). These sets of experiments implied that there was net active reabsorption of oxalate across the rabbit PSE. To further study this possibility, the apical to basolateral oxalate flux, Jox, was measured at room temperature (20°C) (n=4) and at 37°C (n=5). The absorptive flux of oxalate noted at 37°C was essentially abolished at 20°C.

Six experiments were done to study the effects of DIDS on J_{ox}. Of note was the observation that with the addition of DMSO to the control solution, there was a reduction in J_{ox} to 4.7±0.41 nanomoles/cm²/hr when compared to a value of 7.1±1.5 nanomoles/cm²/hr when DMSO was absent. Thus there appeared to be a reduction in J_{ox} just by the addition of DMSO. In four experiments, J_{ox} was further inhibited by the addition of DIDS, only when the control J_{ox} (without DIDS) was above 4.0 nanomoles/cm²/hr. In the remaining two experiments, the absorptive flux of oxalate was low (<4.0 nanomoles/cm²/hr), and further reduction of oxalate transport could not be demonstrated.

As DIDS is not a specific inhibitor of oxalate membrane transport mechanisms, ionic substitution experiments were conducted in an attempt to further characterize transepithelial PSE oxalate transport. The experiments demonstrated similar amounts of oxalate absorption with or without the presence of sodium in the apical bathing solution. These results suggest the absence of the sodium dependent oxalate absorptive mechanism on the PSE apical surface. However, when chloride was replaced with gluconate in the perfusion solutions, there was an approximately 50% increase in J_{ox}. Thus, it appears that the presence of a net chloride/oxalate exchange mechanism at the PSE apical surface leads to net secretion of oxalate, thereby limiting the total amount of oxalate that can be absorbed by the PSE.

DISCUSSION

Understanding the nature of the anatomical sites and mechanisms of initial stone formation within the renal pelvis is fundamental to our understanding of urolithiasis and in determining subsequent therapeutic intervention. The present work was undertaken to begin a study of the sites and mechanisms of initial stone formation within the mammalian kidney. The only systematic evidence for the site of initial stone formation in the human kidney has been provided by Carr[21], who used microradiographic techniques to demonstrate that the initial stone formed within the fornix and in time grew large enough to be detected by conventional radiographic techniques. Carr hypothesized that the mechanism of initial stone formation was renal lymphatic obstruction, but no evidence to support this theory has ever been published. Alternatively, it may be hypothesized that transepithelial transport within the poorly mixed regions of the pelvic recesses, e.g., across the epithelium lining the fornix and secondary pouches, is responsible for initial stone formation by altering the local state of supersaturation of crystal components.

The subepithelial location of Randall's plaques just below the papillary surface further substantiates the concept that transepithelial transport across the papilla could be important for crystal formation within the renal tissue. It seems reasonable, therefore, to study the transport function of the papillary surface epithelium, not only because this transport could favorably or unfavorably affect conditions for stone formation in the renal pelvis, but also because these processes may be involved in the formation of Randall's plaques.

We chose to study transepithelial oxalate transport because of its major role in calcium oxalate urolithiasis. It has been shown that calcium oxalate supersaturation is more sensitive to changes in urinary oxalate concentration than an equivalent percentage change in urinary calcium concentration[13]. The concentration of oxalate used in our experiments was 1.7×10^{-5}M such that the solubility product of calcium oxalate, as determined by the computer program EQUIL[11], was not exceeded.

The major finding of the present study is that the rabbit papillary surface epithelium absorbs oxalate. This conclusion is based on the observation that there was net uptake of oxalate when both the apical and basolateral PSE surfaces were exposed to the same concentration of oxalate and this transport was inhibited when the experiments were repeated at 20°C. As degradative enzymes that metabolize oxalate to formic acid and carbon dioxide do not exist in humans, nor are known to exist in other mammalian tissue[12], the disappearance of oxalate from the apical droplet is unlikely to be due to cellular uptake and metabolic degradation. Interestingly, the apical to basolateral flux of oxalate across the PSE (7.1±1.5 nanomoles/cm^2/hr) is comparable to the secretory flux determined across the rabbit proximal tubule (2.6 nanomoles/cm^2/hr, assuming a diameter of 26.5 microns for the proximal tubule)[14]. Although the PSE cuboidal epithelium has some morphological features suggestive of a brush border, it is not as extensive as that of the proximal tubule[4]. This perhaps suggests a higher density of oxalate transporters on the PSE apical surface than that found in the brush border of the rabbit proximal tubule.

Aronson's group[15,16] has extensively studied the transport of oxalate not only in membrane vesicles from the proximal tubule, but also from the rabbit ileum. Whereas the ileum absorbs oxalate, the proximal tubule secretes it; however, the transporters are localized in a similar fashion at the apical and basolateral membranes and obviously must work in both directions across their respective membranes. Karniski and Aronson[15] suggest that chloride transport from the proximal tubule is initially mediated by formate (FOR-) being exchanged for chloride (Cl-) across the brush border. Formate in turn is recycled back into the cell on a transporter that exchanges oxalate for either formate or chloride. The net result is chloride absorption and oxalate secretion in the proximal tubule.

In the present study, we studied selected oxalate transport mechanisms located on the epithelial apical surface. Although there was no overall statistically significant difference in apical to basolateral oxalate flux in the presence or absence of DIDS, in four experiments where J_{ox} was greater than 4 nanomoles/cm^2/hr, an inhibition of oxalate transport did occur. Thus, given the inherent limitations of the *in vitro* technique to study the effects of DIDS in the intact PSE epithelium, the experiments overall suggest that anion transport mechanisms, as found in the proximal tubule and ileum, are probably operative in the rabbit PSE as well.

The mechanism of oxalate transport is likely to be anion exchange driven either by secondary or tertiary active transport [17]. As removal of sodium from the apical surface did not alter J_{ox} this suggests that sodium sulfate/oxalate cotransport is not responsible for the observed oxalate transport. Results of chloride substitution experiments would suggest that an apical chloride/oxalate exchanger operates in a secretory fashion (a process similar to that found in the proximal tubule), as the addition of chloride resulted in a 50% decrease of net absorptive oxalate flux.

As a similar cuboidal epithelium lines the papilla, fornix, and the inner wall of the pelvic septum (pelvic recesses), absorption of oxalate from the fluid bathing the pelvic recesses will reduce the concentration of oxalate within this space. This process will thus serve as an inhibitory mechanism of initial kidney stone formation within the pelvic recesses. Alternatively, the absorbed oxalate may contribute to the high concentrations of oxalate found within the renal tissues, as reported by Hautmann et al[18]. According to the intranephronic theory of initial kidney stone formation, these high oxalate concentrations could subsequently lead to nucleation within the renal tissue itself, and may serve as nidus for initial kidney stone formation.

Increased red cell oxalate transport has been reported in stone forming patients compared to normals, which is thought to be an inherited characteristic[12]. It is thought that this generalized abnormality in stone formers is present in the gastrointestinal tract and the kidney, such that these individuals absorb and secrete more oxalate from their gastrointestinal tract and kidneys, respectively. One may speculate that if this abnormality was also present in the papillary surface epithelium, where the transport process seems to be similar to the ileum, increased absorption of oxalate from the PSE would increase papillary tissue oxalate concentration in stone formers. If true, this finding would support the intranephronic theory of initial calcium oxalate stone formation.

REFERENCES

1. GW Drach, Urinary Lithiasis, *in* "Campbell's Urology". PC Walsh, RF Gittes, AD Perlmutter, and TA Stamey, eds. fifth edition. Philadelphia, Saunders, (1986).

2. RJ Carr, Aetiology of renal calculi: micro-radiographic studies, *in* "Proceedings of the Renal Stone Research Symposium", Editors Hodgkinson, A, and BEC Nordin, Churchill Ltd., London, (1968).
3. B Schmidt-Nielsen, The renal pelvis, *Kidney Int* 31:621-628, (1987).
4. JM Sands, MA Knepper, and KR Spring, Na-K-CI cotransport in apical membrane of rabbit renal papillary surface epithelium, *Am J Physiol* 251 (*Renal Fluid Electrolyte Physiol* 20): F475-484, (1986).
5. HL Sheehan and JC Davis, Anatomy of the pelvis of the rabbit kidney, *J Anat* 93(4):499-502, (1959).
6. PA Narath, Renal Pelvis and Ureter. New York: Grune and Stratton p 9, (1951).
7. E Schultz, Studies on the influence of the flow field in the pelvi-calyceal system on the formation of urinary calculi, *Urol Res* 15:281-286, (1987).
8. FL Coe and JH Parks, Defenses of an unstable compromise: crystallization inhibitors and the kidney's role in mineral regulation, *Kidney Int* 38:625-631, (1990).
9. RJ Riese, JW Riese, JG Kleinman, JH Weissner, GS Mandel, and N.S. Mandel, Specificity in calcium oxalate adherence to papillary epithelial cells in culture, *Am J Physiol.* 255 (*Renal Fluid Electrolyte Physiol* 24): Fl1025-F1032, (1988).
10. JM Sands and MA Knepper, Urea permeability of mammalian inner medullary collecting duct system and papillary surface epithelium, *J Clin Invest* 79:138-147, (1987).
11. B Finlayson, Calcium stones: some physical and chemical aspects. *in* "Calcium metabolism in renal failure and nephrolithiasis" Edited by David S David, John Wiley and Sons, New York, (1977).
12. HE Williams and TR Wandzilak, Oxalate synthesis, transport, and hyperoxaluric syndromes, *J Urol* 141 (3): 742-747, (1989).
13. WG Robertson, and M Peacock, The cause of idiopathic calcium stone disease: hypercalciuria or hyperoxaluria?, *Nephron* 26:1055, (1980).
14. HO Senekjian and EJ Weinman, Oxalate transport by proximal tubule of the rabbit kidney, *Am J Physiol* 243:F271-F275, (1982).
15. LP Karniski and PS Aronson, Anion exchange pathways for Cl transport in rabbit renal microvillus membranes, *Am J Physiol* 253:F513F521, (1987).
16. RG Knickelbein, PS Aronson and JW Dobbins, Oxalate transport by anion exchange across rabbit ileal brush border, *J Clin Invest* 77:170175, (1986).
17. G Burckhardt and KJ Ullrich, Organic anion transport across the contraluminal membrane -dependence on sodium, *Kidney Int* 36:370377, (1989).
18. R Hautmann, A Lehmann and S Komor, Calcium and oxalate concentration in human renal tissue: The key to the pathogenesis of stone formation, *J Urol* 123:317, (1980).

OXALATE TRANSPORT IN LLC-PK1 CELLS: EVIDENCE FOR OXALATE TRANSPORT BY ANION EXCHANGE

S. Ebisuno[1], H. Koul[2], L. Renzulli [2], M. Menon[2] and C. Scheid[3]

[1]Division of Urology
Wakayama National Hospital
Wakayama, 646, Japan
[2]Division of Urology and Transplantation Surgery and
[3]Department of Physiology
University of Massachusetts Medical School
55 Lake Avenue North
Worcester, MA 01655

INTRODUCTION

Previous studies of renal oxalate handling have demonstrated that oxalate can be transported by a number of anion transport systems within the kidney including the SO_4^{2-}/oxalate exchanger on basolateral (abluminal) and apical (luminal) membranes[1,2] and the Cl^-/oxalate exchanger on apical membranes[3,4]. These studies demonstrated that pathway(s) exist for transcellular oxalate flux, but technical limitations with studies on membrane vesicles (limited timecourse for transport due to the small intravesicular volumes) and with micropuncture studies (limited control of the composition of the extracellular space) have made it difficult to predict the magnitude or direction of this flux under normal and pathological conditions. Thus the present studies have examined LLC-PK1 cells, a renal epithelial cell line with many characteristics of proximal tubular cells[5] as a possible model system for assessing renal oxalate handling.

METHODS

Cell Culture

LLC-PK1 cells (CRL 1392, American Type Culture Collection) were employed for uptake experiments between passage 206 and 225. The cells were maintained in Dulbecco's modified Eagle's minimal essential medium (DMEM) supplemented with 10% fetal bovine serum, glucose and antibiotics. For uptake studies, cells were subcultured into six well culture dishes and used at confluence (3-5 days after seeding).

Uptake Studies

Cultured cells were incubated in serum-free DMEM for at least 3 h prior to uptake studies. Cells were then washed with Hank's balanced salt solution (HBSS) and incubated for an additional 30 min with test buffer ±1 mM 4,4'-diisothiocyanostilbene-2,2-disulfonic acid (DIDS). After this preincubation period, the buffer was replaced with fresh buffer

Urolithiasis 2, Edited by R. Ryall *et al.*,
Plenum Press, New York, 1994

(±DIDS) containing ^{14}C oxalate at a final concentration of 100 µLM. Test buffer contained (in mM): NaCl (138), KCL (5.8), Na_2HPO_4/NaH_2PO_4 (0.78), $CaCl_2$ (0.25), $MgCl_2$ (1.0), HEPES (N-2-hydroxyethyl piperazine-N'-2-ethanesulfonic acid, 10), pH 7.4. For studies of the influence of SO_4^{2-} on oxalate uptake, KCl was replaced by K_2SO_4. For experiments to assess the effects of Cl- on oxalate uptake, NaCl, $MgCl_2$ and KCl were replaced by Na gluconate, $MgSO_4$ and K gluconate, respectively.

Cells were allowed to accumulate label at 25°C in room air for varying periods. The test buffers were then removed, cells were washed twice with ice cold isotonic saline, harvested and counted. Total ^{14}C associated with cells was corrected for trapped label (measured by assessing the carryover of ^3H-inulin). The DIDS-sensitive and hence carrier-mediated portion of uptake was calculated by subtracting the amount of label accumulated in the presence of 1 mM DIDS from the total uptake. Data were then normalized for the specific activity of the labelled oxalate and oxalate uptake was expressed as pmoles/10^6 live cells (cell density and viability determined separately).

Statistical Analysis

The data are the means±SEM for 3 - 9 separate experiments with 3 to 6 measurements in each experiment. Statistical analysis employed Statview (Brainpower Inc., Calabasas, California). ANOVA was used for multiple comparisons with a single control, Scheffe's test for all possible comparisons. The differences between groups were considered significant when the probability of chance occurrence was <0.05.

RESULTS

Timecourse and Concentration Dependence of Oxalate Uptake

LLC-PKl cells accumulated ^{14}C oxalate in a time- and concentration-dependent manner. Uptake plateaued between 2 and 3 h and a significant fraction of the total uptake was inhibited by DIDS (Figure 1, left panel). The DIDS-sensitive (and presumably carrier-mediated) portion of uptake exhibited saturation at oxalate concentrations ≥ 200 µM (Figure 1, right panel).

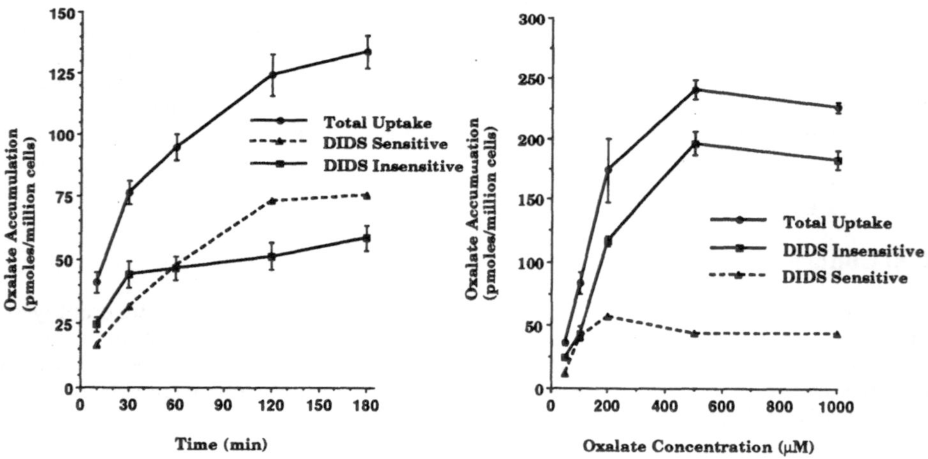

Figure 1. Timecourse and concentration dependence of oxalate uptake in LLC-PK1 cells.
Left panel: Cells accumulated label in a time-dependent manner. A large fraction of total uptake (filled circles) was inhibited by DIDS (squares). The residual DIDS-sensitive uptake (triangles) plateaued within 2 h. Data are the means±SEM for 9 measurements at each time point.
Right panel: Accumulation of label after 30 min at varying concentrations of oxalate. DIDS-sensitive uptake (triangles, dashed line) plateaued at ~100 µM oxalate. Data are means±SEM for 3-6 measurements.

Effect of Sulfate and Chloride on Oxalate Uptake

Previous studies using purified brush border and basolateral membrane fractions provided evidence for at least two types of oxalate carriers: one which exchanges oxalate for sulfate and one which exchanges oxalate for chloride[1,4]. Both transporters exhibit *cis* inhibition, eg, reduced uptake of oxalate when the counterion is added to the medium. Thus we determined whether or not oxalate uptake in LLCPKl cells would exhibit similar characteristics.

The results from these studies (Figure 2) indicated that DIDS-sensitive oxalate uptake was significantly reduced in the presence of 5 mM sulfate (left panel) and in the presence of varying concentrations of chloride (25-140 mM, right panel). These additions had no effect on DIDS-insensitive uptake (hatched bars), suggesting that the anions compete with oxalate for carrier-mediated transport.

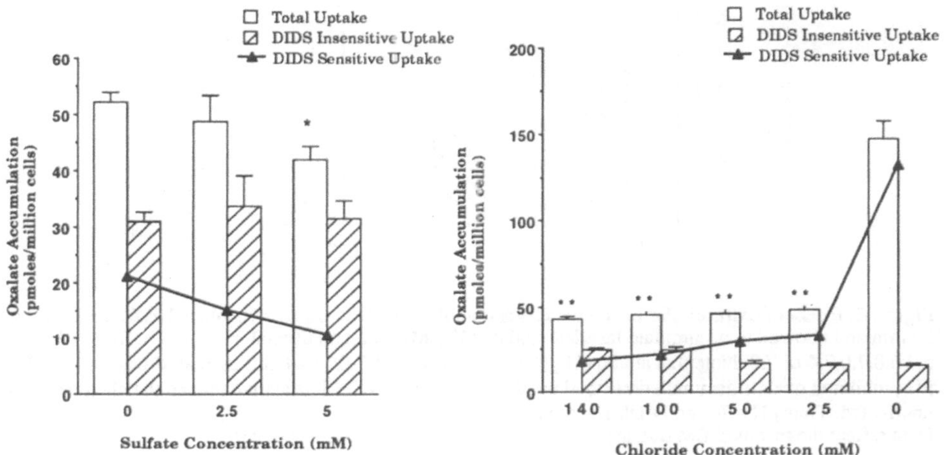

Figure 2. Effect of sulfate and chloride on oxalate accumulation.
Left panel: Oxalate uptake was examined after 30 min in the presence of 0, 2.5 or 5 mM sulfate. Note that 5 mM sulfate produced a significant inhibition of total uptake (open bars) but had no effect on DIDS-insensitive uptake (hatched bars) such that the DIDS-sensitive component of uptake (triangles) was reduced. Data are means±SEM from 6 measurements. * $p < 0.01$.
Right panel: Oxalate accumulation after 30 min was also inhibited in the presence of chloride. In chloride free buffer (chloride replaced by gluconate), total uptake (open bars) was significantly increased; DIDS sensitive uptake (hatched bars) was unaffected. Thus chloride appears to compete with oxalate for a DIDS-sensitive transporter (triangles).
Data are means±SEM from 6 determinations. ** $p < 0.005$.

Effect of pH on Oxalate Uptake

Previous studies on the chloride/oxalate exchanger of brush border membranes indicated that this transport system is affected by pH, with alkaline pH inhibiting and acidic pH promoting oxalate accumulation[3,4]. To determine the pH sensitivity of oxalate exchange in LLC-PKl cells, oxalate accumulation was assessed after 30 min in buffers adjusted to pH 6.6,7.0,7.4 or 7.8. Alkaline pH_o(7.8) produced a small but significant inhibition while acidic pH_o (pH 6.6 or 7.0) produced a significant stimulation of oxalate uptake. DIDS-insensitive uptake was unaffected by these manipulations (Figure 3).

Effect of Organic Acids and Diuretics on Oxalate Uptake

In the intact kidney, infusion of organic acids reduces oxalate clearance[6,7], and it was suggested that these organic acids may compete for a common carrier. In LLC-PKl cells, however, addition of various organic acids (0.1 to 5 mM malate, succinate, phenylsuccinate, urate or para-aminohippurate) to the preincubation and uptake buffers had no significant effect on oxalate uptake (data not shown). Other studies examined the effects of diuretics on oxalate accumulation[3]. Mersalyl, furosemide and chlorothiazide all produced a significant inhibition of oxalate uptake at mM concentrations; the apparent K_i for inhibition of oxalate uptake was 2.1, 1.75 and 4.33 mM for mersalyl, furosemide and chlorothiazide, respectively. Exposure to probenecid also produced a marked inhibition of oxalate accumulation in LLC-PKl cells, with half maximal inhibition seen at 1.48 mM. The carbonic anhydrase inhibitor acetazolamide had relatively little effect on oxalate uptake (Table 1).

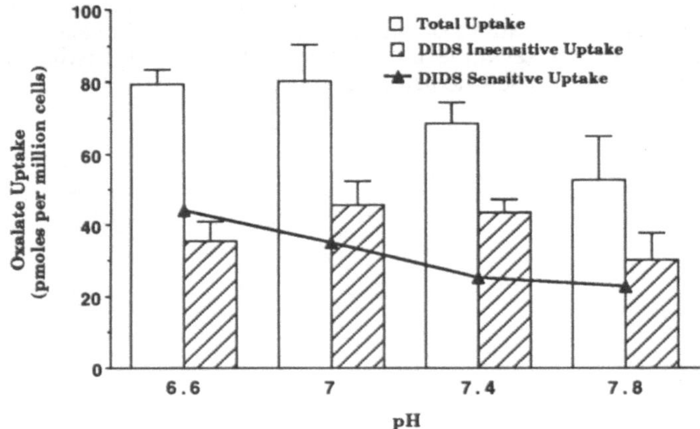

Figure 3. Effect of extracellular pH on oxalate accumulation. Cells were preincubated ±DIDS at pH 7.4 for 30 min and allowed to accumulate labelled oxalate (50 µM) for an additional 30 min in buffers adjusted to pH 6.8,7.0,7.4 or 7.8. Note that acidic pH_o (outwardly directed OH^- gradient) increased whereas alkaline pH_o inhibited oxalate accumulation (total uptake, open bars). DIDS insensitive uptake (hatched bars) was unaffected. Thus pH affected DIDS-sensitive uptake (triangles).
Data reflect the means±SEM obtained for 9 separate wells for each level of pH_o.

Table 1. Effect of Transport Inhibitors on Oxalate Uptake

	Per Cent Inhibition [a]	K_i (mM) [b]
DIDS	67.7±2.7	0.8
Furosemide	43.2±3.2	1.75
Probenecid	39.4±4.9	1.48
Mersalyl	42.8±1.8	2.10
Chlorothiazide	30.8±1.5	4.33

[a] Cells were preincubated with the transport inhibitors at a concentration of 1 mM for 30 min prior to the uptake studies. Labelled oxalate (50 µM) was then added and the amount of ^{14}C oxalate accumulated after 30 min was compared in treated and untreated cells. Data are the average inhibition observed in 6 separate wells. $p < 0.05$ for all treatments.

[b] Cells were preincubated for 30 min with varying concentrations (50 µM to 5 mM) of transport inhibitors and uptake was carried out (in the continued presence of drug and in the presence of 50 µM oxalate) for an additional 30 min. K_i values were determined using a Dixon plot.

DISCUSSION

Several lines of evidence from these studies indicate that oxalate is transported by LLC-PKl cells: (1) the accumulation of labelled oxalate occurred in a time- and concentration-dependent manner with uptake plateauing at ~200 μM oxalate (2) this accumulation was blocked by inhibitors of anion transport (DIDS and probenecid) (3) the accumulation was sensitive to inorganic anions (chloride, sulfate) and pH.

Indeed the present studies indicate that LLC-PKl cells express at least two DIDS-sensitive transport systems. One resembles the SO_4^{2-}/oxalate exchanger seen in luminal and basolateral membrane fractions from rat and rabbit renal cortex[1,2]. Transport on this carrier was inhibited in the presence of sulfate, but uptake was unaffected by Na^+ removal (data not shown) or by the addition of other anions (malate, succinate, urate, phenylsuccinate or para-aminohippurate). The other transport system resembles Cl^- (formate)/oxalate exchanger in rat and rabbit renal microvillus membranes[3,4]. Transport on this carrier was increased by removing chloride from the extracellular medium (eg. when there was an outwardly directed chloride gradient) and reduced by agents (furosemide or chlorothiazide) that inhibit chloride cotransport systems (eg, when the transmembrane chloride gradient was reduced). Moreover, oxalate uptake was stimulated when the external medium was acidic. This finding was of interest for several reasons. For one, oxalate uptake rates (at least in red cells) are increased in stone forming patients[8]. Secondly, aciduria increases the incidence and severity of calcium oxalate stones in experimental animals[9]. Finally many patients with stone disease exhibit defects in urinary acidification[10]. Thus it is interesting to speculate that an oxalate transport system of this type might be involved in calcium oxalate stone disease.

These studies clearly demonstrate that LLC-PKl cells express transport system(s) for oxalate with features similar to anion transporters in normal renal tubular cells and suggest that this cell line will serve as a valuable model for assessing the factors influencing renal oxalate transport.

ACKNOWLEDGEMENTS

Supported by NIH grant DK43184.

REFERENCES

1. C Bastlein and G Burckhardt. Sensitivity of rat renal luminal and contraluminal sulfate transport systems to DIDS, *Am J Physiol* 250:F226 (1986).
2. SM Kuo and S Aronson. Oxalate transport via the sulfate/HCO_3 exchanger in rabbit renal basolateral membrane vesicles, *J Biol Chem* 263:9710 (1988).
3. PS Aronson. The renal proximal tubule: a model for diversity of anion exchangers and stilbene-sensitive anion transporters, *Ann Rev Physiol* 51:419 (1989).
4. LP Karniski and PS Aronson. Anion exchange pathways for Cl- transport in rabbit renal microvillus membranes, *Am J Physiol* 253:F513 (1987).
5. RM Hull, WR Cherry and GW Weaver. The origin and characteristics of a pig kidney cell strain, LLC-PK1, *In vitro* 12:670 (1976).
6. R Greger F Lang, H Oberleithner and P Deetjen. Handling of oxalate by the rat kidney, *Pflugers Arch.* 374:243 (1978).
7. TF Knight, HO Senekjian and EJ Weinman. Effect of para-aminohippurate on renal transport of oxalate, *Kidney Int*15:38 (1979).
8. B Baggio, G Gambaro, F Marchini, E Cicerello, R Tenconi, M Clementi and E Borsatti, An inheritable anomaly of red-cell oxalate transport in primary calcium nephrolithiasis correctable with diuretics, *N Engl J Med* 314:599 (1986).
9. S Khan and RL Hackett. Calcium oxalate urolithiasis in the rat: is it a model for human stone disease? A review of recent literature, *Scanning Electron Micros* II:759 (1985).
10. N Tessitori, V Ortalda, A Gabris et al, Renal acidification defects in patients with recurrent calcium oxalate nephrolithiasis, *Nephron* 41:325 (1978).

POLARIZED DISTRIBUTION OF OXALATE TRANSPORT SYSTEMS IN A LINE OF RENAL EPITHELIAL CELLS (LLC-PK1)

H. Koul[1], L. Renzulli[1], G. Nair[1], C. Scheid[2] and M. Menon[1]

[1]Division of Urology and Transplantation Surgery
[2]Department of Physiology
University of Massachusetts Medical School
55 Lake Avenue North
Worcester, MA 01655

INTRODUCTION

Evidence from clinical[1] and experimental[2] studies demonstrated a significant increase in cellular oxalate handling associated with urolithiasis. These findings led to the speculation that altered oxalate handling within the kidney may contribute to stone formation. However, our current understanding of renal oxalate handling has been limited by a variety of technical difficulties. Studies on intact kidneys are compromised by limited access to various regions of the nephron, difficulty in regulating the ionic composition in these different regions, regional heterogeneity, etc. Biochemical studies on isolated membrane fractions from the kidney have other inherent difficulties (contamination of membrane fractions, small intravesicular volumes and hence limited timecourse for transport studies, etc). Thus our laboratory has explored the possibility of using a simpler model system, LLC-PKl cells, for studies on renal oxalate handling.

This cell line expresses many of the characteristics of normal renal epithelial cells including the formation of a polarized epithelium[3,4]. Moreover, our initial studies demonstrated that LLC-PKl cells express at least two different transport systems for oxalate, one that resembles the SO_4^{2-}/oxalate exchanger on luminal and abluminal membranes[5,6] and one that resembles the Cl^-/oxalate exchanger on luminal membranes from the renal cortex[7,8]. The precise cellular location, eg luminal and abluminal, could not be determined in earlier studies on LLC-PKl cells. Thus for the present studies LLC-PKl cells were grown on permeable membrane supports such that we could independently examine oxalate handling by the luminal and abluminal membrane surfaces. Such studies suggest a polarized distribution of oxalate transport systems in LLC-PKl cells.

METHODS

Cell Culture

LLC-PKl cells (CRL 1392, American Type Culture Collection) were employed for uptake experiments between passage 206 and 225. The cells were maintained in a humidified 37°C incubator in flasks containing Dulbecco's modified Eagle's minimal essential medium (DMEM) supplemented with 10% fetal bovine serum, glucose and antibiotics. For transport studies, cells were subcultured onto 25 mm Nunc membrane

Urolithiasis 2, Edited by R. Ryall *et al.*,
Plenum Press, New York, 1994

filters where they establish a functional epithelium[4]. The ability of these epithelial sheets to exclude ^{3}H-inulin was taken as evidence that the cells had achieved confluence. Only data from membranes with confluent monolayers were included in subsequent analyses.

Influx Studies

Cultured cells were preincubated for 3 h in Earle's balanced salt solution (EBSS). During the last 30 min, 4,4'-diisothio-cyanostilbene-2,2'-disulfonic acid (DIDS) or solvent vehicle was added to the medium. Transport studies were carried out at room temperature in a modified EBSS containing (in mM): NaCl (138), KCL (5.8), Na_2HPO_4/NaH_2PO_4 (0.78), $CaCl_2$ (0.25), $MgCl_2$ (1.0), HEPES (N-2-hydroxyethyl piperazine-N'-2-ethanesulfonic acid, 10), pH 7.4. Where indicated, sodium and potassium chloride were replaced by equimolar amounts of sodium and potassium gluconate. In other experiments, sulfate (10 mM) was added to the incubation medium.

For the uptake experiments described here, the solutions bathing both surfaces of the monolayer were identical. Uptake was initiated by the addition of ^{14}C-oxalate (50 μM) at either the luminal or abluminal surface. At various times after tracer addition, the membrane inserts containing the cells were removed, washed twice with ice cold buffer, drained on paper towels and counted using a liquid scintillation counter. Separate studies determined the fraction of trapped label associated with the monolayers by measuring the fraction of ^{3}H-inulin that remained associated with the membranes after washing and processing for counting. This fraction proved to be negligible; thus uptake data were not corrected for trapped label. Uptake was expressed as nmoles oxalate/10^6 cells (cell counts determined separately on comparable monolayer cultures).

Efflux Studies

Monolayer cultures were incubated with ^{14}C-oxalate for 2 h at 37°C and then transferred to unlabelled room temperature buffer. The loss of radioactivity from the cells was followed as a function of time by removing two or three membrane filters with adherent monolayers at each time point. Filters were then washed and counted. The rate of loss of ^{14}C oxalate from the cells was expressed as a fractional loss/min as follows:

$$\text{fractional loss} = A_t/A_o$$

where A_o = cellular label at time zero.
A_t = cellular label at time "t"

RESULTS

Effect of extracellular chloride on oxalate uptake

In one series of studies, monolayers were preincubated in chloride-containing buffer and oxalate uptake was assessed after acute removal of chloride from the extracellular medium. Under these conditions the chloride gradient was outwardly directed (eg $Cl^-_i > Cl^-_o$. Oxalate uptake at the luminal membrane surface was significantly greater in chloride-free buffer than in chloride-containing buffer (eg uptake was enhanced in the presence of an outwardly directed chloride gradient). This increase in oxalate uptake was inhibited by the anion transport inhibitor DIDS (Figure 1, right panel). Imposing a transmembrane chloride gradient had no effect on oxalate uptake when ^{14}C oxalate was added at the abluminal membrane surface, however (Figure 1, left panel).

Other studies determined whether the luminal oxalate uptake was sensitive to alteration in resting membrane potential. Abolition of the resting membrane potential by treatment with valinomycin (10 μM) in the presence of elevated K^+ (NaCl or KCl replaced isosmotically by Na gluconate or K gluconate) had no effect on luminal oxalate uptake either in the presence or absence of an outwardly directed chloride gradient (data not shown).

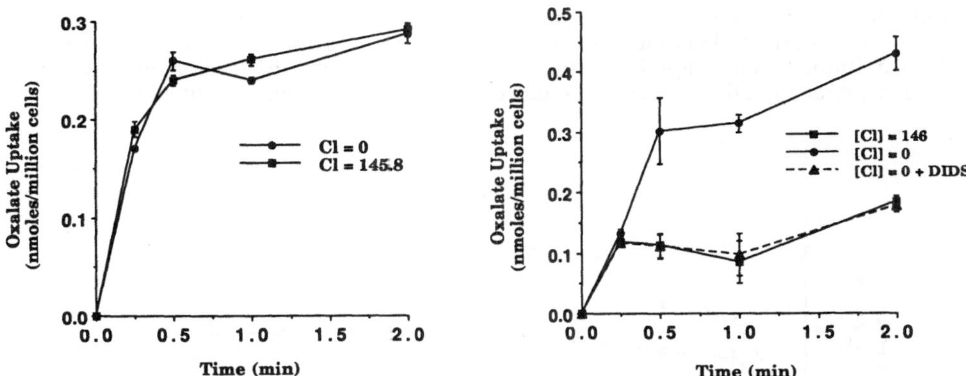

Figure 1. Effect of an outwardly directed chloride gradient on oxalate uptake in LLC-PKl cells. Cells were preincubated ± DIDS for 60 min in a chloride-containing buffer and then transferred to chloride-free buffer. Note that uptake at the luminal membrane surface (right panel) was greater in chloride-free buffer (circles) than in chloride-containing buffer (squares) and that preincubation with DIDS blocked this stimulatory effect (triangles, dashed line). Oxalate uptake at the abluminal membrane surface was unaffected by chloride removal (left panel).

Effect of extracellular chloride on oxalate release

Other experiments examined the release of ^{14}C oxalate from LLC-PKl cells that had been incubated for 60 min in chloride-free buffer containing ^{14}C oxalate (1 μCi/mL). The rate of release of label from the luminal membrane surface was then monitored in chloride-containing buffer and in chloride-free buffer. As is evident in Figure 2, the loss of label occurred more quickly when chloride was present in the washout buffer (squares, dashed line), eg. chloride produced a *trans* stimulation of oxalate efflux. This finding is consistent with the operation of a reversible oxalate/chloride exchanger at the luminal membrane surface.

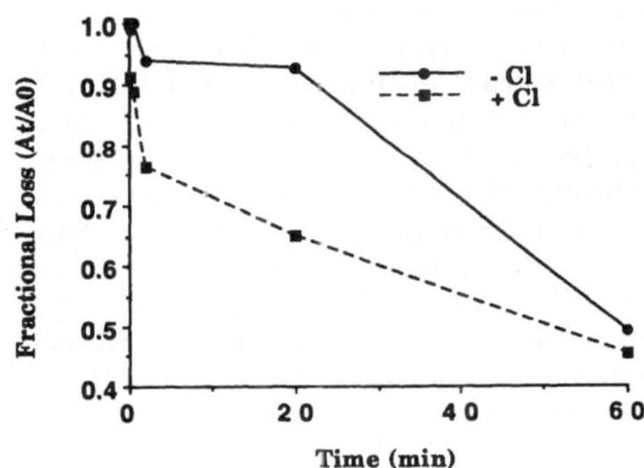

Figure 2. Chloride-induced increase in oxalate release from the luminal surface of LLC-PKl cells. Monolayers were prelabelled with ^{14}C oxalate and the loss of radioactivity from the cells was followed in chloride-containing (squares) and in chloride-free buffer (circles). Oxalate eflux was markedly increased in chloride-containing buffer (eg. when there was an inwardly directed chloride gradient).

Effect of extracellular sulfate on oxalate uptake

In another series of studies, monolayers were preincubated in buffer containing sulfate (10 mM), and oxalate uptake was examined in the presence and absence of an outwardly directed sulfate gradient. As is evident in Figure 3, oxalate uptake at the abluminal surface (left panel) was markedly increased when $[SO_4^{2-}]_i > [SO_4^{2-}]_o$ (eg. in sulfate free buffer). This increase in oxalate uptake was inhibited by DIDS (triangles, dashed line). Oxalate uptake at the luminal surface was unaffected by altering the transmembrane gradient for either sulfate or bicarbonate (Figure 3, right panel).

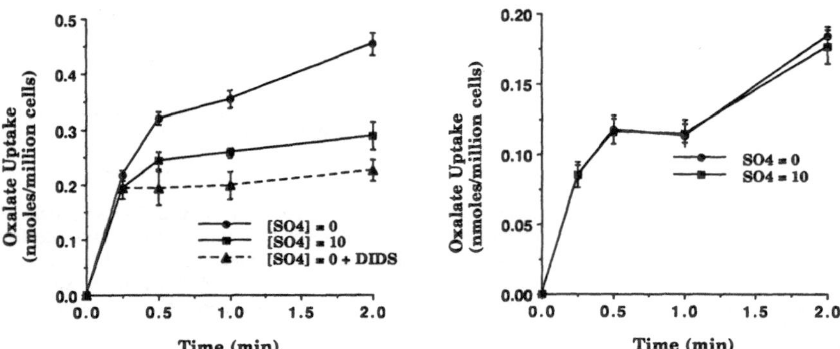

Figure 3. Effect of extracellular sulfate on oxalate uptake. Left panel: oxalate uptake at the abluminal surface. Note that uptake was significantly greater in sulfate-free buffer (circles) than in buffer containing 10 mM sulfate (squares) and that pretreatment with DIDS (squares, dashed line) abolished this stimulatory effect. Right panel: Oxalate uptake at the luminal surface. Note that uptake was virtually identical in sulfate-containing and in sulfate-free buffer.

DISCUSSION

The results from the present studies indicate that monolayer cultures of LLC-PKl cells express different oxalate transport systems at the luminal and abluminal membrane surfaces: a chloride-sensitive transport system at the luminal membrane surface and a sulfate-sensitive transport system at the abluminal surface. Moreover, the characteristics of the transport systems at the different membrane domains resemble those seen in purified membrane fractions from normal rat and rabbit kidneys where brush border membranes express a chloride/oxalate exchanger[7,8] and basolateral membranes express a SO_4^{2-} (HCO_3^-)/oxalate exchanger[5,6]. Specifically, transport on these carriers was sensitive to the anion transport inhibitor DIDS and insensitive to variation in membrane potential. In addition transport by these systems appeared to be reversible since the addition of chloride promoted oxalate release from the luminal membrane surface. Finally these transport systems exhibited both *cis* inhibition (eg, decreased uptake when both oxalate and the other transported anion were present on the same side of the membrane) and *trans* stimulation of oxalate uptake (eg, increased uptake when the counter ion was present on the opposite side of the membrane).

These observations clearly demonstrate that LLC-PKl cells possess a transcellular route for oxalate flux. Further studies using this cell line should prove useful for understanding the various factors that determine the magnitude and direction of net oxalate flux under physiological and pathological conditions.

ACKNOWLEDGEMENTS

Supported in part by NIH grants DK43184 (M Menon) and DK20586 (R Hackett, program project grant, University of Florida at Gainesville).

REFERENCES

1. B Baggio, G Gambaro, F Marchini, E Cicerello, R Tenconi, M Clementi and E Borsatti. An inheritable anomaly of red-cell oxalate transport in primary calcium nephrolithiasis correctable with diuretics, *N Engl J Med* 314:599 (1986).
2. D Sigmon, S Kumar, B Carpenter, M Menon and CR Scheid. Oxalate transport in renal cortical and papillary cells from normal and stone forming animals, *Am J Kidney Dis* 17:376 (1991).
3. RM Hull, WR Cherry and GW Weaver, The origin and characteristics of a pig kidney cell strain, LLC-PKI, *In vitro* 12:670 (1976).
4. HF Cantiello, JA Scott and CA Rabito. Polarized distribution of the Na^+/H^+ exchange system in a renal cell line (LLC-PKI) with characteristics of proximal tubular cells, J *Biol Chem* 261:3252 (1986).
5. C Bastlein and G Burckhardt. Sensitivity of rat renal luminal and contraluminal sulfate transport systems to DIDS, *Am* J *Physiol* 250:F226 (1986).
6. SM Kuo and PS Aronson. Oxalate transport via the sulfate/HCO_3 exchanger in rabbit renal basolateral membrane vesicles, J *Biol Chem* 263:9710 (1988).
7. PS Aronson. The renal proximal tubule: a model for diversity of anion exchangers and stilbene-sensitive anion transporters, *Ann Rev Physiol* 51:419 (1989).
8. LP Karniski and PS Aronson. Anion exchange pathways for Cl- transport in rabbit renal microvillus membranes, *Am J Physiol* 253:F513 (1987).

ENDOGENOUS SYNTHESIS OF OXALATE IN MAGNESIUM DEFICIENT WEANLING RATS

V. Rattan[1], S.K. Thind[1], R.K. Jethi[2], and R. Nath[1]

[1]Department of Biochemistry
Postgraduate Institute of Medical Education and Research
[2]Department of Biochemistry
Panjab University
Chandigarh, India

INTRODUCTION

Magnesium is a risk factor in urolithiasis. Studies have demonstrated the depressive effect of magnesium on endogenous oxalic acid formation[1]. Hypomagnesuria is a very common finding amongst stone formers[2]. A subclinical magnesium deficiency has been reported amongst general population from affluent countries, as the dietary intake of magnesium from the prepared meals is far below the recommended dietary allowance[3]. Although several experimental studies have shown a positive relationship between dietary and urinary magnesium and stone formation, the detailed biochemical mechanisms underlying magnesium deficiency has not been worked out. Therefore, the present study was carried out to clearly elucidate the effect of magnesium deficiency on endogenous synthesis of oxalate in liver as well as kidney in male weanling rats.

MATERIALS AND METHODS

Male weanling rats (Wistar strain) weighing 40-50g were randomly divided into magnesium-deficient and pair-fed control groups. The rats were fed a synthetic diet for 30 days, which consisted of (g/kg) casein 200, sucrose 326, dextrin-white 332, refined groundnut oil 80, salt-mixture (magnesium free) 40 and vitamin mixture 22. The salt-mixture and vitamin mixture were prepared as described previously[4]. The composition of the diets was essentially the same except for the incorporation of magnesium sulphate in the control diet. Magnesium content determined by analysis was 70mg/kg in deficient diet and 722mg/kg in the control diet.

Blood and 24 h urine samples were collected at the end of the experimental period. The rats were sacrificed under ether anesthesia. Liver and kidneys were removed, washed with cold normal saline and appropriately processed.

Blood and urine samples were analysed for calcium and magnesium. Urine samples were also analyzed for creatinine and oxalate. Magnesium was estimated by atomic absorption spectrophotometer (Perkin Elmer, 4000). Calcium, oxalate and creatinine were estimated as described previously [4].

The activities of various enzymes i.e. glycolic acid oxidase (GAO), glycolic acid dehydrogenase (GAD), alanine transaminase (ALT), lactate dehydrogenase (LDH) were measured as described previously[5]. GAO, GAD and ALT were assayed in liver, while

Urolithiasis 2, Edited by R. Ryall *et al.*,
Plenum Press, New York, 1994

LDH was assayed in both liver and kidney. Procedures followed for liver and kidney mitochondrial preparations, oxidation of [1-^{14}C]-glyoxylate to [^{14}C]-CO_2, α-ketoglutarate:glyoxylate (α- KG:GA) carboligase were as described earlier[6]. Horizontal cellulose acetate electrophoretic separation of kidney LDH isoenzymes was done by the method of Barnett[7]. The intensity of individual bands was read on a linear scanner and areas under the peaks were measured graphically. Individual isoenzyme activities were expressed as a percentage of total LDH activity. Statistical analysis of the data was done by using Student t test.

RESULTS

Feeding of magnesium-deficient diet for 30 days led to a significant ($p<0.001$) decrease in the body weights of experimental animals, despite pair-fed feeding, thereby suggesting that magnesium is indispensable for growth. Magnesium deficiency led to a significant ($p<0.001$) hypomagnesemia, hypercalcemia, hypomagnesuria, hyperoxaluria and hypocalciuria in the deficient rats compared to the pairfed control rats (Table 1).

Table 1. Effect of magnesium deficiency on plasma and urine magnesium and calcium and urinary oxalate in male weanling rats.

	Pair-fed control	Magnesium-deficient
Plasma magnesium (mmol/L)	0.83±0.01	0.40±0.01***
Plasma calcium (mmol/L)	2.44±0.02	3.07±0.02***
Urinary magnesium (mmol/mmol Cr)	0.87±0.08	0.22±0.04***
Urinary calcium (mmol/mmol Cr)	0.40±0.03	0.17±0.02***
Urinary oxalate (mmol/mmol Cr)	0.19±0.01	0.70±0.03***

All the values are mean ± SEM of 14 rats *** p <0.001 compared to pair-fed control rats

Magnesium deficiency caused significant increases in the specific activities of GAO ($p<0.001$) and GAD ($p<0.01$), and a significant decrease ($p<0.01$) in the specific activity of transaminase (ALT) in the liver of deficient rats compared to pair-fed control rats. The kidney LDH activity was found to be almost twice that of liver. The total LDH levels in the liver and kidney remained unaltered in the experimental group (Table 2).

The glyoxylate oxidation to CO_2 and α-KG:GA carboligase enzyme activity were manifested both in liver and kidney mitochondria. However the level of glyoxylate oxidation and enzyme activity was almost twice and three times more, respectively, in kidney mitochondria compared to liver mitochondria. In magnesium deficient rats, oxidation of glyoxylate was significantly decreased in both liver ($p<0.001$) and kidney ($p<0.01$) mitochondrial preparations. Similarly enzyme activity was also found to be significantly decreased ($p<0.01$) in both liver and kidney mitochondria compared to the pair-fed control group (Table 3).

Cellulose acetate electrophoresis of kidney LDH demonstrated the presence of five isoenzyme peaks LDH-I(H4) to LDHV (M4). Magnesium deficiency caused significant alterations in the relative distribution of LDH isoenzymes. A significant ($p<0.01$) increase in the relative abundance of LDH-I isoenzyme and a decrease ($p<0.01$) in the LDH-V isoenzyme activity in magnesium-deficient rat kidney compared to pair-fed control rats was observed (Table 4).

Table 2. Effect of magnesium deficiency on the specific activities of GAO, GAD, ALT and LDH in the liver and kidney of male weanling rats.

	Pair-fed control	Magnesium-deficient
Liver GAO[1] (U/mg protein)	4.12±0.06	4.69±0.08**
Liver GAD[2] (U/mg protein)	0.18±0.005	0.22±0.01
Liver ALT[3] (U/mg protein)	1.39±0.04	1.21±0.02**
Liver LDH[4] (U/mg protein)	1.15±0.03	1.18±0.04
Kidney LDH[4] (U/mg protein)	2.38±0.07	2.50±0.05

All the values are Mean±SEM of 6 rats ** $p < 0.01$, *** $p < 0.001$ compared to pair-fed control rats
One unit of GAO[1], GAD[2] and ALT[3] is defined as 1 nanomole of glyoxylate[1], oxalate[2] and 1 μmole of pyruvate[3] produced per minute at 37°C. One unit is defined as change in 0.1 OD/min at 340 nm and at 25°C

Table 3. Effect of magnesium deficiency on conversion of ^{14}Cglyoxylate to $^{14}C\text{-}CO_2$ and α–C-KG:GA carboligase activity in liver and kidney of male weanling rats.

Parameter	Liver		Kidney	
	Pair-fed control	Magnesium deficient	Pair-fed control	Magnesium deficient
Nanomoles of glyoxylate oxidised/min/mg protein x10^{-3}	2.25±0.05	1.65±0.07***	4.51±0.17	3.72±0.13**
α - KG:GA carboligase[1] (U/mg protein)	0.09±0.04	0.64±0.06**	2.97±0.16	2.14±0.13**

All the values are Mean±SEM of six rats ** $p < 0.01$, *** $p < 0.001$ as compared to pair-fed control rats
[1]One unit is defined as one nanomole of ^{14}C-glyoxylate decarboxylated per minute at 37°C.

Table 4. Effect of magnesium deficiency on percent distribution of kidney LDH isoenzymes in male weanling rats.

LDH - Isoenzyme	Pair-fed control	Magnesium-deficient
	(Percent distribution)	
LDH-I	26.34+0.46	29.60+0.69
LDH-II	23.61±0.81	23.62±0.79
LDH-III	12.99±0.87	14.50±1.14
LDH-IV	15.82±0.75	17.13±0.96
LDH-V	21.18±0.86	15.10±1.05

All the values are Mean±SEM of 5 rats **$p < 0.01$ compared to pair-fed control rats

DISCUSSION

Glycolate and glyoxylate are the two immediate precursors of endogenous oxalate. Glyoxylate is a very reactive compound. It can be converted to oxalate by GAO and LDH or can be reduced to glycolate, which is further converted to oxalate by GAD. Mitochondrial membrane is permeable to glyoxylate, where it is rapidly converted to CO_2 in the liver and kidney by TPP-dependent α-KG:GA carboligase[8] as well as via the glyoxylate oxidation cycle[9].

In the present study, glyoxylate decarboxylation in liver and kidney mitochondria was significantly decreased. The glyoxylate oxidation cycle is blocked due to both lack of TPP (coenzyme of α-KG dehydrogenase) and inhibition of key enzymes of this cycle by glyoxylate, which has been shown to be a potent inhibitor of 4-hydroxy-2-ketogluturate-aldolase[10] and 2-KG dehydrogenase[11]. The decarboxylation of glyoxylate by α-KG:GA carboligase was also significantly decreased in both liver and kidney mitochondria. This enzyme is associated with the α-KG decarboxylase moiety of the α-KG dehydrogenase complex[8]. A significant decreased thiamine content of liver and kidney has been reported in magnesium deficiency[12]. Thus due to lack of TPP, the activities of mitochondrial TPP-dependent enzymes are decreased. Thus decarboxylation of glyoxylate in mitochondria is significantly decreased, which leads to an increased pool in the tissues.

Glyoxylate is permeable to mitochondrial membrane and enters the cytosol. LDH-V isoenzyme (M4) of hepatic LDH (the predominant form) has a greater affinity towards glyoxylate and efficiently converts it into glycolate, which can enter peroxisomes. In the peroxisomes, glycolate is either directly converted to oxalate by GAD or is converted into glyoxylate. A significant decrease in the peroxisomal ALT was also observed in the present study, which also leads to increased glyoxalate production by D-amino acid oxidase. Glyoxylate produced from glycolate or glycine is efficiently converted into oxalate by GAO, which explains the increased activities of GAO and GAD in the present study.

A significant increase in the LDH-I isoenzyme with a significant decrease in the LDH-V isoenzyme was observed in the present study. Studies by Everse and Kaplan[13] suggested that H-LDH is responsible for the conversion of lactate or hydrated glyoxylate to pyruvate or oxalate and is distributed in tissues with predominantly aerobic metabolism, whereas, the M-subunit, which preferentially converts pyruvate or unhydrated glyoxylate to lactate or glycolate is prevalent in anaerobic tissue such as kidneys. Studies have shown increased permeability of protons of the inner membrane from mitochondria during magnesium deficiency[14], thus disrupting ATP production, thereby implying that magnesium-deficient rats use glycolytic pathways as the primary source of ATP leading to decreased ATP pool. The conversion of pyruvate to lactate and glyoxylate to glycolate is diminished as NADH produced during glycolysis is preferentially used to generate ATP via the electron transport chain. Hence the M-LDH unit is no longer required. The H and M subunit of LDH are synthesized by separate genes[15]. The shift in LDH distribution pattern suggests the repression of the M-gene with a concomitant induction of the H-gene by pyruvate or glyoxylate. This explains the decrease in LDH-V and increase in LDH-I distribution in deficient rat kidney. Studies have suggested that kidney peroxisomes contain significant amounts of L-α-hydroxy acid oxidase activity, which can act on thiol-glyoxylate adducts and oxidize them to oxalyl thioesters that get hydrolyzed to produce free oxalate.

Thus, it can be concluded that magnesium deficiency in rats leads to hyperoxaluria by significantly altering the endogenous production of oxalate both in the liver and kidney.

REFERENCES

1. P Brundig, W Berg and HJ Schneider, The influence of magnesium chloride on blood and urine parameters in calcium oxalate stone patients, *Eur Urol* 7: 97 (1981).
2. D Wangoo, V Rattan, SK Thind, GS Gupta and R Nath, Circadian rhythmicity in the urinary excretion of metallic ions (Mg, Cu, Fe and Zn) amongst stone formers (SF) and non-stone formers (NSF) in North -Western India (Abstract), *Magnesium Res* 4:211 (1991).

3. M Abdulla and F Reis, How adequate is the dietary intake of magnesium from a global point of view? (Abstract), *Magnesium Res* 4:248 (1991).
4. V Rattan, D Wangoo, HK Koul and SK Thind, Urinary excretion of lithogens and inhibitory activity towards calcium oxalate monohydrate crystallization in vitamin A deficient rats, *in*: "Urolithiasis Research," R Nath and SK Thind, ed, Ashish Publishers, New Delhi, India (1989).
5. S Sharma, H Sidhu, R Narula, SK Thind and R Nath, Comparative studies on the effect of Vitamin A,B, and B_6 deficiency on oxalate metabolism in male rats, *Ann Nutr Metab* 34:104 (1990).
6. H Sidhu, R Gupta, SK Thind and R Nath, Oxalate metabolism in thiamine-deficient rats, *Ann Nutr Metab* 31:354 (1987).
7. H Barnett, The staining of lactate dehydrogenase isoenzymes: Electrophoretic separation on cellulose acetate, *J Clin Path* 17:567 (1964).
8. MA Schlossberg, RJ Bloom, DA Richert and WW Westerfield, Carboligase activity of α-KG dehydrogenase, *Biochemistry* 9:1148 (1970).
9. EE Dekker and SC Gupta, Oxidation of L-2 keto-4hydroxyglutarate to malyl-CoA by α-ketoglutarate dehydrogenase and its role in a pyruvate catalyzed glyoxylate oxidation cycle (Abstract), *Fed Proc* 38: 2339 (1979).
10. SR Grady, JK Wang and EE Dekker, Steady state kinetics and inhibition studies of aldol condensation reaction catalyzed by bovine liver and E.Coli 2-keto-4hydroxyglutarate aldolase, *Biochemistry* 20:2497 (1981).
11. A Adinolfi, S Oilezza and A Ruffo, Inhibition of oxoglutarate dehydrogenase by glyoxylate and its condensation compounds, *J Biochem* 104:50p (1967).
12. Y Itokawa, K Inoue, Y Notori, K Okazaki and M Fujiwara, Effect of thiamine on growth, tissue magnesium, thiamine levels and transketolase activity in magnesium - deficient rats, *J Vitaminol* 18:159 (1972).
13. J Everse and NO Kaplan, Lactate dehydrogenase: Structure and function, *Adv Enzymol* 37:61 (1973).
14. FW Heaton and JP Elie, Metabolic activity of liver mitochondria from magnesium-deficient rats, *Magnesium Exp Clin Res* 3:21 (1984).
15. EJ Brush and GA Hamilton, Thiol-glyoxalate adducts as substrates for rat kidney L- α-hydroxy acid oxidase, *Biochem Biophys Res Commun* 103:1194 (1981).

INCREASED HEPATIC OXALATE PRODUCTION IN RATS TREATED WITH CLOFIBRATE

A.M. Rofe[1], R.A.J. Conyers[2] and R. Bais[1]

[1]Division of Clinical Chemistry
Institute of Medical & Veterinary Science
Adelaide, Australia
[2]Department of Biochemistry
Alfred Hospital
Prahran VIC Australia

INTRODUCTION

Peroxisomes are thought to play an important role in hepatic metabolism[1,2]. In oxalate stone disease where urinary oxalate is derived mainly from endogenous metabolic processes[3], peroxisomes are important in that they are a site of oxalate production[4] with glycollate oxidase, amino acid oxidase, glyoxylate:alanine aminotransferases and catalase being present in these organelles[2,5-7]. The hypolipidaemic drug, clofibrate, causes both peroxisomal proliferation and alterations in the activity of enzymes located in these organelles, particularly catalase[1,8,9]. Clofibrate therefore offers a means of better understanding the regulation of hepatic oxalate production. In this study we describe the effect of clofibrate administration on the changes in enzyme activities and the corresponding effects on oxalate production in hepatocytes isolated from clofibrate-treated rats.

MATERIALS AND METHODS

Clofibrate (ethyl, 2[4-chlorophenoxy], 2-methyl propionate) was purchased from ICI Australia, [U-^{14}C] glyoxylate, [l-^{14}C] glycollate and [U-^{14}C] oxalate from Amersham, Australia and all other biochemicals from Calbiochem or Sigma, USA.

Male Porton derived rats (300-330 g) were given daily subcutaneous injections of clofibrate (500 mg/kg body wt) for 10 days. These and the control animals were fed a standard rat diet during this period. Following an overnight fast, rats were anaesthetised with pentobarbitone (50 mg/kg body wt ip) and hepatocytes were prepared by collagenase digestion[10,11]. Hepatocytes were suspended in Krebs-bicarbonate buffer at a concentration of 5 x 10^6 cells/mL and incubated in a final volume of 1 mL with 0.5 µCi/2µmol [l-^{14}C] glycollate; (0.5 µCi/5µmol [U-^{14}C] glyoxylate, or 0.5 µCi/5µmol, [U-^{14}] glycine. After 60 minutes at 37° in a shaking water bath, the reactions were terminated with 1.5 mL of 1.0 M potassium citrate, pH 3.0. One mL of this mixture was assayed immediately for labelled CO_2 and oxalate, using oxalate decarboxylase[10,11].

A portion of the hepatocyte suspension was homogenised (Sorvall Omnimix) in 10 volumes of ice-cold 10mmol/L Na-phosphate buffer, pH 7.5 containing 0.25M sucrose (referred to as sucrose/phosphate buffer). Enzymes were assayed by established methods;

glycollate oxidase[5], catalase and succinate oxidase[6], alanine:α-ketoglutarate aminotransferase[12], urate oxidase[13], and lactate dehydrogenase, acid phosphatase and γ-glutamyltranspeptidase by automated procedures on a Technicon SMAC analyser. To partially purify the oxalate synthesising activity, the precipitate from 40-60% ammonium sulphate fractionation of liver homogenate containing mitochondria, peroxisomes and cytosol in sucrose/phosphate buffer was chromatographed on CM sephadex. The protein peak contained the major oxalate synthesising activity with glycollate oxidase and catalase as the only significant enzyme activities. Oxalate and CO_2 assay conditions were as described for hepatocyte suspensions with assays being performed in duplicate.

RESULTS

Clofibrate treatment significantly decreased the hepatic activity of urate oxidase and increased that of catalase, succinate oxidase, acid phosphatase and γ-glutamyl transpeptidase (fig 1). The activities of glycollate oxidase, alanine:α-ketoglutarate aminotransferase and lactate dehydrogenase were not significantly changed. Clofibrate also caused a 30% increase in liver weight, together with an increase in total protein concentration per g wet weight of liver. The enzyme activities in this study have been expressed per mL of hepatocyte suspension as these then relate directly to the conditions under which the changes in oxalate and CO_2 production were observed.

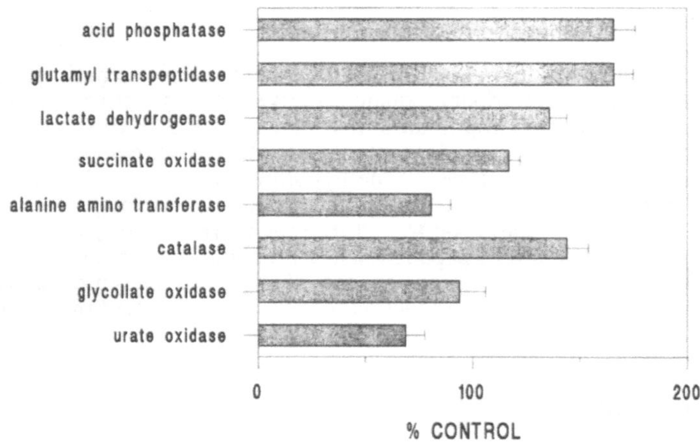

Figure 1. Effect of clofibrate treatment on liver enzyme activities. The results are shown as the mean±SEM (n = 4), * $p < 0.05$ compared to control (100%).

Effect of clofibrate treatment on oxalate and CO_2 production

Oxalate production from glyoxylate increased by 50% in hepatocytes prepared from clofibrate treated rats, while that from glycollate was not significantly changed (+19% compared to untreated rats) (fig 2). The production of CO_2 from both glyoxylate and glycollate was significantly decreased (-34% and -40%, respectively) in hepatocytes from clofibrate-treated rats.

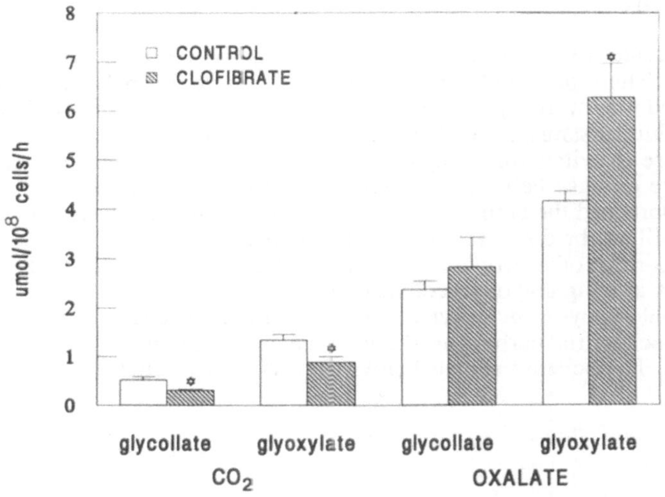

Fig 2. Effect of clofibrate treatment on oxalate and CO_2 production from glycollate and glyoxylate in hepatocytes. The results are shown as the mean±SEM (n = 4), * $p < 0.05$ compared to control.

Inhibition of catalase: Effect on oxalate and CO_2 production

The increased oxalate and decreased CO_2 production observed following clofibrate treatment were associated with increased hepatic catalase activity. To examine further the role of catalase in oxalate production, catalase was inhibited in a purified liver preparation which contained glycollate oxidase and catalase as the major enzyme activities, and therefore excluded other confounding factors, including carboligase and aminotranferase activity. Inclusion of azide (1 mM), or heating this preparation to 56° for 15 min abolished catalase activity without affecting glycollate oxidase. Both treatments resulted in a marked increase in CO_2 production and decreased oxalate production from glycollate and glyoxylate (fig 3), the reverse of that seen with clofibrate treatment where catalase activity was increased.

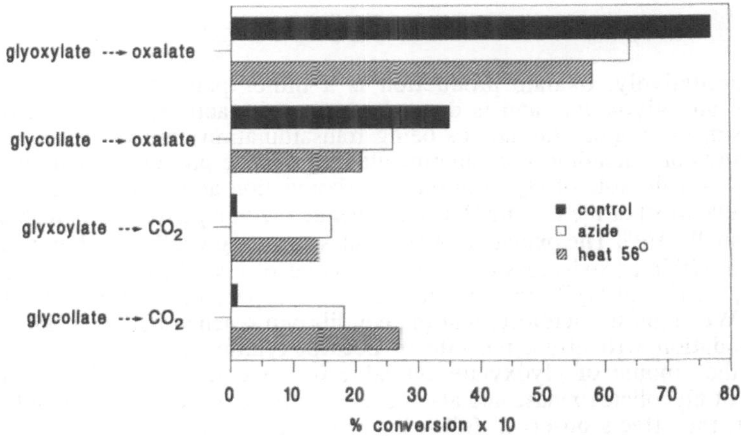

Figure 3. Effect of inhibiting catalase activity on oxalate and CO_2 production from glycollate and glyoxylate in a purified liver fraction containing glycollate oxidase and catalase.

DISCUSSION

Clofibrate administration can alter the activities of a wide spectrum of liver enzymes[14]. The reported effects of clofibrate on peroxisomal enzyme activities appear variable, and may well depend on the route of administration, the type of rat used, its age and its nutritional status. An early report[8] showed an increase in catalase and a decrease in urate oxidase activities following clofibrate treatment. Both enzymes are peroxisomal in origin, urate oxidase being associated with the peroxisomal core. In the light of these varying reports and the comparative nature of our study, the changes observed in enzyme activities will not be discussed further. The important observation, and one in agreement with the reports of other workers[1,8,9] was that clofibrate increased catalase activity without significantly altering that of glycollate oxidase.

Ultimately, cytosolic enzymes such as xanthine oxidase and lactate dehydrogenase, and the mitochondrial carboligase[15] should be considered in the overall control of oxalate production. The scheme shown below (fig. 4) is restricted, however, to peroxisomal enzymes.

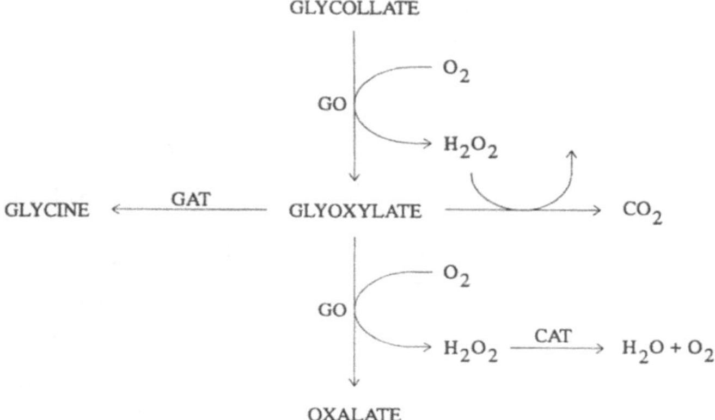

Fig. 4. Scheme for oxalate production in peroxisomes where hydrogen peroxide catalyses the non-enzymatic decarboxylation of glyoxylate. GO, glycollate oxidase; AT, aminotransferase.

Quantitatively, oxalate production is a minor pathway in the metabolism of glycollate and glyoxylate and is dependent upon the activity of alternative routes of metabolism, the major alternatives being transamination and decarboxylation. In this context, any consideration of compartmentalised oxalate production within peroxisomes must allow for the role of H_2O_2 in the decarboxylation and oxidation of glycollate and glyoxylate[16]. Most of the hepatic H_2O_2 is generated within peroxisomes with little leaving these organelles[17,18]. The oxidation of substrates such as glycollate and urate causes H_2O_2 generation within peroxisomes due to the activities of glycollate and urate oxidases[1,2,17]. However, the rate of H_2O_2 accumulation will also depend on the rate of its utilisation by catalase. We suggest, therefore, that the equilibrium reached between H_2O_2 production and degradation will affect the rate of decarboxylation of glyoxylate and therefore regulate the amount of glyoxylate available for oxidation to oxalate. Given that the activities of glycollate oxidase and alanine aminotransferase were not altered by clofibrate treatment, the effects observed following clofibrate treatment are consistent with this mechanism. The increased CO_2 and decreased oxalate production from glycollate and glyoxylate following the inhibition of catalase further support the involvement of catalase and H_2O_2 in hepatic oxalate production.

The use of drugs which alter peroxisomal enzyme activities may provide a fruitful area for further research into the control of oxalate production, particularly with regard to regulating the alternative pathways of glyoxylate metabolism[19]. The clinical use of clofibrate may also need to be considered as an additional risk factor in oxalate stone disease.

ACKNOWLEDGEMENTS

This work was supported by the National Health and Medical Research Council of Australia.

REFERENCES

1. PB Lazarow, Peroxisomes *in*: "The Liver: Biology and Pathology", 2nd Edit. Eds IM Arias, WB Jakoby, H Popper, D Schachter, DA Shafritz, Raven Press Ltd NY. p241 (1988).
2. NE Tolbert, Metabolic pathways in peroxisomes and glyoxysomes, *Ann Rev Biochem* 50:133 (1981).
3. R Bais, AM Rofe and RAJ Conyers, The inhibition of metabolic oxalate production by sulfhydryl compounds, *J Urol* 145:1302 (1991).
4. DA Gibbs, S Hauschildt and RWE Watts, Glyoxylate oxidation in rat liver and kidney, *J Biochem* 82:221 (1977).
5. LL Liao, and KE Richardson, The inhibition of oxalate biosynthesis in isolated perfused rat liver by DL-phenyllactate and n-heptanoate, *Arch Biochem Biophys* 154:68 (1973).
6. E McGroarty, B Hsieh, DM. Wied, R Gee, and NE Tolbert, Alpha-hydroxy acid oxidation by peroxisomes, *Arch Biochem Biophys* 161:194 (1974).
7. A Hand, Ultrastructural localisation of L-hydroxy acid oxidase in rat liver peroxisomes, *Histochem* 41:196 (1975).
8. R Hess, W Staubli and W Riess, Nature of the hepatomegalic effect produced by ethyl-chlorophenoxyisobutyrate in the rat, *Nature* 208:856 (1965).
9. H Hayashi, T Suga and S Niinobe, Studies on peroxisomes, V Effect of ethyl-p-chlorophenoxyisobutyrate on the centrifugal behaviour of rat liver peroxisomes, *J Biochem* 77:1199 (1975).
10. AM Rofe, AH Chalmers and JB Edwards, [^{14}C] oxalate synthesis from [U^{14}C] glyoxylate and [1^{14}C] glycollate in isolated rat hepatocytes, *Biochem Med* 16:277 (1976).
11. AM Rofe, HM James, R Bais, JB Edwards and RAJ Conyers, The production of [^{14}C] oxalate during the metabolism of [^{14}C] carbohydrates in isolated rat hepatocytes, *Aust J Exp Biol Med Sci* 58:103 (1980).
12. HU Bergmeyer, and E Bernt, *in* "Methods of Enzymatic Analysis", HU Bergmeyer, ed, *Acad Press* NY. p 846 (1963).
13. F Leighton, B Poole, H Beaufay, P Baudhin, JW Cofey, S Fowler & C DeDuve, The large scale separation of peroxisomes, mitochondria and lysosomes from the liver of rats injected with triton WR-1339, *J Cell Biol* 37:482 (1968).
14. D Zakim, RS Paradini and RH Herman, Effect of clofibrate (ethylchlorophenoxyisobutyrate) feeding on glycolytic and lipogenic enzymes and hepatic glycogen synthesis in the rat, *Biochem Pharmacol* 19:305 (1970).
15. JV O'Fallon and RW Brosemer, Cellular localization of ketoglutarate: glyoxylate carboligase in rat tissues, *Biochim Biophys Acta* 499:321 (1977).
16. B Halliwell and VS Butt, Oxidative decarboxylation of glycollate and glyoxylate by leaf peroxisomes, *Biochem J* 138:217 (1974).
17. A Boveris, N Oshino and B Chance, The cellular production of hydrogen peroxide, *Biochem J* 128:617 (1972).
18. B Poole, Diffusion effects in the metabolism of hydrogen peroxide by rat liver peroxisomes, *J Theor Biol* 51:149 (1975).
19. R Bais, AM Rofe and RAJ Conyers, Investigations into the effect of glyoxylate decarboxylation and transamination on oxalate formation in the rat, *Nephron* 57:460 (1991).

HYPEROXALURIA IN NUTRITIONAL DEFICIENCIES OF VITAMINS AND MAGNESIUM IN IDIOPATHIC CALCIUM OXALATE UROLITHIASIS

S.K. Thind and R. Nath

Department of Biochemistry
Postgraduate Institute of Medical Education and Research
Chandigarh, India

INTRODUCTION

In developed/still developing countries, the syndrome of calcium oxalate (CaOx) urolithiasis prevails in 70-80% of stone formers (SF)[1]. Hyperoxaluria occurs in 15-50% in CaOx SF[2-4]. Marginal deficiencies of magnesium[5] and vitamins[6] in calculous patients is very common in developing countries. The present study was conducted to assess the effect on the endogenous biosynthesis, absorption and excretion of oxalate in vitamin (A/B1/B6) and Mg deficient rats. The clinical status of these deficiencies in CaOx SF and the beneficial effects of combined oral administration of pyridoxine-HCl and magnesium oxide in lowering hyperoxaluria in CaOx SF was also evaluated.

MATERIALS AND METHODS

Experimental Study

Male wistar rats of 50-60g BW were divided into 8 groups of 8-10 rats each and fed *ad libitum* respective diets for specified periods as follows:

Group I	-	Vitamin A-deficient diet, for 5 weeks
Group II	-	Pair-fed with Group I +150IU retinyl acetate/rat/day by gastric intubation (g.i.)
Group III	-	Vitamin B_1-deficient diet, for 4 weeks
Group IV	-	Pair-fed with Group III +100μg thiamine HCl/rat/day by g.i.
Group V	-	Vitamin B_6 deficient diet, for 4 weeks
Group VI	-	Pair-fed with Group V +100 μg pyridoxine/HCl/rat/day by g.i.
Group VII	-	Mg-deficient diet (Mg 70mg/kg diet/day), for 4 weeks
Group VIII	-	Pair-fed with Group VII (Mg 722mg/kg diet/day)

Food consumption and weight gain of individual rats were measured every alternate day till the end of the experimental period when clinical symptoms of dietary deficiency became apparent. Before sacrificing, 24 h urine samples were individually collected and analysed for calcium, phosphorus, oxalate and uric acid, according to the routine methods of this laboratory. The animals were sacrificed under light ether anaesthesia. The vitamin A, B_1 and B_6 status was ascertained biochemically[7]. Magnesium was measured by Atomic Absorption Spectrophotometry (Perkin Elmer, 4000). The activities of oxalate-biosynthesizing enzymes viz. glycolate oxidase (GAO), glycolate dehydrogenase (GAD) and lactate dehydrogenase (LDH) were assessed as reported

Urolithiasis 2, Edited by R. Ryall *et al.*,
Plenum Press, New York, 1994

earlier[7]. GAO, GAD and LDH were estimated in the liver only, estimated by Lowry's method as described earlier[7].

Clinical Study

Vitamin (A, B_1,B_6) and magnesium status was biochemically assessed in the blood of 50 idiopathic CaOx SF (30-50 years) visiting the Department of Urology, Nehru Hospital, Chandigarh and 25 healthy subjects/cohorts. The assay methods were same as stated earlier except blood vitamin A which was measured by the method of Neeld and Pearson[9]. Combined supplementation of pyridoxine-HCl (10mg/day) and MgO (300mg/day) was administered periodically (0, 30, 60, 90 and 120 days) to CaOx SF and its effect on 24 h urinary excretion of calcium, oxalate, phosphorus, citrate, uric acid, creatinine, GAGs and plasma Mg was assayed by routine methods.

RESULTS AND DISCUSSION

The highly significant ($p< 0.001$) decrease in the body and liver weights in vitamin B_1 and Mg deficient rats and lesser ($p< 0.05$) in vitamin A and B_6 deficiencies, could be postulated due to "metabolic starvation" and lesser assimilation/utilisation of administered food as well as their respective pair-fed controls.

The vitamin and magnesium deficiencies biochemically assessed in the rats at the end of the experimental period were manifested by 95-96% depletion of hepatic vitamin A reserves, significant decreases ($p< 0.001$) both in erythrocyte transketolase (in vitamin B_1 deficiency) and alanine transaminase activities (in vitamin B_6 deficiency) compared with their respective controls.

Significant hyperoxaluria was observed in all these deficiencies in the order of Mg > vitamin B_6 > vitamin A > vitamin B_1) (Fig.1), though their mode of action on oxalate metabolism seems to be quite different.

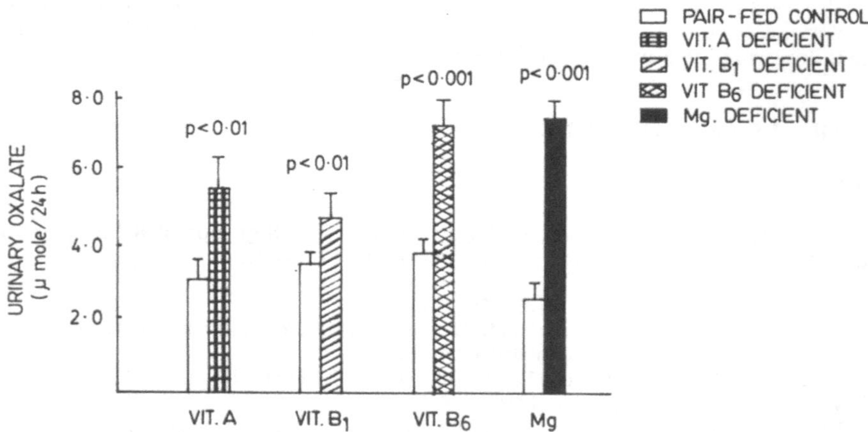

Fig.1 Urinary excretion of oxalate in vitamin A, B_1 and B_6 and magnesium deficient rats Values are Mean ±SEM of 8-10 animals.

The effect of vitamin or magnesium deficiency on oxalate metabolism in rats was seen by markedly enhanced hepatic activities of GAO in all deficiencies ($p< 0.001$), GAD in vitamin A ($p< 0.001$), vitamin B6 and Mg ($p< 0.01$) deficiencies, while LDH remained unaltered in both liver and kidney (Fig.2).

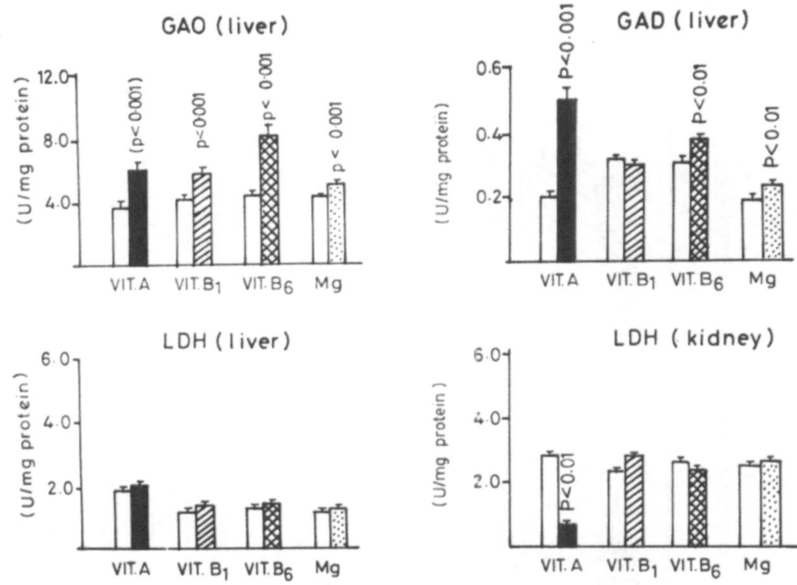

Fig.2 Activities of GAO, GAD and LDH in vitamin (A,B1 B6) and magnesium deficient rats and their pair-fed controls. Each value is mean±SD of 6-8 observations. One unit of GAO is defined as 1 nmol of glyoxylate produced per minute at 37°C; one unit of GAD is defined as 1 nmol of oxalate produced per minute at 37°C; one unit of LDH is defined as change in 0.1 optical density per minute at 340nm at 25°C.

Glycine and presumably other amino acids are oxidized for energy by thiamine-deficient rats[10], leading to increased glyoxylate formation which can induce synthesis of GAO to produce less harmful oxalate. The activity of α-ketoglutarate: glyoxylate carboligase, a thiamine-pyrophosphate-dependent enzyme, is significantly decreased in thiamine deficiency, thus blocking the conversion of glyoxylate to CO_2 whereas, excess glyoxalate is converted to oxalate by LDH and GAD or excreted as such[11]. In thiamine-deficient rats, GAD does not play any significant role, because glyoxylate is a potent inhibitor of GAD. The unaltered behaviour of the kidney and liver LDH supports our earlier reports[7], thus leading to the conclusion that hyperoxaluria in this deficiency is of endogenous origin, as indicated by similar intestinal oxalate uptake rates in thiamine-deficient and pair-fed control groups[7]. The alterations in the uptake rate of oxalate in vitamin B_6 deficiency are attributable to the induction of a new biphasic transport carrier, which facilitates its passage across the enterocyte microvillus membrane[12]. The accumulation of glycine, tyrosine and tryptophan in vitamin A deficiency which are important endogenous precursors of oxalate may be inducing the activity of GAO and GAD (Fig.2), leading to hyperabsorption of oxalate[7] and inhibition of the activity of kidney LDH, since some isoenzymes of LDH are known to be sensitive to the higher concentrations of its product[13]. In the clinical study, no differences in plasma vitamin A levels were noted, although vitamin B6 status (%PALP) stimulation) was significantly ($p< 0.05$) higher in SF. The 24 h urinary excretion of lithogenic constituents (calcium, oxalate, uric acid) were each significantly ($p< 0.001$) increased, while the inhibitory substances (citrate, Mg and GAGs) were significantly ($p< 0.001$) decreased in SF. Pyridoxine and magnesium supplementation have been reported[14,15] to be beneficial in the prevention of recurrent formation of CaOx renal stones. Combined supplementation (MgO 300mg) pyridoxine-HCl 10mg/d administered for 120 days had additive beneficial effect compared to only pyridoxine supplementation[14] by significantly ($p< 0.001$) decreasing

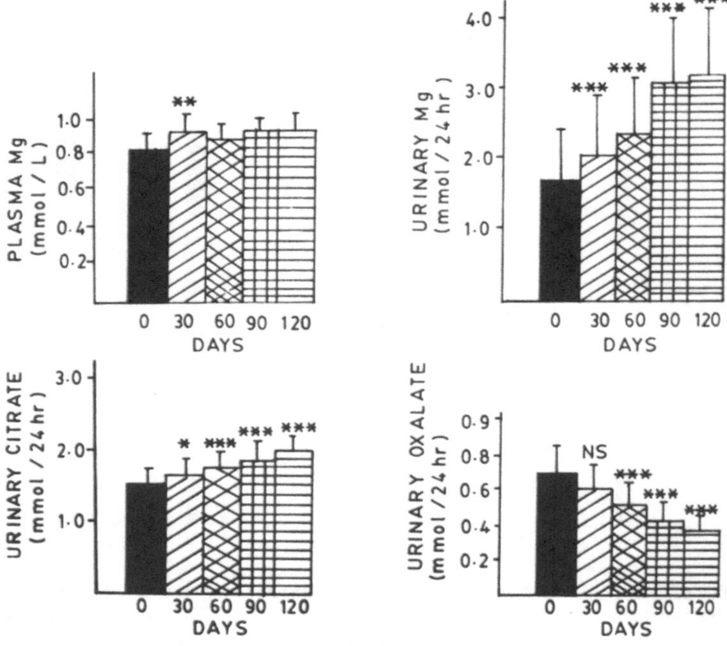

Fig.3 Effect of magnesium oxide (300mg/day) and pyridoxine-HCl (l0mg/day) administered to stone formers (n=16) on the plasma magnesium and urinary magnesium, citrate and oxalate. All the values are Mean±SD. *$p < 0.05$, ** $p < 0.01$, *** $p < 0.001$ compared to 0 day

oxalate excretion and concomitantly increasing ($p < 0.001$) Mg, GAGs and citrate excretion (Fig.3) in SF, thereby lowering risk factors of CaOx stone formation.

Thus the present study emphasizes that the deficiencies of these micronutrients may increase oxalate biosynthesis thereby promoting hyperoxaluria.

REFERENCES

1. LH Smith. Urolithiasis, *in.*"Diseases of the kidney", R Schrier and C Gottschalk, Little Brown, Boston (1983).
2. WG Robertson and M Peacock, The cause of idiopathic calcium stone disease: hypercalciuria or hyperoxaluria? ,*Nephron* 26:105 (1980).
3. LH Smith, PG Werness, SB Erickson et al, Postprandial response to a normal diet in patients with idiopathic calcium urolithiasis, *in* "Urinary stone", RL Ryall, JG Brockis, VR Marshall et al Churchill Livingstone, New York (1984).
4. L Larsson and H-G Tiselius, Hyperoxaluria, *Mineral Electrolyte Metab.*13:242 (1987).
5. M Abdulla and R Fatima, How adequate is the dietary intake of magnesium from global point of view?, *Magnesium Res.* 4:248 (1991).
6. A Anasuya, Nutritional factors in urolithiasis: Recent studies, *in* "Multidimensional Approach to Urolithiasis" PP Singh and AP Pendse, Agarwal Printers, Udaipur (India) (1987).
7. S Sharma, H Sidhu, R Narula, SK Thind and R Nath, Comparative studies on the effect of vitamin A, B_1 and B_6 deficiency on oxalate metabolism in male rats, *Ann Nutr Metab* 34:104 (1990).
8. R Gupta, H Sidhu, V Rattan, SK Thind and R Nath, Oxalate uptake in intestinal and renal brush border membrane vesicles (BBMV) in vitamin B_6 deficient rats, *Biochem Med and Metabol Biol* 39:190 (1988).
9. JB Neeld and WN Pearson, Macro and micromethods for the determination of serum vitamin A using trifluoroacetic acid, *J Nutr* 79:454 (1963).

10. CC Liang, Alternative metabolic pathway of rats suffering from thiamine deficiency, *J Nutr Sci Vitamino* l.22 (Suppl.):47 (1976).
11. H Sidhu, R Gupta, SK Thind and R Nath, Oxalate metabolism in thiamine-deficient rats, *Ann Nutr and Metabol* 31:354 (1987).
12. S Farooqui, R Nath, SK Thind and A Mahmood, Effect of pyridoxine deficiency on intestinal absorption of calcium and oxalate, Chemical composition of brush border membranes in rats, *Biochem Med* 32:34 (1984).
13. SK Thind and R Nath, Lactate dehydrogenase isoenzymes in the rat kidney and bladder in experimental urolithiasis, Int Res Comm Services (IRCS) *Med Sci* 5:429 (1977).
14. MSR Murthy, S Farooqui, HS Talwar, SK Thind, R Nath, L Rajendran and BC Bapna, Effect of pyridoxine supplementation on recurrent stone formers, *International J Clin Pharmacol Therapy and Toxicol* 20:434 (1982).
15. K Jarrar, V Graef and RH Boedeker, Magnesium therapy in calcium oxalate stone patients, *in*: "Urolithiasis" VR Walker, RAL Sutton, ECB Cameron, CYC Pak, WG Robertson, Plenum Press, New York (1989).

REDUCTION OF URINARY OXALATE EXCRETION BY THE ORAL ADMINISTRATION OF THE CYSTEINE PRECURSOR, 2-OXOTHIAZOLIDINE-4-CARBOXYLIC ACID

P. Baker, M.H. Tilley, A.M. Rofe and R. Bais

Division of Clinical Chemistry
Institute of Medical and Veterinary Science
Frome Road
Adelaide, Australia

INTRODUCTION

Calcium oxalate is a major component of over 70% of renal stones[1-3] and small changes in urinary oxalate concentration significantly change the risk of stone formation [1]. Although the diet contributes to the formation of such stones, greater than 80% of urinary oxalate derives from endogenous metabolism [2]. Studies in our laboratory suggest that the control of oxalate production is in the common pathway where the immediate precursors of oxalate, glycollate and glyoxylate are metabolized[4,5]. We have recently shown that a number of sulfhydryl compounds including cysteine, N-acetyl cysteine, dithiothreitol, thioglycollate, 2-mercaptoethanol and captopril inhibit CO_2 and oxalate production from glyoxylate by rat liver homogenates and hepatocytes[6]. The most significant inhibition occurred with cysteine. In rats made hyperoxaluric by administering ethylene glycol in their drinking water, daily intraperitoneal injections of (L)-cysteine rapidly and markedly decreased urinary oxalate excretion[6]. The level of oxalate excretion in these ethylene glycol-treated rats was reduced to that of the controls. We have postulated that this decrease is due to the formation of a (L)-cysteine-glyoxylate adduct, thiazolidine-(L and D)2,(L)4-dicarboxylic acid which prevents glyoxylate being further oxidized to oxalate. One aim of this study was therefore to prepare and characterize this putative adduct.

Although (L)-cysteine caused the greatest effect on oxalate production both *in vivo* and *in vitro,* it can cause necrosis of neurones and weight loss when taken orally, and disseminated neuronal degradation, brain atrophy and lethargy when injected subcutaneously. However, these effects all appear to be due to extracellular cysteine. (L)2-Oxothiazolidine-4-carboxylic acid (OTC) is metabolized to (L)-cysteine *in vivo* by the intracellular enzyme, 2-oxoprolinase. OTC has been shown to significantly increase intracellular levels of (L)-cysteine[7,8]. We have therefore examined the ability of this compound to alter urinary oxalate excretion by increasing the intracellular concentration of (L)-cysteine sufficiently to inhibit glyoxylate metabolism through to oxalate.

MATERIALS AND METHODS

Preparation of Adduct

The (L)-cysteine-glyoxylate adduct was prepared using the following procedure. Equal volumes of 8 mM (L)-cysteine and glyoxylic acid in glass distilled water were mixed and de-aerated with nitrogen. The mixture was incubated for 2.5 hours at 80°C in a shaking water bath. After 2.5 hours the solution was centrifuged to collect the precipitated adduct. The adduct was washed with ethanol and collected by centrifugation. This procedure was repeated three times. The adduct was then dried by lyophilization and stored undesiccated at room temperature. The yield was approximately 40%.

In vivo Experiments

A 15 day study was used to investigate the effect of OTC on urinary oxalate excretion. Six male Porton rats were fed a standard diet whilst the remaining six were fed a standard diet containing 1.2% OTC. All rats were housed individually in metabolic cages for the duration of the experiment. Water was given *ad libitum* and food intake was restricted to 17g per day. Urines were collected daily for the determination of oxalate, cyst(e)ine, calcium, magnesium, creatinine, phosphate and pH. The first five days of the experiment were used to collect baseline data before commencing the appropriate diets.

RESULTS AND DISCUSSION

Characterisation of Adduct

Proton NMR spectroscopy, mass spectroscopy, infrared spectroscopy, gas chromatography and melting point determinations were used to confirm the identity and purity of the adduct. In the reaction of (L)-cysteine and glyoxylic acid to make adduct, two optical centres form and thus a pair of diasteriomers are produced with different NMR spectra. Hamilton et. al[9] have shown that these diasteriomers form in an approximate 1 to 1 ratio. The predicted NMR splitting pattern for the adduct in D_6-DMSO agreed with the observed splitting pattern (figure 1).

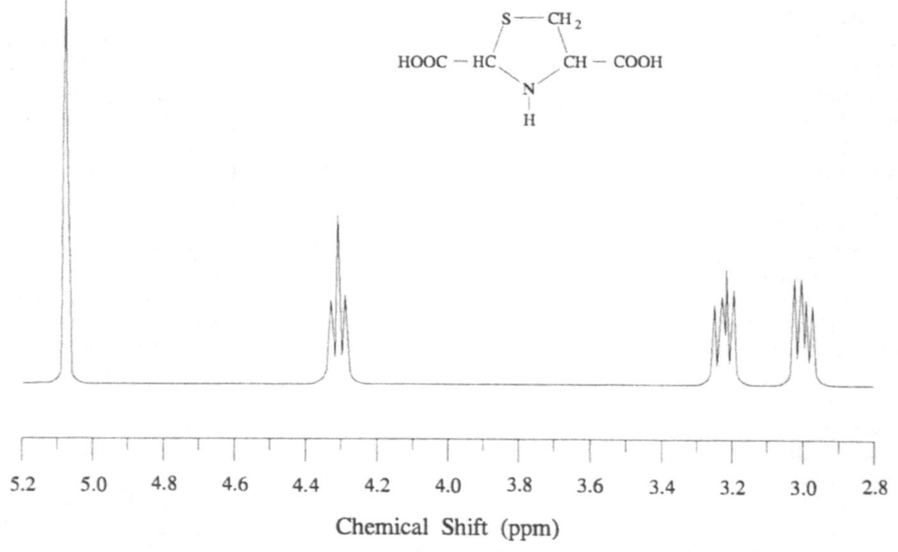

Figure 1. Proton NMR spectrum of adduct in D_6-DMSO (Day 1).

However, it was found that on leaving the adduct in D_6-DMSO for 12 days and re-running the NMR spectrum an identical splitting pattern appeared, only displaced by a few ppm δ (figure 2). The change to the NMR spectrum is due to the epimerisation of one chiral centre on the isolated adduct to form the other diasteriomer. Therefore, only one diasteriomer of a possible pair was isolated during the preparation.

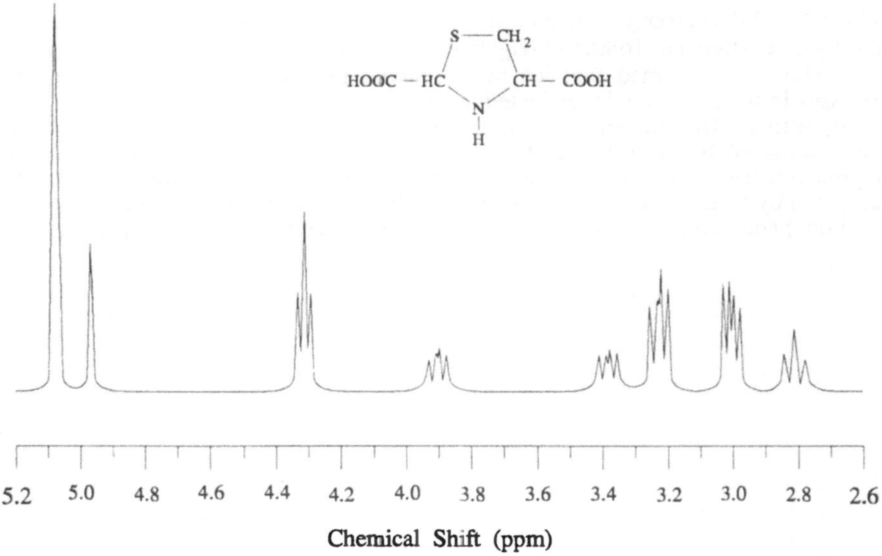

Figure 2. Proton NMR spectrum of adduct in D_6-DMSO (Day 12).

A tetramethylsilyl derivative of the adduct was prepared to enable simultaneous gas chromatography and mass spectroscopy. A single sharp peak was observed in the chromatograph indicating that the sample was of high purity. All major peaks in the mass spectrum of the derivative can be explained by logical fragmentation patterns. The presence of the molecular ion at 321 amu confirmed the empirical formula of the derivative (figure 3).

The presence of both the 2° amine and the carboxylic acid functional groups were confirmed by infrared spectroscopy. The 2° amine is seen as a week N-H stretch at

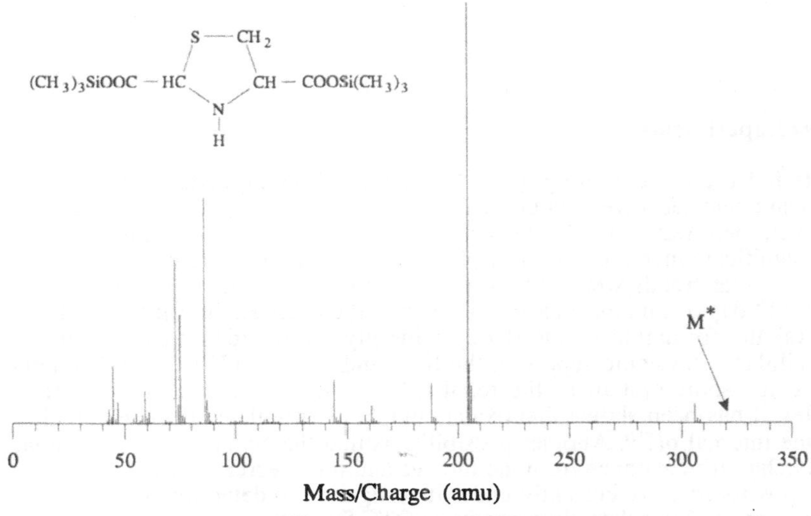

Figure 3. Mass spectrum of the tetramethylsilyl derivative of adduct.

3400cm^{-1} and the carboxylic acid groups are seen as a strong O-H stretch at 3100cm^{-1} and a strong C=O stretch at 1680cm^{-1} (figure 4).

Having confirmed that we had prepared pure adduct, its effect on metabolic processes in the liver was investigated as the liver is the site of major oxalate production and therefore also the site for adduct formation. Addition of the adduct up to a concentration of 10 mmol/L to hepatocytes had no effect on cell viability and intracellular enzyme release, or on the fundamental cellular processes including gluconeogenesis (measured by lactate conversion to glucose), glycogenolysis (measured by glucose release in fed rats) and ketogenesis (measured by hydroxybutyrate/acetoacetate production).

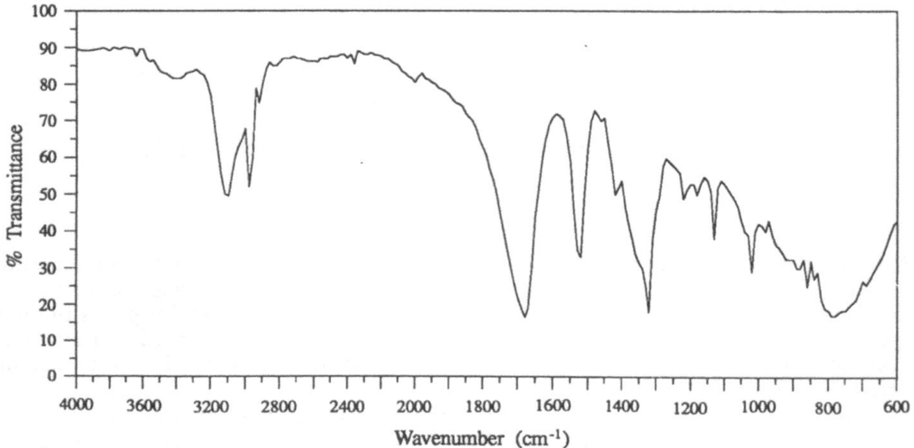

Figure 4 . Infrared spectrum of adduct in a KBr disk.

In vivo Experiments

In rats fed a diet containing 1.2% OTC, the following observations were made. No significant changes were seen in the urinary excretion of magnesium or calcium (table 1). However, there was a significant decrease in urinary oxalate, creatinine, uric acid and pH and a significant increase in urinary phosphate and cyst(e)ine (table 1).

Similar trends were seen in another study where OTC was fed for a prolonged period (27 days). The mechanism underlying the decrease in urinary oxalate may result from (a) the formation of the (L)-cysteine-glyoxylate adduct as a result of increased intracellular (L)cysteine supply in the liver and/or (b) a pH dependent alteration in the oxalate transport system in the renal tubules. In rabbit renal basolateral membrane vesicles, it has been shown that oxalate uptake is stimulated if the external pH is lower than the internal pH[10]. Another possibility is that the adduct, being an anion, competes with oxalate for clearance from the tubules and thus decreases urinary oxalate excretion. These possibilities are currently under investigation to determine the correct mechanism for the decrease in oxalate clearance upon OTC feeding.

Table 1. Effect of OTC on Urinary Parameters. Baseline data, day 1 to 5, and feeding data, day 6 to 10. Significance p <0.05 Control vs OTC for that time period. Values are given as Mean±SEM for n=6. Significance was determined using the Students t test.

	Control	O.T.C.	
Day	**24Hr. Oxalate (umol/L)**	**24Hr. Oxalate (umol/L)**	**Significance**
1 to 5	749 ± 25	729 ± 29	N.S.
6 to 10	641 ± 31	549 ± 27	P < 0.05
Day	**24Hr. Cyst(e)ine (mmol/L)**	**24Hr. Cyst(e)ine (mmol/L)**	**Significance**
1 to 5	4.62 ± 0.28	4.34 ± 0.27	N.S.
6 to 10	4.31 ± 0.26	7.89 ± 0.17	P < 0.001
Day	**24Hr. Phosphate (mmol/L)**	**24Hr. Phosphate (mmol/L)**	**Significance**
1 to 5	68.0 ± 1.6	68.9 ± 7.0	N.S.
6 to 10	67.6 ± 2.7	78.4 ± 3.1	P < 0.05
Day	**24Hr. Creatnine (mmol/L)**	**24Hr. Creatnine (mmol/L)**	**Significance**
1 to 5	6.40 ± 0.21	5.65 ± 0.37	N.S.
6 to 10	6.28 ± 0.30	5.18 ± 0.17	P < 0.01
Day	**24Hr. Uric Acid (mmol/L)**	**24Hr. Uric Acid (mmol/L)**	**Significance**
1 to 5	473 ± 56	415 ± 23	N.S.
6 to 10	588 ± 52	414 ± 33	P < 0.02
Day	**24Hr. Calcium (mmol/L)**	**24Hr. Calcium (mmol/L)**	**Significance**
1 to 5	1.11 ± 0.18	1.15 ± 0.17	N.S.
6 to 10	0.61 ± 0.10	1.04 ± 0.17	N.S.
Day	**24Hr. Magnesium (mmol/L)**	**24Hr. Magnesium (mmol/L)**	**Significance**
1 to 5	39.6 ± 1.2	37.4 ± 3.7	N.S.
6 to 10	38.4 ± 1.3	36.6 ± 1.7	N.S.
Day	**24Hr. pH**	**24Hr. pH**	**Significance**
1 to 5	6.44 ± 0.03	6.43 ± 0.08	N.S.
6 to 10	6.55 ± 0.04	6.10 ± 0.03	P < 0.001
Day	**24Hr. Urine Volume (mL)**	**24Hr. Urine Volume (mL)**	**Significance**
1 to 5	12.3 ± 0.5	13.3 ± 0.5	N.S.
6 to 10	13.4 ± 0.8	14.2 ± 0.4	N.S.

REFERENCES

1. WG Robertson, Urinary Tract Calculi *in* "Metabolic Bone and Stone Disease", Editor BEC Nordin, Churchill Livingstone, Edinburgh, 271, (1984).
2. A Hodgkinson, Oxalic Acid in Biology and Medicine, *Academic Press*, New York, (1984).
3. AM Rofe, RAJ Conyers and DW Thomas, Renal Stone Disease in South Australia, *Med J Aust*, 2,158, (1981).
4. R Bais, AM Rofe and RAJ Conyers, Inhibition of endogenous oxalate production: Biochemical considerations of the role of glycollate oxidase and lactate dehydrogenase, *Clin Science*, 76, 303-309, (1989).
5. R Bais, AM Rofe and RAJ Conyers, Investigations into the effect of glyoxylate decarboxylation and transamination on oxalate formation in rats, *Nephron* 57, 460-469, (1991).
6. R Bais, AM Rofe and RAJ Conyers, The inhibition of metabolic oxalate production by sulphydryl compounds, *J Urol*, 145, 1302-1305, (1991).
7. ME Anderson and A Meister, Marked increases of cysteine levels in many regions of the brain after administration of 2-oxothiazolidine-4-carboxylate, *FASEB J*, 3, 1632-1636, (1989).
8. TK Chung, MA Funk and DH Baker, (L)-2-Oxothiazolidine-4-carboxylate as a cysteine precursor: Efficacy for growth and hepatic glutathione synthesis in chicks and rats, *J Nutr*, 120,158-165, (1990).
9. CL Burns, DE Main, DJ Buckthal and GA Hamilton, Thiazolidine-2-carboxylate derivatives formed from glyoxylate and (L)-cysteine or (L)-cysteinylglycine as possible physiological substrates for D-aspartate oxidase, *Biochem Biophys Res Comm*, 125, 1039-1045, (1984).
10. SM Kuo and PS Aronson, Oxalate transport via the sulphate/HCO_3 exchanger in rabbit renal basolateral membrane vesicles, *J Biol Chem*, 88, 199-204, (1985).

ACUTE EFFECTS OF ALKALI AND EARTH ALKALI CITRATES IN HUMANS - A SYNOPSIS OF PRELIMINARY DATA IN URINE

P.O. Schwille, U. Herrmann, J. Fan, C.H. Schick,
M. Manoharan and G. Hoffmann

Mineral Metabolism and Endocrine Research Laboratory
Departments of Surgery and Urology
University of Erlangen
Germany

INTRODUCTION

Reduction of oxaluria is important in the treatment of idiopathic renal calcium stone disease. Oral calcium supplementation was found to be effective[1,2], but detailed information is lacking. We tested the relative potency of citrates in reducing oxaluria and supersaturation, and increasing inhibitor activity, against breakfast-induced mild hyperoxaluria[3].

PROCEDURES

Male healthy volunteers, age 25-56 years, fasted overnight (12-14 h), mildly hydrated orally with distilled water, underwent three consecutive 2 h creatinine clearance periods between 8:00 am and 2:00 pm (basal, i.e. before oral load at 10:00 am; 1,2 thereafter). Load types: breakfast (2 sandwiches, 20 g butter, 20 g bees honey; I), I + acidic potassium sodium citrate (Oxalyt-C; II), I + neutral potassium citrate (Kalinor; III), I + acidic calcium sodium citrate (Acetolyt; IV), I + neutral magnesium citrate (V). Citrate dosages were 3.41-5.0 g, supplying 40-53 net base equivalents. Established methodologies (oxalate: ion chromatography) were employed, except measurement of calcium oxalate crystal growth and aggregation[4].

RESULTS

Cumulative postprandial oxaluria was highest with load I (breakfast), unchanged with loads II, III (alkali citrates), but substantially lower with loads IV, V (earth alkali citrates); when comparing oxaluria in the combined group IV + V versus I, the difference approached significance ($p < 0.06$). On a molar base, median reduction in oxaluria was similar with both calcium and magnesium citrate (3.3 versus 3.1 µmol oxalate per mmol urinary cation). Citrates (II - V) tended to inhibit nucleation and growth (II vs I: $p < 0.02$ each), and to prolong aggregation time (ns) of calcium oxalate. Thus, citrate medication needs to be re-visited; tailoring a reduction in oxaluria and/or an increase in inhibitors should be feasible by selecting the citrate with the appropriate cation(s).

Urolithiasis 2, Edited by R. Ryall *et al.*,
Plenum Press, New York, 1994

Group code	n	Median change from basal; [(1 - basal) + (2 - basal)] pH[a]	Ca*,a	Ox*	Cit*,a	Mg*,a	CaOx**	Bru**	HAP**,a
I	5	-0.17	74	8.7	78	38	0.93	-1.37	-0.40
II	5	2.86[b]	12	8.0	551	-0.5[b]	1.60	4.28[b]	8.24[b]
III	4	1.82[b]	4.1	7.9	647	-20	-0.51	-0.01	5.72[b]
IV	5	0.49	165[b]	3.4	238	57	-0.13	-1.98	0.12
V	5	0.92	96	2.32	225	116	-1.20	1.44	4.16

*Values are mg per g creatinine; **Relative Supersaturation Product (Δ G, EQUIL-II), for calcium oxalate, brushite, hydroxyapatite; [a]$p < 0.05$ (ANOVA); [b]$p < 0.05$ vs I (U-test) Data are x±SEM

REFERENCES

1. DL Earnest, HE Williams and WH Admirand. Treatment of enteric hyperoxaluria with calcium and medium chain triglycerides. *Clin Res* 23: 130 (1974).
2. DE Barilla, C Notz, D Kennedy and CYC Pak. Renal oxalate excretion following oral oxalate loads in patients with ileal disease and with renal and absorptive hypercalciuria. *Am J Med* 64: 579 (1978).
3. PO Schwille and U Herrmann. Environmental factors in the pathophysiology of recurrent idiopathic calcium urolithiasis (RCU), with emphasis on nutrition. *Urol Res* 20: 70 (1992).
4. J Fan, PO Schwille, M Manoharan and A Wenig. Evaluation of precipitation, growth and aggregation of calcium oxalate in a self-nucleating whole urine system - Comparison between renal calcium stone patients and controls, without and with treatment. In preparation.

GLYCOLATE AND OXALATE PLASMA LEVELS AND RENAL HANDLING IN PATIENTS WITH TYPE 1 PRIMARY HYPEROXALURIA

M. Marangella, M. Petrarulo, C. Vitale, D. Cosseddu and F. Linari

Renal Stone Laboratory
Mauriziano Hospital
Turin, Italy

INTRODUCTION AND METHODS

Detection and differentiation of primary hyperoxaluria type: (PH1) is based on the finding of high urine levels of both oxalate and glycolate. However, these assays may produce misleading results when PHl is associated with chronic renal failure (CRF). We have recently developed accurate procedures for the determination of oxalate (ion chromatography, IC) and glycolate (HPLC) in both plasma and urine. We have used these techniques to detect and study PH1.

Ten patients (6 males) were considered, with presumed PHI and varying degrees of CRF. They were compared to 19 healthy controls (7 males), aged 32.5±12.9 years. Ten parents from 5 PH1 patients were also enrolled. All were studied for oxalate and glycolate plasma levels and renal clearances. GFF was assessed by means of the creatinine clearance. Five PH1 patients were also studied while on 300 to 900 mg/day pyridoxine.

RESULTS

The results are listed in the Table: means±(SD) are shown and expressed as µmol/L (plasma), µmol/24 h (urine), mL/min (clearances).

	P-Gly	U-Gly	C-Gly	P-Ox	U-Ox	C-Ox
Controls	7.9 (2.4)	422 (137)	40 (16)	2.4 (.6)	276 (82)	91 (51)
PH1	199 (147)	3375 (4070)	21 (27)	59 (72)	1360 (786)	35 (26)
Parents	9.5 (4.3)	717 (311)	61 (30)	2.3 (.4)	345 (69)	110 (30)

Glycolate and oxalate plasma levels were closely related ($p < 0.001$). Plasma values from PH1 patients did not overlap with those from controls, whereas urine values did. The decrease in net renal clearances of both glycolate and oxalate was dependent on the decrease of GFR. Instead, fractional renal clearances remained stable in spite of CRF, being 0.48±0.21 vs 0.46±0.18 for glycolate, and 1.21±0.70 vs 1.12±0.81 for oxalate in PH1 and controls respectively. Pyridoxine reduced oxalate levels in 2/5 patients and these changes were faithfully paralleled by glycolate changes. Mild derangements in glycolate levels were observed in 30% of the PH1 parents.

EFFECT OF HIGH DOSE VITAMIN C ON URINARY OXALATE LEVELS

P.A. Davis, T.R. Wandzilak, S.D. D'Andre and H.E. Williams

University of California
Davis
California

INTRODUCTION

Urinary oxalate arises from two sources: endogenous metabolism of glyoxylate and ascorbic acid (Vit C), and dietary intake. Although the RDA for Vit C is 60 mg/day, much larger doses are consumed by a portion of the population. Controversy exists as to whether an increased ingestion of Vit C can significantly increase the urinary excretion of oxalate. We undertook this study to re-evaluate the relationship between Vit C and oxalate by using a modified ion-chromatography method (0.5M borate diluent) to quantitate urinary oxalate.

MATERIALS AND METHODS

Nine male and 6 female subjects, ages 20-55 years, with no history of renal stone disease were enrolled in the study. Subjects were given 1g, 5g and 10g/day of Vit C for five days, each separated by a five day period of no Vit C. 24 h urinary Vit C, oxalate and creatinine were determined. Various amounts of Vit C (100mg-20.0g) added to a pooled urine sample with a known oxalate concentration were analysed for oxalate.

RESULTS AND CONCLUSION

A statistically significant increase in oxalate (36 µM) in the pooled sample was observed at 1g/L of added Vit C. In subjects taking 1g/day Vit C urinary Vit C levels were approximately 3.0mM, and there was a 21 µM increase in oxalate relative to the basal oxalate level. This increase in oxalate (8.1%) was slightly lower than the 8.8% percent increase noted for the direct addition of 1g/L ascorbate to a pooled urine sample. At the 5 and 10g/day ascorbate intake level, this discrepancy became even more pronounced as the 5g/day and 10g/day dose showed an apparent increase of only 45 and 71 µM of oxalate respectively.

Our data show that the ingestion of megadose quantities of Vit C for five days by normal subjects does not produce a physiologically significant increase in the urinary excretion of oxalate.

PATHOPHYSIOLOGY OF INCOMPLETE RENAL TUBULAR ACIDOSIS. EVALUATION OF BONE TURNOVER

P.J. Osther[1], J. Bollerslev[2], A.B. Hansen[3], K. Engel[3] and P. Kildeberg[4]

Departments of [1]Urology, [2]Medical Endocrinology
[3]Clinical Chemistry and [4]Pediatrics
University Hospital
DK-5000
Odense C, Denmark

INTRODUCTION

The fact that no cases of osteomalacia have been reported in stone formers with incomplete renal tubular acidosis (iRTA) indicates that mechanisms other than acidosis are involved in the lithopathogenesis of iRTA[1]. In previous studies bone metabolism has been rather neglected, however, and the existence of subclinical bone lesions cannot be excluded.

MATERIALS AND METHODS

Ten recurrent stone formers with iRTA, 10 recurrent stone formers with normal urinary acidification (NUA), and 10 normal controls (NC) participated in the study. Urinary acidification ability was evaluated by a short ammonium chloride loading test and 24 h urinary excretion of titratable acid and ammonium. Bone formation was evaluated by fasting serum osteocalcin levels and bone resorption by 24 h urinary excretion of hydroxyproline. Values are presented as medians and 95% confidence intervals of medians.

RESULTS AND CONCLUSION

Serum osteocalcin levels were elevated in iRTA, 7.3 (6.2-16.2) μg/L, compared to NUA, 6.1 (3.8-7.4) μg/L ($p < 0.01$); as was 24 h urinary excretion of hydroxyproline, 340 (217-1365) μmol/24 h, compared to NUA, 204 (127-357) μmol/24 h ($p < 0.05$), and NC, 185 (140-227) μmol/24 h ($p < 0.01$), indicating increased bone turnover in iRTA.

Stone formers with iRTA had increased bone turnover compared to other stone patients and healthy subjects.

REFERENCE

1. VM Buckalew Jr and RJ Caruana, *in* "Renal tubular disorders", HC Gonick, VM Buckalew Jr, eds, Marcel Dekker, New York, 357 (1985).

ENZYME CHANGES IN EXPERIMENTAL MODELS ON ADMINISTRATION OF IP OXALATE

C. Aravindakshan, N. Sylaja, N. Jayanthi Bai[1], S. Varghese[1] and Y.M. Fazil Marickar

Departments of Surgery and [1]Biochemistry
Medical College
Trivandrum, India

INTRODUCTION AND RESULTS

Reports of the tissue damage and enzymatic changes taking place in the kidney of man and animals exposed to risk factors of urolithiasis are not many. This study was undertaken to identify the enzymatic changes produced in the kidney tissue of experimental animals receiving intraperitoneal sodium oxalate injections for inducing lithogenesis.

A test group and control group of 6 male Wistar albino rats of approximately 200 g weight were utilised for the study. Both groups were given standard pellet feed and deionised water *ad libitum*. The test group was given in addition, daily injections of sodium oxalate solution intraperitoneally (6 mg/100 g rat weight). Urine was collected periodically for microscopy and biochemical study. After two months, all the animals were sacrificed and kidneys collected for oxalate determination and enzyme studies, namely alanine amino transferase (AAT), alkaline phosphatase (ALP), acid phosphatase (ACP) and lactate dehydrogenase (LDH). Soon after collection, 10% tissue homogenates in normal saline were prepared in cold condition, centrifuged in a cold centrifuge and the supernatant analysed immediately for the biochemical parameters mentioned above. The differences between the groups were statistically analysed.

There was considerable decrease in the mean AAT level in the experimental group (45 iu/L) compared to the control group (665 iu/L), ALP (40 KAU in experimental group and 200 KAU in controls) and ACP (19.8 KAU in experimental and 28.6 KAU in control group). The experimental group rat kidneys had higher levels of LDH (116 iu/L) and oxalate (977 mg%) compared to the control group, (68 iu/L and 788 mg% respectively). All the differences were statistically significant ($p < 0.001$) in all the parameters studied.

Regulation of oxalate synthesis in mammalian systems involves the enzymes glycolate oxidase and LDH. The significant increase of LDH and oxalate in the kidney of experimental rats in our study explains the formation of oxalate through the oxidation of glyoxylate by LDH. A decrease in AAT, ALP and ACP in the kidneys of experimental rats indicates their involvement in the process of urolith formation.

CONCLUSION

It is concluded that enzyme changes are significant in rats given intraperitoneal oxalate.

URIC ACID METABOLISM IN CALCIUM OXALATE STONE DISEASE

N.E. Thomas, H.K. Moorthy, S.V. Roshni, N. Sylaja, C. Aravindakshan and Y.M. Fazil Marickar

Department of Surgery
Medical College Hospital
Trivandrum, India

INTRODUCTION AND RESULTS

The concept of a direct relationship of uric acid to oxalate metabolism in the formation of the calcium oxalate crystals in urine is becoming well recognised. This work was done to identify the importance of uric acid abnormalities in patients who develop pure calcium oxalate stones. Seventy-nine patients who passed stones or had surgery done for the removal of pure calcium oxalate or predominantly calcium oxalate stones were included. They were classified into whewellite and weddellite stone groups. Fifty age and sex matched normals formed the control group. A detailed study of the biochemical parameters of these patients was done to look for metabolic abnormalities. The number of biochemical assessments ranged from one occasion to 13 with a mean of 3.2 in the patients. The uric acid levels were compared between patients who formed pure calcium oxalate monohydrate and pure weddellite. All patients who showed high uric acid level in urine or blood were treated with allopurinol in a dose ranging from 100-300 mg/day. Associated oxalate problems were corrected with pyridoxine 40-240 mg/day.

Forty-seven percent of the patients showed abnormalities of uric acid metabolism. The normal range for the laboratory was worked out as 400-600 mg/day. The mean urine uric acid level was significantly higher in the study group than controls ($p < 0.001$). The serum uric acid was significantly different between patients and controls. Whewellite stone formers had higher urine uric acid than weddellite stone formers ($p < 0.01$), but there was no significant difference in serum uric acid levels between the two. In patients with recurrent urinary stone disease with formation of pure calcium oxalate monohydrate stones, the control of the uric acid levels in the urine or blood was successful in preventing further stone formation. No such difference was seen in the serum uric acid level.

CONCLUSION

It is concluded that a significant number of patients who form pure calcium oxalate stones have errors in uric acid metabolism. Correction of such uric acid abnormalities is helpful in preventing the recurrence of stone disease in these patients. This is over and above the involvement of uric acid metabolic pathology in patients who form pure uric acid stones and patients who form mixed stones of oxalate and uric acid. Long term prophylaxis of the uric acid problem will benefit the patient in preventing recurrence.

EFFECT OF CAFFEINE-LOADING ON URINARY CALCIUM IN CONTROL SUBJECTS AND RECURRENT CALCIUM OXALATE STONE FORMERS WITH HYPERCALCIURIA

R.W. Norman, S.J. Whiting and J.N. Hughes

Dalhousie University and
Mount Saint Vincent University
Nova Scotia, Canada

INTRODUCTION

Dietary moderation plays an important role in reducing the risk of recurrent calcium oxalate stone disease. Most emphasis has been placed on managing hypercalciuria by limiting dietary calcium, sodium and protein but xanthine derivatives, such as caffeine and theophylline may play a significant role in some patients.

METHODS AND RESULTS

In order to assess the effect of caffeine-loading on the urinary excretion of calcium we studied 12 control subjects (4 males, 8 females) and 9 patients (4 males, 5 females) with recurrent calcium oxalate stone disease and hypercalciuria (4 with an elevated and 5 with a normal Ca/Cr ratio). Each was given 5 mg/kg body weight of caffeine after an overnight fast and the excretions of calcium, sodium and creatinine were monitored over the next 3 h and values were compared to similar collections without the drug.

In control subjects the fasting urinary Ca/Cr rose from 0.24±0.04 (Mean±SEM) to 0.52±0.05 after caffeine (p <0.0001); Na/Cr increased from 8.1±1.2 to 17.2±2.5 (p <0.001). In all patients the Ca/Cr ratio increased from 0.51±0.09 to 10.75 ±0.09 (p <0.007) and Na/Cr was unchanged at 0.7±2.5 without caffeine and 16.9±2.9 (n.s.) with caffeine. In the subgroup with fasting normocalciuria the Ca/Cr jumped from 0.28±0.02 to 0.61±0.09 (p <0.01) and Na/Cr increased from 6.6±0.9 to 17.1±3.7 (p <0.03). In the subgroup with fasting hypercalciuria there was no change in the calcium, 0.80±0.04 compared with 0.93±0.10 (n.s.), or the sodium, 15.7±4.7 compared with 16.6±5.2 (n.s.), after caffeine.

CONCLUSION

Control subjects and recurrent calcium oxalate stone formers with hypercalciuria, and normal fasting Ca/Cr have a calciuric effect from caffeine but patients with high fasting Ca/Cr do not. This implies that the defect in the latter group involves a mechanism that caffeine can turn on in the fasting normocalciuric patients and controls but which cannot be aggravated further by caffeine-loading in the fasting hypercalciuric group.

TEA - A RECOMMENDABLE BEVERAGE IN CALCIUM OXALATE UROLITHIASIS?

R. Siener and A. Hesse

Division of Experimental Urology
Department of Urology
University of Bonn, Germany

INTRODUCTION

Various studies have revealed the importance of a high fluid intake to an effective metaphylaxis in recurrent stone formation due to the increase in urinary volume and the consequent decrease in urinary supersaturation. Since beneficial effects of fluid therapy will be negated by the increased excretion of lithogenic components, the avoidance of ingesting oxalate-rich beverages, such as tea and chocolate, was recommended. The present study was performed in order to evaluate the influence of an excessive tea load on the risk of calcium oxalate stone formation.

MATERIALS AND METHODS

Ten healthy male subjects with a mean age of 26 years participated in the study. The investigation was conducted in two consecutive phases, each of them lasting 5 days. In the course of phases I and II the subjects received a standard diet with a constant fluid intake of 2.5 L/d.

In phase I, 1.5 L/d of the total fluid intake consisted of an oxalate-free fruit tea as control beverage. During phase II, fruit tea was substituted for an equal volume of tea (21 g tea steeped in 1.5 L hot water for 5 min), containing 86.3 mg oxalate. The amount of tea used for the loading test was in excess of that usually consumed per day, but not unreasonable for an occasional daily consumption.

RESULTS AND DISCUSSION

Ingestion of tea caused no significant delayed rise in urinary oxalate from 0.309 mmol/d to 0.340 mmol/d, so that only 3.2% of the ingested oxalate was excreted. On the other hand urinary citrate excretion increased significantly from 2.793 mmol/d to 3.387 mmol/d (21% increase) during tea load. Although urinary oxalate excretion and concentration increased during ingestion of tea, relative supersaturation of calcium oxalate remained unchanged due to the significant increase in urinary citrate excretion and concentration.

The results derived from this study in normal subjects may not apply entirely to calcium oxalate stone formers, since the available oxalate may be more prominent in patients with absorptive hyperoxaluria.

BONE MINERAL CONTENT IN RENAL STONE FORMER (RSF) PATIENTS WITH IDIOPATHIC HYPERCALCIURIA

I.P. Heilberg, L.A. Martini, V.L. Szejnfeld,
A.B. Carvalho, S.A. Draibe and N. Schor

Nephrology Division
Escola Paulista de Medicina
São Paulo, Brazil

INTRODUCTION

The association between idiopathic hypercalciuria and decreased bone content has been recognized by many authors[1]. It is not very well established whether or not calcium intake plays a critical role in loss of bone mass [2]. The purpose of the present study was to evaluate bone mineral density (BMD) of white adult RSF.

PATIENTS AND METHODS

Fifty-five RSF with either absorptive (AH), n=35 (22 Female/13 Male) or renal (RH) hypercalciuria, n=20 (7F/13M) were submitted to dual photon absorptiometry in lumbar spine and femoral sites. Only premenopausal women were included in the study. Calcium intake was assessed by 72 h dietary record. Results were compared to age, sex and weight matched controls.

RESULTS AND DISCUSSION

Osteopenia (OP) was detected in 11 of 55 patients (20%) being more common among men, 9/26 (35%) than in women, 2/29 (7%), $p < 0.05$. Apparently, OP was more frequent in RH, 7/20 (35%) than in AH group 4/35 (11%), but the difference was not statistically significant. BMD measurements did not correlate with calcium intake or excretion.

Preliminary data showed an elevated prevalence of OP among RSF with idiopathic hypercalciuria. Albeit not significant, RH tended to present more in OP than in AH patients. It remains to be clarified why the prevalence of OP was higher among male RSF.

REFERENCES

1. S Lawoyin, S Sismilich, R Browne, C Pak, Bone mineral content in patients with calcium urolithiasis *Metabolism* 28(12): 1250 (1979).
2. P Bataille, JM Achard, A Fournier, et al, vitamin D and vertebral mineral density in hypercalciuric calcium stone formers, *Kidney Int* 39:1193 (1991).
3. R Marcus, Calcium intake and skeletal integrity: is there a critical relationship? *J Nutr* 117:631 (1987).

SENSITIVE HIGH-PERFORMANCE LIQUID CHROMATOGRAPHIC MICROASSAY FOR HUMAN LIVER L-GLUTAMATE:GLYOXYLATE AMINOTRANSFERASE ACTIVITY

M. Petrarulo, S. Pellegrino, M. Marangella, D. Cosseddu and F. Linari

Renal Stone Laboratory
Mauriziano Hospital
Turin, Italy

INTRODUCTION AND RESULTS

The specificity of the assay of liver alanine:glyoxylate aminotransferase (AGT) activity is affected by the presence of glutamate:glyoxylate aminotransferase (EC 2.6.1.4) (GGT), which cross-reacts with alanine and glyoxylate. As a consequence, the assay for GGT activity must be coupled to that of the "raw" AGT activity to obtain an indirect estimate of the "true" AGT activity.

We propose a rapid and sensitive liquid chromatographic technique to determine GGT activity in human liver. Samples from percutaneous needle biopsies taken during laparoscopy have been analySed. The extract of the tissue homogenate is incubated for 60 min in the presence of the substrates. The generated 2-oxoglutarate, left to react with phenylhydrazine, is converted into the corresponding phenylhydrazone which is determined using reverse-phase high-performance liquid chromatography. The procedure allows the detection of the enzyme activity expressed by 5 µg of liver proteins and is more sensitive and less time-consuming than the previous spectrophotometric assay. Samples from nine normal individuals, three patients with primary hyperoxaluria type 1 (PHI), one with hyperoxaluria type 2 (PH2) were analyzed for both GGT and AGT activities.

Reference values for GGT activity are in the 0.82 to 1.45 µmol/h/mg protein (U) range (1.08±0.21, mean±SD, n = 9). The PH2 patient, a 54 year-old man with systemic oxalosis and chronic renal failure on regular dialysis treatment, had GGT of 0.57 U. Two PHI patients had normal GGT, whereas in a third, an eight-months old infant, the enzyme activity was 16% the mean from control values.

CONCLUSION

We conclude that the more sensitive and rapid HPLC method could represent a suitable alternative to the spectrophotometric assay, especially when limited amounts of sample are available.

URINARY EXCRETION OF CITRATE AND FEMALE STEROID HORMONES

P.J. Osther[1], H. Mathiasen[2] and A.B. Hansen[3]

Departments of [1]Urology, [2]Gynecology and Obstetrics and [3]Clinical Chemistry
University Hospital
DK-5000 Odense C, Denmark

INTRODUCTION

Women usually have higher urinary citrate excretion than men. The reason for this is not fully understood. Early observations on variations in urinary citrate excretion during the menstrual cycle have, however, indicated that it might be due to effects of female steroid hormones[1].

MATERIALS AND METHODS

Twenty healthy female and 20 healthy male volunteers aged 26-58 and 28-56 years, respectively, participated in the study. Citrate was determined in 24 h urine samples by a method using citrate lyase. Serum samples were collected in the morning on the day of urine sampling. Serum estradiol, progesterone and follicle stimulating hormone (FSH) were determined by radio-immunoassay.

RESULTS AND CONCLUSION

Twenty four hour urinary citrate excretion and citrate/creatinine ratios were significantly correlated ($r=0.9$, $p< 0.001$). Urinary citrate/creatinine ratios were significantly lower in healthy men, (0.21±0.02 [±SE]) than in healthy females (0.32±0.04), ($p< 0.01$). There were no significant correlations between urinary citrate/creatinine and the serum levels of estradiol ($r=0.05$), progesterone ($r=0.28$), and FSH ($r=0.09$).

Urinary citrate excretion was lower in healthy men than in healthy females. A relationship between serum levels of female steroid hormones and urinary citrate excretion could not be demonstrated.

REFERENCE

1. E Shorr, AR Bernheim, H Taussky, Regulation of urinary citric acid excretion to menstrual cycle and steroidal reproductive hormones, *Science* 95: 606 (1942).

DISTAL RENAL TUBULAR ACIDOSIS: A CAUSATIVE FACTOR OF NEPHROLITHIASIS IN NORTHWESTERN INDIA

S. Sharma[1], S. Vaidyanathan[2], S.K. Thind[1] and R. Nath[1]

Departments of [1]Biochemistry and [2]Urology
Postgraduate Institute of Medical Education and Research
Chandigarh, India

INTRODUCTION

The importance of distal renal tubular acidosis (dRTA) as a cause of recurrent stone formation has been emphasised in recent years. To evaluate its signficance in stone formers in northwestern India, we evaluated renal acidification ability in 69 renal stone patients (55M/14F) ranging in age from 18-70 years.

RESULTS AND DISCUSSION

All patients admitted to the study had normal serum and urine creatinine levels and sterile urine cultures. No patients were taking any drug that could affect renal function. In 64 patients, the pH of fasting morning urine was <6.0, thus eliminating the possibility of dRTA in these patients. In the remaining five patients in which fasting urine pH was >6.0, an ammonium-chloride-load test was performed. Following 6 h of fasting, ammonium chloride was administered orally in solution (0.1 g/kg body weight). To ensure an adequate diuresis, 100 mL of water was given every hour during the test, and urine pH was measured hourly for four hours. Plasma bicarbonate was measured prior to and at 2 and 4 h after ammonium-chloride loading. In one patient, urine pH was <5.4, but plasma bicarbonate was >16 mmol/L, indicating that the test was inconclusive. Therefore, a diagnosis of dRTA could not be confirmed in any of the patients evaluated in the study. Thus, dRTA appears an uncommon cause of stone formation in northwestern India.

SEX-DEPENDENT UROLITHIASIS IN THE PORTOCAVAL SHUNT RAT

U. Engelmann, A. Heidenreich, P. Schramek, G. Haupt and T. Senge

Department of Urology
Ruhr-University
Bochum, Germany

INTRODUCTION AND METHODS

The portocaval shunt rat represents an interesting experimental model of endogenous stone formation.

Two hundred and twenty Sprague-Dawley rats (127 males, 93 females) weighing 250-350g entered the study. Fifty eight underwent sham operation while 162 had portocaval shunting (end to side, pancreatico-duodenal vein included, 10-0 nylon sutures). Animals were fed commercially available rat diet. Urine was collected and analysed for pH, sodium, potassium, urea, uric acid and creatinine. The animals were sacrified at set intervals (mean 49.7 weeks, range 3 to 69 weeks). The kidneys, adrenal glands, ureters and bladder were investigated histologically, stones were analysed by scanning electron microscopy, energy dispersion spectrometry and X-ray diffraction.

RESULTS

Sixty eight of 97 (70.1%) male shunted rats developed urolithiasis; 88.2% of these were bladder stones. The incidence in female shunted rats was only 7.5%. In the control group no female and 20% of the males had urolithiasis. The majority of evaluable stones were potassium-hydrogen-urate (23/39); 9 rats had struvite stones without urinary tract infection and 7 rats had composite stones. Organ changes (atrophy of the liver, the testes and ovaries, hypertrophy of the kidney and the adrenal glands) were significant and differed according to sex. Postoperatively the rats developed polyuria, elevated potassium and uric acid excretion (all sex-dependent). Urinary pH levels did not change significantly.

DISCUSSION

Although uric acid in rats is further metabolised (unlike humans) uric acid stone formation in shunted rats was surprising when first recognised in 1972. Diminished hepatic clearance of uric acid, metabolism of uric acid in other organs and increased production due to postoperative catabolism and weight loss may account for the stone formation. However sex difference in stone formation cannot yet be adequately explained. Investigations on the relevant hormonal changes in shunted rats may contribute to explaining the sex difference.

SCANNING ELECTRON MICROSCOPIC STUDIES OF ERYTHROCYTES FROM VITAMIN B_6 DEFICIENT RATS AND VITAMIN B_6 DEFICIENT RATS FED WITH GALACTOSE

P. Kaul[1], S. Vaidyanathan[2], S.K. Thind[1] and R. Nath[1]

[1]Departments of Biochemistry and [2]Urology
Postgraduate Institute of Medical Education and Research
Chandigarh, India

INTRODUCTION

The role of the red blood cell membrane in urolithiasis has recently been investigated, mainly for modification of its ion-transporting capacity or structure and composition of its individual constituents [1,2]. However, the implication of any modifications on the physical state or the morphology of cells has not been investigated. In the present study scanning electron microscopy and osmotic fragility tests were performed in erythrocytes from vitamin B_6 deficient rats and vitamin B_6 deficient rats fed with galactose (well established models of hyperoxaluria).

RESULTS

The majority of erythrocytes from the vitamin B_6 control rats were discoid (i.e. normal), whereas, a number of stomatocytes, leptocytes and a few pitted cells were observed in the vitamin B_6 deficient group. Vitamin B_6 deficient rats fed with galactose showed a number of stomatocytes, surface pits and plastic flow, and the galactose control group exhibited a number of spherocytes and plastic flow. These shape changes were in concurrence with red cell osmotic fragility, which decreased both in the vitamin B_6 deficient group and the group fed galactose (19% and 33% hemolysis at 4g/L NaCl respectively) compared with the vitamin B_6 control (55% hemolysis at 4g/L NaCl), whereas osmotic fragility increased (73% hemolysis at 4g/L NaCl) in the galactose control group.

DISCUSSION

These findings suggest changes in red cell membrane composition, which in turn would alter their function.

REFERENCES

1. A Borsatti, *Kidney Int* 39:1283 (1991).
2. R Selvam and V Ravichandran, *Biochem Int* 23:1007 (1991).

IN PRIMARY HYPERPARATHYROIDISM DUE TO ADENOMAS, POST-OPERATIVE HYPOCALCEMIA IS NOT INFLUENCED BY HISTOLOGY OF REMAINING GLANDS

A. Khan, H. Sheikh[1] and J. Talati

Departments of Surgery (Urology) and [1]Pathology
The Aga Khan University Medical Center
Karachi, Pakistan

INTRODUCTION AND METHODS

Between April 1988 and September 1991, 35 patients were diagnosed as having primary hyperparathyroidism (pHPT). Nineteen patients had surgery. One was a repeat exploration for mediastinal adenoma. Fifteen patients had adenomas, two had carcinomas and 2 had glands that were hyperplastic. In adenomatous patients, of a total of 57 expected glands, 15 adenomatous and 27 non-adenomatous glands were identified. The wedge biopsy of the non-adenomatous glands was reviewed by one histopathologist.

RESULTS

In 27, non-adenomatous glands, 22 were class I (fat content greater than 30% Cusumano's classification[1]), four were class II (less than 30% fat), none were class III (hyperplastic nodule) and one was class IV (totally hyperplastic). Two patients in class I had markedly suppressed glands evident by more than 70% fat (suggested classification:I, suppressed [Is]). The fat content of the glands did not correlate with lowest post-operative calcium or duration of hypocalcemia (mean and S.D. of serum calcium in group Is : 7.5±0.30, entire group I : 7.76±0.67 and in groups II and IV : 7.6±0.51 respectively).

CONCLUSION

Histopathology of remaining glands is useful in ensuring that all additional parathyroid tissue is identified and reduces the chance of missing a second adenoma, but is of no value in predicting post-operative serum calcium decrease.

REFERENCE

1. RJ Cusumano, P Mahadevia and CE Silver, Intra-operative Histologic Evaluation in Exploration of the Parathyroid Glands, *Surg Gynaecol Obstet* 169:506-510 (1989).

RENAL TUBULAR ACIDOSIS IN URINARY STONE DISEASE

Y.M. Fazil Marickar, N.E. Thomas, S.V. Roshni, H.K. Moorthy and C. Aravindakshan

Department of Surgery
Medical College Hospital
Trivandrum, India

INTRODUCTION AND METHODS

Renal tubular acidosis resulting in metabolic acidosis is rarely reported in India. We present 13 cases related to urinary stone disease. This study is based on analysis of the data of 138 patients with hypercalciuria out of 2005 stone patients who attended the stone clinic at the Medical College Hospital, Trivandrum over a period of 12 years (1980-1991). Of these patients, 69 with bilateral renal stones or recurrent stone disease with more than three incidents or hypercalciuria without evidence of hyperparathyroidism were submitted to an acid load test. All patients had sterile urine and a complete stone metabolic work up for stone disease.

The test was performed on three consecutive days by administering ammonium chloride orally in a dose of 150 mg/kg body weight. The serum bicarbonate level and the urine pH were estimated prior to the administration of the ammonium chloride and at intervals of 1 h, 2 h, 3 h and 4 h after the ammonium chloride administration. Diagnostic criteria were a low serum bicarbonate < 22 mEq/L and failure to acidify even one sample of urine to pH less than 5.5 after acid load.

RESULTS AND CONCLUSION

It was observed that amongst the 69 patients, a positive ammonium chloride load test was present in 13. Of these, seven were diagnosed during the first ammonium chloride load test. The remaining six patients needed repeated tests to establish a diagnosis. None of the patients showed evidence of severe metabolic acidosis or hypocalcemia and needed hospital admission. All the patients were given sodium bicarbonate (3 g/day). Temporary symptomatic relief and reduction in calculogenic propensity were achieved in all patients. Subsequent biochemical study showed that five of these patients had other biochemical abnormalities namely hyperoxaluria and hyperuricosuria. A clinical follow up of the patients for a period from two years to 14 years showed recurrence of stone disease in eight. The apparent reason for the recurrence had been cessation of treatment during asymptomatic periods. It is concluded that renal tubular acidosis is a rare but significant cause for recurrent urinary stone formation in India. It may be associated with other biochemical abnormalities. Distal acidification defect (Type - 1) is the only form of renal tubular acidosis associated with renal stones.

The acid load test is simple and effective in diagnosis. Meticulous treatment and follow up are necessary to prevent stone recurrence.

CASE REPORT: HYPEROXALURIA AND DIETARY CALCIUM

V.R. Walker[1], R.A.L. Sutton[1] and S. Chan[2]

[1]Department of Medicine and
[2]Department of Surgery
Vancouver General Hospital
910 W.10th Avenue
Vancouver, BC Canada

INTRODUCTION AND RESULTS

Idiopathic hyperoxaluria in CaOx stone formers may be dietary in origin. We have such patients who can reduce their urinary oxalate to within the normal range (0.18-0.56 mmol/day) on a low oxalate or oxalate-free diet. These patients may have low dietary calcium intakes, associated with low urinary calcium (1-3 mmol/day vs normal, 2.5-7.5 mmol/day) and hyperoxaluria. We report here the effect of increased dietary calcium intake in a typical case.

The patient, a 49 year old male of East Indian extraction, has a 4 year history of recurrent bilateral CaOx stones and has undergone multiple nephrolithotomies, basket retrievals, and ESWL. His diet is low in animal protein and, since the time of his first stone episode, he has consumed virtually no dairy products. Baseline urinary calcium excretion was 1.4 mmol/day and urinary oxalate was 1.54 mmol/day. To determine the contribution of dietary oxalate to his hyperoxaluria, he was given 0-1.5 g/day of calcium in the form of Calcium Sandoz effervescent tablets, or milk, in divided amounts with each meal. Urine samples were collected after 4 days on each calcium regimen:

24 h Urine (mmol/mmol)	Ca/Creatinine	Ox/Creatinine
No supplements	0.18 (1.4)	0.194 (1.54)
Ca Sandoz (1g Ca/day)	0.26	0.065
No supplements	0.16	0.158
Milk (1g Ca/day)	0.37	0.093
Milk (1.5g Ca/day)	0.29	0.066

Note: values in brackets are baseline outputs in mmol/day.

Increased Ca intake reduced urinary oxalate to values close to the normal range (Ox/creatinine: <0.040) while urinary Ca remained low. For the past 1.5 years, he has continued milk with his meals. His general health has improved and his stone frequency has been reduced. The provision of dietary Ca to bind dietary oxalate, thus rendering oxalate unavailable for absorption, has markedly reduced his urinary oxalate excretion.

SECTION II: OXALATE METABOLISM AND TRANSPORT

VECTORIAL OXALATE TRANSPORT ACROSS A MONOLAYER OF RENAL EPITHELIAL CELLS (LLC-PK1)

H. Koul[1], M. Yanagawa[2], L. Renzulli[1], M. Menon[1] and C. Scheid[3]

[1]Division of Urology and Transplantation Surgery
[3]Department of Physiology
University of Massachusetts Medical School
55 Lake Avenue North
Worcester, MA 01655
[2]Department of Urology
Mie University School of Medicine, Mie, Japan

INTRODUCTION

Studies on erythrocytes from stone forming patients[1] and on renal papillary cells from stone forming rats[2] demonstrated abnormally high rates of oxalate transport, and it has been suggested that increased oxalate uptake plays an important role in urolithiasis. Understanding how a change in oxalate uptake affects oxalate excretion is difficult, however, because information on renal oxalate handling is limited. Available data from micropuncture studies indicate that oxalate undergoes both reabsorption and secretion within the kidney with net secretion occurring in the proximal tubule[3]. Such studies are technically quite difficult and provide only limited control of the composition of extracellular space such that it is difficult to predict the magnitude and direction of oxalate flux under physiological and pathophysiological conditions. These difficulties led us to examine vectorial oxalate flux using a simpler model system, LLC-PKl cells. These cells retain many of the characteristics of normal renal epithelial cells and form intact monolayers that can be used for examination of vectorial transport[4]. Moreover, studies in our laboratory have demonstrated that these cells express different transport systems for oxalate at the luminal and abluminal membrane surfaces: a Cl^-/oxalate exchanger at the luminal surface and a SO_4^{2-}(oxalate)/HCO_3^- exchanger at the abluminal surface. Thus, the present studies examined the magnitude and direction of vectorial oxalate flux across confluent monolayers of LLC-PKl cells.

METHODS

Cell Culture

LLC-PK1 cells (CRL 1392, American Type Culture Collection, passage 206 and 225) were employed for uptake experiments. The cells were maintained in Dulbecco's modified Eagle's minimal essential medium (DMEM) supplemented with 10% fetal bovine serum, glucose and antibiotics. For assessment of transcellular oxalate flux, cells were subcultured onto 25 mm Nunc membrane filters (Thomas Scientific Co., Swedesboro, NJ) and grown to confluence. The ability of the monolayers to exclude ^{3}H-inulin was taken as evidence that the cells had achieved confluence; only data from membranes with confluent monolayers were included in subsequent analyses.

Uptake Studies

Cultured cells were preincubated for 3 hrs in Earle's balanced salt solution. During the last 30 min, 4,4'-diisothio-cyanostilbene-2,2'-disulfonic acid (DIDS) or solvent vehicle was added to the medium. Transport studies were carried out at room temperature in buffer containing (in mM): Na (143), K (5.8), Ca (0.25), Mg (1), Cl (125), SO_4 (2), HEPES (N-2-hydroxyethyl piperazine-N'-2-ethanesulfonic acid, 10), pH 7.4. Uptake was initiated by the addition of ^{14}C oxalate (50 μM) to the incubation medium at either the luminal or abluminal surface. At various times after tracer addition, aliquots (100 μl) were removed from both sides of the monolayer and counted. Transcellular flux of oxalate was assessed by examining the disappearance of label from one surface and the appearance of label at the opposite membrane surface. Flux rates were calculated as follows:

Oxalate Influx Rate (Rate of Disappearance of Label) = $[(A_o - A_t) \times V]/t \times SA$
Oxalate Efflux Rate (Rate of Appearance of Label) = $[(A_t - A_o) \times V]/t \times SA$

where: A_o = label in the extracellular space at time zero
A_t = label in the extracellular space at time "t"
V = volume of the extracellular compartment (luminal or abluminal)
SA = surface area of the monolayer

RESULTS AND DISCUSSION

Figure 1 illustrates vectorial oxalate flux in LLC-PKI cells (Left panel: changes in radioactivity at both surfaces of the monolayer when label is added at the luminal surface. Right panel: changes in radioactivity when label is added at the abluminal surface).

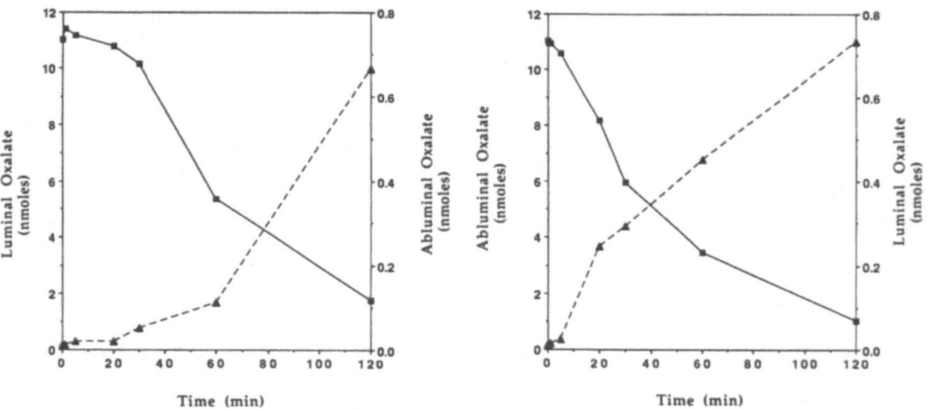

Figure 1. Timecourse of transcellular oxalate flux in LLC-PKI cells. Data show the disappearance of label at one surface (left hand axes, solid lines) and the appearance of label at the opposite surface (right hand axes, dashed lines) when label was added to the luminal (left panel) or abluminal (right panel) compartment.

Several interesting findings emerge from these studies. For one, these monolayers are clearly capable of transcellular oxalate flux. The appearance of label at the opposite surface of the monolayers could not be explained by leakage through the filters since the monolayers were capable of excluding ^{3}H inulin and since the appearance of label was sensitive to DIDS, an inhibitor of oxalate transport (data not shown). Moreover this transcellular flux was bidirectional, ie. label added to either cell surface rapidly appeared at the opposite surface. Secondly, the rate of oxalate influx into the cells from either surface was considerably higher that the rate of oxalate efflux. When label was added to the luminal surface, the rate of oxalate disappearance from the medium averaged 4.15 nmoles.hr^{-1}.cm^{2}. The rate of oxalate appearance at the contraluminal surface averaged 0.20 nmoles.hr^{-1}.cm^{2}. When label was added to the abluminal surface the rate of oxalate disappearance from the medium averaged 24.7 nmoles.hr^{-1}.cm^{2}. The rate of oxalate appearance at the contraluminal surface averaged 1.4 nmoles.hr^{-1}.cm^{2}. Thus LLC-PKl cells accumulate oxalate intracellularly under these experimental conditions. A comparison of the net fluxes across the monolayers indicate that transcellular flux from the abluminal to the luminal compartment is more than 5 fold higher than the transcellular flux from the luminal to the abluminal compartment, eg. vectorial flux across LLC-PKl cells leads to a net secretion of oxalate under these conditions.

These studies indicate that LLC-PKl cells should prove useful for investigating the factors affecting the magnitude and direction of oxalate flux under physiological and pathophysiological conditions.

ACKNOWLEDGEMENTS

Supported in part by NIH grants DK43184 to M.Menon and DK20586 to R Hackett (Program project to the University of Florida at Gainesville).

REFERENCES

1. B Baggio, G Gambaro, F Marchini, E Cicerello, R Tenconi, M Clementi and E Borsatti, An inheritable anomaly of red-cell oxalate transport in primary calcium nephrolithiasis correctable with diuretics, *N Engl J Med* 314:599 (1986).
2. D Sigmon, S Kumar, B Carpenter, M Menon and CR Scheid, Oxalate transport in renal cortical and papillary cells from normal and stone forming animals, *Am J Kidney Dis* 17:376 (1991).
3. R Greger, F Lang, H Oberleithner and P Deetjen, Handling of oxalate by the rat kidney, *Pflugers Arch.* 374:243 (1978).
4. HF Cantiello, JA Scott and CA Rabito, Polarized distribution of the Na^+/H^+ exchange system in a renal cell line (LLC-PKl) with characteristics of proximal tubular cells, *J Biol Chem* 261:3252 (1986).

ISOLATION AND CHARACTERIZATION OF OXALATE DECARBOXYLASE FROM HUMAN INTESTINE BACTERIA

H. Ito[1], H. Hayashi[1], T. Kotake[1], Y. Yokoo[2], K. Yamamoto[2], T. Hara[2], T. Furukawa[3] and Y. Nakagawa[4]

[1]Department of Urology
Teikyo University School of Medicine
Ichihara Hospital, Ichihara
[2]Tokyo Research Laboratories
Kyowa Hakko Kogyo Co Ltd
Machida
[3]Nutri-Quest, Inc
Chesterfield
[4]Nephrology Program
University of Chicago, Chicago

INTRODUCTION

Allison et al reported the presence of an oxalate degrading microorganism, *Oxalobacter formigenes*, in the human colon[1]. They suggested that this microorganism decomposes oxalate in intestine and consequently decreases the amount of oxalate available for absorption. In order to study the degradation and absorption of oxalate in the intestine of human subjects, we isolated the oxalate decarboxylase from anaerobic bacteria obtained from human feces, and examined the biochemical properties of this enzyme.

MATERIALS AND METHODS

Isolation of Oxalate Degrading Bacteria

Human feces were incubated in Barber culture medium using oxalate as a carbon source at 37°C under anaerobic conditions. Oxalate degrading bacteria were obtained after successive cultivations. These bacteria were used for the purification of oxalate decarboxylase. The amount of oxalate consumption in culture medium was measured by an HPLC method.

Ammonium Sulfate Precipitation

Bacteria cells were collected by centrifugation at 10,000g for 20 min, then suspended in 0.05M sodium phosphate buffer, pH 7.0 and sonicated three times for one min at 50W. The suspension was centrifuged at 12,000 rpm for 25 min, and the supernatant was collected. The supernatant was made 80% ammonium sulfate saturated and kept stirring at 4°C overnight. After centrifugation at 12,000 rpm for 25 min, the precipitate was collected

Urolithiasis 2, Edited by R. Ryall *et al.*,
Plenum Press, New York, 1994

and dissolved in 5mL of 0.05M sodium phosphate buffer, pH 7.0. The solution was dialyzed against two liters of the same buffer to remove the remaining ammonium sulfate and kept at -80°C. Protein was determined using a modified Lowry method[2]. Bovine serum albumin 1mg/mL, was used as a standard with the range from 10μg to 50μg.

Enzyme Assay

Oxalate decarboxylase activity was determined using a coupled enzyme system by measuring the increase in absorbance at 340nm due to the reduction of NAD to NADH. The assay mixture contained the following:

S-[N-morpholino] propanesulfonic acid (MOPS) buffer, 0.1 mM, pH 7.0	2.5mL
NAD 10 μM	0.1mL
sodium oxalate 1 μM	0.1mL
formate dehydrogenase 8 units	0.05mL
sample solution	0.05mL

The absorbance was measured at 0 time and after 30 min incubation at 37°C, then the background reading was corrected to calculate the enzyme activity using a Perkin-Elmer Lambda 7 spectrophotometer. For the optimum pH determination, the following 0.05M buffer solutions were used: pH 4 and 5, sodium acetate; pH 6, MES; pH 7, MOPS; pH 8 and 8.8, Tris-HC1; and pH 9.5, sodium bicarbonate.

Spectrophotometric Measurements

UV absorption of the enzyme in 0.05M sodium phosphate buffer, pH 7.0, was measured by a Beckman DU-40 spectrophotometer between 220nm and 320nm with a scanning speed of 500nm/min.

Circular dichroism was measured using a JASCO-600 circular dichroic spectrometer. The enzyme was dissolved in 0.05M sodium phosphate buffer, pH 7.0, and a 1mm light path thickness cylindrical cuvette was used for the far UV region (200~250nm), while 10mm light path thickness cylindrical cuvette was used in near UV region (250~300nm) .

Amino Acid Analysis

In order to determine the amino acid composition of the isolated oxalate decarboxylase, it was hydrolyzed in 6N HCl at 110°C, for 24, 48 and 72 h. The hydrolysates were dried in vacuo to remove HCl, and dissolved in pH 2 sodium citrate buffer. The samples were analyzed by Beckman amino acid analyzer, Model 119C.

SDS-Polyacrylamide Gel Electrophoresis

A stab gel, 7.5% cross-linking, was prepared as described by Andrews[3]. Running conditions were: running buffer, 0.05MTris-0.38M glycine containing 0.1% SDS, pH 8.3, constant voltage at 100V for 4 h. After the gel was fixed and washed, protein was stained by Coomassie Blue R250. A low molecular weight standard purchased from Gibco-BRL was used.

RESULTS

1. Human feces were incubated in Barber culture medium using oxalate as a carbon source at 37°C under anaerobic conditions.
2. Bacteria which could consume oxalate were used for isolation of oxalate decarboxylase.
3. Oxalate decarboxylase was purified by ammonium sulfate precipitation and moleculur sieve chromatography.
4. The isolated oxalate decarboxylase had optimum pH at 7.0.
5. The molecular weight of this enzyme was 50KD determined by a gel filtration column chromatography (Fig. 1) and PAGE.
6. K_m and V_{max} for oxalate were $1.2x10^{-3}$M and 5.0 at 27°C, respectively.

7. Conformational changes occurred after the enzyme bound NAD.
8. The amino acid composition was found to be relatively rich in glycine, hydrophobic and aromatic amino acid residues compared with average proteins (Table 1).

Table 1. Amino Acid Composition

	mole%
Glycine	43.47
Alanine	13.15
Glutamic acid	10.02
Aspartic acid	7.12
Serine	5.41
Proline	3.70
Arginine	3.21
Valine	2.82
Lysine	2.69
Leucine	2.39
Threonine	2.25
Phenylalanine	1.34
Isoleucine	1.12
Histidine	0.90
Tyrosine	0.41

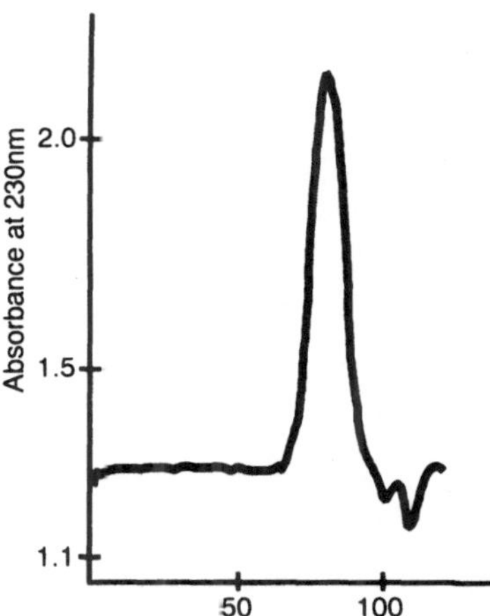

Figure 1. Molecular sieve chromatography

CONCLUSION

Oxalate decarboxylase was isolated from anaerobic bacteria of human intestine origin and the properties were investigated. This enzyme might be different from those isolated from *Pseudomonas oxalaticus*, *Oxalobacter formigenes*, and *Collybia velutipes*.

REFERENCES

1. MJ Allison, HM Cook, DB Milne, S Gallagher and RV Clayman, Oxalate degradation by gastrointestinal bacteria from humans, *J Nutr* 116: 455 (1986).
2. OH Lowry, NJ Rosebrough, AL Farr and RJ Randall, Protein measurement with the Folin phenol reagent, *J Biol Chem* 193: 265 (1951).
3. AT Andrews, *in:* "Electrophoresis", 2nd Ed. Chapter 5 Clarendon Press, Oxford (1986).

OXALATE CONCENTRATION IN RAT PLASMA DETERMINED BY *IN VIVO* ISOTOPIC DILUTION AND ENZYMIC ANALYSIS

J.F Costello[1,2], M. Smith[2], and C. Stolarski[2]

[1]Allegheny General Hospital
Medical College of Pennsylvania and
[2]Allegheny-Singer Research Institute
Pittsburgh, PA

INTRODUCTION

The rat is the animal most widely used for the study of oxalate metabolism and as a model for human renal stone disease. However, few, if any, values have been reported for the oxalate concentration of rat plasma. In the study presented here, we determined plasma oxalate concentration using a double enzymic method as well as by an *in vivo* isotopic dilution procedure.

METHODS

Enzymic Method

Rats were anesthetized with sodium pentothal, 60mg/kg, and a midline incision was made. One hundred units of heparin were injected into the heart and, after one minute, the rats were exsanguinated by cardiac puncture. The plasma, approximately 6 mL, was ultrafiltered and assayed for oxalate as previously described[1].

In Vivo Isotopic Dilution Method

^{14}C-oxalic acid, 2 µCi/day (4.6mCi/mmol), was infused subcutaneously over four days by mini osmotic pump (Alzet). Daily blood samples were withdrawn from the orbital sinus for ^{14}C determination. Twenty-four h urine collections were made into 5.0 mL of 3.5N HCl over the four days of the study. Urinary oxalate was determined using our enzymic procedure[2] and aliquots were counted for ^{14}C determination. Plasma oxalate was calculated from the formula:

$$\text{plasma oxalate concentration} = \frac{^{14}\text{C activity per unit volume of plasma X concentration of oxalic acid in urine}}{^{14}\text{C-oxalate activity per unit volume of urine}}$$

Urolithiasis 2, Edited by R. Ryall *et al.*,
Plenum Press, New York, 1994

RESULTS AND DISCUSSION

Plasma oxalate concentrations in separate groups of male Wistar rats were virtually identical when determined by *in vivo* isotopic dilution or by an enzymic assay method, Table 1. This confirms the accuracy of this enzymic method for determining plasma oxalate. Using this method, we have previously obtained plasma oxalate concentrations for human subjects, 1.25±0.47 µmol/L (mean±SD)[1], which are identical to values reported by *in vivo* isotopic dilution procedures[3-5]. Male Wistar and Sprague Dawley rats had similar plasma oxalate concentrations, Table 1. Further studies are required to determine if there is a sex related difference in the plasma oxalate concentration of these animals. Plasma oxalate concentration in the rat is approximately twice that found in healthy human subjects. Studies carried out in our laboratory have shown that the oxalate concentration in rat urine is also about twice the concentration found in human urine (unpublished data). The *in vivo* isotopic rat model used in the studies presented here is a useful model to monitor changes in blood oxalate concentration when studying oxalate metabolism.

Table 1. Rat Plasma Oxalate Concentrations

Rat Strain	Enzymic Analysis	*In Vivo* Isotopic Dilution
	µmol/L	
Male Sprague Dawley	2.93±0.52 (mean of 5 rats)	
Male Wistar	2.85 (pooled plasma from 5 rats)	2.18 ± 0.22 (n=2)
		2.63 ± 1.71 (n=2)
		3.31 ± 0.64 (n=2)
	3.27±1.10 (mean of 5 rats)	2.83 ± 0.15 (n=2)
		3.93 ± 0.18 (n=2)
		3.41 ± 0.30 (n=3)
		2.69 ± 0.16 (n=3)
		4.12 ± 0.36 (n=3)
		3.07 ± 0.69 (n=3)
Female Wistar	3.69 (pooled plasma from 6 rats)	
		3.13 ± 0.63 (mean ±SD of isotopic values)

n = The number of consecutive daily isotopic determinations on each rat. Isotopic studies were carried out on male Wistar rats. Values are mean ±SD.

REFERENCES

1. J Costello and DM Landwehr, Determination of oxalate concentration in blood, *Clin Chem* 34:1540 (1988).
2. J Costello, M Hatch and E Bourke, An enzymatic method for the spectrophotometric determination of oxalic acid, *J Lab Clin Med* 87:903 (1976).
3. A Hodgkinson, R Wilkinson, and BEC Nordin, The concentrations of oxalic acid in human blood, *in:* "Urinary Calculi", L Cifuentes Delatte, A Rapado, A Hodgkinson, eds, S Karger, Basel (1973).
4. AR Constable, AM Joekes, GP Kasidas, P O'Regan and GA Rose, Plasma level and renal clearance of oxalate in normal subjects and in patients with primary hyperoxaluria or chronic renal failure or both, *Clin Sci* 56:299 (1979).
5. P Boer, JAC Prenen, HA Koomans and EJ Dorhout Mees, Fractional oxalate clearance in subjects with normal and impaired renal function, *Nephron* 41:78 (1985).

ABNORMAL ERYTHROCYTE OXALATE SELF-EXCHANGE IN IDIOPATHIC STONE FORMERS IS NOT LINKED WITH CHROMOSOME 17

G. Gambaro[1], A. De Bortoli[1], G.A. Danieli,[1]
A. Borsatti[1], H.E. Williams[2], F. Marchini[1] and B. Baggio[1]

[1]Institute of Internal Medicine and Department of Biology
University of Padova
Italy
[2]School of Medicine
University of California
Davis, USA

INTRODUCTION

A faster trans-membrane erythrocyte self-exchange of oxalate is frequently observed in idiopathic calcium-oxalate stone formers[1]. This functional anomaly, which seems to be pathogenetically related to lithiasis, is associated with an abnormal phosphorylation rate of band 3 protein, the anion carrier[2], and is transmitted as an autosomic dominant trait[1]. A primary structural defect in band 3 might constitute the possible mechanism underlying the origin of these anomalies. We addressed this aspect by evaluating a possible linkage between increased red blood cell oxalate flux and chromosome 17q21-qter, where band 3 gene is located[3]. We reasoned that if such a linkage does not exist, this would exclude band 3 primary anomalies as the cause of abnormal oxalate self-exchange.

METHODS

We studied two families, in which the oxalate transport anomaly was detected. The largest pedigree is one of the five families described in our first description of the autosomal dominant nature of abnormal oxalate self-exchange[1]. This family includes 26 living subjects of whom 25 from three generations were available for the study. The second family has not been described before; it includes 7 subjects with one abnormal RBC oxalate exchange subject in each generation.

DNA was extracted from peripheral blood by standard phenol/chloroform methods. The long arm of chromosome 17 was screened by different DNA polymorphic markers: HHH202 (17q11.2-12); MPO (17q21-23); GIP (17q21.3-22); and GH (17q22-24). PCR analysis and polyacrylamide or agarose gel electrophoresis were performed as suggested for each probe. Two point linkage analysis was done by the LINKAGE program.

RESULTS

No evidence of linkage with these markers was found. The total lod score (logarithm of the odds for linkage versus non-linkage) for HHH202 was 0.105, for MPO -1.464, for GIP -2.190, and for GH 0.246.

DISCUSSION

The very low or negative lod scores suggest that abnormal erythrocyte self-exchange of oxalate is not linked with chromosome 17, and therefore not with band 3 gene, which is located in this chromosome. This finding supports the idea that a primary anomaly of this protein is not crucial for the expression of the red cell oxalate self-exchange anomaly observed in idiopathic calcium renal stone formers[1].

In the light of this observation, and our previous report of an abnormal phosphorylation status also of band 2 in nephrolithiasic patients[2], we are now addressing the possible existence of a band 2 protein (spectrin) primary abnormality as the cause of abnormal erythrocytic transport of oxalate, and a number of other cellular anomalies[4]. Since spectrin-ankyrin-anion transporter interaction is highly cooperative[5], a primary band 2 structural abnormality could affect the phosphorylation levels of associated proteins, specifically of band 3, thus leading to an abnormal transport function.

However, at the moment, we cannot exclude the existence of a primary defect leading to an imbalance between protein kinases and phosphatases and altering a common energizing step between band 3 and band 2 proteins. This mechanism would also explain abnormal phosphorylation of both membrane proteins.

REFERENCES

1. B Baggio, G Gambaro, F Marchini et al, An inheritable anomaly of red blood cells in idiopathic calcium oxalate nephrolithiasis correctable with diuretics, *New Engl J Med* 314: 599 (1986).
2. B Baggio, G Clari, G Marzaro et al, Altered red blood cell membrane protein phosphorylation in idiopathic calcium oxalate nephrolithiasis, *IRCS Med Sci* 14: 368 (1986).
3. LC Showe, M Ballantine and K Huebner, Localization of the gene for erythroid anion exchange protein, band 3 (EMPB3) to human chromosome 17, *Genomics* 1: 71 (1987).
4. G Gambaro, B Baggio, Idiopathic calcium oxalate nephrolithiasis: a cellular disease, *Scann Microsc* 6: 247 (1992).
5. CD Cianci, M Giorgi, JS Morrow, Phosphorylation of ankyrin down-regulates its cooperative interaction with spectrin and protein 3, *J Cell Biochem* 37: 301 (1988).

THE EFFECT OF DIETARY PROTEIN AND GLUCAGON ON THE URINARY EXCRETION OF OXALATE IN THE GUINEA PIG

R.P. Holmes, C.H. Hurst and D.G. Assimos

Department of Urology
Bowman Gray School of Medicine
Wake Forest University
Winston-Salem, NC USA

INTRODUCTION

Urinary oxalate is supposedly derived from three sources: the diet (10-20%), ascorbate breakdown (40-50%) and endogenous metabolism (40-50%)[1]. Factors influencing oxalate excretion remain poorly defined. Endogenous metabolism, which appears to occur predominantly in the liver, is thought to involve amino acid catabolism[1]. Therefore, dietary protein as a source of amino acid may influence urinary oxalate excretion. However, studies with humans have not been conclusive. We have used a guinea pig model to further study the effects of dietary protein on urinary oxalate excretion. This animal model was chosen since the guinea pig, like humans, has a key biosynthetic enzyme, alanine-glyoxylate aminotransferase type I (AGT), located only in peroxisomes[2]. Guinea pigs were fed semi-purified diets containing 10, 20% or 40% of casein based protein. Urinary oxalate was monitored and hepatic AGT, glycolate oxidase (GO) and lactate dehydrogenase (LDH) levels were measured. Since glucagon is a known mediator of dietary protein metabolism, the effects of pharmacologic doses of this hormone on urinary oxalate excretion and hepatic enzyme levels were also examined.

MATERIALS AND METHODS

Animals and Synthetic Diets: Male Hartley guinea pigs (250-300 g on purchase) were fed a regular chow diet for 1 week to permit adjustment to the new environment. The diet was then changed to a semi-purified diet modeled on the formulation developed by Reid and Briggs[3]. The diet (Ralston-Purina) contained 20% protein (casein), 21% dextrin, 10% sucrose, 7.8% glucose, 13% cellulose, 10% alfalfa meal, 7% corn oil, 7% mineral mix, 2% vitamin mix, 0.25% L-arginine, 0.2% L-methionine, 0.2% choline chloride, 0.2% ascorbic acid, 0.4% magnesium oxide, and 1.3% potassium acetate. Animals underwent a one week period of acclimatisation to the diet and metabolic cage environment. For baseline measurements on this 20% protein diet, 24 h urine samples were collected for one week in vessels containing 1.1 mL of conc. HCl to ensure that the pH of the urine remained below pH 2. Animals were then switched to either 10% or 40% protein diets for two weeks with urine samples on the second week obtained for analysis. In these diets sucrose and dextrin content were adjusted, maintaining a ratio of 3:1, sucrose to dextrin.

Glucagon Injection: Animals were fed commercial chow diets containing 18% protein. Three 24 h urine samples were collected in metabolic cages before injecting

Urolithiasis 2, Edited by R. Ryall *et al.*,
Plenum Press, New York, 1994

animals subcutaneously three times daily (8:00 am, 1:00 pm and 6:00 pm) with either glucagon (4 mg/kg body wt/day) or saline. Injections were performed for four days.

Urinary Assays: Oxalate was measured by a modification of the Sigma 591 oxalate oxidase method where the charcoal step was replaced by a precipitation of oxalate at pH 5.5-6.0 with ethanolic $CaSO_4$ with isotopic addition to monitor recovery of oxalate. Creatinine was measured on a centrifugal analyzer by a kinetic picric acid procedure using a kit supplied by the manufacturer (IL). To characterize each animal's excretion on a particular diet the mean excretion over a one week period was calculated.

Liver Assays: Animals were sacrificed by anesthetizing with rompum/ketamine (5/30 mg/kg body wt) and severing the abdominal aorta. Liver tissue was removed and frozen at -70°C. For enzymatic assays tissue was gently homogenized, 10% (w/v), in 0.1 M Tris-HCl (pH 7.5). This produced much better reproducibility compared with homogenizing in buffers containing 0.25 M sucrose. AGT activity was measured as the serine-pyruvate aminotransferase activity using incubation conditions described by Noguchi and Fujiwara[4]. The hydroxypyruvate produced was measured as the phenylhydrazone derivative by reversed phase HPLC. GO activity was measured using the incubation conditions described by Lindquist and Branden[5] and the glyoxylate produced measured as the phenylhydrazone derivative by reversed phase HPLC.

RESULTS AND DISCUSSION

The level of dietary protein had no effect on the urinary excretion of oxalate in male guinea pigs (Table 1). We have observed a similar response in humans on controlled diets where males excreted similar amounts of oxalate on diets containing 0.6 g protein/kg body wt/day or 1.8 g protein/kg body wt/day[6]. There was no significant effect of dietary protein on hepatic AGT or LDH levels. Hepatic levels of GO, however, were twice as high on 40% protein diets compared with 10% protein diets. To date, we have been unable to determine whether this difference was associated with a change in urinary glycolate excretion due to problems associated with glycolate measurements in guinea pig urine. The large change in hepatic GO levels in response to dietary protein, in contrast to the lack of change in AGT levels, indicated that the levels of these two enzymes are regulated differently. It is surprising that despite the increased GO and an expected increase in glyoxalate synthesis, oxalate excretion was unaffected.

Table 1. Effect of dietary protein on urinary oxalate excretion and liver enzyme levels.

20% Protein	10% Protein	40%Protein	
Oxalate-males (g/g Cr) (n=6)	0.19±0.04	0.20±0.06	0.16±0.06
AGT (μmol/min/g)		0.65±0.12	0.58±0.13
GO (μmol/min/g)		0.57±0.09[a]	1.05±0.11

[a] Significantly different; $p < 0.001$

Increases in dietary protein result in increased circulating levels of glucagon, insulin and other hormones in rodents and other species[7]. Glucagon has been identified as important in stimulating the liver to convert amino acids to glucose[8]. To determine its effect on urinary oxalate excretion and the levels of hepatic enzymes we injected male guinea pigs with pharmacologic doses of glucagon. The level of urinary oxalate increased on the second day of injections and had increased by 40% after four days. Significant changes were observed in both GO and AGT levels with GO decreasing by 20% and AGT by 40% (Table 2). This greater decrease in AGT is apparently responsible for the increased synthesis of oxalate because of the decreased capacity of peroxisomes to convert glyoxylate to glycine. This highlights the pivotal role of AGT in oxalate synthesis. Such a role for AGT has been identified in humans with primary hyperoxaluria type I where a deficiency in AGT or its aberrant localization in mitochondria has been identified as a causative agent in the pathogenesis of the disease[9]. Further support for this role of AGT was obtained with preliminary experiments with female guinea pigs (n=2). Females excreted more oxalate than males on 10% and 40% protein diets, and they had lower hepatic AGT levels than males.

The decrease in hepatic GO levels with glucagon injection, in contrast to its increase with increased dietary protein under conditions where glucagon could be expected to be elevated, suggests that the regulation of the enzyme is complex. The ratio of glucagon to insulin may be an important factor, or alternatively, the effects of insulin which is also elevated on high protein diets, may override those of glucagon. Similarly, the regulation of AGT levels is apparently complex with dietary protein exerting little effect whereas a significant reduction in activity occurred with glucagon treatment. The different responses of AGT and GO to glucagon injection further support a different regulation of the levels of these two enzymes.

Table 2. Effect of glucagon on urinary oxalate excretion and hepatic enzyme levels.

	Saline	Glucagon	P
Oxalate (g/g Cr)	0.13±0.03	0.18±0.05	0.03
AGT(μmol/min/g)	0.51±0.02	0.32±0.03	<0.001
GO (μmol/min/g)	1.07±0.07	0.88±0.14	0.02

A previous report that hepatic AGT levels are unaffected 36 h after glucagon injection[10] is consistent with our observation that an increase in urinary oxalate excretion was not observed until 48 h. There appears to be a time delay before levels of AGT fall. This may be related to the turnover of the enzyme, the stability of the AGT mRNA, or a slow increase in a regulatory metabolite or signal that decreases mRNA synthesis. The effect of glucagon may also be indirect and result from a time-dependent change in some regulatory agent directly responding to glucagon. Our results indicate that the guinea pig may be a useful model for identifying systemic factors that modify hepatic oxalate synthesis.

REFERENCES

1. HE Williams and TR Wandzilak, Oxalate synthesis, transport and the hyperoxaluric syndromes, *J Urol* 141:742-747 (1989).
2. S Hayashi and T Noguchi, Alanine: glyoxylate aminotransferase 1 is present in the peroxisomes of guinea pig kidney, *Biochem Biophys Res Comm* 166:1467-1470 (1990).
3. MA Reid, and GM Briggs, Development of a semi-synthetic diet for young guinea pigs, *J Nutr* 51:341-354 (1953).
4. T Noguchi and S Fujiwara, Identification of mammalian aminotransferases utilizing glyoxylate or pyruvate as amino acceptor, *J Biol Chem* 263:182-186 (1988).
5. Y Lindquist and CI Branden, Preliminary crystallographic data for glycolate oxidase from spinach. *J Biol Chem* 254:7403-7404 (1979).
6. RP Holmes, LJ Hart and DG Assimos, The effects of protein intake on oxalate excretion, *J Urol* 147:329A (1992).
7. J Peret, S Foustock, M Chanez, B Bois-Joyeux and R Assan, Plasma glucagon and insulin concentrations and hepatic phosphoenolpyruvate carboxykinase and pyruvate kinase activities during and upon adaptation of rats to a high protein diet, *J Nutr* 111:1173-1184 (1981).
8. G Boden, L Tappy, F Jadali, RD Hoeldtke, I Rezvani and OE Owen, Role of glucagon in disposal of an amino acid load, *Am J Physiol* 259:E225-E232 (1990).
9. CJ Danpure and PR Jennings, Peroxisomal alanine: glyoxylate aminotransferase deficiency in primary hyperoxaluria type I, *FEBS Lett* 201:20-24 (1986).
10. S Hayashi, H Sakuraba and T Noguchi, Response of hepatic alanine: glyoxylate amino transferase I to hormones differs among mammalia, *Biochem Biophys Res Comm* 165:372-376. (1989).

MACRO AND MICRO AUTORADIOGRAPHIC STUDIES ON OXALATE IN NORMAL AND HYPEROXALURIC RATS

T. Kanazawa[1], T. Sugimoto[1], S. Kamikawa[1], H. Iimori[1], K. Yamamoto[1], T. Nakatani[1], H. Oßwald[3], Y. Funae[2] and T. Kishimoto[1]

[1]Department of Urology
[2]Department of Chemistry
Osaka City University Medical School
Osaka, Japan and
[3]Department of Pharmacy
Tübingen University, FRG

INTRODUCTION

It is known that the deposition of calcium oxalate crystals in the kidney is the first stage of stone formation. However, the exact location of this deposition in the kidney has not been determined. Hautmann and Oßwald[1] pointed out that there was a concentration gradient of oxalate and calcium from cortex to medulla of the kidney, and that oxalate and calcium contents in the papilla were hundred times higher than those in urine. These findings indicated that the renal papilla, especially the interstitium of papilla might be the site of the calcium oxalate crystalluric deposition. In the present study, we performed autoradiographic studies on oxalate in hyperoxaluric rats using ^{14}C-oxalate, in order to elucidate the distribution of oxalate in renal tissue and the site of the deposition of calcium oxalate crystals along the nephron.

MATERIALS AND METHODS

Sprague-Dawley rats weighing 180-200g were used. They were divided into two groups. In one group, rats were fed on a control diet and in the other group a vitamin B_6-deficient diet. Four weeks later, urinary excretion and renal content of oxalate were measured by high-performance liquid chromatography[2]. Renal clearance of oxalate was studied, using ^{14}C-oxalate, and these values were compared with those for insulin. We also studied the distribution of oxalate in vitamin B_6-deficient rats by autoradiography, using ^{14}C-oxalate and ^{45}Ca.

RESULTS AND DISCUSSION

Hyperoxaluria was induced in rats after feeding them a vitamin B_6 deficient diet for four weeks. The urinary excretion, renal clearance, and renal content for oxalate are summarized in Fig 1. Urinary excretion of oxalate was 448±129 µg/day/100g body weight in hyperoxaluria rats, about 3 times higher than that in control rats. In this model, hyper-

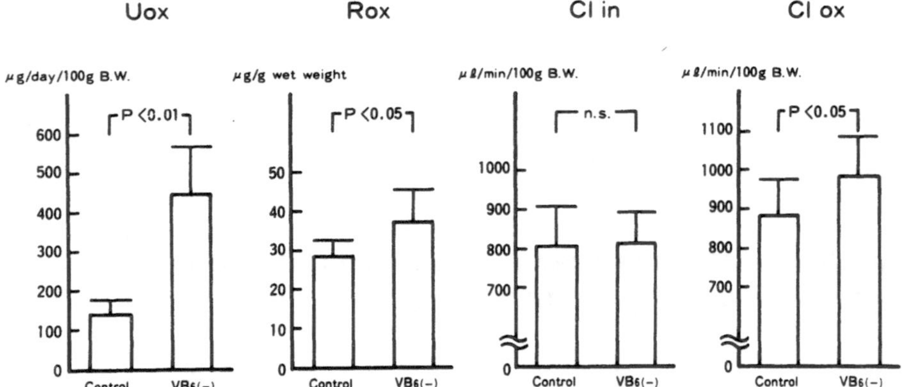

Fig. 1 Comparison of urinary excretion of oxalate (Uox), renal content of oxalate (Rox) and renal clearance for insulin and oxalate (Cl in, Cl ox) between control and VB_6-deficient rats.

oxaluria is mild which is often seen among human stone formers[3]. Renal clearance of oxalate was higher than that of insulin in control and hyperoxaluric rats, indicating that oxalate is secreted at the proximal tubules[4]. On the other hand, fractional excretion of oxalate in hyper-oxaluria rats was higher than that found in control rats. This result confirmed that the secretion of oxalate increases in hyperoxaluric rats. Renal content of oxalate in hyperoxaluric rats was 381±79 μg/g wet weight, about 1.3 times higher than that in control rats.

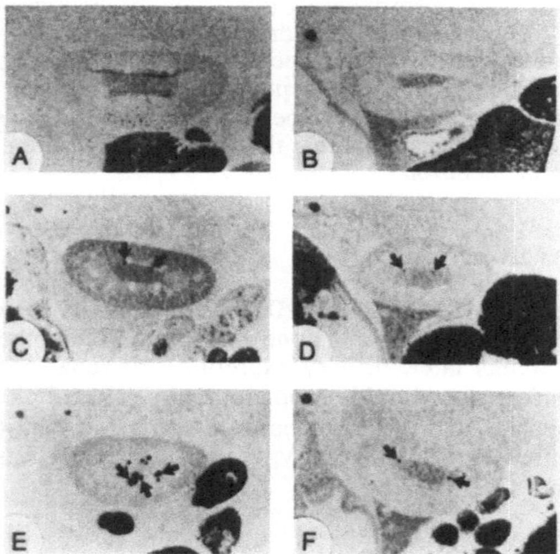

Fig. 2 Distribution of ^{14}C-oxalate and ^{45}Ca in the kidney 8 hours after both radioisotope injection (20 μCi/kg body weight) in control rats (A and B), those in rats feeding VB_6-deficient diet for 4 weeks (C and D), and those in rats feeding VB_6-deficient diet for 8 weeks (E and F), respectively. A, C and E show the autoradiograms of ^{14}C-oxalate, and B, D and F, those of ^{45}Ca.

The autoradiographic study showed that focal depositions of the radioisotope, not found in control rats (Fig 2-A,B), were observed in the renal papilla of vitamin B_6 deficient rats not only after injection of ^{14}C-oxalate but also after injection of ^{45}Ca (Fig 2-C, D) The deposits shown by the autoradiograms of vitamin B_6 deficient rats, fed on the diet specified above for eight weeks, increased in number and size over time (Fig 2-E, F).

CONCLUSION

The first site of calcium oxalate crystalluric fixation might be the interstitial tissue of renal papilla in which fixation might be first stage of stone formation.

REFERENCES

1. Hautmann R, Oßwald H Concentration profile of calcium and oxalate in urine, tubular fluid and renal tissue - some theoretical considerations, *J Urol* 129: 433-436 (1983).
2. Imaoka S, Funae Y, Sugimoto T, Hayahara N, Maekawa M, Specific and rapid assay of urinary oxalic acid using high-performance liquid chromatography, *Anal Biochem* 128: 459-464 (1983).
3. Robertson WG, Peacock M, Ouimet D, Heyburn PJ, The main risk factor for calcium oxalate stone disease in man: Hyperoxaluria or mild hyperoxaluria? *Urolithiasis Plenum* 3-12 (1980).
4. Knight TF, Senekjian HO, Weinman EJ, Effect of para-amino-hippurate on renal transport of oxalate *Kidney Int* 15: 38-42 (1979).

OXALATE-INDUCED CHANGES IN INTRACELLULAR CALCIUM LEVELS IN RENAL PAPILLARY CELLS

M. Yanagawa[1], H. Koul[2], T. Honeyman[3], R.Malhotra[4], C. Scheid[3] and M. Menon[2]

[1]Department of Urology
Mie University School of Medicine
Mie, Japan
[2]Division of Urology and Transplantation Surgery
[3]Departments of Physiology and [4]Pathology
University of Massachusetts Medical School
55 Lake Avenue North
Worcester, MA USA

INTRODUCTION

Alterations in renal tubular and cellular function have been reported in clinical and experimental nephrolithiasis[1-3], and it has been suggested that the presence of calcium oxalate stones within the kidney elicits these changes as a result of mechanical injury to the tubular cells. Evidence in support of this possibility was provided by studies in experimental animals where the induction of calcium oxalate crystal formation led to renal tubular damage (enzymuria, proteinuria, etc,[4,5]). The present studies assessed the possibility that oxalate produces alterations in cell function in the absence of overt crystal formation. Specifically we determined whether or not exposure to physiological levels of oxalate would elicit changes in intracellular calcium levels in isolated papillary cells from normal rat kidneys.

METHODS

Cell Dispersion

Papillary cells from normal male Wistar rats were dispersed enzymatically as described previously[3]. Briefly, rats were anesthetized, kidneys were excised and flushed with buffer A containing in mM: NaCl 118, KCl 5, $MgSO_4$ 1, $NaHCO_3$ 24, KH_2PO_4 1.2, $CaCl_2$ 0.25, glucose 10, pH 7.4. The papillary tips were minced and digested in buffer B containing in mM: NaCl 138, KCl 5, $MgSO_4$ 1, KH_2PO_4 0.3, $CaCl_2$ 0.04, glucose 5.6, HEPES 20, pH 7.4 and 2 mg/mL collagenase D (240 U/mg, Sigma Chemical Co.). The tissue was digested in a shaking 37°C water bath for 60 min (tissue was also triturated with a Pasteur Pipette at 15 min intervals). Digestion was halted by the addition of 4 volumes of collagenase-free buffer B, cells were then centrifuged 10 min at 32 x g and resuspended in enzyme-free buffer B.

Urolithiasis 2, Edited by R. Ryall *et al.*,
Plenum Press, New York, 1994

Intracellular Ca^{2+} Determination

Dispersed papillary cells were incubated for 30 min at room temperature with the membrane permeant, acetomethoxy ester form of fura 2 (fura 2 AM, 5 µM final)[6]. Dye was then removed by centrifugation and cells were resuspended in fresh buffer B. Dyeloaded cells were then transferred to the stage of an Attofluor Digital Imaging Microscope for calcium imaging. This involved collection of emitted light above 500 nm as the excitation wavelength was rapidly alternated between 340 and 380 nm. After appropriate calibration, the level of ionized calcium could be monitored "on line" in individual papillary cells, and the effects of various treatments on ionized calcium levels could be assessed. Usually 6-10 cells/microscope field were monitored in each experiment.

At the end of each experiment cells were exposed to a Ca^{2+} ionophore (bromo-A23 187). Data from cells which failed to respond to the Ca^{2+} ionophore and cells with low light intensities (eg. low concentrations of dye) were excluded from subsequent data analysis. Note that the data are the means±SEM from 3-6 different experiments using freshly dispersed cells from at least three different experimental animals.

RESULTS

In bicarbonate-free buffer containing 10 mM HEPES (pH 7.4) and 40 µM calcium, intracellular calcium levels averaged 113.7±6.5 nM (n=20) in untreated papillary cells. This value increased slightly with time such that intracellular calcium averaged 122.3±6.9 nM after 60 min.

Exposure of isolated papillary cells to sodium oxalate (400 µM) for 60 min produced a small but significant decline in intracellular calcium to a level of 96.3±5.2 nM ($p < 0.01$ relative to the 0 time controls, Table 1). Lower concentrations of oxalate had no significant effect on ionized calcium levels (data not shown).

To determine whether or not this effect of oxalate was specific or simply a consequence of intracellular acidification in the presence of a weak acid, we also examined intracellular calcium levels after 60 min exposure to 400 µM sodium acetate. This treatment produced no significant changes in resting calcium levels, however.

To determine whether oxalate was lowering intracellular calcium levels by complexing extracellular calcium, we monitored intracellular calcium levels after 60 min exposure to 400 µM EGTA. This treatment had no significant effect on ionized calcium levels in dispersed papillary cells (Table 1) despite the fact that this concentration of EGTA would lower extracellular calcium more than 1000 fold (using the EQUIL program to calculate extracellular calcium levels[7] we estimate that ionized calcium levels were approximately 37 µM in the presence of 400 µM oxalate and 7.9 nM in the presence of 400 µM EGTA).

Table 1. Effect of various agents on ionized calcium levels in renal papillary cells.

Treatment	Ionized Calcium Levels (nM) t = 0	t = 60	delta	number of cells
Control	113.7 ± 6.5	122.3 ± 6.9	+8.5 ± 4.9	20
NaOx (400 µM)	122.4 ± 8.0	96.3 ± 5.2	-26.1 ± 2.4*	20
NaAc (400 µM)	102.3 ± 3.8	115.7 ± 6.5	+13.4 ± 6.1	20
EGTA (400 µM)	116.0 ±11.2	101.8 ± 9.4	-14.2 ± 9.0	10

* $p < 0.01$ by paired t-test.

DISCUSSION

A 60 min exposure to sodium oxalate at a concentration (400 μM) that occurs normally in human renal papilla led to a small but significant reduction in the levels of ionized calcium in freshly dispersed papillary cells from normal rat kidneys. This decline in intracellular calcium could not be explained by intracellular acidification on exposure to a weak acid since exposure to sodium acetate had no effect. Nor could it be explained by extracellular complexation of calcium with oxalate since addition of EGTA (400 μM), a calcium chelator, also failed to lower ionized calcium levels. Since papillary cells possess a DIDS-sensitive transport system for oxalate[1], it seems likely that oxalate lowers intracellular calcium by entering the papillary cells and complexing with intracellular calcium. Moreover, it is possible that this process contributes to the formation or retention of calcium oxalate crystals and hence to the pathogenesis of stone disease.

ACKNOWLEDGMENTS

Supported by a grant from the NIH (DK43184).

REFERENCES

1. R Sutton and V Walker. Responses to hydrochlorothiazide and acetazolamide in patients with calcium oxalate stones. Evidence suggesting a defect in renal tubular function, *New Eng J Med* 302:709 (1980).
2. N Tessitore, V Ortalda, A Gabris et al. Renal acidification defects in patients with recurrent calcium oxalate nephrolithiasis, *Nephron* 41:325 (1985).
3. D Sigmon, S Kumar, B Carpenter, M Menon and CR Scheid. Oxalate transport in renal cortical and papillary cells from normal and stone forming animals, *Am J Kidney Dis* 17:376 (1991).
4. SR Khan, B Finlayson and RL Hackett. Experimental calcium oxalate nephrolithiasis in the rat, *Am J Path* 107:59 (1981).
5. SR Khan, PN Shevock and RL Hackett. Acute hyperoxaluria, renal injury and calcium oxalate urolithiasis, *J Urol* 147:226 (1992).
6. G Grynkiewicz, M Poenie and R Tsien. A new generation of Ca^{++} indicators with greatly improved properties, *J Biol Chem* 260:3440 (1985).
7. P Werness, C Brown, L Smith and B Finlayson. EQUIL2: A BASIC computer program for the calculation of urinary saturation, *J Urol* 134:1242 (1985).

OXALATE:BICARBONATE EXCHANGE ON THE BRUSH BORDER MEMBRANE OF RAT PROXIMAL TUBULES

K. Yamakawa, T. Kato and J. Kawamura

Department of Urology
Mie University School of Medicine, Tsu
Mie, Japan

INTRODUCTION

Recently, Na-oxalate co-transport across the brush border membrane of the rabbit renal proximal tubule[1] and oxalate(sulfate):bicarbonate exchange across the basolateral membrane of the rabbit renal proximal tubule[2] have been demonstrated. Although we reported oxalate:OH exchange across the brush border membrane of the rat renal proximal tubule[3], the physiological role of the oxalate:OH exchanger remains unclear. A high-animal-protein diet increases the urinary excretions of calcium, oxalate and uric acid and simultaneously produces acidosis[4]. Therefore, the purpose of this study was to examine the presence of oxalate:bicarbonate exchange across the brush border membrane of the rat proximal tubules.

MATERIALS AND METHODS

Brush border membrane vesicles were prepared from male Wistar rats (weighing 190-300 g), using the $MgCl_2$ precipitation method as described previously[3]. In some experiments, the basolateral membrane was isolated, using a self-orienting Percoll (Pharmacia) gradient by the method of Sactor et al[5]. In brush border membrane vesicles, the enrichment of alkaline phosphatase and Na-K-ATPase was 18.0±1.0 and 1.2±0.5-fold (means±SD), respectively, in comparison with that of the starting homogenate. By contrast, in the basolateral membranes the enrichment of alkaline phosphatase and Na-K-ATPase was 1.5±0.8 and 23.0 ±2.0-fold (means±SD), respectively.

Uptakes of radioisotopes were assayed using a slightly modified version of the rapid filtration technique[6]. Usually, 10 µL of membrane suspension were added to 90 µL of pre-warmed reaction medium at 30°C, containing radioisotopes. At appropriate intervals after the addition of the reaction medium, the reaction in a 10-µL aliquot was terminated with 3 ml of ice-cold stop solution. The mixture was immediately poured on to a 0.45 µm Millipore filter (HAWP; Millipore Corp, Mass., USA) pre-wetted with distilled water, then the tube and filter were washed twice with 3 mL of ice-cold stop solution, which contained 20 mM Tris-Hepes, pH 7.5, and K gluconate at a concentration resulting in an osmolarity equal to that of both the preincubation and reaction medium. The filters were dissolved in 5 mL Emulsifier 299™ (Packard Instruments, Ill, USA), and the radioactivity that had remained on the filters was measured with a beta scintillation counter.

Urolithiasis 2, Edited by R. Ryall *et al.*,
Plenum Press, New York, 1994

RESULTS

As shown in Fig. 1, uptakes of 0.95 μM labeled oxalate were assayed in the presence or absence of an outwardly directed bicarbonate gradient (3 mM, pH 6.5, outside/30 mM, pH 7.5 inside) across the brush border membrane. Oxalate uptake at 2 seconds by the bicarbonate gradient was 2.80±0.45 (means±SE) pmoles/mg protein. In the absence of the bicarbonate gradient (pH 6.5, outside, pH 7.5, inside), oxalate uptake at 2 seconds was 0.93±0.11 (means±SE) pmoles/mg protein. Oxalate uptake was significantly greater in the presence of the bicarbonate gradient compared to the pH gradient ($p < 0.05$). This result suggested oxalate:HCO_3 exchange occurred on the brush border membrane.

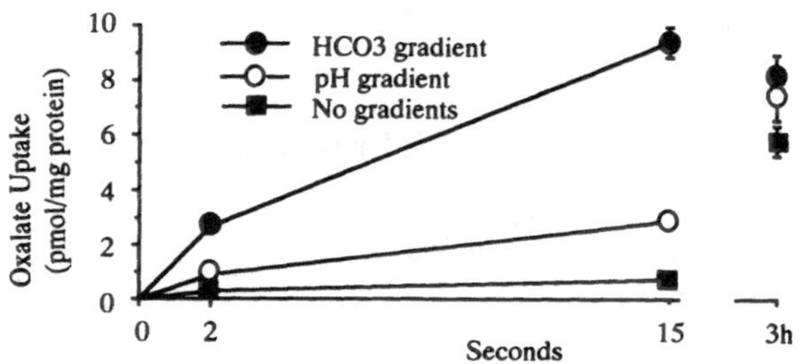

Figure 1. Effect of bicarbonate and pH gradient on oxalate uptake by brush border membrane. Membrane vesicles were preincubated with either 120 mM tetramethylammonium (TMA) gluconate, 30 mM $KHCO_3$, 54 mM Tris, and 86 mM Hepes (pH 7.5) or 120 mM TMA gluconate, 30 mM K gluconate, 56 mM Tris, and 84 mM Hepes (pH 7.5). They were then gassed with either 100% N_2 (no outward bicarbonate gradient) or with 5% CO_2, 95% N_2 (an outward bicarbonate gradient), and mixed with either 120 mM TMA gluconate, 30 mM K gluconate, 33 mM Tris, and 70 mM Hepes, 37 mM 2-(N-morpholino)ethanesulfonic acid (MES) (pH 6.5) or 120 mM TMA gluconate, 30 mM K gluconate, 56 mM Tris, and 84 mM Hepes (pH 7.5). Uptake of 0.95 μM labeled oxalate was assayed. Results are means±SEM for 3 vesicle preparations.

There was no effect of valinomycin-induced inside positive or negative potassium diffusion potentials on oxalate uptake (data not shown). Thus, it seemed likely that the outwardly directed bicarbonate gradient oxalate uptake was not due to electrical coupling, but rather to oxalate:HCO_3 exchange. The existence of oxalate:HCO_3 exchange across the basolateral membrane has been examined since Kou and Aronson observed oxalate:HCO_3 exchange across the basolateral membrane of the rabbit renal proximal tubule[2]. Both an outwardly directed bicarbonate gradient and an inside alkaline pH gradient stimulated oxalate uptake across the basolateral membrane (data not shown). It was therefore important to elucidate whether the oxalate:OH exchange mechanism was present on the brush border or basolateral membranes, since cross-contamination by the basolateral membrane might have caused the oxalate:HCO_3 exchange observed in this study. Therefore, to test this possibility, we examined the sensitivities of oxalate:HCO_3 exchange on both basolateral and brush border membrane to DIDS or 4-acetamido-4'-isothiocyanostilbene-2, 2'-disulfonic acid (SITS). Oxalate:bicarbonate exchange on the basolateral preparation was significantly more sensitive to 10 and 100 μM DIDS and 100 μM SITS than was that on the brush border preparation (Fig. 2 A & B). These results indicate that oxalate:HCO_3 exchange occurs on the brush border membranes of the rat proximal tubules.

DISCUSSION

Oxalate:OH exchange may play a minor role in transporting oxalate across the brush border membranes of rat proximal tubules since there is little difference between intracellular pH and luminal pH in proximal tubules. Bicarbonate is freely filtered from the glomerulus and undergoes net absorption in the proximal convoluted tubule[7] . In this study, oxalate was exchangeable for bicarbonate across the brush border membranes of rat proximal tubules. Therefore, oxalate:bicarbonate exchangers on both membranes may play a major role in secretion of oxalate from proximal tubules. Acidosis may increase

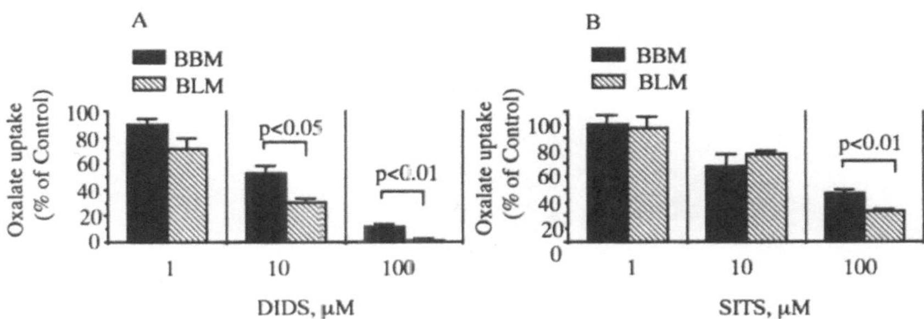

Figure 2. Effect of anion transport inhibitors on bicarbonate gradient-stimulated oxalate uptake by the preparations of basolateral membrane and brush border membrane. A; oxalate uptakes were assayed in the presence of various concentrations of 4, 4'-diisothiocyanostilbene-2, 2'-disulfonic acid (DIDS). B; oxalate uptakes were assayed in the presence of various concentrations of 4-acetamido-4'-isothiocyanostilbene-2, 2'-disulfonic acid (SITS). Membrane vesicles were preincubated with either 120 mM TMA gluconate, 30 mM $KHCO_3$, 54 mM Tris, and 86 mM Hepes (pH 7.5) or TMA gluconate, 30 mM K gluconate, 56 mM Tris, and 84 mM Hepes (pH 7.5). They were then gassed with either 100% N_2 (no bicarbonate gradients) or with 5% CO_2, 95% N_2 (an outward bicarbonate gradient). Uptake of 0.95 μM labeled oxalate was assayed at 2-s in the presence of either 120 mM TMA gluconate, 30 mM K gluconate, 12 mM Tris, 68 mM Hepes 60 mM MES (pTI 5.5), or 120 mM TMA gluconate, 30 mM K gluconate, 56 mM Tris, and 84 mM Hepes (pH 7.5). Valinomycin (60 μg/mL) was preincubated with vesicles for 30 min. Ice-cold stop solution contained 1 mM $HgCl_2$. Outwardly directed bicarbonate gradient-stimulated oxalate uptakes were obtained by subtracting oxalate uptake in the absence of bicarbonate gradient from that in the presence of bicarbonate gradient. Results are represented as % of control uptakes containing no inhibitors, and means±SEM for 3 vesicle preparations.

bicarbonate absorption and oxalate secretion whereas alkalosis may decrease bicarbonate absorption and oxalate secretion across the proximal tubules. Oxalate(sulfate):bicarbonate exchange exists across the basolateral membrane of the rat proximal tubules[2].

In our preparation of basolateral membrane, oxalate(sulfate) was exchangeable for bicarbonate (data not shown). The oxalate:bicarbonate exchange on the brush border membrane was distinguishable from that on the basolateral membrane, using sensitivities of those exchanges to DIDS and SITS. The magnitude of exchange of those ions across the basolateral membrane was greater than that on the brush border membrane. Therefore, there is the possibility that oxalate:bicarbonate exchange on the brush border membrane might be the rate determining step in oxalate transport across proximal tubules. Oxalate secretion via this oxalate:bicarbonate exchange may be related to the pathogenesis of calcium oxalate nephrolithiasis since an intracellular pH may change the intracellular transport of oxalate. Further experiments will be required to elucidate the pathogenesis of calcium oxalate nephrolithiasis.

In conclusion, these results indicates the presence of oxalate:bicarbonate exchange across the brush border membrane of the rat proximal tubules.

REFERENCES

1. C Bastlein and G Burckhardt, Sensitivity of rat renal luminal and contraluminal sulfate transport systems to DIDS, *Am J Physiol* 250:F226 (1986).
2. SM Kuo and PS Aronson, Oxalate transport via the sulfate/HCO_3 exchanger in rabbit renal basolateral membrane vesicles, *J Biol Chem* 263:9710 (1988).
3. K Yamakawa and J Kawamura, Oxalate:OH exchange across rat renal cortical brush border membrane, *Kidney Int* 37:1105 (1990).
4. WG Robertson, Epidemiology of urinary stone disease, *Urol Res*, 18:53 (1990).
5. B Sactor, IL Rosenbloom, CT Liang, and L Cheng, Sodium gradient- and sodium plus potassium gradient-dependent L-glutamate uptake in renal basolateral membrane vesicles, *J Membr Biol* 60:63 (1981).
6. PS Aronson and B Sactor, The Na^+ gradient-dependent transport of D-glucose in renal brush border membranes, *J Biol Chem* 250:6032 (1975).
7. WF Ganong, *in* "Review of Medical Physiology", Appleton & Lang, Connecticut (1989).

OXALATE UPTAKE IN A HUMAN INTESTINAL EPITHELIAL CELL LINE, CACO-2

T.R. Wandzilak[1], L. Calo[2], D.W. Bowyer[1], P.A. Davis[1], A. Borsatti[2] and H.E. Williams[1]

[1]University of California at Davis
California, USA
[2]University of Padova, Italy

INTRODUCTION

Urinary oxalate arises from both endogenous production and dietary intake. To reach the urine, dietary oxalate must pass through the gastrointestinal cells. The mechanisms of oxalate absorption by the gastrointestinal tract have been investigated using a variety of techniques. These include mucosal tissue accumulation[1], everted gut sacs[2], Ussing chamber short-circuited segments[3], membrane vesicles[4] and cell suspensions[5]. A number of tissue types (duodenum, jejunum, ileum, colon) from a variety of species (rat, rabbit, human) have been used to investigate gastrointestinal oxalate absorption[1-5].

Our laboratory was the first to use continuous cell culture to study oxalate uptake in renal cells. In contrast with other methods, the technique of cell culture allows oxalate uptake to be investigated in a defined intact viable cell type. Using this approach, we have recently characterized oxalate uptake in the porcine kidney epithelial cell line, LLCPK[6].

The human colonic cell line, Caco-2, has been used to investigate a number of transport systems. These include bile acids[7], amino acids[8], chloride[9] and potassium[10], folate[11], sodium-dependent sugar[12], and inorganic phosphate[13] transport . We now report studies on the *in vitro* uptake of oxalate in Caco-2 cells.

METHODS AND MATERIALS

Caco-2 cells (passage #16), obtained from American Type Culture Collection, were grown in DMEM/F12 containing 10% fetal calf serum with antibiotics (Gibco), and maintained by serial passage in 75 cm^2 tissue culture flasks (Falcon). Cells were grown as monolayers on 35 x 10 mm tissue culture dishes and refed every other day until confluency was reached (7-12 days), at which time they were used for experiments. All experiments were carried out at 37°C, under an atmosphere of 95% air/5% CO_2.

On the day of the experiment, growth media was aspirated and cells were washed twice in buffer A consisting of 265 mM mannitol, 5 mM Na^+ and K^+, 10 mM Ca-EGTA, 25 mM Hepes/Tris, pH = 7.4. The level of free calcium (1.8 mM) was adjusted using a Ca-EGTA buffer system and therefore the level of free oxalate could be calculated. After the initial washing, cells were preincubated for 10 minutes in buffer A containing particular experimental agents, after which uptake was evaluated.

^{14}C labeled oxalate (specific activity 103 mCi/mmol; Amersham) was used to determine cellular uptake. Non-specific trapping of oxalate was monitored using ^{3}H-

mannitol (specific activity 30 Ci/mmol; NEN). For a typical experiment, cells were incubated with oxalate for various times and the incubation terminated by washing five times with ice-cold phosphate buffered saline. 3 mL of lN NaOH was added to solubilize the cells and 1 mL aliquot was added to 10 mL scintillation fluid (Ready Safe; Beckman). Net oxalate uptake (total-trapped) was determined and expressed as total nanomoles per culture dish.

The diuretics bumetanide and furosemide in DMSO were used at a final concentration of 5 mM/1% DMSO. DIDS (4,4'-diisothiocyanostilbene-2,2'-disulfonic acid) was solubilized in buffer A. For oxalate loading experiments, cells were preincubated for 24 h in DMEM/F12 media containing 500 μM sodium oxalate.

RESULTS

Figure 1 shows uptake of oxalate over time. Uptake slowly increased to reach a maximum of 10 nmoles by 90 minutes and remained level for 3 h. Figure 1 also shows that the anion exchange inhibitor DIDS at 100 μM decreased oxalate uptake 85%.

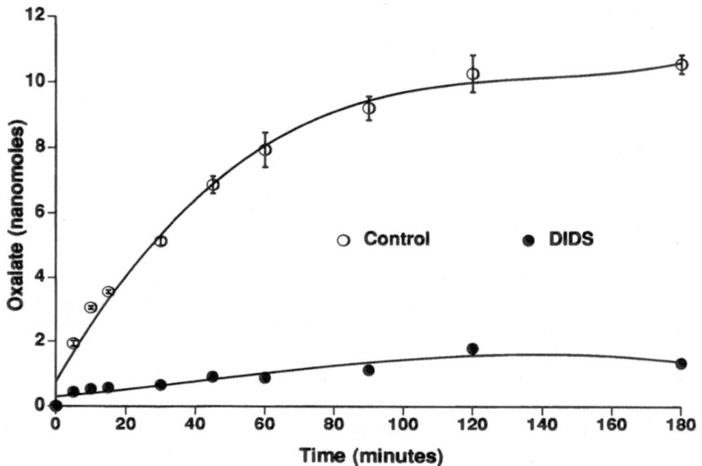

Figure 1. Oxalate uptake time course with and without DIDS.

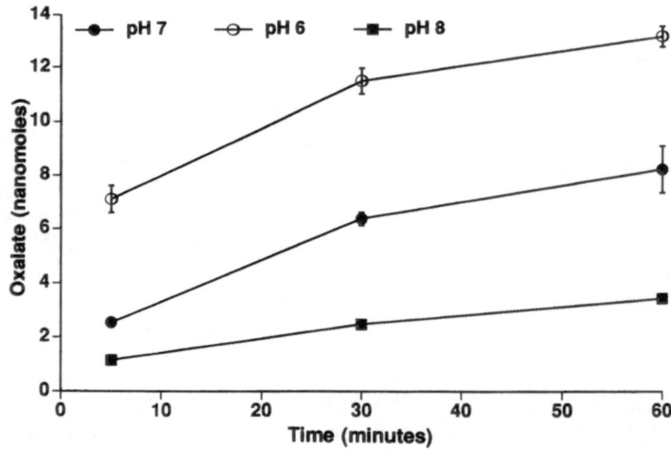

Figure 2. pH Effect.

Figure 2 shows the effect of varying the external pH on the uptake of oxalate at 5, 30 and 60 minutes. Acidification of the external media (pH = 6) increased oxalate uptake, while alkalinization of the media (pH = 8) decreased oxalate uptake.

Figure 3 shows the effect of the diuretics bumetanide and furosemide on 15 and 45 minute oxalate uptake. Oxalate uptake was decreased 90% at 15 minutes and 73% at 45 minutes with 5 mM bumetanide. Furosemide (5 mM) led to a 93% and 68% reduction of oxalate uptake at 15 and 45 minutes, respectively. Cells which had been preincubated (loaded) with 500 µM oxalate for 24 h exhibited an increase in oxalate uptake at both 30 and 60 minutes (48% and 35% respectively). The amount of trapped oxalate, estimated by using ^{3}H-mannitol, averaged 32% with a range of 21-43%. Caco-2 cell viability in buffer A, determined by Trypan blue exclusion, was 94% at 30 min; 96% after 1 h, 89% at 2 h, and 93% after 3 h.

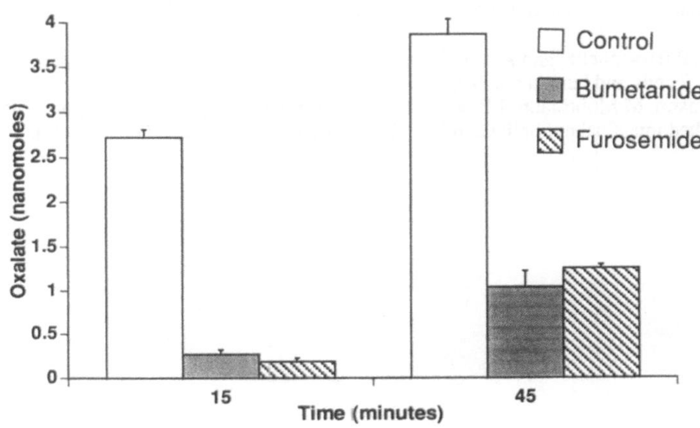

Figure 3. Effect of diuretics

CONCLUSIONS

Caco-2 cells grown in monolayer can be used as a model system to study *in vitro* the basic mechanisms of oxalate absorption by the gastrointestinal tract. Oxalate uptake in this human colonic epithelial cell line occurred in a time dependent manner, with a maximum uptake of 10 nmoles reached at 90 minutes. DIDS at a concentration of 100 µM, decreased oxalate uptake by 85%. Oxalate uptake in the Caco-2 cell line was also affected by the pH of the external medium with greater uptake at a lower pH (pH6 > pH7 > pH8). Oxalate uptake was also reduced by the diuretics, bumetanide and furosemide. These effects of DIDS, pH, and diuretics have also been reported in membrane vesicle preparations[4] and *in vitro* renal epithelial cell studies[6]. The parallel effects of DIDS, pH and diuretics on oxalate uptake in Caco-2 cells and renal epithelial cells in culture suggest that the mechanism for uptake in this cell line may be analogous to the anion exchange system of renal epithelial cells, and possibly that of red blood cells.

Caco-2 cells which had been incubated with oxalate for 24 h (oxalate loaded) demonstrated an increase in oxalate uptake. Possible explanations for this effect are: (1) oxalate loading may cause an increase in oxalate self-exchange, (2) oxalate may up-regulate its own transport, and (3) oxalate may affect other intracellular processes, such as calcium mediated second messenger systems.

REFERENCES

1. HJ Binder, Intestinal oxalate absorption, *Gastro* 67:441 (1974).
2. SE Schwartz, JQ Stauffer, LW Burgess, and M Sheney, Oxalate uptake by everted sacs of rat colon: Regional differences and the effects of pH and ricinoleic acid, *Biochim Biophy Acta* 96:404 (1980).

3. M Hatch, RW Freel, AM Goldner, and DL Earnest, Oxalate and chloride absorption by the rabbit colon:sensitivity to metabolic and anion transport inhibitors, *Gut* 25:232 (1984).
4. RG Knickelbein, PS Aronson, and JW Dobbins, Oxalate transport by anion exchange across rabbit ileal brush border, *J Clin Invest* 77:170 (1986).
5. B Pinto, and JL Paternain, Oxalate transport by the human small intestine, *Invest Urol* 15:502 (1978).
6. TR Wandzilak, L Calo, S D'Andre, A Borsatti, and HE Williams, Oxalate transport in cultured porcine renal epithelial cells, *Urol Res* 20:341 (1992).
7. IJ Hidalgo, and RT Borchardt, Transport of bile acids in a human intestinal epithelial cell line, Caco-2, *Biochim Biophys Acta* 1035:97 (1990).
8. IJ Hidalgo, and RT Borchardt, Transport of a large neutral amino acid (phenylalanine) in a human intestinal epithelial cell line: Caco-2, *Biochim Biophys Acta* 1028:25 (1990).
9. DB Burnham, and JD Fandacaro, Secretagogue-induced Fotein phosphorylation and chloride transport in Caco-2 cells, *Am J Physiol* 256:G808 (1989).
10. JA McRoberts, G Beuerlein, and K Dharmsathaphorn, Cyclic AMP and Ca^{2+} activated K^{+} transport in a human colonic epithelial cell line, *J Biol Chem* 260:14163 (1985).
11. JB Mason, R Shoda, M Haskell, J Selhub, and IH Rosenberg, Carrier affinity as a mechanism for the pH-dependence of folate transport in the small intestine, *Biochim Biophys Acta* 1024:331 (1990).
12. A Blais, P Bissonnette, and A Berteloot, Common characteristics for Na^{+} -dependent sugar transport in Caco-2 cells and human fetal colon, *J Memb Biol* 99:113 (1987).
13. I Mohrmann, M Mohrmann, J Biber, and H Murer, Sodium-dependent transport of Pi by an established intestinal epithelial cell line (Caco-2), *Am J Physiol* 250:G323 (1986).

CONTROL OF ENDOGENOUS OXALATE FORMATION IN RATS ADMINISTERED IMMOBILIZED GLYOXYLATE REDUCTASE

K.G. Raghavan, K.M. Lathika, V. Ramakrishnan and U. Tarachand

Radiation Biology and Biochemistry Division
Bhabha Atomic Research Centre
Trombay
Bombay, India

INTRODUCTION

The major part of oxalate excreted in the urine by man is synthesised endogenously from glyoxylate[1], the immediate precursor of oxalate. Therefore, conditions that lead to expansion of glyoxylate pool in the body are likely to favour its oxidation to oxalate. Glyoxylate, an extremely versatile metabolite, gets acted upon by several enzymes in human liver. These alternative pathways of glyoxylate metabolism are important in genetic disorders in which the deficiency of glyoxylate metabolising enzymes results in hyperoxaluric condition. In man, lactate dehydrogenase is the major enzyme which oxidises glyoxylate to excess oxalate, leading to hyperoxaluric status[2]. In the present study, the feasibility of converting glyoxylate to glycolate by administering polyethylene glycol-immobilized glyoxylate reductase to rats, thereby depriving the availability of glyoxylate to lactate dehydrogenase for oxalate formation, has been demonstrated.

MATERIALS AND METHODS

Immobilization of Glyoxylate Reductase

A purified preparation of glyoxylate reductase was immobilised with monomethoxy polyethylene glycol (5,000 m.wt) at pH 9.2[3].

Implantation of Osmotic Pumps

Alzet osmotic pump (model: 2001, capacity: 200 µL, release rate: 1 µL/hr) was filled with a solution of sodium glyoxylate (2 mmole/mL). Laparotomy was performed on male Wistar rats (200-250 g) under mild ether anesthesia by making a 1 cm dorsal incision. One glyoxylate osmotic pump was inserted into the peritoneal cavity of each animal and the incision sutured[4]. To animals implanted with glyoxylate pump, 10 U of polyethylene glycol glyoxylate reductase (PEG-GR) was administered intraperitoneally at 24 h post surgery. Control animals received injections of distilled water. Normal rats were divided into two groups and the first group, receiving injection of PEG-GR, served as normal treated rats. The second group, not receiving any treatment, served as normal controls. Animals were housed in metabolic cages for urine collection. Urinary oxalate was determined using oxalate oxidase[5].

Urolithiasis 2, Edited by R. Ryall *et al.*,
Plenum Press, New York, 1994

RESULTS

The native glyoxylate reductase as well as the PEG immobilised glyoxylate reductase both were found to be active over a broad pH range of 6.0 - 7.6 (Fig. 1).

The effect of administering glyoxylate and the enzyme glyoxylate reductase to rats is shown in Fig. 2.

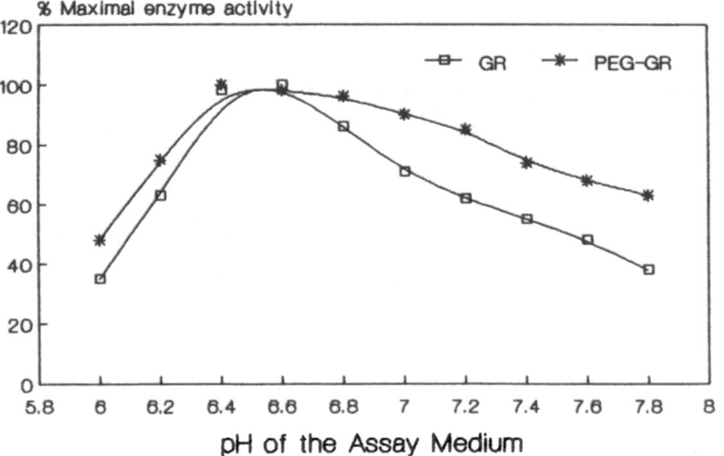

Fig. 1. The pH profile of native and immobilised glyoxylate reductase.

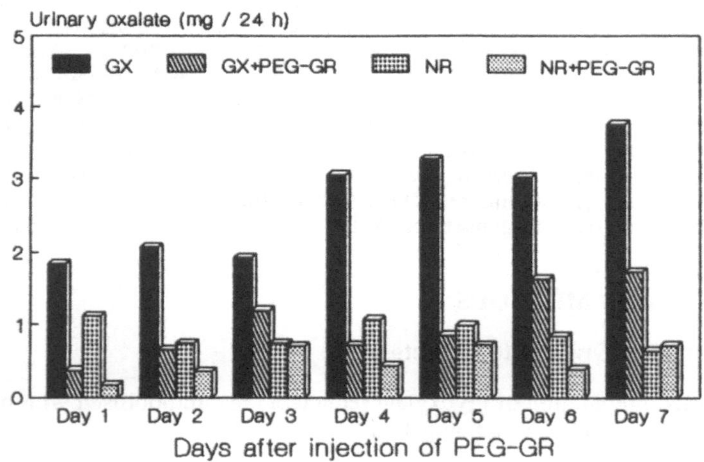

Fig. 2. Urinary oxalate excretion by rats administered with immobilised glyoxylate reductase.

Control animals implanted with the glyoxylate pump excreted a higher amount of urinary oxalate during seven days following implantation of the pump. Animals receiving glyoxylate reductase however failed to excrete such high amounts of urinary oxalate. This effect was more pronounced up to 48 h following enzyme administration. At 72 h the difference between glyoxylate administered control rats and enzyme treated animals in terms of urinary oxalate excretion was minimal. Following a second intraperitoneal injection of glyoxylate reductase on the third day after implantation, a sharp reduction in the excretion of oxalate was observed on subsequent days. Similarly, normal animals on administration of immobilized glyoxylate reductase excreted considerably lower amounts of oxalate than those not receiving the enzyme.

DISCUSSION

Since the majority of calcium oxalate is synthesized endogenously from glyoxylate, inhibiting the endogenous synthesis of oxalate by introducing competitive enzymes that can act on glyoxylate would be the most rational means of controlling hyperoxaluric conditions. However, enzymes administered in their native form get inactivated, provoke an immune response and fail to act at pH other than their own activity range. In addition, evaluation of the efficacy of enzyme therapy is limited by the non-availability of suitable animal models. In the present study, some of these limitations have been overcome. The release of glyoxylate from an osmotic pump in the animal system mimics primary hyperoxaluric condition in which such accumulation of glyoxylate has been reported. The rat, like man, is incapable of metabolising oxalate and therefore serves as a good model for studies simulating genetic disorders of oxalate metabolism. Glyoxylate reductase in native form as well as in immobilized form is active at the physiological pH of the animal. The covalent coupling of enzymes to polyethylene glycol is known to reduce the immunogenicity of the proteins and renders it suitable for repeated administrations.

The enzyme glyoxylate reductase, prevalent in the plant kingdom, reduces glyoxylate to glycolate in the presence of NADH and has a low Km value for glyoxylate. Occurrence of such specific glyoxylate reductase is not known in the mammalian system. In the present study, it is shown that animals administered glyoxylate reductase excreted lesser amount of urinary oxalate compared to animals not receiving the enzyme. This effect was more pronounced in the first two days following the injection of the enzyme. The synthesis of oxalate in the hyperoxaluric condition is a chronic process and therefore for enzyme therapy to be effective, it is essential that the administered glyoxylate reductase sustains its activity for longer duration. A second injection of the enzyme on the third day after implantation of glyoxylate pump ensured reduced excretion of oxalate by these animals on subsequent days. Thus the feasibility of manipulating the enzymes acting on glyoxylate by enzyme therapy as a means of controlling hyperoxaluria is demonstrated for the first time.

ACKNOWLEDGMENTS

The authors wish to express their gratitude to Dr B B Singh, Head, Radiation Biology and Biochemistry Division, for his keen interest, encouragement and involvement in this project.

REFERENCES

1. HE Williams, Oxalic acid and the hyperoxaluric syndromes, *Kidney Int.* 13:410 (1978).
2. HE Williams and LH Smith, Jr. Possible pathogenic mechanism for hyperoxaluria in L-glyceric aciduria, *Science* 171:390 (1971).
3. A Abuchowski, T van Es, NC Palczuk and FF Davis, Alteration of immunological properties of bovine serum albumin by covalent attachment of polyethylene glycol, *J Biol Chem* 252:3578 (1977).
4. KG Raghavan and U Tarachand, Degradation of oxalate in rats implanted with immobilized oxalate oxidase, *FEBS* 195:101 (1985).
5. KV Inamdar, KG Raghavan and DS Pradhan, Development of a procedure for the evaluation of ascorbate interference in the enzymatic assaying of urinary oxalate, *Clin Chem* 37:864 (1991).
6. JS Holcenberg, Enzyme therapy : problems and solutions, *Ann Rev Biochem* 51:795 (1982).

ENHANCEMENT OF URINARY OXALATE EXCRETION BY VITAMIN C: FACT OR ARTIFACT ?

M. Butz[1], M. Kaiser[1] and R. Fitzner[2]

[1]Urology Dept
Joseph Hospital
Paderborn, Germany
[2]Inst. Clinical Chemistry
Klinikum Steglitz, FU Berlin

INTRODUCTION

Oxalate is considered to be a major end-product of ascorbic acid metabolism. Self administration of gram amounts of vitamin C for prophylaxis and treatment of various diseases is very common. There are conflicting reports of hyperoxaluria due to the ingestion of megadoses of vitamin C[5]. Part of the confusion seems to be due to interferences of ascorbate with assay procedures for oxalate[1-4]. In the present study with high intake of vitamin C we compared urinary oxalate values before and after enzymatic elimination of ascorbate.

MATERIALS AND METHODS

Twelve healthy volunteers and 10 recurrent oxalate stone formers (RSF) ingested 6 g of vitamin C daily during five days. Twenty four h urine samples were collected twice before and during the intake of vitamin C. The vessels contained 10 mL of concentrated HCl. Urine samples (50 mL) were adjusted to pH 2.5 and an aliquot was diluted threefold with phosphate buffer (100 mmol/L, pH 7.4) 20 µL of ascorbate oxidase suspension (100 kU/L) were added to 0.21 mL diluted urine and incubated at 37°C in a vibrating water bath for 1 h. Oxalate was determined with a modified oxalate oxidase method (oxalate assay kit, Sigma Chemical Co Ltd, before and after incubation of urine with ascorbate oxidase. For comparison of oxalate excretion without and with ascorbate oxidase incubation the Wilcoxon signed rank test for paired observations was used.

RESULTS AND DISCUSSION

Oxalate assay without ascorbate oxidase

Control values of oxalate excretion in the healthy group were 240 µmol/24 h and 203 µmol/24 h in the RSF group. During intake of vitamin C oxalate excretion increased to 472 µmol/24 h in the healthy group and 633 µmol/24 h in the RSF group (fig 1.)

Urolithiasis 2, Edited by R. Ryall *et al.*,
Plenum Press, New York, 1994

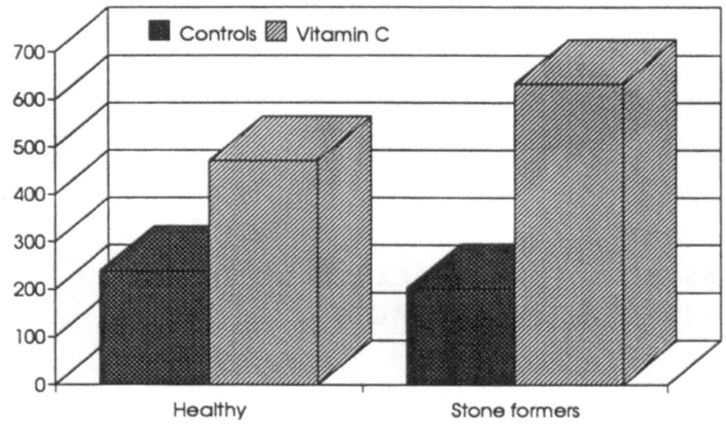

Fig. 1. Urinary Oxalate (µmol/day, median values)
Without Ascorbate-Oxidase Incubation

Oxalate assay with ascorbate oxidase incubation

Control values of oxalate excretion in the healthy group were 212 µmol/24 h and 218 µmol/24 h in the RSF group. During intake of vitamin C oxalate excretion did not increase significantly, in contrast to the untreated urine samples (fig 2). There was no different response to vitamin C loading test between the healthy group and the RSF group.

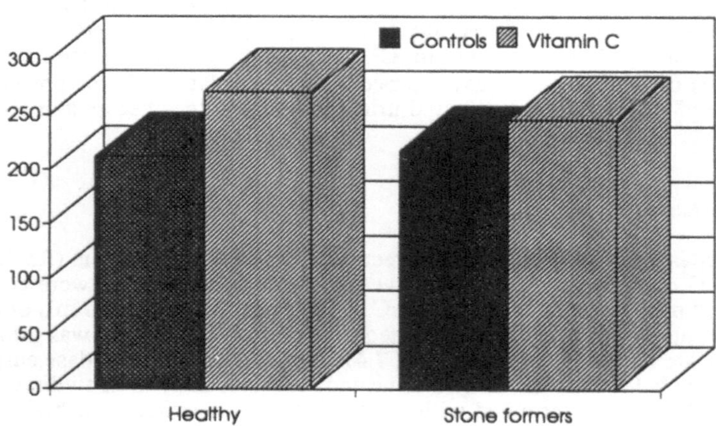

Fig. 2. Urinary Oxalate (µmol/day, median values)
Incubation With Ascorbate-Oxidase

High urinary ascorbate concentrations interfere with oxalate determination by the oxalate oxidase method. The following precautions in sample collection and sample treatment are essential for precise measurement of oxalate in case of consumption of vitamin C:

1. Acidification of urine during collection
2. Degradation of ascorbate by incubation of urine with ascorbate oxidase before analysis of oxalate
3. Adjustment of urinary pH to 2.5 before analysis of oxalate

With these precautions high intake of vitamin C does not result in increased oxalate excretion in healthy persons and in the RSF group. The potential for metabolic conversion of ascorbate is limited and that for renal excretion high. Post-micturition artifacts seem to fake hyperoxaluria when ascorbate is excreted excessively in urine[2,4]. It seems safe to conclude that ingestion of megadoses of vitamin C does not constitute a risk factor for calcium oxalate stone formation in healthy persons and stone formers.

REFERENCES

1. CS Tsao, SL Salimi, Effect of large intake of ascorbic acid on urinary and plasma oxalic acid levels. *Internat J Vit Nutr Res* 54, 245-249 (1984).
2. RAJ Conyers, R Bais, AM Rofe, N Potezny, DW Thomas, Ascorbic acid intake, renal function, and urinary oxalate excretion. *Aust NZ J Med* 25, 353-355 (1985).
3. KV Inamdar, KG Raghavan, DS Pradhan, Five treatment procedures evaluated for the elimination of ascorbate interference in the enzymatic determination of urinary oxalate. *Clin Chem* 37, 384-868 (1991).
4. BC Mazzachi, JK Teubner, RL Ryall, Factors affecting measurement of urinary oxalate, *Clin Chem* 30, 1339-1343 (1984).
5. IS Parkinson, WL Sheldon, MF Laker, PA Smith, Critical evaluation of a commercial enzyme kit (Sigma) for determining oxalate concentrations in urine. *Clin Chem* 33, 1203-1207 (1987).
6. JM Rivers. Safety of high level vitamin C ingestion, *Ann NY Acad Sci* 498, 445-454 (1987).

ERYTHROCYTE BAND 3 PROTEIN PHOSPHORYLATION MODULATES ANION TRANSPORT IN NEPHROLITHIASIC PATIENTS

B. Baggio, L. Bordin, G. Gambaro, M. Vincenti,
M. Nassuato and G. Clari

Institutes of Internal Medicine and Biochemistry
University of Padova
Italy

INTRODUCTION

Higher oxalate[1], and urate[2] transmembrane self-exchange rates, and a faster endogenous phosphorylation of band 3 protein[3], the anion carrier, were found in red cells from idiopathic calcium-oxalate renal stone formers. That these two cellular-abnormalities might be linked is an attractive hypothesis, but it is still debated whether anion transport modulation involves the phosphorylation of band 3 and other membrane proteins. To investigate this problem further, we studied the effect of agents known to interfere with membrane protein phosphorylation on anion transport.

METHODS

This study was carried out on a pool of blood obtained from three blood group compatible "idiopathic" CaOx renal stone formers with erythrocyte oxalate self-exchange rates between 1 and 1.5 x 10^{-2} min^{-1}. Erythrocytes were incubated at 37°C for 30 min (control), or under the following conditions :

(1) with 40 µmol/L Forskoline (FK) and 10 µmol/L Theophylline;
(2) with 100 nmol/L Phorbol Myristate Acetate (PMA);
(3) with 2 µg/mL *Clostridium welchii* Phospholipase C (PhC) and 1 mmol/L $CaCl_2$;
(4) with 0.6 µmol/L of the ionophore A23187 and 100 µmol/L $CaCl_2$;
(5a) and (5b) with 200 µmol/L Oleil-2Acetyl-rac-Glycerol (OAG) also for 16 h;
(6) with 10 mg/L of a Low Molecular Weight Heparin (LMWH).

After incubation the red cells were washed once in the same buffer, recovered by centrifugation, and subdivided into two fractions for determination of transmembrane oxalate self-exchange[1], and membrane protein endogenous phosphorylation[3].

RESULTS

We observed that agents able to modify ^{32}P-labelling of band 3 protein induced a concomitant modification in anion transport. Red cell preincubation with PMA, a potent activator of protein kinase C, was associated with increased ^{32}P- labelling of bands 4.1 and 4.9, while bands 3 and 2, as well as anion transport were unchanged. The same pattern was

Urolithiasis 2, Edited by R. Ryall *et al.*,
Plenum Press, New York, 1994

seen following PhC-induced increase in endogenous diacylglycerol (DAG), incubation with ionophore A23187 and Ca^{2+}, and preincubation with exogenous DAG as oleil-2-acetyl-glycerol.

Erythrocyte preincubation with LMWH (a known inhibitor of protein kinases) promoted modifications in the endogenous phosphorylation level of bands 3 and 2, and in anion transport. These observations demonstrate: 1) a close link between the band 3 phosphorylation state and its anion transport function; 2) cyclic AMP- and phospholipid-sensitive Ca^{2+}- independent protein kinases seem to be critical modulators of band 3 function.

REFERENCES

1. B Baggio, G Gambaro, F Marchini et al, An inheritable anomaly of red blood cells in idiopathic calcium oxalate nephrolithiasis correctable with diuretics. *New Eng J Med* 314: 599 (1986).
2. G Gambaro, F Marchini, M Vincenti et al, Erythrocyte urate self-exchange in idiopathic calcium nephrolithiasis. In this volume.
3. B Baggio, G Clari, G Marzaro et al, Altered red blood cell membrane protein phosphorylation in idiopathic calcium oxalate nephrolithiasis. *IRCS Med Sci* 14: 368 (1986).

CALCIUM PHOSPHATE UROLITHIASIS IN THE MALE RAT - EVIDENCE FOR ENTERIC HYPEROXALURIA AS ONE ASSOCIATED ANOMALY

E. Voss and P.O. Schwille

Mineral Metabolism and Endocrine Research Laboratory
Departments of Surgery and Urology
University of Erlangen
Germany

INTRODUCTION

Calcium phosphate stones in the rat, normally occurring sporadically[1], were observed regularly by us in males of the widely used Sprague-Dawley strain. Retrospective examination of the diet fed after weaning to the "non-stone" rats revealed a lower content of carbohydrates and crude fat, a higher overall ratio of mono-/polyunsaturated fatty acids (0.41 vs 0.39), as well as an imbalance of these [mono-unsaturated fatty acids (ratio none-stone rats/stone rats): eicosa penteneic 2.2, oleic 0.85, eicosenic 1.2; poly-unsaturated fatty acids (ratio non-stone rats/stone rats): linolic 0.79, linoleic 0.79, arachidonic 2.2, icosapenteneic 2.2]. We report on stone frequency, serum lipids, minerals and oxalate in renal tissue, urine, of the two groups of rats.

METHODS

Over 17 weeks, the stone forming MURA (n = 8) and non-stone forming CD (n = 8) animals were *ad lib* fed tap water and normal rat chow; the composition of the latter was close to the one fed by the breeder to CD in the post-weaning period. Analyses included calcium, magnesium, phosphorus, oxalate, all in renal cortex, medulla, papilla, serum cholesterol, beta-carotene (indicating intestinal fat malabsorption). Established procedures and analyses were employed throughout.

RESULTS

Pure calcium phosphate stones were present in 7 MURA and mixed stones (10% calcium oxalate) in two CD rats. MURA developed decreased beta-carotene (5±6 vs 75±6 µg/dL; $p < 0.05$) and hypercholesterolemia (106±7 vs 75±3 mg/dL; $p < 0.001$). Urine calcium, oxalate, calcium oxalate supersaturation, and renal papillary calcium were all increased (table). In MURA signs of steatorrhea and elevated unrecovered (= intake minus urinary) oxalate prevailed, so their metabolic state is consistent with enteric hyperoxaluria, probably initiated by early feeding a diet with non-optimal lipid composition[2]. Renal tissue calcium accumulation may reflect some tissue damage related to the above inadequate diet, more specifically the resulting hypercholesterolemia[3], or hyperoxaluria, or a combination of these, and ultimately leading to calcium phosphate precipitation and nephrolithiasis.

Urolithiasis 2, Edited by R. Ryall *et al.*,
Plenum Press, New York, 1994

	Renal papilla[1]				Urine			
	Calcium	Magnesium	Phosphorus	Oxalate	Calcium[2]	Oxalate[2]	RSP-CaOx[3]	RSP-HAP[3]
CD	2.5±0.1	10.7±0.3	32.7±7	217±38	0.38±0.00	0.047±0.0020	71±0.31	6.13±0.44
MURA	5.4±0.6[b]	13.7±2.5	36.0±5.6	203±39	0.07±0.0[illegible][b]	0.066±0.004[c]	2.73±0.19[a]	7.10±0.35

values expressed as x±SEM
[1]$mmol.kg^{-1}$ wet tissue; [2]$mg.mg^{-1}$ creatinine; [3]Relative supersaturation (Δ G;EQUIL-II)
[a]$p<0.05$; [b]$p <0.01$; [c]$p <0.001$

REFERENCES

1. IC Woodward, SR Khan. Phosphate urolithiasis in the rat, *in*: "Urinary System - Monographs on Pathology" ed Springer, Heidelberg (1986).
2. KM Hendricks, SH Badruddin. Weaning recommendations: The scientific basis, *Nutr Rev* 50: 125 (1992).
3. Q Zhou, S Jimi, TL Smith, FA Kummerow, The effect of cholesterol on the accumulation of intracellular calcium. *Biochim Biophys Acta* 1085: 1 (1991).

OXALATE CLEARANCES IN CALCIUM OXALATE STONE FORMING PATIENTS

D.M. Wilson, R.R. Liedtke and L.H. Smith

Mayo Clinic
Rochester
Minnesota, USA

INTRODUCTION AND RESULTS

A primary defect in membrane transport of oxalate across red blood cells has been reported in patients with idiopathic renal lithiasis (IRL); corrected by thiazide diuretics (TZ). To assess whether this is reflected in abnormal renal tubular handling of oxalate, we evaluated renal handling of oxalate in normals, patients with RL and primary hyperoxaluria (1°ox) where the load is increased. Using an enzyme assay for oxalate (DM Wilson and RR Liedtke, *Clin Chem* 37:1229 [1991]) with sensitivity of 0.5 µM/L, we established a normal value for plasma oxalate of 1.88 µM/L (N=33); range 1.11-5.15 µM/L. Since Pox is dependent on GFR, rising to 5 µM/L at GFR of 10 mL/min and rapidly below this, to 53 and 17 µM/L pre-and post-dialysis, we studied normals and stone subjects with GFR (iodothalamate clearance method) >70 mL/min. Data were developed on renal handling as follows:

Pts	N	GRF mL/min	Pox µM/L	UVox µM/min	Cox mL/min	Feox Acute	Uox mM/24	Feox '24'
Normal	(11)	109	2.7	0.17	98	0.90	0.22	0.81**
IRL	(17)	104	2.3	0.31	167	1.66*	0.28	1.57*
IRL/TZ	(4)	108	1.8	0.17	104	0.95	0.34	1.86
1° ox	(14)	100	5.9*	1.21*	261*	2.73*	1.95	3.33*

* $p < 0.05$ cf normal **Estimated from normal 24° urine oxalate n=100 and plasma oxalate (n=33)

It is apparent that Feox is sensitive to the pre-renal load of oxalate in view of the increased Pox and increased Feox in 1° ox. In the IRL group this is less clear. Feox doubles while Pox is actually lower than normal, suggesting altered renal handling independent of Pox, although the load is increased (UVox). The thiazide group is small. However, Feox and UVox in acute studies, returned to normal in this group. Conversely, Feox from 24 h oxalate data increased, commensurate with the higher 24 h oxalate load (excretion rate).

The data are consistent with net oxalate reabsorption in the renal tubule in normals with sensitive load dependent increase in tubular secretion in the 1° ox group. In IRL, Feox is doubled while Pox is not increased. The increased Feox could reflect either an altered membrane transport or increased load. Early data on thiazides do not yet confirm whether the altered Feox in IRL is corrected by TZ, as seen in the red blood cells.

TRANSEPITHELIAL OXALATE TRANSPORT IN CULTURED RENAL CELLS

T.R. Wandzilak, L. Calo, D.W. Bowyer, P.A. Davis,
A. Borsatti and H.E. Williams

University of California at Davis
Davis, California

INTRODUCTION

We have previously described the uptake of oxalate in continuous cell culture monolayers of the kidney epithelial cell line LLCPK. We now report the transepithelial transport of oxalate in the apical to basolateral direction.

MATERIALS AND METHODS

The LLCPK cells were plated at a density of 2.5×10^5 cells/mL on polycarbonate membranes (Costar) in Eagle's Minimal Essential Media supplemented with 10% fetal calf serum. Transport studies were performed using a defined mannitol/Ca-EGTA buffer system. After 5 days of continuous culture, oxalate transport in confluent cell monolayers was measured using ^{14}C-oxalate over a 2 hour incubation period. The paracellular flux of oxalate was monitored using ^{3}H-mannitol movement.

RESULTS AND CONCLUSIONS

Net oxalate transport (total - paracellular flux) increased over a 2 hour period to 10 nmoles per membrane. The paracellular movement of oxalate averaged 53±8 % over this time. The net transport of oxalate was inhibited by 40±10% using 4,4'-diisothiocyanostilbene-2,2'-disulfonic acid (DIDS), an anion exchange inhibitor, at a concentration of 100 µM. Preliminary studies in this cell line also demonstrate oxalate transport in the basolateral to apical direction, which is inhibitable by DIDS.

This system can be used as a model to study the mechanism of transepithelial oxalate transport in cultured renal cells, with the added advantage of allowing evaluation of transport in both directions.

PLASMA AND URINARY OXALATE AND GLYCOLATE IN NORMAL SUBJECTS AND IDIOPATHIC CALCIUM STONE FORMERS

R.A.L. Sutton, V.R. Walker and L. Hagen

Department of Medicine
University of British Columbia
Vancouver, Canada

INTRODUCTION

We have developed high pressure liquid chromatographic (HPLC) methods for the measurement of oxalate in plasma, and for glycolate (a metabolic precursor of oxalate) in plasma and urine. The present study reports fasting plasma and 24 h urinary oxalate and glycolate in idiopathic CaOx stone formers (SF) and normal subjects.

Ninety-three SF and 39 normal subjects were studied while consuming their normal diets and following 13 h overnight fast, urine samples (both 24 h and fasting) and fasting plasma samples were collected. A single plasma sample was used for both oxalate and glycolate assays and the ultrafiltrate was immediately cooled to 4°C and acidified with a cation-exchange resin. For the oxalate assay, Cl was removed, boric acid was added, and elution took place on a Dionex AS10 anion-exchange column using Na borate as the buffer.

	Stone Formers		Normal Subjects	
	Male (68)	Female (25)	Male (14)	Female (25)
Age±SD	49±12	35±10	41±10	39±10
Plasma, µmol/L				
Oxalate	2.1±0.1	1.9±0.1	2.2±0.2	1.9±0.1
Glycolate	4.6±0.1	4.1±0.1*	4.7±0.4	4.6±0.4
Urine, µmol				
Oxalate	392±23	294±26*	368±26	269±17*
Glycolate	550±31	399±32**	601±111	546±47

Values shown are Means±SEM, n in parentheses. * Signf. diff. from same counterpart, $p < 0.001$, ** vs normal females, $p < 0.05$.

Urinary glycolate correlated positively with urinary urea (r=0.68), urate (r=0.67), and sulphate (r=0.67), all of which were significantly lower in female SF than NS ($p < 0.05$). The lower urinary glycolate in female SF is most likely attributable to dietary factors since following fast, their urinary glycolate, urea, sulphate, and urate excretions were similar to normals.

OXALATE UPTAKE IN RIGHT-SIDE-OUT ERYTHROCYTE VESICLES OF CONTROL AND PYRIDOXINE DEFICIENT RATS

P. Kaul, S. Vaidyanathan, S.K. Thind, and R. Nath

Departments of Biochemistry and Urology
Postgraduate Institute of Medical Education and Research
Chandigarh, India

INTRODUCTION AND RESULTS

Evidence of perturbations in erythrocyte oxalate handling of stone formers has recently been reported[1,2], which strengthen the consideration that stone disease is characterized by a defect in cellular oxalate transport. Speculating that this defect may be because of aberrations at the membrane level, oxalate uptake was investigated in right side-out vesicles (ROV) prepared from erythrocytes of pyridoxine deficient (hyperoxaluric model) and control rats.

The preparation of membrane vesicles on an average contained about 70-75% ROVs as assessed by marker enzymes acetylcholinesterase and glyceraldehyde-3-P dehydrogenase (present on the external and internal side of the red cell membrane respectively). Both the pyridoxine deficient and control rats exhibited time dependent uptake of oxalate which plateaued after 15 minutes. However, pyridoxine deficient rats showed significantly higher oxalate uptake at all incubation times (160 minutes) and oxalate concentrations (0.2-l0mM) studied. Kinetic analysis revealed an increase in both Km (from 13.5 to 20mM) and Vmax (from 100 to 250 nmoles /mg protein) in ROVs of pyridoxine deficient rats as compared to controls, which suggests that either the number of transporting carrier molecules is greatly enhanced and/or the capacity of existing carriers is increased in pyridoxine deficiency.

REFERENCES

1. B Baggio, G Gambaro, F Marchini, E Cicerello, R Tenconi, M Clementi, and A Borsatti, *New Engl J Med* 314:599 (1986).
2. A Borsatti, *Kid Int* 39:1283 (1991).

PRIMARY HYPEROXALURIA TYPE 2: SPECIFIC AND SIMPLE HIGH-PERFORMANCE LIQUID CHROMATOGRAPHIC DETERMINATION FOR L-GLYCERIC ACID IN BODY FLUIDS

M. Petrarulo, M. Marangella, D. Cosseddu and F. Linari

Renal Stone Laboratory
Mauriziano Hospital
Turin, Italy

INTRODUCTION

Urinary L-glycerate was formerly determined by an isotope-dilution colorimetric method coupled with both the chiralselective enzyme reaction and evaluation of the optical rotatory dispersion curve. More recently, gas-liquid chromatography has been used for the screening of glyceric aciduria, but this technique requires a further enantio-selective gas-chromatographic separation for characterizing the L-(S)-configuration. These methods require liquid-liquid extraction, evaporation to dryness of the extract, chromatographic purification of the reconstituted extract, lyophilization of the eluate and derivatization of the residue. The resulting procedure is therefore complex and time consuming.

We describe a reversed-phase high performance liquid chromatographic (RP-HPLC) technique for determining L-glyceric acid in body fluids. The system is based on the derivatization of the acid by means of incubation with lactate dehydrogenase and NAD in the presence of phenylhydrazine. The enzyme allows the oxidation of L-glycerate into ß-hydroxypyruvate which is converted in turn into the related phenylhydrazone. The UV-absorbing derivative is determined using RP-HPLC. The sensitivity is at the µmol/L level and few microliters of sample are required. The imprecision relative standard deviation is 4.0% and the recovery is 94.5±8.8%. The procedure, based on the highly enantio-selective enzymatic conversion, does not require further confirmation of the configuration.

RESULTS

L-Glycerate plasma concentration in six healthy subjects and six patients on dialysis for oxalosis-unrelated CRF was <5 µmol/L. Similar results were found in five dialysed patients with the glycolic variant of primary hyperoxaluria (PH). One patient with systemic oxalosis had 887 µmol/L plasma level, and this established a diagnosis of PH type 2.

TRANSMEMBRANE OXALATE FLUX STUDIES IN INTACT ERYTHROCYTES OF MAGNESIUM-DEFICIENT RATS

V. Rattan[1], S.K. Thind[1], R.K. Jethi[2] and R. Nath[1]

[1]Department of Biochemistry
Postgraduate Institute of Medical Education and Research and
[2]Department of Biochemistry, Panjab University
Chandigarh, India

INTRODUCTION

Defective membrane function has been implicated to be the primary lesion underlying the cellular disturbances in magnesium deficiency[1]. Hyperabsorption of oxalate in intestinal brush border membrane vesicles prepared from magnesium-deficient rats suggested that it might involve a defect in the cellular transport of oxalate[2]. To further test this hypothesis, transmembrane oxalate flux rate constant 'K'[3] was measured in intact erythrocytes from rats fed a magnesium deficient diet (70 mg/kg diet) and from pair-fed controls fed a magnesium-supplemented diet (722 mg/kg diet) for thirty days.

RESULTS AND DISCUSSION

Magnesium deficiency in rats led to a significant ($p < 0.001$) hypomagnesemia, hypomagnesuria and hyperoxaluria in the experimental animals. Oxalate flux rate constant 'K' was -0.53±0.03 (Mean±SEM x 10^{-2} min^{-2}) in the pair-fed control group and -1.07±0.04 (Mean±SEM x 10^{-2} min^{-1}) in the magnesium-deficient group with a significant ($p < 0.001$) difference between the groups. The study indicates that hyperoxaluria also increases the transmembrane oxalate flux in erythrocytes, suggesting a generalised defective membrane function in magnesium deficiency.

REFERENCES

1. S Tongyai, Y Rayssiguier, C Motta, E Gueux, P Maurois, and FW Heaton, *Am J Physiol* 257:C 270 (1989).
2. V Rattan, SK Thind, RK Jethi, and R Nath, *Magnesium Res* 4:227 (1991).
3. B Baggio, G Gambaro, F Marchini, E Cicerello, R Tenconi, M Clementi and A Borsatti, *N Engl J Med* 314:599 (1986).

ERYTHROCYTE OXALATE SELF-EXCHANGE RATES IN JAPAN

T. Kato, K. Yamakawa and J. Kawamura

Department of Urology
Mie University School of Medicine, Tsu
Mie, Japan

INTRODUCTION

Hyperoxaluria is one of the risk factors in idiopathic calcium oxalate nephrolithiasis, but detail of oxalate metabolism is still unclear. Baggio et al[1] reported that the abnormal oxalate influx rate of erythrocytes suggests a defect in cellular transport of oxalate resulting in higher intestinal absorption and urinary excretion of oxalate. In order to clarify the effect of an abnormality in oxalate transport in erythrocytes in calcium oxalate nephrolithiasis, we determined erythrocyte oxalate self-exchange rates in 30 patients with recurrent calcium oxalate nephrolithiasis and 20 healthy controls.

MATERIALS AND METHODS

Subjects consisted of 30 patients with recurrent calcium oxalate nephrolithiasis and 20 healthy controls. The median age was 39.6 (range 29-67) years in the patients, and 30.4 (range 25-45) years in controls. All patients had normal renal function. The controls had normal renal function and did not have past or family histories of urolithiasis. The suspension of erythrocytes was prepared by the method of Baggio et al[1].

RESULTS AND CONCLUSION

Since the erythrocyte oxalate self-exchange rate at 4°C did not proceed under an initial condition rate, experiments were performed at 0°C at which erythrocyte oxalate self-exchange rate could be determined. The erythrocyte oxalate self-exchange rates in patients and in controls were $-1.11 \times 10^{-2} \pm 0.22 \times 10^{-2}$ per min (mean±SD), and $0.77 \times 10^{-2} \pm 0.18 \times 10^{-2}$ per min (mean±SD), respectively. The erythrocyte oxalate self-exchange rate was greater in the patients as compared to the controls ($p < 0.001$). Urinary oxalate excretions in patients and controls were 33.8±11.1 and 24.6±4.2 mg/g creatinine (mean±SD) respectively, and this was not significantly related to the erythrocyte oxalate self-exchange rate in both controls and patients.

Although the erythrocyte oxalate self-exchange rate was elevated in Japanese patients with recurrent calcium oxalate nephrolithiasis, we could not find what the erythrocyte oxalate self-exchange rate indicated in these patients.

REFERENCES

1. Baggio B, Gambaro G, Marchini F, Cicerello E, Tenconi R, Clementi M and Borsatti A, An inheritable anomaly of red-cell oxalate transport in "primary" calcium nephrolithiasis correctable with diuretics, *New Engl J Med* 314:599-604 (1986).

SECTION III: PHYSICOCHEMISTRY, PROMOTERS AND INHIBITORS

CRYSTAL GROWTH AND NUCLEATION RATES FOR CALCIUM OXALATE IN 92% FRESH URINE IN A CONTINUOUS CRYSTALLISER

J.P. Kavanagh[2], N.J. Blacklock[2], S. Nishio[2] and J. Garside[1]

[1]Department of Chemical Engineering
UMIST
Manchester, UK
[2]Department of Urology
University Hospital of South Manchester, UK

INTRODUCTION

Mixed suspension, mixed product removal (MSMPR) continuous crystallisation has been applied in many studies of calcium oxalate crystallisation. It is particularly appropriate for urolithiasis research, as Finlayson has shown that crystallisation within the urinary tract can be modelled effectively by a series of MSMPR continuous crystallisers[1]. The MSMPR method gives a better representation of the renal environment than others because it reaches a steady state with a constant supersaturation and operates with continuous flow. As a result, individual crystals only remain in the suspension for a relatively short time, not much different from transit times through the kidney. Another advantage of the MSMPR technique is that nucleation and growth rates can be measured simultaneously, but independently, in units of no/min/mL and µm/min.

A major limitation of the MSMPR approach has been the difficulty of applying it to whole urine. Most studies have used artificial urine, sometimes with up to 5% urine included[2]. The main reason for using dilute urine is the volume required; in a conventional system between 2 and 4 litres of undiluted urine would be needed[2]. We have developed a small scale crystalliser[3] with a volume of 0.02 litres and hence a total feed solution requirement of about 0.2 litres. This system has been used to study the effect of γ-carboxyglutamic acid[4] in artificial urine and here we describe some applications to fresh urine.

MATERIALS AND METHODS

Twenty two urine samples were obtained from healthy men with no history of urolithiasis and from twenty one male recurrent renal stone formers. Specimens were adjusted to pH 6.0 with HCl or NaOH and analysed for calcium and oxalate. Ten samples from each group were centrifuged (3,000 x g for 5 minutes at 37°C) and the remainder were filtered at 37°C using an 8 µm cellulose nitrate prefilter (Sartorius) and 0.45 µm membrane filter (Millipore).

Urine (at 37°C) was used as one of three feed solutions to the crystalliser, pumped at 92% of the total flow rate. Calcium chloride and sodium oxalate were each fed in at 4% of the total flow. The concentrations of these latter two solutions were adjusted at the start of each experiment to give concentrations in the crystallisation chamber (if no precipitation occurred) of 12mM calcium and 2.4mM oxalate. The experiments were completed about 90 minutes after the sample was passed. For details of the crystalliser see reference 3, and reference 5 for details of the particle size analysis and calculations.

The statistical significance of results was assessed by analysis of variance (ANOVA). Logarithmic transformations were used for growth and nucleation rates to give normally distributed data. Including other factors, such as urinary ionic strength, in the statistical treatment of results allows their influence to be taken in account. Multiple linear regression was used to examine which additional factors were significant and independent, and should therefore be included in calculations. Tukey's critical range test was used to identify significant effects ($p<0.05$) in ANOVAs with multiple groups.

RESULTS

Growth rates were almost identical in control and stone forming groups (Table 1). Nucleation rates were higher in the control samples compared to the stone formers, while the supersaturation within the crystalliser was lower with control samples than with urines from stone formers (Table 1). The supersaturation was estimated from the calculated calcium and oxalate remaining in solution, taking complexation by other ions into account[6].

Table 1. Comparison of controls and stone formers.

	Controls	Stone Formers	p
Growth rate (µm/min)	0.64	0.66	>0.1
Nucleation rate x 10^{-3} (no./min/mL)	95.0	43.0	0.003
Supersaturation index	11.0	14.5	0.003

Filtered samples had higher nucleation rates and lower supersaturations than those that had been centrifuged (Table 2).

Table 2. Comparison of filtered and centrifuged samples.

	Filtered	Centrifuged	p
Growth rate (µm/min)	0.66	0.64	>0.1
Nucleation rate x 10^{-3} (no./min/mL)	83.0	48.0	0.003
Supersaturation index	11.7	13.6	0.01

An inverse relationship between the growth rates and log nucleation rates was noted. Plots of Ln nucleation rate against 1/growth rate were significantly correlated. Analysis of variance of the regression data also showed the relationship was highly significant ($p=0.00001$), with the intercepts ($p=0.0006$), but not the slopes ($p>0.05$) of the regression lines being influenced by the type of sample. The intercept of the control filtered samples was significantly greater than the other three samples and the intercept of the stone formers/centrifuged urine was significantly lower than the others ($p<0.05$) (Table 3).

DISCUSSION

Although we were able to avoid significant dilution of the fresh urine used in this study, some preliminary treatment to remove any crystals or cellular debris that could act as sites for heterogeneous nucleation was considered necessary. In order to compare samples from different patients we adjusted all samples to the same starting concentrations of calcium and oxalate. To bring about sufficient crystallisation, these concentrations were necessarily somewhat higher than physiological. In the absence of crystallisation this would bring about very high supersaturation, which provides the thermodynamic driving force for the crystallisation. Nevertheless, the steady state supersaturation actually achieved within the crystalliser was not much higher than is typical of whole urine.

Table 3. The relationship between Ln nucleation rate and 1/growth rate.

	Ln nucleation rate intercept	Nucleation rate relative to A
A Controls/filtered	7.40	100%
B Controls/centrifuged	6.87	59%
C Stone formers/filtered	6.77	53%
D Stone formers/centrifuged	6.08	27%

Slope = 2.72 ANOVA, p(slope) >0.05; p(intercepts) =0.0006
Tukey critical range test, A > B,C,D; D < A,B,C

The response of urines to this challenging load of calcium and oxalate differed between controls and recurrent stone formers and also depended to a lesser extent on whether the urines had been pre-treated by filtration or centrifugation. Differences were observed in nucleation rate (higher in control samples compared to stone formers urine) but not in growth rate. As a result the supersaturation was lower in the control group. This reproduces observations with whole urine in which stone formers have often been reported to have higher supersaturations than controls eg[7].

The thermodynamic driving force for each sample tested would be very similar, as they all had the same starting concentrations of Ca and oxalate. Their response can be viewed as a balance between the two competing mechanisms of supersaturation relief, nucleation and growth. This balance is reflected in the linear relationship between Ln nucleation rate and 1/growth rate. As the slopes of the lines for the different groups are the same but the intercepts are different, the proportional difference in nucleation rate will be the same for any given growth rate. Taking the nucleation rate of the control/filtered samples as 100%, then for all values of G, the nucleation rates of the other groups will be about 59% (control centrifuged), 53% (stone formers/ filtered) and 27% (stone formers/ centrifuged) (Table 3). The effect of the lower nucleation rate in the stone formers is to bring about a higher steady state supersaturation and this may be a crucial factor distinguishing stone formers from controls. Any crystals which become temporarily lodged within the kidney will have a greater chance of growing and developing into a clinically significant stone when growing in a higher supersaturation environment.

REFERENCES

1. B Finlayson, The concept of a continuous crystalliser. Its theory and application to *in vivo* and *in vitro* urinary tract models, *Invest Urol* 9: 258 (1972).
2. KE Springmann, GW Drach, B Gottung & AD Randolph, Effect of human urine on aggregation of calcium oxalate crystals, *J Urol* 135: 69 (1986).
3. S Nishio, JP Kavanagh & J Garside, A small-scale continuous mixed suspension mixed product removal crystalliser, *Chem Eng Sci* 46: 709 (1991).
4. S Nishio, JP Kavanagh, EB Faragher, J Garside & NJ Blacklock, Calcium oxalate crystallisation kinetics and the effects of calcium and γ-carboxyglutamic acid, *Br J Urol* 66: 351 (1990).
5. JP Kavanagh, Methods for the study of calcium oxalate crystallisation and their application to urolithiasis research, *Scan Microsc* 6: 685 (1992).
6. PG Werness, CM Brown, LH Smith & B Finlayson, EQUIL2: a basic computer program for the calculation of urinary supersaturation, *J Urol* 134: 1242 (1985).
7. WG Robertson, DM Peacock, RW Marshall, DH Marshall & BEC Nordin, Saturation-inhibition index as a measure of the risk of calcium oxalate stone inhibition in the urinary tract, *New Engl J Med* 294: 249 (1976).

INHIBITORY ACTION OF CITRATE ON CALCIUM OXALATE MONOHYDRATE CRYSTAL NUCLEATION AND GROWTH

D.L. Purich, P. Antinozzi and C.M. Brown

Department of Biochemistry and Molecular Biology and
Center for Urolithiasis and Pathological Calcification
University of Florida College of Medicine
Gainesville, Florida

INTRODUCTION

Because calcium oxalate monohydrate is an important urinary stone salt, considerable attention has been devoted to the analysis of COM crystallization in an attempt to understand the associated processes of nucleation, crystal growth, aggregation, and breakup. Considering the paracrystalline dendritic nature of oxalate stones, many urinary agents may bind to and thereby alter crystal growth and/or adhesion to other components in the stone lamina. In addition to calcium oxalate hydrates, other carboxylic ligands that may interact with COM crystals include citrate[1], phosphocitrate[2], the aspartate-rich uropontin[3], γ-carboxyglutamate-containing nephrocalcin[4], and the cytoplasmic membrane component phosphatidylserine (F Novin, CM Brown, and DL Purich, unpublished findings). These agents each contain one or more carboxyl groups which potentially could substitute for oxalate carboxyls within the crystal lattice. Recently, our laboratory conducted crystal growth and nucleation rate measurements to investigate how agents might promote or inhibit COM crystallization[5]. Our work has centered on the inhibitory action of citrate, a prominent urinary metabolite that can sequester calcium ions and thereby reduce the relative supersaturation of urine with respect to calcium oxalate monohydrate[1]. In the present study we describe nucleation and growth kinetics in experiments for which we used the EQUIL speciation software to maintain the free, uncomplexed calcium ion concentration and hence avoid effects of changes in relative supersaturation. We also explored the relationship of these kinetic data to crystal morphology changes arising in the presence of citrate, and we have rationalized structural changes in terms of crystal face-specific interactions with citrate and/or calcium citrate complex. This has required the application of molecular recognition theory[6,7] to calculate the electrostatic surface potential of stable calcium oxalate monohydrate crystal surfaces and to characterize stereochemical interactions of carboxylic ligands with COM crystal faces. Our work supports the conclusion that citrate binds selectively to the (010) face, thereby blocking further crystal growth on that face.

RESULTS

By conducting COM nucleation and growth measurements under conditions of comparable initial supersaturation, we obtained the parameters listed in Table 1.

Urolithiasis 2, Edited by R. Ryall *et al.*,
Plenum Press, New York, 1994

Table 1 Influence of citrate on calcium oxalate monohydrate crystal nucleation and growth at constant relative supersaturation[a].

	(-) Citrate	(+) Citrate
A. Gibbs-Thomson Nucleation Kinetics $[J = A \exp(B \ln^{-2}(RS))]$		
ΔE (erg-cm^{-2})	28.9	26. 8
Particle Production (number-L^{-1})	3.46×10^7	7.14×10^7
B. Parabolic Growth Kinetics $[-dRS/dt = kst\,(RS_t - RS_{oo})^2]$		
k_{growth} (S^{-1}-cm^{-2})	2.36×10^{-6}	0.66×10^{-6}

[a]Experiments were conducted at RS 19.7 using the EQUIL speciation software to maintain uncomplexed calcium ion at identical concentrations for samples prepared in the absence or presence of 3.5 mM citrate.

We found that the presence of citrate had little or no effect on the thermodynamic barrier to nucleation. Overall particle production, however, was higher in the presence of citrate, a finding that might appear to reflect a promotion of nucleation. The origin of these apparently conflicting observations can be traced to differences in crystal growth rate behavior. We found that 3.5 mM total citrate (*ie.*, calcium-citrate and uncomplexed citrate) reduced growth rates by 3- to 4-fold, thereby extending the time period over which such samples remained supersaturated. Net particle formation is the multiplicative product of nucleation rate and this time period; thus, any restrained growth by citrate will necessarily lead to increased particle production, even though citrate has no significant effect on interfacial free energy of nucleation.

We next conducted scanning electron microscopy studies of crystals grown in the absence or presence of citrate to search for any distinctive changes in crystal morphology (Fig 1AB). We found the typical elongated, hexagonal geometry of the (010) face formed in the absence of citrate; however, we observed more equilateral crystals with obviously rounded edges arising in the presence of citrate. In addition, crystals grown with citrate tended to be flatter, indicating that the dimension perpendicular to the (010) face is substantially smaller. Determination of the major and minor axial dimensions for individual crystals resulted in the findings presented in Fig lC. These pairwise crystal measurement data are consistent with altered morphology that occurs in a manner independent of COM particle size. This suggests that citrate binds to one or more crystal faces and alters the growth behavior at specific regions. The resulting crystal geometry can be generated by restricting growth on the (010) face. The diagrams in Fig. lD and lE depict crystal morphology in the absence and presence of citrate, assuming normal growth and no growth onto the (010) plane, respectively.

To understand the nature of face-specific ligand interactions, we considered COM crystal growth from the perspective of molecular recognition theory. The basic idea is that ligands can often bind in a stereospecific (or stereoselective) manner, thereby blocking further accretion on one or more crystal faces[6]. Briefly, this approach involves (a) generating an infinite series from surface lattice vectors with a physical model several layers deep, (b) evaluating the so-called Madelung potential with appropriate summation, and (c) calculating the local electrostatic potential above the surface using a probe proton to estimate attraction and repulsion. Our calculations provided strong evidence that the surface charge and spacing is such that carboxylic ligands can be expected to show preferential interactions with particular faces (CM Brown, EA Colbourn, and DL Purich, unpublished findings). For the present, however, one can obtain a preliminary indication of how citrate may bind to the (010) crystal plane by relying on other modelling methods (Chem3D Molecular Modeling Software, Cambridge Scientific, Inc, Cambridge, MA). In the case shown schematically in Fig. 2, the oxygen atoms of the citrate molecule are illustrated with a special pattern to aid the reader's viewing of citrate's interaction with the (010) face.

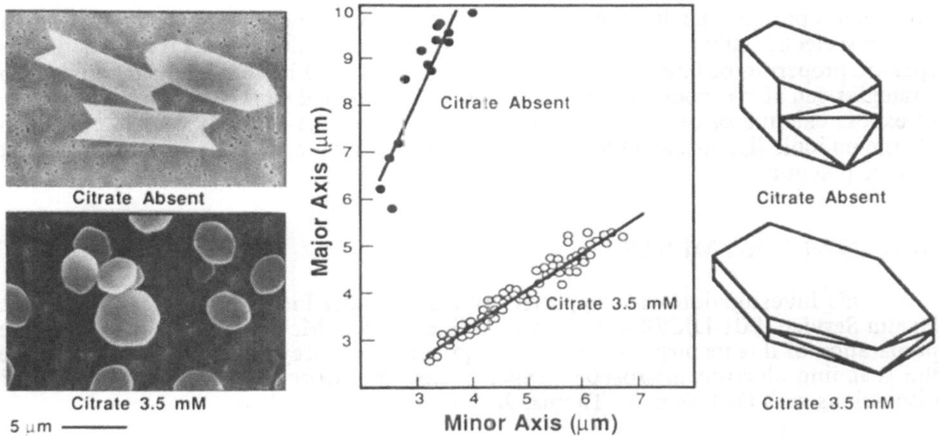

Figure 1. Characterisation of COM crystals formed in the absence or presence of citrate.
A.(upper left) Scanning electron micrographs of control crystals from the lag-phase nucleation studies showing typical COM morphology. Crystals were collected about 10 min after combining calcium chloride and sodium oxalate solutions. Initial relative supersaturation was 19.7 for both control and citrate-containing systems; complexation of calcium was compensated for by increases in calcium suggested by the EQUIL speciation program, with background electrolytes: 0.1 M NaCl and 0.01 M HEPES buffer.
B.(lower left) Scanning electron micrographs (SEM) of crystals grown in the presence of 3.5 mM citrate showing flattened morphology. **C.**(center) Comparison of control and citrate-grown crystals by numerous measurements of prominent hexagonal crystal face axes with crystals from SEM data in Fig. 1. **D,E**(upper right, lower right) diagrammatic illustration of typical morphologies and faces of COM crystals nucleated in the presence and absence of citrate, respectively.

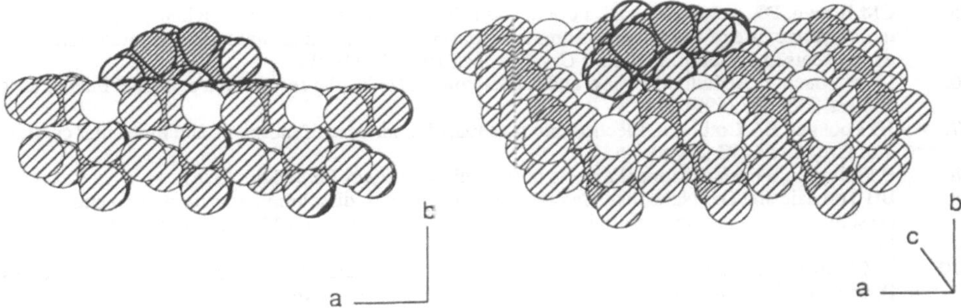

Fig. 2. Space-filling representations of calcium • citrate complex interacting with the (010) calcium oxalate monohydrate crystal face.

DISCUSSION

COM nucleation and growth have been extensively analyzed using such parameters as interfacial free energies of nucleation and rate constants for crystal growth and dissolution[8-10]. Nonetheless, these macroscopic characteristics are insufficient to permit comprehension of the microscopic nature of interactions at all surfaces, edges, and defect sites. Indeed, they necessarily represent properties averaged over all geometries, and this is true for electrophoretic mobility *(ie,* Zeta potential) data as well as the double-layer behavior of crystal-solute-solvent interactions. The only exception may be relative supersaturation, which measures the equilibrium constant for solute coexistence with the most stable crystal face. Cursory examination of the calculated electrostatic potential maps for various stable faces of crystals reinforces the view that each surface is certainly not geometrically equivalent[6], and such structural anisotropy will influence kinetic and thermodynamic properties at the microscopic level. In this respect, molecular recognition approaches promise the opportunity for translating crystal morphology and lattice structure data into electrostatic potential maps that may provide a rational basis for predicting face-specific properties of interactions between crystal salts and inhibitors. Such computational strategies can also account for crystal surface rumpling and surface charge in the presence of excess calcium or oxalate ions. Likewise, effects of pH and temperature can alter the charge on ionic ligands, and these effects can be interpreted as changes in the electrostatic surface potential.

ACKNOWLEDGEMENTS

This investigation was supported by the National Institutes of Health, US Public Health Service (P01-DK20586). We would like to thank Ms Melissa Grace for her careful preparation of this manuscript, and we are grateful to Jeffrey Opalko for his assistance in the scanning electron microscopy experiments. Our work also benefited from critical discussions with Professors W Thomas Jr, S Khan, and R Hackett.

REFERENCES

1. PA Antinozzi, CM Brown, and DL Purich, Calcium oxalate monohydrate crystallization: citrate inhibition of nucleation and growth steps, *J Cryst Growth*, in press (1992).
2. CF Richardson, M Johnsson, FK Bangash, VK Sharma, JD Sallis, and GH Nancollas, The effects of citrate and phosphocitrate on the kinetics of mineralization of calcium oxalate monohydrate, *Mater Res Soc Symp Proc* 174, 87 (1990) .
3. H Shiraga, W Min, WJ VanDusen, MD Clayman, et al, Inhibition of calcium oxalate crystal growth in vitro by uropontin: another member of the aspartic acid rich protein superfamily, *Proc Natl Acad Sci USA* 89: 426 (1992).
4. Y Nakagawa, S Deganello, C Chou, M Ahmed, and FL Coe, Isolation from human calcium oxalate renal stones of nephrocalcin, a glycoprotein inhibitor of calcium oxalate crystal growth, *J Clin Invest* 79: 1782 (1987).
5. CM Brown, DK Ackermann, DL Purich, and B Finlayson, Nucleation of calcium oxalate monohydrate: use of turbidity measurements and computer-assisted simulations in characterizing early events in crystal formation, *J Cryst Growth* 108: 455 (1991).
6. I Weissbuch, L Addadi, M Lahav, and L Leiserowitz, Molecular recognition at crystal interfaces, *Science* 253: 637 (1991).
7. JD Foot and EA Colbourn. Electrostatic potentials for surfaces of inorganic and molecular crystals, *J Mol Graphics* 6: 93 (1988).
8. B Finlayson, Physicochemical aspects of urolithiasis, *Kidney Int* 3: 344 (1978).
9. BB Tomazic and GH Nancollas, Calcium oxalate hydrates, dissolution, transformation and crystallization studies in "Urolithiasis: Clinical and Basic Research," LH Smith, WG Robertson and B Finlayson, eds., Plenum, New York, 392 (1981).
10. CM Brown and DL Purich, Physical-chemical processes in kidney stone formation in: "Disorders of Bone and Mineral Metabolism," FL Coe and MJ Favus, eds., Raven Press, New York, 613.

A NEW CONTINUOUS FLOW MICROSYSTEM OF CRYSTALLIZATION IN GELS AS A MODEL OF URINARY STONE FORMATION

W. Achilles, R. Kothe, U. Jöckel, C. Schalk and H. Riedmiller

Philipps-Universität Marburg
Urologische Universitätsklinik
D-3550 Marburg/Lahn, Germany

INTRODUCTION

Urinary stone formation, including nucleation, growth and agglomeration of crystals, occurs in a fixed, gel-like state from a flow of supersaturated urine. In general, these facts have not been taken into account in conventional models of crystallization applied in urolithiasis research. Therefore, we developed a new flow model of crystallization in gels (FCMG), which is briefly described in this paper. The model represents a dynamic constant composition method which is aimed at simulating physiological conditions of urinary stone formation as far as possible.

METHODS

Two solutions (S_A, S_B), each containing another component (A or B) of a sparingly soluble binary electrolyte AB, are mixed in a flow system. The resulting mixture S_{AB}. which is supersaturated with respect to AB, is conducted over a gel layer. Details of solution compositions can be found in the paper immediately following this one in this volume. Crystallization takes place within the gel matrix or on its surface and can be qualitatively or quantitatively followed by microscopy and/or microphotometry (Figure 1).

The solutions to be mixed (S_A, S_B), are transported by a 24-channel peristaltic pump from their reservoirs to a special reaction plate. This plate, which is located in a thermostatter, contains 12 channels with gel. Gel matrices may be used with or without seed crystals. The mixing process of solutions is achieved at an influx tip just above the gel layer. After passing the gel surface, the mixture S_{AB} is removed by another pump.

The following general conditions were applied: a) gel matrix: 1% (w/w) agar-agar; b) gel volume per channel: 0.75 mL; c) mixing ratio of S_A and S_B: 1:1 (optimal); d) flow rate of S_{AB}: 0.5 mL/min; e) duration of experiment: 4-40 h; e) temperature: 37±1°C.

APPLICATIONS AND DISCUSSION

Using a gel matrix seeded with calcium oxalate monohydrate (COM), growth curves of this crystal phase in gel could be registered microphotometrically by discontinuous measurement of scattered light intensity (dark field mode) as a function of time.

Experiments on the effects of different urinary constituents were carried out with artificial urine at pH 6.0. A measuring example is shown in Figure 2.

It was shown, on the basis of normalized physiological concentrations, that the influence of variation of oxalate on the crystal growth rate of CaOx is much more pronounced than that of a corresponding variation of calcium. Chondroitin-4-sulfate did not show a remarkable effect on inhibition of COM growth up to a concentration of 100 mg/L.

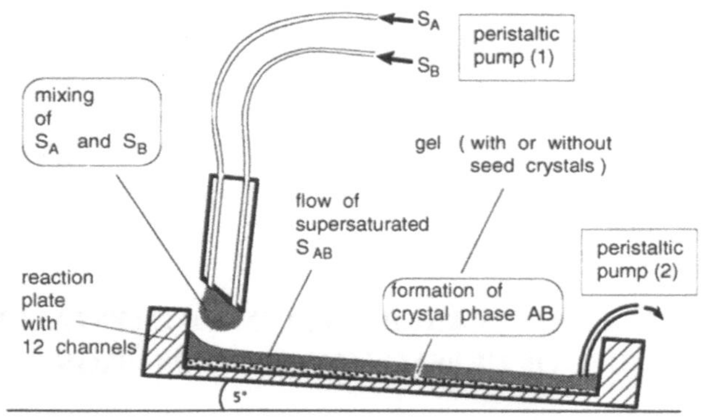

Figure 1. Schematic diagram of the Flow Model of Crystallization in Gels

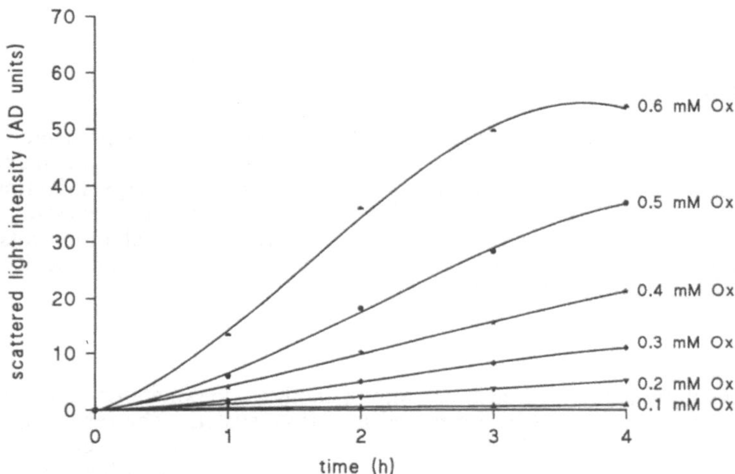

Figure 2. Optically registered crystal growth curves of COM in gel from artificial urine at pH 6: effect of variation of total oxalate concentration

From solutions supersaturated with respect to calcium phosphate crystal phases (CaP), spherulites of CaP could be generated within the gel matrix, which is described in more detail in the following paper in this volume.

The flow model of crystallization in gel matrices demonstrated here simulates physiological conditions of stone formation to a large extent. Formation of crystal phases relevant in urolithiasis (including nucleation, growth, agglomeration and adhesion) can be qualitatively or quantitatively evaluated. Crystal habits formed in the gel, like CaP-spherulites, resemble those observed in urinary stones or human kidneys. The technique provides new insights into the different processes of urinary stone formation.

ACKNOWLEDGEMENTS

This work was supported by the Commission of the European Communities, DG XII (grant no. CI1*/0346). We are indebted to Dr A Schaper, Dipl.-Min. M Burk and Dipl.-Ing. B Kiss for their assistance in this work.

FORMATION OF URINARY STONES *IN VITRO* : GROWTH OF CALCIUM OXALATE ON SPHERULITES OF CALCIUM PHOSPHATE IN GEL

W. Achilles[1], U. Jöckel[1], A. Schaper[2], B. Ulshöfer[1] and H. Riedmiller[1]

Philipps-Universität Marburg
[1]Urologische Universitätsklinik
[2]Fachbereich Geowissenschaften
D-3550 Marburg/Lahn, Germany

INTRODUCTION

Urinary stones are commonly of mixed crystalline composition. Calcium phosphate (CaP), mostly as hydroxyapatite (HAP), has been found to occur in the majority (>75%) of stones containing calcium oxalate (CaOx) as the main component, eg [1, 2]. This observation led several authors to suggest that heterogeneous nucleation of CaOx crystal phases on CaP may be an important etiological factor in the genesis of calcium containing stones[2-4]. Meyer et al [5], Koutsoukos et al [6] and Campbell et al [7] demonstrated epitaxial growth of CaOx on HAP in inorganic solutions *in vitro* .

In vivo, CaP very often has the shape of small spherulites of varying crystalline structure. Spheroid particles have been demonstrated to occur in the central cores of urinary concrements[2,3,8,9], as well as in the tissue of human kidneys [9-11], and spheroidal forms of amorphous CaP can also be produced by spontaneous precipitation of calcium phosphate *in vitro,* eg[12].

The aim of this work was to evaluate appropriate conditions for the formation of CaP-spherulites in gel matrices, in order to investigate their ability to nucleate CaOx crystal phases and to gather preliminary information on potential factors affecting their development. For this purpose, we developed a new flow model of crystallization, which was intended to simulate the formation of urinary stones.

METHODS

The experimental model for studying crystal formation used here is described briefly elsewhere in this volume[13]. Crystallization was carried out in a gel matrix of agar-agar (1% w/w) at 37°C from a supersaturated, metastable solution (S_{AB}) continuously flowing over the gel surface. The constant flow rate was 0.5 ml/min. S_{AB} was generated by a mixture of two solutions, S_A and S_B, each containing one component of the sparingly soluble crystal phase (AB) to be formed.

Crystallization of CaP was generated from the following mixed solution I (final concentrations): 200mM NaCl, 10mM Na_2HPO_4, 10mM NaH_2PO_4, 1.7mM $CaCl_2$; pH 6.8-6.9. Subsequent growth of CaOx on CaP was achieved from the following solution II (final concentrations): 200mM NaCl, 2mM $CaCl_2$, 0.3mM Na_2Ox, 10mM n-morpholino-ethanesulphonic acid (MES; pH = 6.0). Experiments were conducted for 4-8 h.

Urolithiasis 2, Edited by R. Ryall *et al.*,
Plenum Press, New York, 1994

All inorganic reagents were of analytical grade. Phosphatidylcholine was purchased from Sigma (DL-alpha-phosphatidylcholine, dipalmityl; product no. P5911).

The reaction plate (size: 85x128 mm) contained 12 channels which were run simultaneously. Light microscopy was carried out directly on the gel plates using an inverted microscope equipped with a computer controlled scanning stage for 96-well microplates (ZEISS, Oberkochen, Germany). SEM micrographs (SEM device: CamScan 4, Cambridge, GB) were obtained after rinsing and air- or freeze-drying of the gel. Crystallized particles were prepared from the dried material and coated with gold or carbon for SEM.

RESULTS

Generation of Calcium Phosphate Spherulites

During a 4-8 h flow of solution I over the unseeded gel, formation of CaP spherulites, which had a size of up to 200 µm in diameter, were observed within the gel matrix. From the light microscopic image, a concentric lamellar structure of amorphous or microcrystalline material was evident (Figure 1).

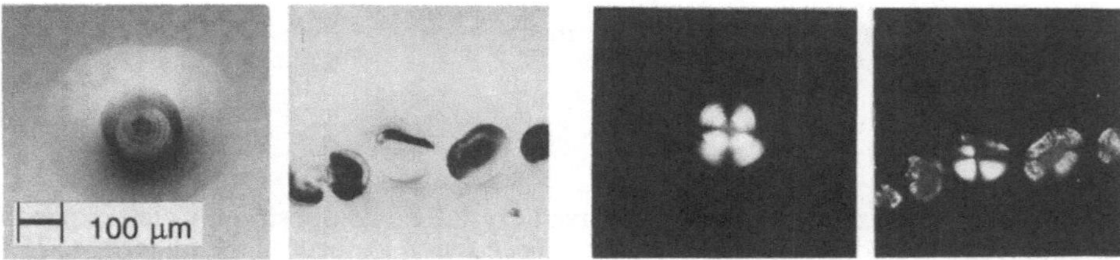

Figure 1. Microscopic image of calcium phosphate spherulites formed in gel. Modes of observation: left bright field; right - polarized light.

The number of spheroid particles formed, as well as their size distribution per volume, were rather different between experiments, thus indicating the obviously predominant role of nucleation for the process under scrutiny. Scanning electron microscopy revealed an unexpected appearance of the particles grown in the gel; spherical cores likely to be microcrystalline CaP were partly surrounded by well crystallized shells, which consisted of radially oriented sheet-like structures (Figure 2). The different crystal phases have not been identified to this time.

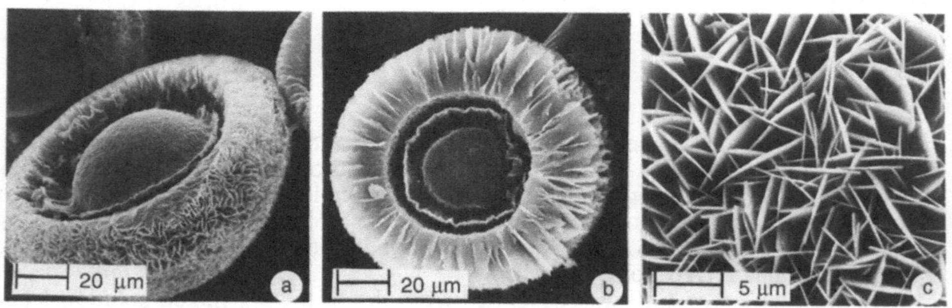

Figure 2. SEM micrographs of calcium phosphate spherulites.
a) mushroom-like whole spherulite with microcrystalline core and incompletely grown shell; b) fractured spherulite showing the radially oriented sheet-like crystals; c) detailed view on outer sheet-like structures.

Growth of Calcium Oxalate on Spherulites of Calcium Phosphate

After rinsing the gel medium with water, a solution supersaturated in calcium oxalate (solution II; see Methods) was passed over the gel for 8 h. As may be seen from the microscopic image (Figure 3), CaOx grew predominantly around the CaP spherulites.

Detailed structures of the crystal phases can be discerned from SEM micrographs (Figure 4) where hexagonal whewellite is seen to be nucleated on the surface of the core, as well as on the outer sheet-like structures. The preformed radial CaP sheets seem to have been continued by a similarly growing CaOx phase. This is especially visible from Figure 4c. Here, hexagonal COM crystals seem to grow directly out of the underlying thinner structures. A borderline which would indicate the transition from the CaP phase to the CaOx phase seems not to be evident from the micrographs.

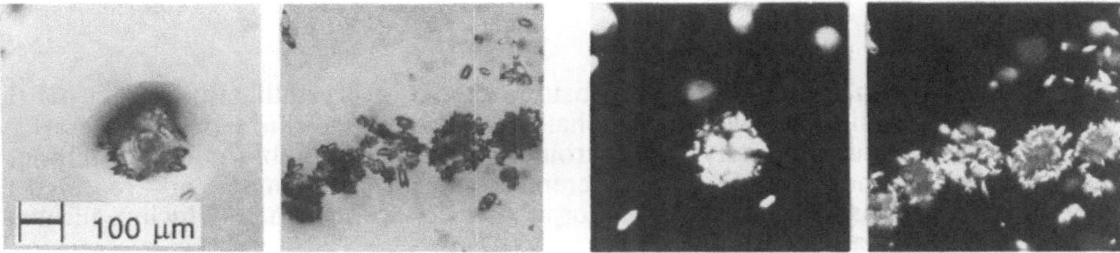

Figure 3. Microscopic image of CaP spherulites overgrown with CaOx. Modes of observation: left - bright field; right - polarized light.

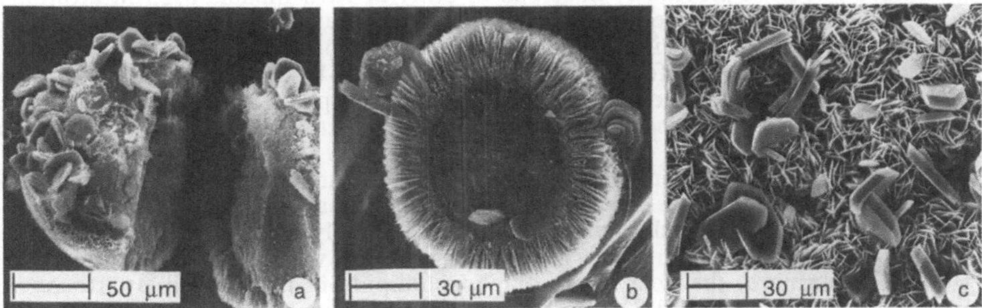

Figure 4. SEM micrographs of CaP spherulites overgrown with CaOx. a) Fractured particle with COM crystals; b) Fractured particle with CaOx growing out of the sheetlike-structured shell. c) detailed view on outer structures with CaOx.

Factors that Influence the Formation of CaP Spherulites

In first experiments on potential effectors of the formation of CaP spherulites, we investigated some phospholipids which are well-known participants in the biomineralization of CaP, eg[14,15]. Phosphatidylcholine was found to act as a very effective nucleator of CaP spherulites in the gel system applied; a physiological concentration of 20 mg per liter in solution I (see Methods) caused a massive nucleation and agglomeration of spherulites, as may be seen from Figure 5. No such effect on nucleation could be demonstrated with phosphatidyl serine using the same concentration.

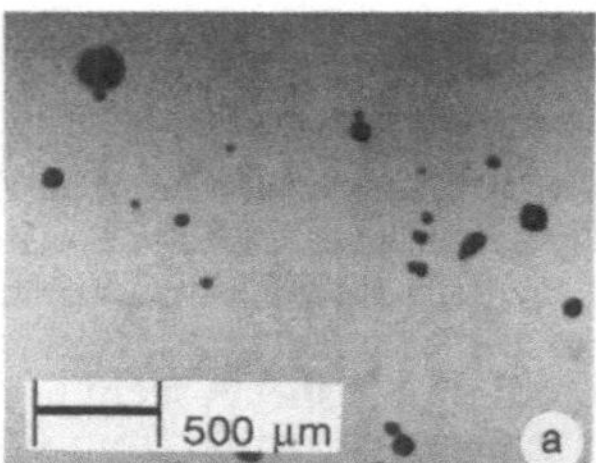

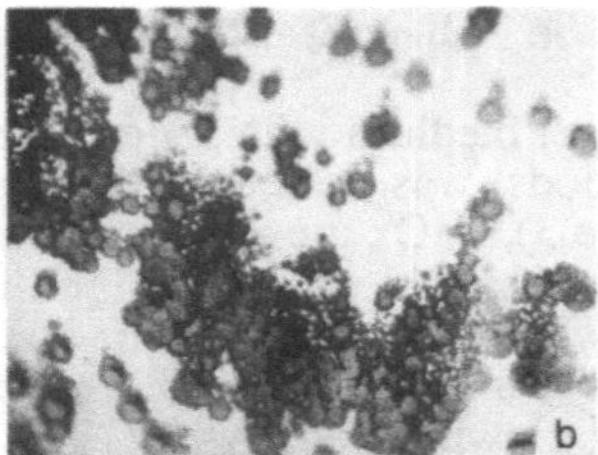

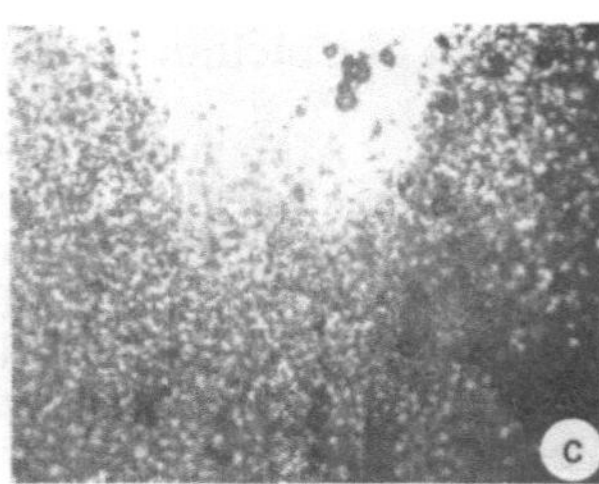

Figure 5. Microscopic demonstration of the effect of phosphatidylcholine on the nucleation of CaP spherulites in gel. a) formation of particles from solution I; b) and c) enhanced nucleation and agglomeration of spherulites in the presence of 20 mg/L phosphatidylcholine.

DISCUSSION

Using the dynamic constant-composition model of crystallization in gel matrices described[13], spherulites of calcium phosphate could be produced *in vitro*. They consisted of crystalline structures radially grown around a core - presumably microcrystalline. At this time we have not identified the different crystal phases. Eanes et al[12] reported that spheroidal CaP consists primarily of amorphous calcium phosphate which is ultimately converted into crystalline apatite.

Apart from their larger size, the particles formed in our gel model bear a strong resemblance to those spheroids found in the core of urinary stones containing calcium[2,3,8,16] and in sections of human kidneys[9,10]. This supports the patho-physiological relevance of the crystallization model used in this study. The findings show directly that the different crystal phases of CaP spherulites nucleate the growth of CaOx which grows as hexagonal COM crystals, and may apparently also form radially oriented structures which have not been previously observed in other *in vitro* models used in urolithiasis research. These findings underline the likely causative role of CaP in the genesis of CaOx formation, since they provide direct evidence to support a role for heterogeneous nucleation of CaOx by pre-formed CaP particles[2,4,8,9,17,18]. Thus, conditions or factors which predispose towards the generation of calcium phosphate spherulites in the urinary tract might be of crucial importance in the formation of stones containing calcium (CaP and/or CaOx) in general.

A possible role for heterogeneous nucleation is also supported by our findings on the nucleation of CaP spherulites by the phospholipid, phosphatidylcholine. Phospholipids or their Ca-P complexes are well known participants in nearly all processes associated with CaP biomineralization and are known to nucleate CaP in various crystallization models, eg[14,15]. Furthermore, the occurrence of phospholipids in human urinary stones was demonstrated by Khan et al[19], and our results support the view of these authors that phospholipids may fulfil a causative role in urinary stone formation.

Though spherulites with radially oriented crystal forms can be generated in aqueous solution with other substances[20], the gel matrix seems to be a stabilizing factor for the formation of CaP having structures which are typically found in urinary stones. This supports the opinion that stone formation occurs in a fixed, gel-like state as supposed by Iwata et al[21]. We hope that the results of this work constitute the first successful step towards simulating the formation of 'urinary stones' under conditions *in vitro*, a step which may provide important insight into the cause of urolithiasis.

ACKNOWLEDGEMENTS

This work was supported by the Commission of the European Communities, Dir. Gen. XII (grant no. CI1*/0346). We are indebted to our colleagues Dipl.-Min. M Burk, Dipl.-Ing. B Kiss and Mrs E Dorr for their assistance in this work.

REFERENCES

1. WH Boyce, Organic Matrix of Human Urinary Concretions, *Am J Med* 45: 673 (1968).
2. DB Leusmann, Erste zusammenfassende Ergebnisse der kombinierten Phasen- und Gefugeanalyse von Harnsteinen mittels Rontgenbeugung und Rasterelektronenmikroskopie, *Fortschr Urol Nephrol* 17: 275(1981).
3. R Blaschke and W Schmandt, Kugelformige Calciumphosphat-Konkremente im Harn und in der Kernzone von Oxalatsteinen, *Fortschr Urol Nephrol* 11: 84 (1978).
4. LH Smith, AD Jenkins, JWL Wilson, and PG Werness, Is Hydroxyapatite Important in Calcium Urolithiasis ?, *Fortschr Urol Nephrol* 22:193 (1984).
5. JL Meyer, JH Bergert, and LH Smith, Epitaxial Relationships in Urolithiasis: The Calcium Oxalate Monohydrate Hydroxyapatite System, *Clin Sci Mol Med* 49: 369 (1975).
6. PG Koutsoukos, ME Sheehan, and GH Nancollas, Epitaxial Consideration in Urinary Stone Formation. II. The Oxalate-Phosphate System , *Invest Urol* 18: 358 (1981).
7. AA Campbell, A Ebrahimpour, L Perez, SA Smesko, and GH Nancollas, The Dual Role of Polyelectrolytes and Proteins as Mineralization Promotors and Inhibitors of Calcium Oxalate Monohydrate, *Calcif Tissue Int* 45:122 (1989).
8. R Blaschke and UB Meyer, Rasterelektronenmikroskopische Untersuchungen zur Rolle steinkernbildender Phosphate in Harn und Harnsteinen, *Beitr elektronenmikrosko Direktabb Oberfl* 12:391(1979).
9. UB Meyer-Jurgens, R Blaschke, and K Maar, Neuere elektronenmikroskopische Untersuchungen an calciumphosphathaltigen Spharolithen in Niere und Harn, *Fortschr Urol Nephrol* 17: 226 (1981).
10. UB Meyer-Jurgens, R Blaschke, DB Leusmann, and K Maar, Analyse von Harnkonkrementen und Mineralstaub im Nierengewebe mittel STEM und ED-RMA sowie SEM, *Fortsch Urol Nephrol* 20:128(1982).
11. F Hering, T Briellmann, G Laond, H Guggenheim, H Seiler, and G Rutishauser, Stone Formation in Human Kidney, *Urol Res 15:* 67 (1987).
12. ED Eanes, JD Termine, and MU Nylen, An Electron Microscopic Study of the Formation of Amorphous Calcium Phosphate and Its Transformation to Crystalline Apatite, *Calc Tiss Res* 12: 143 (1973).
13. W Achilles, R Kothe, M Burk, and H Riedmiller, A New Continuous Flow Microsystem of Crystallization in Gels as a Model of Urinary Stone Formation, *Paper immediately preceding in this volume* (1993).
14. RE Wuthier, The Role of Phospholipid-Calcium-Phosphate Complexes in Biological Mineralization, in The Role of Calcium in Biological Systems, LJ Anghileri and AM Tuffet-Anghileri, eds., CRC Press, Inc, Bacon Raton, Florida (1982).
15. A Boskey, The Role of Calcium-Phospholipid-Phosphate Complexes in Mineralization, *Metab Bone Dis Rel Res 1:* 137 (1978).
16. SR Khan and RL Hackett, Role of Scanning Electron Microscopy and X-ray Microanalysis in the Identification of Urinary Crystals, *Scanning Microsc I:* 1405 (1987).
17. A Randall, Papillary Pathology as a Precursor of Primary Renal Calculus, *J Urol* 44: 580 (1940)
18. L Cifuentes Delatte, JA Medina, J Bellanato, and M Santos, Papillensteine und Randallsche Plaques, *Fortschr Urol Nephrol* 22: 240 (1984).
19. SR Khan, PN Shevok, and RL Hackett, Lipids of Calcium Oxalate Urinary Stones, *in* "Urolithiasis" VR Walker, *et al.*, eds., Plenum Press, New York and London (1989).
20. H Hommel, Phanomen der spharolithischen Poly-Kristallisation und technische Kahlkristallisation. Ein Modellsystem spharolithischer Kristallisation. Die Spharolith-Bildung als elektrochemisches Phanomen, *CT-Bericht* 3185: 1 (1985).
21. H Iwata, S Nishio, A Wakatsuki, K Ochi, and M Takeuchi, Architecture of Calcium Oxalate Monohydrate Urinary Calculi, *J Urol* 133: 334 (1985).

CALCIUM OXALATE CRYSTALLIZATION IN LONG, THIN TUBES

A.S. Bramley[1], M.J. Hounslow[2], R.L. Ryall[3] and V.R. Marshall[3]

[1]University of Adelaide
[2]University of Cambridge
[3]Flinders Medical Centre
Bedford Park, South Australia

INTRODUCTION

The formation of kidney stones relies on two events, the nucleation of crystals and more importantly, the enlargement of these crystals, which may occur as a result of crystal growth and aggregation. Both of these mechanisms have been widely studied; however the geometry of the tubules in the kidney through which both the urine and crystals pass, and its possible effect on crystal enlargement have largely been ignored. The geometry of the tubules determines the velocity profile of the urine which will affect how long the crystals take to pass through the tubules, or their residence time.

The aims of this work were to determine the residence time distribution of the crystals in long, thin tubes with two dimensional fluid flow, and to investigate the effect a tubular geometry has on crystal growth and aggregation.

EXPERIMENTAL

Apparatus: The experimental apparatus consisted of 262 Amicon H1P10-20 hollow porous fibre tubes each with an internal diameter of 410µm, and 45.5cm long, contained as a bundle in a perspex tube 1.9cm in diameter. The porous tubes were held in place at each end of the perspex tube by epoxy resin to a length of 1.5cm. The column was manifolded with collars at both ends that allowed fluid to be fed to the hollow section of each porous tube (the lumen), or to the region surrounding them in the perspex tube (the jacket). Liquid was introduced to the lumen and jacket by centrifugal pumps, via head tanks and rotameters, thus ensuring constant flow rates to both sections of the apparatus during operation.

Method: Residence time distributions (RTD's) were determined by pulse-input tracer responses. A 0.15M saline solution was fed to both the lumen and jacket, with the jacket outlet closed. Thus the fluid fed to the jacket passed through the walls of the porous tubes to mix with that in the lumen. After the addition of a 0.04mL pulse of particle suspension to the fluid entering the lumen, all the fluid leaving the tubes was collected in separate samples. The number of particles per unit volume in 100µL of each sample was then measured using a Coulter Multisizer II particle size analyser fitted with a 70µm orifice. At least five counts were taken from each sample and the average of these counts was used to calculate the RTD.

Urolithiasis 2, Edited by R. Ryall *et al.*,
Plenum Press, New York, 1994

The rate of fluid injection into the lumen is an important parameter for the experimental system. This parameter is called the dilution factor, a, and is the ratio of the total flow rate of fluid leaving the tubes to the fluid flow rate of fluid fed to the lumen, or

$$a = (Q_L + Q_J) / Q_L \qquad (1)$$

In the experiments, the value of the dilution factor was varied from 1 to 5. For each value of the dilution factor two different lumen flow rates were used and particles of two sizes were added. The particles used were callibration standard latex (obtained from Coulter) with number modes of 13.7μm and 8.8μm.

To investigate the effect of tubular geometry on crystal growth and aggregation, experiments were conducted in which calcium oxalate (CaOx) crystals were fed to the tubes in two different solutions, at different flowrates and dilution factors. A well mixed seed suspension, (1g =CaOx/L 0.15M saline) was fed to the tubes by a peristaltic pump at rates of 0.22mL/min and 0.25mL/min for lumen flow rates of 4mL/min and 8mL/min respectively. The two solutions used were a 0.15M saline solution that had been saturated with CaOx and a metastable solution as used by Ryall et al[1] in their study of the batch crystallization of CaOx.

With the system at steady state, samples were collected at the inlet and outlet of the tubes. The time taken to reach steady state was estimated from the results of the RTD experiments, with steady state taken to correspond to F=1. The samples were analysed with a Coulter Multisizer II to give the number and volume of crystals in each sample.

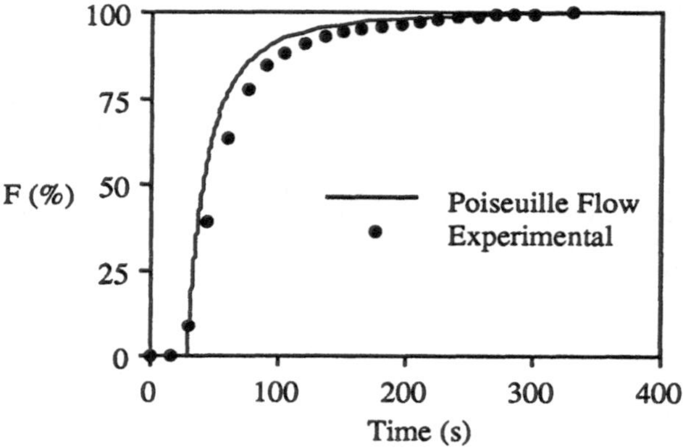

Figure 1. Comparison of the RTD for Poiseuille flow, equation 3 and the experimental RTD when there is no fluid injection through the tube walls.

Analysis: The RTD for a pulse response test such as those employed in these experiments may be calculated from the following formula[2]:

$$F = \sum_{i=1}^{k} N_i / N_t \qquad (2)$$

This is the sum of the number of particles in all the samples up to and including sample k divided by the total number of particles in all of the samples. It follows then that equation 2 gives the fraction of particles that have passed through the tubes at that time t_k.

To determine a theoretical expression for the RTD of the particles in the tubes we must know the fluid velocity profile in them. When there is no fluid injection through the walls of the tubes the velocity field is described by the usual results for Poiseuille flow. In this case[3] the RTD is given by:

$$F = 1 - 1/4\,(\bar{t}/t)^2 \quad \text{for } (\bar{t}/t) < 2 \qquad (3)$$

where $\bar{t}$ is the average time the particles spend in the tube, calculated by dividing the tube length by the average velocity and F is the fraction of particles that has passed through the tubes at time t.

Figure 1 shows the RTD calculated from equation 3 and the experimental RTD when there is no fluid injected through the tube walls. The experimental data are described very well by the RTD for Poiseuille flow. Both the break-through time, *ie* the time at which the first particles leave the tubes, as well as the form, or shape, of the RTD are correct.

When there is fluid injection through the tube walls the above equation cannot be used, as the fluid flow is now two-dimensional, and has both axial and radial components. To obtain the velocity field under these conditions we use the results of Yuan and Finkelstein[4].

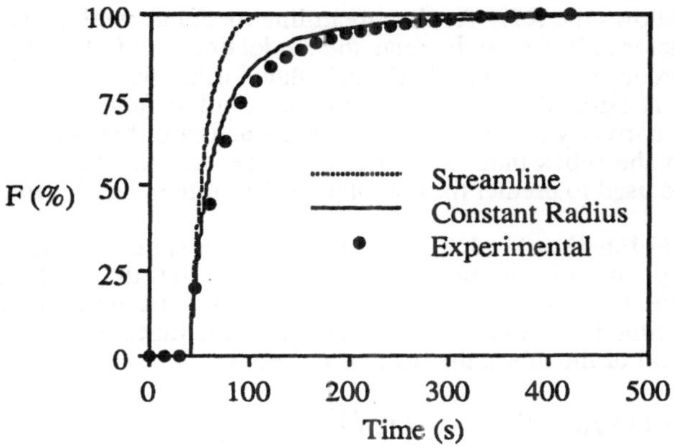

Figure 2. RTD's predicted by the constant radial position model, the streamline model and the experimental RTD for a value of the dilution factor of 2.

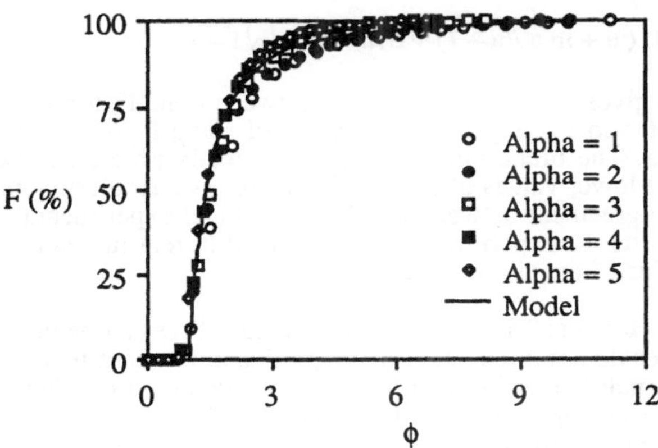

Figure 3. Theoretical RTD of equation 6 and experimental RTD's for different values of the dilution factor.

Streamline Model: In this model it is assumed that the particles follow the fluid streamlines and travel at the local fluid velocity in both axial and radial directions. Or more simply, the particles are swept towards the middle of the tubes as they pass through them by the fluid which is injected through the walls.

By considering the time taken to traverse the length of the tube along a streamline, using the stream function given by Yuan and Finkelstein[4], the residence time for the particles in the section with fluid injection can be determined. No fluid is injected in the sections at the beginning and the end of the tubes, where they are held in the perspex tube by the resin. In these regions we may use the RTD equation for Poiseuille flow. The complete residence time is then:

$$t = \frac{e\,L}{u_o\sqrt{1-F}} + \frac{L}{(a-1)\,u_o}\ln\left[\frac{a - F/2 + \sqrt{a^2 - aF}}{1 - F/2 + \sqrt{1-F}}\right] + \frac{e\,L}{a\,u_o\sqrt{1 - F/(2a-1)}} \tag{4}$$

where e is the fraction of each tube length embedded in epoxy at each end and is equal to 1.5/42.5 or 0.03529.

Figure 2 shows the RTD for the streamline model as well as the experimental RTD for a dilution factor of 2. As can be seen, the model does not fit the experimental data at all well. The difference between the RTD's calculated using this model and those determined experimentally is similar for all values of the dilution factor considered. The break-through time is correctly predicted, but a small fraction of the particles take much longer to pass through the tubes than is predicted by the model. Thus we conclude that this model cannot be used to predict the complete RTD for the particles in the tubes.

Constant Radial Position Model: In this model it is assumed that the particles maintain a constant radial position in the tubes as they pass through them. That is, the fluid being injected through the tube walls does not cause the particles to move radially. By considering the time taken to traverse the length of the tubes, including the sections at the beginning and end of the tubes, the residence time is:

$$t = L\,(e + \ln a/(a-1) + e/a)\ /\ u_o\sqrt{1-F} \tag{5}$$

Figure 2 shows the RTD calculated from equation 3 as well as the experimental RTD for a dilution factor of 2. As can be seen, the constant radial position model fits the experimental data very well. The following simplification allows us to compare the accuracy of the model for all values of the dilution factor. Equation 5 may be re-written as

$$f = t\,u_o\ /\ L\,(e + \ln a/(a-1) + e/a) = 1\ /\ \sqrt{1-F} \tag{6}$$

Equation 6 gives a unique relationship between f and F, for all values of a. Figure 3 shows equation 6 and the experimental values of f and F for the values of the dilution factor considered. The break-through time is correctly predicted and although there is some deviation at lower values of the dilution factor as a result of some particles sticking to the tubes, the agreement between the theoretical and experimental results is excellent. Also, this model fits the experimental data for the different lumen flow rates and particle sizes for each value of the dilution factor.

Observation on RTD's: The fact that the particle RTD's can be predicted by a model in which the particles do not move radially is significant. It means that a small fraction of the particles spend much longer in the tubes than the average time; thus these particles have more time to grow and aggregate and form crystals of pathological significance. The only other work that has considered the residence time of particles[5] assumed that all particles take the average time to pass through the tube. This is clearly untenable, as some particles take much longer, and so this must lead us to question, for example, the conclusions of Finlayson and Reid[5] regarding the expectation of the formation of stones from free particles.

Analysis: Growth and Aggregation: Growth and aggregation will give rise to changes in the crystal number and volume between the inlet and outlet of the tubes; growth causes the crystal volume to increase, aggregation causes the number of crystals to decrease. In saturated saline both crystal number and volume should be similar at the inlet and outlet, as there is no driving force for growth or aggregation. In the metastable solution, which is supersaturated, we expect both growth and aggregation to occur, resulting in an increase in crystal volume and a decrease in number.

If we are to compare accurately the crystal number and volume between the inlet and the outlet of the tubes their values must be corrected to take into account the dilution caused by the fluid injection from the jacket to the lumen. This is done by multiplying the number and volume of crystals at the outlet by a modified version of the dilution factor. The modification to the dilution factor already defined for the RTD experiments is that the flow rate of seeds is included, so the dilution factor for these experiments is:

$$a' = (Q_L + Q_S) / (Q_L + Q_J + Q_S)$$

We may then calculate the following parameters that are the ratios of the corrected crystal number and volume between the inlet and outlet,

$$f_N = (a' N_o) / (N_i|_{a'=1}) \ , \ f_V = (a' V_o) / (V_i|_{a'=1})$$

If no growth or aggregation occurs as the crystals pass through the tubes then both the number ratio, f_N, and the volume ratio, f_V will be one. If growth occurs the crystal volume will be greater at the outlet than at the inlet and the volume ratio will be greater than 1, if aggregation occurs the number of crystals at the outlet will be lower than at the inlet and the number ratio will be less than 1.

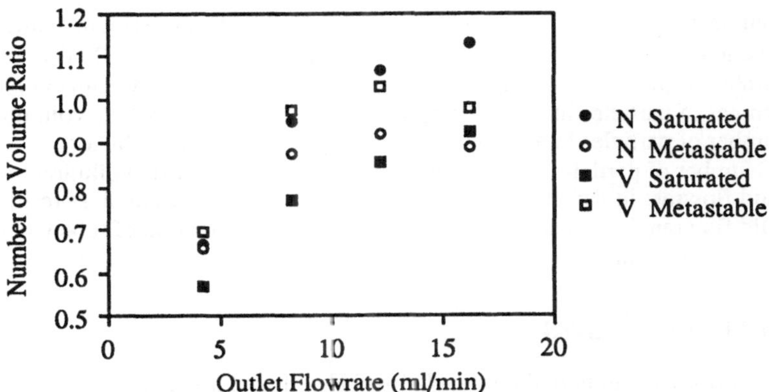

Figure 4. Number and volume ratios for saturated and metastable solutions at various column outlet fluid flowrates.

Figure 4 shows the number and volume ratios plotted against the flow rate of fluid at the outlet of the tubes. From figure 4 we can see that for the saturated saline the volume ratio is less than 1 for all the flow rates considered. This implies that volume is being lost as the crystals pass through the tubes. This can only occur if the crystals are dissolving in the fluid or sticking to the tubes. As the saline has been saturated with CaOx the crystals will not dissolve, so they must be sticking to the tubes. This makes the analysis of the data more difficult, but we can still gain some insight into the extent of growth and aggregation in the tubes from these results. At all the flow rates used the volume ratio is higher for the metastable solution than it is for the saturated saline. Further the difference between them is greater at lower flow rates, when the residence time will be longer. Thus we can conclude that the amount of growth is dependent on the residence time in the tubes.

From figure 4, for the saturated saline at high flow rates the number ratio is actually greater than 1, so more crystals are leaving the tubes than entering them, despite the fact that crystals are sticking to the tubes. This can only occur if the crystals are disaggregating as they pass through the tubes. Further, the amount of disaggregation depends strongly on the outlet flow rate, and this can be explained by the fact that the shear rate the crystals experience as they pass through the tubes will be determined by the flow rate. For the metastable solution the number ratio is less than 1 for all the flow rates, so the the crystals are no longer disaggregating, but probably aggregating, and again the extent of aggregation is dependent on the fluid flow rate and thus shear rate, in the tubes.

CONCLUSION

We conclude that the time a particle spends in a tube is (i) dependent only on the fluid velocity field, (ii) can be predicted accurately by a simple model in which the particles do not move radially and (iii) may be very much greater than the average time previously proposed. It follows that tubular geometry has a significant effect on the time particles spend in the tubes. Further effects on crystal growth and aggregation have been identified. The amount of growth observed is found to depend on the residence time, and the extent of aggregation on the shear rate in the tubes. As a tubular geometry determines both residence time and shear rate, this may be a significant factor in stone formation and certainly one that should be further investigated.

Nomenclature

F	Residence time distribution function	Q_S	Seed flowrate
f_N	Number ratio	t	Time
f_V	Volume ratio	$\bar{t}$	Mean residence time
L	Tube length	u_o	Maximum fluid velocity
N_i	Number of particles in sample i	V_o	Particle volume in outlet sample
N_o	Number of particles in outlet sample	$V_{i\|a'=1}$	Inlet particle volume when a'=1
N_t	Number of particles in all samples	a	Dilution factor
$N_{i\|a'=1}$	Number of particles at inlet when a'=1	a'	Modified dilution factor
Q_J	Fluid flowrate in jacket	f	Dimensionless residence time
Q_L	Fluid flowrate in lumen injection	e	fraction of tube without fluid

ACKNOWLEDGEMENTS

This work was supported by the National Health and Medical Research Council of Australia.

REFERENCES

1. RL Ryall, RM Harnett and VR Marshall, The effect of urine, pyrophosphate, citrate, magnesium and glycosaminoglycans on the growth and aggregation of calcium oxalate crystals *in vitro*, *Clin Chim Acta* 112:349 (1981).
2. O Levenspiel, Chemical Reaction Engineering, 2nd ed., Wiley, New York, (1972).
3. RH Perry, and D Green, Section 5-23, in: Perry's Chemical Engineers' Handbook, 6th ed., McGraw-Hill, New York (1984).
4. SW Yuan and AB Finkelstein, Laminar pipe flow with injection and suction through a porous wall, *Trans ASME* 78:719 (1956).
5. B Finlayson and F Reid, The expectation of free and fixed particles in urinary stone disease, *Invest Urol* 15: 442 (1978).

A GENERAL AGGREGATION MECHANISM FOR CALCIUM OXALATE

M.J. Hounslow[1], G.J. McLaughlin[1], J.E. Olley[1], A.S. Bramley[2] and R.L. Ryall[3]

[1]Department of Chemical Engineering
The University of Cambridge, UK
[2]Department of Chemical Engineering
University of Adelaide, South Australia
[3]Department of Surgery,
Flinders Medical Centre
Bedford Park, South Australia

INTRODUCTION

It has been possible for many years to describe in a semi-quantitative way the mechanisms active in the aggregation of colloidal particles in ionic solutions. These mechanisms are summarised by the theory of Derjaguin, Landau, Verwey and Overbeek (DLVO theory), as presented, for example by Hiemenz[1].

Finlayson et al[2] have proposed that DLVO theory may be used to describe the factors affecting the aggregation of calcium oxalate crystals — it predicts that crystals will aggregate rapidly in urine-like systems as a consequence of the effect of elevated ionic strength on the electrical double layer of each crystal.

However, DLVO theory takes no account of the effects of supersaturation during the aggregation process, so while the results of Sarig et al[3] which were obtained in saturated solution may be accounted for in this way, the more relevant issue of how do crystals aggregate while they are growing, is not addressed.

We have reported elsewhere[4,5,6] our observation and calculation that aggregation rates are directly proportional to supersaturation.

In the current paper we explore the apparent contradiction between these two sets of results. Conventional theory — and some observations — indicate that aggregation should proceed rapidly in saturated solutions. Whereas, in the past, we have observed that no aggregation takes place in saturated solutions but that it does occur in supersaturated solutions, at a rate proportional to the supersaturation.

MATERIALS AND METHODS

We prepared two stock solutions: a supersaturated, metastable solution according to the protocol of Ryall et al[7], and a calcium-oxalate-saturated, saline solution. The saturated solution was prepared from 0.15 M NaCl solution by the addition of 5 g/L of calcium oxalate mono-hydrate crystals. The crystal-saline slurry was allowed to equilibrate for a minimum of 24 h, at the temperature of intended use. Immediately before each experiment, the crystals were removed by 0.22 μm filtration.

Urolithiasis 2, Edited by R. Ryall *et al.*,
Plenum Press, New York, 1994

Each of the experiments reported below entails the addition of a small volume of a 1 g/L seed suspension to a variable volume of either the saturated or supersaturated solution. At various times during each experiment the particle-size distribution (PSD), the total number and the total volume of crystals per unit volume of suspension were determined by the PDI Elzone 280PC.

Experiments were either conducted at room temperature (17.5±2.5° C) or in a Grant Instruments SS40-D shaking water bath at 37±0.5° C.

RESULTS

We conducted four types of experiment using either the supersaturated or saturated solutions described above, with agitation either by the propeller-stirrer supplied with the Elzone 280PC or by Grant shaking water bath. We divide the presentation, here, according to the method of agitation.

Preliminary Studies — Stirred Vessel

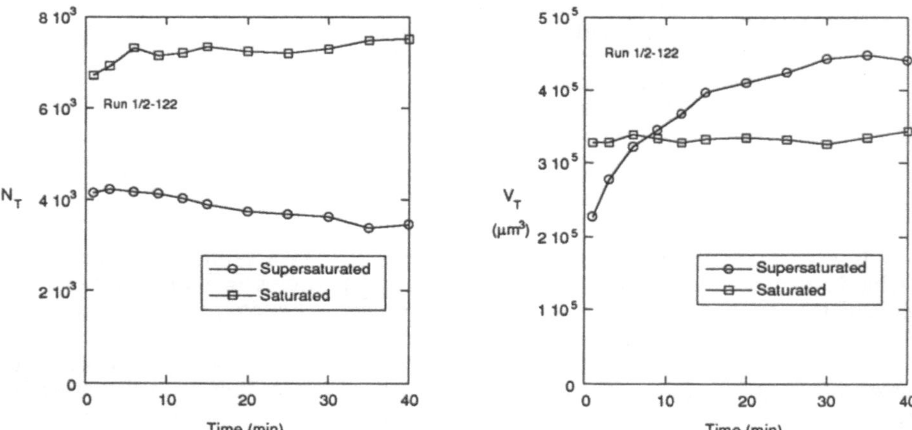

Figure 1. Total number and total volume of crystals in supersaturated and saturated solutions: agitation by stirring.

The Standard Result

As described in the introduction, our experience and expectation is that in the presence of supersaturation, calcium oxalate monohydrate seed crystals will grow and aggregate. Further, in the absence of supersaturation, the seed crystal population will remain unchanged.

In figures 1 and 2 we present results for a pair of parallel experiments conducted in either saturated or supersaturated solution. In the first figure it may be seen that in supersaturated solution the total volume, V_T, of crystals rises, clearly identifying the presence of growth, and the total number, N_T, of crystals falls, identifying aggregation. Conversely, in saturated solution, the volume of crystalline matter remains constant and the numbers slowly rise.

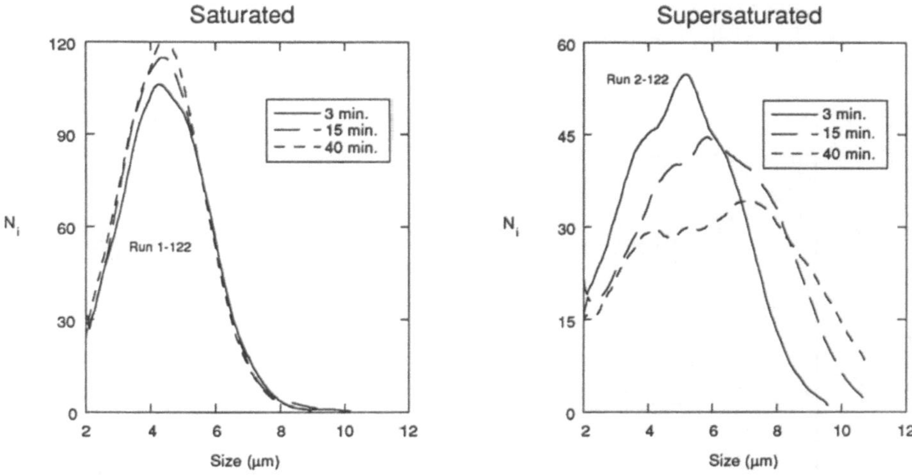

Figure 2. Particle-size distributions in saturated and supersaturated solutions: agitation by stirring.

The second figure shows the PSDs for the same experiments in which the number of particles in each size interval is plotted against the mean size of each interval. We see that in saturated solution the distribution remains static, with some disaggregation possibly identified by the gradual narrowing of the peak. In supersaturated solution the effects of growth and aggregation are clearly visible: the mean size increases and the distribution broadens.

These figures show that under circumstances where aggregation does not occur in saturated solution, it does occur in the presence of supersaturation.

Aggregation In The Absence Of Supersaturation

We turn now to whether, with less vigorous agitation, it is possible to induce aggregation in calcium oxalate crystals in the absence of supersaturation. To explore this prospect we conducted a two-stage experiment over 100 minutes with a medium to fast stirrer speed transition at 60 minutes.

The data displayed in figure 3 show that at the slower stirrer speed aggregation does occur: the total number of crystals decreases while the total volume remains constant. It would appear that when the stirrer speed is increased, the aggregation process is reversed. This observation is confirmed in figure 4, in which it may be seen that the evolution of the size distributions after the increase in stirrer speed is the reverse of that before the increase.

These figures show that aggregation can be induced in saturated solutions, and that it is reversible.

Controlled Studies — Shaking Water bath

We turn now to a series of experiments conducted under better controlled, though less well-defined, hydrodynamic conditions. We show first, in figure 5, that in saturated solution the seeds used are stable at 3 Hz oscillation but disaggregate at 3.5 Hz. Although not show, the crystal number and volume behave as would be expected: both are constant at 3 Hz while only the numbers increase at 3.5 Hz.

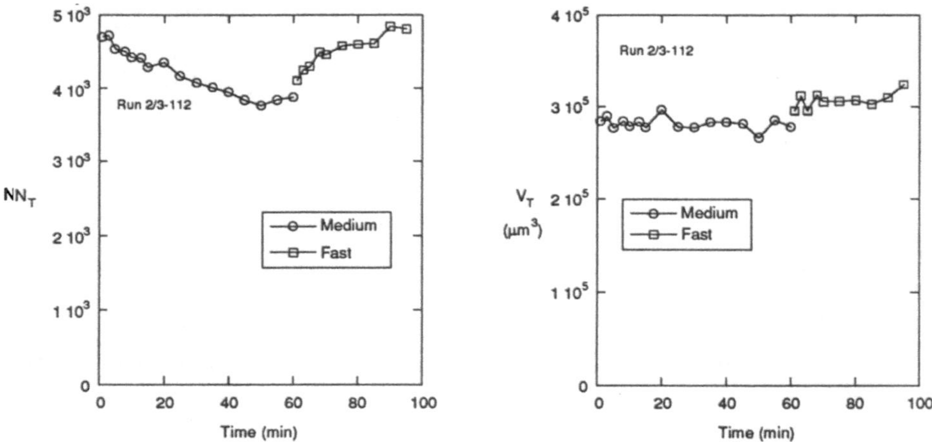

Figure 3. Total number and total volume of crystals in saturated solution: medium to fast stirrer transition.

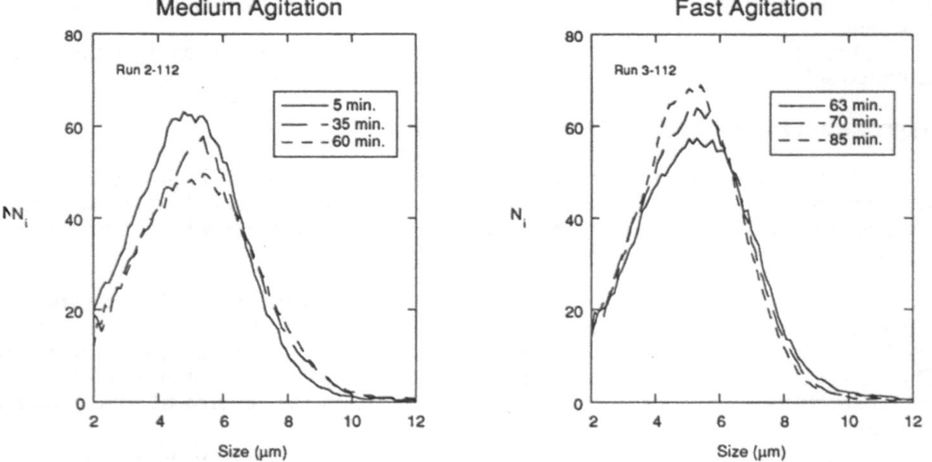

Figure 4. Particle-size distributions in saturated solution: medium to fast stirrer transition.

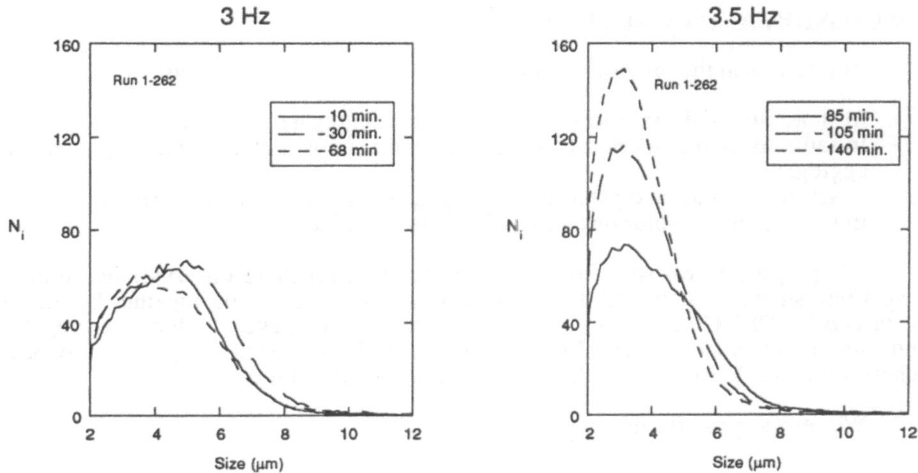

Figure 5. Particle-size distributions for seeds in saturated solution: the effect of shaking rate.

The data shown in figure 5 are a manifestation of the point made in the previous section: in the absence of supersaturation, aggregates (as these seeds apparently are) are free to disaggregate.

Figure 5 shows that at a sufficiently high agitation rate, seeds in saturated solution fall apart. We consider now an analogous experiment, conducted in supersaturated solution. In figure 6 we observe that at the original 3 Hz agitation the crystals grow and aggregate as expected. During this phase the total numbers decrease by *ca* 40% and the volume increases by *ca* 60%. However, when the agitation rate is increased to 3.5 Hz, *no dis-aggregation is observed.* This is in marked contrast to the results shown in figure 5.

We conclude that in the presence of supersaturation, aggregation is irreversible.

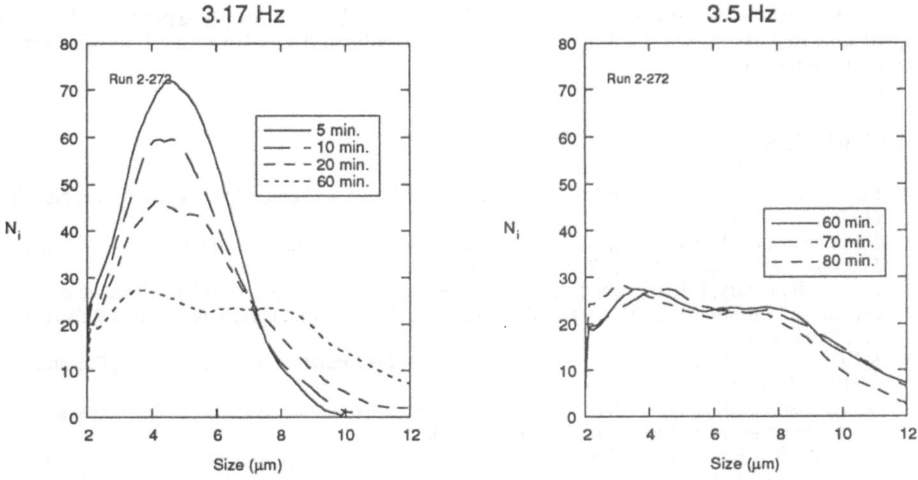

Figure 6. The effect of agitation rate on seed crystals in supersaturated solution.

DISCUSSION AND CONCLUSION

The results in the previous section may be summarised as follows:

- In supersaturated solution, crystals grow and aggregate.
- In saturated solution crystals do not, of course, grow, but they may aggregate or dis-aggregate.
- In saturated solution crystals can aggregate and disaggregate reversibly.
- In supersaturated solution aggregation is irreversible.

We propose a two-stage mechanism to account for these observations: in the first, reversible, stage, crystals collide and form weak aggregates held together by the forces described by DLVO theory. In the second, irreversible, stage, the loose aggregates are cemented together by the deposition of new material. The second stage can only occur in supersaturated solutions, while the first can occur in any solution.

We may summarise this mechanism by:

$$\textit{CRYSTALS} \underset{\text{Agitation, Ionic Strength, Etc.}}{\rightleftharpoons} \textit{LOOSE AGGREGATES} \underset{\text{Super-saturation}}{\longrightarrow} \textit{STRONG AGGREGATES}$$

By means of this mechanism we are able to explain the observations of Sarig et al[3] — that aggregation occurs reversibly in saturated solutions — and our own earlier observations [4,5,6] that aggregation depends only on supersaturation. Clearly Sarig et al were observing the first stage of the mechanism, while the second stage was not active. Conversely, we were operating with a range of particle sizes and agitation rates that moved the "equilibrium" position of the first stage far to the left, causing us to observe no aggregation at zero supersaturation. Only in the presence of supersaturation was it possible for short-lived loose aggregates to be cemented together to form the aggregates we observed.

ACKNOWLEDGEMENT

This work was supported by the SERC of the UK under its Specially Promoted Programme in Particulate Technology and by the National Health and Medical Research Council of Australia.

REFERENCES

1. PC Hiemenz "Principles of Colloid and Surface Chemistry", 2nd edn, Marcel Dekker, Inc, New York (1986).
2. B Finlayson, SR Khan and RL Hackett, Mechanisms of stone formation - an overview, *Scanning Electron Microscop*, 3, 1419, Chicago (1984).
3. S Sarig, R Azoury, E Lerner and F Kahana, Assessment of aggregation of calcium oxalate-containing crystals in "Urolithiasis", VR Walker, RAL Sutton, ECB Cameron, CYC Pak and WG Robertson, eds, Plenum Press, New York (1989).
5. MJ Hounslow, RL Ryall and VR Marshall. Modelling the formation of urinary stones, Presented at CHEMECA 88 Sydney, Australia (1988).
6. MJ Hounslow, RL Ryall and VR Marshall. A discretised population balance for nucleation growth and aggregation. *AI CH EJ* 34, no 11, 1821 (1988).
7. RL Ryall, RM Harnett and VR Marshall, The effect of urine, pyrophosphate, citrate, magnesium and glycosaminoglycans on the growth and aggregation of calcium oxalate crystals *in vitro*. *Clin Chim Acta* 112, 349 (1981).

URINARY CITRATE AND URATE MODULATE CALCIUM OXALATE MONOHYDRATE SOLUBILITY AND AGGLOMERATION

D.T. Erwin, J. Alam, D.J. Kok, A. Annaloro, Jr, J. Vaughn, O. Coker, B. Carriere, J. Lindberg and F.E. Cole

Alton Ochsner Medical Foundation
New Orleans, LA, USA

INTRODUCTION

The exact nature of the chemical abnormalities in the urines of patients who form calcium-oxalate-monohydrate (COM) stones has not been defined. Stone formation could result from the absence of urinary inhibitors of the stone forming process, in conjunction with increased urine concentrations of calcium and oxalate, which are the dominant constituents of these stones, or from some combination of both circumstances. In either case, stone forming events in these patients, when it has been documented that COM crystalluria occurs even in normal persons, dictate that the abnormal stone-forming process must be rapid[1]. Recent attention has turned to physico-chemical processes that affect the agglomeration of pre-existing smaller COM stones and *in vitro* methods developed to evaluate the kinetics of COM crystal growth, agglomeration, and solubility[2,3]. In this study we examine the relationships between urinary urate, citrate, volume and these COM kinetic parameters in stone formers [4,5].

METHODS

Twenty-four hour patient urine collections were performed at room temperature using thymol crystals as a preservative. Oxalate measurements in 24 h urine collections from stone forming patients were performed by oxidizing oxalate enzymatically with oxalate oxidase followed by measurement of H_2O_2 stoichiometrically produced with a peroxidase-coupled colorimetric reaction by the absorbance at 590 nm (Sigma Chemical Co., St. Louis, MO).

Determination of citrate concentration in patient 24 h urine samples was performed by a procedure in which citrate is enzymatically converted using citrate lyase concurrent with the stoichiometric conversion of reduced to oxidized forms of nicotinamide-adenine dinucleotide which is monitored spectrophotometrically by the decrease in absorbance at 340 nm (Boehringer Mannheim Corp., Indianapolis, IN). Uric acid was determined by the stoichiometric production of H_2O_2 by uricase cleavage, with the absorbance read at 700 nm after derivatization with a phenol derivative.

Linear regression relationships

The effect of various patient 24 h urine parameters on COM kinetic measurements was examined. The solubility of COM ($\sqrt{LC}$) is the lowest concentration in mM at which growth of added crystals begins to occur; percent crystal growth inhibition (%GI) is the growth constant which represents the net addition of crystal material at the crystal surface per unit of time; and an agglomeration parameter [tm] is measured in units of time (min). Agglomeration reduces the rate at which mineral components are taken up by the crystals by diminishing the surface area for ^{45}Ca uptake. Measurements were performed in 0.15M NaCl solution with 1:5 diluted urine at pH 6.0. Flasks containing ^{45}Ca, supersaturated with COM, were incubated at 37°C and kept in suspension. At varying times, the crystals were collected on a Millipore filter (0.451μm), solubilized in HCl, and measured for uptake of ^{45}Ca into the crystal mass.

CONCLUSIONS

This is the first report that urinary volume was inversely related to [tm] and %GI. This study showed that *in vitro* COM √LC and [tm] tended to be affected by urinary citrate. This provides further support for the suggestion that calcium-oxalate renal stone formation is modulated by urinary citrate concentrations. Furthermore, these data indicated that urinary urate also affected COM √LC and [tm].

REFERENCES

1. B Finlayson, Where and how does urinary stone disease start? *in*: "Fogarty Int. Cent. Proc. Washington, DC," R van Reen, (ed), US Gov. Printing Office, Washington, DC, 37:7 (1977).
2. DJ Kok, SE Papapoulos, LJMJ Blomen, and OLM Bijvoet, Modulation of calcium oxalate monohydrate crystallization kinetics *in vitro*. *Kid Int* 34:346 (1988).
3. DJ Kok, SE Papapoulos, and OLM Bijvoet, Crystal agglomeration is a major element in calcium oxalate urinary stone formation. *Kidney Int.* 37:5156 (1990).
4. DJ Kok, SE Papapoulos, and OLM Bijvoet. Excessive crystal agglomeration with low citrate excretion in recurrent stone formers. *Lancet*, i:1056 (1986).
5. J Fegan, R Khan, J Poindexter and CYC Pak. Gastrointestinal citrate absorption in nephrolithiasis. *J Urology* 147:1214 (1992).

HETEROGENEOUS DEPOSITION OF CALCIUM PHOSPHATES AT THE SILICON (HYDROUS) OXIDE - WATER INTERFACE

J.H. Adair[1,2], T. Nagira[2], C.M. Brown[1], S.R. Khan[1] and W.C. Thomas Jr[1]

[1]Center for the Study of Lithiasis and Pathological Calcification
University of Florida
Gainesville, Florida
[2]Some of the reported work was performed at
Materials Research Laboratory
Pennsylvania State University
University Park, PA

INTRODUCTION

Calcium phosphates (CP) are found in significant quantities in almost all kidney stones. With a relatively high surface free energy, but only a moderate supersaturation in human urine, it is likely that nucleation of CPs requires a pre-existing surface. As a consequence, there have been a number of studies that have focused on the heterogeneous nucleation of CPs on a variety of materials, from calcium oxalate monohydrate to uric acid and the urate salts [1-3]. However, there have also been studies that have implicated silicon and its oxide compounds, hydrous oxides, or soluble silicate ions, in the nucleation and growth of CPs and hydroxyapatite, the thermodynamically stable form, in particular. These past studies tended to focus on organic silicon or soluble silicates [4,5]. However, silicon has a concentration in human urine in the range of 6 to 17 mg per 24 hours [6]. With a solubility of about 100 ppm in water, solid forms of silicon such as silicon dioxide or the more thermodynamically stable solid silicon (hydrous) oxide are likely to be present in urine as particles.

The objective of the present work was to establish whether the surface *vis-a-vis* soluble silicate species can promote CP deposition through heterogeneous nucleation. The research approach has incorporated the preparation of well-defined spherical, sub-micron silicon (hydrous) oxide particles, such that changes in particle shape or surface chemistry after exposure of the particles in aqueous suspension to calcium and/or phosphate ions as a function of solution pH can be easily followed with various characterization techniques.

MATERIALS AND METHODS

The spherical silicon (hydrous) oxide particles were prepared using a technique previously described by Stober and Bohn[7]. Tetraethoxy silane at 0.25M in ethanol was mixed with ammoniated (4M NH_3) ethanol and deionized water (10M) and allowed to react at 25°C for 24 hours with agitation. The particles were collected by centrifugation, washed repeatedly in ethanol, dried at 25°C, and finally dried at 200°C to remove residual H_20 and NH_3. The recovered particles were exposed to Ca^{2+} as $Ca(NO_3)_2.4H_2O$ and/or PO_4^{3-} as NaH_2PO_4 in aqueous suspensions at various pH values. Suspension pH values were adjusted with either HNO_3 or KOH with 10^{-3} M KNO_3 used as an indifferent electrolyte. The work in this preliminary study was conducted at 25°C. The heterogeneous deposition of CPs at physiological temperatures will be the subject of future investigations.

The surface chemistries of the recovered particles in aqueous suspension, and silicon (hydrous) oxide particle suspensions exposed to Ca^{2+}, PO_4^{3-}, and mixtures of these ions were characterized using particle electrophoretic mobility (EM) measurements (Rank Brothers, Mark II electrophoresis apparatus) to establish if the Ca^{2+}, PO_4^{3-} or CP deposits find heterogeneous sites on which to absorb or deposit on the silicon (hydrous) oxide particle surfaces. Samples were also collected from the suspensions by filtration, dried at 25°C, and selected samples characterized using scanning electron microscopy (SEM) with energy dispersive spectroscopy (EDS; ISI 130 with EDAX) to provide surface chemical analyses of the particles as a function of exposure condition.

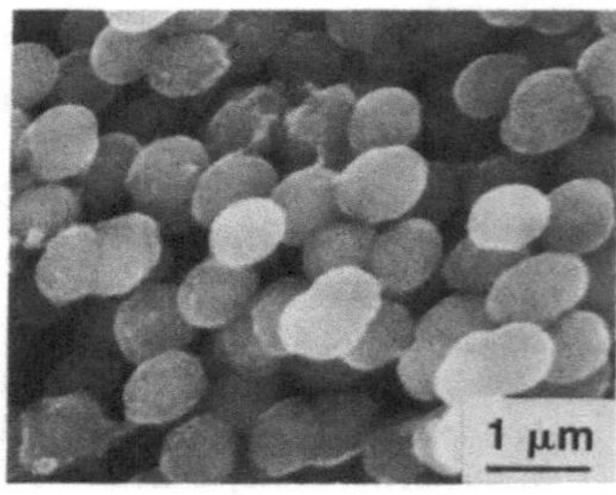

Figure 1. SEM photomicrographs of: (A) 0.025 g/L silicon hydrous oxide particles at pH 7.5, and (B) 0.1 g/L silicon hydrous oxide particles at pH 8.2. Both suspensions contained 10^{-3}M Ca^{2+} and 10^{-4}M PO_4^{3-}. See Table 1 for the EDS analyses on these specimens.

RESULTS AND DISCUSSION

The SEM photomicrographs in Figure 1 demonstrate the physical nature of the CP deposits on the surfaces of the silicon (hydrous) oxide particles. The physical and chemical characteristics of the freshly prepared particles have been presented by Stober and Bohn [7]. The deposits present on the surfaces of the particles in the SEM photomicrographs and the EDS analyses summarized in Table 1 indicate that CP deposits preferentially on the surfaces of the silicon (hydrous) oxide particles and not homogeneously in the solution. Optical density measurements not presented in detail in this report indicate that CP deposition from homogeneous solutions takes place at much greater supersaturation than the supersaturation required for the CP deposition on the silicon (hydrous) oxide particles.

Electrophoretic mobility determinations are a useful characterization technique to study the deposition of materials on the surfaces of fine particles [8]. The EM of the silicon (hydrous) oxide particles in the simultaneous presence of Ca^{2+} and PO_4^{3-} are presented in Figure 2. In data not shown, Ca^{2+} and PO_4^{3-} acting alone did not affect the EM of the silicon (hydrous) oxide as a function of solution pH. When both are present, Ca^{2+} and PO_4^{3-}have a strong influence on the EM of the silicon (hydrous) oxide particles, supporting the hypothesis that CP is deposited on particle surfaces rather than in the solution phase.

In conclusion, the results of the morphological and surface chemical analyses, when combined with EM measurements strongly support the hypothesis that CP preferentially deposits on the surfaces of silicon (hydrous) oxide particles. This conclusion is consistent with other investigations that have demonstrated that the presence of silicon in some form promotes the formation of CPs [4,5], but the present study directly implicates the surfaces of silicon (hydrous) oxides in the heterogeneous deposition of CPs rather than soluble silicate species. Furthermore, the CPs heterogeneously deposit on the surfaces of the silicon (hydrous) oxide particles at relative supersaturations considerably less than the supersaturation required for homogeneous deposition from solution.

Table I. Energy dispersive spectroscopy analyses of silicon (hydrous) oxide particles subjected to aqueous solutions as a function of Ca^{2+}, PO_4^{3-}, and solution pH at various solid loadings.

Solids Loading (g/L)	Ca^{2+} (M)	PO_4^{3-} (M)	Solution	pH	EDS Analysis Atomic Percent SiO_2	CaO	P_2O_5
0.025	10-3	10^{-4}		7.5	12.8	51.7	35.5
0.025	10-3	10^{-4}		10.7	29.5	55.2	15.3
0.1	10^{-3}	10^{-4}		8.2	84.6	11.8	3.6
0.1	10^{-3}	10^{-4}		11.8	69.9	19.6	10.5
0.5	10^{-3}	10^{-4}		6.8	100.0	0.0	0.0
0.5	10^{-3}	10^{-4}		8.9	97.7	2.3	0.0
0.5	10^{-3}	10^{-4}		11.8	92.7	6.1	1.2

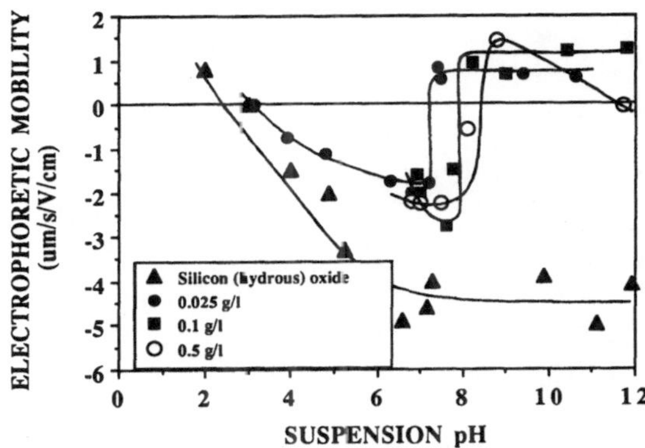

Figure 2. Electrophoretic mobility as a function of suspension pH for different solid loadings of silicon (hydrous) oxide particles in the presence of 10^{-3}M Ca^{2+} and 10^{-4}M PO_4^{3-} with 10^{-3}M KNO_3 present as the supporting electrolyte.

REFERENCES

1. PG Koutsoukos, CY Lam-Erwin, and GH Nancollas, *Invest Urol*, 18178-184 (1980).
2. NS Mandel and GS Mandel, in Urolithiasis: Clinical and Basic Research, LH Smith et al. (eds), Plenum Press, New York, NY, 1981, pp.469-480.
3. JR Burns and B Finlayson, *Invest Urol*, 18 133-136 (1980).
4. LL Hench and G Orcel, *J Non-Cryst Solids*, 82: 1-10 (1986).
5. JJM Damen and JM Ten Cate, *J Dent Res*, 2: 1355-1359 (1989).
6. WC Thomas, Jr, "Urinary Silicates in Calculous Patients," this proceedings.
7. W Stober, A Fink, and E Bohn, *J Colloid Interface Sci* 26 62-69 (1968).
8. RO James, and TW Healy, *J Colloid Interface Sci* 40 42-81 (1972).

THE INHIBITORY EFFECT OF URINARY HEPARAN SULFATE AND CHONDROITIN SULFATE ON CALCIUM OXALATE CRYSTAL GROWTH

H. Iwata, S. Nishio and M. Takeuchi

Department of Urology
Ehime University School of Medicine
Shigenobu
Ehime, Japan

INTRODUCTION

Though the inhibitory effect of glycosaminoglycans (GAGs) on calcium oxalate crystal growth is well known, most experiments have been performed using commercial preparations obtained from animal tissues or crude extracts of urinary GAGs. We have purified heparan sulfate (U-HS) and chondroitin sulfate (U-ChS) from human urine and compared their inhibitory activity with commercial preparations of each.

MATERIALS AND METHODS

Isolation and Purification of U-HS and U-ChS

Urine from a healthy adult male was collected in the presence of sodium azide (0.5 g/L). Batches of 3-5L of urine were filtered through 0.22 μm Millipore filters, concentrated by ultrafiltration (MW cut-off 30,000) and equilibrated with 0.05M sodium phosphate buffer, pH 6.8 (Na = 150mM). The concentrated urine sample was applied to a DEAE-cellulose chromatography column (1.4 x 25 cm, Whatman DE-52) that had been equilibrated with the same buffer. Elution was carried out using a linear NaCI gradient from 0 to lM in the same buffer (total elution volume = 300 ml, elution speed = 18 mL/h). Urinary GAGs eluted between 0.55M and 0.7M Na were collected[1].

The GAGs solution obtained from about 35 L urine was applied to an anion exchange chromatography column (1.4 x 25 cm, Dowex 1X2) that had been equilibrated with 0.5M NaCl. After washing with 0.9M NaCl, elution was carried out using a linear NaCl gradient from 0.9 to 2.5M (total elution volume = 300 mL, elution speed = 36 mL/h)[1]. HS-rich fractions eluted from 0.9 to 1.2M NaCI and ChS-rich fractions eluted from 1.2 to 1.7M NaCl were collected separately, dialyzed against pure water and lyophilized.

The lyophilized materials were dissolved in 0.5M acetic acid containing 5% calcium acetate. Ethanol was added to the solutions to a final concentration of 30%. The precipitates were collected by centrifugation. Stepwise precipitations were further carried out at ethanol concentrations of 35, 40 and 60%. Two-dimensional electrophoresis[2] showed that the 30% ethanol precipitate of the HS-rich fractions contained only HS (Fig 1, left) and the 60% ethanol precipitate of the ChS-rich fractions contained only ChS (Fig 1, right). The yields of U-HS and U-ChS from 100 L urine were 30 mg and 140 mg, respectively.

Urolithiasis 2, Edited by R. Ryall *et al.*,
Plenum Press, New York, 1994

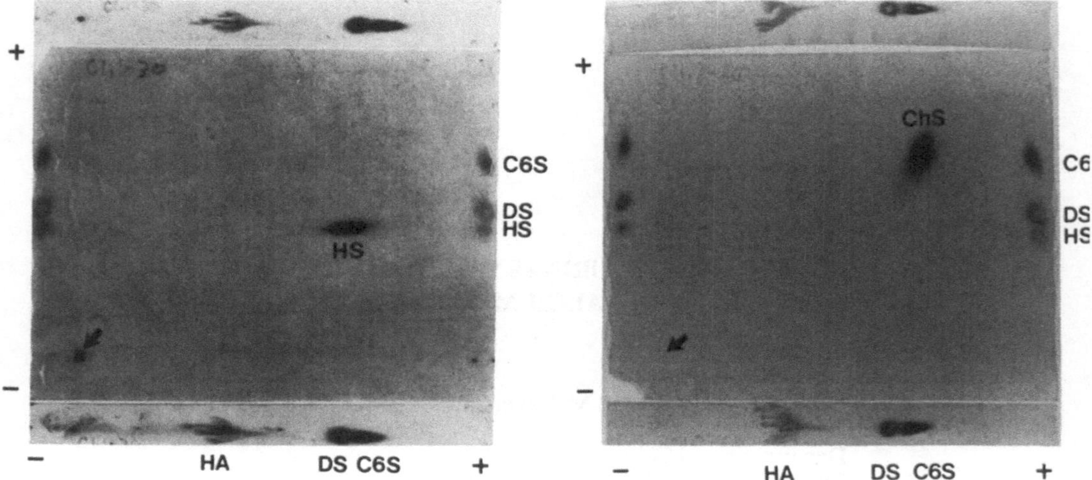

Figure 1. Two-dimensional electrophoresis of purified U-HS (left) and U-ChS (right). HA: hyaluronic acid. DS: dermatan sulfate. C6S: chondroitin-6-sulfate.

Gel-Filtration Chromatography of GAGs

The molecular weight (Mr) of the urinary GAGs was compared to that of the commercial chondroitin-6-sulfate (C6S; obtained from shark cartilage, Seikagaku Kogyo, Mr 40-80kDa) by gel-filtration chromatography (2.6 x 90 cm, Sephacryl S300HR). Elutions were carried out using 0.05M sodium phosphate buffer, pH 6.8, containing 0.5M NaCl (elution speed = 27 mL/h).

Measurement of the Inhibitory Activity of GAGs

The inhibitory activity of the urinary and commercial GAGs on calcium oxalate crystal growth was measured by continuous flow system using a small-scale crystallizer[3] and an artificial urine[4] having a final composition of 8mM calcium and 0.6mM oxalate. The measurements were performed at four different concentrations (1, 5, 10 and 20 µg/mL) of each GAG.

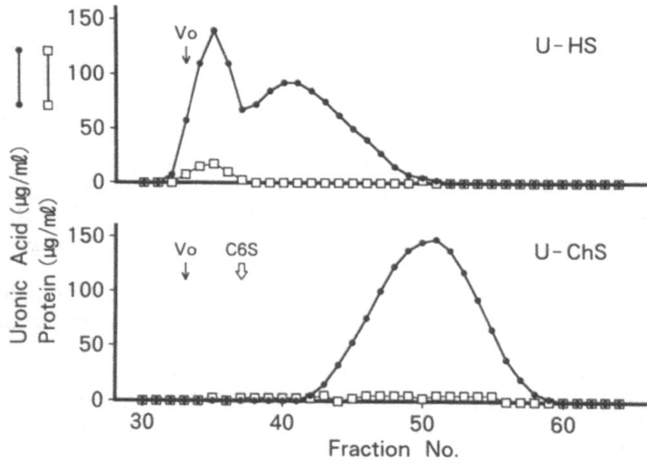

Figure 2. Gel-filtration patterns of U-HS (top) and U-ChS (bottom). Elution peak of C6S (Mr 40-80kDa) is indicated by white arrow.

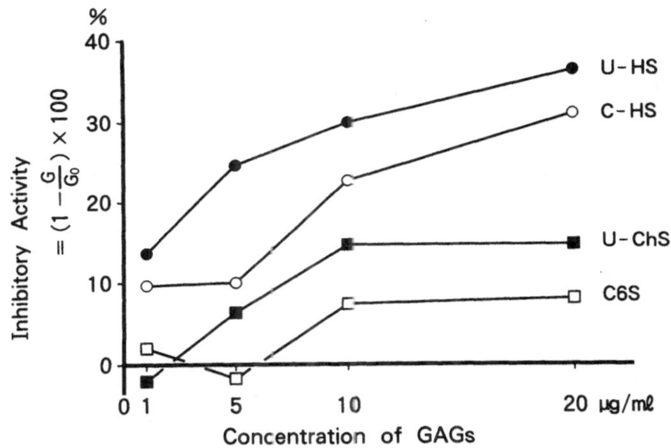

Figure 3. Inhibitory activity of GAGs on calcium oxalate crystal growth. G = crystal growth rate (µm/min) with GAG. Go = crystal growth rate without GAG. C-HS: commercial HS obtained from bovine kidney (Seikagaku Kogyo, Mr unknown).

RESULTS AND DISCUSSION

Gel-Filtration Chromatography of GAGs

U-HS eluted as two peaks, the first one at the void volume and the second one slightly behind C6S. Only the first peak contained detectable amounts of protein (Fig 2, top). These findings suggest that part of U-HS exists in the form of a proteoglycan. U-ChS was eluted well behind C6S (Fig 2, bottom). Therefore the Mr of U-ChS is likely to be much less than that of C6S.

Measurement of the Inhibitory Activity of GAGs

All the GAGs examined showed an inhibitory effect. U-HS showed strong inhibitory activity even at the lowest concentration of 1 µg/mL, though U-ChS showed no inhibitory activity at the same concentration. There was no significant difference between the inhibitory activities of the urinary and commercial GAGs (Fig 3).

In conclusion, native urinary HS and ChS inhibit calcium oxalate crystal growth, and the inhibitory activity of HS is stronger than that of ChS. This study shows that commercial GAGs, at least HS and ChS, can be used instead of native urinary GAGs for evaluating the inhibitory effect of GAGs on calcium oxalate crystal growth.

REFERENCES

1. H Iwata, T Terado, M Kin, S Nishio, M Takeuchi and A Matsumoto, Isolation of urinary heparan sulfate and chondroitin sulfate and their inhibitory activity of calcium oxalate crystal growth. *Jpn J Urol,* 82:405 (1991).
2. R Hata and Y Nagai, A rapid and micromethod for separation of acidic glycosaminoglycans by two dimensional electrophoresis. *Anal Biochem,* 45:462 (1972).
3. S Nishio, JP Kavanagh and J Garside, A small-scale continuous mixed suspension mixed product removal crystallizer. *Chem Eng Sci,* 46:709 (1991).
4. WG Robertson and DS Scurr, Modifiers of calcium oxalate crystallization found in urine. I. Studies with a continuous crystallizer using an artificial urine. *J Urol,* 135:1322 (1986).

THE EFFECT OF GLYCOSAMINOGLYCANS ON CALCIUM OXALATE MONOHYDRATE CRYSTAL MORPHOLOGY

Y. Shirane, A. Yamamoto and S. Kagawa

Department of Urology
Tokushima University
School of Medicine
Tokushima, Japan

INTRODUCTION

Glycosaminoglycans (GAGs) are well-known inhibitors of crystallization of calcium oxalate monohydrate (COM). We have previously reported that GAGs affect the morphology of COM crystals formed in supersaturated calcium oxalate (CaOx) solution *in vitro*, showing characteristic shapes which vary with GAG species. With increasing GAG concentrations, the crystals decrease in thickness - suggesting that GAGs inhibit crystal growth.

The above mentioned differences in morphology of crystals may be related to their binding affinities to the COM crystals and their resulting degrees of crystal growth inhibition. To test this hypothesis, using X-ray microanalysis, we determined whether the crystals contained sulfur (S) which was derived from sulfate residues of GAGs.

METHODS

Supersaturated CaOx solutions, initially containing 4mM calcium chloride, 0.2mM sodium oxalate and 0.1M sodium chloride buffered at pH 5.7 with 50mM sodium acetate, were incubated with 10 μg/mL of GAG in plastic wells at 37°C for 24 h under gentle shaking. Crystal formation was initiated by adding buffered sodium oxalate solutions. The crystals were examined using a HITACHI S-800 SEM and by a Kevex 7000 energy dispersive X-ray microanalyzer attached to a HITACHI H-500 SEM. Quantitative analysis was performed by a Kevex Quantex-Ray analysis using ZAF corrections.

RESULTS

1. The COM crystals formed in the presence of heparin (HP) appeared as oval (crystal upper face) and dumb-bell (side face) shapes consisting of stacks of lamellar crystallites (Figure 1A) and X-ray microanalysis of the crystals revealed the presence of S on the surfaces (Figure 2A, C). The mean atomic percentage of S was 2.16±0.77 ($p < 0.001$).
2. The morphology of crystals formed in the presence of dermatan sulfate (DS) showed hexagonal plates with slightly rounded tops and several lamellae (Figure 1B), and a significant quantity of S (1.25±0.17 %, $p < 0.001$) was detected on the surfaces.

Urolithiasis 2, Edited by R. Ryall *et al.*,
Plenum Press, New York, 1994

3. The crystals formed in the presence of chondroitin-6-sulfate (ChS) had freely-grown thick and elongated hexagonal plates (Figure 1C) similar in morphology to the crystals formed in the absence of GAGs, and microanalysis of the crystals showed the absence of S (Figure 2B, D) with a percentage value (0.47±0.31, ns) similar to that of control crystals (0.58±0.26) formed in the presence of hyaluronic acid, which is a non-sulfated GAG.

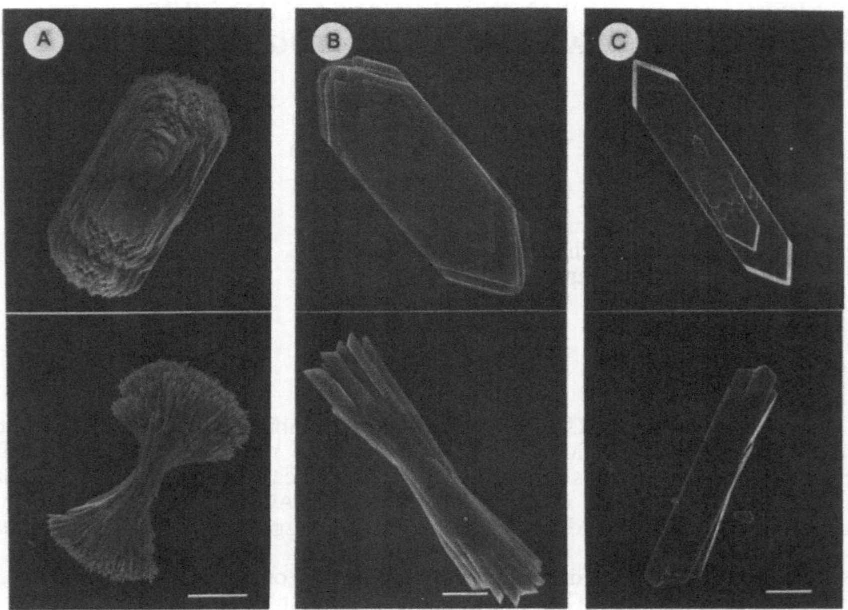

Figure 1. **SEM micrographs of COM crystals formed in supersaturated CaOx solutions coexistent with individual GAGs (10 μg/mL). A, HP; B, DS and C, ChS.**
The top of each panel shows an upper face of crystal and the bottom, a crystal side face. Bar=10μm.

DISCUSSION

These results suggest that HP and DS bind tightly to COM crystals, resulting in the inhibition of continuous crystal growth on their surfaces, and thereby altering crystal morphology. On the contrary, ChS is not retained on the crystals. Thus, the weakly-bound ChS may be easily displaced by the Ox^{2-} lattice ions. Therefore, ChS does not inhibit crystal growth and as a result there is little change in the crystal morphology. Additionally, the results may explain why ChS, the most prominent urinary GAG, is notably absent from urinary stone matrix.

CONCLUSION

The differences in morphology of COM crystals precipitated in the presence of individual GAGs are due to the differences in their binding affinities to the COM crystals and their resulting degrees of crystal growth inhibition.

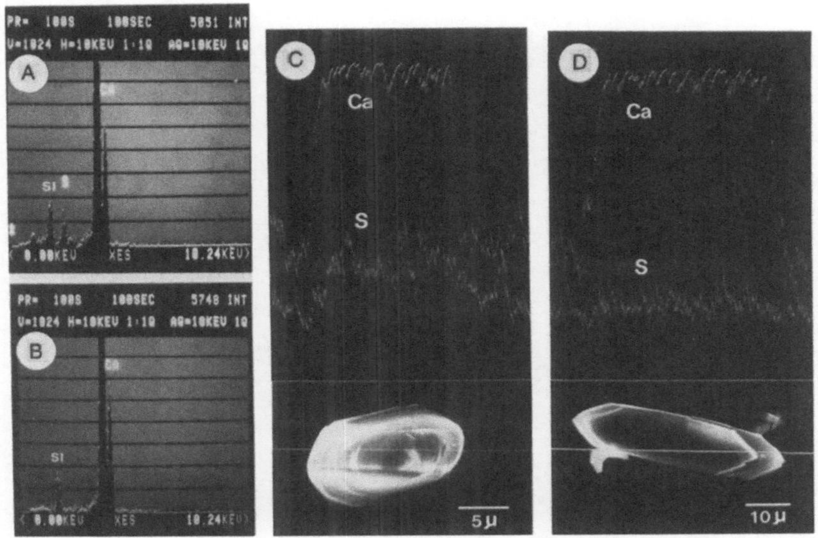

Figure 2. Microanalysis spectrum (A, B) and scanning profiles for Ca (30 eV) and S (70 eV) (C, D) of COM crystals formed in the presence of HP 3 μg/mL with 2mM Mg (A, C) or ChS 10 μg/mL (B, D).

ACKNOWLEDGMENTS

The authors would like to thank Mr Akira Kanaya for his kind technical assistance of X-ray microanalysis.

URINARY GLYCOSAMINOGLYCANS DIFFER IN THEIR CALCIUM OXALATE CRYSTALLIZATION ACTIVITIES

M.D.I. Gohel and D.K.Y. Shum

Dept. of Biochemistry
Faculty of Medicine
The University of Hong Kong
Hong Kong

INTRODUCTION

Various studies have indicated that urinary macromolecules exhibit diverse activities towards calcium oxalate crystallization[1,2]. One group of these macromolecules, the glycosaminoglycans (GAGs), has been shown to affect calcium oxalate crystallization[3,4]. Our interest in GAGs stems from previous evidence that urinary polyanions exhibit differences towards crystallization of urinary calcium oxalate between normal controls and stone-formers[5] and further to this, reference standard GAGs (obtained from non-urinary sources) appear to behave differently from urinary GAGs[6]. The present study indicates that urinary GAG classes from normals and stone formers influence crystallization differently, though electrophoretically they appear similar.

MATERIALS AND METHODS

The preparation of urinary GAGs, urinary ultrafiltrates (10kDa) and the calcium oxalate crystallization test have been described previously[5]. Preparative agarose gel electrophoresis of papainized urinary polyanions was performed using a microcontroller-based apparatus with 0.05M barium acetate buffer (pH 5.8), constant voltage[7]. The eluted fractions were collected at half-hour intervals. The fractions collected were then monitored by (a) cetylpyridinium chloride turbidimetry, (b) cellulose acetate electrophoresis, (c) enzymatic/chemical treatment specific for certain GAG classes and (d) carbazole reaction for uronic acid[8]. All statistical analyses were performed using the Mann Whitney pair test.

RESULTS

The preparative electrophoretic fractions of urinary GAGs were pooled in order of decreasing mobilities into NC1 to NC4 for normal controls and SF1 to SF4 for stoneformers; Chondroitin/dermatan sulphate (CS/DS) and heparan sulphate (HS) contained in these pools were identified by their respective removal subsequent to treatments with chondroitinase-ABC and nitrous acid (pH 4). In general, the fast-moving fractions consisted of CS and DS and the slow-moving fractions, of tailing CS/DS and largely HS. After adjustments to similar uronic acid contents, the pools were tested for crystallization of urinary calcium oxalate. The results are presented in Fig 1. In the normals the higher crystal nucleating activity appears to reside in the electrophoretically slow fractions (NC3 & NC4) which consist mainly of HS, whereas in the stone formers the electrophoretically fast fractions (SFl & SF2), which consist of CS/DS, appear to be more active. The HS of stone formers were also crystal-active but the quantity recovered was not sufficient for a comparison with those of the normals. With regard to inhibition of growth of crystals, it was generally observed that factors that enhance crystal nucleation also inhibit crystal growth.

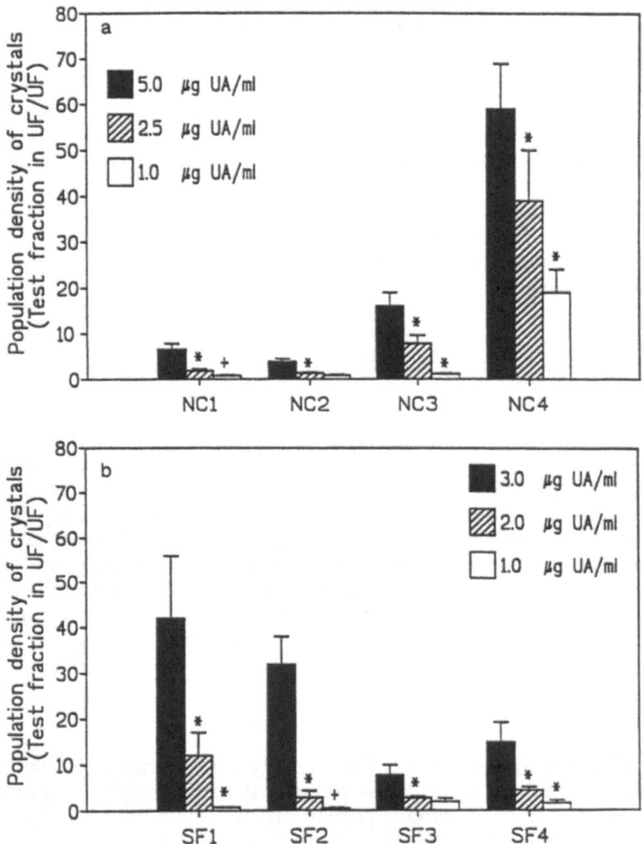

Figure 1. Calcium oxalate crystal-nucleating activities of electrophoretic fractions of urinary GAGs from (a) normals and (b) stone formers. The activities are indicated by the ratio of population density of calcium oxalate crystals formed in solutions of test fractions in urine ultrafiltrate (UF) to that formed in urine ultrafiltrate alone. The significant decrease in crystal-nucleating activities from samples of higher uronic acid content for each pool is indicated. *(n=12; * = P < 0.001; + = P < 0.02)* [Mean±SEM]

DISCUSSION

Urinary GAGs were separated by electrophoresis at a preparative level into different classes of chondroitin sulphates and heparan sulphates. Although the GAG classes excreted by normals and stone formers were presently seen to be similar, we observed for the first time that the influence of these GAG classes on the crystallization of urinary calcium oxalate differed between normals and stone formers.

Heparan sulphate appeared to be a strong promoter of crystal nucleation but an inhibitor of crystal growth. Our earlier study[6] of commercial HS indicated activities similar to urinary HS when considered at similar concentration ranges but considerably higher than heparin, which is absent from urine. Commercial preparations of heparin and CS have, however, been used as reference to reflect the behaviour of urinary GAGs towards calcium oxalate crystallization[3,4,9], notwithstanding that these references may not be comparable to urinary GAGs in such activity. In the stone-formers we found the electrophoretically fast CS to be very active towards crystallization, unlike electrophoretically fast CS fractions of normals or commercial CS, which showed low activity, if at all. Further work will be performed to elucidate the structural basis of these behavioural differences.

ACKNOWLEDGEMENTS

Wong Ching Yee Medical Postgraduate Scholarship for Mr MDI Gohel's attendance at the symposium.

REFERENCES

1. T Koide, M Takemoto, H Itatani, M Takaha and T Sonoda, Urinary macromolecular substances as natural inhibitors of calcium oxalate crystal aggregation, *Invest Urol* 18(5):382 (1981).
2. GH Nancollas, SA Smesko, AA Campbell, M Coyle-Rees, A Ebrahimpour, M Binette and JP Binette. Mineralization inhibitors and promoters, *in:* "Urolithiasis" VR Walker, RAL Sutton, ECB Cameron, CYC Pak and WG Robertson (ed) Plenum Press, New York (1989).
3. K Kohri, J Garside and NJ Blacklock, The effect of glycosaminoglycans on the crystallization of calcium oxalate, *Br J Urol* 63:584 (1989).
4. AH Angell and MI Resnick, Surface interaction between glycosaminoglycans and calcium oxalate, *J Urol* 141:1255 (1989).
5. MD Gohel, DKY Shum and MK Li, The dual effect of urinary macromolecules on the crystallization of calcium oxalate endogenous in urine, *Urol Res* 20:12 (1992).
6. DKY Shum, KF Wu and SW Wong, Crystallization of urinary calcium oxalate. Effects of added calcium, oxalate or commercial glycosaminoglycans (GAGs), in: Book of Abstracts of the 9th Symposium of the Federation of Asian and Oceanian Biochemists, Hong Kong Biochemical Association (Organizers), Hong Kong (1991).
7. LS Cornish, TRC Boyde, YC Chung and CS Ng, A microcontroller-based flat bed preparative electrophoresis apparatus, *Lab Practice* 36:35 (1987).
8. DKY Shum and E Liong, Calcium oxalate crystallizing properties of macromolecules released by renal tubular cells *in vitro*, in *this volume*.
9. RL Ryall, RM Harnett and VR Marshall ,The effect of monosodium urate on the capacity of urine, chondroitin sulphate and heparin to inhibit calcium oxalate crystal growth and aggregation, *J Urol* 135:174 (1986).

THE EFFECT OF URATE ON CALCIUM OXALATE PRECIPITATES

K. Kleboth[1] and J. Joost[2]

[1]Institute of Inorganic and Analytical Chemistry
University of Innsbruck,
Austria
[2]Department of Urology, Hall i.T
Austria

INTRODUCTION

Despite numerous investigations, the role of uric acid in calcium oxalate monohydrate (COM) lithiasis is still speculative. Heterogeneous nucleation and epitaxial growth, specific adsorption of COM growth inhibitors, and salting out effects have been suggested to explain the apparent influence of uric acid and urates on the precipitation of COM[1]. Coprecipitation and mixed crystal formation of COM with urates have also been proposed, based upon the comparison of crystallographic data[2]. Evidence for epitaxial growth of COM on uric acid and sodium hydrogen urate (NaHU) has been presented[3]. In this paper we describe co-precipitation and mixed crystal formation experiments of COM with $Ca(HU)_2$.

EXPERIMENTAL

1. Reagent grade chemicals were used throughout. All experiments were carried out at 37° C. $Ca(HU)_2$ was prepared by the "complex acidification " method as described later.
2. Heterogeneous nucleation was investigated by the constant composition method[4]. $Ca(HU)_2$ was used as seed crystal and the pH was kept constant at pH = 7.0 by means of a PIPES buffer.
3. Equilibrium experiments: Co-precipitation was studied by incubating solid COM in Na(HU) solutions and solid $Ca(HU)_2$ in $Na_2C_2O_4$ solutions until equilibrium was achieved (12 - 20 h).

   Co-precipitation of COM and $Ca(HU)_2$ was also achieved in non-equilibrium systems from clear solutions by homogeneous precipitation using the "complex acidification" method: In a system containing Ca^{2+} and complexing agent (eg, NTA/H_3Y), the Ca^{2+}-activity (pCa) depends on pH, total concentration of H_3Y and total concentration of Ca:

$$Ca^{++} + HY^{--} \qquad CaY^{-} + H^{+}$$

$$K(CaY^{-}) = \frac{[CaY^{-}]\,[H_3O^{+}]}{[Ca^{++}]\,[HY^{--}]}$$

   A wide range of Ca^{2+} activities may be obtained at a certain pH by adjusting the total concentration of Ca and the total concentration of H_3Y (Fig 1).

4. Analytical methods: All precipitates were analysed for cations by atomic absorption spectroscopy and HU was determined by UV absorption measurements. All precipitates were characterized by X-ray powder diffraction and IR spectroscopy.

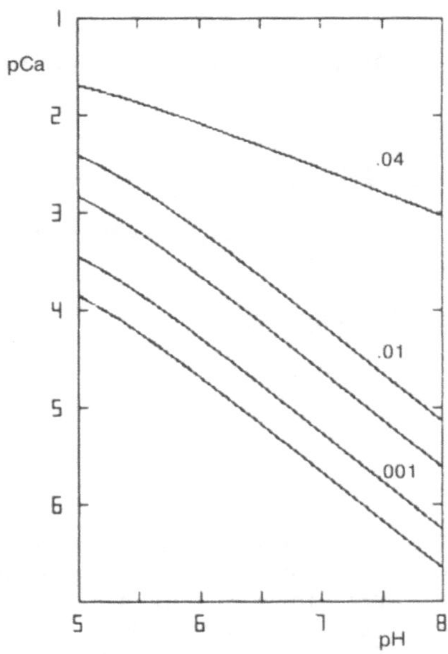

Fig.1. Plot of pCa vs. pH; c(NTA)=0.04, c(Ca)= 0.04, 0.01, 0.004, 0.001, 0.0004 moles/L
Calculated amounts of nitrilo-triacetic acid (NTA), uric acid (H_2U), oxalic acid and $CaCl_2$ were suspended in 150 mL water; LiOH solution was added, until a clear solution was obtained due to the formation of soluble lithium urate and due to the complexation of Ca^{2+} by NTA (∫H3Y): The pH was lowered by slow addition of HCl over several hours; it was monitored by a glass electrode and not allowed to drop below 6,5. By this method, pure Ca-salts (COM and $Ca(HU)_2$) as well as mixed precipitates were prepared.

RESULTS

1. Nucleation experiments indicated that COM crystallization from supersaturated solutions may be induced by the addition of $Ca(HU)_2$ and NaHU seed crystals with an induction period of up to several hours. The crystal growth rate was low at the beginning of the crystallization process, but increased as the crystallization proceeded.

2. The solubility product of $Ca(HU)_2$ was found to be 9.1×10^{-9} at 37°C. The solubility, therefore, is three times the solubility of COM. At physiological oxalate concentrations (3×10^{-4} moles/L) the HU^- concentration must be ten times that of oxalate in order to achieve simultaneous precipitation COM and $Ca(HU)_2$ at the same Ca^{2+} activity.

3. Equilibrium co-precipitation experiments gave different results depending on the $[HU^-]$: $[C_2O_4^{2-}]$ ratio: As expected, a ratio of 1:1 yielded precipitates that were identified as COM, while a ratio of 10:1 precipitated material that gave x-ray powder diffraction diagrams different from those of COM as well as $Ca(HU)_2$, indicating that a new, though unidentified, phase had been formed.

4. Co-precipitation by the complex acidification method confirmed these results. Precipitates from solutions with ratios of $[HU^-]$: $[C_2O_4^{2-}]$ = 10:1 again gave

precipitates that contained calcium, oxalate and urate; again, their powder diffraction patterns did not match those of the pure components COM and $Ca(HU)_2$ respectively. All powder diffraction patterns exhibited numerous but weak reflections, indicating a low degree of crystallinity (Fig. 2). Reproducibility was generally poor.

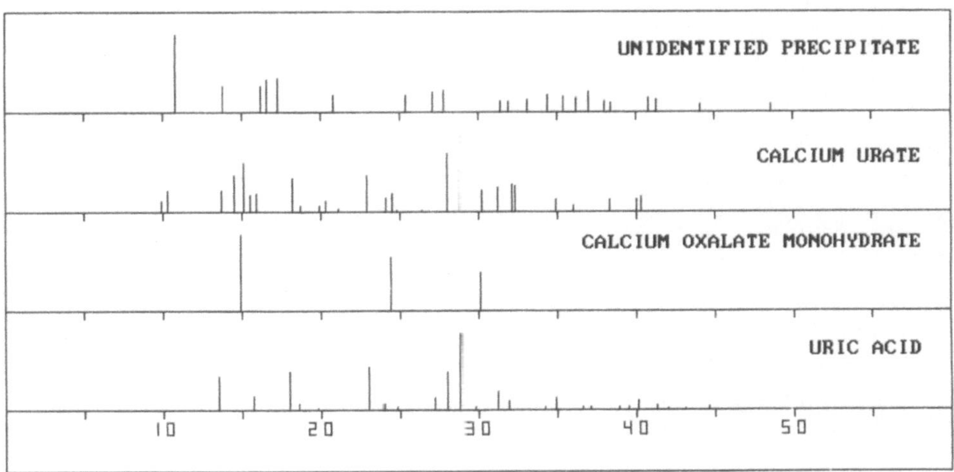

Fig. 2 X-ray powder diffraction patterns of unidentified precipitate and related compounds

CONCLUSIONS

1. NaHU as well as $Ca(HU)_2$ may be used as seeding crystals for the precipitation of COM indicating the possibility of epitaxial growth; however, both showed long induction periods and smaller growth rates compared to COM seed crystals.

2. The co-precipitation experiments indicated that new, but unidentified phases are formed by equilibrium co-precipitation as well as by non-equilibrium co-precipitation. These experiments require more elaborate methods such as electron microscopy to give conclusive evidence of co-precipitation.

3. Precipitation conditions in urine may be widely different from conditions *in vitro* ; these experiments, however, indicate that urates such as Na(HU) or $Ca(HU)_2$ do not play a major role in COM lithiasis.

REFERENCES

1. LJMJ Blomen and AAB Lycklama á Nijeholt, Die Bedeutung der Harnsaure und der Urate für die Kristallisation von Kalziumoxalatsteinen, in: "Harnsäure und Urolithiasis", GIT Verlag, Darmstadt (1984).
2. NS Mandel and GS Mandel, Epitaxis between stone forming crystals at the atomic level, in: "Urolithiasis, Clinical and Basic Research", Plenum Press, New York and London (1981).
3. H-G Tiselius, Effects of sodium urate and uric acid crystals on the crystallization of calcium oxalate, *Urol Res* 12:11-15 (1984).
4. ME Sheehan and GH Nancollas, Calcium oxalate crystal growth, *Invest Urol* 17:446.

THE PROMOTION OF CALCIUM OXALATE CRYSTALLIZATION BY URATE: CAN IT BE THAT EPITAXY IS NOT THE CAUSE AFTER ALL?

P.K. Grover, V.R. Marshall and R.L. Ryall

Department of Surgery
Flinders Medical Centre
Bedford Park, South Australia

INTRODUCTION

In the early 1970's Kallistratos and his colleagues[1] reported that the addition of a solution of urate to human urine prompted the precipitation of calcium oxalate (CaOx). In an extension of this work we recently reported that addition of dissolved urate promoted the nucleation, bulk deposition and aggregation of CaOx crystals generated from undiluted urine in response to increasing the oxalate concentration[2]. Although we have since demonstrated that adding the urate solution to the urine did not cause the formation of particles of urate[3], it is still possible that small nuclei of urate may have formed in the urine samples after addition of the oxalate load used to induce crystallization. If these were to act as epitactic centres for the deposition of CaOx, then scanning electron microscopic examination of the material deposited at the end of the experiment would reveal crystals which, though exhibiting the external morphology of pure CaOx, would in fact contain a core of urate.

The aim of the present investigation was to determine, using chemical analysis, and re-ultrafiltration combined with scanning electron microscopy, whether the process of epitaxy is responsible for the promotion of CaOx crystallization in urine by the addition of dissolved urate.

MATERIALS AND METHODS

Collection of urine samples and measurement of urate concentration before and after CaOx crystallization: 24 h urine samples were collected from 13 healthy men and processed as described earlier[2]. Three of the specimens were used for the re-ultrafiltration studies; the remainder were used for the crystallization experiments as described previously[2]. The urate concentration was measured in control and experimental portions of each urine before and after crystallization. The precipitated crystalline material was 0.22μm filtered and examined by light microscopy.

Analysis of the crystals: The crystalline material isolated from four of the ten urine samples was dried overnight at 37°C and analysed for urate by infra-red spectroscopy, X-ray powder diffraction and ultraviolet wet chemical analysis. For comparison, standards consisted of uric acid, monosodium urate and CaOx monohydrate.

Urolithiasis 2, Edited by R. Ryall *et al.*,
Plenum Press, New York, 1994

Re-ultrafiltration studies Three urine samples were 10kDa ultrafiltered using an Amicon hollow fibre bundle (HIP10-20), and their urate concentration was increased, and CaOx crystallization induced, as described above. After incubating for 5 min at 37°C, duplicate 10mL samples were re-ultrafiltered using Diaflo ultrafiltration membranes (molecular weight cut off 10kDa) of 13mm in diameter. The membranes were then examined for the presence of particles using scanning electron microscopy as described earlier.

The Wilcoxon signed rank sum test was used for all statistical comparisons.

RESULTS

Addition of dissolved urate increased ($p < 0.01$) the median urate concentration from 2.2 to 5.6 mmol/L. No reduction in urate concentration was observed following the crystallization of CaOx. Examination of the infra-red spectra of standard uric acid, monosodium urate, crystals precipitated from one of the urines to which no urate was added (control) and those from the sample of the same urine enriched with urate showed that the absorption maxima of the control and experimental samples were identical, and did not demonstrate peaks specific to sodium urate and/or uric acid (Fig 1).

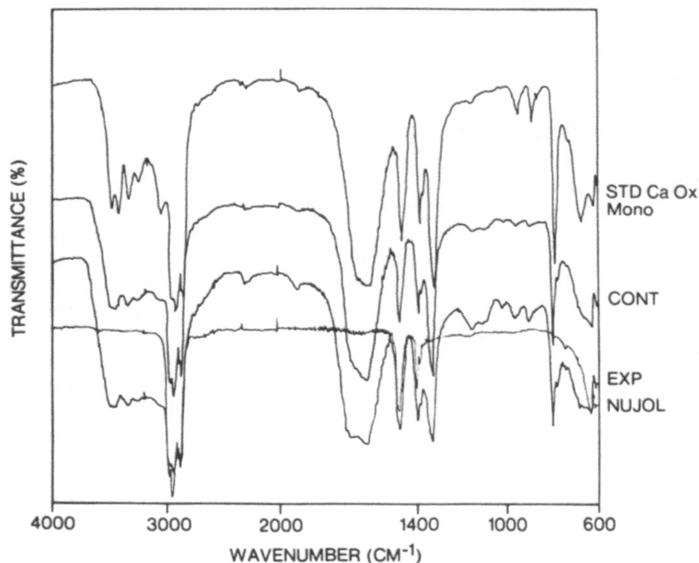

Figure 1. Infra-red spectra of standard calcium oxalate monohydrate and crystals isolated from one of the controls (CONT)and a portion of the same urine spiked with urate (EXP)

The x-ray diffraction patterns of standard uric acid, sodium urate crystals and crystals isolated from the control and test urine samples were identical and did not exhibit contamination with sodium urate and/or uric acid (Fig 2).

Urate was not detected in the crystalline CaOx by wet chemical analysis using ultraviolet absorption spectroscopy. In all three urines examined, scanning electron microscopic examination of the Diaflo membranes used for re-ultrafiltration 5 min after the addition of the oxalate load, did not reveal any colloidal or crystalline material.

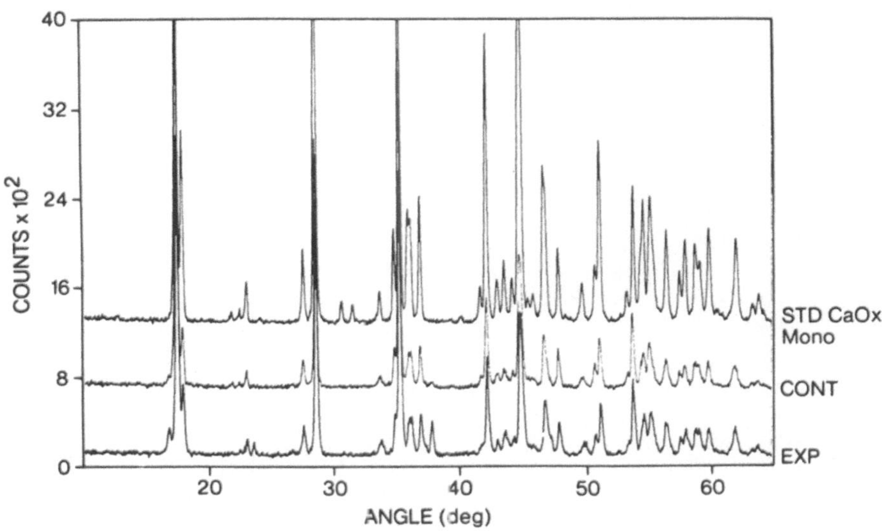

Figure 2. X-ray diffraction patterns of standard calcium oxalate monoydrate and the crystals isolated from one of the controls (CONT) and the same urine to which urate had been added (EXP).

DISCUSSION

If epitaxy is responsible for urate's promotory effect then we might have expected the following results in the experiments reported here:

(1) a decrease in urate concentration after the induction of CaOx crystallization -no such decrease was noted.

(2) the detection of urate as a (minor) contaminant of the CaOx crystals - urate was not detected in the crystals.

(3) colloidal or crystalline particles of sodium urate and/or uric acid on the membranes after re-ultrafiltration of the urine samples early in the crystallization experiments - the filtration membranes were devoid of any particles.

Therefore the evidence overall indicates that no urate was present inside the crystals. Epitaxial nucleation of CaOx by urate crystals can therefore be discounted as a mechanism to explain the promotion of CaOx crystallization invoked by dissolved urate.

ACKNOWLEDGEMENTS

Figures 1 and 2 are reproduced with permission from *Clinical Science.*

REFERENCES

1. G Kallistratos, A Timmerman and O Fenner. Zum Einfluss des Aussalzeffektes auf die Bildung von Calciumoxalat-Kristallen im menschlichen Harn, *Naturwissenschaften* 57: 198 (1970).
2. PK Grover, RL Ryall and VR Marshall. The effect of urate on calcium oxalate crystallization in human urine: Evidence for a promotory role of hyperuricosuria in urolithiasis, *Clin Sci* 79: 9 (1990).
3. PK Grover, RL Ryall and VR Marshall. Calcium oxalate crystallization in urine: The role of urate and glycosaminoglycans, *Kidney Int* 41: 149 (1992).

EFFECTS OF URINARY MACROMOLECULES ON CALCIUM OXALATE CRYSTALLIZATION STUDIED BY CONTINUOUS FLOW SYSTEM AND FRESH UNDILUTED URINE

S. Nishio, T. Terado, H. Iwata and M. Takeuchi

Department of Urology
Ehime University School of Medicine
Ehime, Japan

INTRODUCTION

Continuous crystallization in a mixed suspension mixed product removal (MSMPR) system has been widely used to study calcium oxalate crystallization as it can give quantitative estimates of nucleation and growth rates. But until now most studies have been performed using artificial urines. We studied the effects of urinary macromolecules using fresh undiluted urine and MSMPR system, and compared the difference between normal subjects and stone formers.

MATERIALS AND METHODS

The system used in our present study has a small scale crystallizer with a 20 mL volume, which is one-tenth that of conventional MSMPR crystallizers in which it is impractical to study crystallization because 2L of fresh urine are required for an experiment. The details of the modified system have been described previously[1]. The system has three feeding tubes and one withdrawal tube and urine is used as one of three feed solutions to the crystallizer, pumped at 92% of the total flow rate. The other feed solutions consist of calcium chloride and sodium oxalate, each flowing at 4% of the total flow. Crystals produced in the miniature crystallizer are counted using a Coulter Counter Multisizer and analyzed by microprocessor (Figure 1).

Early morning urine samples were obtained from 29 normal male controls and 13 male stone formers. Specimens were collected into a pre-warmed flask. Before the experiment began, each urine was adjusted to pH 6.0 with hydrochloric acid or sodium hydroxide and analyzed for calcium and oxalate. Samples were then filtered first through a 5 μm, and subsequently through a 0.45 μm membrane filter, steps which are known to remove Tamm-Horsfall mucoprotein from urine, but not other urinary macromolecules[2]. In order to achieve valid comparison between samples, urine was adjusted to 500mOsm/KgH_2O with distilled water. Each urine sample was then divided equally. One half was used for the filtered urine experiment; the other half was ultrafiltered using a nominal Mr cut-off of 10,000 Da in order to remove the urinary macromolecules and used for the ultrafiltered urine experiment. The concentrations of calcium and oxalate solutions were adjusted to give initial concentrations in the crystallizer of 8mM calcium and 0.8mM oxalate, taking into account the concentrations of calcium and oxalate in each urine sample. Crystallization was stable about 50 min after the MSMPR experiment was started. Crystals between 5 and 30 μm in size were analyzed by the Coulter Counter. Crystal growth, nucleation rate and suspension density were calculated.

Urolithiasis 2, Edited by R. Ryall *et al.*,
Plenum Press, New York, 1994

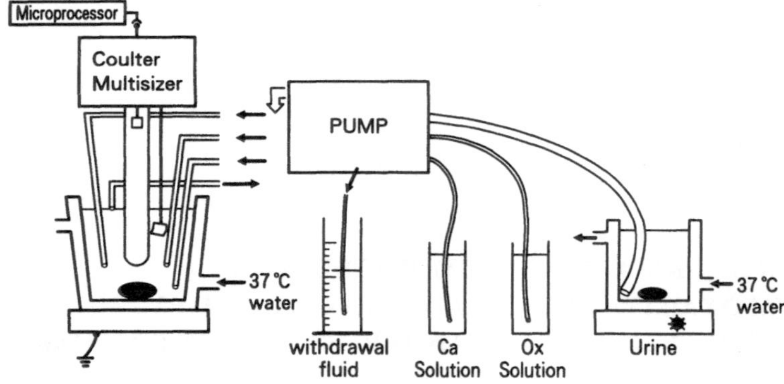

Figure 1. New continuous flow system with fresh undiluted urine

RESULTS

Filtered Urine Experiments

The growth rate of calcium oxalate was lower in the urine samples from stone formers than those from normal subjects ($p<0.1$), whereas the nucleation rate was significantly higher ($p<0.05$). There was no difference in suspension density between the two groups (Figure 2).

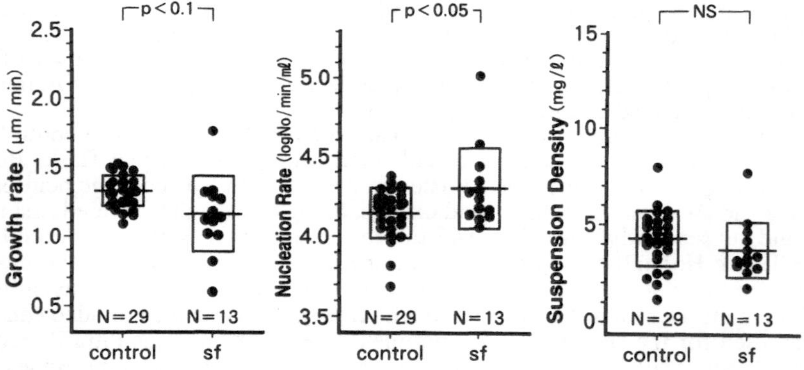

Figure 2. The results of growth rate, nucleation rate and suspension density in undiluted filtered urine from normal subjects (control) and stone formers (sf). The growth rate was lower and the nucleation rate higher in stone former urine than in normal urine.

Ultrafiltered Urine Experiments

In ultrafiltered urine, the growth rate was higher and the nucleation rate was lower than was observed in urine. Although there were some individual differences in filtered whole urine between control subjects and stone formers, there was no overall difference in growth rate, nucleation rate and suspension density between them (Figure 3).

These results indicate that urinary macromolecules have strong inhibitory effects on calcium oxalate crystallization and there is a quantitative or qualitative difference in the urinary macromolecules between normal subjects and stone formers.

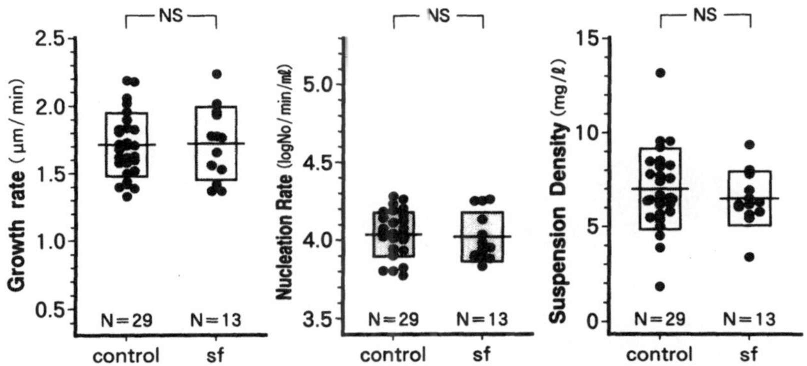

Figure 3. The results of growth rate, nucleation rate and suspension density in ultrafiltered urine from normal subjects (control) and stone formers (sf). There was no difference between the two groups.

DISCUSSION

The present study showed that it is possible to perform crystallization experiments in an MSMPR system with fresh undiluted urine, and to compare the results of each sample by adjustment of conditions such as urinary pH, calcium, oxalate, osmotic pressure and so on. Furthermore, the effect of urinary macromolecules (UMMs) on calcium oxalate crystallization can be directly examined in each sample by using urines with and without (ultrafiltered) their component macromolecules.

The differences of crystallization using filtered urine between normal subjects and stone formers disappeared in ultrafiltered urine. This result indicates that there is a quantitative or qualitative difference in UMMs between them. In whole urine, the growth rate was lower in stone formers' urines than in those from normal subjects. Drach et al also reported a lower growth rate and higher nucleation rate with UMMs from stone formers than with UMMs from normal subjects using an MSMPR system based on artificial urine[3].

It is speculated that UMMs of stone formers may accelerate heterogeneous crystallization and thereby decrease the urinary metastable limits in comparison with those from normal subjects.

REFERENCES

1. S Nishio, JP Kavanagh and J Garside, A small-scale continuous mixed suspension mixed product removal crystallizer. *Chem Eng Sci* 46:709 (1991).
2. IR Doyle, RL Ryall, and VR Marshall, The effect of low-speed centrifugation and Millipore filtration on the urinary protein content. In "Urolithiasis", VR Walker, RAL Sutton, ECB Cameron, CYC Pak, WG Robertson, eds, Plenum Press, New York and London, p 593, (1989).
3. GW Drach, S Thorson, and A Randolph, Effects of urinary organic macromolecules on crystallization of calcium oxalate: Enhancement of nucleation, *J Urol* 123:519 (1980).

DISCUSSION

REFERENCES

STUDIES OF URINARY MACROMOLECULES: AN URGENT APPEAL FOR A STANDARD REFERENCE CRYSTALLIZATION MODEL

A.L. Rodgers[1] and D. Ball[2]

[1]Chemistry Department
[2]Physics Department
University of Cape Town
South Africa

INTRODUCTION

Despite an enormous amount of research on the topic, the role of urinary macromolecules (UMM) in urolithiasis has not been satisfactorily characterised. The literature abounds with numerous examples of conflicting findings. Workers have reported that UMM are promoters[1,2] and inhibitors[3,4,5] of calcium oxalate (CaOx) nucleation[1,3], growth[2,4] and aggregation[5]. The present study was undertaken to gain some insight into the causes of these discrepancies and how they might be addressed in future investigations.

MATERIALS AND METHODS

15 male controls (N) and 10 male CaOx stone formers (SF) provided 24 h urines which were sieved (74 μm) and filtered (0.45 μm) to give the FILTD fraction. Urines were ultrafiltered (Amicon, 10 kDa cut-off) to give ultrafiltrate (UF) and retentate (CONC) fractions. Metastable limits were determined by titration with sodium oxalate; CaOx crystallization kinetics were followed by using turbidimetry and a Malvern particle size analyzer. The Wilcoxon signed rank test and the Wilcoxon rank sum tests were used to analyze the data.

RESULTS

Crystallization rates, particle sizes and particle volumes in the UF fraction of N and SF were lower than those in the corresponding FILTD and CONC fractions ($p<0.05$) but there were no significant differences between the 2 groups of subjects themselves. Figure 1 shows the variation of turbidity vs time for the various fractions obtained from one urine sample. The gradient of the linear portion was taken as a measure of crystallization rate.

Urolithiasis 2, Edited by R. Ryall *et al.*,
Plenum Press, New York, 1994

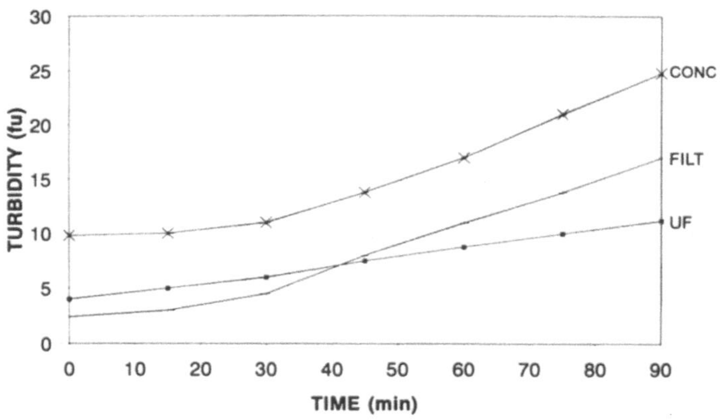

Figure 1. Crystallization Kinetics - Turbidity vs Time plots for different urine fractions after dosing with sodium oxalate.

DISCUSSION

The data indicate that UMM are promoters of calcium oxalate nucleation and growth. It is of concern to us that while several reported studies agree with our findings, there is an equal number which contradict them. It is apparent that the activity of UMM is highly sensitive to the nature of the crystallizer as well as the additive, synergistic and competitive effects of all components in the test solutions. We suggest that different conditions induce a "ripple effect" in which processes involving inhibition and promotion of nucleation, growth and aggregation compete with and support each other to produce a nett, *system dependent* mechanism. It is therefore meaningless to compare experiments which have been conducted in different systems under different operating conditions. It is clear that a standard reference crystallization model is urgently required. Such a model should define the crystallizer itself, the crystal monitoring technique and the procedure for isolating or concentrating the UMM. Terminology too needs to be clarified. Until researchers adopt a standard approach, it is unlikely that consensus on the role of UMM in calcium oxalate urolithiasis will be reached.

ACKNOWLEDGEMENTS

We would like to thank the South African Foundation for Research and Development, the Medical Research Council and the University of Cape Town (Research Committee and Human and Caporn Bequest) for generous awards of research funds.

REFERENCES

1. MD Gohel, DK Shum and MK Li, The dual effect of urinary macromolecules on the crystallization of calcium oxalate endogenous in urine, *Urol Res* 20: 13 (1992).
2. WB Gill, JW Karesh, L Garsin and MJ Roma, Inhibitory effects of urinary macromolecules on the crystallization of calcium oxalate, *Invest Urol* 15: 95 (1977).
3. K Kohri, J Garside and NJ Blacklock, The effect of glycosaminoglycans on the crystallization of calcium oxalate, *Br J Urol* 63: 584 (1989).
4. RL Ryall, RM Harnett and VR Marshall, The effect of urine, pyrophosphate, citrate, magnesium and glycosaminoglycans on the growth and aggregation of calcium oxalate crystals in vitro, *Clin Chim Acta* 112: 349(1981).
5. WG Robertson and DS Scurr, Modifiers of calcium oxalate crystallization found in urine. I. Studies with a continuous crystallizer using an artificial urine, *J Urol* 135: 1322 (1986).

STUDIES OF CALCIUM OXALATE STONE FORMATION: INTRODUCING A MULTI - FACETED CRYSTALLISER

I. Muller[1], P.W. Linder[1], A. Rodgers[1], D. Ball[1] and J. Knight[2]

[1]Department of Chemistry
University of Cape Town
[2]Department of Physics
University of Cape Town
South Africa

INTRODUCTION

The inclusion of various substances ranging from elements to organic ligands and macromolecules in urinary calculi, has prompted numerous studies to assess their possible role as inhibitors or promoters of crystal and stone formation. We have undertaken a study which attempts to address the effects of urinary trace metals on calcium oxalate (CaOx) crystallisation. Our system combines features of a batch crystalliser with the capability of simultaneously monitoring several physico-chemical properties of the reaction solution/suspension: These include [Ca^{2+}], temperature, pH, and particle size distributions.

EXPERIMENTAL

Experiments are performed in a specially designed reaction vessel. It is constructed of teflon since we have shown that it is virtually impossible to obtain good reproducibility between replicate experiments in glass reaction vessels. The vessel is water-jacketed to provide thermal control of the reaction solution. As shown in the diagram, the vessel incorporates two parallel quartz-glass windows (inter-window distance=14.3mm). This facilitates the monitoring of particle size distributions by a laser light-scattering technique using a Malvern 2600 Particle sizer. The laser light (He-Ne 0.1mW laser) is focused by a 100mm focal length lens and the scattered light intensity is measured by a 32 channel annular diode array, and converted to size distributions by a dedicated processor, using an iterative least squares refinement technique.

[Ca^{2+}] is monitored using a Radiometer F2112 Ca Calcium selective electrode. A remote calomel reference electrode with an agar bridge is used. Both are interfaced to an IBM compatible PC via a high precision mV meter, designed and constructed in our laboratories. As indicated in the diagram, two Dosimat 665 burettes are also interfaced to, and controlled by the PC and specifically written software.

In a typical experiment, 9.46 mL of Ca^{2+} solution are buretted into the reaction vessel. The calcium electrode and temperature are allowed to equilibrate for approximately 10 minutes. At this point, 0.542 mL of oxalate solution (concentration adjusted to give the desired initial supersaturation of CaOx) is added at a sufficiently slow rate to avoid local supersaturation maxima. The solution is stirred magnetically. The various properties mentioned above are monitored by the PC and stored on disc. Typical duration of an experiment is two hours.

Urolithiasis 2, Edited by R. Ryall *et al.*,
Plenum Press, New York, 1994

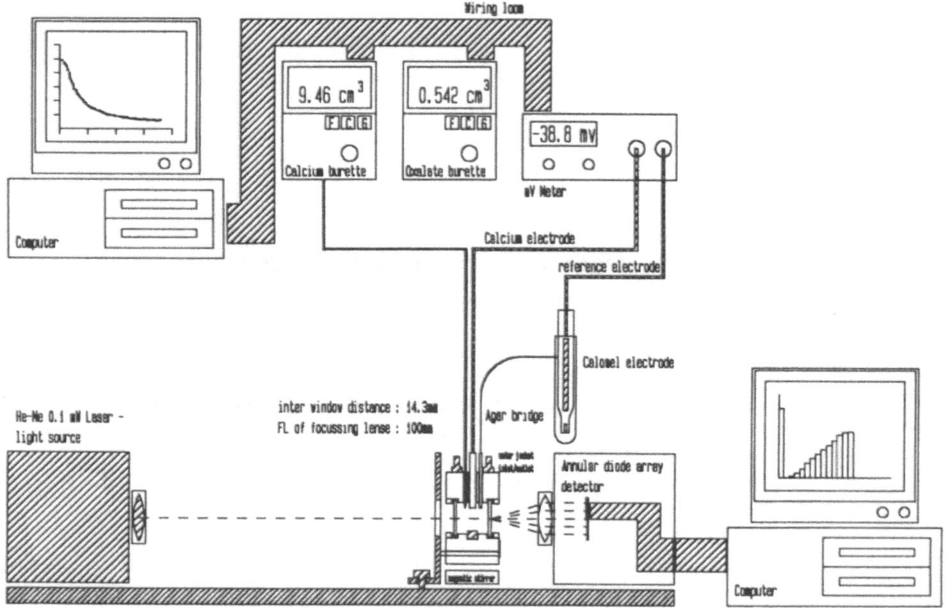

Figure 1. Block diagram of Crystalliser

RESULTS

By way of testing the crystalliser, CaOx crystallisation has been monitored repeatedly and yielded good reproducibility between replicate experiments. Research by Eusebio and Elliot[1] on urinary trace metals indicates that Pb^{2+} is a good inhibitor of CaOx crystallisation. We have performed runs at various concentrations of Lead(II) ranging from the physiologically observed Pb^{2+} concentration in urine of 5×10^{-7}M to higher levels of 1×10^{-5}M. Preliminary data appear to support previous results in that we have observed varying degrees of inhibition of CaOx crystallisation, depending on the concentration. After each experiment the crystals are routinely analysed thermogravimetrically and by scanning electron microscopy. So far both techniques have confirmed the exclusive presence of CaOx monohydrate at equilibrium, under present reaction conditions.

CONCLUSIONS

The usefulness of this crystalliser is not limited to the study of trace metals, but has the potential to be used in the study of the inhibitory or promotory role of virtually all urinary stone constituents. The ability to monitor several properties of the reaction simultaneously and the non-invasive method of particle size distribution measurements (compare Coulter Counter measurements), allow the researcher to make meaningful comparisons between successive experiments, and eliminate the important question about the validity of comparing results obtained from different crystallisers. Investigators are thus urged to adopt a similar approach when undertaking studies of this nature.

REFERENCE

1. E Eusebio and J S Elliot, Effect of trace metals on the crystallisation of calcium oxalate, *Invest Urol* 4: 431 (1967).

EFFECTS OF CITRATE AND URINARY MACROMOLECULES ON CRYSTAL AGGREGATION

H-G. Tiselius, A.M. Fornander and M.A. Nilsson

Department of Urology and Clinical Research Center
University Hospital
Linköping, Sweden

INTRODUCTION

Aggregation of crystals is assumed to be an important risk factor in the formation of calcium renal stones and it is generally believed that urinary macromolecules are efficient in modifying the aggregation of calcium oxalate (CaOx) crystals. It has also been reported that citrate, in addition to its effect on supersaturation and crystal growth, is an inhibitor of CaOx crystal aggregation[1], although this view is not supported by others[2].

METHODS

In an attempt to increase our knowledge of the factors influencing crystal aggregation we measured the rate of crystal sedimentation following a period of stirring, as described by Hess and coworkers[2]. The sedimentation rate was measured in suspensions of CaOx monohydrate (3mg/mL) in saline, with addition of solutions of sodium citrate and dialysed urine, either alone or in combination. The exclusion limit of the dialysis tubing used was 3,000Da. The absorbance was recorded in a Perkin-Elmer Lamba 2 spectrophotometer at 690 nm during 350 seconds after a stirring period of 3 minutes. The inhibition of aggregation brought about by the different solutions was calculated from the absorbance readings after 300 seconds in comparison with that in pure saline suspensions.

RESULTS AND DISCUSSION

Dialysed urine in a concentration of 3.3% had a pronounced inhibitory effect on the rate of crystal sedimentation and such an effect was also recorded following a 10-fold dilution of the urine. This indicates that urine contains macromolecules that strongly counteract crystal aggregation even at very low concentrations. When the citrate concentration in the suspension was increased in the range 0.3 - 3.3 mmol/L the rate of sedimentation decreased significantly; citrate at urinary concentrations is therefore apparently an efficient inhibitor of crystal aggregation. When a mixture of dialysed urine (3.3%) and citrate in the concentration range 0.03 to 3.3 mmol/L was added to the crystal suspension the highest rate of crystal sedimentation was recorded with citrate concentrations in the interval 0.3 to 1.7 mmol/L. The latter observation might indicate either an interaction between urinary macromolecules and citrate in their binding or adsorption to CaOx crystals, or a direct effect of citrate in a certain concentration range on the macromolecules. The clinical importance of this finding is, however, difficult to assess

Urolithiasis 2, Edited by R. Ryall *et al.*,
Plenum Press, New York, 1994

because the concentration of dialysed urine used was low, whereas the citrate concentration corresponded to that in whole urine. Furthermore the crystal concentration necessary for these measurements was far above that expected to occur under physiological conditions.

REFERENCES

1. DJ Kok, SE Papapoulos and OLM Bijvoet, Crystal agglomeration is a major element in calcium oxalate urinary stone formation, *Kidney Int* 37:51(1990).
2. B Hess, Nakagawa Y and FL Coe, Inhibition of calcium oxalate monohydrate crystal aggregation by urine proteins, *Am J Physiol* 257:F99 (1989).

INFLUENCE OF IONIC STRENGTH ON CRYSTAL ADSORPTION AND INHIBITORY ACTIVITY OF MACROMOLECULES

M. Utsunomiya[1], T. Koide[1], T . Yoshioka[1], S. Yamaguchi[1], H. Itatani[2], A. Okuyama[1] and T. Sonoda[1]

[1]Department of Urology
Osaka University Medical School and
[2]Sumitomo Hospital
Osaka, Japan

INTRODUCTION

We investigated the influence of ionic strength on both the ability of heparin to bind to calcium oxalate crystals and to inhibit calcium oxalate crystal growth and aggregation *in vitro*. The effect of ionic strength on the inhibitory activity of urinary macromolecules (UMMS) was also studied.

METHODS

The adsorption of heparin sodium salt was studied both in a seeded (A) and a non-seed crystallization system (B), at various physiological ionic strengths obtained using 0.05, 0.15, 0.25, 0.35 or 0.45 mol/L NaCl in 10 mmol/L sodium cacodylate buffered solutions (pH=6.0) by a solution depletion technique using [^{3}H]-heparin.

RESULTS

In both crystallization systems, the % adsorption of heparin was greater at lower ionic strength than at higher values [System (A):0.05M, 90.4%; 0.15M, 89.0%; 0.25M, 87.8%; 0.35M, 87.0%; 0.45M, 87.1%]. The inhibitory activity of heparin, which was determined by using Coulter Multisizer in system (B), increased in accordance with the increased adsorbed dose (Figures 1,2).

UMMs showed a similar change in inhibitory activity of calcium oxalate crystal growth and/or aggregation in accordance with the change in the ionic strength. These results suggest that the inhibitory activity of UMMS would be greater in urines of lower ionic strength than in those of higher ionic strength, perhaps because adsorption of macromolecules on the surface of calcium oxalate crystals is enhanced under these conditions. Our results have important clinical implications, since they suggest that macromolecular inhibitors may be more potent in human urines of low ionic strength than in those of high ionic strength.

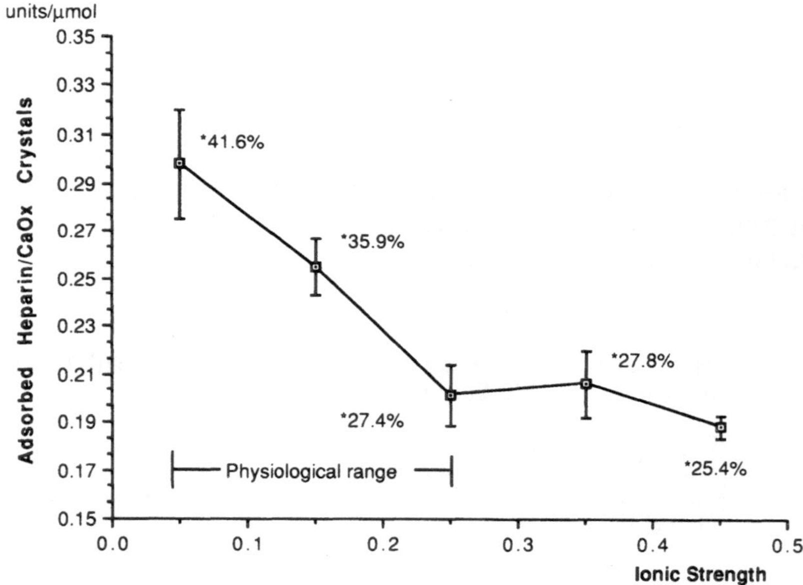

Figure 1. Amount of heparin adsorbed in buffered solutions of various ionic strength.

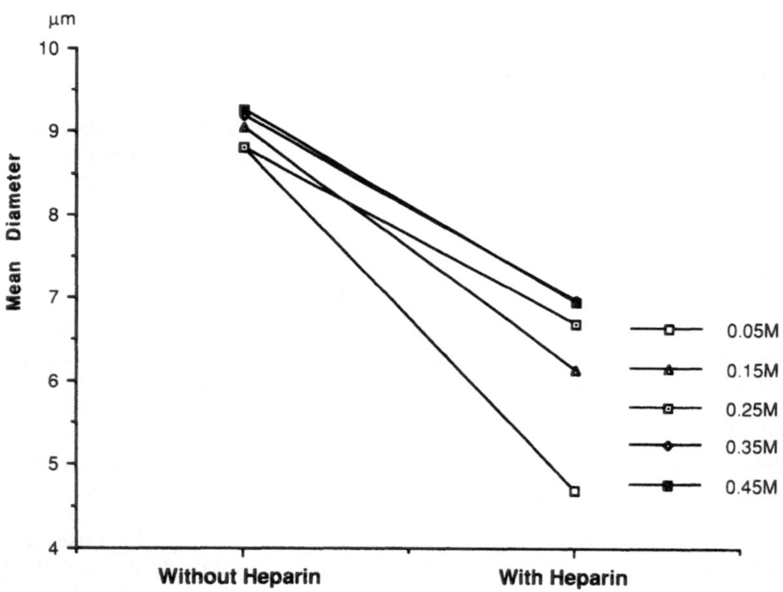

Figure 2. Mean crystal size of calcium oxalate crystals with and without heparin.

INFLUENCE OF AMMONIUM CONCENTRATION AND ANION GAP ON RELATIVE SUPERSATURATIONS IN THE URINE OF STONE FORMING PATIENTS

D.K. Ackermann[1], B. Hess[2], Th. Krebs[1], E. Peheim[3], R. Takkinen[2] and Ph. Jaeger[2]

[1]Department of Urology and
[3]Hospital Laboratories
University of Berne
Berne, Switzerland

INTRODUCTION

For the computation of the relative supersaturation (RS) for struvite, the concentration of ammonium must be known. To estimate ammonium concentration, the use of the anion gap has been proposed[1]. The purpose of this study was to compare the influence of ammonium concentration and anion gap on the RS of struvite, calcium oxalate monohydrate (COM), brushite, and uric acid.

METHODS

Twenty-five sterile 24 h urine samples from 23 stone forming patients were analysed. Ammonium concentration was measured colorimetrically. The anion gap was calculated from the equation: Na + K - Cl. The estimate of ammonium[1] was derived from the equation: ammonium concentration = - 0.8 (anion gap) + 82. The RS values were computed with the EQUIL program of Finlayson (version 1988)[2] in three ways. In a first step, ammonium concentration was set as zero (0), in a second step, it was replaced by the value derived from the anion gap (ag), and in a third step, the measured value (m) was taken for calculation.

RESULTS

The following correlations were found:

-RS (m) versus RS (ag) struvite: $y = 0.06 + 0.15x$ ($R = 0.48$, $p = 0.0147$)
-RS (ag) versus RS (0) COM: $R = 1.00$; RS (m) versus RS (0) COM: $R = 0.99$
-RS (ag) versus RS (0) brushite: $R = 1.00$; RS (m) versus RS (0) brushite: $R = 1.00$
-RS (ag) versus RS (0) uric acid: $R = 1.00$; RS (m) versus RS (0) uric acid: $R = 1.00$

DISCUSSION

Since the inclusion of the ammonium concentration in the computation with EQUIL has only a minor influence on the RS for calcium oxalate monohydrate, brushite, and uric

Urolithiasis 2, Edited by R. Ryall *et al.*,
Plenum Press, New York, 1994

acid, substitution of ammonium concentration - not routinely measurable everywhere - by an estimate derived from anion gap is not necessary. However, urinary anion gap might be used as an indirect substitute for ammonium concentration in the computation of the relative supersaturation with respect to struvite in 24 h urines not infected with urea-splitting organisms.

ACKNOWLEDGEMENTS

This study was supported in part by the Swiss National Science Foundation (Grant No. 32-26428.89)

REFERENCES

1. MB Goldstein, R Bear, RMA Richardson, PA Marsden, and ML Halperin, The urine anion gap: a clinically useful index of ammonium excretion, *Am J Med Sci* 292(4): 198 (1986).
2. B Finlayson, Calcium stones: some physical and clinical aspects, in "Calcium Metabolism in Renal Failure and Nephrolithiasis", DS David, ed., John Wiley & Sons, New York (1977).

CALCIUM CARBONATE CRYSTALS PROMOTE EPITAXIAL CALCIUM OXALATE CRYSTAL GROWTH

J.M. Verdier[1], B. Dussol[2], S. Nitsche[3], P. Berthézène[1], P. Dupuy[1], R. Boistelle[3], Y. Berland[3] and J.C. Dagorn[1]

[1]INSERM U315
[2]Hôpital Sainte-Marguerite
[3]CRMC2 du CNRS
Marseille, France

INTRODUCTION

Calcium oxalate (CaOx) urolithiasis is the most common urinary stone disease in the occident. The physiopathology of this kind of stone is not well understood because crystallization occurs *in vivo* in a very complex environment whose parameters are difficult to control. Calcium oxalate is the major mineral phase of these stones, but it is generally associated with minor amounts of other calcium salts, including carbonate and phosphate.

Recent data showed that fluid in the thin segment of Henle's loop of the rat is supersaturated with respect to calcite and brushite. This suggested that they could crystallize in certain pathological conditions and eventually promote CaOx nucleation. If this were the case, physiological control of crystal growth in the early stages of urine formation would be of considerable importance in the prevention of urinary stone formation. We have demonstrated the presence of a renal protein related to pancreatic lithostathine, the pancreatic inhibitor of calcium carbonate ($CaCO_3$) crystal growth. This protein, which we called renal lithostathine, is also an inhibitor of $CaCO_3$ crystal growth *in vitro* and is localized in the thick ascending limbs of the loops of Henle[2]. Taken together these data prompted us to look for the nucleation of CaOx crystals onto $CaCO_3$ seeds.

METHODS

Seeds of vaterite or calcite were added to a metastable solution of sodium oxalate and calcium chloride containing traces of ^{45}Ca. The accretion of CaOx was monitored by ^{45}Ca-radioactivity incorporation.

RESULTS

We found that (i) seeds of vaterite and calcite may initiate nucleation of CaOx, this nucleation being more prominent with vaterite than with calcite, (ii) the incorporation is closely related to the amount of seeds added (not shown). Moreover, using scanning electron microscopy we clearly demonstrated an epitaxial nucleation of CaOx crystals onto calcite and probably a heterogeneous nucleation onto vaterite (figure 1).

These results show that calcite may act as a heterogeneous nucleus for calcium crystals. As a consequence, renal lithostathine, an inhibitor of $CaCO_3$ crystal growth could secondarily prevent CaOx stone formation, since a defect in the control of CaP or $CaCO_3$ crystallite formation in that segment of the nephron would generate seeds on which CaOx crystals would grow when urine becomes supersaturated with that salt.

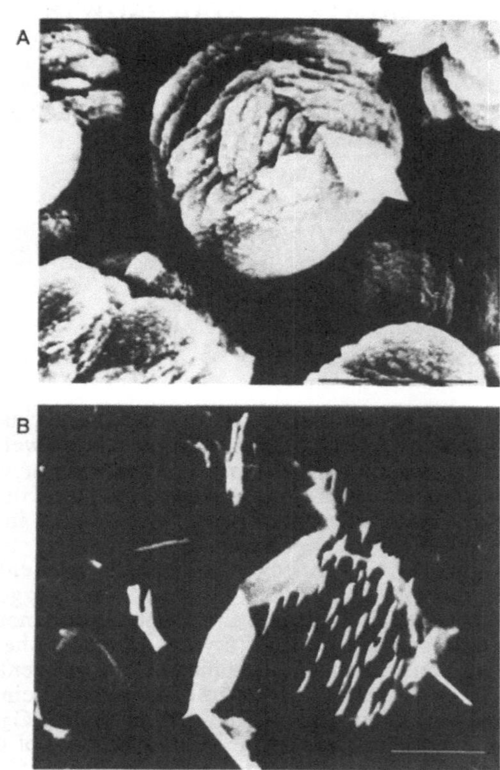

Figure 1. Scanning electron micrographs of crystallization of CaOx on $CaCO_3$ seeds. **A** CaOx growth on vaterite (x 3000). **B** CaOx growth on calcite (x 2400). Horizontal bars correspond to 5 µm.

REFERENCES

1. FL Coe, JH Parks, and Y Nakagawa, Protein inhibitors of Crystallization, *Sem Nephrol* 11:98 (1991).
2. JM Verdier, B Dussol, P Casanova, M Daudon, P Dupuy, P Berthezene, R Boistelle, Y Berland, and JC Dagorn, Evidence that human kidney produces a protein similar to lithostathine, the pancreatic inhibitor of $CaCO_3$ crystal growth, *Eur J Clin Invest* 22:469 (1992).

EFFECTS OF GLYCOSAMINOGLYCANS ON EXPERIMENTAL UROLITHIASIS IN RATS

M.A. Boim[1], C.T. Bergamaschi[1], Y.M. Michelacci[2] and N. Schor[1]

[1]Nephrology Division
[2]Department of Molecular Biology
Escola Paulista de Medicina, SP Brazil

INTRODUCTION

It has been suggested that the glycosaminoglycans (GAGs) heparan sulfate (HS) and chondroitin sulfate (CS) inhibit calcium oxalate (CaOx) crystallization *in vitro*[1]. Moreover we have previously found that stone forming subjects excrete lower levels of GAGs compared to non-stone formers[2]. Thus, since GAGs may inhibit calculi growth *in vivo*, we tested the effect of administering CS to rats using an experimental model of lithiasis.

METHODS

Lithiasis was induced by placing a CaOx stone into rat bladders[3]. Control and test animals were treated intraperitoneally either with 0.3mL saline (S), or CS in doses of 1, 5 or 10 mg/kg/day, for 42 days (n=5 for each group). Urinary excretion of total GAGs, HS and CS were identified by agarose gel electrophoresis and measured by densitometry[2].

RESULTS

As shown in the table, the urinary excretion of GAGs increased in control rats in a dose-dependent fashion, along with a corresponding increase in CS. In contrast, the lithic rats showed a marked reduction in urinary CS excretion. HS was unchanged in all groups. Moreover, the calculi from rats receiving CS showed an elevation in their GAG content, as well as an increase in weight ($p<0.05$ vs 5).

	CONTROL Urinary excretion (mg/g creat)			LITHIC Urinary excretion (mg/g creat)			Calculi	
	GAG	CS	HS	GAG	CS	HS	GAG µg	Δ %
S	5.7	2.6	3.1	2.7	0.3	2.5	<1	9 ± 2
1 mg	9.5	6.0	3.6	2.9	1.0	3.9	2.4	12 ± 2
5 mg	13.2	9.8	3.4	4.6	0.0	4.6	28.9	19 ± 2*
10 mg	17.9	14.5	3.5	4.1	0.0	4.1	29.0	31 ± 3*

Urolithiasis 2, Edited by R. Ryall *et al.*,
Plenum Press, New York, 1994

From these results, it would appear that the effect of CS administration upon the growth of CaOx crystals *in vivo* is opposite to the inhibitory effect observed *in vitro*. It is possible that *in vivo*, GAGs could act as a promoter of nephrolithiasis by its incorporation into previously formed calculi. Thus GAGs may prevent nucleation, but should this occur, they may enhance further crystallization.

REFERENCES

1. WG Robertson, M Peacock, BEC Nordin, Inhibitors of the growth and aggregation of calcium oxalate crystals in vitro, *Clin Chim Acta* 43:31 (1973).
2. YM Michelacci, RQ Clashan, N Schor, Urinary excretion of glycosaminoglycans in normal and stone forming subjects, *Kidney Int* 36:1022 (1989).
3. JL Meyer, LH Smith, Growth of calcium oxalate crystals. I. Model of urinary stone growth. *Invest Urol* 13:31 (1975).

ORAL CIMETIDINE IN HUMANS - EVIDENCE FOR URINE ACIDIFICATION AND INHIBITION OF CALCIUM OXALATE CRYSTALLIZATION

U. Herrmann, P.O. Schwille, M. Manoharan, H. Gruber and A. Wenig

Mineral Metabolism and Endocrine Research Laboratory,
Departments of Surgery and Urology
University of Erlangen
Germany

INTRODUCTION

Medication of subgroups of idiopathic calcium urolithiasis often aims at reducing an alkaline or insufficiently acidic urinary pH. Since (exogenous) acid loading is effective but instantly attacks the body's buffer reservoirs, we tested the hypothesis that inhibition of postprandial gastric proton release into the gastric lumen may lead to proton diversion to kidney via blood, and to subsequent excretion via urine.

SUBJECTS AND PROCEDURES

Ten healthy volunteers (5 males, 5 females; age 22-39 years), fasted overnight (12 -14 h) and mildly hydrated orally with water, underwent three consecutive 2 h creatinine clearance periods at 0800 (basal), 1000 (I) and 1400h (II). Breakfast consisted of 2 sandwiches, 20 g butter, 20 g bees honey, either alone, or in combination with either 400 or 800 mg cimetidine (CM). Reliable laboratory technology was used throughout, including high pressure liquid chromatography for measurement of CM in blood and urine and "tolerable oxalate" in urine (A Wenig, PO Schwille, J Fan, H Gruber. Tolerable oxalate in undiluted urine of renal calcium stone patients and controls - Methodology and effects of oral alkali citrates; unpublished method).

RESULTS AND COMMENTS

Postprandially, the capillary blood acid-base status in both sexes was not altered by breakfast alone, or by additional CM. By contrast, data from postprandial period II (table; values are mean ±SEM) reveal that serum CM and gastrin increased dose-dependently, as did urinary net acid (NAE), whereas urinary pH and hydroxyapatite supersaturation (RSP-HAP) decreased markedly under CM. In addition, serum calcium decreased and the associated mean intact PTH was higher, but oxaluria was unchanged. Sulphoxy-CM, but not CM itself, reduced calcium oxalate nucleation in terms of tolerable oxalate, and also crystal aggregation. Considering these results, a hitherto overlooked gastro-renal functional axis, elicited by the histamine-H_2-receptor antagonist CM, may exist and further more detailed study is justified. Clinical trials in renal stone patients appear warranted, with special emphasis on urinary acid(s), pH, inhibitors, and calcium metabolism[1].

Urolithiasis 2, Edited by R. Ryall *et al.*,
Plenum Press, New York, 1994

SERUM				URINE					
CM $\mu g.mL^{-1}$	T-Calcium[1] $md.dL^{-1}$	Gastrin $g.mL^{-1}$	PTH $g.mL^{-1}$	pH	NAE mmol	Calcium $mg.g\ Cr^{-1}$	Oxalate $g.g\ Cr^{-1}$	RSP-CaOx ΔG	RSP-HAP ΔG
Breakfast alone									
-	9.52±0.06	44±12	39±5	6.49±0.30	-3.9±4.0	100±13	22±2	1.6±0.5	5.8±0.9
Breakfast + 400 mg CM									
1.2±0.2	9.38±0.06	112±29*	44±4	6.01±0.26*	-4.7±9.1	122±25	24±2	1.7±0.5	4.2±0.7*
Breakfast + 800 mg CM									
2.3±0.3	9.21±0.11*	119±32*	49±5	5.89±0.19*	3.4±1.8*	119±25	23±2	1.7±0.5	3.7±0.7*

[1]Total calcium, corrected for serum total protein; *: $p<0.05$ vs breakfast alone; Cr: creatinine; Δ G (free energy; EQUIL II)

REFERENCE

1. GA Williams, RS Longley, EN Bowser et al, Parathyroid hormone secretion in normal man and in primary hyperparathyroidism: Role of histamine H2 receptors. *J Clin Endocrinol Metab* 52:122 (1981).

COMPUTER MODEL OF COMPLEXATION INHIBITORS OF CALCIUM OXALATE STONE FORMATION

D.G. Alto, P.A. Rock and H.E. Williams

Department of Chemistry and the School of Medicine
University of California
Davis
Davis, CA

INTRODUCTION

We have carried out a comprehensive analysis of the known complexation and precipitation equilibria involved in kidney fluids, with the objective of assessing the potential efficacy of various non-protein inhibitors of calcium oxalate stone formation.

METHODS

The model utilizes existing software run on a Macintosh CI computer and involves at least 88 equilibria, 20 independent variables and 105 dependent variables. The model incorporates the temperature dependence of the equilibrium constants and of the mean ionic activity coefficients, and the potential formation of stones containing phosphate.

Provided the relevant equilibrium constants are known or can be estimated reliably, the model permits the rapid, quantitative assessment of the potential effectiveness of proposed complexation based inhibitors of stone formation. The results of the computer analysis are portrayed graphically in readily interpreted 2D or 3D color plots.

RESULTS AND DISCUSSION

At 37°C and a pH of 6.25 the relative effectiveness of naturally occurring inhibitors follows the established order citrate > Mg^{2+} > pyrophosphate > orthophosphate, whereas at a pH of 4.50 the order is Mg^{2+} > citrate > pyrophosphate > orthophosphate. As the pH decreases from 9.0, the driving force for precipitation of calcium oxalate reaches a maximum at a pH of about 4.6 (as a result of protonation of Ca^{2+} complexing agents), and then decreases as the pH is further decreased, in general agreement with earlier studies. The effect of temperature on the driving force for stone formation is significant in the physiological range. The model predicts that succinate and hippurate are not effective complexation inhibitors.

DUAL CONSTANT COMPOSITION STUDIES OF THE SIMULTANEOUS GROWTH AND DISSOLUTION OF CALCIUM OXALATE AND CALCIUM PHOSPHATE

G.H. Nancollas

Departments of Chemistry and Biomaterials
State University of New York at Buffalo
Buffalo, New York

INTRODUCTION AND RESULTS

Human urines are usually supersaturated with respect to calcium oxalate phases but in many instances, the liquid phase may also be supersaturated with respect to the calcium phosphates. Urinary stones frequently contain both calcium oxalate and phosphate phases as evidenced by the layer by layer mixture of these salts in many urinary stones. Constant composition studies have previously demonstrated the nucleation and growth of calcium oxalate monohydrates (COM) on hydroxyapatite (HAP) but this technique enables only a single phase to be investigated. In contrast, the dual constant composition method (DCC) enables the crystal growth and dissolution kinetics of mixtures of sparingly soluble electrolytes to be investigated simultaneously. The method has been used to study the transformation of dicalcium phosphate dihydrate (DCPD) to octacalcium phosphate (OCP) as well as the transformation kinetics from one phase to the other. In the present work, the nucleation kinetics of COM on HAP crystals and the concomitant growth of these two phases have been investigated using the DCC method. It has been shown that the rate of COM growth decreases as the phosphate concentration is increased. However, at constant calcium and oxalate concentrations, the apparent concomitant rate of HAP growth was unaffected by changes in phosphate concentration in the concentration range typical of that in urine.

CONCLUSION

This method allows simultaneous studies to be made of the influence of urinary inhibitors on the growth of mixtures of urinary stone components.

CURRENT CONCEPTS AND FUTURE TRENDS IN UROLITHIASIS RESEARCH WITH SPECIAL REFERENCE TO MOLECULAR BASIS OF DISEASE

R. Nath

Department of Biochemistry
Postgraduate Institute of Medical Education and Research
Chandigarh, India

INTRODUCTION

The major emphasis of research in the last decade has been to delineate the etiopathogenesis, epidemiology and environment factors, nutritional imbalances, inhibitors and promoters, surgical and medical management and non-surgical techniques like PCNL and lithotripsy. Several studies have been undertaken in this laboratory to understand the molecular mechanisms of hyperoxaluria, one of the primary risk factor in stone disease along with the role of inhibitors and promoters of renal stones[1]. The roles of citrate, glycosaminoglycans (GAGs) and nephrocalcin have recently attracted special attention. The hereditary nature of idiopathic hypercalciuria has opened a new area for future research[2]. Use of new salts of citrate i.e. potassium citrate (10mEq tablet), effervescent calcium citrate and potassium-magnesium citrate has led to improved therapeutic regimes for the patients of urolithiasis. New insights in citrate research include origin of citraturic response to these supplements. Future direction of investigations in the function of GAGs research includes, to quantitate individual GAGs in normals and patients of urolithiasis and establish their specificity as promoter or inhibitor of calcium oxalate stone formation. At the molecular level, elucidation of genomic structure, characterization of regulatory elements, mechanism of translocation to nucleus and mechanism by which GAGs regulate gene expression need to be studied. Finally, the role of newly discovered zinc finger proteins which regulate steroid hormones and related thyroid and parathyroid hormone to bind to their respective receptors (which contain DNA-binding domain of zinc finger loops) needs to be discovered to unravel some of the upcoming areas of future research.

REFERENCES

1. S Sharma, S Vaidyanathan, SK Thind, and R Nath, *Ind J Urol* 8: 25 (1991).
2. FL Coe and JH Parks, *Adv Nephrol* 21:31 (1991).

INHIBITORS OF URINARY STONE FORMATION IN CONTROLS AND STONE FORMERS

A. Trinchieri, A. Mandressi, P. Luongo, F. Rovera, P. Longo and G. Zanetti

Istituto di Urologia
Università degli Studi di Milano
Milan, Italy

INTRODUCTION

Citrate, magnesium and zinc are known to be powerful inhibitors of crystal formation while glycosaminoglycans (GAGs) and citrate are effective inhibitors for crystal aggregation[1].

MATERIALS AND METHODS

We have set out to study the urinary excretion of citrate, magnesium, zinc and GAGs and to correlate the results with sex and age in 197 healthy subjects and 104 patients with idiopathic calcium stone disease.

RESULTS

In normal subjects the daily excretion of citrate, magnesium and zinc increased with age to a maximum during the fifth decade and remained relatively constant until the eighth decade when they decreased.The daily excretion of magnesium (105±33 vs 83±46 mg/dL p >0.005) and zinc (0.4±0.2 vs 0.2± 0.1 mg/dL p <0.001) were higher in men than in women, which was attributed to the higher body weights of the men.There was no significant difference between men and women for daily citrate excretion (465±181 vs 463±168 mg/dL p <0.001), but the citrate/creatinine ratio was significantly higher in women than men. The highest GAGs excretion rates were seen during the first two decades. Urinary citrate (342±198 vs 465±173 mg/dL p <0.001) and magnesium (58±30 vs 93±42 mg/dL p <0.01) were lower and GAGs (30±13 vs 19±7 mg/dL p <0.001) and zinc (0.8±0.3 vs 0.3±0.2 mg/dL p <0.001) were higher in stone formers than in controls.

DISCUSSION

It seems that a decreased excretion of citrate and magnesium, together with an increased excretion of calcium, may contribute to the formation of calcium stones. The role of urinary GAGs and zinc remains uncertain.

REFERENCE

1. H Fleisch. Inhibitors and promotors of stone formation, *Kidney Int* 13: 361 (1978).

PHOSPHOCITRATE AS AN INHIBITOR OF CALCIUM OXALATE CRYSTALLIZATION *IN VIVO*

S.R. Khan[1], P.N. Shevock[1], J.D. Sallis[2] and R.L. Hackett[1]

[1]University of Florida
Gainesville, Florida and
[2]University of Tasmania
Tasmania, Australia

INTRODUCTION

Phosphocitrate (PC) is a naturally occurring inhibitor of calcification and has been demonstrated to be a potent inhibitor of hydroxyapatite (HAP) crystal growth and the transformation of amorphous calcium phosphate to HAP. It has also been shown to affect the crystallization of struvite and newberyite. However its impact on calcium oxalate (CaOx) crystallization is not well documented. We studied the effect of PC on CaOx crystallization *in vivo* in a rat model where crystallization is induced by low level hyperoxaluria in the presence of proximal tubular epithelial injury, since PC has also been demonstrated to provide protection against crystal-induced damage to the cell membranes.

METHODS

Male Sprague-Dawley rats weighing approximately 250-300g were given 0.5% ethylene glycol (EG) in their drinking water. PC was administered once a day by I.P. injection at a dose of 112 μmol/kg rat body weight. Injury was induced by a single daily injection of gentamicin sulfate (GS) at a dose of 40mg/kg of rat weight. PC was started on day 1, gentamicin on day 2, and EG on day 3. All rats were sacrificed on day 11. Urines, bladder aspirates, and kidneys were examined for CaOx crystals.

RESULTS AND CONCLUSIONS

EG administration resulted in increased urinary oxalate. EG administration to rats receiving GS was associated with increased levels of renal oxalate and deposition of CaOx crystals in the kidneys of some animals. Increase in renal oxalate and crystal deposition in kidneys was not affected by PC administration. Apparently PC is not protective against GS induced epithelial injury and is not inhibitory against the CaOx crystallization which occurs in this animal model.

ACKNOWLEDGEMENTS

This work was supported by NIH grants #R01DK41434 and #P01DK20586.

RELATIONSHIPS BETWEEN INHIBITOR STRUCTURE AND ITS EFFECTS ON CALCIUM OXALATE MONOHYDRATE CRYSTALLIZATION KINETICS

D.J. Kok, I. Que and S.E. Papapoulos

Dept of Endocrinology
University Hospital
Leiden, The Netherlands

INTRODUCTION

Using a group of model inhibitors, bisphosphonates, relationships between the inhibitor structure and its effects on calcium oxalate crystallization kinetics were studied. The bisphosphonates were tested in a seeded crystal growth system in which the effects on the solubility, the growth and the agglomeration of calcium oxalate monohydrate (COM) crystals are measured as three separate parameters. Changes in structure consisted of variations in the bisphosphonate moiety (containing hydrogen, a hydroxyl group or an amino group), and changes in the remainder of the molecule (rest group), using linear carbon chains of varying length, adding (methylated) amino groups, or including cyclic structures.

RESULTS AND DISCUSSION

As expected, the bisphosphonates increased COM solubility, presumably by complexing calcium. All bisphosphonates tested inhibited crystal growth. The efficacy (based on the concentration at which 50% growth inhibition is obtained) depended on the bisphosphonate moiety; the inhibitory potency increased when the hydrogen was replaced with an hydroxyl group, and even more when substituted with an amino group. No relation with the structure of the rest group was found. The effects on crystal agglomeration, on the other hand, were quite divergent and strongly related to the nature of the rest of the inhibitor molecule.

Bisphosphonates with a small linear rest group stimulated crystal agglomeration, while those with a large rest group either had no effect or inhibited agglomeration. Some having a (methylated) amino group even showed a biphasic effect, namely, inhibition at low and stimulation at high concentrations.

Stimulation of crystal agglomeration seems to depend largely on the ability to form polynuclear calcium bisphosphonate complexes. This in turn depends on the size of the rest group (causing steric hindrance and thus disrupting the complexes) and on the affinity of the bisphosphonate group for calcium. The relationships between bisphosphonate structure and its effects on crystal growth and crystal agglomeration may help in designing bisphosphonates with specific desired properties.

MICRODETERMINATION OF CRYSTAL GROWTH RATES OF CALCIUM OXALATE IN GEL AT INVERSE DISTRIBUTION OF COMPONENTS

W. Achilles[1], C. Lescher[1], M. Burk[1] and H. Füredi-Milhofer[2]

[1]Urologische Universitatsklinik
D-3550Marburg/Lahn, Germany
[2]Rudjer Boskovic Institute
Zagreb, Croatia

INTRODUCTION AND RESULTS

The Gel Crystallization Method (GCM; Achilles et al) has been applied up to now predominantly to the determination of relative crystal growth rates (Vcr) of calcium oxalate (CaOx) using oxalate (Ox) as the soluble component within the gel phase at a defined constant concentration. Though this principle has been useful in most applications of the method, the effect on Vcr of inhibitors at varying [Ox], and of Ox itself, could not be investigated. This paper describes the GCM for CaOx at inverse distribution of the reaction components Ca and Ox (ie with Ca within the gel).

At optimized conditions (gel phase: 0.1 mL 0.5 wt% agar-agar per well of a microplate; 8 mM $CaCl_2$, 0.2 mM sodium oxalate to produce seed crystals, and 10 mM MES (pH 6), relative crystal growth rates of CaOx were determined at different ionic strength (NaCl: 50-400 mM) and total oxalate (0.1-1.0 mM) in solution, in the absence and presence of several effectors of crystal growth.

48 crystal growth curves could be registered simultaneously within 15 minutes by automated scanning microscopy using 96-well microplates. Crystal growth was best detected by scattered light (dark field mode of microphotometric measurement). Vcr values were reproducible within 1-3% (CV) under standard conditions).

The following effectors of CaOx were investigated (total concentrations at 50% inhibition at standard conditions given in parentheses): magnesium (2.710^{-3} M), citrate (1.510^{-3} M), meta-polyphosphate (0.810^{-6} M), chondroitin-4-sulfate (0.310^{-3} M).

Chondroitin-4-sulfate at a physiological concentration (<100 mg/L) produced practically no inhibitory effect. This result was also found when the gel phase had been incubated with this substance prior to the crystal growth procedure, which indicates that the lack of an inhibitory effect could not be ascribed to a decreased diffusion of the macromolecule in the gel phase. Taking into account complex chemical interaction in the gel/solution system, thermodynamic and kinetic effects of the inhibitors of crystal growth could be quantitatively differentiated. The results of this study demonstrate that the GCM is applicable to the highly efficient microdetermination of relative growth rates of sparingly soluble crystal phases independent of the distribution of their phase components.

ACKNOWLEDGEMENTS

This work was supported by the Deutsche Forschungsgemeinschaft (Ac52/2) and the Commission of the European Communities, DG XII (grant no. CI1*/0346).

PHOSPHATE CRYSTAL GROWTH *IN VITRO* - THE MIX UP

S. Sindhu, T.G. Dhanalekshmy, R.K. Vathsala, K.V. Kurien[1], C. Aravindakshan and Y.M. Fazil Marickar

Department of Surgery
Medical College Hospital and
[1]Department of STEC
Trivandrum, India

INTRODUCTION AND RESULTS

The formation of urinary crystals is an essential step in stone disease. Hence proper understanding of these crystals and their accurate identification in tissues and body fluids is of great importance. Growth of urinary crystals *in vitro* has become an integral part of the study of the nucleation and growth characteristics of crystals. It is known that human urinary deposits sometimes contain struvite crystals, and study of urinary stones produced in the body has also shown newberyite crystals. We have been able to grow both struvite and newberyite in test tubes in silica gel medium. This paper details the problems encountered in growing crystals of struvite and newberyite.

Modifications of conventional silica gel medium were utilised for growing newberyite and struvite crystals in Hane's tubes using single diffusion technique. 1.03 density sodium meta silicate was prepared and the pH of the system adjusted to 7 with 3M acetic acid. Gelatification was initiated by adding 0.5M magnesium acetate and 1M ammonium dihydrogen orthophosphate was added on top for growing newberyite crystals. For growing struvite crystals, 0.5M ammonium dihydrogen orthophosphate solution was added to the sodium meta silicate and incorporated into the gel. After gelatification 1M magnesium acetate was added on top. After 30 days, the crystals grown were washed clear of the gel and studied for morphological appearance using light microscopy and scanning electron microscopy and the purity of crystals was assessed by energy dispersive microanalysis and X-ray diffraction.

It was observed that incorporating magnesium acetate into the gel produced orthorhombic newberyite crystals, while the incorporation of magnesium acetate in the top solution produced newberyite crystals. However, in certain situations the procedure for growing struvite crystals produced newberyite crystals, and *vice versa.* Incorporation of ammonium or diammonium hydrogen phosphate in the gel initiated growth of struvite. It will be interesting to study further why newberyite crystals are encountered only in stones formed in urine, but not in the urine of humans and experimental rats. From the present findings, it is concluded that the amount of magnesium in the environment dictates the type of phosphate crystal grown.

GROWTH STUDIES OF CALCIUM OXALATE IN PRESENCE OF INHIBITOR

S. Sindhu, N.E. Thomas, N. Sylaja, C. Aravindakshan, S. Jayadevan, K.V. Kurien[1] and Y.M. Fazil Marickar

Department of Surgery
Medical College Hospital and
[1]Dept. of STEC
Trivandrum, India

INTRODUCTION AND RESULTS

Human urine is is known to be a heterogeneous medium. However, although growth of urinary crystals *in vitro* is usually performed in what is assumed to be a homogeneous solution the presence of various agents in the environment *in vitro* will render solution conditions heterogeneous. The present work was undertaken in order to study the effect of such agents on the process of crystallisation specifically and to assess their effects on nucleation, growth rate and size of crystals. The effects of pyridoxine allopurinol, and a combination of these, on crystal growth *in vitro* were tested.

Calcium oxalate monohydrate crystals were grown by single diffusion method using a modified conventional silica gel medium. The test agents were directly added to the top solution of the crystal growth medium, with distilled water being used as control. The thickness of the crystal column and size of the crystals in the test samples were compared with the findings obtained in the control. Analysis of variance was used for statistical comparison.

The thickness of the crystal columns in the experiments incorporating pyridoxine, allopurinol, a combination of the two, and control were 4.58, 4.48, 4.22 and 5.06 cm respectively on the thirtieth day of the study, and the corresponding crystal sizes were 155.6, 202, 181.2 and 229.6 µm. The thickness of the crystal column and the size of the crystals were significantly less in the presence of pyridoxine, allopurinol and a combination of the two in relation to the control (F ratio = 8.96, $p<0.05$ for thickness; F ratio = 16.98, $p<0.05$ for size of crystals).

Crystal formation from a saturated solution may be retarded or prevented by dopants in solution. These prevent the nucleation of crystal growth or aggregation. *In vitro* studies are useful for testing the ability of substances to inhibit crystal nucleation and growth. The findings *in vitro* correlate well with the effects of these substances *in vivo*. Reduction in crystal size was more significant than the alteration in the thickness of the crystal column. It is noteworthy that strong inhibitors also do not cause a significant reduction in thickness of crystal column. It is concluded from the study that pyridoxine and allopurinol have an inhibitory effect on calcium oxalate crystal formation in the technique used, since they altered the onset of nucleation, the size of the initial crystals, the rate of growth and the morphological appearance of the crystals.

NATURAL URINE VERSUS SYNTHETIC URINE FOR STUDYING INHIBITORS

T.G. Dhanalekshmy, C. Aravindakshan, S. Sindhu, S.V. Roshni,
R.K. Vathsala and Y.M. Fazil Marickar

Department of Surgery
Medical College Hospital
Trivandrum, India

INTRODUCTION AND RESULTS

Although there are many known inhibitors in urine they have not been able to account for the total inhibitor capacity of natural urine. The present study was undertaken to determine whether synthetic urine is useful for studying inhibitor properties and whether it can replace natural urine for such studies. The common crystals seen in human urine were grown in Hane's tubes using a modified conventional silica gel medium. The effect of natural urine was tested by incorporating it into the crystal growth system. Synthetic urine was prepared by the procedure of Henry[1]. The ingredients known to have inhibitory activity, namely, tartrate, citrate and magnesium were added to the synthetic urine in concentrations of 630, 315 and 168 mg% respectively. The changes produced in the rate of growth of crystals, and the size of crystals grown under different environmental conditions in synthetic urine were compared with the findings produced with natural human urine.

In the first instance, the mean size of the crystals in the synthetic urine was significantly higher from day 1 to day 30 in the presence of synthetic urine compared with normal urine. The difference between these values was statistically significant. The rate of growth and size of the oxalate crystals were reduced by addition of known inhibitors to the synthetic urine, compared to the control synthetic urine to which no inhibitors were added, but this reduction was less than the inhibition produced by addition of natural human urine. The inhibitors had a maximum effect in concentration of 630 mg%. This finding proved that the concentration of inhibitors also plays an important role in crystal formation. Among the three different inhibitors tested, tartrate was found to have the most potent inhibitory effect. The difference between the mean thickness of the crystal column in the presence of the three inhibitors was not statistically significant (analysis of variance: ANOVA). The mean size of the crystals in the experiments using synthetic urine containing tartrate, citrate, magnesium, control synthetic urine and whole human urine were 44, 48, 60, 80 and 40 μm respectively. When these data were analysed using one way ANOVA the effects produced by addition of synthetic urine, normal urine, control synthetic urine, and synthetic urine supplemented with tartrate, citrate and magnesium, were significantly different ($p<0.001$). It is concluded that natural urine contains certain unknown, unidentified inhibitors of crystallization and therefore that synthetic urine cannot be used as a substitute for natural urine in the crystallization technique reported here.

REFERENCE

1. JB Henry, Clinical Diagnosis and Management, WB Saunders Co, Philadelphia, pp368-387 (1989).

GROWTH OF URIC ACID CRYSTALS *IN VITRO*

C. Aravindakshan, T.G. Dhanalekshmy, S.V. Roshni, P. Koshy[1], H.K. Moorthy, S. Sindhu, N.E. Thomas and Y.M. Fazil Marickar

Department of Surgery
Medical College Hospital and
[1]Regional Research Laboratory
Trivandrum, India

INTRODUCTION AND RESULTS

Very few reports are available on the growth of uric acid crystals *in vitro*. In this study an attempt was made to grow uric acid crystals in Hane's tubes in modified conventional silica gel medium. 1.03g/mL density sodium meta silicate was prepared in double distilled water. Different concentrations of uric acid at 3 mg%, 4 mg%, 5 mg% and 6 mg% each were incorporated separately into the gel. The pH of each set of experiments was adjusted with lM acetic acid to 10.5, 10, 9.5, 9, 8.5, 8, 7.5, 7, 6.5, 6 and 5.5 for each of the different concentrations of uric acid. The solutions were set aside overnight to gel. After gelatification, 5.25M (1.05g/mL density) acetic acid was added to the top of the gel. The experiment was carried out at room temperature, fridge temperature and body temperature. The tubes were examined for the time of onset of appearance of the crystals, and the size of the crystals and the rate of growth during the different times of observation, namely days 1, 3, 7, 14, 21 and 30 were assessed. The crystals were then washed, dried and subjected to light microscopy, infrared (IR) analysis and Scanning Electron Microscopy (SEM).

Crystal precipitation was observed at concentrations above 5mg% and at pH values below 9.5, even before gelatification occurred. At concentration of 5mg%, precipitation occurred at pH 7 and below before the commmencement of gelatification. At concentrations of uric acid below 5mg% no precipitate occurred at pH 7, and even after gel formation crystal growth did not occur on standing. Thus a concentration of 5mg% and pH 7.5 were found to be the ideal conditions for crystal growth. Uric acid crystals started appearing from day 3 at a size of 1 mm and grew gradually to 8 mm by day 30. IR analysis of the crystals collected at the end of this period exhibited the features of pure uric acid and SEM confirmed the morphology of uric acid crystals. At fridge temperature, crystal growth was greater and crystals were branched, while at body temperature growth was less. The success of the technique depended upon the use of different concentrations of uric acid in the gel in different pH ranges. Uric acid was soluble only in alkaline solution and precipitated as pH became acidic. The solute concentration had to be reduced considerably to avoid precipitation while lowering the pH in order to facilitate its incorporation into the gel at lower pH values. The procedure successfully and reproducibly grew uric acid crystals.

PURITY OF CRYSTALS GROWN *IN VITRO* IN THE PRESENCE OF DOPANTS

S. Sindhu, K.V. Kurien[1], G.N. Subhanna[2], P. Koshy[3], S.R. Khan[4] and Y.M. Fazil Marickar

Department of Surgery, Medical College Hospital
Trivandrum, India
[1]Department of STEC, Kerala; [2]IISc, Bangalore
[3]RRL, Trivandrum and
[4]Department of Pathology
University of Florida, Gainesville, USA

INTRODUCTION AND RESULTS

The addition of impurities to a crystal growth environment to alter the process of nucleation, growth and morphology of crystals is known as doping. The present study was undertaken to assess the purity of calcium oxalate monohydrate (COM) crystals when dopants are added to growth media *in vitro*. COM crystals were grown in Hane's tubes using modifications of the conventional silica gel media. Pyridoxine (0.13%), allopurinol (0.33%), magnesium (5%), citrate (5%) and tartrate (5%) were added as dopants to the top solution of the crystal growth system and the rate of growth and size of crystals determined. On the thirtieth day the crystals were washed from the gel and analysed by X-ray diffraction (XRD) using an X-ray powder diffractometer (PW 1140/90 with CuK a-radiation), Scanning Electron Microscopy (SEM) and energy dispersive x-ray microanalysis (EDAX). One hundred samples were studied.

The chemical composition of the crystals on XRD as shown by the d values (distance between parallel planes of the atoms) was COM. SEM showed a variety of morphologies; they were mainly monoclinic and most of them showed interpenetrating twinning. EDAX spectra showed calcium as the main peak. Silica and gold peaks were also noted. In some specimens amorphous, flaky and often solid glassy material, positive for silica, was seen associated with the crystals. Some plate-like COM crystals were seen to have nucleated on the glassy material in which the silica was detected. Some crystals had minor amounts of magnesium, probably deriving from the magnesium added as dopant. Sometimes strange crystals, but mostly COM and large rosettes of block-shaped crystals and plate-like COM crystals growing out of the glassy silicate material were also seen.

Crystal habit can be influenced both by the crystal's internal structure and by a variety of physicochemical properties of the solution in which the crystals develop. It can also vary in response to different growth conditions, and variation in the number of constituent atoms and nature of any foreign material. Doping affects many physical properties of crystals, including the lattice parameter, which can be used to characterise crystals. XRD can be used for positive identification of the crystals, while SEM is useful for studying crystal habit. We conclude that addition of dopants to the crystal growth medium can alter the morphology of calcium oxalate crystals, but has little influence on chemical composition.

DOES OSTWALD RIPENING OF CALCIUM OXALATE CRYSTALS OCCUR *IN VITRO*?

S. Sindhu, T.G. Dhanalekshmy, P. Koshy[1], N. Sylaja, C. Aravindakshan, K.V. Kurien[2] and Y.M. Fazil Marickar

Department of Surgery
Medical College Hospital
[1]RRL, Trivandrum and
[2]Dept of STEC
Trivandrum, India

INTRODUCTION AND RESULTS

Ostwald ripening occurs when the smaller crystals in a crystal suspension dissolve and the released solute ions are deposited upon the larger crystals, so that the average crystal size of the suspension increases. The present study was undertaken to determine whether this process occurs in a calcium oxalate monohydrate (COM) crystal growth system *in vitro*. Chemically pure COM crystals were grown in Hane's tubes in modified conventional silica gel medium utilising oxalic acid and calcium chloride as reactants, one incorporated into the gel and the other added to the top. The crystals that formed inside the gel were removed using micropipettes at different time intervals, namely, 1, 2, 3, 4, 5, 6 and 24 h. The crystals were washed and their pattern, number and size were studied under the light microscope at magnifications of 70 and 280.

Crystals started developing after one hour and continued to increase in size up to 24 hours. The number of crystals per high power field reduced from the first hour to the 24th hour, reducing from 72/high power field to 20 per high power field at 24 h. The mean size of the crystals was 32μm at 1 hour and 102μm at 24 h. It is apparent from these observations that some of the small crystals disappeared after a period of time while a number of the larger crystals continued to grow still larger.

In order to form crystals in a supersaturated environment, an energy barrier has to be overcome. This barrier stems from the fact that for very small crystals, the Gibbs free energy gain for crystal growth is outweighed by the energy loss due to extension of the crystal surface, which demands a corresponding amount of surface free energy. Consequently, in theory, the solubility of small crystals is greater than that of a sufficiently large ideal crystal. This implies that small crystals below a critical size, which is inversely proportional to the supersaturation, will dissolve at that supersaturation. It is possible that this process of Ostwald ripening takes place in the human urinary system also and is influenced by inhibitors and promoters in the environment. The presence of promoters of Ostwald ripening or the absence of inhibitors may increase the size of the larger crystals, and ultimately, in association with aggregation may initiate the process of stone formation.

CITRATE - A STRONG INHIBITOR OF UREASE-INDUCED CRYSTALLIZATION IN URINE

Y. Wang, A. Edin-Liljegren, L. Grenabo, H. Hedelin and S. Pettersson

Department of Urology
Sahlgren's Hospital
Gothenburg, Sweden

INTRODUCTION AND RESULTS

To study the effect of citrate on crystallization induced by urease we have used the Coulter Counter technique, a method which makes it possible to study different aspects of the crystallization process.

Morning urine specimens from five healthy persons were incubated with citrate-lyase until all the citrate was degraded. Its subsequent replacement with exogenous citrate provides a unique opportunity to study urine containing different concentrations of citrate. Tri-sodium citrate dihydrate was thus added to achieve final concentrations of 1, 2, 3, and 4 mM citrate. The urines were incubated with Jackbean urease in glass reactors with glass rods immersed into the urine. The different processes of the crystallization induced by the urease were monitored using the Coulter Multisizer and light microscopy. After 5 hours incubation, the material precipitated on the glass rods was analyzed for calcium and magnesium by atomic absorption spectrometry.

Although the addition of citrate caused a concentration-related increase in pH, the initiation of the crystallization process (nucleation) was markedly delayed and aggregation strongly inhibited by citrate in a concentration-dependent manner. The crystal configuration was also influenced by citrate. The amount of calcium and magnesium precipitated onto the glass rods was also negatively correlated to the concentration of citrate.

CONCLUSION

Citrate has a distinct effect on crystallization induced by urease in undiluted human urines. Citrate may be one of the factors which influences such crystallization *in vivo* and consequently may play a role in the formation of infection stones.

It is possible that citrate can be used as a prophylactic agent for infection stones in the same way as it is used to prevent calcium oxalate stone formation. This requires further studies.

URINE POLYANIONIC INHIBITORS IN SAUDI STONE FORMERS AND NORMAL CONTROLS

H. Hughes, C.L. Reynolds and W.G. Robertson

King Faisal Specialist Hospital and Research Centre
Department of Biological and Medical Research
Riyadh, Saudi Arabia

INTRODUCTION AND RESULTS

Methods for measuring urinary glycosaminoglycans (GAGS) and ribose nucleic acid (RNA) were developed using internal standards and high performance liquid chromatography (Waters Chromatography Division of Millipore and Dionex Ltd). Detection was by UV spectrometry at two wavelengths (235 and 240 nm for GAGS, and 252 and 260 nm for RNA) to take advantage of wavelength ratioing for specificity. The internal standards, dermatan sulphate (for GAGS) and polyinosinic acid (for RNA), were added to the urine. An aliquot of the purified solution was treated with chondroitinase AC or chondroitinase ABC, and an aliquot was injected on to a Supelco C18 column eluted at 1.5 mL/min with a solution of 0.5-M sodium acetate and 2-mM hexadecyltrimethy ammonium bromide at pH 5. A further aliquot of the purified solution was hydrolysed for 15 h at 37°C with 0.35-M sodium hydroxide (100 μL). An aliquot was injected on to a Dionex AS4A column and eluted with 12-mM sodium dihydrogen phosphate at pH 4.5, using a flow rate of 1.0 mL/min for 12 min which was then raised to 2 mL/min over 30 sec and held for 25 min. Recoveries of added analytes varied from 80-102%, and reproducibility was excellent.

	Di-4S*	Di-6S*	Total GAGS	RNA
		mg		
Expatriates	2.6±1 (19)	1.6±0.8 (19)	4.2±1.3 (19)	0.4±0.6 (21)
Normal Saudis	1.6±1.5 (20)	2.5±1.4 (20)	4.1±2.7 (20)	0.7±1.4 (43)
Saudi SF	3.2±1.9 (21)	2.5±2.1 (21)	5.6±2.0 (21)	0.5±1.2 (43)

Values shown are Means±SD, n in parenthesis * Di-4S and Di-6S are the disaccharides from chondroitin-4- and -6- sulphate.

Data from Western expatriates, normal Saudis, and Saudi stone formers for 24 h urinary GAGS and RNA excretion are shown in the table. This shows that the cause of the high incidence of stone disease in Saudi Arabia cannot be attributed to lower concentrations of macromolecular inhibitors in the urine of stone formers.

THE EFFECT OF KANPOU MEDICINES ON THE GROWTH AND AGGREGATION OF CALCIUM OXALATE CRYSTALS *IN VITRO*

T. Yoshioka, M. Utsunomiya, S. Yamaguchi, T. Koide and A. Okuyama

Department of Urology
Osaka University Medical School
Osaka, Japan

INTRODUCTION AND RESULTS

Although several medicines are administered to prevent new stone formation or stone growth, we have not obtained a satisfactory result. The Kanpou medicines, whose safety has become well established during a long history of use, may be useful as new prophylactic agents if they exhibit strong inhibitory activity. In this study, we examined the inhibitory effect of Kanpou medicines on calcium oxalate crystallization *in vitro*.

The Kanpou medicines studied were Takusha, Akyou, Chorel Bukuryou, Kinsensou and Kagosou, which are commercially available in Japan. They were prepared by boiling the plants for 30 minutes, freeze-drying the extracts and storing the product until required. The Kagosou extract was divided into two molecular weight fractions using a membrane (cut-off Mr of 10kDa).

The inhibitory activities of the reconstituted extracts were measured according to the methods of Robertson et al[1] and Ryall et al[2], and results calculated from the change of the number (Ia), total volume (Ig) and fractional volume (I) of crystals determined by the Coulter® Multisizer in a seeded crystallization system. The concentrations of alcian blue precipitable polyanions (ABPP) of each samples were measured according to Whiteman[3].

Among the samples tested, Takusha and Kagosou strongyl inhibited crystal growth and aggregation at the concentrations of 5 μg/mL and 10 μg/mL (Ia >80%, Ig, I >90 %). However, Kagosou contained large quantities of ABPP, while Takusha contained only small amounts. Moreover, the inhibitory activity of Kagosou was shown to exist predominantly in the molecular weight fraction greater than 10kDa.

It was concluded that some Kanpou medicines may have the ability to prevent stone formation and this is at least partly attributable to macromolecular substances like ABPP.

REFERENCES

1. WG Robertson, M Peacock and BEC Nordin, Inhibitors of the growth and aggregation of calcium oxalate crystals *in vitro*, *Clin Chim Acta* 43:31 (1973).
2. RL Ryall, CJ Bagley and VR Marshall, Independent assessment of the growth and aggregation of calcium oxalate crystals using the Coulter Counter, *Invest Urol* 18: 401 (1981).
3. P Whiteman, The quantative determination of glycosaminoglycans in urine with alcian blue 8GX, *Biochem J* 131:351 (1973).

CONTROLLED RELEASE OF PHOSPHOCITRATE INHIBITS MIXED STONE GROWTH

H. Kamperman and J.D. Sallis

Department of Biochemistry
University of Tasmania
Hobart, Tasmania, Australia

INTRODUCTION AND RESULTS

We have previously reported that phosphocitrate (PC), a potent inhibitor of extraskeletal calcification, can also limit the formation and growth of mixed struvite ($MgNH_4PO_4.6H_2O$) and newberyite ($MgHPO_4.3H_2O$) infection stones in short-term rat models. To date, PC has mainly been administered via intraperitoneal injection because in its present form, oral absorption characteristics are poor. Hence, our current studies have concentrated on developing an injectable sustained release formulation for PC that will efficiently deliver a bioactive dose of the compound over extended periods of time.

A novel technique for incorporating PC into microspheres has emerged. Isotonic NaPC is initially entrapped in glutaraldehyde cross-linked albumin microspheres. Addition of a polymeric poly-D,L-lactide coat which includes CaPC, yields particles of sub-micron dimensions and a greater than 80% overall encapsulation efficiency of PC. *In vitro,* the preparation displays biphasic release kinetics characterised by a fast initial burst, followed by sustained first-order liberation of PC over several hours. This is in sharp contrast to dissolution *in vitro* of PC from simple albumin spheres, which is complete within 10 min.

The composite PC microspheres were tested in rats which develop mixed magnesium salt bladder stones in response to implanted epoxy hemispheres. Over a 15 day period the rats were injected intramuscularly at 5 day intervals with PC at 112 µmol/kg body wt, administered either as a microsphere formulation or as the free Na or Ca salt. Compared to controls, PC in microspheres on average inhibited stone formation by 29%. Neither of the free salts of PC showed any significant response for an equivalent PC dose.

The current advances against mixed magnesium salt urinary stones can be attributed to controlled release of PC from the spheres (as confirmed *in vitro)* which *a priori,* must result in sustained clearance of the compound via the bladder site. In related studies we have also noted that injected microsphere-entrapped PC can significantly reduce subcutaneous calcergy[2]. With the possibility existing of variations in the microsphere coat and composition, greater flexibility in the prophylactic management of a number of crystal deposition diseases should be attainable.

REFERENCES

1. JD Sallis, R Thomson, B Rees and R Shankar, Reduction of infection stones in rats by combined antibiotic and phosphocitrate therapy, *J Urol* 140:1063 (1988).
2. J Sallis, J Meehan, H Kamperman and M Anderson, Chemically modified phosphocitrate and entrapment in microspheres for sustained inhibition of biomineralisation, *Phosphorus, Sulfur & Silicon and the Related Elements* (1993, in press).

DETERMINATION OF RARE EARTH ELEMENTS IN URINARY STONES BY INSTRUMENTAL NEUTRON ACTIVATION ANALYSIS (INAA)

K. Höbarth[1], Ch. Koeberl[2], J. Hofbauer[1] and M. Marberger[1]

[1]Department of Urology
[2]Institute of Geochemistry
University of Vienna Medical School
Vienna, Austria

INTRODUCTION AND RESULTS

Although rare earth elements (REE) are of some biological importance, their concentrations are not frequently assessed in human samples. Although elements have been studied in human urinary stones using various methods to obtain information on the process of calculus formation, the studies were concerned with minor and trace elements. In the present study, urinary stones were examined for REE.

Ten kidney stones (6 oxalate and 4 phosphate stones) were analyzed. Two biological standard materials (NBS SRM 1571 {orchard leaves} and IAEA H-5 {animal bone}) were used for comparison of REE data.

Oxalate stones showed distinctly higher concentrations of REE compared with animal bone but did not reach the REE levels of plants. Phosphate stones revealed elevated levels of high REE (Eu, Tb, Dy, Ho, Er, Tm, Yb, Lu), as did non-biogenic apatites, indicating a related concentration mechanism. Both stone types tested least positive for Sm, which is unusual if compared with non-biogenic samples. Phosphate stones revealed significantly higher levels of Eu and Gd, while oxalate stones showed a lower concentration of Lu ($p<0.005$).

The distribution of minor and trace elements, as well as of REE, reflects the composition of urine over the period of stone formation. By using INAA, it is possible to assess the elemental composition of individual stones. Little is known about the metabolism of REE in the human body. Intake occurs via ingestion and inhalation, and to a smaller extent via intestinal resorption. Urinary stones, being traps for some trace elements, can concentrate REE due to their slow accumulation. Some fractionation of the high REE *versus* the low REE appears to take place during metabolism and deposition.

CONCLUSION

The determination of REE in urinary calculi may shed some light on the issue of stone initiation, by addressing the question of whether the nucleation of urinary stones is heterogeneous. It is not clear whether REE play a role in the growth of deposits of crystalline calcium compounds. Further studies will need to be undertaken to elucidate the function of REE and to determine whether they interact with other factors in stone formation.

ANTICALCULOGENIC PROPERTIES OF *SCOPARIA DULCIS LINN*

P.J. Alphonsa[1] , S. Sindhu, V.V. Nair[1], J. Kurien[1], P. Koshy[2] and Y.M. Fazil Marickar

[1]Dept of Dravya Guna Vijnan
Ayurveda College
[2]RRLTrivandrum & Department of Surgery
Medical College Hospital
Trivandrum , India

INTRODUCTION AND METHODS

Various plant preparations have been used for treatment of urinary stone disease, but their efficacy has not been proved. This paper attempts to scientifically study the anticalculogenic property of one such medicinal plant - *Scoparia dulcis Linn.* The plant extracts were prepared in acetone, petroleum ether and chloroform in concentrations of 20 mg%, 40 mg% & 80 mg% respectively. The inhibitory effects of these plant extracts were studied in calcium oxalate crystal growth in silica gel medium. The different test agents (drug in concentrations of 500 mg and 1 g/100 g rat weight), standard (hydrochlorothiazide 0.25 mg/100 g rat weight) and control (normal saline 0.91% - 2.5 mL/100 g rat weight) were administered orally to male Wistar albino rats. The urine output was recorded at 5 hours and 24 hours to study the diuretic property. Another group of rats was administered the drug in the above concentrations daily for one month and the blood was collected for biochemical estimations and compared with controls.

RESULTS AND DISCUSSION

The petroleum ether extract at a dose of 80 mg% maximately inhibited the growth of calcium oxalate crystals. At a concentration of 500 mg/100 g rat weight, *Scoparia dulcis Linn* produced maximum diuretic effect (67.39% of the total fluid intake compared with 46.06% of the hydrochlorothiazide group and 32.6% of the control group). The extract did not induce any type of crystalluria. Administration of the plant extract lowered the calcium and uric acid level in the urine and blood by 5 days to 20 days. The extracts of the plant produced significant inhibition of calcium oxalate crystals grown *in vitro* and scanning electron microscopy of the crystals showed significant alteration in the surface characteristics.

CONCLUSION

It is concluded that *Scoparia dulcis Linn* has definite anticalculogenic properties.

SECTION IV: CRYSTALLIZATION AND PROTEINS

POSSIBLE ROLE OF STONE MATRIX IN CALCIUM OXALATE STONE FORMATION

S. Yamaguchi[1], T. Yoshioka[1], M. Utsunomiya[1], T. Koide[1], A. Okuyama[1] and T. Sonoda[2]

[1]Department of Urology
Osaka University Medical School
Osaka, Japan
[2]Department of Urology
Osaka Prefectural General Hospital
Osaka, Japan

INTRODUCTION

The role of the organic stone matrix in stone formation is controversial. Some investigators have reported that the organic matrix is selectively incorporated into urinary calculi by promoting crystal growth and aggregation[1,2]. Recently, Resnick et al reported that some particular proteins, which were adsorbed onto crystals, may promote calcium oxalate crystallization[3,4]. Others, however, have suggested that matrix formation is the result of non-specific adsorption of urinary macromolecules onto urinary crystals and does not act as a promoter[5-7]. Moreover, still other researchers have shown that the soluble portion of stone matrix inhibits crystal growth and aggregation *in vitro* [8].

We have previously reported that urinary macromolecules (UMMs) have a strong inhibitory activity[9] and that part of the UMMs are adsorbed as inhibitors onto the surface of calcium oxalate crystals (crystal surface binding substances; CSBS)[10]. The aim of this study was to determine the characteristics of GAGs in the soluble stone matrix as well as in CSBS and UMMs, in order to clarify the role of these GAGs in calcium oxalate stone formation.

MATERIALS AND METHODS

Preparation of pooled urine

A large volume of urine was collected from 10 healthy males between 24 and 37 years old. The urine was collected in sterile glass containers with 0.02% sodium azide as an antibacterial agent and stored at 4°C. After warming to 37°C with shaking, the urine was filtered through a Whatman No 1 filter paper and a 0.22 μm Millipore filter prior to the experiments.

Preparation of UMMs

After filtration, the pooled urine was ultrafiltered with a Labomodule (cut-off: 3kDa). The concentrated and desalinated urinary macromolecular solution was lyophilized and stored at -20°C until use.

Urolithiasis 2, Edited by R. Ryall *et al.*,
Plenum Press, New York, 1994

Preparation of CSBS

We obtained CSBS by the following procedures reported previously[10]. Spontaneous crystallization of calcium oxalate was induced by adding $CaCl_2$ and $Na_2C_2O_4$ to the pooled urine (Urine: 1M $CaCl_2$: 0.1M sodium oxalate = 1000:32:320) followed by incubation for six hours at 37°C in a shaking water bath. After incubation, the samples were centrifuged at 2000 rpm for 10 min. The crystal pellets which were precipitated during centrifugation were washed thoroughly with a volume of saturated calcium oxalate solution sufficient to prevent contamination from the remaining urinary constituents. The crystals were then suspended in a 10% EDTA.4Na solution (pH 8), and dissolved within 30 min. Ultrafiltration was performed with a Labomodule (3kDa cut-off) to remove EDTA.4Na and to concentrate the solution. Finally, the concentrated and desalinated CSBS solutions were Iyophilized and stored at -20°C until use.

Preparation of seed-crystal surface binding substances (Seed-CSBS)

50g of calcium oxalate monohydrate crystals were added to 3 L of pooled urine and incubated for 6 hours at 37°C in a shaking water bath. After incubation, the solution was centrifuged at 2000 rpm for 10 min. The remainder of the procedure was the same as for the preparation of CSBS[10].

Preparation of soluble stone matrix

The calcium oxalate stones were crushed into a powder, placed in a dialysis sac (3.5 kDa cut-off) and dialyzed against an abundant volume of 5% EDTA.4Na solution (pH 7.8) for 2 weeks. The completely decalcified material was dialyzed against distilled water for 72 hours in order to remove EDTA. The dialyzed solution was centrifuged at 3000 rpm for 10 minutes. The supernatant, which contained the soluble stone matrix, was lyophilized and stored at -20°C until use.

Proteolysis

Proteolysis of UMMs, Seed-CSBS, CSBS and soluble stone matrix was performed using the method of Nishio et al.[1]. Uronic acid was measured with the method of Bitter and Muir[11]. Each protein concentration of UMMs, Seed-CSBS, CSBS and soluble stone matrix was also measured with the Bradford protein assay method[12] before proteolysis.

Two-dimensional cellulose acetate membrane electrophoresis

Two-dimensional electrophoresis was performed with the method of Hata and Nagai[13] on a Sepraphore III instrument (Gelman Sciences) using 0.1M pyridine and 0.47M formic acid (pH 3.0) as a solvent in the first dimension for 60 minutes, and 0.1M barium acetate (pH 8.0) in the second dimension for 4.5 hours. UMMs, Seed-CSBS, CSBS and soluble stone matrix were subjected to electrophoresis after proteolysis. Commercially available GAGs (chondroitin sulfate A, dermatan sulfate, hyaluronic acid and heparin), were run simultaneously as standards. Staining was done with 0.1% alcian blue in 0.1% acetic acid for 10 minutes. The Sepraphore III instrument was washed with 300 mL of 0.1% acetic acid for 20 minutes.

Inhibition assay

The inhibitory effect of UMMs, CSBS and soluble stone matrix on calcium oxalate crystal growth and aggregation was measured with the modified method of Robertson et al[14]. Crystal distribution was measured with the Coulter Multisizer (Coulter Electronics Limited, England). Calculation of percent inhibition was based on determination of the number of particles of greater than 9 μm in diameter before incubation and after incubation for 4 hours. The effect of adding various concentrations of UMMs, CSBS and the soluble stone matrix on the growth and aggregation of calcium oxalate crystals was studied. Percent inhibitory activity was expressed as a function of uronic acid and protein content.

RESULTS

Two-dimensional electrophoresis revealed that UMMs contained various GAGs which are usually seen in normal human urine. These included chondroitin sulfate A and C, dermatan sulfate, keratan sulfate, hyaluronic acid and heparan sulfate (Fig 1-A). Seed CSBS, the substances adsorbed onto the seeded calcium oxalate monohydrate crystals, showed a similar pattern to UMMs but the heparan sulfate spot was stained more strongly (Fig 1-B). The GAGs in the CSBS, the substances adsorbed onto calcium oxalate crystals during crystallization in whole urine, consisted of heparan sulfate and a small amount of dermatan sulfate (Fig 1-C). Furthermore, the only GAG in the soluble stone matrix was heparan sulfate; other GAGs could not be found (Fig 1-D).

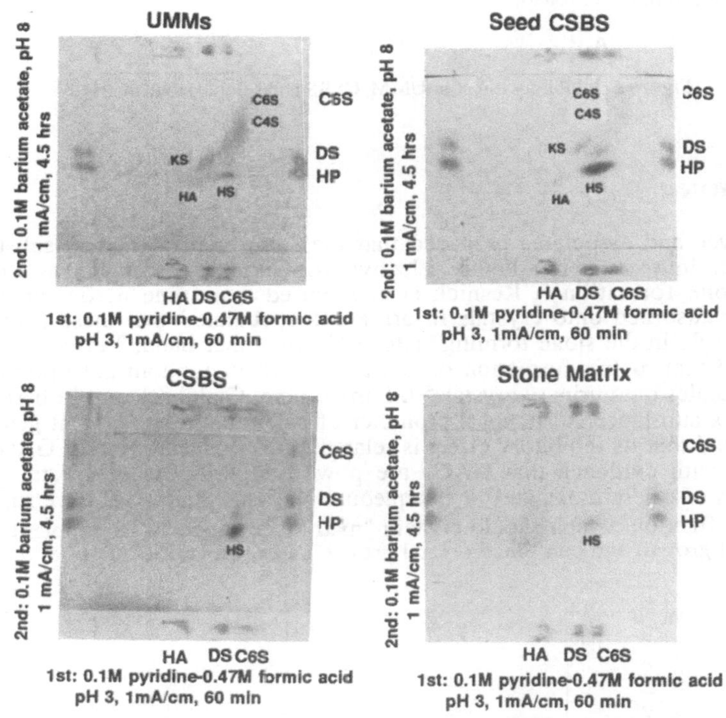

Figure 1. Top row 1-A, 1-B; Bottom row 1-C, 1-D. Two-dimensional cellulose acetate membrane electrophoresis. C6S:chondroitin-6-sulfate, C4S:chondroitin-4-sulfate, KS:keratan sulfate, DS:dermatan sulfate, HS:heparan sulfate, HA:hyaluronic acid, and HP:heparin

Comparison of the inhibitory effect of UMMs, CSBS and the soluble stone matrix on calcium oxalate crystal growth and aggregation is shown in Figs 2-A and B. Although the inhibitory effect of soluble stone matrix, CSBS and UMMs did not differ greatly from each other when the inhibitory effect was expressed in terms of protein concentration (Fig 2-A), there was a major difference when they were expressed in terms of uronic acid concentration. Percentage of inhibition in terms of uronic acid was strongest in the soluble stone matrix and CSBS exhibited a fairly strong inhibition (Fig 2-B).

The strong inhibitory effect of the soluble stone matrix, however, was weakened by protein digestion (Figure 3.).

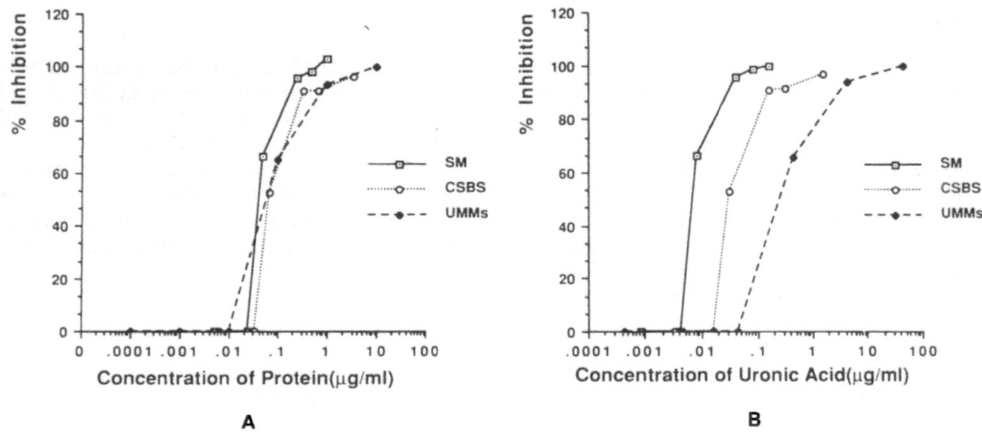

Figure 2. Inhibitory effect of UMM, CSBS and soluble stone matrix

DISCUSSION

Boyce and associates proposed that the stone matrix (matrix substance A) is incorporated into stones in a highly selective manner and thereby acts as a promoter of urinary stone formation[15]. Resnick et al reported that some particular proteins are selectively adsorbed onto crystals in urine and speculated that these proteins play a promotive role in the stone forming state[3,4]. On the other hand, Vermeulen and others have stated that matrix formation occurs as a result of random adsorption of urinary macromolecules onto urinary crystals[5-7]. Furthermore, Gjaldback and Robertson showed that "matrix substance A" is not a promoter of calcium oxalate crystal growth, but an inhibitor, and that its inhibitory effect is related to the concentration of GAGs[16]. This is consistent with evidence that GAGs are powerful inhibitors of crystal growth and aggregation of calcium oxalate[17,18] in aqueous solutions. Scurr et al have suggested that the soluble fraction, which should contain "matrix substance A", is a potent inhibitor of both crystal growth and combined crystal growth and aggregation[8].

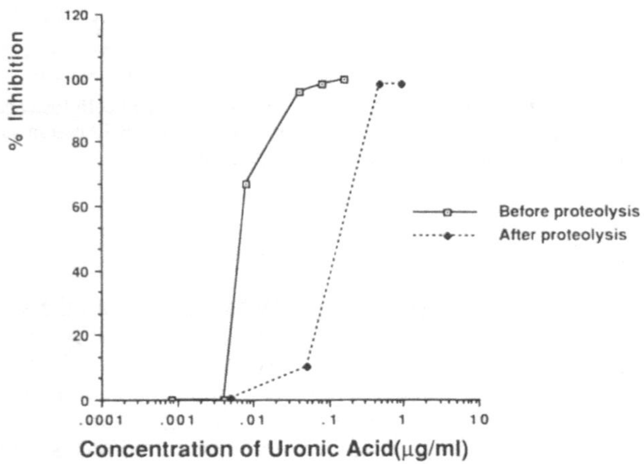

Figure 3. Inhibitory effect of the soluble stone matrix before and after proteolysis

With respect to the GAGs in the stone matrix, Nishio and colleagues showed that heparan sulfate is the sole GAG in calcium oxalate monohydrate stone matrix and speculated that this GAG is a promoter rather than an inhibitor of urinary stone formation, enhancing stone growth by affecting crystal growth rates and crystal aggregation[1]. In this study, however, we proved that soluble stone matrix shows a very strong inhibitory effect on calcium oxalate crystal growth and aggregation, although the composition of GAGs in the stone matrix confirms the findings of Nishio et al[1].

In a previous report, we proved that CSBS, the material adsorbed onto crystals, was not a promoter, but a strong inhibitor of crystal growth and aggregation[10]. Moreover, Doyle et al, in a recent study, proved that crystal matrix protein exhibits a marked affinity for calcium oxalate crystals and is a potent inhibitor of crystal aggregation[19]. When we examined the composition of GAGs, we found that the proportion of heparan sulfate was the highest in the stone matrix, the second highest in the CSBS, the third highest in the Seed-CSBS, and the lowest in the UMMs. These results indicate that with respect to GAGs, Seed-CSBS appears to derive from the UMMs, the CSBS constitute part of the Seed-CSBS, and soluble stone matrix a portion of the CSBS. The difference between CSBS and Seed-CSBS is that the CSBS was formed under more active crystallizing conditions than Seed-CSBS. It is our hypothesis that when rapid crystallization is occurring, stronger inhibitors are adsorbed more actively onto crystals than weaker inhibitors, in accord with the finding that weaker inhibitors bind with less affinity than do stronger inhibitors[20]. It follows from this hypothesis that heparan sulfate may have the strongest affinity for the crystals and the strongest inhibitory effect on crystal growth and aggregation. GAGs other than heparan sulfate might not contribute significantly to this inhibitory effect. Therefore, the stone matrix, in which heparan sulfate was the predominant GAG, exhibits the strongest inhibitory effect.

Nonetheless, the inhibitory activity cannot be ascribed solely to GAGs; it must be stressed that the inhibitory strength of the soluble stone matrix was markedly reduced by protein digestion. There are two possible explanations. One might be that the heparan sulfate in the stone matrix usually exists in the form of protein-heparan sulfate complex, so that its strongest inhibitory activity appears in the form of proteoglycans and consequently, this inhibitory power is weakened by proteolysis. This notion is supported by our inability to obtain a strong heparan sulfate spot during two-dimensional electrophoresis without proteolysis. This finding indicates that heparan sulfate is incorporated into the calcium oxalate stone in the form of proteoglycan. Gjaldback et al reported that the strong inhibitory effect of the soluble stone matrix was associated with the uronic acid-rich fraction[16], suggesting that GAGs components play an important role in inhibition, albeit in the form of proteoglycans. Therefore, it may be correct to assume that heparan sulfate exists in the form of a proteoglycan, and exhibits its strongest inhibitory activity as a form of the protein-heparan sulfate complex. Nonetheless, another explanation for the inhibitory effect of soluble stone matrix might be that some matrix proteins, such as nephrocalcin[21], are strong inhibitors and that proteolysis reduces their potency. Since CSBS comprise various substances as reported previously[10], the latter possibility cannot be ignored in relation to the soluble stone matrix.

In conclusion, it was found that the soluble portion of stone matrix is composed partly of CSBS, at least with regard to GAGs, and has a very high affinity for the crystals. The GAGs in soluble stone matrix consist principally of heparan sulfate, which may exist in the form of a protein-heparan sulfate complex, and exhibit a strong inhibitory effect on calcium oxalate crystal growth and aggregation. Stone formation might thus start when the ambient conditions overwhelm the inhibitory effect of heparan sulfate crystallization, after which heparan sulfate might be incorporated into the stones and form part of the stone matrix. The insoluble fraction of the stone matrix should be investigated further to determine whether the stone matrix has any promotive function in itself.

REFERENCES

1. S Nishio, Y Abe, A Wakatsuki, H Iwata, K Ochi, M Takeuchi and A Matsumoto, Matrix glycosaminoglycans in urinary stones, *J Urol* 134:503 (1985).
2. SD Roberts and MI Resnick, Glycosaminoglycans content of stone matrix, *J Urol* 135:1078 (1986).

3. RM Morse and MI Resnick, A new approach to the study of urinary macromolecules as a participant in calcium oxalate crystallization, *J Urol* 139:869 (1988).
4. MI Resnick, ME Sorrell, JA Bailey and WH Boyce, Inhibitory effect of urinary calcium-binding substances on calcium oxalate crystallization, *J Urol* 127:568 (1982).
5. B Finlayson, CW Vermeulen and EJ Stewart, Stone matrix and mucoprotein from urine, *J Urol* 86:355 (1961).
6. BT Murphy and LM Pyrah, The composition, structure, and mechanisms of the formation of urinary calculi, *Br J Urol* 34:129 (1962).
7. OW Vermeulen and ES Lyon, Mechanism of genesis and growth of calculi, *Am J Med* 45:684 (1968).
8. DS Scurr, CM Bridge and WG Robertson, In: *Urolithiasis.* Studies on inhibitors and promoters of the crystallization of calcium oxalate in urine and in matrix from calcium oxalate stones. Edited by LH Smith, WG Robertson and B Finlayson, Plenum Press, New York p601 (1980).
9. T Koide, M Takemoto, H Itatani, M Takaha and T Sonoda, Urinary macromolecular substances as natural inhibitors of calcium oxalate crystal aggregation, *Invest Urol* 18:382 (1981).
10. T Koide, T Yoshioka, S Yamaguchi, S Hosokawa, M Utsunomiya and T Sonoda, Urinary crystal surface binding substances on calcium oxalate crystals, *Urol Res* 18:387 (1990).
11. T Bitter and HM Muir, A modified uronic acid carbazole reaction, *Anal Biochem* 4:330 (1962).
12. MM Bradford, A rapid and sensitive method for the quantitation of microgram quantities of protein utilizing the principle of protein-dye binding, *Anal Biochem* 72:248 (1979).
13. R Hata and Y Nagai, A rapid and micromethod for separation on acidic glycosaminoglycans by two dimensional electrophoresis, *Anal Biochem* 45:462 (1972).
14. WG Robertson, M Peacock and BEC Nordin, Inhibitors of the growth and aggregation of calcium oxalate crystals *in vitro*, *Clin Chim Acta* 43:31(1973).
15. WH Boyce, Organic matrix of human urinary concretions, *Am J Med* 45:673 (1968).
16. J Chr Gjaldbaek and WG Robertson WG, Does urine from stone-formers contain macromolecules which promote the crystal growth rate of calcium oxalate crystals *in vitro*? *Clin Chim Acta* 108:75 (1980).
17. RC Bowyer, JG Brockis and RK McCulloch, Glycosaminoglycans as inhibitors of calcium oxalate crystal growth and aggregation, *Clin Chim Acta* 95:23 (1979).
18. JD Sallis and MF Lumley, On the possible role of glycosaminoglycans as natural inhibitors of calcium oxalate stones, *Invest Urol* 16:296 (1979).
19. IR Doyle, RL Ryall and VR Marshall, Crystal matrix protein - An Inhibitor of calcium oxalate stone formation? *Br J Urol* 67:399 (1991).
20. AH Angell and MI Resnick, Surface interaction between glycosaminoglycans and calcium oxalate, *J Urol* 141:1255 (1989).
21. Y Nakagawa, M-A Ahmed, SL Hall, S Deganello, FL Coe, Isolation from human calcium oxalate renal stones of nephrocalcin, a glycoprotein inhibitor of calcium oxalate crystal growth, *J Clin Invest* 79:1782 (1987).

ROLE OF UROPONTIN IN URINARY CALCIUM STONE FORMATION

J.R. Hoyer

Departments of Pediatrics and Medicine
University of Pennsylvania and
The Children's Hospital
Philadelphia, PA, USA

INTRODUCTION

Most stones formed in the urinary space are mineralized with calcium salts; the majority of these urinary calculi are composed principally of calcium oxalate[1]. Although normal urine is frequently supersaturated with respect to calcium oxalate, most humans do not form stones. The precise role of urinary inhibitors in the complex process of urinary stone formation that involves multiple factors is uncertain. The inhibitors of calcium oxalate crystal growth, nucleation and aggregation present in normal urine[2-6] slow crystal growth and raise the concentration of calcium oxalate required for the spontaneous formation of new crystals. Protein macromolecules, rather than lower molecular weight moieties account for the majority of the inhibition of crystal growth observed in normal urine[3]. Even highly diluted normal urine will markedly decrease the growth of crystals in calcium oxalate solutions. The growth and aggregation of calcium oxalate crystals *in vitro* are significantly more inhibited by urine from normal subjects than from stone formers[7,8]. New information concerning the molecular features of inhibitory proteins will provide the basis for better understanding their role(s) in stone formation.

Isolation and characterization of Uropontin

We have used an immunologic approach to the problem of identifying and characterizing crystal inhibitor proteins. Hybridoma cells producing monoclonal antibodies were obtained after immunization with the main inhibitory peak of human urine protein[3]. These monoclonal antibodies were used to purify an inhibitor of calcium oxalate monohydrate (COM) crystal growth from human urine by immunoaffinity chromatography[9]. The major band of this inhibitory protein was detected in Western blots and by silver staining of 16% SDS-PAGE gels at a position of ~50 kDa. Comparison of its N-terminal sequence with known protein sequences in the Swissprot computerized protein database showed that its sequence was identical to that of human osteopontin[10]. Further evidence of the relationship of this urinary protein to osteopontin was provided by amino acid composition showing a very high percentage of aspartic acid residues[10] and an overall distribution of amino acids comparable to that determined or predicted for human osteopontin from bone[12,13]. Since multiple tissues can produce pontin proteins, it seems unlikely that this urinary protein is derived exclusively from bone. Thus, to denote its source, we refer to it as uropontin, rather than using the term urinary osteopontin. In any case, it is the urinary form of the protein encoded by the OPN gene located on chromosome 4 of the human genome[11].

Urolithiasis 2, Edited by R. Ryall *et al.*,
Plenum Press, New York, 1994

Role in Biomineralization

On the basis of the biological distribution of aspartic acid-rich proteins (AARP) in close association with mineralization events in a broad range of organisms and tissues, and our studies showing very potent inhibition of calcium oxalate crystal growth *in vitro* by uropontin (the Kd for inhibition is 2.5 x 10^{-8}M [14]), we identified a potential common functional role[10]. This hypothesis concerning the nature of the interaction of uropontin with crystals and a role of pontin proteins in biomineralization is derived from a series of elegant studies *in vitro* demonstrating tissue specific patterns of modulation of crystal growth by aspartic acid rich proteins from other mineralizing tissues. For example, solutions of an AARP from mollusk shells interact specifically with selected faces on calcium dicarboxylic acid crystals and thereby slow the overall rate of crystal growth[15]. While inhibitory in the fluid phase, adsorption of this AARP onto a solid substrate induced calcite crystal formation in an orientation characteristic of biological mineralization[15]. As discussed in detail elsewhere in this volume (Purich et al), the very potent inhibition of crystal growth appears to require a specific structural conformation that produces a complementary pattern of charge density. Although the concentration of uropontin required for inhibition of COM crystal growth is well within the range present in normal human urine[16], the precise nature of the interaction of uropontin with crystals remains to be defined.

Several of the known features of the family of pontin proteins would support their role in the regulation of mineralization within the fluid spaces that are supersaturated with respect to their mineral constituents. These features would allow either restrictive or permissive effects on the formation or remodelling of crystal surfaces in a variety of body fluids. The responsiveness to vitamin D and parathyroid hormone demonstrated for pontins and the multiplicity of forms resulting from several post-translational modifications have been previously summarized[10]. In addition, responsiveness to estrogen and progesterone[17] and the regulated temporal and spatial expression of OPN during pregnancy[18] are conducive to regulatory roles for these pontin proteins.

Structural Features

The multiple bands observed on SDS-PAGE gels may reflect differences in the extent of post-translational modifications or may be due to alternative RNA splicing. The heterogeneity detected in human mRNA transcripts supports this possibility. An exon of the mouse OPN gene[17,19] encodes the sequence of 14 amino acids shown in Figure 1 that is highly conserved and deleted in one of the isoforms encoded by cDNAs from human kidney, bone and decidua [10,11,13]. Factors responsible for differential expression of the isoforms and functional differences among isoforms are not yet known.

HUMAN	N	A	V	S	S	E	E	T	N	D	F	K	Q	E
RAT	x	S	x	x	x	x	x	x	D	x	x	x	x	x
MOUSE	x	x	x	x	x	x	x	K	D	x	x	x	x	x
PIG	x	T	I	x	x	x	x	x	D	x	x	x	x	x
COW	x	S	x	x	x	x	x	x	D	x	N	x	x	N

Figure 1. Aligned amino acid sequences of pontin proteins showing the high degree of homology to the 14 amino acid sequence deleted in some human cDNA clones. The amino acids in the other mammalian sequences[20-23] that are identical to those in the human sequence are represented by an x. The amino acids immediately preceding and following this sequence (glutamine and threonine, respectively) are identical in the 5 species. By contrast, only 5 amino acids of the corresponding chicken sequence[24] are homologous to the mammalian sequences. The single letter codes for the amino acids used in Figures 1-3 are: A = alanine, D = aspartic acid, E = glutamic acid, F = phenylalanine, G = glycine, H = histidine, I = isoleucine, K = lysine, L = leucine, M = methionine, N = asparagine, P = proline, Q = glutamine, R = arginine, S = serine, T = threonine, V = valine, Y = tyrosine.

Regions within proteins that are highly conserved among species may provide important clues to their potential functional roles. The term osteopontin was originally applied to an acidic protein isolated from the mineralized matrix of bone since it contained a cell binding (RGD) sequence and would bind to hydroxyapatite; it was postulated that this protein provided a bridge between the mineral and cells within bone[20]. Subsequent studies have amply demonstrated that it effectively promotes cell attachment (summarized in 25). As shown in Figure 2, the domain containing the RGD sequence is highly conserved in each of the species studied. More recent studies suggest that the role of this protein in attachment is not limited to providing a physical bridge, since the binding of osteopontin or uropontin to the $\alpha_v\beta_3$ integrin on the surface of osteoclasts stimulates intracellular signal transduction[27,28].

HUMAN	D	G	R	G	D	S	V	V	Y	G	L	R	S	K	S	K
PIG	x	x	x	x	x	x	x	x	x	x	x	x	x	x	x	x
RAT	x	x	x	x	x	x	L	A	x	x	x	x	x	x	x	R
MOUSE	N	x	x	x	x	x	L	A	x	x	x	x	x	x	x	R
COW	x	x	x	x	x	x	x	A	x	x	x	K	x	R	x	x
CHICKEN	A	x	x	x	x	x	x	A	x	x	F	x	A	x	A	H

Figure 2. Aligned amino acid sequences of integrin-binding (RGD) domain and the thrombin cleavage site[26] of pontin proteins. Amino acids in the other 5 sequences that are identical to those in the human sequence are represented by an x.

The aspartic acid-rich sequence of human osteopontin is another highly conserved domain among species. Since the sequences of amino acids shown in Figure 3 appeared to be strong candidates for binding to calcium oxalate crystal surfaces, we synthesized a series of peptides containing these sequences in order to evaluate their functional capacity. A dose dependent inhibition of COM crystal growth by short peptides (MW <4000) of the human protein was observed at concentrations less than 10 µg/mL (data not shown). Thus, this domain appears to be involved in the interaction of the protein with crystal surfaces.

MOUSE	N	E	S	H	D	H	M	D	D	D	D	D	D	D	D	D	D
RAT	x	x	x	x	x	x	x	x	x	x	x	x	x	x	x	x	G
HUMAN	x	x	x	x	x	x	x	x	x	M	x	x	E	x	x	x	x
PIG	x	x	x	P	E	Q	T	x	x	V	x	x	x	x	x	E	x
COW	x	x	x	P	E	Q	T	x	x	L	x	x	x	x	x	N	S
CHICKEN	S	K	x	Q	E	A	V	x	x	x	x	x	x	x	N	x	S

Figure 3. Aligned amino acid sequences in the aspartic acid-rich domain of pontin proteins. Amino acids identical to those in the mouse sequence are represented by an x.

Analysis of urinary stones

Since uropontin inhibits COM crystal growth *in vitro,* we tested the hypothesis that it specifically interacts with calcium crystals *in vivo* by examining the content of uropontin and other urinary proteins in 0.5M EDTA extracts of human urinary stones by ELISA. These studies showed that uropontin is a major matrix component of COM stones[29] with a mean content of ~800 µg of uropontin/100mg of stone[30]. The contents of albumin, urinary α-1-inter-trypsin inhibitor (UITI)[31] and Tamm-Horsfall protein (TH) detected by ELISA were all <10µg/100 mg of stone in each of the 10 COM stones[30]. By contrast, calcium oxalate dihydrate (COD) stones contained much less uropontin (mean of 16µg/100mg of stone, n=10) with more albumin (>20µg/100mg of stone) than uropontin, being present in 4 of these COD stones. The content of UITI in all COD stones, and TH in 9/10 of COD stones was <10µg/100 mg of stone[30]. Thus, nearly 1% of the weight of COM stones was uropontin, an amount substantially greater than the quantity of nephrocalcin recovered from calcium oxalate stones[32]. Our findings are consistent with previous studies of the amino acid composition of calcium oxalate stones[33-35]. The presence of uropontin within calcium oxalate stones has been confirmed by cloning studies using a polyclonal antiserum to calcium oxalate stone protein[36]. We further demonstrated specificity of the interaction of uropontin with COM crystals by showing that the uropontin content of other types of stones was very low. Hydroxyapatite, uric acid, cystine and struvite stones (at least 5 stones in each category) all contained <10 µg of uropontin/100mg of stone.

However, the uropontin content of 7 brushite stones was relatively high (mean of ~92μg per 100mg of stone) while the content of albumin, UITI and TH in brushite stones was not increased (all <10μg per 100 mg of stone). In view of the roles in bone mineralization that have been postulated for osteopontin (summarized in [25,37]), the virtual absence of uropontin from urinary hydroxyapatite stones was not expected. Differences in the extent or type of post-translational modifications of pontin proteins occurring at various sites within the body may be important determinants. For example, the sulfation of osteopontin appears to be closely related to mineralization in bone[38]. Although all tissue sources for uropontin are not known, available evidence strongly supports renal synthesis of uropontin, a possibility that has been previously summarized[10]. More recently, OPN expression in normal adult mouse kidneys has been shown by studies combining *in situ* hybridization and immunohistochemical analysis to be primarily localized in thick ascending limbs of Henle's loop and distal convoluted tubules[39].

CONCLUSIONS

Within the concepts of mineralization outlined above, a pivotal role in the pathogenesis of calcium nephrolithiasis for urinary proteins sharing molecular characteristics with uropontin is quite plausible. In the fluid phase, such molecules have a net inhibitory effect on crystal growth. Our studies of uropontin excretion by normal adults showed that uropontin is present in normal human urine[16] at concentrations that will effectively slow the rate of COM crystal growth *in vitro*. The uropontin concentration varied inversely with urine volume[16], a physiological characteristic that favors its prevention of urinary crystallization of calcium oxalate. The protective effects of inhibitors may be particularly important during times of physiologically increased calcium excretion, such as during pregnancy. In this regard, the relatively low incidence of stone formation during pregnancy may be at least partially explained by the increased OPN expression observed during pregnancy[18].

It appears likely that quantitative deficiencies in uropontin synthesis would favor stone formation by decreasing the inhibitory capacity of urine. Molecular abnormalities resulting from defects in the primary sequence or in post-translational modifications could result in alterations in the conformation or patterns of charge density of uropontin that would alter the interaction with crystals so as to cause less effective inhibition of crystallization. Alternatively, molecular abnormalities of uropontin could also lead to a net promotion, rather than inhibition of crystallization. The high content of uropontin in COM stones strongly suggests a specific interaction of this aspartic acid-rich protein with COM crystals and supports the concept that this protein has an important role in calcium oxalate stone formation. The elucidation of the primary structure of uropontin and its structural relationship to other pontin proteins provides the basis for promising new investigative approaches at the molecular level. These should lead to a better definition of the role of urinary proteins in stone formation.

ACKNOWLEDGEMENTS

These studies were supported by grants from the National Institutes of Health (DK33501) and the University of Pennsylvania Research Foundation.

REFERENCES

1. LC Herring, Observations in the analysis of ten thousand urinary calculi, *J Urol 88:* 545 (1962).
2. JL Meyer and LH Smith, Growth of calcium oxalate crystals. II. Inhibition by natural urinary crystal growth inhibitors, *Invest Urol 13:* 36 (1975).
3. Y Nakagawa, V Abram, FJ Kezdy, ET Kaiser and FL Coe, Purification and characterization of the principal inhibitor of calcium oxalate monohydrate crystal growth in human urine, *J Biol Chem* 258: 12594 (1983).
4. J Asplin, S Deganello, Y Nakagawa and FL Coe, Evidence that nephrocalcin and urine inhibit nucleation of calcium oxalate monohydrate crystals, *Am J Physiol* 261: F824 (1991).

5. B Hess, Y Nakagawa and FL Coe, Inhibition of calcium oxalate monohydrate crystal aggregation by urine proteins, *Am J Physiol* 257: F99 (1989).
6. KA Edyvane, CM Hibberd, RM Harnett, VR Marshall and RL Ryall, Macromolecules inhibit calcium oxalate crystal growth and aggregation in whole urine, *Clin Chim Acta* 167: 329 (1987).
7. WG Robertson and M Peacock, Calcium oxalate crystalluria and inhibitors of crystallization in recurrent renal stone formers, *Clin Sci* 43: 499 (1972).
8. A Ligabue, M Fini and WG Robertson, Influence of urine on *in vitro* crystallization rate of calcium oxalate; determination of inhibitory activity by a ^{14}C-oxalate technique, *Clin Chim Acta 98:* 39 (1979).
9. H Shiraga, MD Clayman, EG Neilson and JR Hoyer, Affinity purification of urinary crystal growth inhibitor (CGI), *Kidney Int* 35: 363 (1989).
10. H Shiraga, W Min, WJ VanDusen et al, Inhibition of calcium oxalate crystal growth *in vitro* by uropontin: another member of the aspartic acid-rich protein superfamily, *Proc Nat Acad Sci USA 89:* 426, 1992.
11. MF Young, JM Kerr, JD Termine, UM Wewer et al, cDNA cloning, mRNA distribution and heterogeneity, chromosomal location, and RFLP analysis of human osteopontin (OPN), *Genomics* 7: 491(1990).
12. LW Fisher, GR Hawkins, N Tuross and JD Termine, Purification and partial characterization of small proteoglycans I and II, bone sialoproteins I and II, and osteonectin from the mineral compartment of developing human bone, *J Biol Chem* 262: 9702 (1987).
13. MC Kiefer, DM Bauer and PJ Barr, The cDNA and derived amino acid sequence for human osteopontin, *Nucleic Acids Res 17: 3306* (1989).
14. J Asplin, JR Hoyer, Y Nakagawa and FL Coe, Uropontin (UP) inhibits calcium oxalate monohydrate (COM) crystal growth and nucleation, *J Am Soc Nephrol* (in press)(1992).
15. L Addadi and S Weiner, Interactions between acidic proteins and crystals; stereochemical requirements in biomineralization, *Proc Natl Acad Sci USA*. 82: 4110 (1985).
16. C Chalko, G Krishna, JR Hoyer and S Goldfarb, Characterization of urinary uropontin excretion in humans, *J Am Soc Nephrol* (in press)(1992).
17. AM Craig and DT Denhardt, The mouse gene encoding secreted phosphoprotein 1 (OPN, osteopontin); promoter structure, activity, and induction *in vivo* by estrogen and progesterone, *Gene* 100: 163 (1991).
18. P Waterhouse, RS Parhar, X Guo, PK Lala and DT Denhardt, Regulated temporal and spatial expression of the calcium-binding proteins, calcyclin and OPN (osteopontin) in mouse tissues during pregnancy, *Mol Reprod Develop* 32: 315 (1992).
19. Y Miyazaki,M Setoguchi, S Yoshida et al, The mouse osteopontin gene; expression in monocytic lineages and complete nucleotide sequence, *J Biol Chem* 265:14432 (1990).
20. A Oldberg, A Franzen and D Heinegard, Cloning and sequence analysis of rat bone sialoprotein (osteopontin) cDNA reveals an arg-gly-asp cell-binding sequence, *Proc Natl Acad Sci USA* 83:18819 (1986).
21. AM Craig, JH Smith and DT Denhardt, Osteopontin, a transformation-associated cell adhesion phosphoprotein, is induced by 12-o-tetradecanoylphorbol-13-acetate in mouse epidermis, *J Biol Chem* 264:9682 (1989).
22. JL Wrana, Q Zhang and J Sodek, Full length cDNA sequence of porcine secreted phosphoprotein-l (SSP-l, osteopontin), *Nucleic Acids Res* 17:10119 (1989).
23. JM Kerr, LW Fisher, JD Termine and MF Young, The cDNA cloning and RNA distribution of bovine osteopontin, *Gene* 108: 237 (1991).
24. P Castagnola, P Bet, R Quarto, M Gennari and R Cancedda, cDNA cloning and gene expression of chicken osteopontin, *J Biol Chem* 266: 9944 (1991).
25. WT Butler, The nature and significance of osteopontin, *Conn Tiss Res* 23: 123 (1989)
26. DR Senger, CA Peruzzi, A Papadopoulos and DG Tenen, Purification of human milk protein closely similar to tumor-secreted phosphoproteins and osteopontin, *Biochim Biophys Acta* 996: 43 (1989).
27. A Miyauchi, J Alvarez, EM Greenfield et al, Recognition of osteopontin and related peptides by an $\alpha_v\beta_3$ integrin stimulates immediate cell signals in osteoclasts, *J Biol Chem* 266: 20369 (1991).
28. Z Zimolo, G Wesolowski, H Tanaka, JR Hoyer and GA Rodan, The $\alpha_v\beta_3$ integrin is a signal transducing receptor in rat osteoclasts and mouse osteoclast-like cells, *J Bone Min Res* (in press) (1992).
29. JR Hoyer and E Daikhin, Uropontin is a major matrix component of urinary calcium oxalate monohydrate stones, *Con Tiss Res* 27: 188 (1992).
30. JR Hoyer, W Wikoff and SC Bock, Selective incorporation of uropontin into urinary calcium oxalate monohydrate stones, *J Am Soc Nephrol* (in press) (1992).
31. K Hochstrasser, P Reisinger, GJ Albrecht, E Wachter and OL Schonberger, Isolation of acid-resistant urinary trypsin inhibitors by high performance liquid chromatography and their characterization by N-terminal amino-acid sequence determination, *Hoppe-Seyler's Z Physiol Chem* 365: 1123 (1984).
32. Y Nakagawa, M Ahmed, SL Hall, S Deganello and FL Coe, Isolation from human calcium oxalate renal stones of nephrocalcin, a glycoprotein inhibitor of calcium oxalate crystal growth, *J Clin Invest* 79: 1782 (1987).

33. J Lian, EL Prien, MJ Glimcher and PM Gallop, The presence of protein-bound γ-carboxyglutamic acid in calcium-containing renal calculi, *J Clin Invest* 59:1151(1977).
34. AR Spector, A Gray and EL Prien, Kidney stone matrix, differences in acid protein composition, *Invest Urol* 13: 387 (1976).
35. MA Warpehoski, PJ Buscemi, DC Osborn, B Finlayson and EP Goldberg, Distribution of organic matrix in calcium oxalate renal calculi, *Calcif Tissue Int* 33: 211 (1981).
36. K Kohri, Y Suzuki, K Yoshida et al, Molecular cloning and sequencing of cDNA encoding urinary stone protein, which is identical to osteopontin, *Biochem Biophys Res Comm* 184: 859 (1992).
37. JP Gorski, Acidic phosphoproteins from bone matrix: a structural rationalization of their role in biomineralization, *Calcif Tissue Int* 50: 391 (1992).
38. S Kasugai, R Todescan, T Nagata, KL Yao, WT Butler and J Sodek, Expression of bone matrix proteins associated with mineralized tissue formation by adult rat bone marrow cells *in vitro*, *J Cell Physiol* 147: 111 (1991).
39. CA Lopez, JR Hoyer, PD Wilson and DT Denhardt, Heterogeneity of osteopontin expression among nephrons in mouse kidneys. (submitted for publication).

DISTRIBUTION AND QUANTIFICATION OF CRYSTAL MATRIX PROTEIN IN THE HUMAN KIDNEY

A.M.F. Stapleton, A.E. Seymour[1] J.S. Brennan[1], I.R. Doyle, V.R. Marshall and R.L. Ryall

Departments of Surgery and [1]Histopathology
Flinders Medical Centre
Bedford Park, South Australia

INTRODUCTION

Throughout all kidney stones there exists a lattice-work of organic material commonly referred to as the matrix[1]. Its importance in the pathogenesis of stone disease must not be overlooked, for it may hold the key that will allow us to manipulate the natural history of this condition in the future. More than 70% of urinary calculi contain calcium oxalate (CaOx) as their principal component [2], and in these stones, the matrix constitutes approximately 2.5% of the dry weight and is distributed throughout the entire structure[3-5].

Chemical analysis of stone matrix has proved difficult because of its poor solubility[6] and the fact that its dissolution and extraction prior to analysis necessitate the use of aggressive procedures, such as acid hydrolysis, which may alter its chemical structure. Furthermore, stones take varying periods to grow before becoming clinically apparent and are often stored for years prior to analysis[7]. As a result, consequent chemical cross-linking and degeneration of proteins may alter the final composition of matrix so that it bears little resemblance to that prevailing when the core of the stone was initially formed. If this were not enough to undermine the validity of analytical findings, the matter is further complicated by the fact that organic material in the stone's crystalline structure may derive from two distinct sources, namely constituents normally present in the urine and those liberated variably from the epithelial lining or the interstitium of the urinary tract as a consequence of trauma induced by the enlarging stone[8].

Thus, recent attention has turned to the extraction and analysis of organic matrix from the precursors of stones, CaOx crystals, freshly precipitated from whole urine[8,9]. Since these crystals are grown in the absence of substances that could be termed opportunistic contaminants of the growing stone's matrix, their organic matrix should be representative of normal urinary macromolecules associated with CaOx crystal formation in stone pathogenesis.

In an extension of the study by Morse and Resnick[9] we have analysed the protein content of CaOx crystals harvested in this manner and have isolated a urinary protein not previously described. A 31kDa glycoprotein, this has been named crystal matrix protein (CMP)[8]. Unpublished studies from our laboratory have shown that it is a potent inhibitor of CaOx crystal aggregation, and is present in CaOx crystals in quantities far in excess of those expected from its concentration in urine. This lends support to previous findings[9] that the incorporation of urinary proteins into CaOx crystals is a highly-selective phenomenon.

Urolithiasis 2, Edited by R. Ryall *et al.*,
Plenum Press, New York, 1994

The anatomical origin of CMP is unknown. Doyle et al[8] did not detect CMP in human serum using an enzyme-linked immuno-absorption assay (ELISA), indicating that it may be produced and secreted within the urinary tract. Since the protein may be important in the control of stone pathogenesis and, perhaps even have a role in renal physiology and homeostasis, the aims of this study were: a) to delineate the distribution of CMP within the human urinary tract; b) to investigate its presence in other human tissues; and c) to assess histologically the quantity of CMP in men, women, and CaOx stone formers.

MATERIALS AND METHODS

Antibodies

Crystal matrix protein (CMP) was isolated from urine collected over a 48 h period from healthy volunteers and processed as described by Doyle et al[8]. Pre-immune plasma was bled from each of three New Zealand white rabbits before they were immunized. The antiserum was pre-absorbed with human serum immuno-absorbent[10] to reduce non-specific background staining. The specificity of this antiserum was assessed by its use as a primary antibody in the immunochemical staining of Western blots[11] of proteins (including Tamm-Horsfall glycoprotein [THG]) separated by sodium dodecy/sulphate polyacrylamide gel electrophoresis (SDS-PAGE) from human urine, serum, and crystal-matrix extract. Epithelial Membrane Antigen (EMA) was detected in kidney specimens using a monoclonal goat anti-human EMA antibody from DAKOPATTS (DK-2600 Glostrup, Denmark).

Tissue

Fresh human kidney was obtained from 10 nephrectomy specimens removed because of renal cell carcinoma or transitional cell carcinoma of the renal pelvis[3]. This was obtained from seven males and three females with a median age of 74 years (range 57-86 years). Macroscopically, normal tissue was sampled away from the site of known pathology and, subsequently, shown by light microscopy to be within normal morphological limits. Each specimen was processed as follows:

a) Tissue blocks not more than 3 mm thick were embedded in cryomoulds containing OCT Compound (Miles Laboratories), quenched in an isopentane bath cooled in liquid nitrogen, and stored at -70°C.

b) Tissue blocks not more than 5 mm thick, were fixed in 10% buffered formalin (pH 7.4) for 12-24 h and embedded routinely in paraffin wax.

Archival renal tissue (formalin-fixed paraffin-embedded) was obtained from the routine tissue blocks of nephrectomy specimens from 35 individuals. Adult human tissue other than kidney was gathered from stored specimens or was obtained from fresh samples following surgical resection.

Immuno-histochemistry

Frozen tissue sections (5 μm thickness) were cut using a cryotome, placed on chrome-alum gelatin-coated slides, air dried, and washed in Tris-buffered saline, pH 7.6 (TBS) prior to immuno-staining. Paraffin sections (5 μm thickness) were cut using a standard microtome, placed on acid-cleaned glass slides, de-paraffinized in xylene, and passed through graded alcohol before placing in TBS prior to immuno-staining. Endogenous peroxidase activity was blocked by incubating sections for 10 minutes in a solution of 0.1% H_2O_2 in 100% ethanol.

Each primary antibody (Ab) was diluted in the normal serum of the species used to generate the secondary Ab, and the sections were incubated in a humid chamber overnight at 4°C. All subsequent incubations were performed at room temperature. The anti-CMP Ab was routinely diluted 1:400 to 1:800, and the anti-EMA Ab diluted 1:1000.

For single staining, the avidin-biotin immuno-peroxidase technique was used as previously described by Hsu et al[13] using a Vectastain ABC Kit (Vector Laboratories, Burlingham, CA). The chromogen used was 3:3-diaminobenzidine in the presence of

0.024% hydrogen peroxide to give a brown, insoluble precipitate which was stopped with water after 3 min. Sections were counter-stained in Mayer's haematoxylin, acid differentiated, blued in lithium carbonate, dehydrated, and mounted in PIX.

Sections were double-labelled using a combination of the immuno-peroxidase described above using an alkaline-phosphatase method. The immuno-peroxidase/DAB technique was used first. After immuno-labelling as detailed above, the tissue was incubated overnight with the second primary Ab. After incubation with the second biotinylated-linking Ab directed against the second primary Ab, 0.5% streptavidin alkaline phosphatase was applied for 30 min. Naphthol AS-TR phosphate was dissolved in dimethyl formamide and added to Gomari's buffer (pH 9.8) containing 0.03% levamisol. The chromogen, Fast Red TR Salt, was added to give a characteristic red reaction which was stopped after 10 minutes by adding excess water. Sections stained by this method were subsequently air dried and mounted in Locktite which was polymerised under ultraviolet light for 15 min. Between each of the incubation steps, sections were washed three times in TBS for 5 min.

Light microscopy was performed using a Leitz-Wetzler-Diaplan microscope, and micrographs taken using a Wild-MPS-45 camera system with Kodak 35 mm 50ASA professional colour film.

Controls

All sections were run in parallel with a negative control of pre-immune rabbit serum (diluted 1:400) for the anti-CMP Ab, and TBS for the anti-EMA Ab. Each batch was accompanied by at least one renal section that had been consistently found to be positive for CMP.

The anti-CMP Ab was assessed by an affinity-absorption test, whereby aliquots of CMP, purified by gel filtration and noted to display a single band on SDS-PAGE, were incubated with the Ab for 2 h at 37°C, and overnight at 4°C. Maximum concentrations of CMP were estimated to be in excess of the anti-CMP Ab by a factor of 50-100x. The samples were then centrifuged at 300 rpm for 15 min, and the supernatant was applied to known positive renal tissue and processed as described above. Included in this batch was a section incubated with an appropriately diluted sample of anti-CMP Ab without added antigen to act as the positive control.

Quantification of CMP

Subjective blended, quantification of intracellular CMP was performed independently by two of the authors (AMFS and AES) examining single immuno-peroxidase labelled sections using a direct visual analogue system from 0 (absence of staining), through 1+, 2+, 3+, and 4+ (indicating heavy staining). Each score represents the sum of assessments of the distal convoluted tubule (DCT) and the thick ascending limb of Henle's loop (TALH). Forty-five kidneys were assessed, and the data were analysed using the Wilcoxon Rank Sum Test to allow comparison between men and women and between healthy subjects and CaOx stone formers.

RESULTS

Immuno-absorption using human serum had no perceptible effect on the titre of the CMP antiserum. However, the specificity of the antiserum was significantly enhanced by the procedure. There was no cross-reactivity with urinary THG.

The sections incubated with pre-immune serum, which were used as negative controls in each batch, were unstained. Absorption of the anti-CMP Ab to purified antigen eliminated labelling.

Immunohistochemistry

The EMA was confirmed to be distributed along the apical membranes of the distal convoluted tubule and collecting duct epithelia, and absent from the glomerulus, proximal tubule, and the loop of Henle, as previously reported[13,14].

At appropriate dilutions of the anti-CMP Ab, background staining was low and the signal high. Staining for CMP was evident in all but one of the kidney specimens (98%). In the renal cortex, CMP was found in the cytoplasm of the epithelial cells of DCT. Double staining with antibodies to EMA and CMP confirmed co-localisation in DCT epithelia. In the medulla, single immuno-peroxidase labelling of CMP showed it to be in TALH, and this was confirmed using double-labelling with antibodies to CMP and EMA to immuno-dissect the tubules. Interestingly, CMP did not stain all nephrons. In fact, staining was heterogeneous amongst the nephron population such that, in any one individual, between 10% and 60% of the nephrons were positive. The proportion of positively-stained nephrons was not changed by manipulating the conditions for immuno-labelling, for example, Ab dilution or time of development of the end product. No one group, such as cortical or juxtamedullary nephrons, was consistently stained. Material within the lumen distal to the TALH was also noted to be positive for CMP, as one would expect with release of CMP into the urinary space. The CMP was consistently demonstrated in the cells of the macula densa in those nephrons that were positively stained elsewhere.

CMP was not detected, despite using varying Ab dilutions, in 27 separate human organs examined: renal pelvis, ureter, bladder, prostate, testis, epididymis, heart, lung, liver, spleen, pancreas, jejunum, uterus, ileum, cerebral cortex, colon, skeletal muscle, skin, thyroid, parathyroid glands, bone marrow, tonsil, stomach, adrenal, breast, smooth muscle, and oesophagus.

Quantification of CMP in Renal Tissue

The quantification data of intracellular CMP are shown in Fig. 1. Significantly less CMP was observed in normal women compared with stone formers ($p< 0.01$), and normal men compared with stone formers ($p< 0.01$). The difference in the amount of CMP detected in kidneys from normal males, as opposed to normal females, failed to reach statistical significance ($p< 0.11$), although the trend suggested less CMP in the latter group.

DISCUSSION

The consistent presence and disproportionate abundance of CMP within the structure of CaOx crystals generated from human urine suggest that the protein may fulfil an active role in CaOx crystallization and perhaps, therefore, in CaOx stone disease. This possibility is supported by other data from our laboratory which show that CMP is a potent inhibitor of CaOx crystal aggregation and growth.

Using a polyclonal antibody, we have demonstrated the presence of CMP in the tubular epithelia of the TALH and the DCT of the human kidney. The CMP was also clearly seen within the cells of the macula densa. The location of CMP within the lumen adjacent and distal to these sites suggests that the protein is produced by and secreted from these cells into the urinary collecting system. Furthermore, the protein appears to be specific to the kidney, with none being detected elsewhere in the urinary tract or in other human tissues.

This very limited and precise location, both within the body and within the kidney, lends credence to the possibility that CMP may fulfil some regulatory role in stone pathogenesis. Given that calcium transport is significantly controlled in this section, and that luminal fluid is progressively concentrated dependent upon hydration both here and distally, calcium and oxalate concentrations may be maximal in the collecting ducts. Therefore, it would be physiologically prudent to minimise the risk of crystal precipitation in the collecting ducts by releasing an inhibitor of CaOx crystallisation into the luminal fluid immediately upstream in the nephron. However, although CMP would appear to fulfil both the anatomical and physicochemical criteria of a critical macromolecular determinant of CaOx stone disease, other urinary proteins can also claim such distinction, and it is salient to compare their properties with those of CMP.

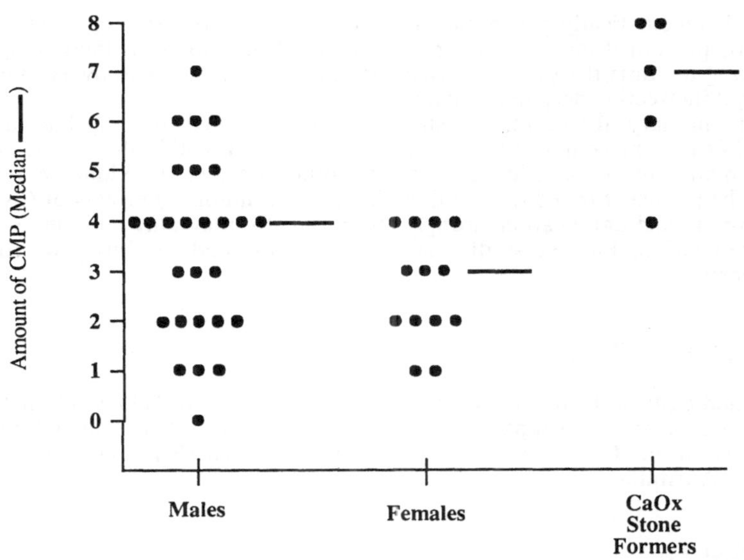

Figure 1. The amount of CMP in kidney tissue derived from normal men and women and from CaOx stone formers.

Nephrocalcin (NC) is a 14-kDa glycoprotein that has been shown to inhibit CaOx crystal growth in inorganic solutions, but whose effects in undiluted urine have not been reported[15]. Using a rabbit polyclonal Ab, NC has been shown to be present in the proximal tubules and the TALH epithelia of both man and mouse kidney[16]. Their dissimilar distributions within the nephron and the disparity in their molecular weights suggest that they are different proteins and probably fulfil different functions.

More recently, another potentially important urinary protein, uropontin, has been described[17]. The macromolecule is rich in aspartate and is an inhibitor of CaOx crystal growth in inorganic solutions. There is indirect evidence that uropontin is synthesised by cells in the proximal and distal tubules, and in the loops of Henle in mice[17,18]. Uropontin has an Mr of 55 kDa and lacks Gla[17], features that clearly distinguish it from CMP.

CMP was detected in significantly greater amounts in sections of kidneys from stone formers than in those from individuals unaffected by the disease and, although the difference failed to attain statistical significance, the data also demonstrated a trend for normal males to have greater quantities of CMP than normal females. The observed quantity of CMP, therefore, mirrors the risk of stone disease, such that an increased amount of CMP correlates with an increased incidence of stone disease. These data appear to conflict with the fact that CMP is a potent inhibitor of CaOx crystal aggregation where one might expect a greater risk of stone disease to result from a decreased level of urinary inhibitors. Several explanations can be proposed for this apparent contradiction.

First, the molecular structure, and therefore the functional activity, of CMP might differ between the three groups. Evidence to support this comes from our recent observations that there are differences in the amino-acid composition of CMP between these groups; in particular, Gla is reduced in the CMP derived from urine of recurrent stone formers. This could alter the binding affinity of CMP for the surface of CaOx crystals and, thereby, lessen its inhibitory efficacy. Second, release of CMP may be regulated by a feedback-control mechanism, such that production and secretion of the protein is stimulated under conditions of high urinary supersaturation which predispose to crystal nucleation. This hypothesis is consistent with a protective function for the protein. The increased quantity of the protein in the renal tissue of normal men may account for their resistance to stone formation in the presence of a degree of urinary supersaturation which is characteristically greater than that in women. In stone formers, on the other hand, it might be presumed that though the production of the protein is increased, the urinary saturation at the time the stone was first formed was so great as to overwhelm the CMP, rendering it ineffective despite its increased production.

In summary, therefore, this study has demonstrated the distribution of a newly described urinary protein, crystal matrix protein, (CMP), within the TALH and the DCT of the human nephron, and its absence from other areas of the kidney and many other organs. The precise, limited distribution, and known inhibitory effects of CMP on CaOx crystal growth and aggregation strongly suggest that it may play a pivotal role in stone-matrix deposition. Further studies are clearly warranted to clarify its role in stone pathogenesis.

ACKNOWLEDGEMENTS

Figure 1 and parts of the text are reprinted with permission of Kidney International (paper in press). This work was supported by grants from the Australian Kidney Foundation, the Royal Australasian College of Surgeons and the National Health and Medical Research Council of Australia.

REFERENCES

1. AJ Butt, Historical survey of etiological factors in renal lithiasis, *in* "Etiological Factors in Renal Lithiasis', AJ Butt, ed, Thomas, Springfield (1962).
2. EL Prien and EL Prien Jr, Composition and structure of urinary stone, *Am J Med* 45:654 (1968).
3. WH Boyce and FK Garvey, The amount and nature of the organic matrix in urinary calculi: a review, *J Urol* 76:213 (1956).
4. MA Warpehoski, PJ Buscemi, DC Osborn, B Finlayson and EP Goldberg, Distribution of organic matrix in calcium oxalate renal calculi, *Calcif Tissue Int* 33:211 (1981).
5. J Stacholy and EP Goldberg, Microstructural matrix-crystal interactions in calcium oxalate monohydrate kidney stones, *Scanning Electron Microscopy* II: 781 (1985).
6. WH Boyce, JS King and ML Fielden, Total non-dialyzable solids (TNDS) in human urine. XI. Immunological detection of a component peculiar to renal calculous matrix and to urine of calculous patients, *J Clin Invest* 41:1180 (1962).
7. H Iwata, O Kamie, Y Abe, S Nishio, A Wakatsuki, K Ochi and M Takeuchi, The organic matrix of urinary uric acid crystals, *J Urol* 139:607(1988).
8. IR Doyle, RL Ryall and VR Marshall, Inclusion of proteins into calcium oxalate crystals precipitated from human urine: a highly selective phenomenon, *Clin Chem* 37:1589 (1991).
9. RM Morse and MI Resnick, A new approach to the study of urinary macromolecules as a participant in calcium oxalate crystallisation, *J Urol* 139:602 (1988).
10. S Avrameas and T Temynck, The cross-linking of proteins with glutaraldehyde and its uses for the preparation of immunoadsorbents, *Immunochemistry* 6:53 (1969).
11. H Towbin, T Staehelin and J Gordon, Electrophoretic transfer of proteins from polyacrylamide gels to nitrocellulose sheets: procedures and some applications, *Proc Natl Acad Sci* 76:4350(1979).
12. SM Hsu, L Raine and H Fanger, Use of avidin-biotin-peroxidase complex (ABC) in immuno-peroxidase techniques. A comparison between ABC and unlabelled antibody (PAP) procedure, *J Histochem Cytochem* 29:577 (1981).
13. S Fleming, GBM Lindop and AAM Gibson, The distribution of epithelial membrane antigen in the kidney and its tumours, *Histopath* 9:729 (1985).
14. JP Sloane and MG Ormerod, Distribution of epithelial membrane antigen in normal and neoplastic tissues and its value in diagnostic tumor pathology, *Cancer* 47:1786(1981).
15. Y Nakagawa, V Abram, FJ Kezdy, ET Kaiser and FL Coe, Purification and characterization of the principal inhibitor of calcium oxalate monohydrate crystal growth in human urine, *J Biol Chem* 258:12594 (1983).
16. Y Nakagawa, M Netzer and FL Coe, Immuno-histochemical localization of nephrocalcin (NC) to proximal tubule and thick ascending limb of Henle's loop (TAHL) of human and mouse kidney: resolution of a conflict, *Kidney Int* 37: A474 (1990).
17. H Shiraga, W Min, WJ VanDusen, et al, Inhibition of calcium oxalate crystal growth *in vitro* by uropontin: another member of the aspartic acid-rich protein superfamily, *Proc Natl Acad Sci, USA* 89:426 (1992).
18. S Nomura, AJ Wills, DR Edwards, JK Heath and BLM Hogan, Developmental expression of 2ar (osteopontin) and sparc (osteonectin) rna as revealed by *in situ* hybridisation, *J Cell Biol* 106:441 (1988).

LOCALIZATION OF NEPHROCALCIN IN THE NEPHRON

M.F. Netzer, Y. Nakagawa and F.L. Coe

Nephrology Program and Department of Biochemistry
and Molecular Biology
University of Chicago
Chicago IL, USA

INTRODUCTION

Nephrocalcin (NC) is known as a major inhibitor of crystal aggregation and growth in the renal tubular system. Therefore, the cellular origin of this protein is of great interest to understand its physiological function in the nephron. Previous attempts at NC immuno-localization have been unsuccessful because of cross-reactivity of the antisera to Tamm-Horsfall protein (THP). The current study utilizes a new antibody to NC which confirms localization of the antigen to the thick ascending limb of Henle's loop (TAHL) and proximal tubule.

MATERIALS AND METHODS

Nephrocalcin was purified from medium conditioned with a human renal-cell line as described previously[1]. New Zealand Albino female rabbits (~ 4 kg body weight) were immunized with 130 μg of NC emulsified in 1 mL of complete Freund's adjuvant (Sigma). The rabbits were boosted using the same amount of antigen emulsified in incomplete Freund's adjuvant (Sigma) at the second and fourth weeks after the first immunization. The sera, 50 mL each, were collected at the sixth week by ear-vein puncture using a bleeding apparatus (Bellco Glass Inc.) with a vacuum pump. All immunological tests used in this report were performed using a direct Enzyme Linked Immunosorbent Assay (ELISA). Standard 96-well microtiter plates (Nunc) were coated overnight at 4° C, with 1.0 μg per well of a test protein in 100 μL PBS (0.01 M sodium phosphate buffer, pH 7.4, containing 0.15 M NaCl), the solution was decanted and wells were blocked with 400 μL of 0.25% gelatin in PBS for 1 h at 37°C. The plates were washed three times with distilled water and PBS, then 100 μL of antiserum, 1:5,000 dilution in Tween/PBS (0.05% Tween 20 (Sigma) in PBS), was added and incubated for 1 h at 37°C. After the plate was washed with distilled water and PBS, 100 μL of biotinylated goat anti-rabbit IgG (Vector Laboratories), 1:1,000 dilution in Tween/PBS, was added, then incubated for 30 min at 37°C. After washing the plate with water and PBS, 100 μL of Avidin-Biotin conjugated to horseradish peroxidase (Vector ABC-kit, 1:500 in Tween/PBS) was added and incubated at 37°C for 10 min. Finally, 100 μL of 0.1% o-phenylenediamine and 0.01% hydrogen peroxide in 0.05 M sodium citrate buffer, pH 4.2, was added to each well. Developed color was measured at 450 nm by a microplate reader (Dynatech MR 700, Dynatech Laboratories). The obtained antisera showed no cross reactivity against purified THP, human albumin (Sigma), chondroitin sulfate (Sigma), hyaluronic acid (Sigma), and β_2-microglobulin (Pharmacia-LKB) (Fig.1). Also plasma α_1-antitrypsin inhibitor (Boehringer) and urinary antitrypsin inhibitor isolated from human urine by the method of Hochstrasser and Wachter in this laboratory were negative against the antibody described in this report.

Immunohistochemistry

Human kidney tissue obtained at surgery was immediately fixed in 10% formaldehyde and embedded in paraffin. Microsections were prepared by a microtome and placed on a glass slide, de-paraffinized in 100% xylene for 10 sec, then rehydrated by successive washing for 10 sec with ethanol solution by decreasing 10% at each step from 100% to 70%. Endogenous peroxidase was blocked using 3% hydrogen peroxide in methanol for 30 min.

Sections were incubated in 10% non-immune goat serum (NGS, Vector Laboratories) in PBS buffer for 90 min to avoid non-specific antibody binding. The samples and controls were incubated in anti-NC antibody solution, 1:1,000 in PBS, containing 1% NGS, for 12 h at 4°C on a shaker, then washed three times with PBS. The slides were incubated in 1:50-diluted biotinylated goat anti-rabbit IgG (Vector) at 37°C for 60 min. After rinsing with PBS three times, the slides were incubated in 1:500 diluted Avidin-Biotinylated horseradish peroxidase complex (ABC kit, Vector) in PBS for 30 min followed by 0.02% 3,3'-diaminobenzidine (DAB, Sigma) and 0.3% hydrogen peroxide in PBS. Color development was stopped by washing with PBS three times. The slides were counterstained by Mayers' hematoxylin (Fisher Scientific) for 2.5 min, then dehydrated by increasing ethanol concentration from 70% to 100% in 10% increments each for 10-sec immersion, and finally in 100% xylene for 10 sec. The slides were mounted with Permont (Fisher).

Transmission electron microscopy

Fresh kidney tissue from the surgical specimen was diced into 0.5 x 0.5 x 0.3 cm in the operating room and immediately frozen in acetone-dry ice mixture and stored at -20°C. These specimens were cut by a microtome and fixed in 10% buffered formalin (pH 7.2-7.4) for 6 h at 4°C, then rinsed and kept in PBS for further staining.

Sections were immersed in 0.05% saponin-PBS (Fisher) for 30 min, then incubated in anti-NC antiserum in 10% NGS (1:1,000) for 12 h at 4°C. The control sections were incubated in 10% NGS and non-immune rabbit serum (Vector) diluted 1:1,000 in PBS for 1 h at 37°C followed by ABC diluted 1:500 in PBS for 30 min. The sections were fixed with 1% glutaraldehyde (Pella Inc.) for 30 min, and incubated in 0.05% DAB in PBS for 20 min, 0.3 mL of 1% hydrogen peroxide in 10 mL of DAB solution was added. After 20 min incubation, specimens were washed in fresh PBS, then fixed in 2.5% glutaraldehyde for 12 h. Sections were rinsed with Mollonig's buffer (sodium phosphate buffer, pH 7.2), and fixed in 1% osmium tetraoxide (Lanum Co.), then washed three times in 0.2 M collidine buffer, and transferred to 50% ethanol. Thereafter, the sections were progressively dehydrated by increasing ethanol concentration by 10% to 90%, then 5% each up to 100%. After 10 min in 100% ethanol, the specimens were washed twice in 100% propylene oxide (Baker Inc.) and immersed in a solution of propylene oxide and plastic (1:1, v/v) mixture (5 mL Medcast resin, 3 mL nadic methyl anhydride and 2 mL dodecanyl sulfinic anhydride, Pella Inc.) for 30 min, and transferred to propylene oxide:plastic mixture for 1 h. Finally, the samples were embedded in 100% plastic mixture, polymerized for 12 h at 60°C, and thick sections (1 μm) were cut into 0.25 μm using a Porter Blum MT 1 microtome (Sorvall).

The slide sections were placed on mesh grids (Lanum), and double stained for 20 min by incubation in a saturated solution of uranylacetate (Fisher) in 5% ethanol, and the grids were washed three times with water. After drying over a filter paper, sections were stained with lead citrate for 2 min. After being rinsed with water and dried over a filter paper, the specimens were examined using an electron microscope (Phillips Model PW 6006).

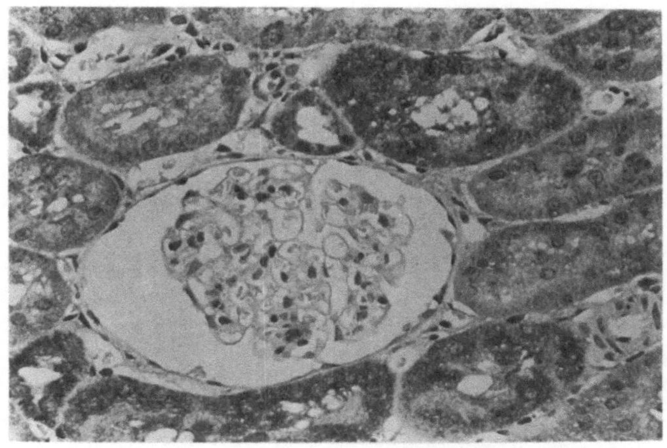

Fig 1. Proximal tubules in a cortical area with positive immunostaining for NC.

RESULTS

The results indicated that the antibody raised in rabbits was specific for NC in human urine but not for other major urinary proteins; thus, we applied this antibody to localize a NC secretion site in kidneys using immuno-histochemical staining techniques. Light microscopy showed brown pigment indicating DAB from a reaction with the anti-NC antibody deposited in the cytoplasm of proximal tubule cells (Fig. 1) and in the TAHL (Fig. 2). Glomeruli, vessels and their lumens, and interstitial tissues remained unstained, but some intratubular debris did stain.

Immuno-electron microscopy showed pigment deposit most strongly on the luminal fringe of the proximal-tubule-cell brush borders, but also small ones between microvilli (Fig. 3). Staining occurred in rough endoplasmic reticulum (Fig. 4) and golgi apparatus (Fig. 5).

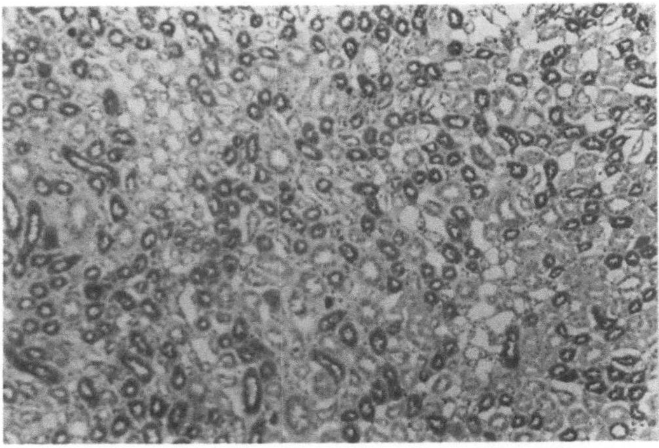

Fig 2. Thick ascending limb of Henle's loop (TAHL) with a strong intracellular antibody reaction.

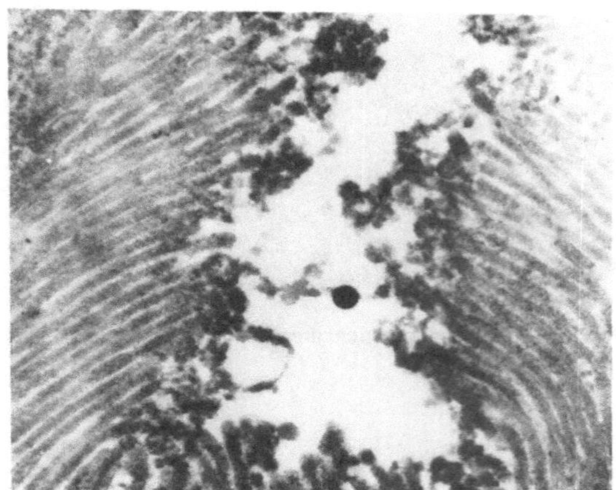

Fig 3. Black dots indicate NC deposits on microvilli.

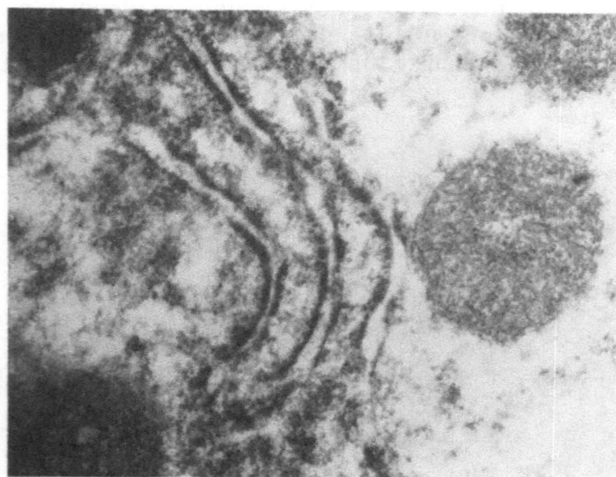

Fig 4. The presence of NC is seen in the rough endoplasmic reticulum.

DISCUSSION

We have immunized rabbits with NC isolated from human renal-cell tissue-culture medium. These antisera obtained from two immunized rabbits were specific to human NC and did not cross-react with THP, albumin, chondroitin sulfate, hyaluric acid, human plasma and urinary antitrypsin inhibitor, and ß-2-microglobulin. We previously reported that NC is of kidney origin[2], and now we are able to demonstrate the localization of NC in the proximal tubule cells and TAHL. Previously, NC was thought to localize in distal tubules[3]. However, the antibody used in that study cross-reacted to THP and the localization of NC was, therefore, uncertain as THP is produced in distal tubule cells. We raised antisera specific to NC but not to major urinary proteins and confirmed the localization of NC in human kidneys. Electron microscopy confirmed that NC-antibody staining was observed in the cytoplasm of the proximal tubule and TAHL. The results of electron microscopy studies showed the presence of NC in the rough endoplasmic reticulum and golgi apparatus. Since NC is found in urine, it is likely that after synthesis in the cytoplasm, it is transported to brush borders and thence, into the tubular lumen as suggested by electron microscopic images.

We conclude that cells of both proximal tubule and TAHL contain NC-immunoreactive material. These cells produce NC and secrete this inhibitor into at least two separate tubule sites as a defense against crystallization.

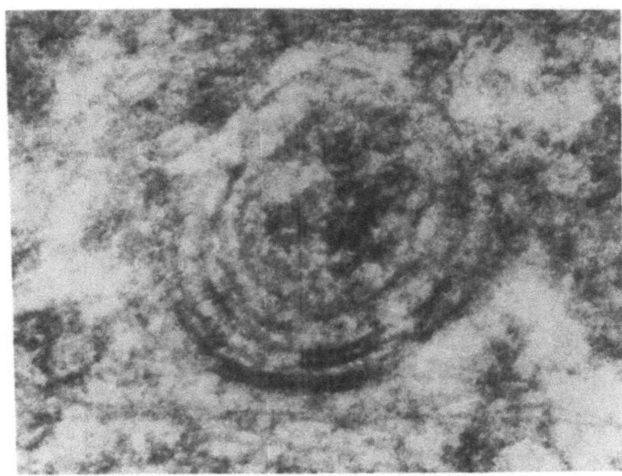

Fig 5. Golgi apparatus with positive immuno-cytochemical reaction.

ACKNOWLEDGEMENTS

The authors thank Dr B Spargo of the Pathology Department for his valuable discussions and critical comments. We also thank Siok Le Dun and Mark P Janulis for their excellent technical assistance. This research was supported by Deutsche Forschungsgemeinschaft grant NE 356/1-1 and National Institutes of Health grant 5P01 AM 33949.

REFERENCES

1. Y Nakagawa, HC Margolis; S Yokoyama, FJ Kezdy, ET Kaiser, FL Coe, Purification and characterization of a calcium oxalate monohydrate crystal growth inhibitor from human kidney tissue culture medium, *J Biol Chem* 256: 3936 (1981).
2. D Sirivong, Y Nakagawa, WK Vishny, MJ Favus and FL Coe, Evidence that mouse renal proximal tubule cells produce nephrocalcin, *Am J Physiol* 257: F397 (1989).
3. M Lopez, Y Nakagawa, FL Coe, C Tsai, AF Michael, JI Scheinman, Immunochemistry of urinary calcium oxalate crystal growth inhibitor, *Kidney Int* 29: 827 (1986).

IN UNDILUTED HUMAN URINE TAMM-HORSFALL MUCOPROTEIN MITIGATES THE PROMOTION OF CALCIUM OXALATE CRYSTALLIZATION INDUCED BY URATE

P.K. Grover, V.R. Marshall and R.L. Ryall

Department of Surgery
Flinders Medical Centre
Bedford Park, South Australia

INTRODUCTION

We have previously reported that dissolved urate promotes calcium oxalate (CaOx) crystallization by enhancing crystal nucleation, bulk deposition and aggregation[1,2]. All these experiments were carried out either with spun and 0.22 μm filtered or 10 kDa ultrafiltered urines which are devoid of Tamm-Horsfall mucoprotein (THM). However, THM is a natural macromolecular component of whole urine and inhibits CaOx crystal aggregation[3]. Thus it is quite possible that the promotory effect of dissolved urate we observed may not necessarily occur *in vivo* where the protein is present in large amounts.

Therefore, the aim of this investigation was to determine whether a physiological concentration of THM mitigates or cancels the promotion of CaOx crystallization induced by dissolved urate.

MATERIALS AND METHODS

THM was prepared from normal human urine by a slight modification of the method of Tamm and Horsfall as described in detail elsewhere[3].

24-hour urine specimens were collected without preservative from 20 healthy men and 10 kDa ultrafiltered as described previously[2]. Each sample was then divided to form experimental and control portions. A solution of THM in distilled water was added to the experimental aliquots to give a final concentration of 35mg/L, while an equivalent volume of distilled water was added to the controls. Both the experimental and control portions of each urine were then spiked with identical amounts of sodium urate solution as described earlier[1] and CaOx crystallization was induced by the oxalate load technique[4]. Because THM interferes with crystal counting using the Coulter Counter[3], the metastable limits of the test samples were determined both by using the Coulter Counter and examining the samples for CaOx crystals under the light microscope. The median urate concentration of urines was increased from 2.3 to 6.1mmol/L ($p < 0.01$).

At the end of each experiment, lmL aliquots were 0.22μm filtered and prepared for scanning electron microscopy as described previously[1]. Data were compared using the Wilcoxon signed rank sum test.

RESULTS

Addition of THM did not significantly change the median metastable limit of urate-enriched ultrafiltered urines. Therefore, to study the effect of THM on particle size and bulk deposition of crystalline material, identical amounts of oxalate were added to each of the control and test urine samples.

Different protein concentrations were incubated with COM crystals under supersaturated conditions in 50 mM HEPES buffer, pH 7.4 at 37°C. The crystals were washed with the buffer and incubated with a monoclonal anti-THP antibody. Crystals with bound protein were examined under scanning electron microscopy using protein A gold detection.

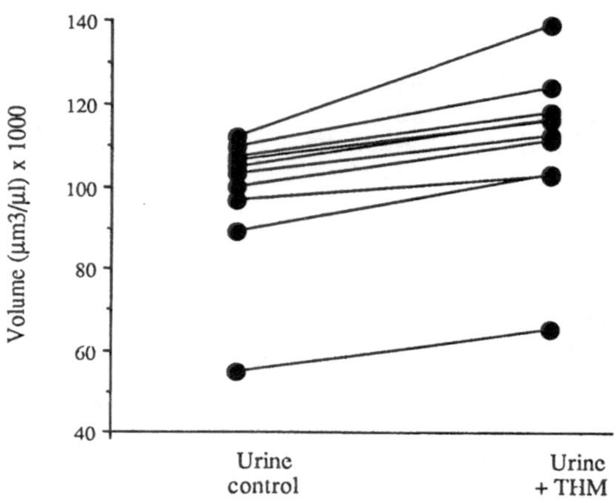

Figure 1. The volume of particulate material deposited following the addition of a standard load of oxalate in control urines and samples of the same urines enriched with THM.

Fig 1 shows the volume of the particulate material deposited in the presence and absence of THM. The median volume increased slightly, but significantly ($p<0.01$) from 104,000$\mu m^3/\mu L$ in the absence of THM to 114,000$\mu m^3/\mu L$ in its presence - a rise of approximately 9.6%. Experiments performed with ^{14}C-oxalate showed that the increase in volume was not a result of increased CaOx deposition.

The presence of THM significantly ($p < 0.01$) reduced the particle size from the median value of 14.3 to 10.7µm (fig 2). These findings were confirmed by scanning electron microscopy.

Figure 3 exemplifies the crystalline material deposited in the absence and presence of added THM. These scanning electron micrographs show the exclusive deposition of CaOx monohydrate crystals in both portions of the urine. The individual crystals precipitated in the presence of THM were large and not as aggregated as in its absence.

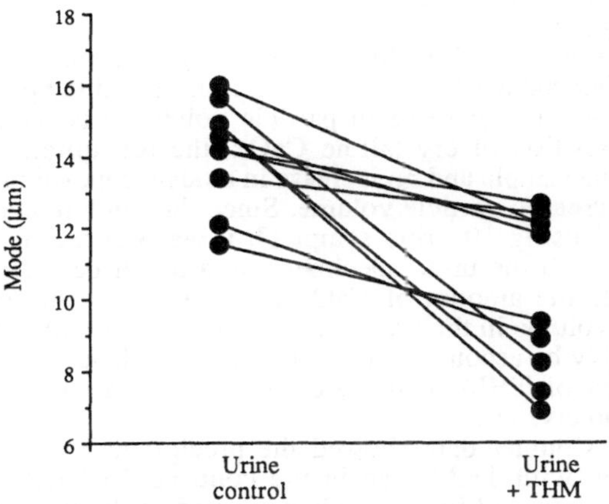

Figure 2. The size of the particles, expressed as the position of the mode of the volume distribution curve in control urines and samples of the same urines which had been supplemented with THM.

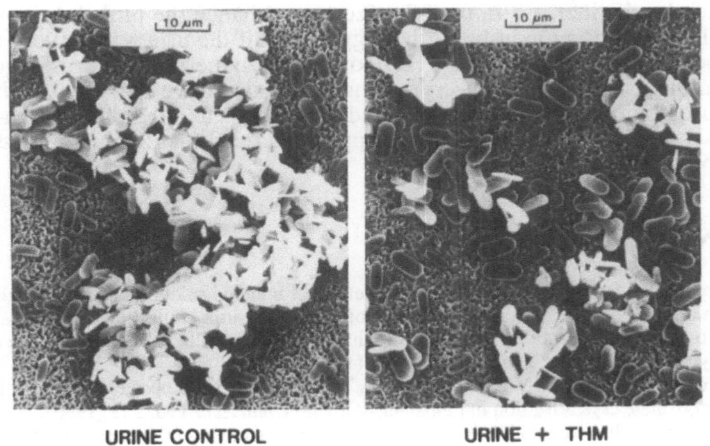

Figure 3. A typical scanning electron micrograph of the material precipitated from one of the urine samples.

DISCUSSION

Results presented earlier revealed that under conditions likely to be operating *in vivo* THM inhibits crystal aggregation[3]. The aim of this study was to determine whether THM can reduce or prevent the promotory effect of urate most importantly, its effect on CaOx crystal aggregation.

Though the presence of THM did not change the minimum amount of oxalate required to induce detectable crystal nucleation, it increased the particle volume by 9.6%. To determine whether this increase in particle volume was reflective of a significant increase in the deposition of crystalline CaOx, the precipitation of ^{14}C-oxalate was measured in one urine sample and no increase in oxalate deposition was observed, despite the finding of an increased particle volume. Since this finding confirmed what we have previously observed using 10 urine samples[3], it was not felt necessary to check ^{14}C-oxalate deposition in all the urine specimens, and it can be reasonably concluded that THM does not alter the amount of CaOx precipitated in the presence of urate. The increase in particle volume in the presence of THM, as observed by the Coulter Counter, can therefore probably be accounted for by the same explanations as offered earlier[3] - an increased deposition of THM in the presence of crystals or detection of composite particles of THM and crystals.

The Coulter Counter data showed the precipitation of smaller particles in the urines supplemented with THM than in the controls. In keeping with this, scanning electron microscopy revealed that crystals deposited in the presence of THM were less aggregated than in its absence. But once again the *individual* crystals deposited in the presence of THM were larger than in its absence.

The results presented here show that addition of THM to urate-enriched ultrafiltered urines did not affect the amount of oxalate required to induce detectable spontaneous crystal nucleation or the volume of crystalline material deposited, but significantly inhibited crystal aggregation. It can be inferred from this that, *in vivo* the presence of THM would not alter the likelihood of CaOx crystal precipitation induced by urate, or the amount of crystalline material deposited. It would reduce but not prevent the formation of large CaOx crystal aggregates. Thus in patients excreting high levels of urate, and urine relatively rich in calcium and oxalate, this increased bulk deposition, accompanied by the increase in particle size may enhance the probability of renal tubule blockage. This situation might be further aggravated if the urines were very concentrated, because under these circumstances THM could be polymerized to form large flocs which would promote the deposition of crystalline material[3] above and beyond the promotion already caused by urate, and might also bind the crystals together into large agglomerates. The whole complex might then become trapped in a renal tubule, thereby setting the stage for stone formation.

REFERENCES

1. PK Grover, RL Ryall and VR Marshall, The effect of urate on calcium oxalate crystallization in human urine: Evidence for a promotory role of hyperuricosuria in urolithiasis, *Clin Sci* 79: 9 (1990).
2. PK Grover, RL Ryall and VR Marshall, Calcium oxalate crystallization in urine: The role of urate and glycosaminoglycans, *Kidney Int* 41: 149 (1992).
3. PK Grover, RL Ryall and VR Marshall, Does Tamm-Horsfall mucoprotein inhibit or promote calcium oxalate crystallization in human urine?, *Clin Chim Acta* 190: 223 (1990).
4. RL Ryall, CM Hibberd and VR Marshall, A method for studying inhibitory activity in whole urine, *Urol Res* 13: 285 (1985).

EVIDENCE THAT URIC ACID ABOLISHES CITRATE-MEDIATED INHIBITION OF CALCIUM OXALATE MONOHYDRATE AGGLOMERATION *IN VITRO*

S.M. Volta, E.E. Fradinger, C.E. Bogado and J.R. Zanchetta

Instituto de Investigaciones Metabolicas
Buenos Aires, Argentina

INTRODUCTION

Calcium oxalate monohydrate (COM) crystal growth and agglomeration in urine are modulated by a complex set of chemical and physical properties of the medium. In such a chemically complex milieu, where many effects of these properties can mask each other, it is virtually impossible to identify and/or quantify them separately. However, considerable advantage can be gained in the understanding of the mechanisms involved by testing the effects of some of the species present in urine in isolation, or in combination with only a small number of other urinary constituents.

Since the theoretical possibility of the epitaxial deposition of COM upon uric acid or sodium urate was first proposed[1], a considerable amount of clinical and laboratory research has been devoted to investigating the phenomenon. From the use of computer programs and direct solubility experiments general agreement has evolved that solid uric acid or urates exhibit a complete lack of any action on the solubility of COM in urine and inorganic solutions[2,3]; however, no conclusive evidence has come to light regarding any possible possible effect of uric acid or urates on the metastable limit of this salt.

A seeded crystal system which uses a tracer as an indicator of crystal growth kinetics was chosen. The method is described in detail elsewhere[4,5]. In our opinion, although somewhat cumbersome, this is one of the most accurate and better described available methods from a mathematical point of view.

For the evaluation of a direct action of urate, a measured quantity of uric acid was added to a mild supersaturated solution of COM which contained only sodium and chloride ions in addition to calcium and oxalate, and physicochemical parameters were compared with those obtained in the absence of uric acid. The indirect action was tested by interaction of uric acid with citrate, a well known potent inhibitor of both crystal growth and aggregation of COM, as well as an enhancer of its solubility.

It follows from the data that uric acid alone has no direct effect on the crystallization kinetics of COM, as evaluated by the present method; on the other hand, it can strongly affect the potency of citrate as an inhibitor of crystal agglomeration, without exerting an appreciable effect on crystal growth.

Urolithiasis 2, Edited by R. Ryall *et al.*,
Plenum Press, New York, 1994

MATERIAL AND METHODS

Experimental System

The experimental design adopted was essentially the same as that described by Blomen[4] and Kok[5] with minor modifications to expression of data and physicochemical symbols used. The system consists of three double-walled vessels which contain supersaturated solutions to be tested, kept at constant temperature by a steady stream of water from a thermostatted water bath. The experiment is initiated by addition to the flasks of COM crystals from a stabilized suspension, to achieve the desired final crystal concentration.

Before addition of the seed suspension a fixed amount of ^{45}Ca is added to each flask to enable measurement of calcium uptake by the crystal mass, which is kept in suspension by a magnetic stirrer. Before, and every ten minutes for 60 minutes after addition of the seed, samples of the suspension are drawn off and immediately filtered by 0.44 μm membrane filters (Micron Separations, Inc, NY) ; an aliquot of the filtered liquid is then used to measure ^{45}Ca activity by liquid scintillation (Beckman Instruments, NY).

Mathematical Treatment of the Data

Activities measured at different times are used to calculate percent Ca uptakes by the seed, assuming equimolar uptake of calcium and oxalate, and absence of any isotope effect on crystal growth kinetics, by means of the following relation:

$$U_t = (A_o - A_t) / A_o \qquad (1)$$

where A_o represents activity at time zero (before the addition of seed), and A_t represents activity at any time, t , after the seed has been added.

Ca uptake by the crystals is known to obey an inverse hyperbolic relation with time elapsed after addition of the seed[6] approaching a limit value, U_{in}, which becomes the ordinate of the null-slope-asymptote of the hyperbola; the description of this function, whose plot is known as a crystal growth curve, can be completed with the constant tm - the time at which calcium uptake by the crystals reaches half the value of U_{in}. This can be represented by the relation:

$$U_t = U_{in} * [\, t/(t + tm)] \qquad (2)$$

In order to find U_{in} and tm from experimental data, the relation (2) is rearranged:

$$t/U_t = (t/U_{in}) + (tm / U_{in}) \qquad (2b)$$

This linear form provides a means of deriving the desired parameters by performing a linear regression of the left-hand side of the equation *vs* time elapsed (t) from addition of the seed.

The derivative of equation (1) constitutes a first approach to the rate of Ca uptake:

$$dU_t/dt = k_1 * [\, (U_{in} - U_t)/U_{in}]^2 \qquad (3)$$

The symbol k_1 represents a parameter which was found by Blomen[4] to depend on experimental conditions as follows:

$$k_1 = k_a * s * [\, U_{in}\,] * [(T_{Ca,i}/T_{Ox,eq}) + (T_{Ox,i}/T_{Ox,eq})]^2 \qquad (4)$$

Here, the rate of calcium uptake by the crystals is explicity dependent on the numerical values defining the chemical composition of the solution with respect to ionic strength and concentrations of calcium and oxalate, as well as to the seed concentration and properties; for this reason, the remaining parameter k_a, is said to be a constant which depends on the kinetic properties imparted to the solution by the chemical species which take part in it, but not on the numerical values which express their concentrations. On the other hand, it was found that U_{in} and tm parameters vary widely with seed concentration, and this variation can be described by two relations:

$$U_{in} = \{[U_{in}]/[tm]\} * tm + [U_{in}] \quad (5)$$

$$\{U_{in}/[U_{in}]\} = s / (s + sm) \quad (6)$$

$[U_{in}]$ was found to be numerically equal to the experimentally found equilibrium uptake in the solubility experiments carried out by the authors of the method[4].

[tm] was found to be strongly dependent on the agglomeration properties of the growing crystals.

[sm] is the seed concentration at which the parameter Uin of the corresponding growth curve equals half the value of $[U_{in}]$; [sm] is used in the calculation of k_a.

Summarizing, this experimental procedure provides two main parameters: k_a and [tm] which allows the evaluation of growth and agglomeration independently; in addition, the parameter $[U_{in}]$ is used to evaluate solubility of COM in a hypothetical solution containing all the chemical species present in the tested solution, but calcium and oxalate.

To study the influence of solution composition on crystal growth kinetics, the physichochemical parameters [tm] and ka are compared to a control solution composed as follows:

0.4mM Na_2Ox + 0.4mM $CaCl_2$ + 0.15M NaCl + 0.45 μCi of $^{45}CaCl_2$.

pH adjusted to 6.0 and regulated by means of 7.5mM sodium dimethyl arsinate.

All the reagents were of analytical grade; the solutions were filtered by 0.45 μm membranes prior to use.

COM crystals were purchased from BDH and exposed during 6 months to a mild supersaturated solution for ageing before use. Test solutions were prepared in concentrations which covered the physiological range ([Cit] = 0.1- 2.0 mM, [Ur] = 0.5 - 6 mM); solutions of uric acid were prepared immediately before use due to their poor stability. Each experiment was performed in duplicate.

RESULTS

Figure 1 shows preliminary results for this work, namely, the effect of increasing citrate concentration on the [tm] parameter.

This parameter (tm) has an inverse relation with the extent of agglomeration of the seed, so the maximum in [tm] produced at 0.4mM citrate corresponds to a minimum in

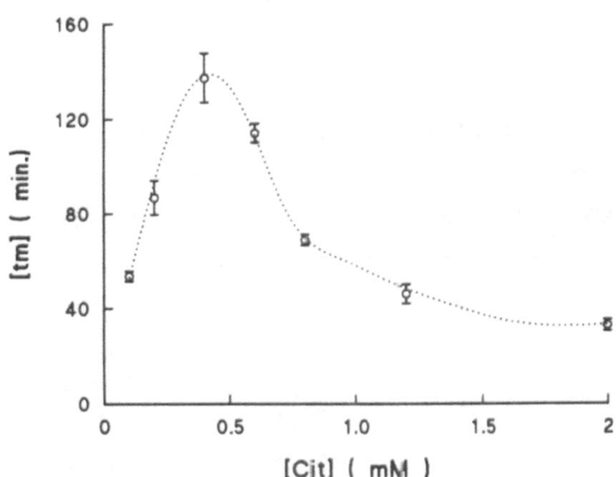

Figure 1. [tm] vs citrate concentration [Cit]

agglomeration; beyond this point the parameter decreases. The crystal growth parameter (graph not shown), which has a direct relation with the calcium uptake rate, decreases linearly with increasing citrate concentration. These results agree with those obtained by Kok[5].

When this scheme was applied to urate, no significant effect was observed in either of the parameters. Two series of experiments were carried out, to obtain the parameters of solutions with two different fixed concentrations of citrate combined with increasing urate concentrations: in the first, a fixed citrate concentration of 0.1mM was used, at which a slight but significant increase is observed (Figure 1) in [tm]; in the second, a fixed citrate of 0.4mM was used, which induced the maximum increase observed in [tm].

Plots of [tm] vs urate concentration for both series are shown in Figure 2, and some selected values are displayed in Table 1.

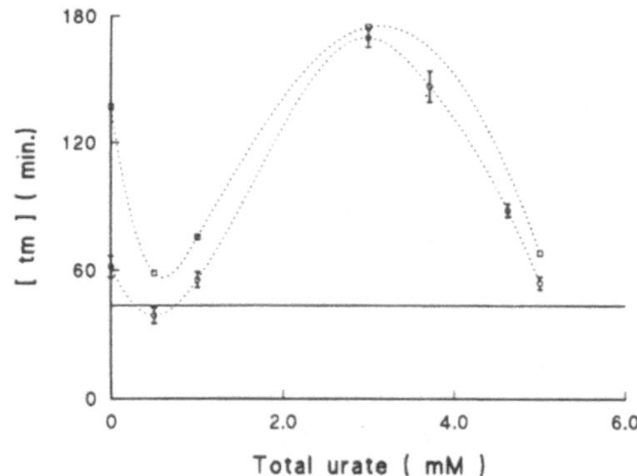

Figure 2. [tm] vs urate concentration ([Ur])
Open circles: citrate 0.1mM; Open squares: citrate 0.4mM. Continuous Line: Control value

A bell-shaped dose-response curve was obtained in both cases. From zero urate concentration to 0.5mM, [tm] decreases in both cases to a local minimum which has a lower numerical value with 0.1mM citrate (Table 2). Beyond this point [tm] rises to a maximum at approximately 3mM and then falls again to an indeterminate value due to the practical impossibility of obtaining higher concentrations of uric acid with the required stability.

Table 1. Physicochemical parameters

[Cit] (mM)	[Ur] (mM)	[tm] (mM)	k_a (lVgr*min)
0.0	0.0	44.27	1.514 e-02
0.1	0.0	55.69	1.206 e-02
0.4	0.0	137.25	0.564 e-02
2.0	0.0	44.14	1.269 e-02
0.1	0.5	45.42	1.531 e-02
0.1	1.0	58.30	1.328 e-02
0.1	5.0	58.55	1.283 e-02
0.4	0.5	56.37	0.721 e-02
0.4	1.0	73.21	0.832 e-02
0.4	5.0	65.23	0.805 e-02

CONCLUSIONS

Data shown account for a strong, direct depletion by urate of the inhibition of the seed agglomeration exerted by citrate when both substances are isolated from other urinary constituents. This effect was shown by the numerical values of the agglomeration parameter, [tm]. This parameter has no direct physical interpretation in view of its units (minutes), but considerable experimental evidence from different sources exists to support its interpretation[4]. As [tm] is just a parameter and not a direct expression of the extent of agglomeration of the crystals, no attempt was made by the authors of the method to estimate how it is influenced by solution conditions; this leads to the possibility that the observed effect could be due to a combination of many factors, including perhaps the method of preparing concentrated uric acid solutions, which is a difficult task requiring extreme conditions, reversed immediately before the crystallization experiment takes place. Nevertheless, when these solutions were tested without citrate, no effect could be found on any of the parameters determined.

On the other hand, no effect was observed in the crystal growth parameter (k_a), which showed no difference between citrate-containing solutions with or without uric acid. Unlike [tm], k_a has a direct connection with the rate of uptake of ^{45}Ca by the seed as it arises from the derivative of the equation governing the crystal growth curve; the units employed may not have a direct physical interpretation in terms of crystal growth due to the mathematical rearrangement necessary to isolate the parameter from its dependence on the solution composition, but this is frequently the rule when working with crystallization kinetics.

No attempt was made at the present to study such an effect in undiluted human urine, or to search for its mechanism. Although these data cannot be directly extrapolated to human urine, in view of the simplification that this kind of system represents, it can be argued that the observed effect may at least partly account for the observation[7] that hyperuricosuria is associated with the development of calcium oxalate kidney stones.

ACKNOWLEDGEMENTS

The authors are indebted to Mr. E. Lores for his competent technical assistance.

REFERENCES

1. K Lonsdale, Epitaxy as a growth factor in urinary calculi and gallstones, *Nature* 217,56 (1968).
2. B Finlayson, Physicochemical aspects of urolithiasis, *Kidney Int* 13,344 (1978).
3. S Volta, E Fradinger, C Bogado, J Zanchetta, Mecanismos de accion de inhibidores y promotores de la cinetica de cristalizacion del oxalato de calcio monohidrato (COM), *Medicina* 51(5),443 (1991).
4. L Blomen, Growth and agglomeration of calcium oxalate monohydrate, *Thesis, Leiden University*, (1982).
5. D Kok, The role of crystallization processes in calcium oxalate urolithiasis, *Thesis, Leiden University* (1991).
6. A Ligabue, M Fini, WG Robertson, Influence of urine on *in vitro* crystallisation rate of calcium oxalate: determination of inhibitory activity by a [^{14}C] oxalate technique, *Clin Chim Acta* 98.39 (1979).
7. FL Coe, Commentary: Allopurinol treatment of uric-acid disorders in calcium stone formers, *J Lithotripsy and Stone Dis* 3(3), 272 (1991).

SAME SEQUENCE BETWEEN OSTEOPONTIN AND URINARY STONE PROTEIN

K. Kohri[1], Y. Suzuki[2], K. Yoshida[2], N. Amasaki[1], T. Yamate[1], T. Umekawa[1], M. Iguchi[1], H. Sinohara[2] and T. Kurita[1]

Departments of Urology[1] and Biochemistry[2]
Kinki University School of Medicine
Osaka 589, Japan

INTRODUCTION

All urinary stones contain an organic matrix which comprises approximately 2 to 5 per cent of the total stone weight. The matrix may be involved in the pathogenesis of urinary stone formation. Despite intensive investigation the role of this material in lithogenesis remains to be understood.

We present the results of the study of the molecular sequencing of the cDNA encoding for urinary stone protein. We also report the purification of osteopontin protein from urinary stone protein. Furthermore, we demonstrate a marked increase in osteopontin mRNA and protein in renal distal tubular cells in a rat model of urinary stones.

MATERIALS AND METHODS

The commercial human kidney cDNA library (Clontech) was screened as described by Sambrook et al[1] with the stone protein polyclonal antibody. Positive clones were isolated and the cDNA inserts were subcloned into M13mpl9 which were used for the construction of a series of deletion mutants containing various lengths of the inserts by the Cyclone System. The cDNA sequence was analyzed by the dideoxynucleotide chain termination method of Sanger et al[2] using the T7 Sequencing Kit. Sequence analysis was performed with a computer program.

The procedure used to extract urinary protein from calcium oxalate urinary stone was similar to that for the preparation of urinary stone antibody. Proteins were analyzed by SDS-PAGE electrophoresis. Samples were dissolved in 10 μL sample buffer containing 1% SDS, 2 M urea and bromophenol blue marker. Digestion was performed at 37°C for 30 min in 10 μL of 10 mM Tris-HCl buffer, pH 8.0 containing 10 mM $CaCl_2$ and 1 unit of thrombin. Following separation, the gels were stained in the dark for at least 48 hours with 0.025% Stains-All.

The experiments were designed to determine if cells expressing osteopontin mRNA could be identified in renal tissues from a rat model of urinary stone induced with glyoxylic acid.

Digoxigenin-labeled single strand RNA probes were prepared using the DIG RNA Labeling Kit (Boehringer). A 984-bp Hind III fragment of 2ar was subcloned into Bluescript pKS(-). Digoxigenin-UTP-labeled sense and antisense RNA probes were prepared by using the DIG RNA labeling kit (Boehringer). Hybridization was carried out as described previously[3].

Total RNA was prepared by the lithium chloride (LiCl)-urea method and electrophoresed in 1.2% agarose-7.0% formaldehyde gel in 1 X MOPS buffer.

Immunodetection of hybridized digoxigenin-labeled RNA probe was performed by using DIG luminescent detection kit. Filters were incubated with 1.0% blocking reagent in DIG buffer 1 at room temperature for 60 min. Then the filters were incubated with 0.1 U/mL of polyclonal sheep anti-digoxigenin Fab fragments conjugated to alkaline phosphatase in DIG buffer 1 at room temperature for 30 min. Excess antibody was removed by washing DIG buffer 1 for 15 min twice. Washed filters were equilibrated twice for 5 min with DIG buffer 3 (100mM Tris-HCl pH 9.5, 100 mM NaCl, 50 mM $MgCl_2$) and assay buffer (100 mM diethanolamine, 2 mM $MgCl_2$) for 3 min. twice. The alkaline phosphatase reaction was measured by incubating filters in 0.1 mg/mL of AMPPD assay buffer at 37°C for 10 min. After removing excess substrate, filters were exposed to X-ray film for 30 min at room temperature and developed.

RESULTS

Initial antibody screening of approximately 50,000 recombinant plaques identified 16 positive clones. Ten of them could be cloned by the second screening, and were subjected to nucleotide sequence analysis. Computer-assisted homology search of the sequence data library established complete identification of 3 kinds of sequences (Fig 1).

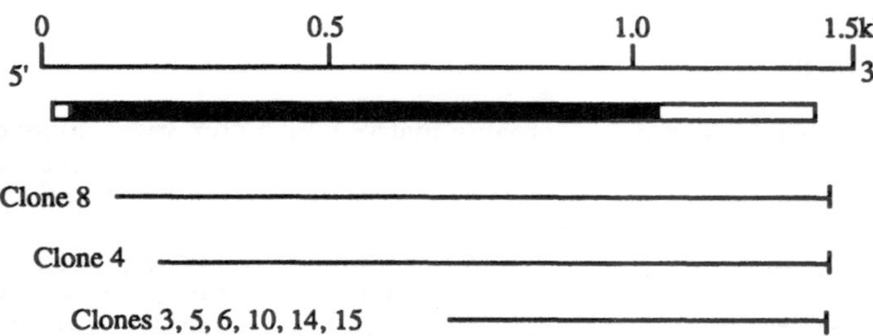

Fig 1. Comparison of the nucleotide sequence of cDNAs encoding osteopontin[5] and calcium oxalate urinary stone protein established in this study. Three types of sequences in 8 clones were completely identical to osteopontin cDNA. The open and solid boxes indicate the non-coding and coding region, respectively.

Analysis of urinary stone protein extracted with 0.5M EDTA and stained with Stains-All revealed three major bands (52kDa, 33kDa, and 23kDa) on 10% cross-linked SDS-PAGE gel. The 52kDa band was quantitatively converted into 33kDa and 23kDa bands following thrombin digestion.

In the control kidney, osteopontin mRNA was detected in a small proportion of distal tubular cells while the renal proximal tubular cells were negative. In contrast, rats with renal stones had a marked increase in osteopontin expression in the distal tubular cells of the medulla. Distal tubular cells in the renal cortex were also sporadically positive 7 days after administration of glycolate although glomeruli and proximal tubular cells still had no detectable osteopontin mRNA. The increase in osteopontin mRNA expression by distal tubular cells and collecting duct cells was associated with the dosage and the duration of administration of glycolate.

We examined the expression of osteopontin gene in control rat kidneys and stone forming rat kidneys by Northern blot analysis using osteopontin cDNA as a probe. A faint signal was observed in the control kidney. The level of osteopontin mRNA in stone forming rats receiving high dosage of glycolate at 7 days was at least twenty-fold higher than that in control rats.

In control rats, epithelial cells of distal tubules were weakly positive for osteopontin protein, and glomeruli and proximal tubular cells were negative. In contrast, rats with renal stones showed a marked increase in intensity of immunoreactive osteopontin in distal tubular cells. The staining pattern for osteopontin protein closely paralleled the increases in expression of osteopontin mRNA.

DISCUSSION

Although a considerable number of reports have appeared in recent years, little is known about the composition of urinary stone matrix. Our previous molecular observation[4] and the further study in this paper have revealed that a cDNA sequence of osteopontin encodes the calcium oxalate urinary stone protein. To confirm that urinary stones also contain osteopontin protein as a matrix, protein extracted with EDTA from urinary stone was stained with Stains-AII, and osteopontin protein was observed in calcium oxalate urinary stone.

This study has also revealed that osteopontin mRNA and protein were sporadically present in renal distal tubular cells in normal rats, and remarkably increased in rats with urinary stones induced with glyoxylic acid as measured with *in situ* hybridization and immunohistochemical staining, respectively. Northern blot analysis also showed a remarkable increase of osteopontin mRNA in proportion to the dosage and the duration of glyoxylic acid. Based on the present data, it is reasonable to presume that osteopontin in renal distal tubular cells plays a principal role as a stone matrix in urinary stone formation. Osteopontin was originally identified in the mineralized matrix of bovine bone, but is now known to be associated with non-mineralized tissue as well. The present study demonstrated that osteopontin is involved as a matrix in stone formation by a similar mechanism to that in bone formation.

REFERENCES

1. Sambrok, J, ZF Fritsch, and T Maniatis, Molecular cloning: a laboratory manual, 2nd ed. Vols. 1-3, Cold Spring Harbor Laboratory, Cold Spring Harbor, NY (1989).
2. Sanger, F, S Nicklen and AR Coulson, DNA sequencing with chain-terminating inhibitors, *Proc Natl Acad Sci* USA 74: 5463-5467 (1977).
3. Nomura, S, AJ Wills, DR Edwards, JK Heath and BLM Hogan, Developmental expression of 2ar (osteopontin) and SPARC (osteonectin) RNA as revealed by in situ hybridization. *J Cell Biol* 106: 441-450 (1988).
4. Kohri, K, Y Suzuki, K Yoshida, K Yamamoto, N Amasaki, T Yamate, T Umekawa, M Iguchi, H Sinohara and T Kurita, Molecular cloning and sequencing of cDNA encoding urinary stone protein, which is identical to osteopontin, *Biochem Biophys Res Commun.*184: 859-864 (1992).
5. Kiefer, MC, DM Bauer and PJ Barr, The cDNA and derived amino acid sequence for human osteopontin. *Nucleic Acids Res.*17: 3306 (1989).

TWO-DIMENSIONAL GEL ANALYSIS OF PROTEINS IN UNPROCESSED HUMAN URINE USING DOUBLE STAIN

P.K. Grover and M.I. Resnick

Division of Urology
University Hospitals of Cleveland
2065 Adelbert Road
Cleveland, Ohio, USA

INTRODUCTION

Analysis of human urine for protein composition is helpful to diagnose various dysfunctions particularly of renal origin. However, in order to detect the excretion of protein(s) in a given disease, it is crucial to identify proteins in urine of healthy subjects which serve as control. Though two-dimensional (2-D) gel electrophoresis was employed in the past to scan urinary proteins, the studies should be regarded with considerable scepticism as there have been serious shortcomings. For instance, despite the inherent protease activity[1] the urines were never collected over protease inhibitors. Only individual urines were analysed, though there were reports of biologic and rhythmic variations in the excretion of proteins[2,3]. The samples were refrigerated during the collection period and centrifuged and filtered. This processing of human urine has now been reported to remove various proteins[4]. Also, though silver stain is 370 fold more sensitive than conventional Coomassie Brilliant Blue[5], proteins in the gels were visualized often by the latter stain.

The aim of the present investigation was to develop double stained 2-D electrophoretograms of pooled unprocessed human urinary proteins, separately for men and women, with samples collected at random times over protease inhibitors.

MATERIALS AND METHODS

Spot mid-stream urines were collected at random times from ten healthy males and ten females directly over a mixture of protease inhibitors in Dewar flasks. Each specimen was 50μm sieved and desalted by extensive dialysis against distilled deionized water. The urines were then freeze-dried and 5mg of the lyophilized protein of each sample was pooled according to sex. The resulting two specimens were dissolved in urea mix containing 9M urea, 5% (v/v) 2-mercaptoethanol, 2% (v/v) nonidet P-40 and 5% (v/v) servalyt 3/10. The protein concentration of the samples was determined and identical protein loads were subjected to 2-D gel analysis. The gels were fixed and double stained: silver stain was followed by Coomassie Brilliant Blue (CBB).

RESULTS AND DISCUSSION

Predominant spots on the double stained profiles of identical amounts of protein from urines of healthy males and females occupied similar relative positions. However,

Urolithiasis 2, Edited by R. Ryall *et al.*,
Plenum Press, New York, 1994

the sample collected from females contained a higher number of spots and a greater proportion of low molecular weight proteins. Nonetheless, the overall protein maps resembled the ones already reported in the literature. The number of protein spots visualized on our electrophoretograms was much higher than in the earlier studies. This could partly be explained by the use of unprocessed urine coupled with high sensitivity of the double stain. To date most of these spots remain unidentified. In the past, profile of human plasma proteins was studied in great detail and most spots recognized. The greater proportion of proteins in urine are contributed by plasma. In an attempt to identify proteins in normal urine, the 2-D electrophoretograms were compared to that of human plasma and spots were recognized by their relative positions. The various proteins discerned included Tamm-Horsfall mucoprotein; albumin; transferrin; fibrinogen alpha chain; fibrinogen beta chain; IgG gamma heavy chain; unknown; Gcglobulin; α-l-antichymotrypsin; α-l-antitrypsin; haptoglobulin beta chain; α-l-acid glycoprotein; G4; IgG light chains; α-l-microglobulin; lipoproteins; unknown; most acid urinary proteins. Most importantly, the proteins known to precipitate during the sample preparation were also visualized. These included Tamm-Horsfall mucoprotein (THM), albumin and transferrin.

In the past, α–1-acid glycoproteins and the most acidic urinary proteins (MAUP) were detected only by electrofocusing the samples in narrow range, very acidic ampholytes i.e., pI 2.5-4.0. However, in the present investigation these proteins were visualized in an ISODALT system with wide range ampholytes i.e., pI 3-10. This difference could probably be attributable to the presence of THM in our samples which being an acidic protein (pI 3-3.5) might have stabilized the pH gradient in that zone thereby making it possible to electrofocus MAUP and α-1-acid glycoproteins. In addition, this might be the result of high sensitivity of the double stain as compared to CBB.

Men are three times more prone to suffer from urinary stones than are women. By corollary of the findings of Hess et al[6], could this be assigned to a defective THM in urine of males? The defect, if any, in THM can arise either from a difference in its amino acid composition or sequence. If THM in urine of males is different in composition, one would anticipate an alteration in its isoelectric point and/or molecular weight. This, in turn, should result in separate spots during 2-D co-electrophoresis of the pooled urines of males and females. However, contrary to our expectation this mucoprotein appeared as a single spot: the spots of THM from the pooled urines of both sexes were superimposed. Its identical charge and molecular weight, irrespective of gender, suggested that the high prevalence of stone formation in males could at least not be ascribed to a compositional difference of THM. Nonetheless, though of identical composition, it is quite possible that THM in urines of males may have a different amino acid sequence. This question should be a matter of priority to address the higher incidence of stone formation in men as compared to women.

In conclusion, the results of this study constitute the first ever developed double stained 2-D maps of proteins in unprocessed human urine collected from healthy subjects and could be used as a data base to identify proteins present under various pathological conditions.

REFERENCES

1. SA Lewis and C Clausen, Urinary proteases degrade epithelial sodium channels, *J Membrane Biol* 122: 77 (1991).
2. N Frearson, RD Taylor and SV Perry, Proteins in the urine associated with Duchenne muscular dystrophy and other neuromuscular diseases, *Clin Sci* 61: 141 (1981).
3. EL Kanabrocki, JA Kanabrocki, RB Sothern et al, Circadian distribution of proteins in urine from healthy young men, *Chronobiol Int* 7: 433 (1990).
4. BM Fraij, Separation and identification of urinary proteins and stone-matrix proteins by mini-slab sodium dodecyl sulfate-polyacrylamide gel electrophoresis, *Clin Chem* 35: 658 (1989).
5. J Guevara, DA Johnston, LS Ramagali et al, Quantitative aspects of silver deposition in proteins resolved in complex polyacrylamide gels, *Electrophoresis* 3: 197 (1982).
6. B Hess, Y Nakagawa, JH Parks and FL Coe, Molecular abnormality of Tamm-Horsfall glycoprotein in calcium oxalate nephrolithiasis, *Amer J Physiol* 260: F569 (1991).

TWO-DIMENSIONAL ANALYSIS OF PROTEINS IN THE URINE OF MALE AND FEMALE STONE FORMERS

P.K. Grover and M.I. Resnick

Division of Urology
University Hospitals of Cleveland
2065 Adelbert Road
Cleveland, Ohio USA

INTRODUCTION

All urinary calculi contain an organic matrix[1]. Chemically, two-thirds of this matrix is protein[2] and this suggests that protein might play a key role as an inhibitor or promotor of stone formation by controlling the initial steps of crystallization of the mineral phase. Urine is the milieu in which stones nucleate and grow. The matrix proteins are therefore thought to arise from proteins present in urine of stone formers. However, a prerequisite to identify any such abnormal protein(s) would be to study proteins in urine of healthy subjects.

In an endeavour to identify proteins that ensue in stone formation, the aim of the present investigation was to study 2-D protein profiles in pooled unprocessed urine of male and female recurrent stone formers, compare them with that of their normal counterparts and discern any abnormal protein spot(s).

MATERIALS AND METHODS

The chemicals used in this study were of analytical grade and purchased either from Fisher Scientific Co or Bio-Rad, USA. Spot mid-stream urine samples were collected at random times from ten male and ten female recurrent calcium oxalate stone formers in Dewar flasks directly over a mixture of protease inhibitors. Each specimen was 50 μm sieved and desalted by extensive dialysis against distilled deionized water. The urines were then freeze-dried and 5 mg of the lyophilized protein of each sample was pooled according to sex. All the individual and the two pooled specimens were dissolved in urea mix containing 9M urea, 5% (v/v) 2-mercaptoethanol, 2% (v/v) nonidet P-40, 5% (v/v) servalyt 3/10. The protein concentration of the samples was determined and subjected to 2-D gel analysis. The gels were fixed and double stained: silver stain was followed by Coomassie Brilliant Blue.

RESULTS AND DISCUSSION

Major protein spots on the urinary electrophoretograms of healthy and stone former males occupied similar relative positions and the overall maps resembled the ones already reported in the literature. However, a visual comparison identified eight spots in the urine of stone formers, seven (protein spots A, B, C, D, E, F, G) of which were missing in their

Urolithiasis 2, Edited by R. Ryall *et al.*,
Plenum Press, New York, 1994

normal counterparts and one (protein spot H) was excreted in higher quantities in the urine of stone formers. The proteins comprising spots A (43 kDa), B (39.5 kDa), C (29 kDa), D (26 kDa) exhibited multiple horizontal spots thereby indicating charge heterogeneity, whereas those of spots E (25.5 kDa), F (26.5 kDa), G (27 kDa) and H (18.5 kDa) did not reveal any charge or molecular weight variants. Similar results were noted with proteins from urines of healthy and stone former women.

The co-electrophoresis study revealed that regardless of sex all these eight proteins had respective identical charge and molecular weight.

One of the primary events in stone formation is the crystallization of mineral phase. Once these crystals have formed they can aggregate into a particle large enough to block a kidney tubule and damage the cells lining the urothelium. In addition to normal urinary proteins this can result in the excretion of serum proteins not present in the urine of healthy subjects. To confirm that the proteins we identified were not secondary to trauma, urines of recurrent calcium oxalate stone formers who had undergone treatment with extracorporeal shock wave lithotripsy (ESWL) and thus had no active stone(s) were also screened. Since ESWL is known to induce a minor hematuria that recedes normally in 5-6 weeks, the urine samples from these patients were collected after 8 weeks of the treatment: all these eight proteins were once again detected (results not shown). This excludes the possibility that the proteins might be a consequence of injury to the urothelial lining by the stone(s).

CONCLUSION

Further studies are required to probe the role of these eight urinary proteins and their likelihood as potential markers of stone formation.

REFERENCES

1. WH Boyce and FK Garvey, The amount and nature of the organic matrix in urinary calculi: A review, *J Urol* 76: 213 (1956).
2. JS King and WH Boyce, Analysis of renal calculous matrix compared with some other matrix minerals and with uromucoid, *Arch Biochem Biophys* 82: 455 (1959).

CRYSTAL MATRIX PROTEIN - SORTING THE SEXES

I.R. Doyle, V.R. Marshall and R.L. Ryall

Department of Surgery
Flinders Medical Centre
Bedford Park, South Australia

INTRODUCTION

The very presence of an organic matrix throughout the mineral structure of all renal calculi implies that it fulfils a decisive role in stone formation, since it may direct the initial steps in the process, either as an inhibitor or a promoter. For this reason, the matrix has been the focus of considerable attention for many years. However, workers have failed to recognize that matrix inclusion may occur as two distinct events: (i) the incorporation of normal macromolecules into crystals comprising the nidus that lodges in a kidney tubule and initiates stone formation, and (ii) secondary deposition of macromolecules as a consequence of tissue injury caused by the growing stone. By studying the organic matrix of calcium oxalate (CaOx) crystals freshly nucleated from human urine, it is possible to overcome the major difficulties associated with the study of stone matrix. It has been shown that the inclusion of urinary proteins within such crystals is highly selective: most proteins are conspicuously absent from the crystal matrix[1,2]. The predominant protein associated with the crystals, crystal matrix protein (CMP), is found in quantities disproportionately greater than its concentration in urine[2]. Because there is no *qualitative* difference in the protein composition of male and female urine[2] and CMP is a potent inhibitor of calcium oxalate crystal aggregation in undiluted urine[3], the aims of this study were to determine whether (i) the urinary concentration of CMP in normal women is greater than that in normal men, and (ii) female CMP binds more avidly to CaOx crystals generated in urine, than does that from men.

MATERIALS AND METHODS

Isolation of crystal matrix extract from male and female urine

24 hr urine specimens were collected without preservative from 10 healthy men and 10 healthy women, and pooled according to sex. They were then centrifuged (10,000x g) and filtered (0.22μm). Preparation of CaOx crystals from the pooled urines and isolation of their organic matrix were performed as detailed by Doyle et al[2], with the urinary metastable limit being determined using a modification[4] of the method of Ryall et al[5].

Enzyme linked immunoabsorption (ELISA) inhibition assay

An antiserum against the crystal extract was raised in rabbits and the IgG fraction isolated by affinity chromatography using Protein A Superose HR. Urinary CMP

Urolithiasis 2, Edited by R. Ryall *et al.*,
Plenum Press, New York, 1994

concentrations were measured using an ELISA inhibition assay system. Briefly, Costar polyvinyl ELISA plates were coated with 10 mg/L of CMP. In a second ELISA plate aliquots of either CMP standards *or* urines were incubated with rabbit anti-CMP IgG. These aliquots were then tranferred to the ELISA plate coated with CMP and, in turn, incubated with alkaline-phosphatase conjugated sheep anti-rabbit IgG.

The CMP concentration was measured in 24 hr urines collected without preservative from 30 men and 30 women and the Wilcoxon signed rank test was used for statistical comparison of the concentrations.

Affinity of CMP from male and from female urine for CaOx crystals

24 hr urine samples were collected from 5 normal men and 5 normal women and pooled so that the final specimen contained equal volumes of male and female urine. After ultrafiltration (10kDa) the specimen was divided and increasing quantities (0 - 10μg/mL) of the male and female CMP isolated as described above was added to each, and CaOx crystallization was induced as before. The CaOx crystals were isolated, 10mg samples demineralised and the component CMP determined using the ELISA inhibition assay

RESULTS

CMP concentration in normal male and female urine

Figure 1 shows the concentration range of CMP in the urines from the normal men and women. The median concentration of CMP in the urine from both males and females was 8.0 mg/L. As expected, there was no difference between the two sexes ($p>0.05$; $n=20$) and this was confirmed when the data were also expressed as daily excretion.

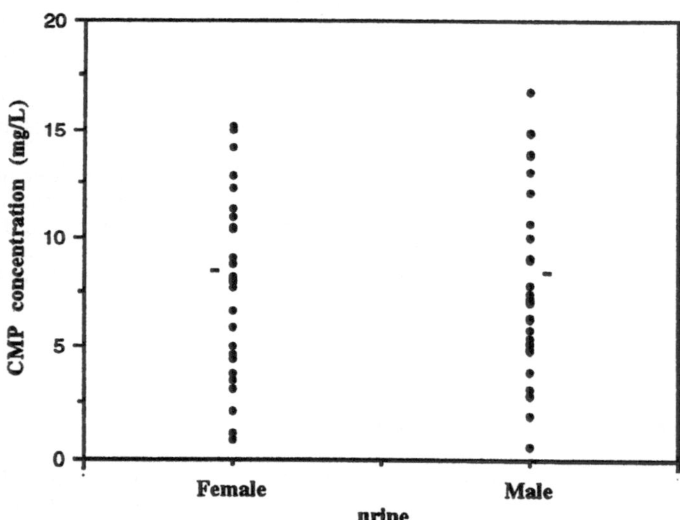

Figure 1. Concentration of CMP in male and female urine.

Affinity of CMP from male and from female urine for CaOx crystals

The amount of CMP isolated from CaOx crystals precipitated from the pooled ultrafiltered urines from men and women, in response to increasing concentrations of added CMP derived from each sex is shown in figure 2. At all concentrations tested, the amount of female CMP in the crystals was approximately double that of the male.

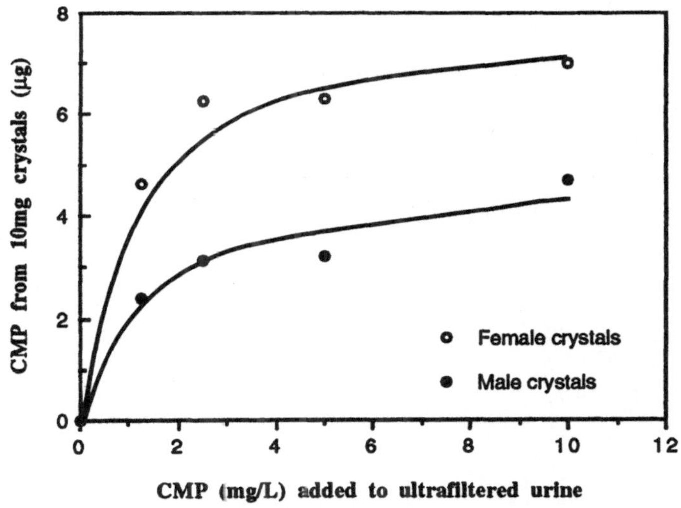

Figure 2. Amount of CMP isolated from CaOx crystals precipitated from ultrafiltered urine.

DISCUSSION AND CONCLUSIONS

These results demonstrate a lack of difference between men and women with regard to the median concentration or range of CMP in their urine, and in the daily amount excreted. However, the amount of female CMP included into CaOx crystals precipitated from urine is approximately double that of male CMP. There are several possible explanations for this disparity: (i) female CMP binds more strongly to CaOx crystals than does male; (ii) women's urine contains some agent(s) which facilitates binding of CMP to the crystals; (iii) men's urine contains some substance(s) which interferes with binding of the protein to the crystals. However, the latter two possibilities may be discounted, because inclusion of CMP into the CaOx crystals occurred in a pooled ultrafiltered urine consisting of equal quantities of urine from each sex.

It is therefore apparent that CMP derived from women possesses a stronger binding affinity for CaOx than does that from men, and this implies a molecular difference between the proteins obtained from each sex. This, combined with the fact that CMP is the most potent inhibitor of CaOx crystal aggregation in undiluted urine that we have yet detected[3], may explain, at least in part, the lower incidence of urolithiasis in women compared with men.

REFERENCES

1. RM Morse and MI Resnick, A new approach to the study of urinary macromolecules as a participant in calcium oxalate crystallization, *J Urol* 139:602-8 (1988).
2. IR Doyle, RL Ryall and VR Marshall, The inclusion of proteins into calcium oxalate crystals precipitated from human urine: A highly selective phenomenon, *Clin Chem* 37:1589-1594 (1991).
3. IR Doyle, VR Marshall and RL Ryall, Crystal matrix protein: A potent inhibitory component of calcium oxalate crystals. This volume.
4. K Suzuki, K Miyazama and R Tsugawa, A Simple method for determining the metastable limit of calcium oxalate. In: *Urolithiasis*, VR Walker, RAL Sutton, ECB Cameron, CYC Pak and WG Robertson (eds). Plenum Press, New York and London, pp 65-66 (1989).
5. RL Ryall, CM Hibberd and VR Marshall, A method for studying inhibitory activity in whole urine, *Urol Res* 13:285-289 (1985).

STUDIES ON URINARY PROTEINS IN CALCIUM OXALATE STONE FORMATION

O. Scharrel and A. Hesse

Division of Experimental Urology
Department of Urology
University of Bonn, Germany

INTRODUCTION

Proteinuria is regarded as one of the main symptoms of renal disease. However, temporary or permanent proteinuria may often occur without renal involvement. Stone disease may be a consequence or cause of renal dysfunction. The subsequent proteinuria may play a role in stone formation. Above all, 2-3 % of renal stone material consists of high molecular weight organic substances, even proteins. The reasons for their appearance not only in the concrements but also in urine, as well as their role in the pathogenesis of urolithiasis, are still not known. Because of the high organisation of these macromolecules in stones, it is very unlikely that they are incorporated non-specifically or randomly. It is important, therefore, to acquire further knowledge of the role of urinary proteins in stone disease.

MATERIALS AND METHODS

Forty-five healthy subjects and 45 calcium-oxalate (CaOx) stone formers were studied. All patients were recurrent pure-CaOx stone formers and none suffered from urinary tract infection or other diseases which might affect protein excretion. Renal function,was normal and all subjects had both kidneys. In the healthy subjects, none had a prior history of stone disease or a positive family history of stone disease.

Twenty four h urine samples were collected from all subjects. Urine samples were analysed for total protein, assayed by the method of Iwata and Nishikaze[1]. For qualitative analysis, SDS-polyacrylamide-gradient-gel electrophoresis was used according to the method of Gorg et al[2] in gels (125 x 225 x 0.36 mm) with T = 4-22.5 % and C = 4 %. Following electrophoresis, the gels were fixed for 30 min in 20 % TCA and stained in 0.5% Coomassie Brilliant Blue R-250 in methanol/acetic-acid/distilled water. Silver staining was carried out according to the method of Heukeshoven and Dernick[3]. Farmer's reducer was used to get clearer bands without background staining.

In normal urine samples, two protein bands (albumin and transferrin) were routinely seen with Coomassie Blue staining; electrophoretograms with more protein bands were regarded as pathological and classified as showing tubular, glomerular or mixed proteinuria. For evaluation of electrophoretograms stained with silver, the number of protein bands with their molecular weight found in urines of the healthy controls were compared with those of CaOx stone formers.

Urolithiasis 2, Edited by R. Ryall *et al.*,
Plenum Press, New York, 1994

RESULTS

Total protein

In urine samples from CaOx stone formers, significantly higher amounts of protein were found, an average of 311 mg/24 h compared to only 131 mg/24 h in healthy subjects. An increased rate of protein excretion (> 150 mg/24 h) was diagnosed, thus, in 8.9% of healthy subjects and in 44.4% of CaOx stone formers. Sex-linked differences were generally observed in the patient group but not in the healthy controls.

Coomassie Blue staining

With Coomassie Blue staining, all forms of proteinuria were observed in the urine samples from CaOx stone formers: 6.7% of the patients showed tubular proteinuria, 6.7% glomerular proteinuria, and 20% mixed proteinuria. No pathological proteinuria was found in the urine samples from the healthy control group.

Silver stain

Figure 1 shows the average number of protein bands for the various ranges of molecular weight in the urines of CaOx stone formers and healthy subjects. In the patient group, we observed significantly more urinary protein patterns, even in those patients with normal total protein values, but not over the entire range of molecular weights. In the range below 20 kD, we found significantly more proteins for the healthy control subjects, whereas in the range of higher-molecular weights, the CaOx stone formers had significantly more protein bands. No sex-linked differences were observed in either group.

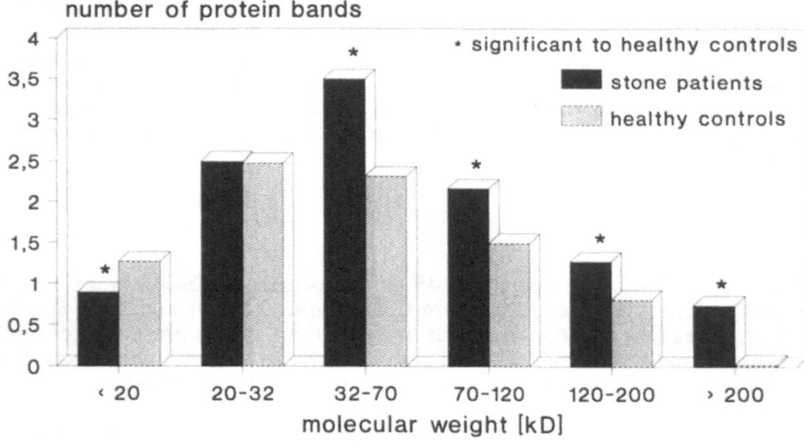

Figure 1. Number of protein bands (M) in silver-stained urine electropherograms of CaOx stone formers (n=45) and healthy subjects (n=45).

DISCUSSION

Significantly higher levels of total urinary protein were found in CaOx stone formers, confirming the data reported by Ibrahim et al[4]. The differences between male and female patients cannot as yet be explained. With Coomassie Blue staining, no homogeneous form of proteinuria and, therefore, no specific renal dysfunction could be diagnosed. Proteinuria in stone patients rather seems to be a consequence than a cause of their disease due to tissue lesions or destruction by the concrements, but there is no real proof because the patients were all recurrent stone formers with a history of several stone episodes.

The results with silver stain demonstrated that CaOx stone formers with high total-protein values showed no non-specific elevation of all protein fractions. In the low molecular-weight range (<20 kD), we observed more proteins in the urines of the healthy control subjects. These findings agree with the data reported by Sofat et al[5] who concluded that these substances may be inhibitors of stone formation which fail in the urine of stone formers, whereas the increased excretion of proteins of high-molecular weight could promote or accelerate urolithiasis. Drach and Randolph[6], who also noted this increased excretion in urines of stone formers, could demonstrate that these high-molecular-weight substances promote nucleation and, therefore, may contribute to stone formation.

CONCLUSIONS

The specific distribution of protein patterns in the urines of CaOx stone formers, even in those with normal total protein values, indicates a special significance of these substances in urolithiasis, especially since proteins can be found in urinary stones themselves.

Monitoring of proteinuria from CaOx stone formers seems to be important and should be integrated in a laboratory program to obtain better knowledge of onset and progression of this symptom. For effective treatment, the proteins involved need to be analyzed in greater detail.

REFERENCES

1. S Iwata and 0 Nishikaze, New turbidimetric method for determination of protein in cerebrospinal fluid and urine, *Clin Chem* 25:1317-1319 (1979).
2. A Gorg, W Postel and J Weser, Horizontale SDS-elektrophorese in ultradunnen gradientengelen ur differenzierung von urinproteinen, *LKB-Note* (1985).
3. J Heukeshoven and R Derniek, Simplified method for silver staining of proteins in polyacrylamide gels.*in:* "Electrophorese Forum '83", BJ Radola, ed, Munehen (1983).
4. AM Ibrahim, YM Shaker, MFS El-Hawary, KJ Fayek, MM Zahran and NK El-Shawarby, Immunochemical studies of serum, urine and calculus proteins in urolithiasis, *Clin Physiol Biochem* 3:16-22 (1985).
5. IB Sofat, MS Rao and RK Jethi, Isolation of polypeptides from the urine of human beings and their role in renal calculi formation, *Urol Res* 16:234 (1988).
6. GW Drach and AD Randolph, Discrimination of the effects of urinary macromolecular inhibitors on nucleation, growth and agglomeration of calcium oxalate, *in:* "Inhibitors of Crystallization in Renal Lithiasis and their Clinical Application", A Martelli, B Buli and B Marchesini, eds, Rome (1988).

ULTRAFILTRATION OF HUMAN URINE AFFECTS CALCIUM OXALATE MONOHYDRATE (COM) CRYSTALLIZATION KINETICS

D.T. Erwin, J. Alam, D.J. Kok, A. Annaloro Jr, J. Vaughn, O. Coker, B. Carriere and F.E. Cole

Alton Ochsner Medical Foundation
New Orleans, LA

INTRODUCTION

The search for reliable, rapid and accurate tools to enable the clinician to predict stone recurrences and prevent patients from forming subsequent calculi has been marginally successful. Chemical measurements of urine constituents dictating the thermodynamics of calcium oxalate monohydrate (COM) stone formation have proved to be of limited use in treating this disease, since crystalluria is as frequent in normal urines as in urines from stone formers[1]. More recently, therefore, attention has turned to methods for evaluating physicochemical processes which dictate rapid growth and agglomeration of COM particle size[2], because the transit time of stones in the urinary tract is so short and the agglomeration process is so rapid[3]. The effect of urine on crystal agglomeration may therefore be a more sensitive index of the likelihood that a patient will form and retain stones[4]. In the present studies we examined the effects of Tamm-Horsfall protein, a urinary glycoprotein which is a variable constituent in renal stones[5]. We quantitatively [6] removed THP from the urines of stone formers by ultrafiltration and assessed its structure, immunologic reactivity and its effects on COM solubility, growth and agglomeration.

METHODS

COM solubility (√/LC) is defined as the lowest concentration (mM) at which growth of added crystals occurs. Percent crystal growth inibition (% GI) is the growth constant which represents the net addition of crystal material to the crystal surface per unit of time. Agglomeration [tm] is measured in units of time (min) and reduces the rate of deposition of mineral components upon the crystals by diminishing the surface area for ^{45}Ca uptake. Measurements were performed in 0.15M NaCl solution with 1:5 diluted urine at pH 6.0. Flasks containing ^{45}Ca, supersaturated with COM, were incubated at 37°C and kept in suspension. At varying times, the crystals were collected on a Millipore filter (0.45µm), dissolved in HCl, and the uptake of ^{45}Ca into the crystal mass measured. THP was removed after measuring the volume of each patient's urine and a portion was filtered through a filter paper (pre), and the remaining portion through an Amicon ultrafilter with a nominal molecular weight cut-off of 30kDa (post). Aliquots (pre and post) were used for THP determinations, and after titration to pH 6.0, their effects on COM kinetics were measured. The ultrafiltration retentate was treated for 24 hours at 5°C with NaCl and NaN_3 (34.0 and 0.25 mg/mL, respectively) and centrifuged at 45,440xg for 20 min at 4°C. The purified THP precipitate was dissolved in distilled, deionized water (DI) and dialyzed (50,000 MW cut-off) for 48 hours with 4 changes of DI using a total volume of 16 liters. The dialyzate was lyophilized (Speed-Vac) and stored at -70°C for future use.

ELISA assay for THP

Materials - Suppliers were as follows: 96-well Immulon #2 plates (Dynatech Labs., Alexandria, VA); anti-human THP monoclonal antibody (Accurate Chemical and Scientific Corporation, Wesbury, NY); antimouse IgG, alkaline phosphatase conjugate developed in rabbits (Sigma Chemical Co, St. Louis, M0); Microwell ELISAmate for phosphatase conjugate (Kirkegaard & Perry Labs, Inc, Gaithersburg, MD). THP isolated from an apparently healthy male donor was used as the standard in the assay.

Methods - In 96-well plates, serial two-fold dilutions were applied in 100µL aliquots with eight dilutions. The THP standard was run in duplicate on every plate. 24 h pre and post urine samples were diluted, applied to the plates, and run in triplicate. Negative controls were included on every plate. The plates were incubated overnight (4°C). BSA diluent/blocking solution was added for ten minutes. THP antibody solution (diluted 1:2000) was applied to each well and reacted for 1 h at room temperature. After 4 washings with phosphate buffered saline Tween, the second antibody (antimouse IgG alkaline phosphatase conjugate, diluted at 1:1000) was added to each well and reacted for 1 h at room temperature. The plates were washed 4 times with PBS-Tween followed by p-nitrophenyl phosphate (1 mg/mL in diethanolamine buffer). The color reaction was stopped after 10 minutes by the addition of 100µl EDTA to each well. The absorbance was read immediately (EIA Reader-Model 2550; Bio Rad, Richmond, CA) at 405 nm.

THP in pre-filtered urines was subjected to SDS/PAGE on 7.5% mini-gels according to Laemmli [7], and transferred to Immun-Lite membranes using a Mini-Trans blot cell (BIO-RAD). The immunoreaction with mouse monoclonal anti-THP antibody (Cedarlane Laboratories Ltd, Hornby, Ontario, Canada), and second antibody, goat anti-mouse IgG-alkaline phosphatase conjugate, were performed using a chemiluminescent assay (BIO-RAD Labs). THP was identified on Western blots, migrating with an apparent molecular weight of ~80kD.

RESULTS AND CONCLUSION

Ultrafiltration of the urine caused a decrease in [tm] from 222.7±22.5 (±SEM) to 81.1±6.6 ($p< 0.001$), in %GI from 84.5±1.7 to 75.9±2.2 ($p<0.0001$), in solubility from 0.22±0.007 to 0.21±0.007 √/LC ($p< 0.005$) and in THP from 28.87±7.76 to 0.00±0.00 mM ($p< 0.0001$).The data in this report suggest that ultrafilterable constituents in human urine, possibly THP, affected COM kinetics, % Gl, solubility, and [tm] *in vitro*. THP from stone formers appeared to be structurally and immunologically identical to that from normals.

REFERENCES

1. WG Robertson, Saturation-inhibition index as a measure of the risk of calcium oxalate stone formation, *in*: Idiopathic urinary bladder stone disease, edited by R van Reen, Fogarty Int Cent Proc Wash. DC, US Gov. Printing Office, 37:55-71 (1977).
2. DJ Kok, SE Papapoulos, LJMJ Blomen, and OLM Bijvoet, Modulation of calcium oxalate monohydrate crystallization kinetics *in vitro*, *Kidney Int* 34: 346 (1988).
3. B Finlayson, Where and how does urinary stone disease start? edited by R. van Reen, Fogarty Int. Cent Proc Wash DC, US Gov. Printing Office, 37:7-31 (1977).
4. DJ Kok, SE Papapoulos, and OLM Bijvoet, Crystal agglomeration is a major element in calcium oxalate urinary stone formation, *Kidney Intl*37:5156 (1990).
5. AMS Grant, LRI Baker and A Neuberger, Urinary Tamm-Horsfall glycoprotein in certain kidney diseases and its content in renal and bladder calculi, *Clin Sci* 44:377 (1990).
6. HH Reinhart, N Obedeanu, D Walz and JD Sobel, A new ELISA method for the rapid quantification of Tamm-Horsfall protein in urine, *Am J Clin Pathol* 92:199 (1989).
7. UK Laemmli, Cleavage of structural proteins during the assembly of the head of bacteriophage T4, *Nature* 227:680 (1970).

NEPHROCALCIN IN SAUDI STONE FORMERS AND NORMAL CONTROLS

D.M. Feuchuk, W.G. Robertson and H. Hughes

King Faisal Specialist Hospital and Research Centre
Department of Biological and Medical Research
Riyadh , Saudi Arabia

INTRODUCTION

Nephrocalcin, a urinary glycoprotein isolated from the urine of normal subjects[1], has been shown to inhibit the nucleation, growth and aggregation of calcium oxalate monohydrate[2]. Analysis indicated that it has a molecular mass of 14.5 kDa and that it contains three γ-carboxyglutamic acid (γ-Gla) residues in addition to a carbohydrate moiety (10%).

A group of glycoproteins possessing a similar amino acid composition but differing in terms of their γ-Gla content was subsequently isolated from normal controls[3]. Glycoproteins with a similar backbone composition isolated from stone formers did not possess γ-Gla residues[3] suggesting that stone formers do not possess the post-translational capability of γ-carboxylating the glutamyl residues of the nephrocalcin backbone. It has to be supposed that this deficiency is specific to the kidneys of stone formers, since, as far as is known, these patients do not suffer from any other disorder traceable to a γ-carboxylation deficiency such as a haemorraghic tendency.

The daily excretion of nephrocalcin in urine is approximately 16 mg/L. At this concentration the protein would be the major γ-Gla-containing macromolecule in urine. Measurement of the γ-Gla concentrations in the molecular weight fraction greater than 5000 Da would thus indicate approximate levels of nephrocalcin in urine samples. In this paper, we have attempted to duplicate these observations in Saudi stone formers and normal subjects.

METHODS

The high performance liquid chromatographic (HPLC) system (Waters Chromatography Division of Millipore Ltd.) consisted of two Model 510 pumps, a model 680 gradient-former, a U6K injector, a model 481 LC spectrophotometer operated at 215 and 280 nm, a Model 490 UV spectrophotometer operated at 254, 280 and 400 nm, a Model 745 data integrator, a Model 740 integrator and an LKB 2-pen recorder. The columns employed were Bio-Sil Sec 125, and Beckman TSK 3000 SW for size exclusion chromatography, a Waters DEAE 8 HR for anion exchange chromatography and a Supelco LC 18 for reverse phase chromatography. For γ-Gla determination, a Waters 710 WISP, a 590 HPLC pump, a Model 470 fluorescence detector and a Model 745 integrator were used. The columns were a Waters C18radial pak and a Beckman C-18 ultrasphere. Spectropor 6000-8000 MW dialysis tubing was purchased from Fisher Scientific Ltd.

Urolithiasis 2, Edited by R. Ryall *et al*.,
Plenum Press, New York, 1994

ß-Carboxyaspartic acid (the internal standard for the γ-Gla assay) was purchased from Sigma Chemical Company. All other chemicals and solvents were Analar or HPLC grade purchased from BDH Ltd. Water was purified using a Millipore Milli-Q system. Authentic nephrocalcin was kindly supplied by Dr. Y. Nakagawa (Chicago).

Urinary Glycoprotein Isolation

The initial stages of the isolation procedure were as previously described[2]. The DEAE extract was chromatographed using a Bio-Sil SEC 125 column (600 x 7.5 mm) and a Beckman TSK 3000 SW (300 x 7.5 mm) column in series. Elution was carried out using dipotassium hydrogen phosphate (0.lM, pH 7.0) at a flow rate of 1 mL/min. Detection was at 280, 254 and 400nm. Fractions were collected and analysed for γ-Gla. These fractions were pooled and further purified using anion exchange chromatography with a Waters AP-1, 100 x 10 mm column (protein pak DEAE 8HR). The proteins were eluted using a linear gradient from 0.1 to 0.4M sodium chloride in 0.05M Tris-HC1 buffer at pH 7.3 (0-10 minutes isocratic, 10-40 min increasing sodium chloride concentration, at a flow rate of 0.5 mL/min). Detection was carried out as above. Alternatively, the fractions were subjected to reverse phase HPLC employing a Supelco LC18 (250 x 4.6 mm 5 μm particle size) column and a Supelco LC18 (75x 4.6mm, 3μm particle size) column in series. The mobile phase consisted of A [0.1% trifluoroacetic acid (TFA)] and B (0.1% acetonitrile). The programme was 0.5 min 100% A, 5-35 min linear gradient to 100% B, 35-45 min 100% B with a flow rate of 1 mL/min. Detection was performed as before. The elution of the proteins was monitored by γ-Gla analysis.

SDS PAGE Separation

Sodium dodecyl sulphate-polyacrylamide gel electrophoresis (SDS-PAGE) was carried out using a Bio-Rad Protean II system. Samples were run on discontinuous polyacrylamide gels (4% stacking gel, 13.5% resolving gel) under denaturing conditions[4].

γ-Carboxyglutamic Acid

Representative samples of urines from Saudi normals and Saudi stone formers were subjected to ultrafiltration through a 5000 Da ultrafilter. Samples of the ultrafiltrate and retentate were hydrolysed with 2M potassium hydroxide and analysed for γ-Gla.

RESULTS

Calibration of the size exclusion columns showed that 17.5 kDa and 1.36 kDa standards had retention times of 29.4 and 36.9 min respectively. Size exclusion chromatography of extracts from normal and stone formers' urines showed a similar series of peaks from 29 min to 34 min. The peak corresponding to the γ-Gla-containing protein (retention 31 min) was taken in each case for further purification. Further chromatography on HP-DEAE and HPRP columns yielded a material with retention times identical to those of authentic nephrocalcin (Rt-DEAE 14.5 min, Rt-RP 4.2 min).

The isoelectric focusing gels indicated that nephrocalcin has an isoelectric point of approximately 3. The protein was visualised with Coomassie Blue and also by Schiff's base formation. The stone formers and non-stone formers had similar electrophoretograms. Polyacrylamide gel electrophoresis yielded similar results.

The chromatograms of γ-Gla from stone formers did not differ appreciably from those of non-stone formers. The excretions of free and 'protein-bound' γ-Gla are shown in Table 1. There was no significant difference between the two groups.

DISCUSSION

The procedures described for the purification of nephrocalcin using various HPLC chromatographic procedures have not indicated any major difference in the nephrocalcin obtained from the urines of Saudi stone formers and non-stone formers. The chromatographic data are supported by the gel electrophoresis patterns of nephrocalcin from the two groups. Further support for these observations was obtained from the estimation of 'protein bound' γ-Gla in urine which showed no significant difference between the two groups.

Table 1. Urinary γ-Gla concentrations

	Stone formers	Non-Stone formers
Free γ-Gla (μmol/day)	38.1±9.8	36.4±9.0
'Protein-bound' γ Gla (μmol/day)	2.56±1.0	2.93±1.2

SUMMARY

In the Saudi stone formers and non-stone formers studied, using HPLC and gel electrophoretic techniques, there is no evidence to substantiate the suggestion that nephrocalcin is different in stone formers and non-stone formers.

REFERENCES

1. H Ito and FL Coe, Acidic peptide and polyribonucleotide crystal growth inhibitors in human urine, *Am J Physiol* 233, F455 (1977).
2. Y Nakagawa, V Abram, FJ Kezdy, EM Kaiser, and FL Coe, Purification and characterization of the principal inhibitor of calcium oxalate crystal growth in human urine, *J Biol Chem* 258, 12594 (1983).
3. Y Nakagawa, V Abram, JH Parks, HS-H Lau, JK Kawooga, and FL Coe, Urine glycoprotein crystal growth inhibitors. Evidence for a molecular abnormality in calcium oxalate nephrolithiasis, *J Clin Invest* 76,1455 (1985).
4. UK Laemmli, Cleavage of structural proteins during the assembly of the head of bacteriophage T4, *Nature* 227, 680 (1970).
5. Y Haroon, Rapid assay for γ-carboxyglutamic acid in urine and bone by precolumn derivatization and reversed-phase liquid chromatography, *Anal Biochem* 140, 343 (1984).

INHIBITION OF CALCIUM OXALATE CRYSTAL AGGREGATION BY TAMM-HORSFALL GLYCOPROTEIN DEPENDS ON CITRATE

B. Hess[1], L. Zipperle[1], R. Takkinen[1], D. Ackermann[2] and Ph. Jaeger[1]

[1]Policlinic of Medicine and
[2]Department of Urology
University Hospital
CH-3010 Berne, Switzerland

INTRODUCTION

Tamm-Horsfall glycoprotein (THP) is the most abundant protein in the urine of healthy individuals, but whether it actually inhibits or promotes calcium oxalate monohydrate (COM) crystallization processes (reviewed in [1]) has been the subject of a great deal of controversy. At higher pH and lower ionic strength, THP primarily inhibits COM crystal aggregation[2,3], whereas at lower pH and higher ionic strength, THP molecules are more polymerized and act as weaker inhibitors of COM crystal aggregation[4]. Viscosity measurements and molecular weight determinations have provided evidence that severely recurrent calcium renal stone formers excrete THPs with an abnormally high tendency to polymerize and thus with reduced inhibitory properties[4].

In this study, we investigated the effects of calcium and citrate on COM crystal aggregation inhibition by THPs purified from urines of healthy men (N) and male idiopathic recurrent calcium renal stone formers (RCSF).

MATERIALS AND METHODS

Individual THPs were purified from pooled 24 h urines of 8 RCSF and 8 N by repeated salt precipitation and passage through a Sepharose 4-B column, as previously described[3,4]. Crystal aggregation in COM crystal slurries (0.7 mg/mL) was measured spectrophotometrically (620 nm) at 37°C by monitoring particle sedimentation[3]. Intrinsic viscosity (VISC) of THPs was measured at 25°C, using a capillary Ubbelohde-type viscometer. All experiments were performed at 200mM NaCl, 5mM $CaCl_2$ and pH 5.70 (buffered with 10mM sodium acetate), either without (CIT-) or with 3.5 mM citrate (CIT+). Values are expressed as means ±SEM; the Mann-Whitney U-test and paired t-test were used for comparisons between and within groups, respectively.

RESULTS

Data from COM crystal aggregation measurements are summarized in Table 1. When no THP was added (control experiments), citrate slightly promoted COM crystal aggregation. In CIT-, 6 out of 8 THPs from RCSF, but only 3 out of 8 normal THPs promoted crystal aggregation; in CIT+, all THPs became inhibitors (inhibition between 4

Table 1. COM crystal aggregation inhibition (% of control in CIT-)

	CIT-	CIT+	p value (vs Cit-)
No THP	0	-14.7±1.8	< 0.001
RCSF-THPs (40 mg/L)	-26.0±12.4	43.4±4.6	< 0.001
Normal THPs (40 mg/L)	22.0±12.9*	49.5±4.6	ns

* , p < 0.025 vs THPs from RCSF in CIT-

and 73%). In CIT-, increasing THP concentration from 16 to 28 and further to 40 mg/L progressively lowered COM crystal aggregation inhibition by stone former THPs (from 19.7±8.0 to 0.1±12.8 and further to -26.0±12.4%, respectively), but not by normal THPs.

In CIT-, VISC of THPs from RCSF was higher (341±58 mL/g) than that of normal THPs (140±29 mL/g, p <0.01); in CIT+, viscosities of THPs from both groups were similar (194±42 in RCSF vs 186±31 mL/g in N; ns). Thus, VISC of stone former THPs was lower in CIT+ than in CIT- (p = 0.029). In CIT-, but not in CIT +, there was a significant negative correlation between crystal aggregation inhibition by 40 mg/L of THP and VISC (r = -0.843, p <0.001).

DISCUSSION

Our study demonstrates that in the presence of calcium at a high physiological concentration, citrate enables promotory THPs to become inhibitors of COM crystal aggregation *in vitro* at pH 5.7 and 200mM sodium chloride. Since citrate alone slightly promotes COM crystal aggregation under these conditions (Table 1), crystal aggregation inhibition in our assay system is apparently not mediated by the formation of calcium-citrate complexes that bind to the crystal surface.

Our viscosity measurements strongly suggest that increased COM crystal aggregation inhibition by THPs is at least partly mediated by the formation of less polymerized THP molecules, which are stronger inhibitors of COM crystal aggregation. Citrate, when added to a solution containing calcium, lowers intrinsic viscosities of strongly polymerizing THPs from recurrent calcium renal stone formers, ie conformational changes of THP molecules (probably depolymerization) occur. Further studies will have to clarify whether this is simply due to chelation of calcium ions in solution, or whether citrate is able to induce conformational changes in THP molecules independently of calcium.

Finally, this study confirms our previous finding that, when compared with proteins from normals, THPs from severely recurrent idiopathic calcium renal stone formers have an increased tendency to polymerize at low pH and high ionic strength[4]. The molecular basis for this abnormal behaviour remains to be elucidated.

ACKNOWLEDGEMENTS

This study was supported by the Swiss National Science Foundation (Grant No. 32-26428.89).

REFERENCES

1. B Hess, Tamm-Horsfall glycoprotein - inhibitor or promoter of calcium oxalate monohydrate crystallization processes? *Urol Res* 20:83 (1992).

2. DS Scurr, WG Robertson, Modifiers of calcium oxalate crystallization found in urine. II. Studies on their mode of action in artificial urine *J Urol* 136:128 (1986).
3. B Hess, Y Nakagawa, FL Coe, Inhibition of calcium oxalate monohydrate crystal aggregation by urine proteins, *Am J Physiol* 257 (Renal Fluid Electrolyte Physiol 26):F99 (1989).
4. B Hess, Y Nakagawa, JH Parks, FL Coe, Molecular abnormality of Tamm-Horsfall glycoprotein in calcium oxalate nephrolithiasis, *Am J Physiol* 260 (Renal Fluid Electrolyte Physiol 29) :F569 (1991).

HUMAN KIDNEY PRODUCES A PROTEIN SIMILAR TO LITHOSTATHINE, THE PANCREATIC INHIBITOR OF $CaCO_3$ CRYSTAL GROWTH

J.M.Verdier[1], B. Dussol[2], P. Casanova[3], M. Daudon[4], P. Dupuy[1], P. Berthézène[1], R. Boistelle[5], Y. Berland[2] and J.C. Dagorn[1]

[1]INSERM U315, Marseille, France
[2]Service de Néphrologie et d'Hémodialyse
Hôpital Sainte-Marguerite, Marseille, France
[3]Service d'Anatomie Pathologique
Hôpital de la Timone, Marseille, France
[4]Laboratoire de Biochimie A
Hôpital Necker, Paris, France
[5]CRMC2 du CNRS, Marseille, France

INTRODUCTION

Pancreatic juice is supersaturated with calcium carbonate. Calcium carbonate ($CaCO_3$) crystal growth is controlled by lithostathine, a secretory protein synthesized by pancreatic acinar cells, first described as a constituent of pancreatic stones[1]. It was recently reported that in the thin descending limb of the Henle's loops of rat kidneys, urine is supersaturated with $CaCO_3$ and calcium phosphate[2]. This observation suggested the presence in kidney of a similar inhibitor.

Antibodies to lithostathine were used to identify a related protein in urine and kidney stones (figure 1). After sodium dodecyl sulphate polyacrylamide gel electrophoresis (SDS-PAGE), Western blot analysis of proteins extracted from concentrated normal urine[3] or kidney stones demonstrated the presence of a protein with an apparent molecular weight of 23,000 Da (figure 1, lane 4). This band was also evidenced in calcium oxalate stones of two patients and in a pool of stones obtained from different patients (not shown). A band was also found in the same position in a gel of urinary proteins from healthy controls (figure 1, lane 3). Taken together, these observations suggest that there is indeed a protein structurally related to lithostathine in the urinary tract. Under the same experimental conditions, lithostathine migrated with a lower Mr, which makes it unlikely that the urinary protein is lithostathine itself or a related degradation product originating from blood by glomerular filtration.

In addition, renal lithostathine did not react with all sera prepared to the pancreatic protein, demonstrating that the proteins differed at least by one epitope (not shown). Furthermore, this protein is not immunologically related to nephrocalcin. Because of its relatively small size, compatible with glomerular filtration, the renal origin of this protein could not be ascertained. Hence, the same antibodies were used in immunolocalization experiments on fresh human surgical nephrectomy specimen cryosections. A strong positive signal was observed in the cells of proximal tubules and thick ascending limbs of Henle's loops (figure 2A). No staining could be visualized in glomeruli, in the descending limbs of the Henle's loops or in distal tubules. Similar results were obtained on all cryosections tested. With IgGs from non-immune serum, no staining was seen (fig. 2B).

Urolithiasis 2, Edited by R. Ryall *et al.*,
Plenum Press, New York, 1994

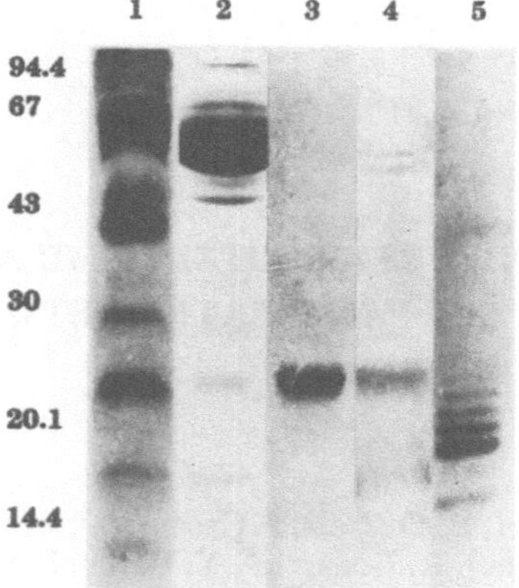

Figure 1. Immunodetection of a protein related to pancreatic lithostathine in urine and renal stone extracts. Proteins extracted from urine (100 µg, lanes 1 and 3) or from stones (100 µg, lanes 2 and 4) were subjected to SDS-PAGE and stained with Coomassie blue (lanes 1 and 2) or transfered to nylon membranes and incubated with antibodies to lithostathine (lanes 3 and 4). As control, lithostathine (0.5 µg) was run and treated in parallel (lane 5). Values of molecular weight standards (in kDa) are indicated on the left. Band(s) revealed by the antibody in lanes 3 and 4 migrate in the 23 kDa region.

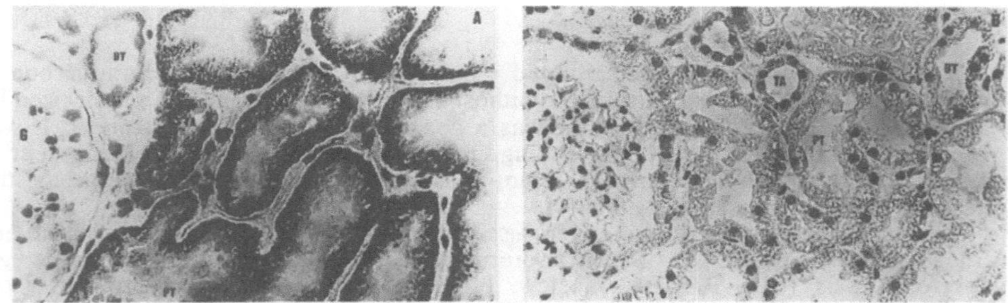

Figure 2. Immunohistochemical localization of renal lithostathine in kidney. With antibodies to lithostathine (A) (Left), immunoperoxidase staining was localized to proximal tubules (PT) and thick ascending limbs of Henle's loops (TA). Only background staining was found in the glomerulus (G) and distal tubules (DT). Non-immune serum used as control (B) (right) gave no signal. Original magnification x250.

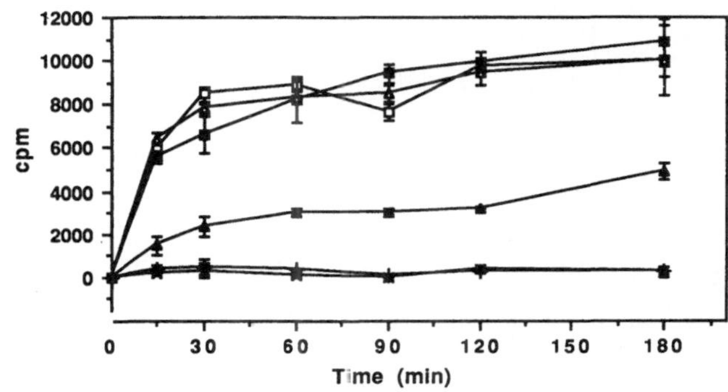

Figure 3. Inhibition of $CaCO_3$ crystal growth. Accretion of ^{45}Ca onto $CaCO_3$ crystals was monitored as a function of time following addition of seed suspension to a solution supersaturated with $CaCO_3$, by counting the radioactivity retained with crystals on a fibreglass filter. In the absence of added protein (-†-), in the presence of albumin (···o···) or after control immunoprecipitation with antibodies to lithostathine and protein A (---Δ---), ^{45}Ca incorporation reached near-maximal values after 30 min. In the presence of 3 μg lithostathine (··· ♦ ···) of 10 μg renal stone protein extract (---•---) or 19 μg urinary proteins (---•---) no significant incorporation occurred. When stone extracts were incubated with antibodies to lithostathine, then with protein A before seeding, incorporation increased to 50 % of control value (---s---). Mean ±SEM (n=3).

The influence of protein extracts from renal stones on $CaCO_3$ crystal growth was tested *in vitro* (figure 3). Using a seeded crystal growth system, we showed that extracts were able to completely prevent crystal growth. When compounds reacting with antibodies to lithostathine were removed from the extracts by immunoprecipitation, inhibition was reduced by more than 50 %.

It is concluded (i) that a protein immunologically related to pancreatic lithostathine is synthesized in the kidney and secreted into urine, (ii) that since the thin descending limbs of Henle's loops might be supersaturated with respect to $CaCO_3$, renal lithostathine might participate in the control of the crystal growth of calcium salts in urine. Hence, the urinary protein immunologically related to lithostathine is also an inhibitor of $CaCO_3$ crystal growth. Because of its structural and functional relationship with the pancreatic protein, it will be henceforth called renal lithostathine.

REFERENCES

1. A De Caro, JJ Bonicel, P Rouimi, JD De Caro, H Sarles, and M Rovery, Complete aminoacid sequence of an immunoreactive form of human pancreatic stone protein isolated from pancreatic juice, *Eur J Biochem* 168:201 (1987).
2. FL Coe, JH Parks, and Y Nakagawa, Protein inhibitors of Crystallization, *Sem Nephrol* 11:98 (1991).
3. JM Verdier, B Dussol, P Dupuy, Y Berland, and JC Dagorn, Preliminary treatment of urinary proteins improves electrophoretic analysis and immunodetection, *Clin Chem* 38:860 (1992).

STUDY OF TAMM-HORSFALL GLYCOPROTEIN BINDING TO CALCIUM OXALATE MONOHYDRATE CRYSTALS

M.F. Netzer, N. Hasenfuß, T.J. Davies and P. Mestres[2]

Departments of Urology and Anatomy[2]
University of Saarland 6650
Homburg/ Saar Germany

INTRODUCTION

Aggregation of crystals is the most important process during stone formation in the kidney and can be influenced by different proteins of renal origin. Tamm-Horsfall glycoprotein (THP) is produced in the cells of the thick ascending limb of Henle's loop and is excreted into the tubular lumen. One of its major physiological roles is the inhibition of calcium oxalate monohydrate (COM) crystal aggregation. However, under specific conditions this protein may also act as a promotor of crystallization. The interaction between THP and the crystal surface during aggregation is not known and is the topic of this study.

METHODS

THP was purified by salt precipitation from healthy individuals' and stone patients' urines. Different protein concentrations were incubated with COM crystals under supersaturated conditions in 50 mM HEPES buffer, pH 7.4 at 37° C. The crystals were washed with the buffer and incubated with a monoclonal anti-THP antibody. Crystals with bound protein were examined under scanning electron microscopy using protein A gold detection.

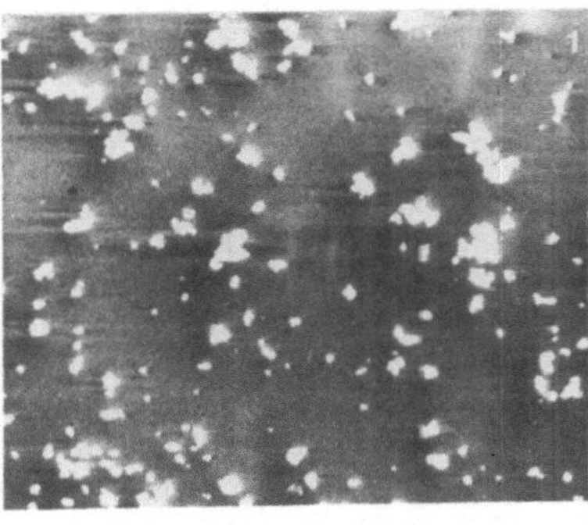

Fig 1. (Left) Crystals coated with THP from healthy individuals (3cm=100μm).
Fig 2. (Right) Same samples as in fig 1 at higher magnification) (2cm=3μm).

RESULTS

The results demonstrate small aggregates and coated crystals for THP isolated from the non-stone formers (fig 1, fig 2) as opposed to weak protein binding and large aggregates (fig 3, fig 4) when using the THP from the stone formers' urines.

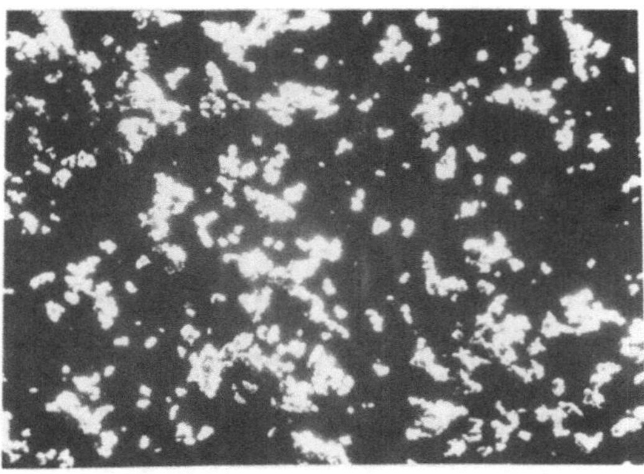

Fig 3. Crystals incubated with stone former THP (3cm= 100μm)

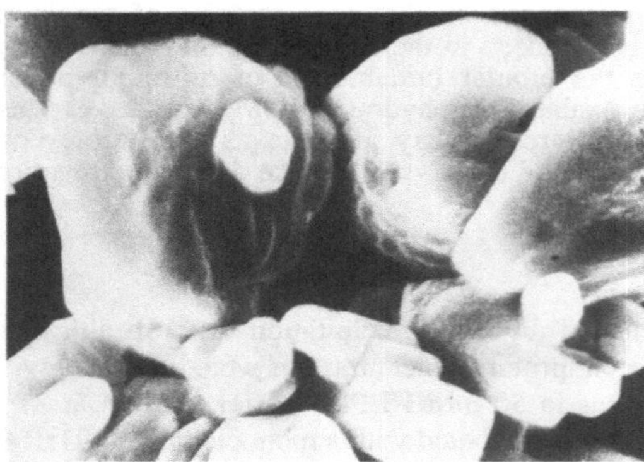

Fig 4. Same sample as shown in fig 3, at higher magnification (2cm=3μm).

DISCUSSION

Our electron microscopic examinations of THP and COM crystal interaction revealed differences between protein isolated from stone formers and healthy controls. We detected only a weak binding of the protein to the crystal surface in stone formers. Extensive washing could remove most protein but we observed different forms of crystal aggregation after this step. It has been reported that the action of THP is influenced by salt concentration and pH, making it either an inhibitor or promotor of aggregation. THP from the normal individual caused small crystal aggregates whereas the stone former's THP induced large aggregates under the same experimental conditions. These findings indicate that crystal interaction and subsequent aggregation may be dependent upon molecular variation in THP produced by individual patients.

CRYSTAL MATRIX PROTEIN: A POTENT INHIBITORY COMPONENT OF CALCIUM OXALATE CRYSTALS

I.R. Doyle, V.R. Marshall and R.L. Ryall

Department of Surgery
Flinders Medical Centre
Bedford Park SA
Australia

INTRODUCTION AND METHODS

Nucleation of crystals is the critical initial step in stone formation; the aims of this study were (i) to analyse the protein content of calcium oxalate(CaOx) crystals freshly precipitated from human urine, and (ii) to ascertain whether their presence results from their acting as inhibitors or promoters.

CaOx crystals precipitated from normal urine by the addition of sodium oxalate were harvested and demineralized, and the proteins in the resulting extract separated by polyacrylamide gel electrophoresis. Most urinary proteins, including Tamm-Horsfall mucoprotein and serum albumin, were absent from the crystals or were present in only minute quantities. However, one protein with a molecular weight of 31kDa, barely detectable in urine itself, was present in the crystal matrix extract (CME) in quantities far exceeding that of all other proteins. Western blotting showed this to be a previously unidentified urinary component which we have named **crystal matrix protein (CMP)**.

To achieve the second aim of this study the effect of CME on CaOx crystallization was tested in an inorganic seeded assay system and in undiluted, ultrafiltered urine.

RESULTS AND CONCLUSION

In the seeded assay CME (0.08-1.25 mg/L) inhibited CaOx crystal growth and aggregation in a dose-dependent manner. Like other macromolecules previously tested, CME inhibited aggregation more potently than growth. Most importantly, its effect on linear growth was at least 10 times that of albumin.

In undiluted urine, CME (1.25-10 mg/L) did not alter crystal morphology or the amount of CaOx deposited. In contrast, it potently inhibited crystal aggregation, reducing the mean size of the crystal particles from a diameter of 14.35 μm in the control containing no CME to 8.00 μm at a CME concentration of 10 mg/L. Scanning electron microscopy confirmed this: crystals precipitated from the ultrafiltered urine were characteristically highly aggregated, while those deposited in the presence of CME were progressively less aggregated with increasing concentrations of protein. At a concentration of 10 mg/L the crystals consisted almost exclusively of singlets or twins of weddellite.

It was concluded that most urinary proteins play no role in CaOx crystal formation. CME (and by implication, CMP) is the most powerful natural urinary inhibitor of CaOx crystal aggregation in urine yet detected and may therefore be important in stone pathogenesis.

TWO-DIMENSIONAL ELECTROPHORETIC ANALYSIS OF UNPROCESSED HUMAN BLADDER WASHINGS: AN EVALUATION OF UROTHELIAL PROTEINS IN PATIENTS WITH NORMAL BLADDER MUCOSA

P. Sweeney, P.K. Grover and M.I. Resnick

Department of Urology
Case Western Reserve University
Cleveland, Ohio

INTRODUCTION

The present investigation was undertaken to define a method of obtaining proteins from normal bladder urothelium and to define a consistent protein profile. Such a method and consistent profile would provide a basis for future studies of pathologic conditions of bladder mucosa, such as bladder cancer and interstitial cystitis.

METHODS

Ten patients with predominantly bladder outlet obstruction symptomatology underwent cystoscopic examination with sterile water irrigation. Each was found to have normal appearing bladder mucosa, was culture-negative at the time of exam, and had normal urine cytologies, which were performed if any irritative symptoms were present. The bladder was drained upon placement of the cystoscope, rinsed free of urine with 100 cc of irrigant, and then washed with 100-150 cc irrigant, which was saved as specimen. The specimen was then dipped to exclude the presence of hemoglobin, sieved (50 µm), dialyzed against distilled water at 4°C and lyophilized. The protein thus obtained was dissolved in urea-mix and subjected to two-dimensional (2-D) electrophoresis utilizing the ISO-DALT system, with gel-staining by silver followed by Coomassie Brilliant Blue (CBB).

RESULTS AND DISCUSSION

The electrophoretic profiles thus obtained were relatively consistent amongst patients, with normal urinary proteins identified. Minor variations in both concentration and number of proteins were noted between individual patient profiles, while some proteins present in normal whole urine electrophoretic profiles such as Tamm Horsfall protein were absent. We conclude that atraumatic bladder washings of documented normal bladder mucosa can be utilized to provide a consistent 2-D electrophoretic profile of bladder urothelial proteins.

ISOLATION AND PARTIAL CHARACTERISATION OF RENAL STONE PROTEINS

A.A. Siddiqui[1], M.A. Waqar[1] and T.S. Talati[2]

Departments of [1]Biochemistry and [2]Surgery
The Aga Khan University Medical College
Stadium Road
Karachi, Pakistan

INTRODUCTION AND RESULTS

The presence of proteins in renal stone matrix is well documented[1]. In order to find promoters and inhibitors of urolithiasis, attempts are being made to isolate and characterize macromolecules from renal stones, which could prove useful in finding clues to the mechanism of renal urolithiasis[2].

Extracts of renal stone proteins[3] were analyzed by SDS-polyacrylamide gradient gel electorphoresis under denaturing conditions. The extracts were further fractionated by gel filtration on Sephadex G-75, in a 2 x 90 cm column equilibrated in phosphate buffer, pH 7.5 and collected in 2 mL fractions.

SDS-PAGE isolated proteins showed at least three species in the M_r range of 12-66kDa. Further fractionation revealed that these proteins are eluted in three major peaks. Peak I, eluted within void volume contained a mixture of proteins in the M_r range of 12-66kDa. Peak II, eluted just after that, contained a protein with an M_r of 15kDa. Peak III also comprised a 15kDa M_r protein along with another band at 12,kDa. Isoelectric focusing showed the presence of bands in the pI range of 4.0 - 5.5, while immunoelectrophoresis using antisera against all human serum proteins did not show a precipitating arc.

Using strong ionic detergent and reducing conditions we have isolated and partially characterized protein species from renal stones, which are acidic in nature and do not appear immunologically identical to normal human serum proteins. Although SDS-PAGE reveals the presence of several bands in the region of 12-66kDa M_r, their elution profile indicates that some of these either exist in strong aggregated forms or as subunits of a larger protein. Attempts are being made to purify them to homogeneity.

ACKNOWLEDGEMENTS

The financial support from NSRDB of Pakistan is gratefully acknowledged.

REFERENCES

1. RM Morse and MI Resnick, Urinary stone matrix, *J Urol* 139:602 (1988).
2. S Sorensen, K Hansen, S Bak and SJ Justesen, An unidentified macromolecular inhibitory constituent of calcium oxalate crystal growth in human urine, *Urol Res* 18:373 (1990).
3. AA Siddiqui, AN Hussain, MA Waqar and J Talati, Isolation of low molecular weight proteins from renal stones, *Biochem Soc Trans* 18:1266 (1990).

FACTORS AFFECTING THE INCLUSION OF CRYSTAL MATRIX PROTEIN INTO CALCIUM OXALATE CRYSTALS: CENTRIFUGATION, SIEVING AND SEX

C.J. Dawson and R.L. Ryall

Department of Surgery
Flinders Medical Centre
Bedford Park, South Australia

INTRODUCTION AND RESULTS

Crystal matrix protein is the most potent inhibitor of calcium oxalate (CaOx) crystal aggregation in undiluted urine yet detected and the principal macromolecular component of CaOx crystals precipitated from human urine. Preliminary empirical observations have demonstrated that such crystals deposited from male urine contain less CMP than do those from female, and those precipitated from samples that have been subjected to preliminary clean-up centrifugation and filtration (C&F) contain less CMP than do those from native urines. The aim of this study was to determine whether urinary CMP concentration is reduced by C&F and if so, whether the magnitude of the reduction is related to the sex of the urine donor.

Fresh random urine specimens were collected from 40 healthy subjects comprising 20 men and 20 women, and the CMP concentration was determined before and after 10,000xg centrifugation and 0.22 µm filtration (Gelman Polysulfone ACRODISC Disposable Filter Assembly), using an enzyme-linked immunosorbent assay (ELISA) based on a polyclonal antibody to CMP raised in New Zealand white rabbits.

Overall, C&F had no effect on CMP concentration: the median values before and after treatment were 10.47 mg/L (range 2.4-29.4 mg/L; n=40) and 10.91mg/L (range 2.1-28.5 mg/L; n=40).

There was no significant difference between the concentration of CMP in the untreated women's urines (median 12.4: range 2.4-27.0 mg/L; n=20) and that in the men's (median 9.9: range 2.5-14.1 mg/L; n=20), confirming other results from our laboratory.

A further 10 male and 10 female urines were treated by 10,000xg centrifugation and 0.22µm filtration using Millipore membrane filters, type GS. There was a significant ($p<0.0002$) loss of CMP from a median value of 4 mg/L (range 0.2 - 12.1 mg/L; n=20) in the centrifuged urines to 0.1 mg/L (range 0.001 - 1.9 g/L; n=20) in those which had also been filtered. However, there was no difference between the extent of this loss in the male and female urines.

It was concluded that filtration through type GS Millipore membrane filters causes an almost complete removal of CMP from urine which does not occur with Gelman Polysulfone ACRODISC Disposable Filter Assemblies. This loss is responsible for the reduced amount of CMP in crystals generated from urine which has been C&F. The discrepant amounts of CMP in crystals derived from male and female urines does not reflect differences in urinary CMP concentration, but may be a consequence of different binding properties of the protein from each source.

QUANTIFICATION OF MINERAL ENCRUSTATION OF CATHETER MATERIALS IN A FLOW SYSTEM OF CRYSTALLIZATION

P. Kollenbach, W. Achilles, C. Haacke and H. Feiber

Philipps Universität Marburg
Urologische Universitätsklinik
D-3550 Marburg/Lahn, Germany

INTRODUCTION AND METHODS

Deposition of mineral phases on urinary catheters during long-term use is a common problem, especially in elderly urological patients. The aim of this study was to quantify the ability of catheter materials to resist mineral encrustation using a flow model of crystallization.

Deposition of calcium phosphate phases and struvite (ammonium magnesium phosphate) were produced by pumping supersaturated urinary solutions through catheters at 37°C (tube length: 10 cm; reaction time: 4 h; flow rate: 0.5 mL/min). Supersaturation was achieved within the tubes by mixing two different solutions (artificial or native urine with aqueous ammonia) from a multi-channel peristaltic pump. The amount of mineralization was quantified by determination of total calcium, magnesium and phosphorus using simultaneous ICP atomic emission spectroscopy, after acidic elution from the tubes.

RESULTS AND DISCUSSION

From the plastic materials under investigation, polyvinyl chloride, polyethylene, and silicon were similar at resisting encrustation *in vitro*. Tubes made of polyurethane showed the lowest rate of deposition in the model applied. Amounts of calcium phosphates and struvite precipitated within the tubes were 20-30% smaller than those deposited in material of standard polyvinyl chloride. Black printing on polyurethane tubes caused an increase of mineral deposits by 20-30%.

Sulfonation of plastic surfaces resulted in a remarkable enhancement of mineralization by more than 100%. This indicates that negatively charged groups, if immobilized on surfaces, induce crystal formation via the primary binding of cations, in contrast to their property of inhibiting crystal growth if present in solution. This phenomenon may be compared with the dual role of polyelectrolytes in inhibiting or promoting crystallization processes depending on their concentration of solute or fixed state.

The flow crystallization model used here was shown to be useful in differentiating between different materials with respect to their ability to deposit minerals from artificial or native urine. It may be successfully applied to control the improvement of the ability of catheter materials (eg by chemical or physical modification of plastic surfaces) to resist natural encrustation of catheters in patients.

NEPHROCALCIN IN CHILDREN WITH CALCIUM OXALATE STONE DISEASE

M.F. Netzer[1], T. Davies[1] and Y. Nakagawa[2]

[1]Urology Clinic
University of Saarland
6650 Homburg/Saar, Germany
[2]Program in Nephrology
University of Chicago
Chicago, IL USA

INTRODUCTION AND RESULTS

Nephrocalcin (NC) is a renal tubular glycoprotein and is a major urinary inhibitor of calcium oxalate monohydrate (COM) crystal growth. It has been isolated from human urine, human kidney tissue culture medium, renal stones and kidney tissue of nine vertebrate species. NC from adult stone-formers is biochemically different from NC produced by healthy adults and inhibits COM crystal growth less strongly. We partially purified NC from urine of children with COM stone disease and healthy children. The NC elution patterns from an anion-exchange chromatography column were compared to those from healthy and stone-forming adults.

Filtered, dialysed urine was applied to a DEAE column and eluted with a linear NaCl gradient (0.05 to 0.55M) in 0.05M Tris-HCl pH 7.3 buffer. Fractions were collected and assayed for COM crystal growth inhibition, absorbance at 230 nm and conductivity.

Table. Summary of NC elution

Urine sample		NC elution fraction (%)			
		A	B	C	D
Adult:	stone (n=18)[1]	24	27	36	13
	healthy (n=20)[1]	22	36	34	8
Children:	stone (n=5)	25	22	13	40
	healthy (n=2)	27	23	27	23

[1]Summary of published data

DISCUSSION

Children with COM stone disease produce more of the most negatively charged, poorly functioning NC (fraction D). On average, 40 % of total inhibitor activity was from this fraction compared with 8-23 % in the other groups examined. The prevalence in urine of this form of NC may contribute to their stone disease.

SECTION V: STONE COMPOSITION, MATRIX AND MEMBRANES

ROLE OF ORGANIC MATRIX IN THE FORMATION AND GROWTH OF CALCIUM OXALATE URINARY STONES

S.R. Khan and R.L. Hackett

Department of Pathology
College of Medicine
University of Florida
Gainesville
Florida, USA

INTRODUCTION

Organic matrix is an integral part of all urinary stones. It accounts for 2-3% of a stone's total dry weight and its presence has long been thought essential for stone formation. It has been proposed that matrix plays an active architectonic role in the morphogenesis of urinary stones. To better understand the role of organic matrix in stone formation we studied calcium oxalate crystal-matrix association in: 1) crystalluria particles obtained from male hyperoxaluric rats and 2) surgically removed human urinary stones.

MATERIALS AND METHODS

Crystalluria was induced in male Sprague-Dawley rats by administration of ethylene glycol through drinking water (0.25 % v/v) and gentamicin sulphate by once a day subcutaneous injection at a dose of 100/mg/kg rat body weight[1]. Crystals were isolated from the bladder aspirate by centrifugation. Crystalluria particles or human stone pieces were embedded in agar and demineralized by treatment with a mixture of ethylenediaminetetraacetic acid (EDTA) and a glutaraldehyde-paraformaldehyde fixative solution[2]. EDTA-insoluble matrix, stones and crystals were examined using light microscopy and scanning and transmission electron microscopic techniques. Natural stones without decalcification were also examined by scanning electron microscopy.

RESULTS

Loss of calcific material but preservation of EDTA insoluble organic matrix, resulted in the formation of crystal ghosts and maintenance of crystal/stone architecture (Fig. 1A) with distinct concentric laminations. All crystal ghosts, irrespective of the crystal source and location in the stones, were limited by a thin organic outer coat and contained organic material inside (Fig. 1B). EDTA insoluble stone matrix (Fig. 1C) contained amorphous, fibrillar and membranous components, some of which could be identified as cellular degradation products[3].

Urolithiasis 2, Edited by R. Ryall *et al.*,
Plenum Press, New York, 1994

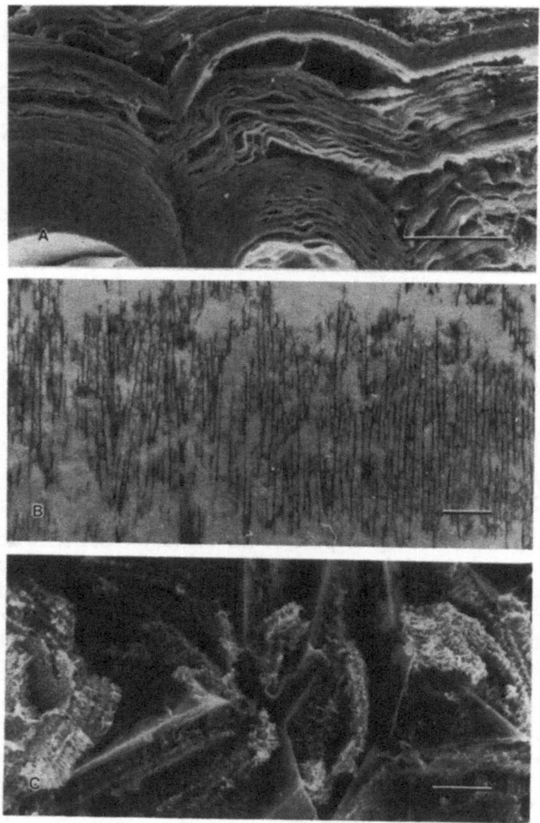

Fig.1. Ultrastructure of decalcified urinary crystals and stones. **A.** Scanning micrograph of a calcium oxalate monohydrate stone showing laminated structure and maintenance of the stone architecture. Bar = 250/μm **B.** Transmission electron micrograph of a calcium oxalate monohydrate stone showing plate-like crystal ghosts. Bar=0.5 μm **C.** Scanning electron micrograph of a calcium oxalate dihydrate stone showing both amorphous and fibrillar components of the EDTA-insoluble stone matrix. Bar=10 μm

In the animal model, ethylene glycol administration at this dosage resulted in low level hyperoxaluria and increased urinary calcium oxalate relative supersaturation[1]. But no calcium oxalate crystals were seen in the urine. However, when the hyperoxaluric rats were made membranuric by additional gentamicin administration, calcium oxalate crystals were present in the urine[1]. The crystals were associated with membranous material (Fig. 1C). In an earlier study[2] we reported that crystal aggregates were surrounded by a coating of cellular degradation products. Matrix of these aggregated crystals was highly electron dense in their nuclear zone and was arranged concentrically in the peripheral zone.

DISCUSSION

Requirement of membranuria for the development of crystals in rat urine metastable for calcium oxalate and the presence of organic material within an organic membrane bound crystal ghosts indicate that involvement of organic material during crystallization in urine starts very early. Presence of a highly dense nucleus within the crystal ghosts also suggests an early participation by urinary macromolecules in crystal nucleation. Calcium oxalate crystals are known to preferentially acquire some proteins from the urine[4]. Organic material found associated with the crystals may be involved in crystal nucleation and may deter the dissolution of incipient nuclei. Membranes isolated from the kidney brush border are capable of inducing calcium oxalate crystallization in a metastable calcium oxalate solution[5] and various components of the cellular membranes have been implicated in the induction of both physiological and pathological calcification processes[6].

Membranous cellular degradation products seen surrounding the crystal aggregates[2] may hold the crystals together in close proximity until they are joined by crystal bridging. Formation of large aggregates facilitates crystal retention within the urinary system[7] and may be the first step towards the formation of a concretion. Thus membranous and macromolecular organic material present in urinary stones as the organic matrix may play an important role in the development of urinary stones.

ACKNOWLEDGEMENTS

This work was supported by NIH grants #ROIDK41434 and #POIDK20586.

REFERENCES

1. RL Hackett, PN Shevock, and SR Khan, Cell injury associated calcium oxalate crystalluria, *J Urol* 144:1535 (1990).
2. SR Khan and RL Hackett, Crystal-matrix relationships in experimentally induced calcium oxalate monohydrate crystals, an ultrastructural study, *Calcif Tissue Intl* 41:157 (1987).
3. B Finlayson, SR Khan and RL Hackett, Mechanisms of stone formation - an overview, *Scan Electr Microsc* 3:1419 (1984).
4. RM Morse and MI Resnick, A study of the incorporation of urinary macromolecules onto crystals of different mineral compositions, *J Urol* 141:641(1989).
5. SR Khan, PN Shevock and RL Hackett, Membrane associated crystallization of calcium oxalate *in vitro*, *Calcif Tissue Intl* 46:116 (1990).
6. AL Boskey, Current concepts of the physiology and biochemistry of calcification, *Clin Orthopedics* 157:225 (1981).
7. SR Khan and RL Hackett, Retention of calcium oxalate crystals in renal tubules, *Scan Microsc* 5:707 (1992).

CALCIUM OXALATE CRYSTALS AND OXALATE IONS ARE INJURIOUS TO RENAL EPITHELIAL CELLS

R.L. Hackett[1], P.N. Shevock[2] and S.R. Khan[1,2]

[1]Department of Pathology
[2]Department of Surgery
College of Medicine
University of Florida
Gainesville, Florida

INTRODUCTION

Human stones and experimentally produced calcium oxalate (CaOx) renal stones are composed of a crystalline and non-crystalline phase, the latter containing a variety of substances, some of which closely resemble cell membranes both structurally and chemically. Experimentally, CaOx crystals can be formed as a result of heterogeneous nucleation in models in which the crystals are induced by administration of a nephrotoxin and an oxalate lithogen. In utilizing *in vivo* animal models it is difficult to determine whether injury precedes or results from exposure to crystals or oxalate. We decided to examine this question using a tissue culture model, the MDCK cell, in order to more closely monitor and examine the interaction between crystals, oxalate and cells. To assess cellular changes, we processed cultures for examination by phase and scanning microscopy and assessed others by trypan blue dye exclusion as well as specific activities of membrane enzymes.

MATERIALS AND METHODS

Cell Culture

MDCK cells (CCL 34) were obtained from ATCC. Cells were plated onto 25 cm^2 flasks in Dulbecco's Modification of Eagle's Medium (DMEM) with 10% newborn calf serum, 1 g/L glucose, antibiotics and 10 mM HEPES, pH 7.40. Cultures were maintained in 5% CO_2 air atmosphere at 37°C. To eliminate interferences from the pH indicator during enzyme analysis, medium without phenol red was replenished on the day of the experiment with or without the addition of KOx or CaOx crystals.

Crystal Experiments

Calcium oxalate monohydrate (COM) was prepared by mixing equimolar solutions of $CaCl_2$ and $K_2C_2O_4$ as described by Brown, et al[1]. Preparation was confirmed by X-Ray diffraction analysis. Calcium oxalate dihydrate (COD) was prepared in a magnesium-citrate buffer, by adding $CaCl_2$ and $K_2C_2O_4$ at a ratio of 0.01 M: 0.002 M, as described by Brown[2]. X-Ray diffraction analysis confirmed a pure COD preparation. Both crystal types were weighed, UV-irradiated and resuspended in medium described above at a 10 mg/mL concentration. Slurry was equilibrated at 37°C and sterility assured before addition to cultures. Slurries were added to monolayers at a final concentration of 500 μg/mL.

Urolithiasis 2, Edited by R. Ryall *et al.*,
Plenum Press, New York, 1994

Oxalate Experiments

$K_2C_2O_4$ stock solution was prepared in normal sterile saline and 0.2 μm-filtered. Stock was standardized using atomic absorption spectrophotometry. KOx was added with fresh medium to monolayers at concentrations ranging from 0.0375 mM to 0.25 mM at the onset of the experiment.

Cell viability as determined by trypan blue exclusion

Monolayers were trypsinized and resuspended in Hank's balanced salt solution. Single cell suspensions were stained with 0.4% isotonic trypan blue and counted with a hemacytometer.

Cell lysis and biochemical analysis

Single cells were re-pelleted and lysed in an EDTA-sucrose buffer with Triton-X 100. Lysis was facilitated by freeze-thawing. Protein was determined using a modified Lowry procedure for membrane bound proteins[3]. Gamma-glutamyl transpeptidase (γ-GT) was determined at pH 9.0 using L-glutamyl-*p*-nitroanilide as a substrate and glycylglycine as an acceptor. Release of *p*-nitroanilide was measured in an azo-dye reaction[4]. N-acetyl-ß-glucosaminidase (NAG) was determined in citrate buffer using p-nitrophenyl-N-acetyl-ß-D-glucosaminide as a substrate. Release of *p*-nitrophenyl was measured[5].

RESULTS

Figure 1a, at 24 h, demonstrates the overgrowth of crystals by cells and resulting detachment. In other reports, endocytosis of crystals has been reported but this mechanism could not be evaluated by SEM. Cell numbers were significantly decreased, up to 40% in crystal studies and up to 25% in oxalate studies. The ability to exclude trypan blue was decreased from 98% to 87% ($p < 0.001$). The cell membrane enzyme γ-GT was consistently increased in the remaining cells in culture as compared to the lysosomal enzyme, NAG. Similar results occurred with COM crystals but the changes appeared to occur at a faster rate.

Figure 1b illustrates the severe cell surface pitting seen in cells exposed to oxalate. Cell death and loss of cells from culture also occurred, similar to the crystal studies. However, the enzyme patterns were different as NAG was consistently elevated compared to γ-GT. These alterations are not specific as they have been described in other studies.

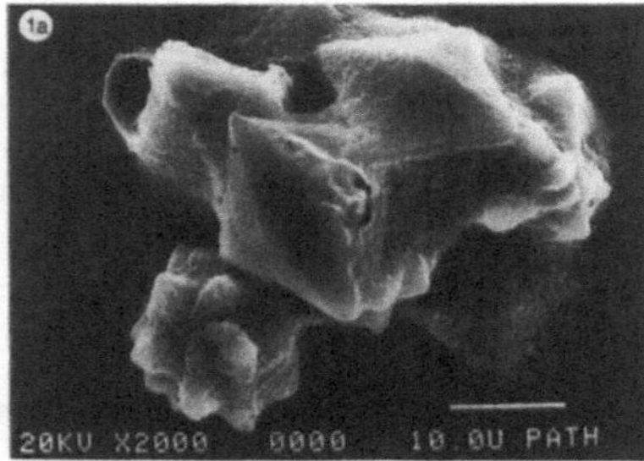

Fig. 1a. (Left) 24 h incubation with COD: Conglomerate of crystals with cells.

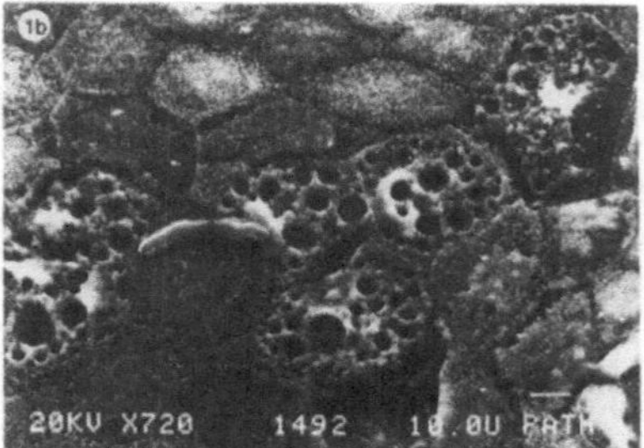

Fig. 1b. (Right) 0.15 mM KOx at 72 h: Pronounced cell surface pitting.

CONCLUSIONS

1. Both CaOx crystals and oxalate are toxic to MDCK cells in culture.

2. Crystal Studies

a. Etching of COD crystals occurs only when crystals are exposed to cells in culture, and not when exposed to media alone: Cells or their products influence COD crystal structure.

b. Calcium oxalate crystals interact with MDCK cells in a step-wise pattern:
1) The crystals are first entangled in the cilia and microvilli of the MDCK cell.
2) Overgrowth of crystals by culture occurs. Endocytosis of crystals by MDCK cells has been reported but SEM could not evaluate that mechanism.
3) Conglomerates composed of crystals and various components of cell structure are released into the medium.

c. The cellular enzyme, γ-GT, usually considered to be a cell membrane enzyme, is consistently more elevated in the remaining cells in culture than is NAG, a lysosomal enzyme.

3. Oxalate Studies

a. Cell damage as demonstrated by severe pitting of cell surface results from exposure to KOx.

b. The cellular enzyme, NAG is consistently more elevated in remaining cells in culture, compared to γ-GT.

c. These changes occurred most dramatically within 24 h regardless of oxalate concentration.

d. Neither change is specific as similar alterations have been described in other studies.

ACKNOWLEDGEMENTS

This work was supported by NIH grant PO1 DK20586.

REFERENCES

1. CM Brown, DK Ackermann, DL Purich and B Finlayson, Nucleation of calcium oxalate monohydrate: use of computer-assisted simulations in characterizing early events of crystal formation, *J Crystal Growth* 108: 455 (1991).
2. P Brown, D Ackermann and B Finlayson, Calcium oxalate dihydrate production. *in:* "Urolithiasis", VR Walker, RAL Sutton, ECB Cameron and CYC Pak (eds.), Plenum Press, New York, 193 (1989).
3. MAK Markwell, SM Haas, NE Tolbert and LL Bieber, Protein determination in membrane and lipoprotein samples: manual and automated procedures *Methods Enzymol* 72: 296 (1981).
4. L Natflin, M Sexton, JF Whitaker and D Tracey, A routine procedure for estimating γ-glutamyl transpeptidase activity, *Clin Chim Acta* 26: 293 (1969).
5. R Kornfeld and C Siemers, Large scale isolation and characterization of calf thymocyte plasma membranes, *J Biol Chem* 249: 1295 (1974).

THE PROTEIN CONTENT OF URINARY CALCULI IN CHINESE PATIENTS

Y. Lee, H. Shen, L.S. Chang, M. Chen, B. Jiaan and J. Huang

Department of Surgery (Urology) and Medical Research
Veterans General Hospital
Kaohsiung and Taipei
Taiwan National Yang-Ming Medical College
Taipei, Taiwan, Republic Of China

INTRODUCTION

In 1990, Jones and Resnick utilized two-dimensional polyacrylamide gel electrophoresis (PAGE) to characterize soluble matrix proteins in human renal calculi[1]. They suggested that different urinary calculi had different matrix protein maps and the majority of the protein maps showed proteins with low molecular weight (MW < 17.5 kDa). By using a similar method, we analyzed the protein map of 36 consecutive urinary calculi. In order to prevent the unavoidable loss of protein during fragmentation of urinary calculi by extracorporeal shock wave lithotripsy, lasertripsy, electrohydraulic and ultrasonic lithotripsy, only stones removed by open surgery were included in this study. The inorganic composition of stones were analyzed by Jasco IR-700 infrared spectrophotometer was described previously[2].

RESULTS

The crystalline compositions of 36 urinary calculi were whewellite in five, uric acid four, struvite two, carbonate apatite two, whewellite and carbonate apatite mixture three, whewellite and weddelite mixture two, struvite and carbonate apatite mixture three, whewellite, weddelite and carbonate apatite mixture fifteen. The molecular weights (MW) and isoelectric points (PI) of each protein spot in different urinary calculi are shown in Table 1. Some protein spots were detected in more than one kind of stone. According to the MW and PI matching, the spot designated as "B" probably represents uromucoid or Tamm Horsfall protein,[3] spot "D" may represent albumin[4], and spot "N" may represent IgG light chain[5]. In the meantime, the results obtained from one - and two - dimensional PAGE were consistent in the same stone.

The weight of protein content in different urinary calculi also varied (Table 2), but because the number of cases for each class was so different, no statistical significance was noted. However, the carbonate apatite stone tended to contain more protein than other stones, followed by uric acid, struvite and carbonate apatite mixture; on the contrary, pure whewellite stone contained the least amount of protein among all the stones analyzed.

Table 1. The protein components in different urinary calculi.

Protein	MW(kDa)	PI range	Whewellite	Apatite	Struvite	Uric acid
A	88	7.4-7.6			+	
B	87	6.7-7.1	+			
C	74	5.8-6.0				+
D	67	6.6-7.0	+		+	+
E	62	5.3-S.5	+			
F	57	7.0-7.2			+	
G	52	5.9-6.2	+		+	
H	49	5.1-5.7		+		+
I	43	6.0-7.2		+		
J	41	5.9-6.6	+	+	+	
K	40	6.8-7.0			+	
L	35	6.4-7.2	+			+
M	31	5.9-6.6		+		+
N	29	6.2-7.7		+	+	+
O	21	5.8-6.3		+		
P	16	5.8-7.6		+	+	
Q	14	5.6-7.6		+	+	+
R	<14	5.6-7.6	+	+	+	+

Table 2. The protein content weight in different urinary calculi.

Stone composition	μg protein / g stone
Struvite	16.50±10.72
Apatite	45.61±17.20
Uric acid	24.94±16.57
Whewellite	14.77±4.30
Struvite + Apatite	21.84±12.03
Whewellite + Weddellite	19.09±3.34
Whewellite + Weddellite + Apatite	18.10±13.44

DISCUSSION

Two-dimensional SDS-PAGE has been widely applied in protein analysis. However, because methods available for protein extraction yielded relatively insoluble material, the application of two-dimensional gel electrophoresis in analyzing matrix proteins of stones had not been reported until 1990 when Jones and Resnick used 10% acetic acid to extract matrix proteins. They concluded that the electrophoretic patterns of matrix proteins were complex and reproducible, similar to the two-dimensional electrophoretic patterns of urinary proteins recovered from urinary crystals, that the patterns were different when based on the crystalline content of the stone and remarkable for the preponderance of low molecular weight (MW < 17.5 kDa)[1]. Results obtained from the present study showed a similar finding that part of the soluble proteins of stone matrix were of low molecular weight (equal or less than 14 kDa). Furthermore, we found that different urinary calculi contained a varied degree of protein content and carbonate apatite contained more soluble proteins than other stones. These findings also suggest that urinary proteins may be actively incorporated within the crystalline structure of the stone during its formation. However, the exact role of protein in urinary calculi formation merits further investigation.

REFERENCES

1. WT Jones and MI Resnick,The characterization of soluble matrix proteins in selected human renal calculi using two-dimensional polyacrylamide gel electrophoresis, *J Urol* 144: 1010 (1990).
2. YH Lee, MT Chen, JK Huang and LS Chang, Analysis of urinary calculi by infrared spectroscopy, *Chin Med J (Taipei)* 45: 157 (1990).
3. N Iwata, S Kamei, Y Abe, S Nishio, A Waikatsuki, K Ochi and M Takeuchi, The organic matrix of urinary uric acid crystals, *J Urol* 139:607 (1988).
4. N Frearson, RD Taylor and SV Perry, Proteins in the urine associated with Duchenne muscular dystrophy and other neuromuscular disease, *Clin Sci* 61:141 (1981).
5. RP Tracy and DS Young, Clinical applications of two-dimensional gel electrophoresis, *in*: "Two dimensional gel electrophoresis of proteins. Methods and applications," JA Clis and R Bravo, eds, Academic Press, Orlando, (1984).

THE PREVENTION OF EXPERIMENTAL NEPHROCALCINOSIS WITH EICOSAPENTAENOIC ACID (EPA) AND EVENING PRIMROSE OIL (GLA)

A.C. Buck, W.S. Smellie, A. James, D. Horrobin and M. Marko

Department of Urology
Glasgow Royal Infirmary
Glasgow, UK

INTRODUCTION

Eicosanoids derived from essential fatty acids comprise the structural components of all cell membranes and determine a variety of membrane functions including fluidity, permeability, ion channels and the behaviour of membrane-associated receptors and enzymes. In a previous report we have shown that the D-3 fatty acid, eicosapentaenoic acid (EPA) derived from fish oil prevents experimental nephrocalcinosis and reduces urinary calcium excretion in an animal model. We have also shown that feeding Evening Primrose oil, a rich source of the D-6 fatty acid, dihomo-γ-linolenic acid (GLA), prevented hypercalciuria in the streptozotocin diabetic rat. Experimental and clinical studies have shown that the effects of D-3 fatty acids (EPA) are potentiated in the presence of the n-6 fatty acid GLA. This study was performed to evaluate the effect of a combination of EPA and GLA on the process of renal parenchymal calcification and solute excretion in an animal model of nephrocalcinosis.

METHODS

Fifty female PVG rats were divided into five groups. Nephrocalcinosis was induced in the control rats (n=10, Group 1) by an intraperitoneal injection (ip) of 10% calcium gluconate for 10 days. Group 2 rats (n=l0) were gavage fed 1 mL of fish oil (EPA 400 mg) for 4 days prior to receiving ip calcium gluconate and the fish oil treatment continued during the 10 day course of calcium gluconate. Group 3 rats (n=10) were gavage fed Evening Primrose Oil (GLA) prior to calcium gluconate. Group 4 rats (n=10) were fed a combination of Evening Primrose oil and fish oil and Group 5 rats (n=10) were fed sunflower oil according to the same regimen. Twenty-four hour calcium excretion was measured before treatment, at days four, eight and fourteen. The animals were killed and the kidneys examined histologically for the presence of calcification and for calcium content by wet chemical analysis.

Urolithiasis 2, Edited by R. Ryall *et al.*,
Plenum Press, New York, 1994

RESULTS

Extensive, dense nephrocalcinosis was present in the kidneys of the control animals. By contrast calcification was absent or only faintly present on histology in the kidneys of the rats treated with the various oils. Tissue calcium concentration was significantly less ($p < 0.005$). Urine calcium was significantly reduced ($p < 0.001$) at day four on oil treatment and did not rise to the levels that occurred in the rats given calcium gluconate alone or sunflower oil, particularly in the groups treated with Evening Primrose oil and the combination of GLA plus fish oil.

INCIDENCE AND FORMATION OF MIXED KIDNEY STONES CONTAINING CALCIUM OXALATE AND URIC ACID

P.T. Cheng[1], K. Tupy[1] and A. Pierratos[2]

[1]Departments of Pathology and Clinical Biochemistry
Mount Sinai Hospital
[2]Division of Nephrology
Wellesley Hospital
University of Toronto
Toronto, Canada

INTRODUCTION

Theoretical consideration of the crystal lattice dimensions of calcium oxalate monohydrate (COM) and those of uric acid (UA) shows that there are dimensionally closely matched crystal surfaces and suggests that epitaxial growth can take place between these two common kidney stone components. *In vitro* demonstration of epitaxial growth of COM crystallites on well developed (100) faces of UA crystals has added support to the theory. Another study has shown that monosodium urate (MSU) is a better heterogeneous nucleator for COM, but MSU crystals are rarely observed in kidney stones. The important question of whether or not epitaxial growth of COM on UA or vice versa actually takes place *in vivo* has not been answered. Obviously, a direct observation *in situ* is not practical. Indirectly, we can look carefully at the end products - mixed kidney stones containing UA and COM or calcium oxalate dihydrate (COD), for clues of such an event.

METHODS

The compositions of all kidney stone fragments recovered from the urines of patients after lithotripsy in Toronto were determined by X-ray powder diffraction. From this data base, the incidences of pure COM or COD or UA, and of COM+UA or COD+UA mixed stones were determined for a period of four years (1988-91). The urinary chemistry results, including pH and soluble UA level, from these identified patients were then analyzed. The structural relationship between COM and UA in the mixed stones was studied by scanning electron microscopy (SEM) and energy dispersive X-ray (EDX) microanalysis.

RESULTS

A total of 3968 stones were analyzed. After excluding stones that contained any component other than COM, COD, UA, the following results were obtained:

Pure stones			Binary mixed stones		
COM	25.5	± 2.1%	COM+UA	1.6	± 0.4%
COD	3.5	± 0.2%	COD+UA	0.0%	(n=1)
UA	1.7	± 0.6%			

Both urine pH and urinary UA results showed significant differences between the average values of pure COM and pure COD patient groups, as follows:

Urine pH			Urinary UA		
COM	5.72	± 0.75	COM	2.9	± 1.2 (n=466)
COD	5.93	± 0.72	COD	3.4	± 2.3 (n=75)
	($p < 0.05$)			($p < 0.05$)	

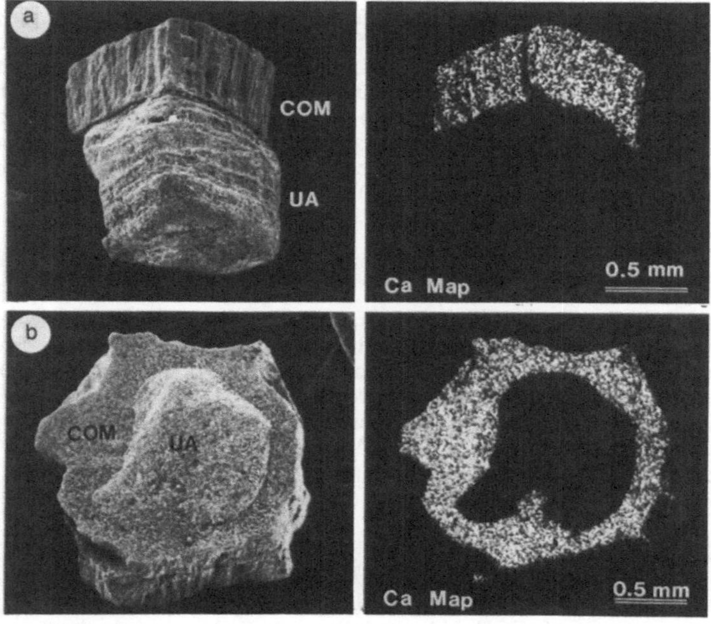

Figure 1. SEM micrographs and EDX Ca maps of COM+UA stone fragments, from two different patients, which have UA cores and COM shells.

Figure 2. Two different ultrastructural relationships between COM and UA: a) tight COM shell/UA core junction; b) UA crystal clusters on COM stone surface.

DISCUSSION

The results show that, for the Toronto lithotripsy patient group, the incidence of COM+UA is as high as that of pure UA. This is high as the conventional value for mixed to pure UA ratio is only 1:4. This probably reflects the selection criteria for lithotripsy patients.

Despite the significantly higher urinary UA level for patients with pure COD stones, the formation of COD+UA stones is inhibited. The significantly higher urine pH for COD patients should contribute to the inhibition.

The SEM and EDX results show that COM crystals do grow on UA stone surfaces, some with tight junctions (Fig. 2a); but we do not know whether the nucleation is epitaxial or non-specific in nature. The same can be said for the growth of UA crystal clusters on COM stone surfaces (Fig. 2b).

We conclude that 1) both COM nucleation by UA and vice versa exist *in vivo*, 2) the exact nucleation mechanism has not been determined here, and 3) COD+UA stones are very rare due to the low incidence of COD stones and to the higher urine pH of COD patients.

ACKNOWLEDGEMENTS

We thank the E C Bovey Lithotripsy Unit, Wellesley Hospital, University of Toronto for collaboration and support.

REFERENCES

1. S Deganello and FL Coe, Epitaxy between uric and whewellite: experimental verification, *N Jb Miner* Mh. H6:270 (1983).
2. K Lonsdale, Human stone, *Science* 159:1159 (1968)a.
3. K Lonsdale, Epitaxy as a growth factor in urinary calculi and gallstones, *Nature* 217:56 (1968)b.
4. CYC Pak, Y Hayashi and LH Arnold, Heterogeneous nucleation between urate, calcium phosphate and calcium oxalate, *Proc Soc Exp Biol Med* 153:83 (1976).
5. EL Prien and EL Prien, Jr, Composition and structure of urinary stone, *Am J Med* 45:654 (1968).

Figure ... COD ... and UV ...

DISCUSSION

The results ... for the ... assessment of ... the incidence of
COD ... patient ...

The ... analysis show ... on growth in DO pond subjected ...

With ... with COD ...

ACKNOWLEDGEMENTS

We wish ... Hospital, University of Toronto.

REFERENCES

PRELIMINARY RESULT OF CALCIUM OXALATE CRYSTAL INTERACTION WITH MADIN-DARBY CANINE KIDNEY CELLS IN CULTURE

S. Ebisuno, T. Yoshida, Y. Kohjimoto, M. Uehara and T. Ohkawa

Department of Urology
Wakayama Medical College
Wakayama, 640, Japan

INTRODUCTION

Recently, interactions of crystals and tubular cell of the kidney have been extensively examined on inner medullary collecting duct epithelial cells in primary culture by Wiessner et al[1] and Riese et al[2,3]. However, the exact and specific processes that mediate attachment of stone crystals to tubular epithelium are not yet known.

Madin-Darby canine kidney (MDCK) cells exhibit many of the characteristics of the cortical collecting tubule. The present investigation is designed to study adhesion of calcium oxalate crystals on the surface of intact MDCK cells, and to measure the effects of glycosaminoglycans (GAGs) on these adhesions quantitatively.

MATERIALS AND METHODS

MDCK cells were obtained from American Type Culture Collection and maintained in Minimum essential medium (MEM) with 10% fetal calf serum. The cells in suspension (2 x 10^5/ml) were cultured in 35 mm culture dish in 3 days, for the estimation of the adhesion of calcium oxalate crystals.

Crystals of calcium oxalate monohydrate (COM) were prepared by adding equimolar amounts of calcium chloride(10 mM) and sodium oxalate (10 mM) solutions according to the method by Pak[4].

After cells reached confluence in 35 mm culture dishes, they were rinsed twice with 5 mL of Hank's balanced salt solution (HBSS). The culture dishes were then dipped vertically into a slowly stirring COM crystal suspension in Tris-HCl buffer containing 0.15 M NaCl (pH 7.4). The crystal-adhesion monolayer MDCK cells were lysed with 10% trichloroacetic acid after washing with 5 mL of HBSS three times, and then scraped using a rubber policeman meticulously. The calcium concentrations of COM on the cells in culture dish were finally measured by an atomic absorption analysis after the complete harvest of the cells.

After the cells were pre-treated with GAGs (2 mL in HBSS) for 1 minutes before exposure of COM crystals, the effects of GAGs on the COM crystal adhesions were immediately measured in an identical manner. In the present experiment, following GAGs were studied; heparin, sodium pentosan polysulphate, chondroitin sulphate A, -B and -C, hyaluronic acid and heparan sulphate.

Urolithiasis 2, Edited by R. Ryall *et al.*,
Plenum Press, New York, 1994

RESULTS

Figure 1 is a phase-contrast micrograph of MDCK cells that have been exposed to COM crystals (2 mg/mL, 2 min). The COM crystals adhered on the surface of the cell bodies predominantly rather than intercellular spaces. The crystal adhesion on the cells seemed to be very strong, and seemed to be actively bound or trapped under a finding of scanning electron microscopy (Figure 2).

The amounts of COM crystals were loosened by washing the cells, however those crystals were not decreased after washing with HBSS three times or more.

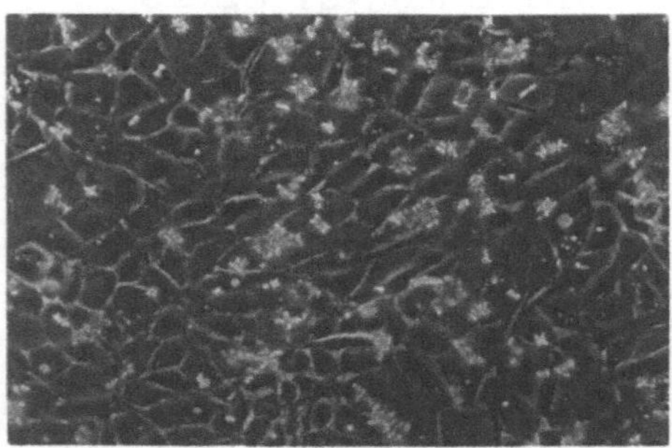

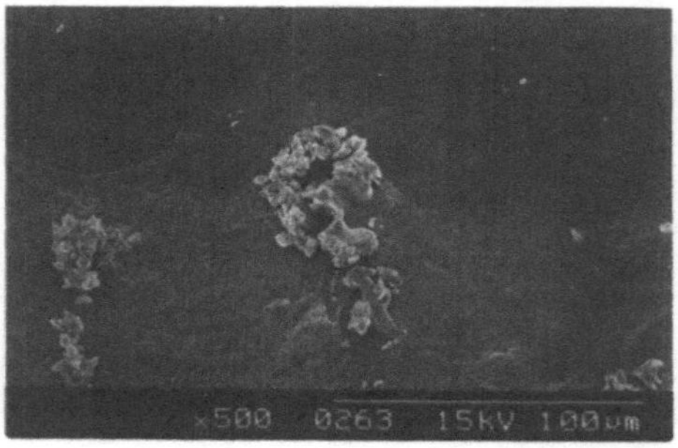

Fig. 1 (Left) Phase-contrast microscopic finding.

Fig. 2 (Right) SEM finding.

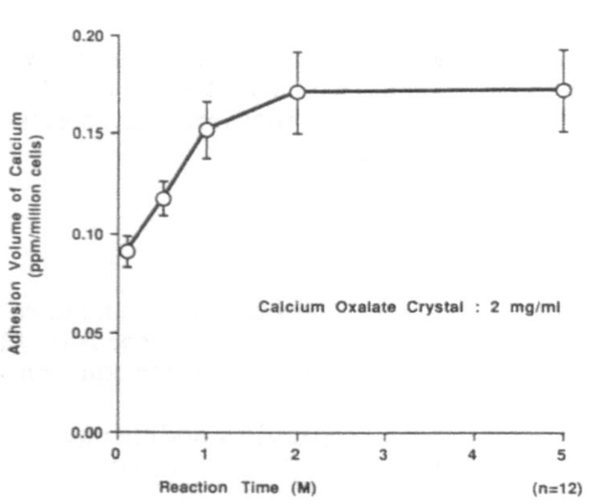

Fig.3. (Left) Time course of adhesion of COM adhesion crystal on the MDCK cells.

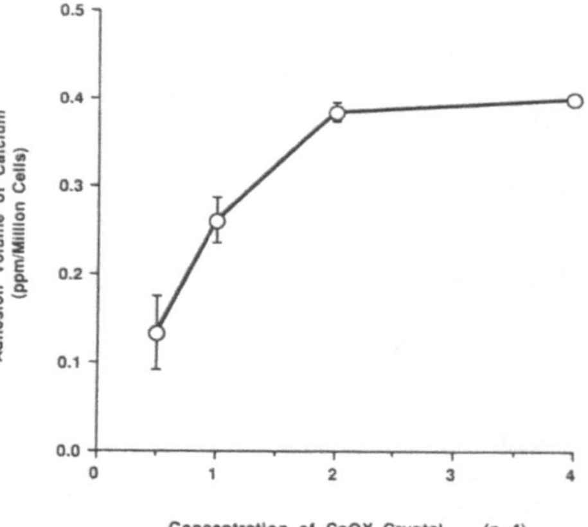

Fig.4. (Right) Concentration dependence of of COM crystal on the MDCK cells.

The COM crystals adhered to MDCK cells in a time dependent manner, with the amount plateauing at 2 minutes (Figure 3).

Increasing the concentration in the suspension of COM crystals from 0.5 mg/mLto 4.0 mg/mL led to a progressive increase in the amount of adhesive crystals plateauing at 2 mg/mL (Fig. 4).

The pre-treatment with low concentration of GAGs (heparin, sodium pentosan polysulphate, chondroitin sulphate A, -B and -C, hyaluronic acid, and heparan sulphate) produced significant reductions of the adhesive amount of COM crystals. And these effects were proportional to these concentrations. In these drugs, hyaluronic acid had the most sensitive effect against the crystal adhesion as compared with other GAGs.

Table 1. Comparison of the inhibitory effects of GAGs on the adhesions of COM crystals.

	Heparin	Sodium	Hyaluronic A	Heparin 5	Chond S-A	Chond S-B	Chond S-C
0.0001mg/mL	22.9	20.7	0	9.2	17.5	0	22.1
0.001mg/mL	30.5	23.5	62.3	16.8	40.5	49.8	49.3
0.01mg/mL	53.7	39.5	74.2	32.7	57.4	65.3	67.6
0.1mg/mL	83.1	79.7	84.1	85.2	86.6	93.4	88.6

The results (% of control) were expressed as means of 4 experiments.

DISCUSSION

Attachment of microcrystals to tubular cell membranes may be an important component in the pathophysiology of renal stones. Particularly a relationship between crystal deposition and tissue injury seems to have great importance for elucidation of the process of stone genesis. Recently, Wiessner and associates have reported an increased calcium oxalate crystal adherence to rounded up cell as compared with flat monolayer cells in rat papillary collecting tubule cells in primary culture[1]. Furthermore in their laboratory, when those cells were treated with EGTA, the crystal adherence to the cells was increased drastically, and the crystal attachment returned to control level after removal of EGTA or readdition of calcium[3]. Thereby they suggested that alteration of cell surface characteristics or loss of epithelial membrane polarity may result in enhanced capacity to bind calcium oxalate.

In the present study, COM crystals adhered to the MDCK cells considerably, although the cells suffered no resulting injuries. We do not currently know if there are specific sites of the cell surface for crystal adherence, and how this phenomenon would affect some substances on the cell surface. Recent study has shown that there are some glycoproteins on the apical or basolateral membrane in MDCK cells[5]. It is assumed that the adherence of crystals to the tubular cells may depend on some specificities of their surface proteins or some GAGs.

The normal urothelium is covered by a GAG layer that acts as a barrier to the adhesion of crystals. Destruction of the GAG layer increased the number of adhered crystals, while treatment by exogenous GAG reduced the adherence[6]. In the present results, all examined GAGs showed significant reductions of the adherent crystals. It is therefore suggested that GAGs could "coat" the cell surface and reduce their ability to adhere the crystals.

Our studies may support the hypothesis of attachment of microcrystals to the cellular membrane, which is one of the most important and earliest processes of the pathophysiology of kidney stones. The current quantitative system on MDCK cells may serve as a useful model for the investigation of interactions with crystals and tubular cells. It is also possible to gain insights into those detail characteristics, and to assess the COM adhesion quantitatively under various conditions.

REFERENCES

1. JH Wiessner, JG Kleiman, SS Blumenthal, JC Garancis, GS Mandel and NS Mandel, Calcium oxalate crystal interaction with rat renal inner papillary collecting tubule cells, *J Urol* 138: 640 (1987).
2. RJ Riese, JW Riese, JG Kleinman, JH Wiessner, GS Mandel and NS Mandel, Specificity in calcium oxalate adherence to papillary epithelial cells in culture, *Am J Physiol* 255: F1025 (1988).
3. RJ Riese, NS Mandel, JH Wiessner, GS Mandel, CG Becker and JG Kleinman, Cell polarity and calcium oxalate crystal adherence to cultured collecting duct cells, *Am J Physiol* 262: F177 (1992).
4. CYC Pak, M Ohata and K Holt, Effect of diphosphonate on crystallization of calcium oxalate *in vitro*, *Kidney Int* 7: 154 (1975).
5. PJ I Salas, DE Vega-Salas, J Hochman, E Rodriguez-Boulan and M Edidin, Selective anchoring in the specific plasma membrane domain: A role in epithelial cell polarity, *J Cell Biol* 107: 2363 (1988).
6. WB Gill, KW Jones and KJ Ruggiero, Protective effects of heparin and other sulfated glycosaminoglycans on crystal adhesion to injured urothelium, *J Urol* 127: 152 (1982).

INHIBITION OF CALCIUM OXALATE MONOHYDRATE SEED CRYSTAL GROWTH IS DECREASED IN RENAL INJURY

R.L. Hackett[1], P.N. Shevock[2], S.R. Khan[1, 2] and B. Finlayson*

[1]Department of Pathology
[2]Department of Surgery
College of Medicine
University of Florida, Gainesville, FL
*Deceased

INTRODUCTION

We have demonstrated previously that sub-lithogenic doses of ethylene glycol (EG), when combined with administration of the proximal tubule nephrotoxin, gentamicin sulfate (GS), cause calcium oxalate (CaOx) crystalluria in association with cellular degradation products[1]. In this model, consistent elevation of CaOx relative supersaturation (RSS) occurs as calculated by the ion speciation program EQUIL[2]. Because EQUIL calculations do not consider large molecules, we examined the growth inhibitory activity of compounds > or < 10,000 mol. wt. (10K MW) using a multi-well crystallizer technique.

MATERIALS AND METHODS

Animal Protocols and Urine Processing

Ten male Sprague Dawley rats, 200 g each, were divided into 2 groups: Group I were normal controls; Group II received daily subcutaneous injections of 100 mg/kgbw of GS for 6 days. Twenty four-hour iced urine collections were collected daily from Day 1 to Day 6. At the end of each collection period, the volume and pH were noted and urines were processed as follows. Urine samples were warmed to 37°C, 5 µm-filtered and an aliquot was applied to an Amicon YM10 membrane to fractionate material greater or less than 10K MW. All crystallizer experiments were conducted on the same day as the urine was collected.

Multi-Well Crystallizer Procedure

All solutions were 0.2 µm-filtered and standardized using atomic absorption spectrophotometry (AA) before use. 1.0mL of 1.2mM $CaCl_2$ was pipetted into each well of the crystallizer, and equilibrated at 37°C. Urine substrates were added at 1:20 and the same volume of buffer was added to controls. At time=0, 1 mL of ^{14}C-$K_2C_2O_4$ was added simultaneously with 10 µL of the seed slurry.and the contents were agitated with micro-magnetic stirring bars. At time=20, on a well-by-well basis, contents were aspirated and filtered through a 0.2 µm syringe filter. The filtrate was assayed for depletion of calcium,

Urolithiasis 2, Edited by R. Ryall *et al.*,
Plenum Press, New York, 1994

potassium and sodium by AA, and ^{14}C by scintillation counting, and the final pH was noted. Results at time=0 and time=20 were evaluated by the EQUIL program for CaOx RSS. An increasing percentage of RSS depletion in the filtrate was assumed to indicate increasing incorporation of Ca and Ox into the crystalline phase, thus reflecting a decreased inhibition of crystal growth.

RESULTS

Table 1 illustrates that the CaOx RSS in the filtrate was depleted about 20% by 5% whole urine from normal animals and 40-50% by urine from GS-treated rats, in comparison to around 75% in the CaOx seed control. Both normal and GS values differed significantly ($p<0.001$) from the seed control. The p values for GS urine relate to comparison with normal urine. These figures indicate that normal rat urine is strongly inhibitory and that urine from GS animals is considerably less potent, falling about half-way between normal and CaOx seed controls.

Table 1. COM seed crystal growth inhibition with 5% GS whole urine Depletion of Filtrate CaOx RSS; Initial RSS=10.

Day	Nornal Urine	GS-Urine	COM Control
1	8.18 ± 0.82	6.66 ± 1.55 (p <0.005)	2.31 ± 0.05
2	7.72 ± 0.49	5.67 ± 1.13 (p < 0.02)	2.41 ± 0.15
3	8.06 ± 1.24	5.53 ± 0.54 (p<0.002)	2.22 ± 0.06
4	8.00 ± 0.48	5.43 ± 0.42 (p<0.001)	2.11 ± 0.08
5	8.32 ± 0.99	5.09 ± 0.33 (p<0.001)	1.93 ± 0.18
6	8.41 ± 1.06	4.08 ± 0.38 (p<0.001)	2.27 ± 0.06

Table 2 shows the data obtained using the <10K MW fraction. Again, a significant loss of inhibitory activity was evident. The p values for GS result from comparison with normal control urine. The >10K MW fraction also demonstrated some decrease in inhibition, but the changes compared to >10K MW normal rat urine showed relatively minor deviations: CaOx RSS of seed control, 2.13±0.17; >10K MW normal urine, 2.74±0.17; and CaOx RSS after 5 doses of GS, 2.51±0.18. Thus, the loss of inhibition was less dramatic than that observed with the < 10K MW fraction.

Both whole urine and the <10K MW fraction from GS-treated animals demonstrated a significant loss of inhibition of COM crystal growth compared to normal controls. These changes resulted after only 1 dose of GS and persisted throughout the duration of the experiment. In our previous work[1], we found significantly lowered urinary citrate levels, but not until 4 doses of GS had been administered; therefore the impact of the <10K MW fraction must occur considerably earlier than the drop in urinary citrate levels.

CONCLUSIONS

1. Previous work has shown that crystalluria occurs when a lithogenic substance is combined with a nephrotoxin; citrate levels decline late in the process.
2. We examined the contribution of renal injury alone and determined that in the urine, species < 10K MW play a role in decreased inhibition of COM growth.
3. This effect occurs earlier than the reduction in urinary citrate levels that have also been observed, but ultimately, loss of both elements probably contributes to COM growth.

Table 2. COM seed crystal growth inhibition with the <10K MW fraction from GS Urine Depletion of Filtrate CaOx RSS.

Day	Normal Urine	GS-Urine	COM Control
1	13.49 ± 0.81	10.51 ± 0.80 (p<0.02	5.72 ± 0.59
2	10.95 ± 0.13	9.32 ± 0.47 (p<0.01)	5.58 ± 0.19
3	11.53 ± 1.01	10.75 ± 2.56	5.92 ± 0.25
4	11.38 ± 1.01	9.71 ± 3.02	5.82 ± 0.37
5	10.04 ± 0.23	6.93 ± 1.62 (p <0.005)	5.98 ± 0.33
6	11.80 ± 0.78	7.66 ± 0.25 (p<0.01)	5.80 ± 0.32

ACKNOWLEDGEMENTS

This work was supported by NIH grant POl DK20586.

REFERENCES

1. RL Hackett, PN Shevock and SR Khan, *J Urol 144:* 1535 (1990).
2. Y Nakagawa, HC Margolis, S Yokoyama, TJ Kezdy, ET Kaiser and FL Coe, *J Biol Chem* 256: 3936 (1981).
3. CM Brown, P Senior, H Charest and B Finlayson, *in*: "Urolithiasis and Related Clinical Research" PO Schwille, LH Smith, WG Robertson and W Vahlensieck (eds), Plenum Press, New York, 843 (1985).

DO BACTERIAL MEMBRANES CONTRIBUTE TO CALCIUM OXALATE (CaOx) CRYSTAL DEPOSITION?

M.S. Cohen[1], R.J. Lobinske[1] and S.R. Khan[2]

[1]Department of Surgery
Division of Urology
[2]Department of Pathology
University of Florida College of Medicine
Gainesville, FL

INTRODUCTION

Clinical studies demonstrate the presence of bacteria in as many as one third of calcium oxalate/phosphate stones[1]. A role for such organisms has not been previously recognized. Bacteria may develop intracellular calcium phosphate crystals (hydroxyapatite) in unique clinical situations (e.g. dental tartar) and when inoculated into calcium phosphate enriched growth media *in vitro*[2]. Previous work in our laboratory suggests that bacteria may form intracellular electron dense deposits containing calcium when inoculated in human urine[3]. Such bacteria demonstrate x-ray powder diffraction patterns consistent with hydroxyapatite[3,4]. Additional *in vitro* studies suggest that bacteria may form microcolonies on tissue surfaces in urines that demonstrate increased levels of calcium[5]. The following study was designed to test the hypothesis that bacteria or bacterial membranes (mineralized and non-mineralized) may form a platform for calcium oxalate crystal deposition.

MATERIALS AND METHODS

Bacteria

Escherichia coli was isolated from a patient with a urinary tract infection and identified using standard biochemical techniques. Serotyping identified the organism as a K-15 strain. Absence of urease activity was confirmed by growth on Christensen's urea media. Organisms were isolated, washed three times with phosphate buffered saline and studied at a concentration of 1×10^{12} organisms/mL.

Chemically Defined Growth Medium (CM)[2]

CM consisted of a buffered metastable calcium phosphate solution known to support microbial calcification, with essential nutrients for bacterial growth. A maintenance medium (MM) was used as a control medium which would support bacterial growth but not calcification. Organisms were incubated in CM and MM at 37°C for two and four weeks respectively, recovered by centrifugation and studied as intact organisms or membrane fragments as prepared below.

Batch Crystallization Studies[6]

A stable supersaturated solution of calcium oxalate (relative supersaturation equal to 6 or 10 as determined by EQUIL) was prepared from calcium chloride and potassium oxalate with PIPES buffer. Fifty mL was maintained at 37°C in a water-jacketed beaker and stirred at a fixed rotational speed. pH was monitored continuously. Following addition of 1 mL of bacterial suspension, crystal formation was monitored by studying 1 mL aliquots over 48 hours by scanning electron microscopy (SEM), electron dispersive spectrophotometry (EDS) and calcium depletion from solution (expressed in parts per million; ppm) by atomic absorption spectrophotometry (AA). The calcium chloride and potassium oxalate consumed in the reaction were not replenished in this study.

Constant Composition Crystallization Studies[7]

Five mL of potassium oxalate and 5mL of calcium chloride solution were carefully combined in a water-jacketed beaker maintained at 37°C (relative supersaturation = 10) and stirred at fixed rotational speed. After five minutes equilibration, 10 µL of bacterial suspension was added. Calcium ion activity and pH were monitored by reference electrodes. Calcium ion activity was kept constant by the potentiostatic controlled addition of calcium chloride and potassium oxalate. Crystal deposition was monitored in terms of moles of titrant added over a 180 minute period.

RESULTS

Transmission electron microscopy confirmed the presence of electron dense intracellular deposits consistent with crystal (hydroxyapatite) formation in *E coli* recovered from CM. Electron microscopy of filtrates from the continuous crystallization chamber revealed calcium oxalate crystal deposition in frequent association with the bacterial biofilm and bacterial matrix. Groups of organisms and "mats" of membranes with CaOx crystals on the margins and surface were seen.

Calcium depletion in the batch crystallization chamber was not significantly different from controls. A drop in calcium concentration (10.8-12.6 ppm) was observed in the first 48 hours.

In the constant composition study a statistically significant difference in moles of titrant added between incubations containing organisms and controls was seen (controls =0; bacteria=0.12-0.24 moles, mean=0.17). Among bacteria, sonicated organisms from supplemented medium required more titrant than unsonicated organisms from the same medium (0.24 moles vs 0.19 moles). Unsonicated organisms from unsupplemented medium gave results which differed little from those obtained with sonicated organisms from the same medium.

CONCLUSIONS

This study shows that bacterial membranes may form a platform for CaOx crystal deposition *in vitro*. The effect is most dramatic when studied in a dynamic system such as that afforded by a constant composition technique which mimics the constant replenishment of the mineral source found clinically in urine. In addition, this study suggests that bacteria and bacterial membranes can form a substrate for CaOx deposition. These findings suggest a role for bacteria in stone formation beyond that traditionally recognized.

REFERENCES

1. J Hugosson, L Grenabo, H Hedlon, S Petterson, and S Seeberg, Bacteriology of upper urinary tract stones, *J Urol* 143:965 (1990).
2. JJ Vogel, The microbial model in the study of calcification. *in:* "Methods of Calcified Tissue Preparation", GR Dickson, ed, Elsevier Science Publishers, Amsterdam (1984).
3. MS Cohen, MD Warren, P Baur, JJ Vogel, and CP Davis, Intracellular crystal formation in bacteria from human urines: A contributing factor in urinary calculi, *Urol Res* 9:55 (1981).

4. MS Cohen, CP Davis, PS Baur, and MM Warren, Papillary necrosis *in vitro* study, *J Urol* 127:184 (1982).
5. MS Cohen, CP Davis, PS Baur, MD Warren, Papillary necrosis *in vitro*: A scanning electron microscopic comparison of *Escherichia coli* and *Proteus mirabilis* infection, *Scanning Electron Microscopy*, III: 65 (1981).
6. JL Meyer and LH Smith, Growth of calcium oxalate crystals: A model for urinary stone growth. *Invest Urol*, 13:31 (1975).
7. ME Sheehean and GH Nancollas, Calcium oxalate crystal growth, a new constant composition method for modeling urinary stone formation, *Invest Urol* 17:446 (1980).

CALCIUM OXALATE CRYSTALLIZING PROPERTIES OF MACROMOLECULES RELEASED BY RENAL TUBULAR CELLS *IN VITRO*

D.K.Y. Shum and E. Liong

Department of Biochemistry
Faculty of Medicine
The University of Hong Kong
Hong Kong

INTRODUCTION

Urinary proteins and glycosaminoglycans (GAGs) have been shown to act diversely on the crystallization of calcium oxalate. In media where these macromolecules coexist as solubilized and immobilized forms, they are dually effective as promoters of nucleation and inhibitors of growth in the process of crystallization[1,2]. Structural modifications of urinary proteins or structural differences of urinary GAGs can also influence crystallization[3,4]. Micropuncture and renal clearance studies have shown that only a small proportion of plasma proteins and GAGs normally enter the tubular fluid via glomerular ultrafiltration, and escape tubular reabsorption to be excreted in the urine[5]. It follows that the majority of urinary GAGs, like the urinary proteins, are of renal origin. It is of interest therefore to develop an *in vitro* system to study renal tubular cell activities that release macromolecules that may influence crystallization of urinary calcium oxalate.

MATERIALS AND METHODS

Suspensions of single cortical tubular cells from kidneys of 14-day-old Sprague-Dawley rats were inoculated at $5x10^5$ cells/cm^2 onto 100x20mm dishes or 12mm collagen-coated Millicell inserts (Millipore) in a 50:50 mixture of Dulbecco's Modified Eagle's Medium (DME[D-Val]) and Ham's F12 Medium (F12) supplemented with 10% foetal calf serum[6]. The kidney cells were routinely incubated in a humidified 5% CO_2/95% air mixture at 37°C. After 16 hours in culture, the monolayers were washed in serum-free (SF), hormone-supplemented DME[D-Val]/F12[7] and subsequently maintained in this medium until confluency. Conditioned media were collected 3 days after SF media had been introduced and every 3-4 days thereafter for 7 days for cells grown on plastic. For cells grown on Millicell inserts, media were collected separately from the apical and basal compartments at similar intervals for 21 days. Each collection was centrifuged to remove cell debris. Protease inhibitors (phenylmethylsulfonylfluoride, N-ethylmaleimide, aminohexanoic acid and benzamidine) were added to the supernatants and the mixtures were stored at -20°C until they were thawed for recovery of macromolecules.

Macromolecules in the different conditioned media were precipitated with ethanol. Proteins therein were digested with papain at 65°C and the GAGs were recovered from the digest by sequential precipitation with cetylpyridinium chloride and sodium acetate-saturated ethanol[5]. The crude GAG extract was assayed for uronic acid by the carbazole reaction[8] and

Urolithiasis 2, Edited by R. Ryall *et al.*,
Plenum Press, New York, 1994

for GAG composition by cellulose acetate electrophoresis (0.05 M barium acetate buffer, pH 5.8) before and after treatments with nitrous acid (pH 4)[9] and chondroitinase ABC[10]. The membranes were stained for GAGs with 0.2% Alcian blue.

Normal human male urine ultrafiltrate (10 kDa cutoff, 1,200 mOsm, pH 5.5)[2] was mixed with test solutions of recovered macromolecules or GAGs (4:1, v/v), incubated at 20°C for 16 hours and then thawed. The mixtures were examined for population density of envelope crystals of calcium oxalate using a haemocytometer. All statistical analyses were performed using the Mann Whitney pair test.

RESULTS AND DISCUSSION

Kidney cells cultured in DME[D-Val]/F12 were exclusively epithelial. The initial supplement of serum was necessary to facilitate attachment to the substratum. As serum contains proteins that can influence calcium oxalate crystallization[11], the cells were subsequently maintained in an SF medium so that macromolecules released by the cells could be distinguished from the native macromolecules of serum. The change to an SF medium did not affect attachment or survival of cells grown on plastic or Millicell inserts. Cells grown on the permeable supports differentiated into apical and basal morphologies, but these were not observable in cells grown on impermeable tissue culture plastic.

Macromolecules recovered from the conditioned media by ethanol precipitation contained both proteins and GAGs. After digestion of the former, the GAGs that remained were separable electrophoretically into Alcian blue-stained bands of mobilities similar to reference chondroitin/dermatan sulphates and heparan sulphates. The identities of the bands were confirmed by their respective removal subsequent to treatments with chondroitinase ABC and nitrous acid (pH 4). This was observed of GAGs recovered from non-polarized cultures and the basal compartment of polarized cultures. Alcian blue-stained material resistant to the treatments was, however, evident in the GAG extract from the apical compartment of polarized cultures; this was also resistant to keratanase.

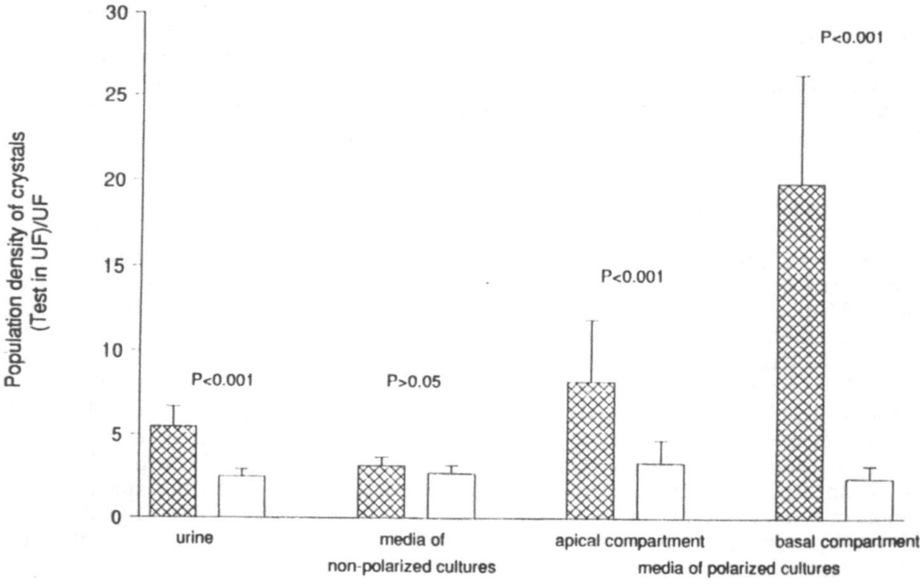

Figure 1. Population density of calcium oxalate crystals formed in solutions of test macromolecules (shaded) or GAGs (blank) in urine ultrafiltrate compared with that formed in urine ultrafiltrate (UF) alone (n=9; mean±SEM). All solutions contained a uronic acid concentration of 1 µg/mL. The test substances were recovered from normal human male urine, media conditioned by non-polarized cultures, media conditioned by the apical and basal compartments of polarized cultures.

This material may well be a heparan sulfate sub-family that is highly N-acetylated and thus resistant to nitrous acid at pH 4[12]. The relative intensity of Alcian blue stained bands suggests a predominance of chondroitin/dermatan sulphates over heparan sulphates in all samples studied.

While urine that has been depleted of macromolecules by ultrafiltration showed minimal crystallization of urinary calcium oxalate[3], introduction of macromolecular or GAG extracts of the conditioned media enhanced nucleation of crystals. At similar uronic acid concentrations of 1 μg/mL, the population density of crystals formed from the incubation of non-papainized macromolecules with ultrafiltrates was consistently higher ($P<0.001$) than that from the papainized counterpart in extracts from polarized cultures (Fig.1). Similar results were observed in urinary macromolecules so treated. The crystallizing properties of a system can thus be modified by the relative concentration of proteins and GAGs. The difference between non-papainized and papainized counterparts was, however, not apparent in macromolecules recovered from non-polarized cultures. Based on the crystal-nucleation-enhancing activities observed, polarized cultures of renal tubular epithelial cells are preferred to non-polarized cultures for the study of turnover of renal tubular macromolecules that can influence crystallization in the urinary environment.

ACKNOWLEDGEMENTS

This work was supported by the Committee on Research and Conference Grants of The University of Hong Kong.

REFERENCES

1. AA Campbell, A Ebrahimpour, L Perez, SA Smesko and GH Nancollas, The dual role of polyelectrolytes and proteins as mineralization promoters and inhibitors of calcium oxalate monohydrate. *Calcif Tissue Int* 45:122 (1989).
2. MD Gohel, DKY Shum and MK Li, The dual effect of urinary macromolecules on the crystallization of calcium oxalate endogenous in urine, *Urol Res* 20:13 (1992).
3. B Hess, Y Nakagawa, J Parks and F Coe, Molecular abnormality of Tamm-Horsfall glycoprotein in calcium oxalate nephrolithiasis, *Am J Physiol* 260:F569 (1991).
4. MD Gohel and DKY Shum, Urinary glycosaminoglycans differ in their calcium oxalate crystallization activities, in: *this volume.*
5. DKY Shum, C Baylis and JE Scott, A micropuncture and renal clearance study in the rat of the urinary excretion of heparin, chondroitin sulphate and metabolic breakdown products of connective tissue proteoglycans, *Clin Sci* 67:205 (1984).
6. PYD Wong, Mechanism of adrenergic stimulation of anion secretion in cultured rat epididymal epithelium, *Am J Physiol* 254:F121 (1988).
7. M Taub and G Sato, Growth of functional primary cultures of kidney epithelial cells in defined medium, *J Cell Physiol* 105:369 (1980).
8. T Bitter and H Muir, A modified uronic acid carbazole reaction, *Anal Biochem* 4:330 (1962)
9. DJ Carey and MS Todd, A cytoskeleton-associated basement membrane heparan sulphate proteoglycan in Schwann cells, *J Biol Chem* 261:7518 (1986).
10. H Saito, T Yamagata and S Suzuki, Enzymatic methods for the determination of small quantities of isomeric chondroitin sulphates, *J Biol Chem* 243:1536 (1968).
11. KA Edyvane, RJ Ryall, RD Mazzachi and VR Marshall, The effect of serum on the crystallization of calcium oxalate in whole human urine: inhibition disguised as apparent promotion, *Urol Res* 15:87 (1987).
12. JE Shively and HE Conrad, Formation of anhydrosugars in the chemical depolymerization of heparin, *Biochem* 15:3932 (1976).

EXAGGERATED RESPONSE OF THE HYPER-OXALURIC STONE FORMER TO AN OXALATE LOAD TEST AT INITIAL METABOLIC EVALUATION

P.S. Chandhoke[1], R. Dunlay[2], P. Stein[2], J. Nitz[2] and K.A. Hruska[2]

[1]Division of Urologic Surgery
[2]Renal Division Jewish Hospital
St Louis at Washington University Medical Center
St Louis, Missouri, USA

INTRODUCTION

Hyperoxaluria is an important risk factor for calcium oxalate stone disease. It may be due to increased metabolic production of oxalate, increased intestinal oxalate absorption, defective renal oxalate transport, or a combination of defects. The available data to date seem to suggest that in the idiopathic calcium stone former, there may be an increased intestinal source of oxalate[1]. However, there are conflicting data as to whether the increased intestinal absorption of oxalate is due to an increased oral intake or to an increased inherent capacity of the intestine to absorb oxalate[1]. Determination of the exact cause of hyperoxaluria in patients with calcium oxalate stones may lead to the development of more specific medical therapeutic strategies.

The objective of the present study was to determine if the recurrent calcium oxalate stone former with hyperoxaluria, defined here as a 24 h urinary oxalate excretion of greater than 40 mg, has an abnormality in oxalate absorption from the intestinal tract.

METHODS

We developed an oral oxalate loading test in an ambulatory setting to evaluate the intestinal absorption of oxalate. Prior to the oral oxalate load test, the recurrent calcium oxalate stone former referred to our stone clinic was placed on a low calcium, low oxalate, low purine, and restricted salt diet for one week. In addition, the patients were instructed not to take any vitamin C. At the end of one week, a 24 h urine sample was collected and analyzed to ensure dietary compliance, and the patient fasted overnight prior to the oxalate load test.

On the day of the oxalate load test, a fasting 2 h urine specimen was collected between 6 and 8 a.m. At 8 am the patient was given an oral oxalate load of 440 mg of sodium oxalate dissolved in 250 cc of distilled water. Urine was then collected over 2 h intervals for the subsequent 6 hours.

Each urine sample collected was analyzed for volume, creatinine concentration, and oxalate concentration. The oxalate concentration was determined using a modified enzymatic assay (Sigma Chemical, St. Louis, MO) with maximal oxalate recovery and minimal interference with ascorbate. An oxalate/creatinine ratio was thus determined for each urine sample.

Urolithiasis 2, Edited by R. Ryall *et al.*,
Plenum Press, New York, 1994

RESULTS AND DISCUSSION

Fifty seven consecutive recurrent calcium oxalate stone formers and three volunteer non-stone formers underwent an oral oxalate load test. Only patients with a 24 h urine creatinine clearance of greater than 75 mL/min were included in the study. The patients were divided into the following four groups: those without hyperoxaluria, those with hyperoxaluria (> 40mg urinary oxalate/ 24 h); those with bowel disease, and non-stone formers.

Table 1. Demographic data of individuals undergoing the oxalate load test.

Patient Group	#of patients	% males	mean urinary oxalate (mg/24 h)
non-hyperoxalurics	38	63%	31.3
hyperoxalurics 16	16	63%	48.6
bowel disease	3	100%	77.0
non-stone formers	3	33%	24.5

Of the 57 patients with recurrent calcium oxalate stone disease, 16 or 28% had hyperoxaluria. The degree of hyperoxaluria in these patients was mild to moderate, with a mean value of 48.6 mg/24 h. There were three patients with hyperoxaluria due to bowel surgery or inflammatory bowel disease, in whom the level of hyperoxaluria was significantly higher (77 mg/24 h).

The results of the oxalate load test are summarized in Figure 1. The ordinate represents the urinary oxalate/creatinine ratios, and the abscissa, the time intervals of urine samples. The first value represents the fasting oxalate creatinine ratio. The following three values correspond to urine samples obtained at 2, 4, and 6 hours after the oxalate load. Note that the peak oxalate/ creatinine ratio occurs during the 4 h urinary collection in all groups.

There are three aspects of the results which are noteworthy in Figure 1. Firstly, in both the non-stone formers and the non-hyperoxaluric stone formers, the fasting and peak urinary oxalate/creatinine ratios are almost identical. Secondly, in the hyperoxaluric stone formers, although the fasting oxalate/creatinine ratio is similar to the non-hyperoxaluric stone former (with a value around 0.02), the peak ratio is significantly higher in the hyperoxaluric stone formers (about 0.08). Thirdly, the stone former with bowel disease has a similar peak value to the hyperoxaluric stone former, but much higher fasting oxalate/creatinine ratios than the other three groups. Altogether, these findings support the hypothesis that the hyperoxaluric stone former has an inherent increased capacity to absorb oxalate from the intestinal tract.

CONCLUSION

In summary, our results show that a subset of recurrent calcium oxalate stone formers, that is patients with hyperoxaluria, have an increased peak urinary excretion of oxalate after an oxalate load test, probably because of increased inherent capacity to absorb oxalate from the intestinal tract. Thus the strategy of decreasing urinary oxalate in these patients should be directed towards decreasing the dietary oxalate or instituting therapies to decrease intestinal absorption of oxalate.

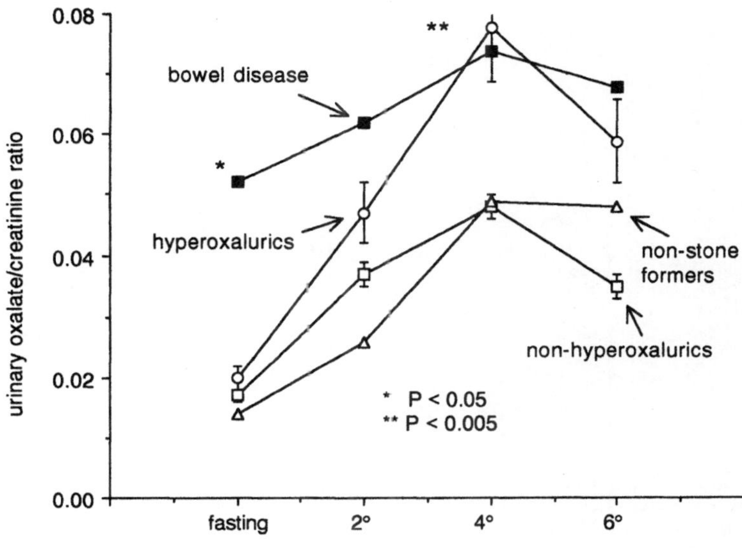

Figure 1. Results of the oxalate load test.

REFERENCE

1. LH Smith, Diet and hyperoxaluria in the syndrome of idiopathic calcium oxalate urolithiasis. *Am Kidney Diseases* 17: 370-375, (1991).

FIBROUS MATRIX IN EXPERIMENTAL UROLITHIASIS

N. Sylaja[1], T.G. Dhanalekshmy[1], C. Aravindakshan[1], M.M. Oommen[2] and Y.M. Fazil Marickar[1]

[1]Department of Surgery
Medical College Hospital and
[2]Department of Zoology
University of Kerala
Trivandrum, India

INTRODUCTION

Fibrous materials[1] and cellular fragments have been implicated as initiators of crystal nucleation in both physiological and pathological mineralization. A high oxalate content in the diet is known to be toxic and to cause renal-tubular damage resulting in stone formation. Renal epithelial injury may induce crystallization in urine by providing membranous nidi for heterogeneous nucleation of crystals, and by changing urinary chemistry[2]. The present work was performed in order to study the relationship between the fibrous matrix and cellular fragments of stones with the deposition of calcium oxalate (CaOx) crystals in experimental rats administered sodium oxalate i.p.

MATERIALS AND METHODS

Two groups of six male albino Wistar rats were selected for study (experimental and control animals). They were fed on normal pelletized rat feed and deionised water *ad libitum*. The experimental group was given sodium oxalate in 0.9% saline i.p. at a dose of 6 mg/100 g body weight. The urine samples of both experimental and control rats were collected daily for microscopic studies. After one month, the rats were sacrificed and the kidneys, ureters and bladders were removed for histopathological studies.

RESULTS

Urinary sediments of the experimental rats showed an initial increase in the extent of oxalate crystalluria which gradually decreased with time. Phosphate crystals, which were present originally, started to reduce in number. The CaOx-monohydrate crystals increased in number as the dihydrate crystals reduced. In the control group, only phosphate crystals were seen under light microscopy. In the urine of experimental rats, in addition to the phosphate crystals, membranous fragments and fibrous structures with branching and entangling patterns were seen. At the end of one week, the early signs of fibrillar materials were noted (Fig. 1). These took up crystal violet and safranin II stain on histochemistry. At the end of two weeks, the extent of fibrous materials increased along with aggregation of the fibres (Fig.2).

Urolithiasis 2, Edited by R. Ryall *et al.*,
Plenum Press, New York, 1994

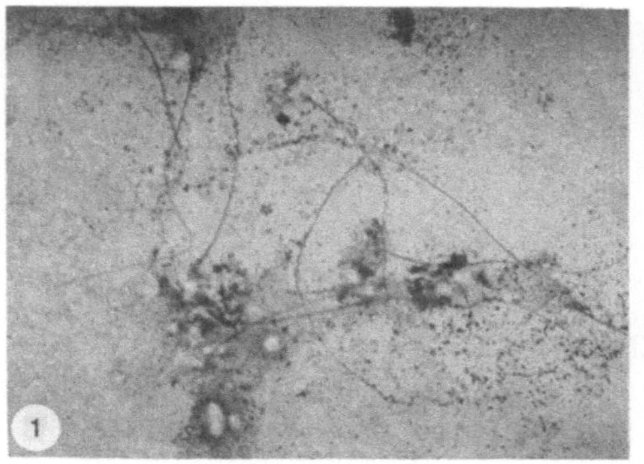

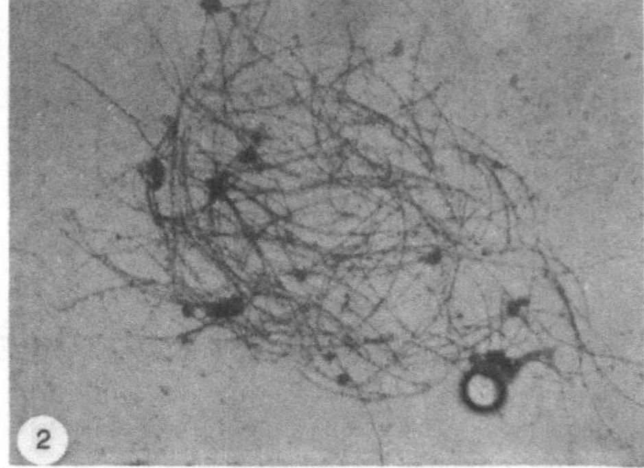

Fig.1 (Left) Early formation of fibrillar material. LMx250. **Fig.2** (Right) Aggregation of fibrillar material. LMx250.

During the still later stages, these fibrous structures were seen to be entangled with different types of crystals (Fig.3). This probably constituted the early microlith. Histopathological examination of the kidneys of the experimental rats showed dilated tubules filled with colloidal and fibrillar materials (Fig.4). These were superimposed on the destructive changes in the cells. No significant changes were noticed in the ureters and bladders.

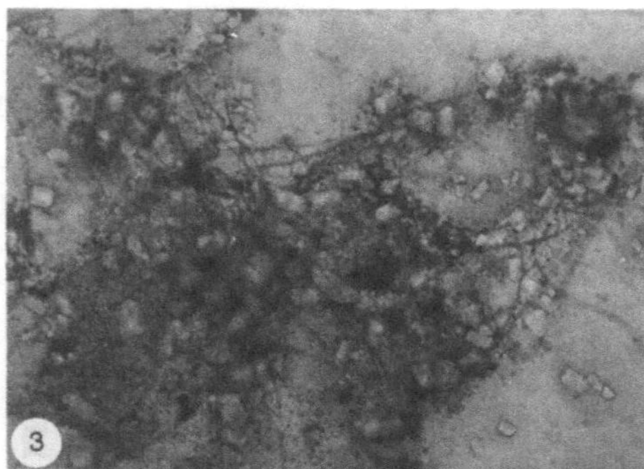

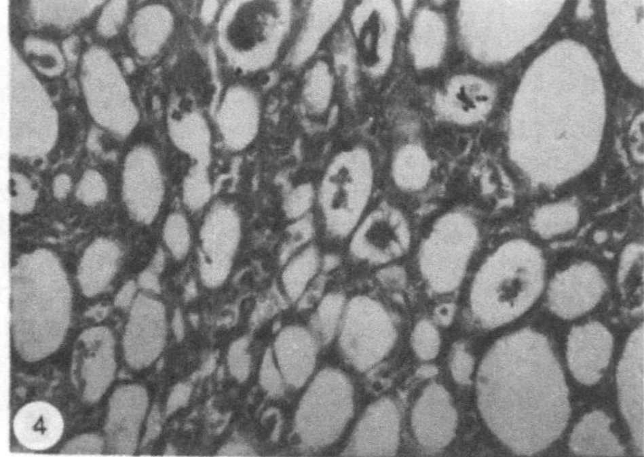

Fig.3 (Left) Crystal fibrillar entangling in urine. LMx450. **Fig.4** (Right) Fibrillar material in renal tubules. LMx250.

DISCUSSION AND CONCLUSIONS

The urine of experimental rats contains different types of crystals such as CaOx and phosphate. However, there have been very few reports of matrix material in the urinary deposits of rats. In the 1950's, investigations using light microscopy showed that the organic matrix had a highly organised architecture composed of fibrils and an amorphous interfibrillar material[3]. The appearances of the fibrillar matrix material identified in the present study are similar to those found in human urinary deposits[1]. Cheng and Pritzker[4] suggested that there are probably two parallel modes of kidney-stone pathogenesis. These include accumulation of organic matrix from cellular degradation, followed by deposition of calcium-apatite and non-apatite crystals, for example, CaOx crystallisation followed by agglomeration of discrete crystals stacked together like masonry and cemented by organic matrix material. The occurrence of cellular debris and colloidal and crystalline aggregates within the dilated renal tubule in the present study supports the findings of Khan et al[5]. In their study in rats given oxalate i.p., they found crystals in the lumen associated with a material that appeared eosinophilic, amorphous, or granular by light microscopy, and tubular, fibrillar, or amorphous by scanning electron microscopy (SEM). SEM examination revealed this material to be cellular debris composed of membranous and vesicular material and cellular organelles in various stages of necrosis[5]. The SEM study[6] of the matrix material in CaOx renal stones revealed that the matrix was composed of amorphous and fibrillar materials and cellular debris.

It is concluded from the study that oxalate toxicity will produce precipitation of fibrous matrix-like materials and colloidal substances and cellular fragments which might form the nidus for heterogeneous nucleation of crystals.

ACKNOWLEDGEMENTS

The authors acknowledge the financial assistance rendered by the Indian Council of Medical Research.

REFEFENCES

1. Fazil Marickar YM, Sachidev K, Joseph T, Sindhu S and Koshy, P, Microlith formation in urine, *in*: "Urolithiasis", Walker VR, Sutton RAL, Cameron BEC, Pak CYC and Robertson WG (eds), Plenum Press, New York (1989). 37-38.
2. Khan SR, and Hackett RL, Renal proximal tubular injury and crystallisation of calcium oxalate in rat urine, Ibid 117-119.
3. Roberts SD and Resnick MI, Urinary stone matrix and urinary macromolecules, *in*: "Urolithiasis and Related Clinical Research", Schwille PO, Smith LH, Robertson WG and Vahlensieck W (eds), Plenum Press, New York (1985). 911-918.
4. Cheng PT and Pritzker KPH, Alternating crystallisation - a proposed mechanism for lamellar structure formation in renal stones, *in*: "Urolithiasis and Related Clinical Research", Walker VR, Sutton RAL, Cameron BEC, Pak CYC and Robertson WG (eds), Plenum Press, New York, (1989) 919-921.
5. Khan SR, Finlayson B and Hackett RL, Experimental calcium oxalate nephrolithiasis in the rat, *in*: "Investigative Urology", The William Wilkin Co (1978), 236.
6. Khan SR, Finlayson B and Hackett RL, A microscopic study of some calcium oxalate renal stones, *in*: "Urolithiasis and Related Clinical Research", Schwille PO, Smith LH, Robertson WG and Vahlensieck W, Plenum Press, New York (1985). 923-925.

ANALYSIS OF THE PROTEIN CONTENT OF FIVE DIFFERENT TYPES OF KIDNEY STONES

B. Dussol[1-3], M. Daudon[2], P. Dupuy[1], R. Michel[1], Y. Berland[3], J.C. Dagorn[1] and J.M. Verdier[1]

[1]INSERM U315, Marseille, France
[2]Hôpital Necker, Paris, France
[3]Hôpital Ste-Marguerite, Marseille, France

INTRODUCTION AND RESULTS

The aim of this study was to characterize the protein matrix of five different types[1] of kidney stones: calcium oxalate monohydrate (type I), calcium oxalate dihydrate (type II), uric acid (type III), struvite (type IV) and cystine (type V). After extraction of calcium salts, the proteins were run on sodium dodecyl sulphate polyacrylamide gel electrophoresis (SDS-PAGE) and analyzed by Western blot experiments using specific antibodies. In all types, the most abundant protein found was albumin (HSA). α-1-acid glycoprotein, α-1-microglobulin, immunoglobulins A, G, M (Igs A, G, M), apolipoprotein Al, transferrin and α-1-antitrypsin were also present in the five types of calculi. ß-2-microglobulin was only found in types I, II and III. Furthermore, using antibodies to pancreatic lithostathine, an inhibitor of $CaCO_3$ crystal growth in pancreatic juice, we were able to detect an immunologically related protein in all types of stones (figure 1). Conversely, we failed to detect caeruloplasmin, haptoglobulin and, surprisingly, the Tamm-Horsfall protein (THP) which is the most abundant protein in normal urine.

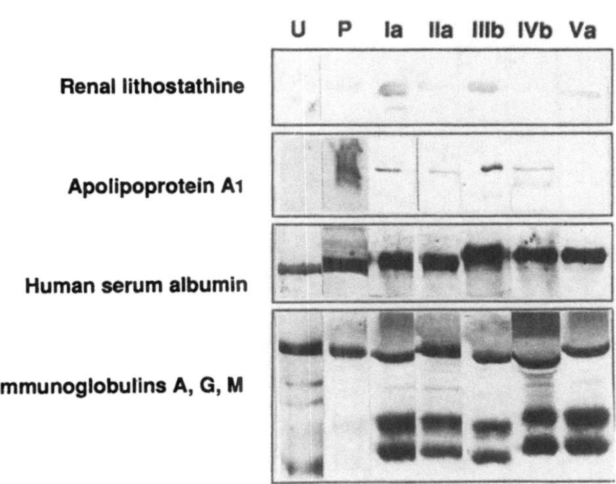

Figure 1. Some of the immunodetection experiments. Crude urine (lane U), proteins extracted from urine[2] (lane P) or from stones (lanes Ia, IIa, IIIb, IVb, Va) were subjected to SDS-PAGE, transferred to nylon membranes and incubated with antibodies to RL, apolipoprotein Al, HSA, Igs A, G, M.

Urolithiasis 2, Edited by R. Ryall *et al.*,
Plenum Press, New York, 1994

From all these results, it is concluded that (i) not all the proteins of normal urine are present in stone protein matrix, (ii) THP is probably not a constituent of kidney stones, (iii) kidney stones contain a protein immunologically related to pancreatic lithostathine which we called renal lithostathine (RL) for reasons described elsewhere[3].

It is therefore possible that the proteins present within stones play a specific role in kidney stone formation particularly with regard to the well known calcium- and protein-binding properties of human albumin.

REFERENCES

1. M Daudon, Methodes d'analyse des calculs et des cristaux urinaires, Classification morpho-constitutionnelle des calculs, *in:* "Lithiase Urinaire," P Jungers, M. Daudon, A Le Duc, eds., Flammarion, Medecine-Sciences (1989).
2. JM Verdier, B Dussol, P Dupuy, Y Berland, and JC Dagorn, Preliminary treatment of urinary proteins improves electrophoretic analysis and immunodetection, *Clin Chem* 38:860 (1992).
3. JM Verdier, B Dussol, P Casanova, M Daudon, P Dupuy, P Berthezene, R Boistelle, Y Berland, and JC Dagorn, Evidence that human kidney produces a protein similar to lithostathine, the pancreatic inhibitor of $CaCO_3$ crystal growth, *Eur J Clin Invest* 22:469 (1992).

LONG TERM STUDIES OF UREASE-INDUCED CRYSTALLIZATION IN HUMAN URINE

A. Edin-Liljegren, L. Grenabo, H. Hedelin, S. Pettersson and Y. Wang

Departments of Urology
Sahlgrenska Sjukhuset
University of Göteborg
Sweden

INTRODUCTION

The aim of this study was to evaluate if urease-induced crystallization is augmented in the presence of viable *Proteus mirabilis*.

METHODS

Urine from three individuals was divided into two parts, one inoculated with *P mirabilis* and the other with Jack bean urease. The growth of crystals and crystal aggregation were followed in a Coulter Counter and the size and shape with polarisation microscopy. Bacterial cultures and pH were continuously monitored. The experiment was performed five times.

Proteus was cultured on almost all 14 days the process was followed. The end pH (around 9.0) was reached after 3-4 days and it was always 0.1 to 0.4 pH-units higher in the portions inoculated with *P mirabilis*.

Nucleation started earlier and at a lower pH in the urines inoculated with *P mirabilis* and the median crystal size was 1.5 - 6.5 µm larger in the presence of the organism. The crystal size was positively correlated to pH but not to time. Small crystals, < 5 µm, thus started to precipitate at pH 7.25 and larger crystals when pH 8.25 was reached. Both crystalline and amorphous material precipitated. The crystals were present in several different shapes and the crystallization pattern differed distinctly between urine samples inoculated with *Proteus* and Jack bean from two of the studied individuals.

DISCUSSION

Our results demonstrate that the urease-induced crystallization is augmented in several ways in the presence of viable *Proteus mirabilis*.

OXALATE CRYSTALLURIA FOR POST-HEPATIC AND RENAL TRANSPLANTATION MANAGEMENT IN PRIMARY HYPEROXALURIA

I.P. Jouvet, A. Valogne, P. Hubert, M.F. Gagnadoux, D. Jan and M. Daudon

Pediatric Department
Hôpital Necker
Paris, France

INTRODUCTION

In primary hyperoxaluria type I (PHI) after hepatic and renal transplantation (HRT), the oxalate pool is slowly removed by urine excretion. To minimize the amount of oxalate deposition in the graft during the first weeks, haemodialysis (HD), hyperdiuresis and diuretics have been suggested. Crystallization risk is usually evaluated by calcium oxalate product (pCaOx) but in PHI it can be inappropriate, so we measured total volume of oxalate crystals.

We studied crystalluria of two children during the first month after HRT for PHI. Renal function was normal in both. Measures were made on fresh urine samples. The oxalate crystal volume (OCV) per µL of urine was determined as: number of whewellite (Wh) x mean volume of Wh + number of Weddellite (Wd) x mean volume of Wd.

The mean volume of Wh and Wd was mathematically calculated from the geometric structure (on 10 urine samples) and size (microscope measurement). We compared oxalate crystal volume to calcium oxalate product for different treatments:

	Diuresis (1.5 to 2L/m^2) n=3	Hyper-diuresis (>2L/m^2) n=5	Hyper-diuresis + thiazide n=6	HD* n=1
OCV (µm^3)	2780	1406	96	34094
(range)	(320-7562)	(24-5903)	(21-287)	
pCaOx(mM/L)^2	0.28	0.45	0.29	5.45

* Data of the 3rd haemodialysis performed on the 3rd day after HRT.

In the first patient with daily haemodialysis, we observed very high OCV as pCaOx associated with obstruction of the ureteral catheter by oxalate deposition. So haemodialysis was stopped on the fourth day and hyperdiuresis alone was continued. Results were improved in the second patient with thiazide addition to hyperdiuresis. We conclude that oxalate crystal volume is an easy test quickly performed. This quantitative information can be helpful for the post-transplantation management of PHI.

IS INFRA-RED (IR) ANALYSIS AN ALTERNATIVE TO CHEMICAL ANALYSIS?

S.V. Roshni, H.K. Moorthy, S. Sindhu, R.K. Vathsala, N. Sylaja, C. Aravindakshan and Y.M. Fazil Marickar

Department of Surgery
Medical College Hospital
Trivandrum, India

INTRODUCTION AND METHODS

Conventional wet chemical analysis of urinary stone fragments gives meagre details and is often not possible due to small sample size. X-ray diffraction is often not feasible, due to non availability of equipment and expertise, so an attempt was made to standardise an alternative investigation for routine stone analysis. Infra-red analysis was done using a Perkin Elmer IR Spectrophotometer. For comparison, ordinary qualitative and quantitative wet chemical methods were utilised in situations where sufficient stone samples were available.

One hundred and fifty stone samples were analysed by IR spectroscopy. The stone samples were powdered and mixed with potassium bromide in the ratio of 1:200 and pelletised. The IR wavelengths were recorded on a computer screen and photographed. The IR-charts were scanned by the computer and peak wavelengths were recorded. Characteristic peaks at definite wavelengths were produced in the finger print region 600-1600 nm. The components of each sample were identified by comparing the peaks in the graph obtained with those of standard graphs.

RESULTS AND CONCLUSION

Characteristic peaks at definite wavelengths were obtained for whewellite (660, 778, 883, 1317 and 1618 nm), weddellite (604, 778, 988, 1046, 1064 and 1322 nm), uric acid (621, 657, 747, 878 and 1028 nm) and different varieties of phosphates namely brushite, apatite, newberyite, struvite and octocalcium phosphate. Thus it was possible to obtain a qualitative/semi quantitative analysis of the stone fragments and which made it possible to distinguish between pure whewellite (60%), pure weddellite (20%), pure uric acid (2%) and mixed (18%) calculi. Such distinction was often not possible by wet chemical analysis due to interference of phosphates.

It is concluded from the studies that IR analysis is less time consuming, less costly, more reliable and accurate than wet chemical analysis for routine qualitative/semi quantitative analysis of urinary stones where full facilities for X-ray diffraction studies are not available. Usefulness of the IR analysis in deciding prophylaxis of the patients is stressed. It is also applicable to assess the purity of crystals grown *in vitro*. Rare elements in stones can also be identified.

CALCIUM OXALATE CRYSTAL MATRIX-WHAT ABOUT THE GLYCOSAMINOGLYCANS?

K. Suzuki, K. Mayne, I.R. Doyle and R.L. Ryall

Department of Surgery
Flinders Medical Centre
Bedford Park, South Australia

INTRODUCTION

The aims of this study were to determine whether GAGs contribute to the organic matrix of CaOx crystals deposited from normal urine, and if so, whether they affect CaOx crystallization in undiluted human urine.

Urine was collected from healthy men: one portion (WU) was sieved (20μm), one (S&F) centrifuged (10k x g) and filtered (0.22μm), and the third (UF) ultrafiltered (10kDa). Separate samples of the UF urine were treated with heparan sulphate (UF+HS; 2μg/mL) and chondroitin sulphate (UF+CS;10μg/mL). CaOx crystallization was induced in the samples by the addition of an oxalate load, the crystals harvested and demineralized, and the GAGs content of the resulting extract analyzed using cellulose acetate electrophoresis.

RESULTS

No GAGs were detected in UF urine. Only one band, identified as HS by digestion with heparitinase, was visible in the electrophoretograms of crystal extracts derived from the WU, S&F and UF+HS urines. CS was not detected in WU or S&F samples, but was visualised in the UF+CS electrophoretograms, suggesting that HS is preferentially bound to CaOx crystals in whole urine. To determine whether HS is likely to act as a promoter or inhibitor of stone pathogenesis, CaOx crystallization was induced by the addition of an oxalate load to additional samples of S&F, UF, and UF+HS (1-2μg/mL) normal male urine. There was no difference between the amount of CaOx deposited from these specimens, but HS reduced the size of CaOx particles from 13μm in the UF sample to 11μm in the UF+HS, though not to the value of 7μm observed in the S&F urine. The reduction in size was shown by scanning electron microscopy to result from a decrease in the degree of crystal aggregation.

CONCLUSION

It was concluded that the inclusion of GAGs in CaOx crystals is selective; despite the fact that CS is the most abundant of the urinary GAGs, only HS was detected in the crystal extracts. Although HS inhibits CaOx crystal aggregation in undiluted human urine, it cannot account for the total inhibitory activity of all urinary macromolecules.

PART 1. A REFINED ETIOLOGICAL CLASSIFICATION OF CALCIUM OXALATE UROLITHIASIS BASED ON CALCULI STUDIES

M. Daudon, C. Bader, R.J. Reveillaud and P. Jungers

Inserm U 90 Hôpital Necker
Paris, France

INTRODUCTION

Current etiological classifications of calcium-oxalate urolithiasis do not reflect the various pathological situations that may be implicated. An etiological classification using morpho-constitutional analysis (MCA) of the calculi was studied. This analysis combines structural and infrared studies of the surface, the section, and the nucleus of the calculi (*Adv. Nephrol.*15:219 (1986)). Using MCA, the etiological classifications of calcium oxalate calculi are:

1° MCA surface type (reflects late metabolic activity)

1-1: Homogeneous MCA surface

Ia	Intermittent mild hyperoxaluria+hyperuricosuria; medullary sponge kidney (MSK)
Ib	Mild hyperoxaluria; stasis
Ic	Primary hyperoxaluria
Id	Hyperoxaluria + multiple stones + anatomical confinement
IIa	Hypercalciuria
IIb	Hypercalciuria + mild hyperoxaluria
IIc	Hypercalciuria + multiple stones + anatomical confinement

1-2: Mixed MCA surface type (metabolic association)

Ia+IIa	Intermittent mild hyperoxaluria + hypercalciuria
Ia+IIa+IVa	Intermittent mild hyperoxaluria + MSK
Ia+IIIb	Hyperuricemia, hyperuricosuria + mild hyperoxaluria
IIa+IVa	Hypercalciuria; hyperparathyroidism

2° Nucleation (nucleus) and growth (section) processes

Ia	Ia, Ib; Ia+IVa	Mild hyperoxaluria; MSK
Ia, Ib	Ia, Ib, Id	Mild hyperoxaluria
Ic	Ic, Ia+Ic	Primary hyperoxaluria
IIa,	IIb IIa	Hypercalciuria
IIb	IIb	Hypercalciuria+Mild hyperoxaluria
IIa, IIb	IIa+IVa	Hypercalciuria, Hyperparathyroidism
IIc	IIc	Hypercalciuria+ Anatomical confinement

This etiological classification is of interest to define both selective biological investigations and specific medical management.

PART II: MORPHO-CONSTITUTIONAL ANALYSIS (MCA) OF CALCULI IN HYPERPARATHYROIDISM, DISTAL RENAL TUBULAR ACIDOSIS, AND PRIMARY HYPEROXALURIA

M. Daudon, P. Jungers, R.J. Reveillaud and C. Bader

Inserm U90 Hôpital Necker
Paris, France

INTRODUCTION

Calculi from patients with primary hyperarathyroidism (PHP, n=55), distal renal tubular acidosis (dRTA, n=19), and primary hyperoxalosis (PHO, n=40) were analyzed by morpho-constitutional analysis (MCA) (*Adv. Nephrol.*15: 219 (1986). The table shows the MCA of calculi in patients with PHP, dRTA and PHO.

	PHP		dRTA		PHO	
Compound (occurrence, % in decreasing order)						
1	Carbapatite	100	Carbapatite	100	Whewellite	100
2	Weddellite	88	Whewellite	63	Carbapatite	23
3	Whewellite	84	Weddellite	21	Weddellite	15
Mean content (%)						
1		51		90		98
2		33		6		1.5-2
3		13		2		<1
Main morphological types (% occurrence)						
Surface						
1 IIa+IVa		60	IVa2	84.2	Ic	60
2 IVa		25.5	IVb	10.5	Ic+Ia	25
Section						
1 IIa+IVa		67	IVa2	89	Ic	87
2 IVa		11	IVd	5	Ic+Ia	13
Nucleus						
1 IVa		47	IVa2	74	Ic	100
2 IIa+IVa		23	IVa	11		

Note: for definition of classification, see Part I. A refined etiological classification of calcium-oxalate urolithiasis based on calcium studies, previous paper in this volume.

In conclusion, MCA of calculi allows a better etiological definition than constitutional analysis alone.

SECTION VI: RISK FACTORS AND DIET

URINARY SILICATE IN CALCULOUS PATIENTS

W.C. Thomas

University of Florida College of Medicine and
VA Medical Center
Gainesville, Florida

INTRODUCTION

Silicate calculi are frequent in herbivorous animals[1] and small amounts of silicate are regularly present in human calculi[2]. Recently it has been reported that the presence of polymerized silicate will decrease the concentrations of calcium and phosphorus required for mineralization of tissues *in vitro*[3]. Previously the author has reported that the addition of small amounts of polymerized silicate to urine of normal subjects, 5 to 20 mg per 100 mL of urine, will overcome the effect of inhibitors of mineralization in such urine[4].

MATERIALS AND METHODS

To investigate the possible contribution of silicates to calculus formation, urinary silicate was measured in 12 healthy control subjects and in three groups of patients with renal calculi[5]. All participants were hospitalized in a metabolic unit and received a diet free of dairy products during the 24 h collections of urine. Values for urinary silicate ranged from 6 to 17 mg per 24 h in the healthy subjects with an average value of 12 mg per 24 h. However, the silicate content (as SiO_2) of urine from 24 patients with idiopathic renal calculi was greater and ranged from 15 to 62 mg per 24 h with an average value of 35 mg per 24 h. Interestingly, the urine silicate in 13 patients with hyperparathyroidism ranged from 4 to 21 mg per 24 h with an average value of 12 mg per 24 h.

RESULTS AND DISCUSSION

Values for urine silicate in 10 patients with medullary sponge kidneys were also similar to the values obtained in normal subjects in that the range was from 7 to 19 mg per 24 h with an average value of 10 mg per 24 h.

During the four-day periods of evaluation and daily collections there was no correlation between the amounts of urinary silicate and either urine volume or urine calcium content. Also, administration of orthophosphate (1500 mg of orthophosphate phosphorus per day) to several patients did not decrease urine silicate. All of the above cited urine values were obtained while the participants were drinking tap water which contained approximately 17 mg of silicate per liter. Three patients were re-evaluated while drinking only distilled water, and within three days there was a marked reduction (to approximately 5 mg per 24 h) in urine silicate.

In the previously cited mineralization studies[3,4], it was noted that only polymerized silicate increased the mineralizing potential of urine or inorganic solutions. Interestingly, and despite the relatively low concentrations of silicate in the urine of the healthy subjects and in two of the groups of stone formers, 20 - 30% of the silicate was in a polymerized state in all of the subjects evaluated.

An explanation is lacking for the increased urine silicate in the patients with idiopathic stone disease, but suggests the possibility of increased absorption which could relate to a previous daily intake of large amounts of water. The regional incidence of renal calculi does not correlate with the silicate content of the water in the states with low or high water silicate.

Table 1.[6,7] Average silicate concentration (mg/L) of water in states with a high and low incidence of calculi.

High Incidence (>20/10,000 Pop)		Low Incidence (<13/10,000 Pop)	
Mississippi	23	26	New Mexico
Georgia	19	22	Washington
South Carolina	17	21	California
North Carolina	15	19	Oregon
Arkansas	14	14	Wyoming
Florida	12	11	Utah
Alabama	10	9	Colorado
Virginia	9	5	New York
Tennessee	7	5	Maine
West Virginia	6		
Oklahoma	6		
Kentucky	4		

The presence of increased silicate in urine of patients with idiopathic renal stones and the potential of polymerized silicate to promote hydroxyapatite formation suggests that a reduction in urine silicate might aid in controlling calculus formation. Thus, the use of demineralized water by patients who prove resistant to usual therapeutic regimens could possibly be beneficial.

ACKNOWLEDGEMENTS

This study was supported in part by NIH grants PO1 DK20586 and RR00082.

REFERENCES

1. CB Bailey, Silica metabolism and silica urolithiasis in ruminants: A review, *Can J Animal Sci* 61:219 (1981).
2. AA Levinson, N Miloslav, M Davidman, EL Prien Sr, EL Prien Jr, and RG Stevenson, Trace elements in kidney stones from three areas in the United States, *Invest Urol* 15:270 (1978).
3. A Anasuya and BS Marasingo Rao, Effect of fluoride, silicon and magnesium on the mineralizing capacity of an inorganic medium and stone formers urine tested by a modified *in vitro* method, *Biochem Med* 30:146 (1983).
4. WC Thomas Jr, Inhibitors of mineralization and renal stones, *in:* "Renal Stone Research Symposium", A Hodgkinson and BEC Nordin, eds, J & A Churchill, Ltd, London (1969).
5. EJ King and BD Stacy, The colorimetric determination of silicon in the micro-analysis of biological material and mineral dusts, *Analyst* 80:441 (1955).
6. US Geological Survey. Water Supply Papers 1299 and 1300, U.S. Government Printing Office, Washington, DC (1954).
7. R Sierakowski, B Finlayson and R Landes, Stone incidence as related to water hardness in different geographical regions of the United States, *Urol Res* 7:157 (1979).

RECURRENT RENAL CALCULI IN PATIENTS WITH MEDULLARY SPONGE KIDNEY

A.M. Meyers, N. Whalley, M. Martins, M. Sonnekus and L.P. Margolius

Metabolic Stone Clinic and Department of Nephrology
University of the Witwatersrand
Johannesburg, South Africa

INTRODUCTION

Medullary sponge kidney (MSK), was first described in 1948. The prevalence of the disorder is unknown but is said to occur in 0.5% of all unselected excretory urograms and is often clinically disregarded. However, we and others have found MSK to be a frequent association in patients with recurrent calcium calculi (RCRC) particularly in females (F). Although 3 large series of MSK and RCRC have been published[1,2,3], more data are required, particularly with respect to stone pathogenesis[4]. This study was designed to address these issues.

PATIENTS AND METHODS

Twenty four patients (15F, 9M) with MSK (Group 1) were studied, 19 of whom had RCRC. Control groups consisted of our last 334 patients with RCRC (Group 2) of whom 105 age and sex-matched stone controls (Group 3) were selected for detailed comparative studies. The patients with MSK were divided into 2 grades according to a novel radiological classification which is diagramatically depicted in Figure 1.

All patients underwent a full metabolic stone work-up in our laboratory as previouly described[5,6]. This included qualitative chemical analysis on the 18 available calculi. The short acid load test was performed on all group 1 and on 22 group 3 patients because their urine pH readings were consistently >5.5. Arterial gas estimations were performed on all. Subjects with pure uric acid calculi, enteric hyperoxaluria, oxalosis, cystinuria, primary hyperparathyroidism without MSK and primary infective calculi or anatomical abnormalities of the lower urinary tract were excluded from this study. Statistical analyses were made using the Chi square test (standard and Fisher exact), Students t-test and correlation co-efficients where indicated.

RESULTS

Six percent of all patients had MSK. The M:F ratio in group 2 was 3.1:1 and in group 1 was 0.58:1 ($p < 0.0001$). Thus of the 353 patients with RCRC, 3% were M and 13% F. Clinical and biochemical findings correlated with radiological grading is shown in Table 1. Of the five grade II patients with hypercalciuria, four had complete RTA and one had complete RTA with "renal" hypercalciuria. Four of the five had renal calculi. Hypocitraturia was present in seven cases with RTA and appeared as an isolated defect in

Urolithiasis 2, Edited by R. Ryall *et al.*,
Plenum Press, New York, 1994

three grade I patients all of whom had RCRC. The Ca:Cr was abnormal in eight and showed an absorptive pattern in five grade I patients but none in grade II. One patient in grade I and two in grade II had a "renal" pattern on the Ca:Cr ratio but only one had persistent hypercalciuria. Of the eighteen calculi analysed, the majority contained calcium and oxalate. Clinical and biochemical findings in group I compared with group III patients are shown in Table 2.

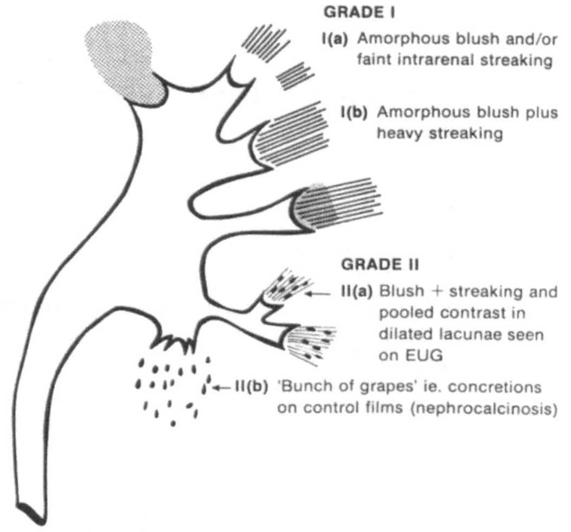
GRADE I
I(a) Amorphous blush and/or faint intrarenal streaking
I(b) Amorphous blush plus heavy streaking
GRADE II
II(a) Blush + streaking and pooled contrast in dilated lacunae seen on EUG
II(b) 'Bunch of grapes' ie. concretions on control films (nephrocalcinosis)

Figure 1. Anatomical classification of MSK via excretory urogram.

Table 1. Clinicobiochemical data correlated with radiological grading.

	Grade 1	Grade II	p value
Sex: F	7	8	<.05
M	8	1	
Calculi	15	4	<.003
UTI	3	7	<.008
RTA: Partial	4	1	<.045
Complete	1	6	
Nephrocalcinosis	0	6	<.05
Hypercalciuria	0	5	<.05
Mild Hyperoxaluria	2	0	
Hypocitraturia	4	6	<.06
Hyperuricosiuria	0	0	
Ca:Cr Normal	9	2	NS
Abnormal	6	2	
Ser Creat (μmol/L)	85(13)	102(23)	NS
GFR (mLs/min)	95(17)	58(23)	<.01

Of 10 group 1 patients with hypocitraturia, 7 had complete RTA-6 in grade II, four of whom had RCRC, and 1 in grade I who also had RCRC. There were no differences in the serum calcium, ionized calcium, phosphorous, PTH, 1,25(OH)2D_3, urinary volume (L/24h), fractional excretion of sodium or U.Ca/24 h between the 2 groups. There were no differences in the serum creatinine or GFR between grade I MSK and controls, but grade II patients exhibited a significantly lower GFR (58±23 vs 98±21 mLs/min/1.73m^2, p <0.01).

Table 2. Clinicobiochemical correlations: MSK vs Group III controls.

	MSK	Controls (III)	p value
Age at Referral	37.7 (11/1)	39/3 (13.1)	NS
Hypertension	1	14	<.004
Family History	1	38	<.001
No. Calculi/pat	9.4 (11.8)	8.9 (7.6)	NS
UTI(F)	8	2	<.002
RTA	12	1	<.03
Hypercalciuria	5	13	NS
Hyperoxaluria	2	7	NS
Hypocitraturia	10	36	NS
Hyperuricosuria	0	19	<.014
Abnormal Ca:Cr	8	46	NS

DISCUSSION

The prevalence of MSK in patients with RCRC has been found to vary between 3.6%[5] and 21%[4]. Parks et al. found the incidence to be 12% for M and 20% for F[6]. Our series lies somewhere between, that is, 6% for the total, 3% in M and 13% in F. Several important differences relating to stone pathogenesis were highlighted by dividing MSK into the two radiological grades. Stone disease was significantly more common in grade I and UTI more in grade II (Table 1). Hypercalciuria was only present in grade II and 4 of the 5 patients developed calculi. Complete RTA was closely related to the development of hypercalciuria. Hypocitraturia was also more common in grade II but did not quite reach significance. Urinary uric acid levels were normal in all our patients. The low incidence of hyperuricosuria has been noted by others[1,3]. Of considerable importance was the frequency of and potential role played by RTA in the genesis of RCRC in both grades of MSK. However, further discussion of the data is beyond the scope of this paper. Grouped as a whole, our MSK patients had a similar pattern of biochemical risk factors for stone production as did our group III controls (Table 2). Corresponding comments have been made by others[1]. Furthermore, we have also shown hyperoxaluria between MSK and stone controls. Previous studies did not estimate these parameters. However, in spite of the similarities, the underlying mechanisms are probably quite different. "Renal hypercalciuria" in patients with MSK might indicate underlying nephron damage and dysfunction, as well as the effects of chronic acidosis. Hypocitraturia relates in 70% of our cases to the presence of RTA but not in the stone controls. The presence of an absorptive pattern to the Ca:Cr ratio is unexplained but has been noted previously[3]. Likewise, the mechanism of hyperoxaluria remains obscure but again may be related to tubular dysfunction in MSK. In summary, we have confirmed and extended previous publications on the frequency and pathogenetic mechanisms of stone formation in patients with MSK. We have not found primary hyperparathyroidism to be associated as have others[3,7]. However, it is important for urologists and nephrologists to recognize that MSK is frequently encountered in patients with RCRC as treatment modalities may differ. The condition is particularly common in F with RCRC as are urinary tract infections, but the two complications occur independently. The role of RTA in the pathogenesis of RCRC in these patients cannot be overstressed and must be looked for assiduously. Finally, the practical importance of dividing patients into two radiological grades has been clearly demonstrated. It is our contention that many adults with RCRC and RTA due to grade II MSK have been previously incorrectly diagnosed as primary distal (Type I) idiopathic RTA.

ACKNOWLEDGEMENTS

This work was supported by a grant from the Medical Research Council of South Africa.We are indebted to Miss Lana Rootenberg for the preparation of this manuscript.

REFERENCES

1. JH Parks, FL Coe, AL Strauss, Calcium nephrolithiasis and medullary sponge kidney in women, *New Eng J Med* 306: 1088 (1982).
2. ER Yendt, S Jarzylo, WA Finnis et al, Medullary Sponge Kidney (tubular ectasia) in calcium urolithiasis. *in:* "Urolithiasis: Clinical and basic research." LH Smith, WG Robertson, B Finlayson, eds, Plenum Press, New York, 105 (1981).
3. M O'Niel, NA Breslau, CYC Pak, Metabolic evaluation of nephrolithiasis in patients with medullary sponge kidney, *JAMA* 245:1233 (1981).
4. ER Yendt, Medullary sponge kidney and nephrolithiasis, *New Eng J Med* 306: 1106 (1982).
5. NA Laminski, AM Meyers, MI Sonnekus et al, Prevalence of hypocitraturia and hypophosphaturia in recurrent calcium stone formers, *Nephron* 56: 379 (1990).
6. NA Laminski, AM Meyers, M Kruger et al, Hyperoxaluria in patients with recurrent calcium oxalate calculi: Dietary and other risk factors, *Brit J Urol* 68: 454 (1991).
7. DS Rao, B Frame, M Block et al. Primary hyperparathyroidism, A cause of hypercalciuria and renal stones in patients with medullary sponge kidney, *JAMA* 237: 1353 (1977).

EFFECT OF VARIOUS TYPES OF ANIMAL PROTEIN IN LITHOGENESIS

R.K. Vathsala, T.G. Dhanalekshmy, N.E. Thomas, S.V. Roshni, H.K. Moorthy, C. Aravindakshan, P.L. Vijayammal[1] and Y.M. Fazil Marickar

Department of Surgery
Medical College Hospital and
[1]Dept. of Biochemistry
University of Kerala
Trivandrum, India

INTRODUCTION

The prevalence of calcium oxalate stone disease is rising in affluent societies. Major dietary changes in affluent societies show so far a high consumption of animal protein[1]. Anand et al[2] have shown that a high animal protein diet increases urinary risk factors namely calcium, oxalate and uric acid. Robertson et al[3,4] advise stone formers to become vegetarians. Marangella et al[5] suggest that a vegetable rich diet may increase the intestinal absorption and urinary excretion of oxalate and that dietary animal protein has a minimal effect on oxalate excretion. This study was carried out to see whether animal proteins play a role in urolithiasis and to find out the influence of different types of animal protein on the biochemistry of the stone patient.

MATERIALS AND METHODS

This study was performed in two parts. In the first part we studied the difference in biochemical values between vegetarians and non vegetarians. Secondly, we analysed the effect of various types of animal protein such as fish, mutton, beef and chicken on lithogenesis. 1,500 patients with urinary stone disease were studied for their dietary habits. Of these, 98% were non-vegetarians and 2% were pure vegetarians. The 2% vegetarians were taken as controls. The non-vegetarians were classified as fish and meat eaters. The fish eaters were further classified as eating small amounts or large amounts of fish. The meat eaters were classified as mutton, beef or chicken eaters. All the non-vegetarians who consumed more than 75g of any type of the above animal proteins were taken into consideration. The biochemical study was performed by analysing the 24 h urine samples of these patients for calcium, oxalate and uric acid, and serum for calcium and uric acid. Comparisons were made using student t test and the variation in the effect of various types of animal proteins in lithogenesis was assessed by ANOVAR.

RESULTS

The non-vegetarians had significantly higher urinary oxalate and uric acid and serum calcium and uric acid levels compared to vegetarians (Table 1).

Table 1. Comparison of the biochemical values of urine and serum in vegetarians and non vegetarians.

Sl. No.	Biochemical parameters	Vegetarians Mean±SE	Non vegetarians Mean±SE	T value	P value
1	Urine calciummg/day	195.63±16.7	219.40±9.6	1.18	0.1
2	Urine oxalate mg/day	49.75±6.2	71.58±4.4	2.8	0.001
3	Urine uric acid mg/day	527.25±32.8	716.85±18.1	4.86	0.001
4	Serum calcium mg%	10.29±0.1	9.89±0.1	2.04	0.05
5	Serum uric acid mg%	4.42±0.2	4.85±0.1	1.97	0.05

The biochemical data obtained in the different types of animal protein eaters are shown in Table 2. The calculated F ratio for comparison between these different types of animal protein eaters is found to be greater than the table F value and hence the deviation due to these three different diets is statistically significant ($p < 0.05$).

Table 2. Comparison of biochemical values of urine and serum in patients consuming different types of animal protein.

Sl. No.	Biochemical parameter	Fish eaters	Large amts of fish	Small amts of fish	Meat eaters	Mutton eaters	Beef eaters	Chicken eaters	F ratio	P value
1	U.Ca. mg/day	239.9	210.1	269.7	222.7	172.9	236.6	258.5	3.92	0.05
2	U.Ox. mg/day	90.6	92.3	89.0	59.5	44.8	76.2	57.2	14.97	0.05
3	U.Ua. mg/day	681.3	716.0	646.7	744.7	693.9	756.7	783.6	8.81	0.05
4	S.Ca. mg%	10.6	9.9	10.1	10.0	9.6	9.7	10.8	4.44	0.05
5	S.Ua. mg%	4.6	5.1	4.1	5.1	4.9	5.1	5.3	4.19	0.05

Fish eaters had high levels of urine and serum calcium and urine oxalate while meat eaters had high levels of urine and serum uric acid. Comparative studies between values in the two groups of animal proteins showed that calcium excretion was significantly higher in the group eating small amounts of fish than in the group eating large amounts of fish ($p < 0.008$). Among meat eaters the excretion of calcium and uric acid was significantly higher in chicken eaters. The oxalate level was seen to be higher in beef eaters. The serum calcium was higher in the group eating small amounts of fish compared to the group eating large amounts of fish but it was not significant. At the same time the uric acid level was high in the group eating large amounts of fish compared to the group eating small amounts of fish and it was statistically significant. Chicken eaters had higher serum calcium than mutton eaters ($p < 0.001$) and beef eaters ($p < 0.001$).

DISCUSSION AND CONCLUSIONS

The influence of animal protein consumption in stone formers is well recognised. Our studies confirm that non-vegetarians had high levels of risk factors in urine and serum compared to vegetarians. Animal protein contains a higher amount of certain amino acids like tyrosine, phenylalanine and tryptophan than vegetables[7]. This aggravates the risk of calcium oxalate stone formation by increasing the urinary excretion of oxalate[6]. Different authors have different opinions on the oxalate response to dietary proteins[8,9]. In our study, eaters of large amounts of fish, and beef eaters, had higher oxalate excretion. Dietary protein and its relation to urinary calcium have received attention in the urolithiasis literature[10]. Among the non-vegetarians, those eating small amounts of fish had high levels of urine calcium and chicken eaters had high levels of serum calcium. Another study showed that a high content of animal protein in the diet caused an increase in the urinary excretion of calcium and a decrease in urinary pH and the excretion of citrate, a pattern which is thought to be unfavourable for renal calcium stone formation[11]. It was seen that among the non-vegetarians, the chicken eaters had maximum serum and urine uric acid. Indulgence in animal protein may cause hyperuricosuria[12,13], especially with meats, which are rich in purines including nucleic acid (RNA, DNA) and mononucleotides (AMP, GMP). It is concluded that the continuous intake of excess animal protein is conducive to kidney stone formation. Among the different types of animal proteins, mutton is the least harmful.

ACKNOWLEDGEMENTS

The authors acknowledge the financial support rendered by the Indian Council of Medical Research, the Kerala State Committee on Science, Technology and Environment and The University of Kerala where this work was performed.

REFERENCES

1. WG Robertson, M Peacock and A Hodgkinson A, Dietary changes and the incidence of urinary calculi in the United Kingdom between 1958 and 1976, *J Chron Dis* 32: 469 (1979).
2. CR Anand and HM Linkswiler, Effect of protein intake on calcium balance of young men given 500mg calcium daily, *J Nutr* 104: 695 (1974).
3. WG Robertson, PJ Heyburn, M Peacock, The effect of high animal protein intake on the risk of calcium stone formation in the urinary tract, *Clin Sci* 57: 285 (1979).
4. WG Robertson, M Peacock, PJ Heyburn, Should recurrent calcium stone formers become vegetarians?, *Br J Urol 51:* 427 (1979).
5. M Marangella, O Bianco, C Martini, M Petrarulo, C Vitale and F Linari, Effect of animal and vegetable protein intake on oxalate excretion in idiopathic calcium stone disease, *Br J Urol* 63: 348 (1989).
6. AJ Clifford and DL Story, Levels of purines in foods and their metabolic effects in rats, *J Nutr* 106: 435 (1976).
7. RL Gambardella and KE Richardson, The pathways of oxalate formation from phenylalanine, tyrosine, tryptophan and ascorbic acid in the rat, *Biochem Biophys Acta* 499: 156 (1977).
8. M Butz, H Hofmann and J Kohlbecker, Dietary influence on serum and urinary oxalate in healthy subjects and oxalate stone formers, *Urol Int* 35: 309 (1980).
9. JG Brockis, AJ Levitt and SM Cruthers, The effects of vegetable and animal protein diets on calcium, urate and oxcalate excretion, *Br J Urol* 54: 590 (1982).
10. HE Harrison, Factors influencing calcium absorption, *Fed Proc* 18: 1085, (1959).
11. B Fellstrom, Urate metabolism and renal calcium stone disease, *Scan J Urol & Nephrol* Supplement 62: 7 (1981).
12. NA Breslau, L Brinkley, KD Hill, CYC Pak, Relationship of animal protein-rich diet to kidney stone formation and calcium metabolism, *J Clin Endocrinol Metab* 66: 140 (1988).
13. N Zöllner, Influence Of Diet On Urinary Uric Acid Excretion. *in*: "Urolithiasis Research" (Eds). H Fleisch, WG Robertson, LH Smith and W Vahlensieck W. Plenum, New York 155 (1976).

EFFECT OF CAFFEINE CONSUMPTION ON URINARY CALCIUM, MAGNESIUM, OXALATE AND CITRATE IN CALCIUM STONE FORMERS

L.K. Massey[1] and R.A.L. Sutton[2]

[1]Food Science and Human Nutrition
Washington State University
Spokane, USA
[2]Department of Medicine
University of British Columbia
Vancouver, Canada

INTRODUCTION

Caffeine has been previously shown to increase urinary Ca for three hours after its consumption in several age and gender groups[1]. Caffeine increases urinary Ca excretion by decreasing reabsorption with little increase in GFR[2]. Caffeine most likely exerts its effects on the kidney via adenosine receptors, as adenosine antagonists can block its effect on Ca excretion[3]. Similarities exist between patterns of mineral excretion occurring after caffeine loading in non-stone formers and the continual elevated calcium excretion in some stone formers (SF) with fasting hypercalciuria of presumed "renal leak" origin. Fasting hypercalciuric SF have abnormal responses to diuretics that are consistent with the presence of a defect in proximal tubular reabsorption[4]. Overall, these data suggest that the effect of caffeine on the renal handling of Ca may be similar to the defect causing renal hypercalciuria.

METHODS

Thirty-nine patients (age 51±12 [±SD] years; M/F, 29/10) with Ca-containing kidney stones were recruited from the Vancouver General Hospital Stone Clinic and from the practices of urologists utilizing the lithotripter. Nineteen of the 39 patients had chemically-confirmed CaOx stones. Thirteen patients had only one prior stone passage, while 26 were recurrent SF. All patients were studied at least one month after ESWL or surgical intervention for stone removal. All were normocalcemic and were free of urinary tract infection at time of study. Forty-eight normal healthy subjects (age 45±15 years; M/F, 33/15) served as controls.

Following a 14 h overnight fast, all subjects supplied a spot urine sample, and fasting blood samples were collected. Each subject then consumed an oral caffeine load of 6 mg/kg lean body mass (range: 250-400 mg caffeine, equivalent to 15 to 25 oz of brewed coffee) in 180 mLs of warm water. Urine was collected for 2 hr following the caffeine load. In addition, 57 subjects supplied a 24 h urine collection from the day prior to the caffeine load while consuming their normal diet.

RESULTS

Fasting total serum ultrafilterable Ca and Mg levels were not different between SF and controls, nor were differences found between the two groups in fasting urinary Ca, Mg, oxalate (Ox), or citrate expressed as a ratio with urinary creatinine (creat). However, both SF and controls showed an increase in urinary Ca and Mg ratios in response to the caffeine load, but only SF showed an increase in urinary citrate. The Tiselius Risk Index [5] increased similarly in SF (+78) and controls (+85), although it was only significant in SF. Since changes in urinary Ox were not observed, this increase in the Risk Index was attributed to the increase in urinary Ca, not offset by beneficial increases in Mg and citrate.

Table 1. Acute effect of caffeine on urine composition 2 h after consumption.

	Stone formers (39)		**Controls (48)**	
	Fasting	Post-caffeine	Fasting	Post-caffeine
Ca/creat	0.24±0.14	0.38±0.21*	0.23±0.14	0.43±0.29*
Mg/creat	0.20±0.09	0.28±0.15*	0.23±0.14	0.33±0.17*
Ox/creat	0.021±0.006	0.021±0.006	0.019±0.011	0.018±0.016
Cit/creat	0.20±0.010	0.29±0.13*	0.24±0.11	0.27±0.14
Risk Index	290±164	368±180*	217±93	302±285

Values shown are Means±SD, n in parentheses. * Significantly different from fasting values, $p< 0.001$.
Note: Urinary Ca, Mg, Ox and citrate (Cit) are expressed as a ratio with urinary creatinine (mmol/mmol).

Table 2 shows fasting and post-caffeine Ca/creat. and Na/creat ratios in normocalciuric and fasting hypercalciuric (CA/creat. >0.311) subjects divided by prior 24 h urinary Na excretion relative to the overall mean of 145 mmol Na/day. A higher Na consumption the day before caffeine testing is associated with lack of a significant calciuretic response in fasting hypercalciuric subjects but not in normocalciuric subjects (Table 2). A blunted or absent hypercalciuric response (less than 10% increase in Ca/creat) after caffeine was found in 6 of 39 SF (5 of 12 with fasting hypercalciuria, 1 of 27 with normocalciuria), and 3 of 48 controls (3 of 13 with fasting hypercalciuria, none of 35 with normocalciuria).

Table 2. Interactions of previous Na consumption and calciuric response to caffeine by fasting Ca/creat.

	Hypercalciurics		Normocalciurics	
	Low Na (10)	High Na (7)	Low Na (20)	High Na (20)
Fasting				
Ca/creat	0.46±0.14	0.38 ±0.05	0.15±0.09	0.16±0.07
Na/creat	11.0±7.2	16.3±6.1*	6.2±3.0	9.5±5.0
Change Post-load				
Ca/creat	0.31±0.26*	0.12±0.22	0.15±0.09	0.11±0.06
Na/creat	14±10.5	7.5±12.2	8.0±5.0	9.3±7.2

Values shows are Means±SD (mmol/mmol), n in parentheses. * Significantly different from all other groups, $p< 0.05$.

DISCUSSION

Consumption of caffeine increased the Tiselius Risk Index for CaOx stone formation in Ca stone formers. A hypercalciuric response to caffeine in fasting hypercalciuric subjects, either SF or controls, was associated with consumption of lower Na diets. Nies et al[6] found that salt loading did not change the intrinsic vasoconstrictor response to an adenosine analog in isolated rat kidneys as it did in intact animals. The extrinsic factor responsible for the change in response after salt loading was not identified, but other studies suggest that changes in either angiotensin II[7] or sympathetic nerve activity[8] may alter renal responses to adenosine.

The differences in the hypercalciuric response to caffeine, an adenosine antagonist, appear to be related to physiological adaptations associated with previous consumption of dietary Na, rather than with adenosine receptor defects in Ca SF.

REFERENCES

1. LK Massey and KJ Wise, Impact of gender and age on urinary water and mineral excretion response to acute caffeine doses, *Nutrition Res* 12:605 (1992).
2. EA Bergman, LK Massey, KJ Wise and DJ Sherrard, Effects of dietary caffeine on renal handling of minerals in adult women, *Life Sci* 47:557 (1990).
3. MD McPhee and SJ Whiting, Effect of adenosine and adenosine analogues on methylxanthine-induced hypercalciuria in the rat, *Can J Physiol Pharmacol* 67:1278 (1989).
4. RAL Sutton and VR Walker, Responses to hydrochlorothiazide and acetazolamide in calcareous renal stone formers: Evidence suggesting a defect in renal tubule function, *NEJM* 302:709 (1980).
5. H-G Tiselius, An improved method for the routine biochemical evaluation of patients with recurrent calcium oxalate stone disease, *Clin Chim Acta* 122:409 (1982).
6. AS Nies, ML Beckmann and JG Berber, Contrasting effects of changes in salt balance on the renovascular response to A1-adenosine receptor stimulation *in vivo* and *in vitro* in the rat, *J Pharmacol Exptl Therapeutics* 256:542 (1991).
7. G Deray, R Sabra, WA Herzer, EK Jackson and RA Branch, Interaction between angiotensin II and adenosine in mediating the vasoconstrictor response to intrarenal hypertonic saline infusions in the dog, *J Pharmacol Exptl Therapeutics* 2562:631 (1990).
8. P Hedqvist, BB Fredholm and S Olundh, Antagonistic effects of theophylline and adenosine on adrenergic neuroeffector transmission in the rabbit kidney, *Circulation Res* 43:592 (1978).

SHOULD HYPERCALCIURIA BE REDEFINED?

G. Mobb[1], J.P. Kavanagh[2] and P.N. Rao[2]

[1]Department of Urology
Stepping Hill Hospital
Stockport, UK
[2]Department of Urology
University Hospital of South Manchester, UK

INTRODUCTION

It is customary to advise idiopathic stone formers to drink large quantities of fluid in addition to dietary modification to reduce their risk of recurrent stone formation. However, neither the ideal daily volume which would provide maximum benefit nor its direct effect upon solute excretion is known.

Hypercalciuria is a common finding in stone formers and is associated with a significantly increased risk of further urinary tract stone formation. It is usually corrected by dietary manipulation, and thiazide diuretics are prescribed to increase the distal tubular resorption of calcium[1].

Although increased fluid intake is always routinely recommended, it has been previously demonstrated that calcium excretion may rise when urinary volume is increased[2], in which case increasing fluid intake may be counterproductive. Alternatively, since concentrations are more important in determining supersaturation and hence the risk of stone formation, total daily excretion values may not be relevant. In addition to change in total solute excretion, any decrease in the urinary concentration of magnesium and citrate which may occur would counteract the beneficial effect of decreasing calcium and oxalate concentrations. The overall effect upon calcium oxalate supersaturation ratios remains unknown.

PATIENTS AND METHODS

The effect of increased fluid intake was investigated in 25 idiopathic urinary stone formers (4 F; 21 M, mean age 52 yr, range 30-74). Of these, nine were hypercalciuric by definition (>7.5mmol/24 h). These patients had been attending an Out Patient Stone Clinic for some years and were already following standard dietary advice.

Each patient made three 24 h urine collections whilst established upon their usual dietary regimen. Seven days later three further 24 h urine collections were made on consecutive days whilst being advised to drink at least 5 litres of fluid per day.

A dietary diary was recorded by each patient for a three day period and a specimen of local tap water was analysed for each subject. Dietary histories revealed a mean daily intake of 803mg of calcium, 92g of carbohydrate, 78g of protein and 20g of fibre. Local tap water was classified as "very soft" (mean of 0.31 mmol/L of calcium).

A blood sample was collected prior to the study and at the end of each three day period and was analysed for serum biochemistry. Serum creatinine concentration was used

Urolithiasis 2, Edited by R. Ryall *et al.*,
Plenum Press, New York, 1994

to calculate both creatinine clearance and glomerular filtration rate using the formula of Gault and Cockcroft[3]. All urine samples were analysed for calcium, oxalate, magnesium, citrate and other electrolytes, and urinary supersaturation ratios were calculated using the computer program EQUIL2[4]. The effect of high fluid intake was examined in the group as a whole and also more specifically in those patients who were hypercalciuric. The Wilcoxon rank sign test was used to compare values measured before and after fluid loading and the chi squared test was used to examine crystallization risk (supersaturation) in samples less than or greater than 3 litres.

RESULTS

Serum creatinine, creatinine clearance and glomerular filtration rate fell within normal accepted ranges and showed no significant change when urinary volume was increased. Serum calcium and phosphate concentrations also remained unchanged throughout the study period (Table 1).

Table 1. Serum parameters and renal function on standard and high fluid intake.

	Standard	High	p
GFR (mL/min)	94.8	93.9	NS
Cl_{Cr} (mL/min)	144.0	141.0	NS
Creatinine (µmol/L)	93.9	98.9	NS
Calcium (mmol/L	2.25	2.23	NS
Phosphate (mmol/L)	1.03	1.00	NS

Analysis of the first three 24 h urine collections revealed that patients were excreting an average of 2.13 L/day (sd 0.72) following standard dietary advice. After deliberately increasing daily fluid intake to a recommended minimum of 5 litres, daily urinary volumes increased significantly to a mean of 4.44 L/day (sd 0.67) ($p <0.001$) (Table 2).

Table 2. Urinary excretion and supersaturation on standard and high fluid intake.

	Standard	High	p
Volume (L)	2.13	4.44	<.001
Calcium (mmol)	6.45	7.51	0.016
Oxalate (mmol)	0.39	0.51	0.036
Magnesium (mmol)	4.41	5.22	0.013
Citrate (mmol)	2.91	3.17	NS
Sodium (mmol)	140	160	0.033
Creatinine (mmol)	13.93	14.16	NS
Supersaturation ratio	6.0	3.1	<.001

It was found that increasing daily fluid intake significantly increased both the total daily excretion of calcium ($p <0.016$) and oxalate ($p <0.036$) in all patients. However, calcium oxalate supersaturation values significantly decreased from 6.0 to 3.1 ($p <0.001$) because of a significant fall in the concentration of all urinary solutes measured. Creatinine output was unchanged (Table 2). Daily urine samples of less than 3 litres had a

78% chance of having a supersaturation >5.0 (high risk), but only 15% of samples more than 3 litres had supersaturations > 5.0 ($p < 0.00001$).

When the hypercalciuric patients were examined more closely, they also demonstrated similar changes. All nine excreted more than 7.5 mmol of calcium per day (mean 9.9 mmol/day, sd 1.38) and this also increased significantly on high fluid intake (mean 11.78 mmol/day, sd 2.60, $p = 0.05$). Once again the concentrations of all urinary solutes decreased significantly. All hypercalciuric patients increased their output to above 3 litres/day and in eight of these patients the supersaturation ratio also changed from above to below our guideline value (5.0) for high risk (mean of 6.6 falling to 3.7) - thus questioning the need for further therapy.

DISCUSSION

In 1980 Pak[5] demonstrated that diluting urine *in vitro* or encouraging patients to drink distilled water reduced the supersaturation of calcium oxalate and reduced the risk of stone formation. In our more natural setting, the substantial increase in daily fluid consumed by all of our stone formers produced a significant increase in the daily urinary excretion of calcium and oxalate. On the other hand, the concentrations of all measured urinary solutes and supersaturation ratios decreased. It has been suggested that increased calcium excretion on high fluid intake results from calcium ingested in water consumed[2], but as our patients reside in a "very soft" water district the changes demonstrated are more likely to reflect reduced calcium reabsorption in the presence of increased urinary flow. Nevertheless, the form in which the extra fluid is taken could be of importance and in other circumstances might lead to even more significant increases in calcium and oxalate excretion. The beneficial dilutional effects of increased fluid intake can overcome the increased output of calcium and oxalate but does not become apparent until urinary volume exceeds 3 litres per day. Using solute-free water, Pak[5] arrived at a similar value (2.5 litres per day).

Hypercalciuric patients on high fluid intake will also benefit by significantly reducing their urinary calcium concentration but, by definition, still remain "hypercalciuric". However, due to the demonstrated significant fall in supersaturation ratios, their risk of stone formation is reduced. The dilutional effect of the large fluid intake in these patients is such that the "hypercalciuria" became insignificant with respect to the risk of stone formation. Therefore, the term "hypercalciuria" may be misleading with regard to stone formers and should perhaps be redefined on the basis of concentration values rather than total daily urinary excretion. In conclusion, the necessity for drug therapy in hypercalciuria needs to be reconsidered whilst the clinical significance of this condition perhaps should be re-examined.

REFERENCES

1. M Ohkawa, S Tokunaga, TM Nakashima, M Orito and H Hisazumi, Thiazide treatment for calcium urolithiasis in patients with idiopathic hypercalciuria, *Br J Urol* 69: 571 (1992).
2. OM Embon, GA Rose and T Rosebaum, Chronic dehydration and stone disease, *Br J Urol* 66: 357 (1990).
3. DW Cockcroft and MH Gault, Prediction of creatinine clearance from serum creatinine, *Nephron* 16: 31 (1976).
4. PG Werness, CM Brown, LH Smith and B Finlayson, EQUIL2: a basic computer program for the calculation of urinary supersaturation, *J Urol* 134: 1242 (1985).
5. CYC Pak, K Sakhaee, C Crowther and L Brinkley, Evidence justifying a high fluid intake in treatment of nephrolilhiasis, *Ann Int Med* 93: 36 (1980)

COMPARISON OF ACUTE AND CHRONIC CALCIUM-LOAD TESTS IN RENAL STONE FORMERS

I.P. Heilberg, L.A. Martini, L. Cuppari, S.A. Draibe,
H.C. Perrone and N. Schor

Nephrology Division
Escola Paulista de Medicina
São Paulo, Brazil

INTRODUCTION

Idiopathic hypercalciuria, a common metabolic disturbance affecting up to 60% of renal stone formers[1], is defined as a urinary Ca excretion higher than 4 mg/kg body weight/day[2]. The enhanced urinary Ca concentration raises the saturation for CaOx and phosphate, thus leading to formation of calcareous stones. Hypercalciuria may be due to a primary increase in intestinal Ca absorption, a renal tubular Ca leak, bone resorption, or combined tubular abnormalities among other mechanisms[3]. The renal or absorptive origin of hypercalciuria can be discerned by means of an acute oral Ca load test (ACLT)[4]. In our population of 86 Ca stone formers who consumed a low or normal Ca intake, the ACLT disclosed the presence of absorptive hypercalciuria (AH) and renal hypercalciuria (RH) in many of them. However, the majority of these patients did not have hypercalciuria, based on 24 h urine Ca excretion, when they were consuming their usual diets. Therefore, in order to evaluate their calciuric response, these patients were challenged to a normal or higher Ca intake by means of a chronic Ca load test (CCLT).

PATIENTS AND METHODS

Eighty-six adult Ca stone formers (40 male/46 female) were given an ACLT described elsewhere[4], but with no prior dietary restrictions. Two 24 h urine samples were obtained from each of them while consuming their normal diets. Seventeen healthy volunteers were submitted to the same protocol. The second part of the study consisted of a CCLT in which Ca stone formers and healthy subjects were given 1g/day of oral Ca in a single dose in the form of gluconolactate + carbonate for seven days in addition to their usual diet. On the seventh day, they again collected 24 h urine samples for Ca determination. Serum levels of $1{,}25(OH)_2D_3$ were also determined (Dr. C Langman, Laboratory of Mineral Metabolism, Children's Memorial Hospital, Chicago, USA). A 72 h dietary record was obtained from all patients to assess their usual Ca intake.

RESULTS

Based on the the results of the ACLT, 51 Ca stone formers (59%) were classified as AH and 35 as RH (41%). Data regarding the number and percentage of subjects with

Table 1. The number and percentage of subjects with normocalciuria or hypercalciuria while consuming their normal diet as indicated by 24 h urinary Ca excretion.

Group	Total n	Normocalciuria	Hypercalciuria
Normal subjects	17	14.82%	3.18%
AH	51	40.78%	11.22%
RH	35	22.63%	13.37%

either normocalciuria or hypercalciuria, based on 24 h urine samples collected while on their normal diet, indicated that 82% of healthy normal subjects, 78% of AH and 63% of RH were normocalciuric (Table 1).

The mean intake of Ca (±SD) during the period of normal dietary intake was 451±257, 520±268, and 573±366 mg/day, for normal subjects, AH and RH, respectively. These values were not statistically different. Table 2 shows that the mean urinary Ca excretion on a normal diet was below 4mg/kg/24 h in all three groups of subjects with RH having the highest output ($p< 0.05$). After the CCLT, the mean urinary Ca excretion was significantly higher than that during the period on a normal diet only in healthy subjects and AH.

Table 2. A comparison of 24 h urinary Ca excretion while on a normal diet and following the CCLT.

Group	Normal diet	CCLT
Normal subjects	2.5±1.1	3.7±1.8*
AH	2.9±1.6	3.7±1.5*
RH	3.7±1.8#	4.1±1.5

Values shown are Means±SD. *Significantly greater than while consuming a normal diet, $p< 0.05$; # Significantly greater than other groups on a normal diet, $p< 0.05$.

Data regarding the number and percentage of patients who developed hypercalciuria or were normocalciuric following the CCLT are shown in Table 3. While 64% of normal subjects, 70% of AH and 55% remained normocalciuric, 36% of normal subjects, 30% of AH and 45% of RH developed hypercalciuria. Serum levels of 1,25 $(OH)_2D_3$ were not different between the groups.

Table 3. The number and percentage of subjects with normocalciuria and hypercalciuria following the CCLT.

Group	Total n	Normocalciuria	Hypercalciuria
Normal subjects	14	9 (64%)	5 (36%)
AH	40	28 (70%)	12 (30%)
RH	22	12 (55%)	10 (45%)

DISCUSSION

Preliminary data showed that despite the initial classification of AH or RH by the ACLT, most of the patients were normocalciuric when evaluated while consuming their normal diet. In these normocalciuric patients, only 30% of the AH group were sensitive to a normal or higher Ca intake induced by the CCLT. In contrast, 45% of the RH group were sensitive to this intake. These data suggest that in this population, the response to the ACLT should be further confirmed by the CCLT. Both AH and RH may have a different sensitivity to dietary Ca intake which makes it difficult to clearly separate them[5]. Levels of serum $1,25(OH)_2D_3$ which are similar in AH and RH do not help in a further sub-categorization of these groups.

REFERENCES

1. CYC Pak, Etiology and treatment of urolithiasis, *Am J Kidney Dis* 18:624 (1991).
2. A Hodgkinson, LN Pyrah, The urinary excretion of calcium and inorganic phosphate in 344 patients with calcium stone of renal origin, *Br J Surg* 16:10 (1958).
3. CYC Pak, Medical management of nephrolithiasis in Dallas: update 1987, *J Urol* 140:461 (1988).
4. CYC Pak, R Kaplan, H Bone, J Townsend, O Waters, A simple test for the diagnosis of absorptive, resorptive and renal hypercalciurias, *N Engl J Med* 292:497 (1975).
5. FL Coe, MJ Favus, T Crockett, AL Strauss, JH Parks, A Porat, CL Gantt and LM Sherwood, Effects of a low calcium diet on urine calcium excretion, parathyroid function and serum $1,25(OH)_2D_3$ levels in patients with idiopathic hypercalciuria and in normal subjects, *Am J Med* 72:25 (1982).

CIRCADIAN RHYTHMOMETRIC STUDIES ON THE LOW MOLECULAR WEIGHT URINARY INHIBITORS OF CALCIUM OXALATE UROLITHIASIS

D. Wangoo[1], V. Rattan[1], S.K. Thind[1], G.S. Gupta[2] and R. Nath[1]

[1]Department of Biochemistry
Postgraduate Institute of Medical Education and Research
[2]Department of Biophysics
Panjab University
Chandigarh, India

INTRODUCTION

The possible significance of low molecular weight inhibitors (Mg, citrate and pyrophosphate) in the prevention of calcium oxalate urolithiasis has been emphasized[1]. Estimation of total 24 h excretion of urinary inhibitors in stone-formers (SF) will not reveal any circadian fluctuations. However, measurements of 3-hourly urine samples collected over a 24 h span should give a better picture of their circadian rhythmicity, which may help to expose any urinary biological abnormalities present therein. Therefore, the circadian rhythm of the 3 h excretion of Mg, citrate and pyrophosphate, and urinary inhibitory activity towards calcium oxalate monohydrate (COM) crystallization in SF and healthy subjects (N), were investigated.

MATERIALS AND METHODS

Three-hourly urine samples (over a 24 h span) beginning at 00:00 h were collected from 25 SF (mean age 39±7 years) and 25 age-sex matched N (mean age 41±6 years). The volume of the samples was recorded and they were analyzed for Mg (atomic absorption Spectroscopy; Perkin Elmer, 4000), citrate[2], pyrophosphate[3] and inhibitory activity[4]. Creatinine was also estimated for the assessment of complete urine collection and kidney function. Details of the conventional methods (mean, t-test) used to describe the circadian rhythm and other changes have been reported previously [2].

RESULTS

The mean 3 h excretion of Mg varied from 0.53 mmoles to 0.90 mmoles amongst SF, whereas it varied from 0.64 mmoles to 1.43 mmoles amongst N. The amplitude in N was found to be double that in SF. Rhythmometric analysis of Mg excretion revealed a significant circadian rhythm amongst N, which was absent from SF (Table 1).

The mean 3 h excretion of citrate varied from 0.21 mmoles to 0.31 mmoles amongst SF and from 0.23 mmoles to 0.39 mmoles in the N group. Although the 3 h urinary excretion of citrate was lower in SF than in N, the difference was not significant.

Cosinor rhythmometric analysis of citrate excretion revealed a significant circadian rhythm amongst N, which was absent from SF (Table 1).

The mean 3 h excretion of pyrophosphate varied from 0.029 to 0.046 mmoles in the SF, and from 0.062 mmoles to 0.123 mmoles amongst the N. The amplitude was quite high amongst N in comparison with SF. Cosinor rhythmometric studies revealed a significant circadian rhythm amongst N which was absent from SF (Table 1).

Table 1. Circadian rhythm of magnesium, citrate, pyrophosphate excretion and inhibitory activity (1 cycle = 360°) in N and SF.

	N		SF
		Magnesium	
Mesor (mmoles)	0.96		0.72
Amplitude	0.218		0.106
Acrophase	29°		148°
		Citrate	
Mesor (mmoles)	0.31		0.26
Amplitude	0.06		0.017
Acrophase	216°		352°
		Pyrophosphate	
Mesor (mmoles)	0.074		0.036
Amplitude	0.049		0.003
Acrophase	49°		11°
		Inhibitory activity	
Mesor (U/μmole Cr)	2.76		1.73
Amplitude	1.70		0.328
Acrophase	215°		212°

Inhibitory activity of the 3 h urine samples (expressed as units)/μmole creatinine) varied from 1.40 to 2.30 amongst SF and from 1.35 to 5.79 amongst N, with the amplitude being about 5 times higher in N than in SF. Cosinor rhythmometric analysis of 3 h urine samples revealed a significant circadian rhythm amongst N, which was absent from SF with uniform and low inhibitory activity (Table 1). The acrophase of inhibitory activity in N was found at 215°, ie at a time which also coincided with that of citrate (216°).

DISCUSSION

The present study demonstrates the existence of a circadian rhythm for Mg, citrate, and pyrophosphate excretion, and urinary inhibitory activity amongst N and the absence of such rhythms from SF. The lack of rhythmicity in the various urinary and serum parameters amongst idiopathic SF has also been reported from various laboratories[5,6]. The occurrence of the acrophase for Mg and pyrophosphate excretion amongst N in the morning hours suggests that maximum protection is provided to this group at this time, when the concentration of stone minerals is highest and the risk of precipitation greatest. The occurence of the acrophase for citrate excretion in the afternoon hours suggests that it significantly contributes to the overall inhibitory activity of urine. This is the precise period during the day when additional protection is required as indicated by the excretion profiles of various lithogenic substances. A significantly lower (negative) inhibitory activity in the afternoon and evening urine samples of SF has also been reported, suggesting an increased tendency towards crystallization at these periods compared to normal subjects[7]. Thus, in contrast to SF, N are protected during the whole 24 h cycle against the risk of calcium and oxalate precipitation.

REFERENCES

1. K Kohri, J Garside and NJ Blacklock, The role of magnesium in calcium oxalate urolithiasis, *Brit J Urol* 61:107 (1988).
2. D Wangoo, SK Thind, GS Gupta and R Nath, Chronobiology of urinary citrate amongst stone-formers and healthy males from North-Western India, *Urol Res* 19:203 (1991) .
3. RCG Russel and A Hodgkinson, Urinary excretion of inorganic pyrophosphate by normal subjects and patients with renal calculus, *Clin Sci* 31:51 (1986).
4. FL Coe, HC Margolis, LH Deutsch and AL Strauss, Urinary macromolecular crystal growth inhibitors in calcium nephrolithiasis, *Mineral Electrolyte Metab* 3:268 (1980).
5. Y Touitou, C Touitou, G Charransol et al, Alterations in circadian rhythmicity in calcium oxalate renal stone formers. *Int J Chronbiol* 8:175 (1983).
6. N Kinoshita, Diurnal variation in plasma oxalate concentration and oxalate clearance in calcium oxalate stone formers with special reference to the effect of oxalate loading, *Hinyakika Kiya* 33:1331 (1987).
7. MK Li, DK Shum and S Choi, Detection of crystallization in inhibitory activity of whole urine with a gel model, *Urol Res* 15:75 (1987).

RELATIONSHIP BETWEEN 24 HOUR URINARY VOLUME, URINARY SPECIFIC GRAVITY, AND RELATIVE SUPER-SATURATIONS IN STONE FORMING PATIENTS

D.K. Ackermann[1], R. Takkinen[2], B. Hess[2] and Ph. Jaeger[2]

[1]Department of Urology
[2]Policlinic of Medicine
University of Berne
Berne, Switzerland

INTRODUCTION

For prophylactic purposes, calcium stone forming patients are generally advised to increase 24 h urinary output to 2 L[1,2] and to lower urinary specific gravity to 1.015 g/mL or less[2,3]. Since urinary supersaturation is the driving force for stone formation[4], its relation to urinary output and specific gravity should be demonstrated.

MATERIALS AND METHODS

Fifty-four sterile 24 h urine samples were analysed from 40 stone forming patients (15 females, 25 males) who collected urine at home on an unrestricted diet. Urinary specific gravity was measured refractometrically.

The relative supersaturations (RS) with respect to calcium oxalate monohydrate, brushite, and uric acid were computed with the EQUIL program of Finlayson (version 1988)[4,5]. For the analyses of regression, linear and non linear curve fitting programs were used.

RESULTS

Specific gravity and 24 h urinary volume were inverseley related according to the equation:

$$y = 1.1119 * X^{-0.011959} \text{ with } R^2=0.79$$

The relative supersaturation for uric acid showed significant correlations with urinary volume:

$$y = 19.306 * e^{-0.00080895x} \text{ with } R^2 = 0.53$$

as well as with specific gravity :

$$y = - 572.73 + 569.17x \text{ with } R^2 = 0.72$$

No significant correlations were detected between relative supersaturations for calcium oxalate monohydrate (RS COM) (Fig. 1) or brushite (Fig. 2) and 24 h urinary volume or between the same relative supersaturations (Figs. 3 and 4) and specific gravity in the observed ranges (24 h urinary volume: 460 - 5660 mL, specific gravity: 1.004 - 1.034 g/mL). Whenever 24 h urinary volume was higher than 3 L or specific gravity was lower than 1.010 g/mL, the relative supersaturations for calcium oxalate monohydrate and brushite were always < 6.0 and < 1.5, respectively. For a 24 h urinary volume of 2 L, the relative supersaturation for calcium oxalate monohydrate ranged from 3 to 14 and for brushite from 0.05 to 3. A similar range of values of these relative supersaturations was found for a specific gravity of 1.015 g/mL.

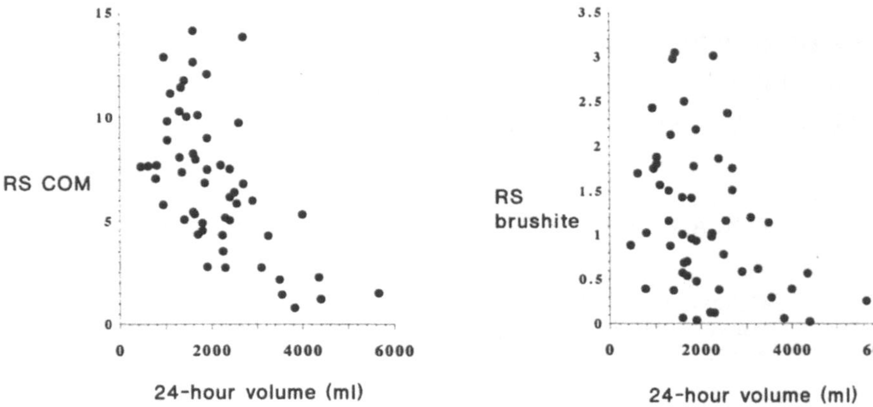

Fig. 1. Relative supersaturation for calcium oxalate monohydrate (RS COM) versus 24 h urinary volume.

Fig. 2. Relative supersaturation (RS) for brushite versus 24 h urinary volume.

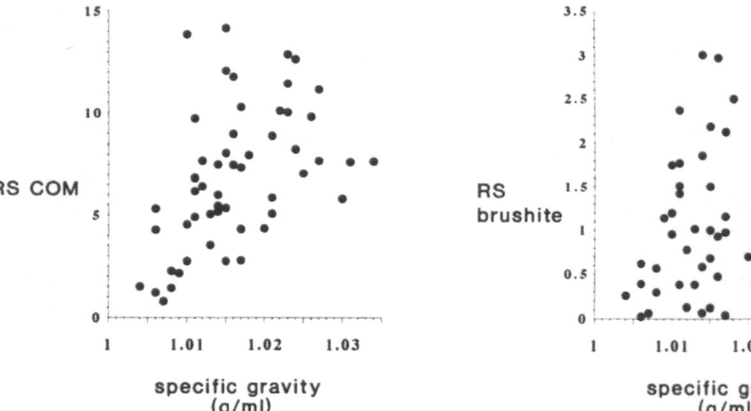

Fig. 3. Relative supersaturation for calcium oxalate monohydrate (RS COM) versus specific gravity.

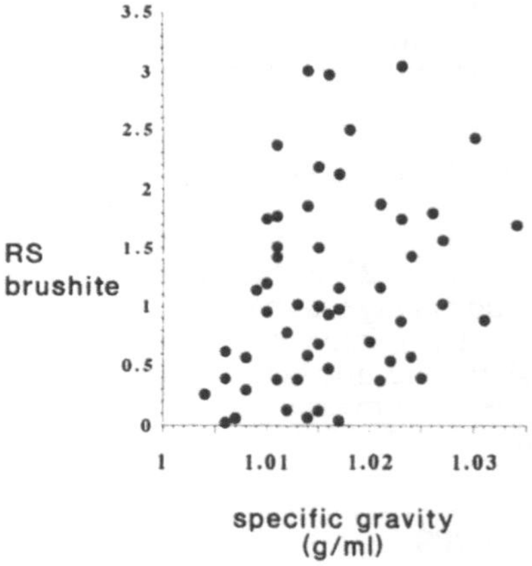

Fig. 4. Relative supersaturation (RS) for brushite versus specific gravity.

DISCUSSION

In the present study, we observed a significant relationship between the RS for uric acid and urinary volume as well as specific gravity. Both dilution of uric acid and some increase in urinary pH owing to water diuresis account for this finding[1].

No significant correlations were detected between the RS for calcium oxalate monohydrate or brushite and urinary volume or between the same RS and specific gravity. These results are in keeping with the findings by other authors[3]. Urinary specific gravity correlates strongly with ionic strength[3] which is governed mainly by sodium and chloride concentrations. Obviously, the concentrations of these ions depend more on urinary output than the concentrations and interactions of oxalate, citrate, phosphate, calcium, and magnesium, ie those substances which are more important with respect to urinary supersaturation for calcium oxalate monohydrate and brushite[5,6]. However, despite the lack of significant correlation over the observed range, some relationship between urinary volume, specific gravity, and RS of calcium salts seems to exist. If urines are highly diluted - 24 h volume higher than 3 L and specific gravity lower than 1.010 g/mL - one may reasonably assume that urinary supersaturations for calcium oxalate monohydrate and brushite are low enough.

Therefore, stone forming patients should be advised to increase 24 h urinary volume to more than 3 L and/or to lower urinary specific gravity to less than 1010 g/mL. Urinary 24 h output and urinary specific gravity are parameters that patients can easily measure at home and are useful in monitoring stone prophylaxis. Measurements of urinary specific gravity allow the assessment of urinary dilution in every single urine portion throughout the day and seem to stimulate the patients to increase fluid intake[7].

ACKNOWLEDGEMENTS

This study was supported in part by the Swiss National Science Foundation (Grant No. 32-33543.92).

REFERENCES

1. GW Drach, Urinary lithiasis, *in* "Campbell's Urology", PC Walsh, RF Gittes, AD Perlmutter, and TA Stamey, eds, WB Saunders Company, Philadelphia (1986).
2. D Kutter, A Kremer, Die Diureseuberwachung bei der Rezidivprophylaxe der Harnsteinbildung, *Urologe* (B) 29: 288 (1989).
3. Hesse, H Wuzel, A Classen and W Vahlensieck, An evaluation of test sticks used for the measurement of the specific gravity of urine from patients with stone disease, *Urol Res* 13: 185 (1985).
4. B Finlayson. Calcium stones: some physical and clinical aspects, *in* "Calcium Metabolism in Renal Failure and Nephrolithiasis", DS David, ed, John Wiley & Sons, New York (1977).
5. PG Werness, CM Brown, LH Smith and B Finlayson, EQUIL 2: a basic computer program for the calculation of urinary saturation, *J Urol* 134: 1242 (1985).
6. WG Robertson and M Peacock, The cause of idiopathic calcium stone disease: hypercalciuria or hyperoxaluria? *Nephron* 26: 105 (1980).
7. M McCormack, J Dessureault and M Guitard, The urine specific gravity dipstick: a useful tool to increase fluid intake in stone forming patients, *J Urol* 146: 1475 (1991).

THE BONN UROLITHIASIS POST-EPISODE CARE PROGRAM - NEW RESULTS

A. Hesse[1], A. Nolde[1], V. Hagmaier[2], O. Scharrel[1] and W. Vahlensieck[1]

[1]Division of Experimental Urology
Department of Urology
University of Bonn, Germany
[2]Medical Research
Madaus AG
Cologne, Germany

INTRODUCTION

Prevalence of urolithiasis in Germany is estimated to stand at 4% and incidence at 0.4 to 0.5%, i.e. 300,000-400,000 patients/per year will develop stones; 60% (200,000 patients) are recurrent stone formers[4,9]. Experience with recurrent stone formers in our hospital over the past 12 years has shown a decrease in the recurrence rate amounting to 20% under a specific therapy[5,6]. Without such therapy the recurrence rate was 60 to 70%.

Therefore, in June 1988 the Bonn Post-Episode Care Program was initiated in order to prevent relapses in patients with recurrent stone formation by propagation of specific diagnostic measures (standard program) and a special metaphylaxis. Over the last three years 1,359 patients have participated in the program.

MATERIALS AND METHODS

In close co-operation with the attending physician, check-ups at four-month intervals were made. 24 h urines were collected and nutrition records taken, once an anamnesis and therapy records at 4 month intervals were set up by the attending physician. Depending on the analysis of stone composition, if available, 24 h urines (standard program: pH value, volume, calcium, magnesium, uric acid, oxalic acid, citric acid, phosphate), serum parameters, nutrition records and accompanying diseases, general as well as specific recommendations were given by the study center concerning diet, fluid intake and drugs. The study center provided patients with collection samplers, information material, reminded them of follow-up checks, evaluated all documentation and entered all data in the Bonn Urolithiasis register.

RESULTS

Evaluation of the first 1,000 patients has shown the following results. Stone analyses were available for 815 patients. Calcium oxalate was the most frequent type of stone (81%), followed by carbonate apatite (5.3%), cystine (4.8%), uric acid (3.9%) and infection stones (2.1%). Brushite and urates were extremely rare (< 1%). No difference in the pattern of the age of first stone manifestation (21 - 30 years) between male and female

Urolithiasis 2, Edited by R. Ryall *et al.*,
Plenum Press, New York, 1994

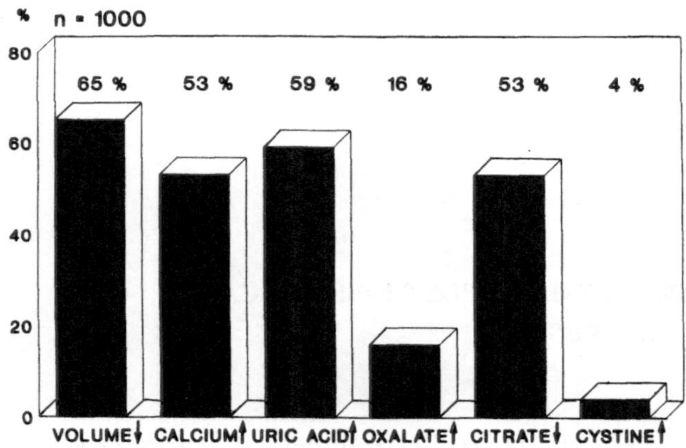

Figure 1. Frequency of pathological findings (initial finding)

patients could be found. The average stone passage rate previous to participation in the program amounted to 9.3 stones in men and 7.6 stones in women. 20.3% of all patients had more than 10 stone passages before they participated in the study.

Specific measurements of 24 h urines (standard program) indicated multiple physicochemical disturbances. In the beginning of our investigations more than 80% of all patients showed three or more abnormal urine findings such as low urine volume (65%), hypercalciuria (53%), hyperuricosuria (59%), hyperoxaluria (16%), hypocitraturia (53%) and cystinuria (4%), (see figure 1 above).

Patients suffering from hypercalciuria, hyperuricosuria, or hyperoxaluria showed a significant decrease in the excretion of calcium, uric acid, and oxalic acid after at least four check-ups had been made. The excretion of citric acid and the volume of urine were significantly increased in those patients who had hypocitraturia and low volumes when they joined the program. In addition there was a significant decrease in urine saturation index of calcium oxalate and uric acid, calculated by the EQUIL program according to Finlayson[3]. The average stone loss rate 12 months previous to the start of the program amounted to 1.69 stones in women and 1.36 stones in men. Twenty four months later this rate decreased to 0.99 stones in women and 0.69 stones in men, i.e., the rate of stone passage per patient per year was reduced by 46%.

DISCUSSION

Introduction and revolutionary development of endoscopic methods (percutaneous nephrolithotomy, ureteroscopy) and extracorporal shock wave lithotripsy (ESWL) have pushed stone metaphylaxis into the background in recent times. But all these methods only treat symptoms and not the cause of stone diseases. Eisenberger et al[2] have shown that only 75% of all patients become stone-free after ESWL treatment. Furthermore there is no exact knowledge of late complications after repeated ESWL treatment. Without specific metaphylaxis the recurrence rate was 50 to 100% as published by Brandl[1] and Resnick[8].

In addition, cost-benefit evaluations emphasize the importance of metaphylaxis. Periods of indisposition, medical treatment, hospitalization etc. burden our public health system. Preminger et al[15] have shown that less than $1,000 per year are calculated to treat and follow patients medically, and calculations in USA have shown that the cost of a metabolic study for a patient with urolithiasis varies between $400 and $600 but treatment with ESWL averages $6,000 to $8,000 as reported by Resnick[16].

Important to successful metaphylaxis are stone analysis by infrared spectroscopy or X-ray diffraction, analysis of lithogenic and inhibitory parameters in 24 h urine (standard program) and serum parameters, nutrition records and anamnesis. Dietary intake and habits of patients are important to know because many risk factors are diet-dependent[12].

The results of our program have shown that effective metaphylaxis on the basis of the above-mentioned parameters could not only be realized in clinic but also in outpatients. The risk of stone formation was reduced significantly after four months as a result of decreased concentration of lithogenic substances like calcium, uric acid and oxalic acid and an increase of citric acid concentration and urine volume. Stone passage rate was reduced by 46%.

CONCLUSIONS

All stone patients, especially those with recurrent stone formation, should participate in a specific diagnostic program. Individual metaphylaxis, including dietary management and medical treatment, can prevent recurrent stone formation. To be effective, the program requires constant surveillance by the physician, good compliance by the patient and close co-operation between physician and laboratory.

REFERENCES

1. H Brandl, B Liedl, and FJ Marx, Nachsorge and Metaphylaxe bei Harnsteinerkrankungen, *Munch med Wschr* 128:422-423 (1986).
2. F Eisenberger, K Miller, and J Rassweiler, "Stone Therapy in Urology", Thieme, Stuttgart, New York (1991).
3. B Finlayson, Calcium stones: some physical and clinical aspects, in: "Calcium metabolism in renal failure and nephrolithiasis", DS David, ed, John Wiley & Sons, New York, London, Sydney, Toronto (1977).
4. A Hesse, Zur Epidemiologie des Harnsteinleidens, *J Clin Chem Clin Biochem* 26: 841-842 (1988).
5. A Hesse, K Klocke, RM Schafer, W Schneeberger and W Vahlensieck, Ergebnisse von Stoffwechseluntersuchungen bei 354 Kalziumoxalat-Rezidiv-Harnsteinpatienten, *Fortschr Urol Nephrol* 25: 15-19 (1987).
6. K Klocke, A Hesse, W Schneeberger, A Nolde and W Vahlensieck, Rezidivierende Kalziumoxalat-Urolithiasis. Ergebnisse von Stoffwechseluntersuchungen, TW *Urol Nephrol* 2:401-408 (1990).
7. GM Preminger, R Peterson, PC Peters and CYC Pak, The current role of medical treatment of nephrolithiasis: the impact of improved techniques of stone removal, *J Urol* 134:6-10 (1985).
8. MI Resnick, Are metabolic studies of urolithiasis necessary? *J Urol* 137:960-961 (1987).
9. EW Vahlensieck, A Hesse and RM Schaefer, Epidemiologische Studien zur Inzidenz, Pravalenz und Mortalitat des Harnsteinleidens in der Bundesrepublik Deutschland 1979 und 1984, *Fortsch Urol Nephrol* 25:1-4 (1987).

THE OXALATE CONTENT OF SESAME SEEDS AND RELATED FOODS

J.F. Costello[1], M. Smith[2], C. Stolarski[2] and N. Fituri[3]

[1]Allegheny-General Hospital/Medical College of Pennsylvania and
[2]Allegheny-Singer Research Institute
Pittsburgh, PA
[3]Department of Nephrology
Cairo University
Cairo, Egypt

INTRODUCTION

The incidence of calcium oxalate stone disease is reported to be higher in the Middle East than in the developed countries. The oxalate content of Middle Eastern diets has not been thoroughly investigated. Foods prepared with sesame seeds are quite common in the Middle East. This study investigated the oxalate and calcium content of sesame seeds, sesame oil, and foods prepared with sesame seeds, namely Tahina and Halava.

METHODS

Ground sesame seeds, as well as Tahina, and Halava (19 of each in duplicate) were defatted four times with 1:1, ether:petroleum:ether mix, lyophilized and extracted three times with 10 mL of 0.5N HCl. Aliquots of each extraction were assayed by atomic absorption for calcium. Extracts of each sample were combined, tracer ^{14}C-oxalate added, the oxalate precipitated with ethanol and calcium sulfate, and the precipitate lyophilized and assayed for oxalate content as previously described[1]. Tracer ^{14}C-oxalate was added to 500 μL of sesame oil (in duplicate), which was then extracted with 10 mL of 0.5N HCl. The extract was assayed for calcium and oxalate as described above.

RESULTS AND DISCUSSION

This study establishes that the oxalate content of sesame seeds 2.8g/100g of seeds, Table 1, is probably the highest of any known food. Foods prepared with sesame seeds namely Tahina and Halava, also contain significant quantities of oxalate (Table 1). These foods are regularly consumed in Middle Eastern countries and may contribute to increased urinary oxalate excretion and a greater incidence of stone disease.

Urolithiasis 2, Edited by R. Ryall *et al.*,
Plenum Press, New York, 1994

Table 1. Calcium and Oxalate Content of Foods

Foods Analyzed	Calcium (mg/100g)	Anhydrous Oxalic Acid (molar ratio)	Calcium/Oxalate
Sesame Seeds	1143.0	2799.0	0.92
Tahina	77.0	117.0	1.46
Halava	43.0	59.0	1.53
Sesame Oil	3.6	0.8	10.00

REFERENCE

1. J Costello and DM Landwehr, Determination of oxalate concentration in blood, *Clin Chem* 34:1540 (1988).

CHANGES OF SERUM AND URINE PARAMETERS IN IMMOBILIZED CHILDREN AND ADOLESCENTS

H. van Ahlen[1], K. Klocke[1], A. Pfeiffer[1], H. Meßler[2] and A. Hesse[1]

[1]Dept of Urology and
[2]Orthopedics
University of Bonn, FRG

INTRODUCTION AND METHODS

Long term immobilization increases the risk of urolithiasis, especially in children. Several factors may be important in the pathogenesis of stone formation. These are changes in urinary composition due to increased metabolism in the growing bone, urinary infections and possible urodynamic disturbances.

To assess the changes in serum and urine composition during complete, long term immobilization we followed 10 children immobilized for various orthopedic diseases over a period of at least five weeks.

Blood and 24 h urine samples were drawn twice a week. The following parameters were measured in serum: calcium, phosphate, uric acid, alkaline phosphatase (AP), parathyroid hormone (PTH) and calcitonine (CT); in urine: pH, calcium, phosphate, oxalic acid, uric acid, citric acid, magnesium, hydroxyproline and GAGs.

In serum during immobilization there was an early increase in PTH activity, CT and AP activities increased in the second half of the observation period. There was a continuous significant increase in phosphate concentration whereas the calcium level remained unchanged.

In 24 h urines no infections were observed. The pH, the excretion of oxalic acid and hydroxyproline as well as GAG and magnesium levels were not altered. A marked increase in calcium excretion was accompanied by an increase in citric acid excretion.

DISCUSSION

In immobilized children without urinary tract infection the increased risk of forming calcium stones is less than in adults because the raised calcium excretion is accompanied by an increase in the excretion of the most potent inhibitor of calcium stone formation - citric acid.

STONE FORMATION RISK INDEX(SFRI): A POSSIBLE PROGNOSTIC FACTOR GOVERNING THE NEED FOR METAPHYLAXY

T. Esen, M. Akinci and S. Tellaloglu

Department of Urology
Medical Faculty of Istanbul
Istanbul, Turkey

INTRODUCTION AND RESULTS

The need for specific metaphylaxis in urolithiasis is controversial. In this study we tried to define a biochemical parameter to differentiate between "severe"and "harmless" disease.

In an effort to find a single reliable parameter differentiating patients with a low and high risk of stone recurrence, we analysed 175 adult patients in a multivariate setting. Actual age, age at the onset of the disease, recurrence rate as well as quotients and equations such as Ca/citrate, Ca/magnesium, Ca x oxalate/Mg x creat $Ca^{4.7}$ x oxalate x $3.8/Mg^{0.8}$ x $citrate^{0.4}$ x $volume^{4.2}$ were calculated and a new equation; Ca x oxalate x uric acid/Mg x citrate x volume, were designed and calculated using average values obtained from two 24 h urinalyses.

The average Stone Formation Risk Index (SFRI)[1] for the whole group was 41.7. No parameter showed a significant difference between the groups of primary and recurrent disease. Yet, defining 50 as a critical value for SFRI we found that 70% of first-time stone formers had an SFRI <50 while 67.7% of the recurrent stone formers had a value >50.

CONCLUSION

We concude that SFRI could be a starting point in deciding the need for metaphylaxy for the individual patient, together with age and nature and severity of the present metabolic disorder.

REFERENCE

1. HG Tiselius and L Larsson, Biochemical evaluation of patients with urolithiasis, *Eur Urol* 7:31 (1981).

UROLITHIASIS IN DIABETES MELLITUS

R.K. Vathsala, H.K. Moorthy, T.G. Dhanalekshmy,
S.V. Roshni, C. Aravindakshan and Y.M. Fazil Marickar

Department of Surgery
Medical College Hospital
Trivandrum, India

INTRODUCTION AND METHODS

Detailed nutritional analysis in stone patients has identified various interesting correlations. This paper attempts to study the relationship between diabetes and urinary stone disease. Fifty patients having diabetes for more than 3 years and urinary stone disease were included in the study. Fifty non-diabetic stone patients were taken as controls. The 24 h urine biochemistry was studied at first visit. Urinary Ca, oxalate, uric acid and citrate levels in the two groups were compared using the unpaired Student's t test.

RESULTS AND CONCLUSION

The mean levels of all risk factors studied namely urinary oxalate, uric acid and calcium levels were significantly higher in the diabetic stone formers as detailed in the table. The mean urinary citrate level was not significantly different (NS).

Table. Biochemical values of diabetic and non diabetic stone patients (SP).

Biochemical parameters mg/day	Diabetic SP mean ± SE	Non diabetic SP mean ±SE	p Value
U.calcium	228.2±17.7	175.7±14.5	<0.03
U.oxalate	114.9±11.5	41.0± 3.7	<0.001
U.uric acid	718.2±39.4	544.8±33.0	<0.001
U.citrate	163.6±42.4	122.8±13.1	NS

Diabetes affects renal tubular function and thereby causes increased excretion of these ions in urine. The unutilised carbohydrates in the diabetic may also be responsible for elevation of urinary calcium and other lithogenic factors. It has been shown that in the diabetic, the presence of a constant raised blood glucose level, and the administration of insulin, result in an exaggerated urinary excretion of calcium compared to the non-diabetic. The finding of the higher uric acid in urine of diabetic stone formers may be due to the fact that the normal diabetic patients consume more wheat than the non-diabetic, this in turn increasing the urine uric acid levels.

RENAL TOXICITY INDUCED IN RATS BY ORAL SODIUM OXALATE

N. Sylaja, T.G. Dhanalekshmy, H.K. Moorthy,
B. Seemanthini Bai[1], M.M. Oommen[2], C. Aravindakshan
and Y.M. Fazil Marickar

[1]Priyadarsini Institute of Paramedical Sciences and
[2]University of Kerala,
Department of Surgery
Medical College Hospital
Trivandrum, India

INTRODUCTION AND RESULTS

Histopathological changes occurring in rat kidneys on administration of oral oxalate were studied. Two groups - one control and one experimental - each consisting of six male albino Wistar rats were studied. The controls were fed on a diet of known composition and deionised water *ad libitum*, and the experimental rats, on a diet containing 3 mg% sodium oxalate and deionised water *ad libitum*. After one month, the kidneys were examined histopathology .

Kidneys of experimental rats showed irregular urothelial thickening. In certain cases the urothelium was torn and crystals were exposed. Cellular erosion was seen in most parts of the renal tubules, which were dilated with prominent basement membrane (fig. 1). In most cases the epithelial cells appeared to be sheared from the tubular basement membrane. Cellular debris and colloidal materials were seen in most of the dilated tubules. Crystal aggregates were also seen in many parts of the cortico-medullary region (fig.2).

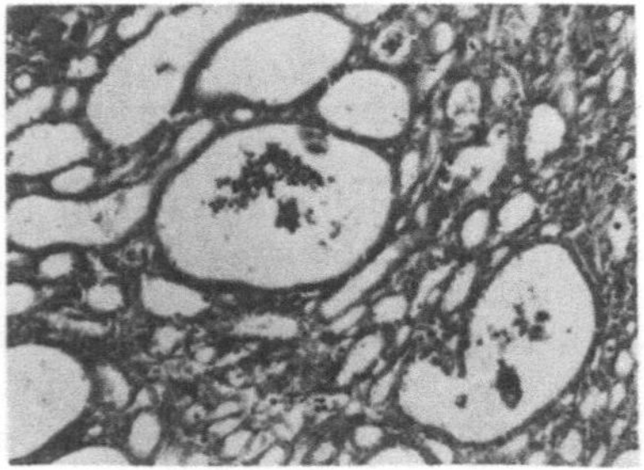

Fig. 1 - Cellular aggregates in tubules X 450.

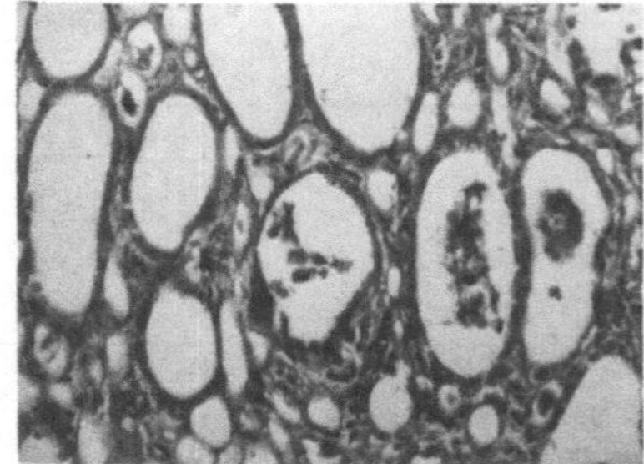

Fig.2 - Crystalline aggregates in tubules X 250.

CONCLUSION

It is concluded that high oxalate content in the diet produces urothelial erosions and deposition of colloid or crystalline materials in the renal cells and the tubules.

NEW INSIGHTS ABOUT DIET AND NEPHROLITHIASIS IN AN OBESE POPULATION

G. Gambaro, G. Bertaglia, E.M. Inelmen, F. Marchini, M.A. Nassuato, G. Enzi and B. Baggio

Institute of Internal Medicine
University of Padova, Italy

INTRODUCTION

Many clinical-epidemiological and experimental observations show that a number of nutrients have an adverse effect on the physical-chemical risk profile in urine, thus suggesting that idiopathic calcium nephrolithiasis (ICN) might be related to diet[1]. However, a number of reports failed to show evidence for different eating habits between ICN subjects and controls[2]. We reasoned that if dietary habits were crucial in determining ICN, then obese patients, in whom western dietary patterns are or were maximized, might be at risk for ICN.

METHODS AND RESULTS

In a series of 920 (875 females) obese subjects and 114 non-obese controls, we evaluated the frequency of nephrolithiasis, and the following parameters: physical and working activity; personal and family history for nephrolithiasis; presence of hyperuricemia; age at onset of obesity; dietary habits concerning meat, cereals, macaroni, rice, refined sugar, pastry, fruits, and alcohol. A metabolic study (24 h oxaluria, calciuria and uricuria; serum PTH, calcium, phosphate, uric acid and oxalate RBC self-exchange) was performed in the 23 obese patients with a personal history of nephrolithiasis, in 20 randomly selected obese subjects, and in 20 non-obese controls. The prevalence of stone disease in controls (1.8%) and in obese subjects (2.5%) was not statistically different, and was similar to that observed in the general population in Italy. In comparing obese patients with and without nephrolithiasis, no difference emerged regarding the various parameters examined. Compared to non-obese stone formers, obese stone formers had a distinct risk profile for calcium stone formation, as genetic, clinical and urinary risk factors were rare.

Our study suggests that diet alone is not sufficient to promote renal stone formation, but we cannot exclude that it might be crucial in the presence of predisposing genetic and/or metabolic factors.

REFERENCES

1. S Goldfarb, Dietary factors in the pathogenesis and prophylaxis of calcium nephrolithiasis, *Kidney Int* 34: 544 (1988).
2. DJP Barker, JA Morris and BM Margetts, Diet and renal stone in 72 areas in England and Wales, *Br J Urol* 62: 315 (1988).

ATP DEPLETION: A POSSIBLE ROLE IN THE PATHOGENESIS OF URIC ACID NEPHROLITHIASIS IN THE ELDERLY

A. Trinchieri, F. Rovera, R. Nespoli, F. Colombo,
A. Guarneri, G. Zanetti and E. Austoni

Istituto di Urologia
Università degli Studi di Milano
Milan, Italy

INTRODUCTION

Undue acidity of the urine attributed to a deficiency in renal ammoniogenesis has been claimed to be prevalent in certain groups of patients who may form uric acid stones even in the absence of hyperuricosuria[1].

METHODS AND RESULTS

We studied 56 patients older than 50 years with idiopathic stone disease. The patients were divided in 2 groups: uric acid stone formers (19 M, 5 F) and calcium stone formers (26 M, 6 F). Serum was analyzed for urate, pyruvate, glucose and creatinine; 24 h urine samples were analyzed for their urate and creatinine contents and fasting 2 h samples were collected for determination of pH, titrable acid and ammonium.

The uric acid stone formers had significantly higher serum urate (5.3±1.5 vs 4.5±1.5 mg/dL) and pyruvate (0.90±0.63 vs 0.55±0.21 mg/dL) levels than the calcium stone formers ($p < 0.05$).The mean urinary pH (5.1±2.8 vs 5.5±0.7) and the mean urate clearance (9.2±5 vs 5.5±2.8 mL/min) were significantly lower ($p < 0.01$) and the mean titrable acid/ammonium ratio (0.74±0.59 vs 0.39±0.23) was significantly higher ($p < 0.053$) in the uric acid stone group than in the calcium stone group.

DISCUSSION

High serum pyruvate levels suggest a disorder of adenosine triphosphate (ATP) synthesis in relation with ageing that may induce hyperuricaemia and a deficiency in renal ammoniogenesis in uric acid stone formers. On the other hand renal ammoniogenesis is tied to the activity of the Krebs cycle which is closely coupled to the generation of ATP in oxidative phosphorylation.

REFERENCE

1. AB Gutman and TF Yu, Uric acid nephrolithiasis, *Am J Med* 45:756 (1968).

RELATIVE HYPOPARATHYROIDISM AND CALCITRIOL UP-REGULATION IN IDIOPATHIC HYPERCALCIURIA

B. Hess[1], J.P. Casez[1], D. Ackermann[2], R. Takkinen and Ph. Jaeger[1]

[1]Policlinic of Medicine and [2]Department of Urology
University Hospital
CH-3010 Berne, Switzerland

INTRODUCTION AND RESULTS

The association of idiopathic hypercalciuria (IH) with elevated intakes of animal protein and of NaCl is well recognized. In order to study the actual conditions of renal stone formation, metabolic evaluation and bone mineral density (BMD) measurements were performed in 61 recurrent male idiopathic calcium renal stone formers on free-choice diets, i.e. while maintaining their individual steady state.

Calcium intake from dairy products (Dairy Ca) was estimated using a questionnaire based on a table of Ca content of Swiss dairy products. Protein intake was quantified from 24 h urine excretion of urea, sulfate, phosphate, and sodium intake derived from $U_{Na} \times V$. Stone formers were divided into hypercalciurics (HCSF) and normocalc-iurics (NCSF); hypercalciuria was defined as $U_{Ca} \times V > 7.5$ mmol/d.

Values are means±SEM. Dairy Ca was identical in HCSF (623±99 mg/d) and NCSF (634±97 mg/d). In HCSF, urinary excretion rates of urea ($p = 0.005$ vs. NCSF), sulfate ($p = 0.004$) and phosphate ($p = 0.0001$), all markers of protein consumption, were higher than in NCSF, as was $U_{Na} \times V$ (223±9 vs 185±12 mmol/d, $p = 0.016$). At identical blood Ca^{2+} levels, intact PTH was lower in HCSF (25.3±1.8 pg/mL) than in NCSF (31.4±1.8 pg/mL, $p = 0.017$). Since neither fasting urinary hydroxyproline nor pyridinoline/deoxypyridinoline excretions nor values of BMD (Hologic QDR 1000) were different between HCSF and NCSF, chronic bone dissolution could be excluded as the cause of relative hypoparathyroidism in HCSF. Despite lower PTH levels and similar blood concentrations of phosphate, Ca^{2+} and IGF-1 (other main regulators of calcitriol production), serum calcitriol levels were slightly higher in HCSF (52.8±3.2 pg/mL) than in NCSF (47.3±2.9 pg/mL, NS), and calcitriol/PTH ratio was elevated in HCSF (2.52±0.29) vs NCSF (1.66±0.15, $p = 0.001$). Although urinary creatinine excretions per kg BW were equal in both groups, creatinine clearance was higher in HCSF than in NCSF (113±4 vs 92±3 mL/min/1.73 m^2, $p = 0.0001$), and was positively correlated with $U_{urea} \times V$ ($r = 0.474$, $p=0.0001$) and $U_{Na} \times V$ ($r = 0.342$, $p=0.007$). Serum calcitriol was positively correlated with creatinine clearance in all stone formers ($r = 0.350$, $p= 0.006$).

In conclusion, IH under free-choice diet is, at least in part, the consequence of protein and NaCI overconsumption, both apparently raising GFR (functional renal mass) with subsequent up-regulation of calcitriol synthesis and relative hypoparathyroidism.

ACKNOWLEDGEMENTS

This study was supported in part by the Swiss National Science Foundation (Grant No. 3226428.89).

CHANGES OF SERUM AND URINE PARAMETERS IN IMMOBILIZED ADULTS

K. Klocke[1], H. van Ahlen[1] C. Pütz[1], H. Meßler[2] and A. Hesse[1]

[1]Dept. of Urology and
[2]Orthopedics
University of Bonn, FRG

INTRODUCTION

Long term immobilization increases the risk of urolithiasis. Several factors may be important in the pathogenesis of stone formation in immobilized patients. These are changes in urinary composition due to increased bone metabolism, urinary infections and possible urodynamic disturbances.

METHODS AND RESULTS

To assess the changes in serum and urine composition during complete, long term immobilization we followed 10 patients (five females, five males, average age 41 years), immobilized for various orthopedic diseases over a period of at least five weeks. Blood and 24 h urine samples were drawn twice/week. The following parameters were measured in serum: calcium, phosphate, uric acid, alkaline phosphatase (AP), parathyroid hormone (PTH) and calcitonine (CT); in urine: pH, calcium, phosphate, oxalic acid, uric acid, citric acid, magnesium, hydroxyproline and GAGs.

In serum during immobilization there were no changes of CT and PTH activity. AP and phosphate showed a temporary increase with partial normalization over the time of evaluation. There was a constant, significant increase in calcium and uric acid concentration. In 24 h urines no infections were observed. The pH, the excretion of oxalic acid and hydroxyproline and GAG levels were not altered. A marked, but only temporary increase in calcium excretion appeared whereas the inhibitors citric acid and magnesium decreased significantly.

DISCUSSION

In immobilized patients that do not suffer from urinary tract infection we were able to show that the risk of forming calcium stones is not only increased by evelated serum and urine calcium levels but also by a significant reduction of inhibitory activity due to a marked decrease of citric acid and magnesium excretion.

EFFECT OF A VEGETARIAN DIET ON URINARY OXALATE EXCRETION IN HEALTHY SUBJECTS

R. Siener and A. Hesse

Division of Experimental Urology
Department of Urology
University of Bonn, Germany

INTRODUCTION AND METHODS

In the past, calcium oxalate stone formers have been repeatedly advised to become vegetarians, since a diet low in animal protein has been suggested to decrease the urinary excretion of oxalate. The reason for increased oxalate excretion in high animal protein intake was seen as the result of the increased metabolism of phenylalanine, tyrosine, tryptophan and hydroxyproline. All are present in higher proportions in animal than in vegetable protein. The aim of the present study was to investigate the effect of a vegetarian diet on oxalate excretion in the urine of healthy subjects.

Ten healthy male individuals were kept on two different standard diets for five days each. Diet 1, the control diet, was established as a normal mixed diet, whereas Diet 2 represented an ovo-lacto-vegetarian meal plan. Diet 1 and 2 were calculated according to the dietary recommendations of the German Society of Nutrition (DGE). Both diets were isoenergetic with equal amounts of the main nutrients and a constant fluid intake. The total protein content of each diet amounted to 65 g/d with an animal protein intake of 37 g/d by Diet 1 and 28 g/d by Diet 2.

Although ingestion of oxalate-rich foods such as spinach, rhubarb, cocoa, beetroot and others was excluded, 129 mg/d oxalate were supplied with Diet 2 vs 74 mg/d with Diet 1. Samples from 24 h urine collections were analysed for oxalate by ion chromatography.

RESULTS

On the normal mixed Diet 1 urinary oxalate excretion was 0.287 mmol/d. Reducing animal protein by 24%, but raising oxalate intake by 74% in the ovo-lacto-vegetarian Diet 2 resulted in a statistically significant increase in oxalate excretion by 31% to a mean value of 0.376 mmol/d.

DISCUSSION

These data indicate that a vegetarian diet with a low proportion of animal protein may increase urinary excretion of oxalate by its high oxalate content. According to these findings stone formers with a mild hyperoxaluria, due to intestinal hyperabsorption of oxalate, should be advised to avoid an excess of fruits, vegetables and cereals.

CALCIUM OXALATE SUPERSATURATION OF URINE: WHICH FACTORS SHOULD BE MEASURED?

J.P. Kavanagh and P.N. Rao

Department of Urology
University Hospital of South Manchester
Manchester, UK

INTRODUCTION AND RESULTS

Estimation of urinary supersaturation with calcium oxalate (CaOx) is a valuable tool in urolithiasis investigations. Methods used to assess supersaturation are usually empirical or dependent on a complete ionic analysis of the urine for computation of the supersaturation ratio. EQUIL2[1] requires analysis of 14 urinary components. In this study CaOx supersaturation ratios (SS-CaOx) were estimated for 97 24 h urines from 67 patients using EQUIL2 and compared with results from three simplified methods: (1) Substitution of 25 μM for the observed pyrophosphate concentration in the EQUIL2 calculation; (2) a regression equation based on calcium, oxalate, citrate and magnesium (equation i); (3) the AP(CaOx)-index of Tiselius[2] modified to give the best numerical agreement with the EQUIL2 results (equation ii).

$$\text{SS-CaOx} = 0.712 + 1.071 \times [\text{Ca}] + 10.89 \times [\text{Ox}] - 0.0022 \times [\text{Mg}] - 0.567 \times [\text{Cit}] \quad \text{(i)}$$

$$\text{AP(CaOx)-index} = (3.8 \times [\text{Ca}]^{0.71} \times [\text{Ox}]/[\text{Mg}]^{0.10}) \times 2.57 + 1.645 \quad \text{(ii)}$$

	correlation mean	% error	% error range	error > 10%
1. PP_i set to 25 μM	1.00	0.01	-0.2 to 0.9	0 out of 97
2. By regression equation	0.93	6.6	-43 to 108	51 out of 97
3. By AP(CaOx)-index	0.94	8.1	-40 to 136	52 out of 97

SS-CaOx estimated by methods 1-3, compared with EQUIL2 values are shown in the table. Excluding pyrophosphate from the EQUIL2 computation made little difference. While both equations (i) and (ii) gave values which correlated well with EQUIL2 and with low average errors, the range of errors with both of these methods was very large.

Pyrophosphate estimations are not essential for calculation of SS-CaOx, but simple formulae based only on calcium, oxalate citrate and magnesium are not likely to be adequate for determining this important risk factor.

REFERENCES

1. PG Werness, CM Brown, LH Smith & B Finlayson, EQUIL2: a basic computer program for the calculation of urinary supersaturation, *J Urol* 134: 1242 (1985).
2. H-G Tiselius, An improved method for the routine biochemical evaluation of patients with recurrent calcium oxalate stone disease, *Clin Chim Acta* 122: 409 (1982).

URINARY FIBRINOLYTIC ACTIVITY OF STONE FORMERS - DOES DIET MATTER?

D.P. Sandhu[1], B.M. Margetts[2], J.P. Kavanagh[3] and N.J. Blacklock[3]

[1]Department of Urology
Leicester General Hospital, UK
[2]Department of Human Nutrition
University of Southampton, UK
[3]Department of Urology
University Hospital of South Manchester, UK

INTRODUCTION AND RESULTS

Evidence suggests that diet may be important in the development and recurrence of renal stones[1]. Urinary fibrinolytic activity (UFA) has been found to be lower in the urine of stone formers (SF) compared to matched controls (N)[2], and deficient UFA may be an important (but correctable) contributor to urinary stone disease[3]. In a small series of subjects we were unable to confirm this[4], although we noted that SF who had reduced their meat intake tended to have higher UFA[4]. We have now studied a larger population, paying particular attention to diet.

The UFA of spot urine samples is representative of 24 h samples and was measured by a fluorometric assay with a specific synthetic substrate[4]. Urine from 105 SF (64 M, 41 F) and 107 N (67 M, 40 F) was analysed for UFA, Ca, oxalate, other electrolytes and creatinine. Each participant completed a dietary questionnaire[5].

There were no significant differences in food and nutrient consumption between SF and N, and UFA was not correlated to any dietary factor, Female SF had a lower UFA (10u/mL) than their corresponding N (19.9u/mL) but this was due to an age effect (age adjusted results, female SF 13.5u/L, female N 16.3u/ml, p=0.37). Comparing SF and N adjusted for age and sex, gave means of 10.2u/mL (SF) and 12.3u/mL (N), p=0.16. Expressing UFA in relation to creatinine did not reveal any significant differences. Age was the only factor found to be associated with UFA. There was a significant decrease in UFA with age (p=0.001), declining from a mean of 21.0u/mL in those less than 30 to 7.3u/mL above 70 years old. Age was a dominant factor affecting UFA. Diet and stone disease were not significantly related to UFA.

REFERENCES

1. NJ Blacklock. Epidemiology, *in:* "Scientific Foundations of Urology", 2nd edition, GD Chisholm & DI Williams, eds, Heinemann, London, (1982).
2. CAC Charlton and C Osmond, Deficient urinary fibrinolysis in renal stone disease, *Br Med J* 292:1239 (1986).
3. CAC Charlton, A urinary detergent and urolithiasis, *Br J Urol* 63:561 (1989).
4. BM Margetts, DJP Barker, JP Kavanagh and NJ Blacklock, Do stone formers have lower urinary fibrinolytic activity than controls? *Br J Urol* 66:581 (1990).
5. JE Cade and BM Margetts, Nutrient sources in the English diet: quantitative data from three English towns, *J Epidemiol* 17:844 (1988).

CHRONOBIOLOGY OF URINARY VOLUME, CALCIUM, OXALATE AND URIC ACID IN STONE FORMERS AND HEALTHY SUBJECTS

D. Wangoo[1], V. Rattan[1], S.K. Thind[1], G.S. Gupta[2] and R. Nath[1]

[1]Department of Biochemistry, Postgraduate Institute of Medical Education and Research and
[2]Department of Biophysics, Panjab University
Chandigarh, India

INTRODUCTION AND METHODS

The increased excretions of lithogenic substances (calcium, oxalate and uric acid) are known to be significant risk factors in stone formation[1]. Twenty five stone formers (SF) and age-sex-matched healthy subjects were asked to collect 3 h urine samples over a 24 h period. These were analyzed for calcium, oxalate, and uric acid by routine procedures. Computer programs using the method of least squares were used to detect a circadian rhythm[2].

RESULTS

A significant rhythm for urine volume, calcium, oxalate and uric acid with their acrophase at 263° (17:31 h), 232° (15:27 h), 242° (16:31 h) and 180° (12:00 h) respectively was detected in healthy subjects, whereas, SF showed a lack of circadian rhythmicity for all of these parameters. In healthy subjects, the pattern for oxalate excretion was found to be similar to that for calcium, suggesting a high risk period for calcium and oxalate precipitation in urine. However, this period was also accompanied by peak excretion of citrate and inhibitory activity towards calcium oxalate crystallization.

DISCUSSION

The present study suggests that while healthy subjects are protected during the high risk period in the day, SF are exposed to high concentrations of lithogenic substances during the same period, thereby increasing their risk of stone formation.

REFERENCES

1. R Nath, SK Thind, MSR Murthy, HS Talwar, and S Farooqui, Molecular aspects of idiopathic urolithiasis, *Molec Aspects Med* 7:1 (1984).
2. W Nelson, YL Tong, JK Lee, and F Halberg, Methods for cosinor-rhythmometry, *Chronobiologia* 6:305 (1979).

INVOLVEMENT OF TUBULAR INJURY IN STONE FORMATION: EXPERIMENTAL INVESTIGATIONS

K-H. Bichler, W.L. Strohmaier and H.J. Nelde

Dept of Urology
Eberhard-Karls-University
Tübingen, Germany

INTRODUCTION

Two theories of stone formation have been presented: the free particle, and the fixed-particle theory. In regard to the incongruity between slow crystal growth and rapid urine flow through the nephron, the free-particle theory does not yield satisfactory explanations for stone formation. Crystal growth presupposes the crystal's fixation to the tubular cell. Such fixations are rare or only transient with uninjured tubular cells. However, lesions of the epithelium may promote them (dystrophic calcification). The present study in two animal models shows the role of tubular stone formation.

METHODS AND RESULTS

A cholesterol diet was fed to 60 male Wistar rats for four weeks. Twenty rats received a control diet. All animals were sacrificed after four weeks and kidneys were taken for histology. Calcifications were found in the cortico-medullary junction. Predominantly, these calcifications were intracellular and interstitial and they showed concentric growth.

1,25-dihydroxycholecalciferol was injected subcutaneously in 20 male rats (125 pmol/day). After six days, kidneys were taken for histology. Calcifications were found in the cortico-medullary junction, mainly intracellularly. Intraluminal concrements were rare.

CONCLUSION

Both supersaturation of the urine and the tubular epithelium of the kidney play major roles in stone formation, either by the presence of pre-existing lesions or as a starting point for crystallization.

CHEMICAL COMPOSITION OF THE URINE IN THE NORMAL BLACK AND WHITE POPULATION

A.M. Meyers, N. Whalley, W.J. Zakolski and T. Shar

Metabolic Stone Clinic
Johannesburg Hospital
University of Witwatersrand and
The South African Institute for Medical Research

INTRODUCTION AND RESULTS

The prevalence of nephrolithiasis in white males and females is 12 and 5% respectively, and in blacks is <1%. In order to further investigate these striking discrepancies we compared dietary intake, serum values, urine solute, inhibitor content and Ca:Creat ratios before and after ingestion of a diet consisting of a standard intake of Na (100 mmol/24 hr) and Ca (400 mg/24 hr), in 31 (19F and 12M) recently urbanized blacks and 29 (17F and 12M) white age-matched volunteers. Glycosaminoglycans were measured on early morning urines by glucuronic acid determination.

Blacks had a higher dietary intake of Na than whites (3840±3387 vs 2482±973 mg/day, $p< 0.08$). There were no apparent differences in Ca or oxalate intakes, or in serum Ca, ionised Ca, PO_4 and 1,25-$(OH)_2D_3$. The table shows the urine results. No differences in 24hr values and creatinine excretion were seen. Mean Ca:Cr pre and post Ca load in blacks vs whites were 0.05±0.03 and 13±0.08 vs 0.05±0.02 and 0.14±0.07 respectively (NS). Mean pyrophosphate: creatinine ratios in blacks vs whites were 2.86±2.36 vs 3.92±3.83 respectively (NS). There was a significant correlation between Na and Ca excretion ($r=0.72$ $p < .002$) in whites. No correlation was found in blacks.

	Na	K	Ca	Pyro[1]	Urate	Ox[1]	Citrate[1]	GAGS[2]
Black mean	172	43	1.5	18.0	1.5	272	1.4	4.28
SD	63	28	1.1	5.7	0.8	127	0.9	1.02
White mean	112	36	3.5	27.0	1.6	236	2.8	2.95
SD	40	32	1.6	7.0	0.9	142	1.2	0.93
p<	.005	NS	.001	.001	NS	NS	.001	.0004

Results in mmol/24 hr; [1]mmol/24 h; [2]μmol/g creatinine

In conclusion, a lower Ca excretion combined with the absence of the usual Na:Ca relationship may be important in preventing calculi in blacks. Low urinary citrate in blacks remains unexplained and should favour calculus formation. However, urinary GAGS levels were significantly higher in blacks. Preliminary studies of inhibition of calcium oxalate crystallization have shown increased inhibition in black F and M compared to white F and M (.3 and .15 vs .24 and .08). Further studies are required to elucidate the mechanisms of low urinary citrate and to identify other possible macromolecular inhibitors in blacks.

EFFECT OF DRASTIC DEHYDRATION ON URINE LITHOGENIC RISK FACTORS IN NORMAL HUMANS

B. Baggio, G. Gambaro, A. Burighel, G. D´Amelio and A. Boninsegna

Institutes of Internal Medicine
Division of Nephrology, and
Clinical Medicine
University of Padova, Italy

INTRODUCTION AND RESULTS

The pathogenesis of idiopathic calcium nephrolithiasis most likely involves an imbalance between factors promoting and inhibiting calcium crystal growth and aggregation in the urine. Among many factors able to cause this disorder, urine concentration induced by chronic dehydration seems to be an important condition that increases the risk of stone formation[1]. To verify the effect of drastic dehydration on lithogenic risk, we carried out this study.

Ten normal subjects who were undergoing mud therapy for osteoarthropathy were investigated. Urines were collected before, during, and after therapy, and the following were evaluated: pH, osmolarity, oxalate (Ox), uric acid, calcium (Ca), citrate (Cit), and glycosaminoglycans (GAGs). The Ox/Cit x GAGs[2] and Ca/Cit[3] ratios were also calculated to estimate the imbalance between some of the promoting and inhibiting factors in urine.

Mud therapy induced a significant increase in urine osmolarity (before: 712±105 mOsm/L; during: 762±93; after: 811±89; F= 54.23; p<0.001}, but no significant variation in urinary uric acid, calcium, oxalate, and GAGs. The lithogenic risk ratios were both unchanged (Ox/Cit x GAGs: 5.32±1.15, 5.49±0.76, 5.57±1.01; F= 0.40, n.s.; Ca/Cit: 0.76±0.11, 0.74±0.13, 0.77±0.11; F= 2.01, n.s.).

These results show that drastic urine concentration following rapid dehydration in normal subjects does not induce an imbalance between a number of factors promoting and inhibiting calcium lithogenesis. We cannot exclude the possibility that stone formers, in the presence of predisposing genetic and/or metabolic factors, behave in a different manner. Future studies will clarify this problem.

REFERENCES

1. VR Walker, S Colleen, M Nisa et al, Volume control in the desert: stone formers in Saudi Arabia, in: *Urolithiasis*, eds VR Walker, RAL Sutton, ECB Cameron, CYC Pak and WG Robertson, Plenum Press, New York (1989), p.769.
2. B Baggio, G Gambaro, O Oliva et al, Calcium oxalate nephrolithiasis: an easy way to detect an imbalance between promoting and inhibiting factors, *Clin Chim Acta* 124:149 (1982).
3. JH Parks and FL Coe, A urinary calcium-citrate index for the evaluation of nephrolithiasis, *Kidney Int* 30: 85 (1986).

DOES SIALIC ACID POSSIBLY INFLUENCE STONE FORMATION?

J. Hofbauer, S. Fang-Kircher[1], K. Höbarth and M. Marberger

Department of Urology
[1]Institute of Medical Chemistry
University of Vienna Medical School
Vienna, Austria

INTRODUCTION AND METHODS

Sialic acids (acylneuraminic acids) are involved in the regulation of a wide variety of biological phenomena. For instance, they appear to facilitate binding of cationic compounds to macromolecules and cells. It is well known that uromucoid, which may promote stone formation, contains sialic acid. The impact of this substance on stone formation is not yet clear and a subject of controversy. In the present study, sialic acids were investigated in the 24-hour urine of recurrent calcium oxalate stone formers (n=26), patients with single stone episodes (n=30), and healthy individuals (n=21). The subjects in each group were sex- and age-matched. Free sialic acid before and after acid hydrolysis was determined using the periodate acid and thiobarbituric acid reaction. To eliminate interference by other subtances, for example 2-deoxyribose, in this assay, measurements were performed using two different wavelengths, and values were corrected.

RESULTS

There was no statistical difference in the concentration of total sialic acids between the three groups, although the excretion in 24 h urines differed significantly ($p<0.0005$) between healthy controls and recurrent stone formers. Concentrations of free sialic acid were significantly lower in recurrent stone formers than in healthy individuals ($p<0.005$). Bound sialic acid, by contrast, was significantly increased in stone formers compared with healthy individuals ($p<0.0005$). The ratio of urinary free:bound sialic acid in healthy individuals, patients with a single stone episode, and recurrent stone formers was 1.87, 1.0 and 0.84, respectively ($p<0.0005$).

DISCUSSION

As a preliminary explanation of our findings, we hypothesize that residues of sialic acid increase the intrinsic viscosity of all glycoproteins and facilitate gel formation in water. Sialoglycoproteins may promote stone formation, whereas free sialic acid appears to have an inhibitory effect.

THE RELATIONSHIP BETWEEN URINARY Ca^{2+}, pH AND STONE DISEASE

S. Langley[1] and C. Fry[2]

[1]Urology Department
[2]Physiology Department
St Thomas' Hospital
London, UK

INTRODUCTION AND RESULTS

The ionized portion of the total urinary calcium content, Ca^{2+}, is an important determinant of stone formation. The concentration of this fraction, $[Ca^{2+}]$, is dependent not only upon the total concentration, but also upon the pH and other urinary components. We have investigated the role of urinary Ca^{2+} and pH in relation to stone disease, using Ca^{2+} selective electrodes to measure directly the $[Ca^{2+}]$[1].

Spot urine samples were obtained from calcium stone formers (n=52) and non-stone formers (n=37), under standardised conditions. Each sample was initially acidified to pH 4.75 and the $[Ca^{2+}]$ measured as the pH was subsequently increased in 0.25 pH unit increments. At high pH values calcium phosphate precipitates were formed, as detected by a light scattering technique. $[Ca^{2+}]$ was calibrated against a standard of 1.0mM $CaCl_2$ in 0.15M NaCl, expressed as a ratio of the urinary creatinine concentration.

The urinary pH of each sample, at the voided PCO_2, was significantly more alkaline in the stone formers than in the normal group (6.64±0.58; 6.12±0.57, respectively, $p<0.05$). The relationship between $[Ca^{2+}]$ and pH was non-linear, with a point of inflection prior to precipitation. This point of inflection was characterised, and showed that non-stone formers were able to maintain Ca^{2+} free in solution over a significantly greater range of pH values than the stone formers ($p<0.001$). Raised urinary calcium, magnesium and phosphate concentrations correlated with the formation of precipitates at a more acidic pH. No correlation was seen with oxalate, citrate or urate. The stone formers exhibited a raised $[Ca^{2+}]$/[urinary creatinine] measured at the voided pH, than the normals, which became more evident when the pH of the urine was standardised to pH 5-7; (i.e. $[Ca^{2+}]$/[urinary creatinine] = 0.28±0.25; 0.11±0.07 at pH 5.0 respectively; $p<0.001$). Stone formers had a greater urinary $[Ca^{2+}]$ than the normal group, and this was also excreted at a higher pH.

Stone formers have a smaller safety factor in the variation of urinary pH before precipitation, and possible stone formation occurs.

REFERENCE

1. SEM Langley and CH Fry, Measurement of the ionic composition of urine with ion-selective electrodes, *J Physiol* 438: 75 (1991).

OXALATE CRYSTALLURIA - A DISEASE

H. K. Moorthy, R.K. Vathsala, N. Sylaja, S. Sindhu, S.V. Roshni, C. Aravindakshan and Y.M. Fazil Marickar

Department of Surgery
Medical College Hospital
Trivandrum, India

INTRODUCTION AND RESULTS

Urinary crystals are the fundamental units of stone formation. Their presence indicates the process of urolithiasis. Calcium oxalate crystalluria can cause low backache, dysuria or 'crystalgia'. This study was intended to assess whether sympotmatic crystalluria can exist alone without obvious urinary stones.

One hundred patients with varying degrees of loin pain or low backache and having oxalate crystalluria, but with no evidence of obvious stones in the urinary system, were studied. After assessing the crystalluric status, the urinary risk factors namely calcium, oxalate and uric acid and serum risk factors, namely calcium and uric acid, were estimated. The patients were given pyridoxine (40 -120 mg/day) for oxalate abnormalities and allopurinol (100 - 300 mg/day) for uric acid abnormalities for one month. The response was assessed by the decrease in the severity of pain and extent of crystalluria and correction of metabolic abnormalities.

Seventy percent of the patients had calcium oxalate dihydrate crystalluria, 10% had calcium oxalate monohydrate crystalluria and 20% had a combination of the two. All these patients were symptomatic with loin pain, low abdominal pain and dysuria. Of the crystallurics 855 had biochemical abnormalities in the urine in the form of hyperoxaluria (15%) and mixed urinary abnormalities (41%). Hyperuricaemia was observed in 30% of cystallurics. The mean values of urinary calcium, oxalate and uric acid were significantly ($p < 0.001$) higher (189, 93 and 476 mg/day respectively), compared to the mean levels of these risk factors in the non crystallurics (172, 84, 427 mg/day respectively). The serum uric acid was also higher (5.9 mg%) compared to the non-crystallurics (4.9 mg%). One month of pyridoxine treatment reduced the risk factors in urine and serum with maximum effect on urine oxalate level. Allopurinol also reduced all the risk factors with significant effect on urine and serum uric acid levels. Both these drugs alleviated the symptoms in all the patients with disappearance of crystalluria.

Thus it is concluded that calcium oxalate crystalluria without obvious urinary stone formation is itself a definite clinical entity. It is often associated with biochemical abnormalities in serum or urine. The biochemical abnormalities and the crystalluria respond well to appropriate medical treatment. Both allopurinol and pyridoxine reduce the oxalate and uric acid levels in urine and cause disappearance of the calcium oxalate crystalluria. The propensity for stone formation in these patients can be assessed only by long term follow up.

URINARY CITRATE IN RELATION TO DIETARY CITRATE

N.E. Thomas, H.K. Moorthy, R.K. Vathsala,
S. Jayadevan, C. Aravindakshan and Y.M. Fazil Marickar

Department of Surgery
Medical College Hospital
Trivandrum, India

INTRODUCTION AND METHODS

Diet plays an important role in urolithiasis due to the presence of various promoters or inhibitors of stone formation. Citrate is very important in inhibiting calculogenesis. Administration of alkalinising drugs are known to increase urinary citrates. However patients with urinary stone disease are confronted with the problem of prolonged treatment with alkali. Since this is generally unpalatable, and citrate-containing fruits and vegetables are available, this study was intended to assess the efficacy of dietary supplementation with citrate containing fruits in patients with hypocitraturia.

Sixty male patients with proven urinary stones were selected randomly and only those with hypocitraturia (citrate levels less than 150 mg/day) were included in the study. The random samples of urine collected from these patients were studied for pH and microscopically examined to assess the crystalluric status. Their 24 h urinary levels of calcium, oxalate, uric acid and citrate were reassessed before the study. They were then asked to consume high citrate-containing fruit, namely lemon, in varying quantities for one month in addition to their normal diets. The patients were reassessed for their urine pH and biochemical and crystalluric status at the end of one month.

RESULTS AND CONCLUSION

The mean levels of calcium, oxalate, uric acid and citrate in the patients at the beginning of the study were 205, 34, 403 and 68 mg/day respectively. At the end of one month of supplementation with high citrate containing diets, the mean levels of these risk factors were 196, 46, 471 and 131 mg/day respectively. The paired t test did not show statistically significant variation in the biochemical parameters, pH of the urine or the extent of crystalluria. Supplementation of the diet with high citrate containing foods in hypocitraturic stone patients will therefore not significantly alter the various urinary risk factors of urolithiasis. The possible reasons for this may be the lack of relationship between the oral citrates and the urinary citrate levels and also presence of other substances like vitamin C in some foods.

Thus it is concluded that high dietary citrate as such has no role in reducing the urinary risk factors of stone formation.

CLINICAL SIGNIFICANCE OF HYPOCITRATURIA

K. Matsushita[1], K. Tanikawa[1], A. Masuda[1], M. Tanaka[1], S. Baba[1], J. Matsunaga[2] and S. Matsuzaki[2]

Department of Urology
Tokai University Tokyo Hospital
[1]Department of Urology
Keio Cijuku University
[2]Department of Urology
Inagi Municipal Hospital

INTRODUCTION AND METHODS

Hypocitraturia is defined as one of the risk factors for calcium oxalate (CaOx) nephrolithiasis because citrate inhibits CaOx crystallization and retards its crystal growth. However, the clinical importance of hypocitraturia is not yet clearly understood.

From the record of 731 patients with urolithiasis seen at our clinic since 1981, 62 cases with hypocitraturia were selected. Hypocitraturia was defined as < 320 mg excretion in two or more 24 h urine collections at outpatient clinic.

RESULTS AND DISCUSSION

In 37 cases having hypocitraturia as a sole abnormality, the mean of the urinary citrate was lower than 320 mg/day. However, in 25 cases having hypocitraturia associated with other stone risks, the standard deviation tended to be great and 13 of them had a mean value over 320. This may be that the latter group of patients with higher risks had more chance to receive supplementary citrate (see below) and they adhered more to a high-fiber and low animal protein diet than the former group. These factors probably enhanced their urinary citrate excretion.

Between the two groups there was no statistical difference in stone recurrence, stone management or stone composition. Only seven (19%) having hypocitraturia alone needed supplementary citrate, while 11 (44%) of 25 hypocitraturics associated with other stone risks had oral citrate but not allopurinol or thiazide. This means that hypocitraturia complicated with other conditions has led us to use citrate more often than hypocitraturia alone.

CONCLUSION

This study failed to show a clinical significance of hypocitraturia, either as a single or associated disorder. However, the presence of hypocitraturia apparently affected our attitude toward medical therapy.

URINARY SATURATION IN PATIENTS WITH INFLAMMATORY BOWEL DISEASE

J.L. Nishiura, L.C. Ferreira, S.S.N. Coelho, R.M. Porto, S. Misputen, H. Ajzen and N. Schor

Nephrology and Gastroenterology Divisions
Escola Paulista de Medicina
São Paulo, Brazil

INTRODUCTION AND RESULTS

Patients with inflammatory bowel diseases (IBD) have an increased incidence of nephrolithiasis (4-19%). Crohn's Disease and ulcerative colitis can modify urinary solute concentration and acid-basic equilibrium. The present study was undertaken in order to evaluate lithogenic factors and renal stone disease in patients with IBD.

Control Group: (Cont) 17 non-stone formers (8M/9F), mean age 37±3 years. Stone Former Group (SF): 10 patients (7M/3F), mean age: 43±3 years, with elimination of calculi. IBD Group: 12 patients (8M/7F), mean age: 40±3 years, with Crohn's Disease (6) and ulcerative colitis (6), out of agudization period and without other systemic diseases. Laboratory investigation: Ca and Mg (atomic absorption spectrophotometry), uric acid (Follin & Dennis method), creatinine (Jaffe method), citrate (citrate-lyase method), oxalate (quantitative enzymatic determination).

	Diuresis (mL/24h)	pH	Ca	UrAc	Oxal mg/24h	Cit	Mg
Cont	1878*	-	159	648	44	342	76
(n=17)	±184		15	49	4	46	6
SF	1855	5.4	192	705	-	463	62
(n=10)	127	0.3	34	43	-	67	7
IBD	1575	5.4	169	639	35	327	63
(n=12)	272	0.2	40	51	3	55	11

* X±SE

IBD patients did not show any significant alterations in the parameters analysed. However, since 25% of the subjects in the present series had renal stones it is important to evaluate and follow these high risk stone patients.

REFERENCES

1. MP Banner, Genito urinary complications of inflammatory bowel disease, *Radiol Clin North Am* 25: 199 (1987).
2. JH Clarck, JF Fitzgennd and JM Bergstein, Nephrolithiasis in childhood inflammatory bowel disease, *J Pediatr Gastroenterol Nutr* 4:829 (1985).

URINARY CALCIUM AND URIC ACID EXCRETION IN PATIENTS WITH MULTIPLE CALCULI ELIMINATION

L.C. Ferreira, R.O. Patino, J.L. Nishiura, S.T.S.N. Coelho and N. Schor

Nephrology Division
Escola Paulista de Medicina
São Paulo, Brazil

INTRODUCTION

Accelerated nephrolithiasis (AN) occurs in patients who have passed more than ten calculi[1]. The lithogenic factor(s) involved in this specific situation have not been identied to date[2]. Thus, the present study was performed in order to evaluate the metabolic alteration(s) in a group of out-clinic patients with accelerated nephrolithiasis.

Seventeen controls (CONT; 8M/9F), mean age: 37±3 years. Ten patients (SF; 7M/3F), mean age: 43±3 years, with elimination of <10 calculi and 10 patients (AN; 6M/4F), mean age: 42±3 years, with elimination of >10 calculi. All subjects had no other systemic disease and normal renal function, and were not taking any drugs at the time of the study. The following investigations were undertaken on 24 h urine samples: Ca and Mg excretion (atomic absorption spectrophotometry), uric acid (UA; Follin & Dennis method), creatinine excretion (Jaffe method), phosphorus excretion (Fiske & Subbarow method), citrate excretion (citrate-lyase method), total GAG: chondroitin sulfate (CS) and heparan sulfate (HS) excretion (Agarose gel electrophoresis).

	Diuresis mL/24h	Ca	UA mg/24h	Cit	Mg	Total GAG	CS mg/g creat	HS
CONT	1878*	159	648	342	76	11.90**	9.58	2.40
	±184	15	49	46	6	0.73	0.64	0.09
SF	1855	192	705	463	62	3.82#	3.02	0.80
	127	34	43	67	7	0.23	0.19	0.04
AN	1867	186	628	281#	104#	1.13#**	1.04	0.09
	±131	36	84	99	8	0.12	0.13	0.01

*X±SE; $p< 0.05$: ** vs CONT; # vs SF

Patients with AN did not have a higher urinary excretion of Ca or UA compared with the CONT or SF groups. Hypocitraturia and lower urinary total GAG excretion were found in patients with accelerated stone formation. These patients will need to be investigated specially for the inhibitor factors of lithogenesis.

REFERENCES

1. FL Coe, JH Parks and AL Strauss, Accelerated calcium nephrolithiasis, *JAMA* 244:809 (1980).
2. FL Coe and MJ Favus, Nephrolithiasis *in*: "The Kidney", ed WB Saunders Co, Philadelphia (1991).

COFFEE AND TEA CONSUMPTION - RISK FOR UROLITHIASIS

R.K. Vathsala, T.G. Dhanalekshmy, N.E. Thomas, S.V. Roshni, S. Jayadevan, C. Aravindakshan and Y.M. Fazil Marickar

Department of Surgery
Medical College Hospital
Trivandrum, India

INTRODUCTION AND METHODS

The importance of nutritional imbalance in the etiology of urolithiasis is well recognised. The aim of the present study was to assess the effect of coffee and tea consumption on the level of urinary oxalate in stone formers (SF) and, thus, on the risk of calculogenesis.

Twenty-four h urinary oxalate excretion was determined in 800 SF who consumed more than 4 cups of tea or coffee a day. Fifty age- and sex-matched SF who drank low quantities of tea or coffee served as controls. During the time of urine collection, these patients did not consume any other oxalate-rich foods. They were classified into strong (> 20 g of tea powder/day), medium (12.5 to 20 g/day) and light tea drinkers (6.25 to 12.5 g/day) and strong (> 20 g of coffee powder/day) and light (6.25 to 12.5 g/day) coffee drinkers. The oxalate values for the different groups were compared and the statistical significance of the difference assessed using analysis of variance.

RESULTS

Among the tea and coffee drinkers, 24.4% were single SF and 75.6% were multiple/ recurrent SF, whereas in the control group the corresponding figures were 62% and 38%. The mean urinary oxalate excretion in tea drinkers was significantly greater than that of coffee drinkers, being 85 and 46 mg/day, respectively, ($p < 0.001$). The difference in oxalate excretion between the tea drinkers and controls (85 and 43 mg/day, respectively) was also highly significant ($p < 0.001$), whereas no difference was found between the the coffee drinkers and controls (46 and 43 mg/day). The mean urinary oxalate excretions in strong, medium and light tea drinkers were 148, 93 and 45 mg/day respectively. In coffee drinkers, the mean urinary oxalate excretions in the strong and light drinkers were 69 and 36 mg/day, respectively. The differences between these values were statistically significant ($p < 0.05$).

The urinary oxalate excretions in SF increased significantly after consumption of more than cups of coffee or tea a day and thus increased the risk of stone formation. Tea was more oxaluric than coffee.

A point to be considered in this aspect of study of SF is the variation in the oxalate content in the different samples of tea available in the different regions, and whether or not the tea is consumed with milk.

COMPUTERISED DIETETIC SURVEY

R.K. Vathsala[1], T.G. Dhanalekshmy[1], C. Aravindakshan[1],
P.L. Vijayammal[2], H.K. Moorthy[1] and Y.M. Fazil Marickar[1]

[1]Department of Surgery
Medical College Hospital and
[2]Department of Biochemistry
University of Kerala
Trivandrum, India

INTRODUCTION AND METHODS

Several nutrients influence the occurrence of urinary stone disease. However, no effective method has been available for a scientific assessment of the nutrient intake of individuals. This work was undertaken to create a computer program for nutritional assessment which could be easily handled by clinicians and laboratory staff.

The program written in turbo C and Clipper languages was designed in three steps. The first was to input the various constituent nutrients in food which included carbohydrates, protein, fats, fibre, minerals and vitamins. The food products from which these were obtained included the various grains, vegetables, meats, oils, eggs, and beverages. As a next step, commonly-consumed meal items such as hamburgers, french fries, bread, biscuits and ice cream, for example, were entered into the program for their contents. The last part of the program was the user part. The patient's dietary intake was recorded in the patient's own words and the computer program was able to analyse the total nutrients consumed by the individual in a day or for a particular meal. Dietary intake was studied in 600 stone patients. During the day of study, a 24 h urine sample and a blood specimen were collected for biochemical assessment and correlation with the dietary intake. Biochemical and crystal-pattern studies proved that diet does alter the urine and blood biochemistry, and crystal nucleation, growth, and aggregation in stone formers.

RESULTS

It was possible to analyse the nutrient content effectively in all the patients analysed. The program has flexibility and can easily be modified as far as expansion of nutrients and food types are concerned.

CONCLUSION

It is concluded that the computer-assisted nutritional survey is a good scientific assessment of nutrient intake of individuals. It is not only useful to clinicians, but also to dieticians and even lay public to formulate dietary patterns for patients with different types of diseases requiring dietary adjustments and corrections.

RISK OF CALCIUM STONE FORMATION IN RELATION TO SEX

A. Trinchieri[1], F. Rovera[1], R. Nespoli[1], G. Zanetti,[1]
E. Austoni[1] and A. Mandressi[2]

[1]Istituto di Urologia dell´ Universita di Milan
Milan, Italy
[2]U.O. di Urologia
Ospedale di Busto Arsizio, Italy

INTRODUCTION AND RESULTS

Idiopathic calcium-stone formation occurs mainly in men, the male/female ratio being about 4:1. The reason for the higher risk of idiopathic calcium stone formation in men than in women is due to a combination of urinary risk factors.

Investigations were carried out in 81 normal women, 70 normal men, 37 female stone formers and 67 male stone formers. Twenty-four h urine samples were collected in order to measure potassium, sodium, calcium, phosphate, oxalate, urate, citrate, glycosaminoglycans (GAGs), magnesium, zinc, creatinine, and pH. In 12 premenopausal women, 24 h urine samples were collected during different phases of the menstrual cycle.

In normal subjects, the daily excretions of potassium, sodium, calcium, magnesium, zinc, phosphate, urate and GAGs were significantly higher in men than in women. Sodium, phosphate, and urate were significantly higher in male stone formers than in female stone formers. These differences were reversed when excretion was related to creatinine. No significant difference was observed between men and women in daily citrate excretion, but citrate/creatinine ratios were significantly higher in females than in males, whether normal subjects or stone formers. The urinary output of electrolytes was unrelated to normal cycles of oestrogen and progesterone and no significant changes were found between postmenopausal women and those using oral contraceptives.

DISCUSSION

The lower biochemical risk of stones in women seems to be due to a better balance between urinary calcium and citrate. The relationship between sex steroids and urinary citrate is still unsettled[1].

REFERENCE

1. ML Hammar, GE Berg, L Larsson, H-G Tiselius, E Varenhorst, Endocrine changes and urinary citrate excretion, *Scand J Urol Nephrol* 21:51 (1987).

EFFECTS OF DIETARY NUTRIENTS UPON URINARY EXCRETION OF CALCIUM, URIC ACID AND CITRATE IN RENAL STONE FORMERS

I.P. Heilberg, L.A. Martini, L. Cuppari, S.A. Draibe and N. Schor

Nephrology Division
Escola Paulista de Medicine
São Paulo, Brazil

INTRODUCTION AND RESULTS

The impact of diet on the pathogenesis of urolithiasis is well recognised[1]. Dietary nutrients such as Ca, Na, carbohydrate (CHO), and animal protein may affect urinary excretion of lithogenic substances[2], but the link between them may depend on which population is being studied[3]. The aim of the present study was to evaluate the correlation between urinary Ca, citrate, and uric acid with dietary Ca, Na, protein, and CHO intakes.

Seventy-nine renal stone formers (SF) were divided into absorptive (AH; n=46) and renal hypercalciuric (RH; n=33). Seventeen normal subjects were also evaluated. A 24 h dietary record, together with a 24 h urine sample for determination of Ca, Na, citrate, and uric acid were obtained from all subjects.

In normals, urinary Ca did not correlate with dietary Ca, Na, protein, and CHO. In AH, urinary Ca correlated with dietary Ca ($r=0.29$) and Na ($r=0.47$; $p< 0.05$. In RH, urinary Ca correlated with dietary Ca ($r=0.52$), Na ($r=0.52$) and CHO ($r=0.52$; $p< 0.05$. In normals, no correlations were found when urinary citrate and uric acid were correlated with dietary intake. In AH, urinary citrate correlated with dietary Na ($p<0.05$). No correlation existed between urinary citrate and uric acid and dietary protein in any group.

DISCUSSION

In normal subjects, Ca, Na, CHO and protein intake did not significantly affect Ca, citrate, and uric acid excretion. The correlations in RH between Ca excretion and CHO intake, and uric acid and Na intake, suggest the presence of competitive absorption along the proximal renal tubule in RH. Unexpectedly, Ca excretion correlated with Ca intake even in RH, who demonstrated fasting hypercalciuria. In RH, the calciuric effects of dietary Ca and Na were more significant than those of either CHO and protein intakes.

REFERENCES

1. PN Rao, V Prendiville, A Buxton, DG Moss, NJ Blacklock, Dietary management of urinary risk factors in renal stone formers, *Br J Urol* 54: 578 (1982).
2. S Goldfarb, The role of diet in the pathogenesis and therapy of nephrolithiasis, *Endocr Metab Clin North Am*, 19:805 (1990).
3. WG Robertson, M Peacock, The pattern of urinary stone disease in Leeds and in the United Kingdom in relation to animal protein intake during the period of 1960-1980, *Urol Int* 37: 394 (1982).

INCREASED RISK OF CALCIUM OXALATE STONES IN YOUNG FEMALES

P.W. Baker, P. Coyle, R. Bais and A.M. Rofe

Division of Clinical Chemistry
Institute of Medical and Veterinary Science
Adelaide, Australia

INTRODUCTION AND RESULTS

A retrospective study of 2,800 renal calculi analysed at the IMVS from 1977 to 1991 was undertaken. The percentages of stone types were calcium oxalate (CaOx), 66%; uric acid (UA), 18%; magnesium ammonium phosphate (MAP), 13%; calcium phosphate (CaP), 3%; and cystine, <1%; which is in accordance with studies from other affluent countries. The frequency of CaOx, UA and CaP stones from males exceeded that of females by 175, 269 and 35% respectively. The frequency of MAP stones from females exceeded that of males by 39%. The combined frequency of predominant stones was 135% greater in males than in females. In addition, a distinct seasonal pattern was observed, with the summer to winter ratio of stones being: UA, 1.86±0.79; CaP, 1.60±0.83; CaOx, 1.21±0.33; and MAP, 0.76±0.30 per year (mean yearly ratio±SD, n=15).

The overall incidence of calcium oxalate stones in males was 275% greater than that in females. Males showed a peak incidence between the ages of 51-55 years followed by a sharp drop in incidence between 56-60 years. In female CaOx stone formers (SF), peak incidences were observed between the ages of 21-25, 31-35 and 51-55 years, these groups representing 9.5, 13.1 and 9.5% of total female SF, respectively. The equivalent values for the male SF in these age groups was 2.4, 9.6 and 14.2%, respectively. The increased incidence of CaOx stones in young females was first noticed in 1980 by Rofe et al[1]. This trend has persisted over the last decade to the extent that females between 21-25 years have a 146% greater stone risk than males in the same age group.

The distribution of the frequency of CaOx stones with age is clearly different between males and females. The maximum female frequency of CaOx stones between 31-35 years of age occurs 20 years earlier than the maximum male frequency. Our observations of a stone peak in females between 51-55 years supports the postmenopausal stone peak observed by other investigators.

DISCUSSION

Reasons for the increased frequency of female CaOx stones may include changes in occupation, diet, drug history, and more importantly, hormonal changes.

REFERENCE

1. AM Rofe, RAJ Conyers, DW Thomas, *Med J Australia* 2:158 (1981).

SECTION VII: EPIDEMIOLOGY AND INFECTION

OXALATE-DEGRADING BACTERIA IN THE GUT - DO THEY INFLUENCE CALCIUM OXALATE STONE FORMATION?

K. Kleinschmidt, A. Mahlmann and R. Hautmann

Department of Urology
University of Ulm
7900 Ulm, Germany

INTRODUCTION

Several hundred milligrams of dietary oxalate pass daily through the human gastrointestinal tract. Nutrition therefore represents the largest potential source of oxalate for the human body. Hodgkinson estimated that approximately half the ingested oxalate is destroyed by bacterial action. Oxalate is absorbed via the intestinal mucosa by an ion exchange process. Two to 20% of the ingested oxalate in absorbed via the ileal or colonic mucosa. The rest is considered thus far to be excreted in the feces[5,6].

Recently a species of anaerobic bacteria (*Oxalobacter formigenes*), living exclusively on oxalate, has been described[2]. In animals up to 600 milligrams per day of the oral oxalate load are metabolized to CO_2 and formate by these bacteria. This species is present in the human gut, too. There is evidence that it is mainly active in the proximal colon.

Based on this knowledge we set up a new hypothesis concerning etiology and pathogenesis of calcium oxalate-urolithiasis: the gastrointestinal tracts of stone formers possibly are not sufficiently colonized by oxalate-degrading bacteria, leaving soluble oxalate available for absorption. The resulting hyperoxaluria in patients will lead to stone formation. In healthy persons exogenous oxalate is microbially decarboxylated to a significant extent. Thus hyperoxaluria and the risk of urolithiasis are prevented.

The purpose of the study was to analyze oxalate-degrading bacteria in human feces with regard to urinary oxalate concentration and excretion comparing stone formers with non-stone formers. We tried to answer the question whether the absence of oxalate degrading bacteria in the gut represents a risk factor for calcium oxalate urolithiasis.

MATERIALS AND METHODS

Fifty-one persons participated at the study after informed consent. Thirty-two were selected patients with calcium oxalate-urolithiasis, 15 patients were free of calculi, 17 patients presented with recurrent disease. Nineteen urological patients without urolithiasis served as controls. Age, sex and body weight of these three subgroups were comparable.

All study participants had no gastrointestinal disorders or bowel resections. No antibiotic therapy or ascorbic acid was allowed at least three weeks before examinations. The stone formers had pure calcium oxalate calculi as shown from stone analysis by X-ray diffraction. As in-patients, all study participants received an individual diet for two

Urolithiasis 2, Edited by R. Ryall *et al.*,
Plenum Press, New York, 1994

days, followed by an oxalate-rich diet for three days. Feces samples were analyzed daily and urine samples were analyzed daily in five fractions per day.

Quantification of anaerobic oxalate-degrading bacteria was achieved using selective media containing 20 mmolar oxalate. Feces samples were incubated for three weeks in Hungate roll tubes at 37°C. Colony forming units (CfU) of oxalate degrading bacteria with clear zones in the medium were counted. Urinary oxalate concentrations were measured by ion-chromatography[3].

RESULTS

We were able to demonstrate anaerobic oxalate degrading bacteria in human feces. If present, we found a number of 10 million CfU per g feces. Colony counts of oxalate degrading bacteria could be stimulated by an oxalate-rich diet. We observed a significant increase in numbers. In highly active stone formers with more than four stone episodes per life, anaerobic oxalate-degrading bacteria were completely absent. In contrast, they were present in comparable numbers in both stone formers with 1 - 3 episodes and control persons (Table 1).

The circadian oxalate concentrations and excretions were generally higher in stone patients compared with the controls. In contrast to the controls stone formers showed a peak concentration of urinary oxalate during the night time (Fig. 1).

Table 1. Correlation of stone frequency and oxalate-degrading bacteria (CfU/g feces).

Patients	Lack of oxalate-degrading bacteria
Controls (n=19)	47%
Stone formers	
- 1 episode (n=9)	33%
- 2 episodes (n = 8)	29
- 3 episodes (n= 8)	50%
- 4-30 episodes (n=7)	100%

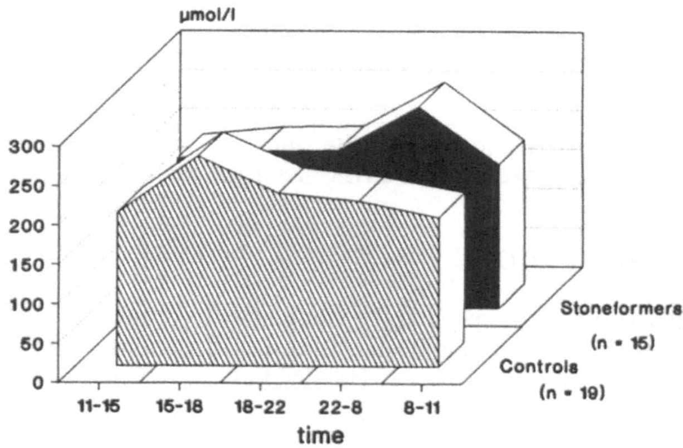

Figure 1. Circadian urinary oxalate concentrations (oxalate-rich diet).

DISCUSSION

As reported by others[1,4] we were able to isolate oxalate-degrading bacteria from human feces. When present, we found about 10 million colony-forming units per g fecal sample. The bacteria were stimulated by an oxalate-rich diet. Although we have only small numbers, there is evidence that in highly active stone formers there is a lack of anaerobic oxalate-degrading bacteria in the gut. These bacteria may be able to prevent critical oxalate concentrations in the urine by metabolizing dietary oxalate to CO_2 and formate. The circadian concentrations of urinary oxalate indicate that during the night time this microbiological system may be active. In controls a concentration peak is prevented. During day time the lack of a concentration peak in stone formers may be explained by a higher fluid intake with a mean of 2.5L vs. 2.0L in control persons.

We conclude that a lack or lowered numbers of oxalate-degrading bacteria in the gut may represent a risk factor for calcium oxalate urolithiasis.

ACKNOWLEDGEMENTS

Sponsored by the Deutsche Forschungsgemeinschaft (DFG) Grant HA 1371/3-1.

REFERENCES

1. MJ Allison, HM Cook, DB Milne, RV Clayman, Oxalate degradation by gastrointestinal bacteria from humans, *J Nutr* 116:455 (1986).
2. MJ Allison, KA Dawson, WR Mayberry, YG Foss, Oxalobacter formigenes gen. nov., sp. nov.: oxalate-degrading anaerobes that inhabit the gastrointestinal tract, *Arch Microbiol* 141 (1985).
3. A Classen, A Hesse, Measurement of urinary oxalate: an enzymatic and an ion-chromatographic method compared, *J Clin Chem. Clin Biochem* 25:95 (1987).
4. LT Doane, M Liebman, DR Caldwell, Microbial oxalate degradation: effects on oxalate and calcium balance in humans, *Nutr Research* 9:957 (1989).
5. M Hatch, RW Freel, AM Goldner, DL Earnest, Oxalate and chloride absorption by the rabbit colon: sensitivity to metabolic and anion transport inhibitors, *Gut* 25:232 (1984).
6. HE Williams, TR Wandzilak, Oxalate synthesis, transport and the hyperoxaluric syndromes, *J Urol* 141:742 (1989).

URINARY ENZYMES : THEIR ROLE IN RENAL STONE FORMATION

C.H. van Aswegen, P.J. du Toit and D.J. du Plessis

Department of Urology
University of Pretoria
Private Bag X169
Pretoria, RSA

INTRODUCTION

Urolithiasis is a multifactorial disorder of some complexity. Various theories have been postulated to explain lithiasis, namely the so called matrix theory; the precipitation theory; and the theory on the lack of inhibition[1]. According to the matrix theory a protein such as uromucoid activates the initial crystallization process by promoting calcium oxalate and calcium phosphate crystal formation and concretion in whole urine[2]. We therefore postulated that stone formation is initiated by uromucoid decomposition when optimal conditions occur[3]. Factors that may influence this process are desialylation of uromucoid and enzyme regulation of urinary uromucoid concentration.

In this respect the activity of proteases, like urokinase, would play an important role. Since it is accepted that urinary urate, calcium, magnesium, certain bacteria, and sex play a role in urolithiasis, these factors have been formally investigated in relation to urolithiasis and enzyme activity.

MATERIALS AND METHODS

All reagents were of Analar grade. First morning urine was obtained from men of different ages with and without renal stones. Spectrophotometric analysis of sialidase[4] and urokinase[5] activities was performed in the presence of urine, urate, minerals or bacteria.

RESULTS AND DISCUSSION

Sialidase

Urolith patients of different ages had significantly higher sialidase activity than healthy men of comparable ages ($p < 0.001$[4]). Urine of renal patients contained significantly lower bound ($p = 0.0012$) and total ($p = 0.003$) sialic acid concentrations than their healthy counterparts[4]. The promoter, calcium, and the inhibitor, magnesium, of urolithiasis had no effect on the sialidase activity [6].

Urolithiasis 2, Edited by R. Ryall *et al.*,
Plenum Press, New York, 1994

Urokinase

An increase in urokinase inhibition was found in the urine of urolith patients ($p < 0.001$), which could be attributed to an increase in urinary urate ($r = 0.762$), an increase in calcium[6] and/or a decrease in testosterone ($r = 0.7305$, ref 7) concentrations. The data presented here concur well with published results and thus, if the postulate is correct, it is expected that bacteria also may affect these enzymes, as the association between urinary tract infection and renal stone formation is well established[1]. It is generally theorised that *Proteus mirabilis* and other urease producing bacteria, are responsible for renal stone formation.

The importance of non-urease-producing bacteria *(Escherichia coli,* etc) in urolithiasis is still unclear. Spectrophotometric studies show that bacteria associated with infection stones, like *P mirabilis* and E *coli,* respectively, inhibit and stimulate urokinase and sialidase activity[8]. In contrast, bacteria not associated with infection stones *(Bacillus subtilis)* affected the urokinase and sialidase activity significantly less[8].

CONCLUSION

It seems highly probable that urinary sialidase and urokinase play a role in renal stone formation. Thus, future prospects will include the determination of bacterial effects on tissue urokinase and sialidase in renal parenchyma.

REFERENCES

1. U Backman, BG Danielson and S Ljunghall, *in* Renal Stones. Almquist and Wiksell, Stockholm (1985).
2. PC Hallson and GA Rose. Uromucoids and urinary stone formation, *Lancet* 1; 1000 (1979).
3. CH van Aswegen and DJ du Plessis. Pathogenesis of kidney stones, *Med Hyp* 36:368 (1991).
4. CH van Aswegen, CA van der Merwe and DJ du Plessis, Sialic acid concentrations in the urine of men with and without renal stones. *Urol Res* 18:29 (1990).
5. CH van Aswegen, AWH Neitz, PJ Becker and DJ du Plessis, Renal calculi-urate as a urokinase inhibitor. *Urol Res* 16:143 (1988).
6. CH van Aswegen, JC Dirksen van Schalkwyk, et al, The effect of calcium and magnesium ions on urinary urokinase and sialidase activity. *Urol Res* 20:41 (1992).
7. CH van Aswegen, P Hurter, CA van der Merwe and DJ du Plessis, The relationship between total urinary testosterone and renal calculi. *Urol Res* 17:181 (1989).
8. PJ du Toit, CH van Aswegen et al, The effects of bacteria involved with the pathogenesis of infection-induced urolithiasis on the urokinase and sialidase (neuraminidase) activity. *Urol Res* (In press).

UROLITHIASIS IN TURKEY: EPIDEMIOLOGICAL FEATURES AND CAUSAL FACTORS OF STONE FORMATION

M. Akinci, T. Esen and S. Tellaloglu

Department of Urology
Medical Faculty of Istanbul
Istanbul, Turkey

INTRODUCTION

Turkey is thought to be an endemic region for urolithiasis yet sound epidemiological and clinical studies were missing at the present time. Most of the data on urinary stone disease in Turkey date back to studies in the 1950's and 1960's[1]. To determine the severity of the disease in the country, we conducted an epidemiological study which included a prospective clinical trial and a retrospective analysis of stone patients' files.

MATERIALS AND METHODS

In a nation-wide study, we interviewed 1,504 patients in 14 provinces, the population of which is representative for Turkey. A total of 780 male and 724 female subjects were randomly selected, and the prevalence of urolithiasis in Turkey was estimated to be 10±5 (±SD)%. In a prospective clinical trial, 230 consecutive stone patients were evaluated metabolically. Investigations included two 24 h urine analyses and X-ray-diffraction stone analysis. In two time periods (1969-79 and 1979-89), we compared the rate of surgical intervention for stones, the average patient age, the ratio of male/female patients, the occurrence of paediatric patients, and the rate of occurrence of lower urinary-tract stones.

RESULTS

This study indicated that the overall prevalence of urolithiasis in Turkey was 14.8%. The male/female ratio was 1.5 to 1. The prevalence did not correlate with habitat or occupation, but significantly differed in various educational and socio-economical sub-groups as shown in Table I. A geographical distribution - though statistically not significant - was observed with higher values in southern and southeastern parts of the country. Causal factors of stone formation are shown in Table 2.

Urolithiasis 2, Edited by R. Ryall *et al.*,
Plenum Press, New York, 1994

Table 1. General epidemiological data on urolithiasis in Turkey.

Overall prevalence	14.8%
Male/female ratio	1.5:1
Prevalence in relation to education	
Illiterates/Primary school	19.0%
University graduates	8.5%
Prevalence in relation to habitat	
Rural areas	14.4%
Cities	15.4%
Prevalence in relation to occupation	
Manual	15.2%
Non-manual	13.7%
Prevalence in relation to socio-economics	
Low	20.8%
Medium	14.2%
High	12.7%

Table 2. Causal factors of stone formation.

Hypocitraturia	22.3%
Hypercalciuria	22.6%
Hypomagnesiuria	43.3%
Hyperoxaluria	10.6%
Infection	10.0%
Hyperuricosuria	8.0%
Urinary anomalies	6.0%
Hyperparathyroidism	6.0%
Renal tubular acidosis	3.0%
Cystinuria	2.0%
Low urinary volume	4.0%
Idiopathic	14.0%

When comparing two time periods, we saw that the rate of stone operations to total urological operations had decreased from 35% to 25.3% from the 1969-1979 decade to the 1979-1989 period. Similarly, the rate of occurrence of pediatric patients and lower urinary-tract stones was also reduced in the latter 10 years compared to the previous decade (29% vs 13.6 % and 24.7% to 14.5%, respectively). The average patient age increased, however, from 28.3 to 35.5 years. The proportion of female patients also rose from 25.9% to 29.3%.

According to Asper[1], these demographic data show the same pattern as is found in high-level European countries. If classified according to the data from the 1969-1979 period, the classification would be a low/medium socio-economical level urolithiasis pattern.

DISCUSSION

The overall prevalence of urolithiasis in Turkey is 14.8% which exceeds that of other European countries[2,3,6,7]. The male/female ratio of 1.5:1 confirms the trends of recent reports[2,5]. Interestingly, people with low educational and socio-economical status had significantly higher prevalence rates. In contrast, profession and habitat did not affect the prevalence rate. This might be the result of mass migration in Turkey from rural areas to the cities. The variability in prevalence in different geographical regions needs further and more detailed studies.

Clinical investigations revealed that a deficiency in inhibitory factors is the leading causal factor of urolithiasis in Turkey. Hypocitraturia and hypomagnesiuria are more commonly found than elsewhere[4,5]. This could be attributed to dietary habits which are maintained irrespective of habitat in the country.

The comparison of various data in two time periods showed that the urolithiasis pattern of Turkey is close to that of high socio-economical countries according to Asper's definition[1]. Yet, more detailed studies are needed to reach conclusions. Meanwhile, urolithiasis remains as a serious health problem for Turkey, which requires special clinical attention.

REFERENCES

1. R Asper, Epidemiology and socioeconomical aspects of urolithiasis. *Urol Res* 12:1 (1984).
2. L Borghi, PP Ferretti, GF Elia et al. Epidemiological study of urinary tract stones in a northern Italian city, *Br J Urol* 65:231 (1990).
3. F Disilverio, M Gallucci, Progressi nella calcolosi renale *in:* Acta Edizioni e Congressi, Roma (1988).
4. CYC Pak, Medical management of urolithiasis, *J Urol* 128: 1157 (1982).
5. CYC Pak, Should patients with single renal stone occurrence undergo diagnostic evaluation?, *J Urol* 127:1855 (1982).
6. WG Robertson, State of the Art Lecture. 2nd European Symposium on Stone Disease. Basle (1990).
7. EW Vahlensieck, D Bach and A Hesse, Incidence, prevalence and mortality of urolithiasis in the German Federal Republic. *Urol Res* 10:161 (1982).

LIQUOR IN LITHOGENESIS

R.K. Vathsala, T.G. Dhanalekshmy, H.K. Moorthy,
C. Aravindakshan, P.L. Vijayammal[2] and Y.M. Fazil Marickar[1]

[1]Department of Surgery
Medical College Hospital and
[2]Dept of Biochemistry
University of Kerala
Trivandrum, India

INTRODUCTION

A marked increase in the incidence of urinary-stone disease has been correlated with the affluence of the population. Affluence has led to an increase in consumption of alcoholic drinks. The present study was performed to find out the variations in the values of biochemical parameters in alcohol-consuming stone formers and non-alcoholic stone formers, and to assess the effect of these on the risk of stone formation.

PATIENTS AND METHODS

Fifty stone formers attending the urinary stone clinic of Medical College Hospital, Trivandrum who had consumed alcohol for more than five years, and 50 non-alcohol-consuming stone formers were studied. The 50 alcohol-consuming patients were classified as heavy drinkers (> 900 mL/week), moderate drinkers (300 - 900 mL/week), and occasional drinkers (< 300 mL/week). Depending on the type of liquor consumed by the patients, they were further classified into whisky/brandy, rum, beer and arrack drinkers. Twenty-four h urine samples from all patients were analysed for Ca, oxalate (Ox), and uric acid.

Biochemical differences between alcohol-consuming and non-alcohol-consuming stone formers were analysed using the Student's *t* test. The comparison between heavy, moderate and occasional drinkers, and whisky/brandy, rum, beer and arrack drinkers was done using ANOVAR.

RESULTS

The mean values for 24 h urinary Ca, Ox and uric acid, and serum Ca and uric acid were higher in the alcohol-consuming patients than in the non-alcohol-consuming patients (Table 1). However, only urinary uric acid was significantly higher ($p < 0.001$).

Urolithiasis 2, Edited by R. Ryall *et al.*,
Plenum Press, New York, 1994

Table l. Biochemical changes in alcohol-consuming and non-alcohol-consuming stone formers.

	Stone formers	
	Alcohol-consuming (50)	Non-alcohol-consuming (50)
Urine (mg/day)		
Ca	212±18	186±17
Ox	65±9	44±13
Uric acid	626±45	474±47***
Serum (mg/dL)		
Ca	9.4±0.4	9.3±0.4
Uric acid	5.0±0.3	4.7±0.3

Values shown are Means±SEM, n in parentheses. *** Significantly different from alcohol-consuming patients, $p < 0.001$.

The values of risk factors varied with the frequency of drinking. The difference in the urine Ca and uric acid levels between the heavy, moderate and occasional drinkers was statistically significant ($p < 0.05$), while the differences in urine Ox, serum Ca, and uric acid did not differ (Table 2).

Table 2. Biochemical variation between heavy, moderate, and occasional alcohol consumers.

	Alcohol Consumption		
	Heavy	Moderate	Occasional
Urine (mg/day)			
Ca	176±17	200±29	300±59*
Ox	90±15	52±14	46±16
Uric acid	747±62	738±91	546±56*
Serum (mg/dL)			
Ca	9.1±0.5	9.3±1	10.0±0.4
Uric acid	5.4±0.4	5.0±0.1	4.1±0.4

Values shown are Means±SEM. * $p < 0.05$.

Table 3 compares urinary Ca, Ox, and uric acid, and serum Ca and uric acid in the alcohol-consuming stone formers divided according to the type of alcohol consumed. The whisky/brandy consumers had a higher urinary Ca excretion than the other alcohol-consuming groups. Their urinary Ox and uric acid excretions were similar to those of the rum consumers.

DISCUSSION

Alcohol consumption has been blamed as a potentiating factor for the risk of stone formation[1]. Consumption of alcoholic beverages has variable effects on urate excretion. During ethanol abuse, the hyper-lactacidemia that can occur may be sufficient to suppress the renal excretion of uric acid by inhibiting its secretion[2]. However, this hypo-uricosuric effect may be opposed by ethanol-induced enhancement of urate synthesis[3], and by provision of exogenous purines (yeast) by certain alcoholic beverages such as beer and wine[4]. Alcohol consumption also increases the *de novo* synthesis of uric acid leading to an increased excretion of uric acid[5]. Our study also shows that the alcohol-consuming stone patients have higher urinary uric-acid levels than non-alcohol-consuming stone patients. Disturbances in uric-acid metabolism have recently been suggested to be important not only for uric-acid stone disease but also for CaOx stone formation[6]. In our study, urinary uric acid and oxalate were higher in the heavy drinkers, and urate was also high in the moderate drinkers compared to the occasional consumers.

Table 3. Biochemical values of risk factors in different types of drinkers.

	Whisky/brandy	Rum	Beer	Arrack
Urine (mg/day)				
Ca	288±35	208±24	209±41	143±25*
Ox	85±18	81±21	46±17	45±11
Uric acid	704±81	705±112	538±71	556±83
Serium (mg/dL)				
Ca	9.4±0.8	10.0±0.3	9.3±0.9	9.1±0.6
Uric acid	5.6±0.9	5.4±0.7	5.4±0.3	5.0±0.6

Values shown are Means±SEM. *$p< 0.05$.

Alcohol is known to suppress intestinal Ca absorption[7] resulting in a calcium-depleted state[8]. Various authors have documented disturbances of Ca absorption and renal reabsorption[9,10]. Our findings also indicate that urinary Ca excretion can be elevated in alcohol-consuming stone formers. The increasing frequency of urate-containing stones in patients drinking alcohol represents an important aspect to be considered in the prophylactic management of urinary stone disease[11].

ACKNOWLEDGEMENTS

The authors acknowledge the financial support of the Indian Council of Medical Research, the Kerala State Committee on Science, Technology and Environment for conducting the project, and the University of Kerala for the Fellowship grant to RKV.

REFERENCES

1. O Zechner, D Lat, H Pftuger and U Scheiber, Nutritional factors in urinary stone disease. *J Urol* 1225: 51, (1981).
2. CS Lieber and CS Davidson, Some metabolic effects of ethyl alcohol. *Am J Med* 33: 319, (1962).
3. J Faller and IH Fox, Ethanol induced hyperuricemia. Evidence for increased urate production by activation of adenine nucleotide turnover. *N Engl J Med* 307: 1598, (1982).
4. CYC Pak, Hyperuricosuric calcium nephrolithiasis. *in*: "Urolithiasis, A Medical And Surgical Reference". MI Resnick and CYC Pak (eds). WB Saunders 79-87 (1990).
5. EW Vahlensieck, A Strenge and A Hesse, Nutritional history of recurrent calcium oxalate stone formers pre and post diet, *in* "Urinary Stone", R Ryall , JG Brockis, VR Marshall and B Finlayson (eds) Churchill Livingstone (1984) 8: 41-46.
6. WG Robertson, Physical chemical aspects of calcium stone formation in the urinary tract. *in*: "Urolithiasis Research", H Fleisch, WG Robertson, LH Smith and W Vahlensieck (eds) Plenum Press, New York (1976) pp 25.
7. EL Krawith, Effect of ethanol ingestion on duodenal calcium transport. *J Lab Clin Med* 85: 665, (1975).
8. MH Criqui, RD Langer, DM Reed, Dietary alcohol, calcium and potassium independent and combined effects on blood pressure. *Circulation*. 80: 609, (1989).
9. S De Marchi, E Ceuhin, A Basile, Fractures and hypercalciuria, two markers of severe dependence in alcoholics. *Br Med J* 288: 1457, (1984).
10. H Heath, PW Lambert, FJ Service, SB Arnaud, Calcium homeostasis in diabetes mellitus. *J. Clin. Endocrinol Metab* 49: 462, (1979).
11. O Zechner and Scheiber, Alcohol as an epidemiological risk in urolithiasis. *in*: "Urolithiasis Clinical And Basic Research" , LH Smith, WG Robertson and B Finlayson (eds) Plenum Press, New York, pp 315-319 (1981).

EPIDEMIOLOGY OF URINARY STONE DISEASE IN SAUDI ARABIA

W.G. Robertson and H. Hughes

Department of Biological and Medical Research
King Faisal Specialist Hospital and Research Centre
Riyadh, Saudi Arabia

INTRODUCTION

Urolithiasis is a disorder which is recognised in every country in the world although its prevalence varies widely from one to another. The pattern of the disorder has altered with time, a process which is continuing even to this day. However, from a knowledge of this changing picture and of the parallel fluctuations occurring in suspected risk factors for stone-formation, it is possible to construct a hypothesis of what external factors are important in the etiology of stone-formation in the urinary tract. In this connection, several studies have been carried out in the past 10 years in some of the affluent states of the Arabian Gulf which have thrown light on some of the more important epidemiological and urinary risk factors which combine to make this region probably the most prolific area for stone formation in the world. In this study, we concentrate on the factors found to be of particular importance in Saudi Arabia.

Stone Prevalence

A study carried out on stone formers who were seen at four hospitals in Riyadh indicated that urolithiasis is an extremely common disorder in Saudi Arabia. The disorder is more common in men than in women (male/female ratio = 3.2:1). In terms of idiopathic stone formation, this ratio increases to well over 4:1. Taking into account the very skewed age-distribution in the population as a whole (50% are aged 15 or under), the age-specific expectancy of stones in men reaching the age of 60 is over 20%. This compares with values of 7.8% and 13.0% reported from the UK and USA respectively (Table 1). A more detailed study indicates that the higher age-specific expectancy of stones in Saudi males extends over the entire age range including that of children. Indeed, stone formation, particularly the idiopathic form, starts at an earlier age in Saudis and is about twice as common as in children from Western countries.

Stone Composition

Table 2 shows that, with the notable exception of infection stones, all other types of stone are more common in Saudi children than in children from other countries. The vast majority consist of calcium oxalate and/or uric acid and appear to be idiopathic in origin. Only 9% are secondary to a urinary tract infection, which is a considerably lower figure than that reported in juvenile stone formers in the West. There are relatively more of the rare genetic and metabolic types of stone such as cystine, xanthine and 2,8-dihydroxyadenine. This is not unexpected in the light of the reports of the relatively high

Urolithiasis 2, Edited by R. Ryall *et al.*,
Plenum Press, New York, 1994

Table 1. Relationship between life-time expectancy of urolithiasis in men and urinary and dietary composition in normal men in various countries.

Country	Life-time stone expectancy in men (%)	Urinary Ox (mmol/day)	Urinary Ca (mmol/day)	Dietary Ox/Ca (mmol/day)	Dietary AP (g/day)
China	1.5	-	3.5	-	21
Japan	5.4	0.31	4.1	-	39
U.K	7.8	0.35	6.0	0.04	53
West Germany	8.0	0.36	5.8	0.06	59
Sweden	8.6	0.35	5.0	0.06	60
Canada	12.0	0.39	5.5	-	66
New Zealand	12.5	0.40	4.3	-	72
U.S.A.	13.0	0.41	5.5	0.10	76
U.A.E.	18.0	0.46	4.2	0.25	'very high'
Saudi Arabia	20.1	0.53	3.2	0.26	82

Note: AP = animal protein; Ox = oxalate; Ca = calcium

Table 2. Composition of stones from children in various countries.

Stone type	UK (%)	Austria (%)	USA (%)	KSA (%)	Saudi/Western* frequency
Uric acid	0	3	4	15	9.0
Calcium	32	37	58	63	2.1
Infected	60	53	28	9	0.3
Rare	8	7	10	13	2.2

*Including the overall Saudi/Western prevalence ratio of 1.4:1. KSA = Kingdom of Saudi Arabia.

incidence of inborn errors of metabolism in this part of the Middle East. Endemic bladder stones do not seem to occur in children from the Riyadh area, presumably because of the relatively high level of nutrition received by most children in that part of the country.

The pattern of stone formation in Saudi adults (Table 3) is not dissimilar from that in Saudi children. Stones which consist predominantly of uric acid, calcium oxalate, or one of the rare constituents are much more common than in the West; phosphatic calculi, on the other hand are much less common.

Diet

Diet histories taken from male Saudi stone formers and normals have shown that Saudis, in general, have a very high intake of animal protein and purine, an extremely high intake of oxalate, and a relatively low intake of calcium and of 'fibre'. This combination provides a very high oxalate/calcium ratio in the intestine (Table 1) and a high acid load to the body. Together these dietary abnormalities explain the more acid urine (acid-ash diet), hyperuricosuria (high purine intake), extensive mild hyperoxaluria (high oxalate/calcium ratio in the diet) and hypocitraturia (acid urine). Furthermore, hypercalciuria is rare because of the combination of the reported low vitamin-D levels and

Table 3. The proportion of calculi according to predominant mineral in adult stone formers from Saudi Arabia, the USA and the UK.

Predominant mineral	KSA	USA (%)	UK	KSA/UK* frequency ratio
Uric acid	14.6	10.1	4.0	5.8
Calcium oxalate	71.3	58.8	53.8	2.1
Calcium phosphate	7.6	20.3	30.9	0.4
MAP	3.7	9.3	9.6	0.6
Rare constituents	2.9	1.5	1.7	2.6

* Including the overall Saudi/UK prevalence ratio of 1.6:1. KSA = Kingdom of Saudi Arabia.

low calcium intake in the Saudi population. When these urinary abnormalities are combined with a tendency to a low urinary volume (because of mild dehydration due to the hot, dry climate), it is not difficult to understand the high prevalence of calcium oxalate and/or uric acid stones, and the relatively low occurrence of phosphatic stones in the region.

Mild Hyperoxaluria

We have previously described the high occurrence rate of mild hyperoxaluria (> 0.5 mmol/day) in the Saudi population and have largely attributed the high prevalence of stones in Saudi Arabia to this phenomenon. Further studies on the life-expectancy of stones in men in various countries, including Saudi Arabia, have shown that there is a strong relationship between the life-expectancy of stones in the male population of a given country and the average urinary oxalate excretion in normal men from the same country (Table 1). This relationship has recently been extended to include data from patients with various forms of gross hyperoxaluria. There is no equivalent positive relationship, however, between the life-expectancy of stones and the mean urinary excretion of calcium in the same populations.

There are also strong relationships between the life-expectancy of stones and (i) the oxalate/calcium ratio in the diet and (ii) the dietary intake of animal protein in the population (Table 1). The former accounts for the high prevalence of mild hyperoxaluria and stones containing calcium oxalate and the latter for the high prevalence of hyperuricosuria, a relatively acid urine and uric-acid-containing stones.

AN EPIDEMIOLOGICAL STUDY OF UPPER URINARY STONES IN KAIZUKA CITY

M. Iguchi, T. Umekawa, Y. Katayama, C. Takamura, K. Kohri and T. Kurita

Department of Urology
Kinki University School of Medicine
Osaka-Sayama
Osaka, Japan

INTRODUCTION

Urolithiasis is as common in Japan as it is in Europe and the United States. Since the Second World War when Japan experienced rapid economic growth and changes in dietary habits, the incidence of urolithiasis in Japan has been steadily increasing[1,5]. Four nation-wide surveys of urolithiasis have been coordinated by the Japanese Urological Association[2-5].

In these surveys, however, hospital statistics which cannot provide sufficient information regarding the true incidence and prevalence of urolithiasis in a population were utilized. There has been no survey of the prevalence of urolithiasis in a Japanese city utilizing postal questionnaires. In the present study in cooperation with the municipal government, we surveyed the prevalence of upper urinary-tract calculus among the residents of Kaizuka City.

MATERIALS AND METHODS

Kaizuka City is located in southern Osaka Prefecture and has a population of about 80,000 people. The subjects were 3,000 residents of Kaizuka City ranging in age from 20-59 years. They were randomly selected (male to female ratio, 1:1) from the census register in numbers consistent with the population distribution in each district. Only one member of a given family was selected. On January 15, 1992, the prospectus which included a questionnaire consisting of 47 items and a return envelope was sent to the 3,000 subjects described. On January 31, 1992, a letter was sent to all subjects to urge them to return the completed questionnaire. Statistical analysis was performed using the chi-square test.

RESULTS

As of March 31, 1992, completed questionnaires had been returned by 1,975 (65.8%) subjects; 1,972 (total, 65.7%; males: 62.9%; females: 68.1%) were considered valid. The mean age of the respondents (41.4±11.5 years [±SD]) was consistent with that of the entire group. Of the respondents, 7.0% had a history of urinary stones, and the prevalence among males (9.6%) was 2.1 times as high as that among females (4.5%). In all of these cases, the history of stones was confirmed by the individual's response that

Urolithiasis 2, Edited by R. Ryall *et al.*,
Plenum Press, New York, 1994

they had been diagnosed by a physician or by their observation of the spontaneous passage of stones. The prevalence of urolithiasis increased linearly with age, and that of respondents in their 50's was over 10% (Fig. 1). Extrapolation from these data suggests that over 10% of the general population (over 13% of males and over 7% of females) can be expected to suffer from urinary calculi at least once in their lifetime. On the other hand, the nation-wide surveys conducted in 1985 indicated that 5.4% of the general population were projected to suffer from urolithiasis at least once. Although the reason for the far higher prevalence in Kaizuka City than that nation-wide was not precisely clarified, it is thought that this difference was due to differences in the methods of epidemiological study used. We consider that the results of the present study more accurately reflect the true prevalence. About 85% of stone formers experienced the first stone episode in Kaizuka City. There was no difference between males and females in the district where the first stone was diagnosed. Over three-quarters of all stone formers were single stone formers, and the rate of recurrent stone formation in males was about twice that in females. The relative distribution of occupations of male Kaizuka residents from 20-59 years old was thought to be representative for a satellite city of Osaka. There was no difference between stone formers and non-stone formers in occupational status except that of an administrative worker which was twice as frequent in the stone formers. The prevalence of urinary stones among administrative workers was 19.6%, which was significantly higher than that in any other occupation (0-9.9%), albeit the mean age was slightly higher than that in other occupations.

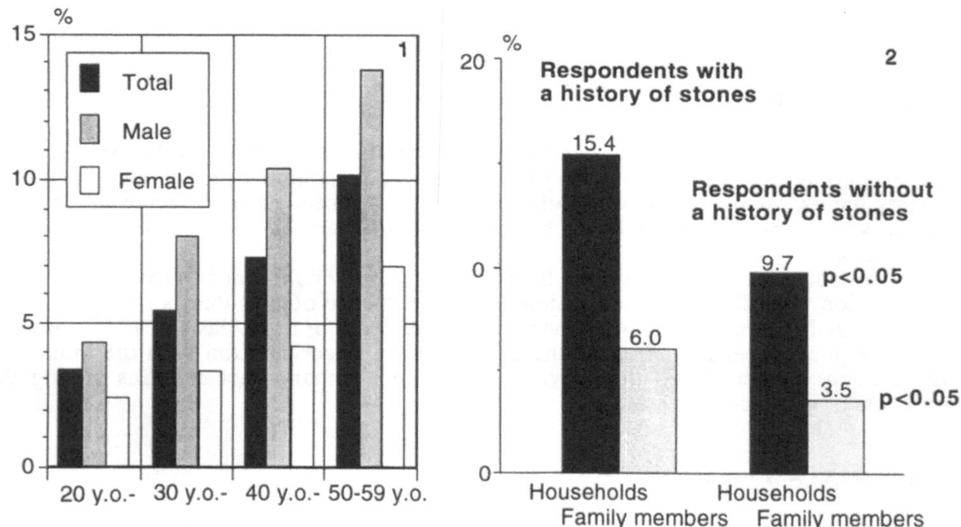

Figure 1.(left) The prevalence of urolithiasis in Kaizuka City according to age.

Figure 2.(right) The prevalence of other stone formers among the family members.

In the survey population of 7,568 subjects which included the respondents and their family members, the overall prevalence of urinary stones in Kaizuka City was 4.5%, and 16.2% of all families had at least one member with a history of stones. Among respondents who had a history of stones, 15.4% of the household had family members who also had a history of stones and 6.0% of family members had a history of stones. On the other hand, among respondents without a history of stones, 9.7% of the households had family members who had a history of stones and 3.5% of family members had a history of stones. There was a significant difference between the two groups (Fig. 2). In the families of stone former respondents, the individuals who had stones were frequently their blood relatives (72.7%).

The incidence of relatives with stones was lower in the families of non-stone former respondents than in those in the families of stone former respondents, and the individuals who had stones were frequently their spouses (46.2%). There was no

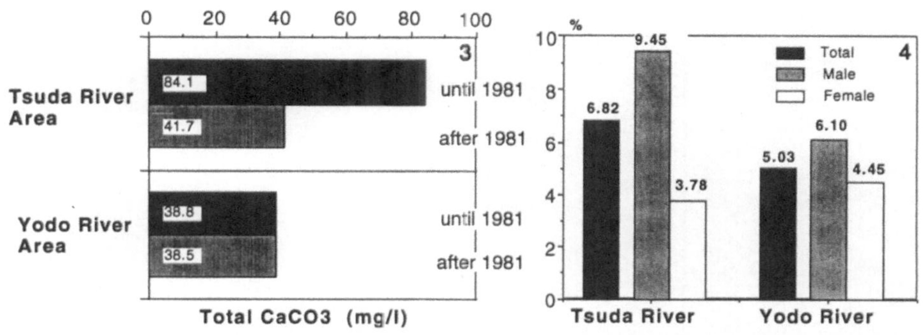

Figure 3. (left) Hardness of tap water. Until 1981, when lime was added during purification of the water from the Tsuda River, the tap water was harder than that from the Yodo River. After 1981, there was no difference between them in hardness.

Figure 4. (right) Prevalence of urolithiasis in the areas near the two rivers from where the tap water was obtained.

difference between the families of stone formers and the families of non-stone formers (including respondents and family members) in annual family income before taxes. Monthly food costs for families of stone formers were significantly greater than those in families of non-stone formers (includes respondents and family members) ($p < 0.05$).

In Kaizuka City, tap water comes from two rivers with different origins (the Yodo River and the Tsuda River). The hardness of tap water from these two rivers differed greatly until 1981 (Fig. 3). For this reason, we also analyzed the prevalence of urinary stones among the respondents living in the two districts differing in tap-water origin. Of all the respondents, 1,136 (including 68 stone formers) who had lived in the same district for more than 20 years and had experienced their first stone episode while residing in that district, were selected. In terms of age, occupation, income and food expenses, no significant difference was observed between the two groups. The prevalence of urolithiasis in the Tsuda River Area was higher than that in the Yodo River Area, although this difference was not statistically significant (Fig. 4). The prevalence of households with stone formers in the Tsuda River Area was higher than that in the Yodo River Area, although this difference was not significant (17.3% and 13.7%, respectively). The prevalence of stone forming family members in the Tsuda River Area was significantly higher than that in the Yodo River Area (5.1% vs 3.8%, respectively, $p < 0.05$).

In summary, hereditary predisposition and the composition of the tap water are thought to be related to the prevalence of urinary stone formation.

REFERENCES

1. M Iguchi, T Umekawa, Y Ishikawa, et al, Dietary intake and habits of Japanese renal stone patients, *J Urol* 143:1093 (1990).
2. T Inada, Statistical study on urolithiasis in Japan, *Acta Urol Jpn.*1: 143 (1955).
3. T Inada, Research on urolithiasis, *Jpn J Urol* 57: 917 (1966).
4. O Yoshida, Epidemiology of urolithiasis in Japan, *Jpn J Urol* 70: 975 (1979).
5. O Yoshida and Y Okada, Epidemiology of urolithiasis in Japan: a chronological and geographical study, *Urol Int* 45: 104 (1990).

EXPERIMENTAL BLADDER-STONE PRODUCTION IN RATS BY *UREAPLASMA UREALYTICUM*

Y. Arai[1], Y. Okada[1], T. Tomoyoshi[1] and H. Takeuchi[2]

Department of Urology
[1]Shiga University of Medical Science and
[2]Kyoto University
Shiga and Kyoto, Japan

INTRODUCTION

Infection stones are thought to develop in the urinary tract infected by urea-splitting bacteria such as *Proteus* species. However, urea-splitting bacteria are often not isolated from the urine or stones of the patients with infection stones. Micro-organisms not recovered by conventional bacterial cultures may be associated with stone formation.

Ureaplasma urealyticum is a unique mycoplasma which possesses urease. This organism is frequently isolated from patients with male infertility and non-gonococcal urethritis and, occasionally, from patients with urolithiasis[1-4]. There have been several experimental studies concerning urolithiasis and *U. urealyticum*[5-7]. In this study, we have attempted to clarify a possible relationship of struvite stone formation and *U. urealyticum in vivo*.

MATERIALS AND METHODS

Ureaplasma urealyticum (T-960) was incubated aerobically in Taylor-Robinson medium for 12 h at 37°C. Male Wistar rats weighing about ~200 g were used. Aliquots of 0.1 mL of *U. urealyticum* suspension containing 10^7 colour-changing units/mL were injected after surgically-implanting a zinc disc into the bladder of the rats. In the control animals, a zinc disc was inserted followed by injection of 0.1 mL of Taylor-Robinson medium without *U. urealyticum*. The animals were housed in stock cages and fed laboratory pellet chow and drinking water *ad libitum*. On days 2 and 6 following surgery, urine was obtained by forced urination for pH and culture of *U. urealyticum* using Taylor-Robinson medium. Half of the rats were sacrificed at 7 days, and the remainder at 14 days after surgery. Urinary tracts were carefully examined for gross pathological lesions. The dry weights of bladder stones, pH of urine, BUN, and serum creatinine were obtained, and bladder histology was performed.

RESULTS

A total of 46 rats, consisting of 10 rats in the control group and 36 rats inoculated with *U. urealyticum* were studied, after excluding those with wound infections and those having positive urine cultures at time of sacrifice.

Urolithiasis 2, Edited by R. Ryall *et al.*,
Plenum Press, New York, 1994

Serial changes in urine cultures for *U. urealyticum* on day 2 , day 6, and at the time of sacrifice are shown in Fig. 1. The urine culture from 10 of the 36 rats inoculated with U. *urealyticum* were negative on the second day following surgery. We concluded, therefore, that *U. urealyticum* infection was not established in these rats. Of the remaining 26 animals, on day 6, 12 rats were found to have negative urine cultures and 14 animals had positive ones. At the time of sacrifice, only six animals had positive urine cultures. This indicates a tendency towards spontaneous elimination of *U. urealyticum*.

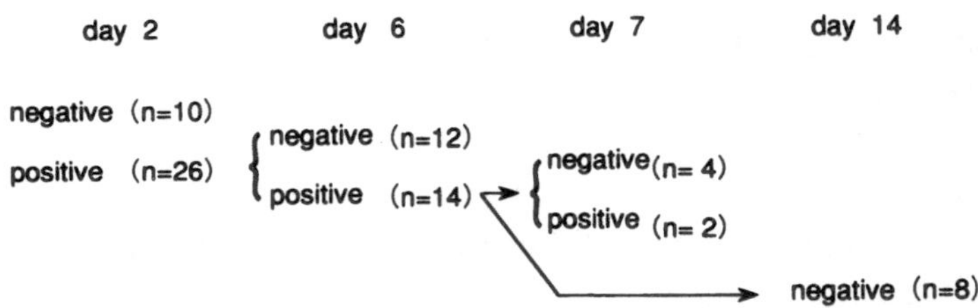

Figure 1. Serial changes in urine culture for *U. urealyticum*.

Serial changes in the urinary pH of the rats after inoculation of *U. urealyticum* are shown in Fig. 2. Compared to pre-operative urinary pH, urinary pH was not increased in association with *U. urealyticum* infection.

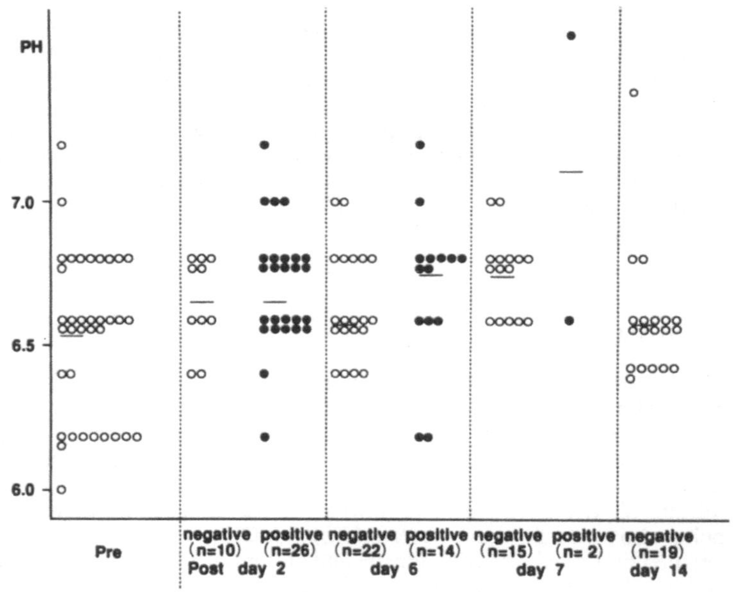

Figure 2. Serial changes in urine pH of rats after inoculation of *U. urealyticum*.

The animals inoculated with *U. urealyticum* were divided into three groups according to the results of urine culture of *U. urealyticum* (Fig. 3). In group 1 (n=10), *U. urealyticum* in the urine disappeared two days after surgery. In group 2 (n=12), the culture of *U. urealyticum* in urine was positive on day 2 but negative on day 6. In group 3 (n=14), the cultures were positive on day 2 and 6.

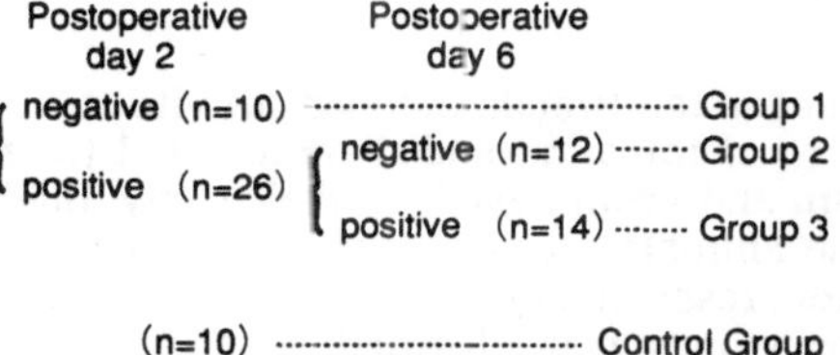

Figure 3. Grouping of rats according to the results of urine culture of *U. urealyticum.*

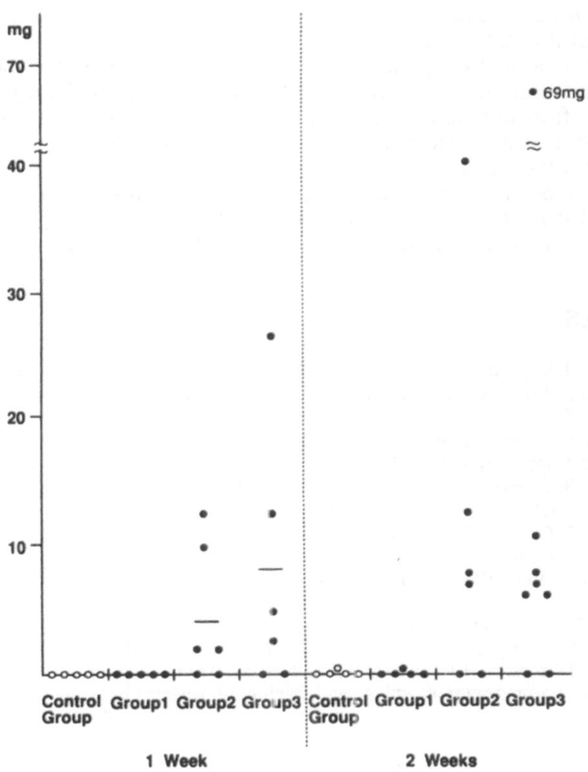

Figure 4. Weight of bladder stones according to the groups and day of sacrifice.

In the control group and group 1, no bladder stones were formed. In group 2, the mean stone weights were 4.5 mg and 11.2 mg, 7 and 14 days after surgery, respectively. In group 3, the mean weights were 8 mg and 13.4 mg, 7 and 14 days after surgery, respectively. However, in several rats in both group 2 and group 3, stones did not develop even in the presence of *U. urealyticum* infection. Crystallographic analysis revealed that the stones formed consisted of pure struvite.

Observation of the urinary tract of the rats at the time of sacrifice revealed dilation of the ureter in those rats having relatively large stones. However, no elevation in BUN or serum creatinine level was observed. In addition, no renal abscesses or renal stones were seen in any of the rats. Histological findings of the urinary bladder revealed slight inflammatory changes in the rats in which stones had formed.

DISCUSSION

In 1974, Friedlander and Braude induced struvite-stone formation in rats by *U. urea Plasma (T. mycoplasma)* infection and demonstrated for the first time an association between this organism and urolithiasis[5]. In their study, bladder stones were observed in 90% and 78%, of the animals six weeks after inoculation of *U. urealyticum* into the bladder and the kidney, respectively. In our study in which a zinc disc was inserted into the bladder, bladder stones were formed in 50% of the rats inoculated with *U. urealyticum,* and in 69% of the animals in which *U. urealyticum* infection was established. This difference may be ascribed to the difference in the strain of the rats or *U. urealyticum* employed, or in the duration of the rearing period.

Repeated culture and pH measurements of urine post-operatively revealed that a) *U. urealyticum* disappeared within ~one week after inoculation, which suggested a considerably weak virulence of the organism, and b) the urinary pH was not markedly elevated by *U. urealyticum* infection. This phenomenon may be related to the optimal pH for the urease of this organism.

We have evaluated bladder-stone formation by the same method using various bacteria. For example, bladder stones with a mean weight of 30 mg were observed in all rats one week after inoculation of *Proteus*. *U. urealyticum* was shown to be a possible cause of infection stones. However, its stone-forming ability is considered to be weaker than that of other bacterial species, including *Proteus*.

There have been several clinical reports regarding urolithiasis and *U. urealyticum*[3,4]. When urea-splitting bacteria are not isolated from patients with infection stones, *U. urealyticum* may be considered as a cause of the infection stone.

REFERENCES

1. WR Bowie, HM Pollock, PS Forsyth, JF Flord, ER Alexander, SP Wang and KK Holmes, Bacteriology of the urethra in normal men and men with non-gonococcal urethritis, *J Clin Microbiol* 6: 482 (1977).
2. CE Swenson, A Toth and WM O'Leary, *Ureaplasma urealyticum* and human infertility: the effect of antibiotic therapy on semen quality, *Fertil Steril* 31: 660 (1979).
3. H Hedelin, JE Brorson, L Grenabo and S Pettersson, *Ureaplasma urealyticum* and upper urinary tract stones, *Brit J Urol* 56: 244 (1983).
4. L Grenabo, G Claes, H Hedelin and S Pettersson, Rapidly recurrent renal calculi caused by *Ureaplasma urealyticum* : a case report, *J Urol* 135: 995 (1986).
5. M Friedlander and AI Braude, Production of bladder stones by human T Mycoplasmas, *Nature*, 247: 67 (1974).
6. S Takebe, A Numata and K Kobashi, Stone formation by *Ureaplasma urealyticum* in human urine and its prevention by urease inhibitors, *J Clin Microbiol* 20: 869 (1984).
7. L Grenabo, J Brorson, H Hedelin and S Pettersson, *Ureaplasma urealyticum* -induced crystallization of magnesium ammonium phosphate and calcium phosphates in synthetic urine, *J Urol* 132: 795 (1984).

COMPOSITION OF URINARY CALCULI IN THE SOUTH OF PORTUGAL

J.M. Reis-Santos

Department Of Urology
Hospital de Curry Cabral and
Faculty of Medical Sciences
New University of Lisbon
Lisbon, Portugal

INTRODUCTION

The incidence and composition of urinary stones vary throughout the world. There are marked differences between stones from inhabitants of under-developed countries and those from industrialized ones. Taking into account intrinsic and, above all, extrinsic factors responsible for stone formation, variations in composition can be expected. The objective of this study was to analyse this variation in southern Portugal.

MATERIALS AND METHODS

Between 1974-1982, (Group I), 1,320 upper urinary tract stones (UUS), were obtained from 664 male and 656 female patients > 20 years of age who had been living in Lisbon or in southern Portugal for more than five years[1]. These stones were obtained surgically (n=730) or following spontaneous passage (n=590). A further group of 1,235 UUS were collected from 1983-1991 (Group II). These were obtained from 650 male and 585 female patients following surgery (n=498), ESWL (n=258), or spontaneous passage (n=498). Stones were analysed using microscopy, spectroscopy, chemistry, and X-ray diffraction. Crystallographic analysis yielded qualitative and quantitative determination of the constituents[2-5].

RESULTS

The nucleus was identified in 55% of Group I stones and 63% of Group II stones. Apatite and calcium-oxalate (CaOx) monohydrate were the most common chemical components of nuclei in both groups of stones (Table 1). The stone most frequently found in both groups was tri-mineralic with a nucleus of apatite and subsequent layers of whewellite (WH) intergrown with weddellite (WHD). In Group I, a tri-mineralic stone was found in 32.8% of the patients, followed by a bi-mineralic struvite/apatite stone in 12.3% and WH/WHD stone in 11.2%. A mono-mineralic stone of WH was found in 5.8% of patients. In Group II, a tri-mineralic stone was found in 36.8% of patients, and a bi-mineralic stone in 26.6%. In both groups, the rarest stones were those with more than three phases.

Table 2 shows the percentage of stone types in Groups I and II.

Urolithiasis 2, Edited by R. Ryall *et al.*,
Plenum Press, New York, 1994

Table 1. Components identified as the nucleus in Group I (1974-82) and Group II (1983-91) stones in southern Portugal.

Components	Group 1 (n = 1,320)	Group II (n=1,235)
Apatite	462	553
Whewellite	210	164
Uric Acid	26	18
Brushite	20	31
Cystine	3	1
Ammonium acid urate	3	3
Others	0	4
Total	724	774

Table 2. Comparison and percentage of stone types in southern Portugal from 1974-82 (Group I) and 1983-91 (Group II).

Stone composition	Group I (%)	Group II
CaOx CaOx + Ca phosphate (CaP) CaP	64.5	82.4
Struvite (S) S + CaP S + CaP/Ox	14.0	3.2
Uric Acid (UA) UA + CaOx UA + CaOx/P	19.0	12.0
Cystine Cystine + CaP	0.9	0.6
Others	1.5	1.8

DISCUSSION AND CONCLUSIONS

Results from the comparison of stones from two groups of patients living in southern Portugal for more than five years (~70% of these in the Lisbon area) between 1974-1982 and 1983-1991 show that there has been a change in the composition of these stones between these time periods. Even when taking into consideration that patients sent to our Department are already pre-selected, the alterations in stone composition with time are highly significant. Stones containing CaOx and/or CaP increased from 64.5% to 82.4% which is close to percentages found in the industrialised countries of Northern Europe and the USA. However, whereas there is a predominance of these stones in male patients in the industrialised countries[6,7], we have found this type of stone equally distributed between the sexes, a fact which we cannot explain. However, a post-mortem survey carried out by Larsen and Phillips[8] also showed these stones equally distributed, irrespective of the sex of the patients. Again, a study by Scott[9] of the prevalence of UUS Ca stones showed both male and female patients equally likely to have this type of stone disease.

Stones with uric acid as the major component decreased from 19% to 11.8%. Small amounts of uric acid were found in the outer layers of other stones (WH + uric acid) but, as the quantity was so small, this was neglected. Struvite stones decreased dramatically in number. They amounted to 14% in Group I and only 3.2% in Group II. Again, struvite was found in the outer layers of some stones of different composition, but this quantity was negligible.

Two hypotheses can be put forward for this reduction in uric acid stones: (a) improved levels of health and hygiene throughout the country, and more comprehensive medical aid to potential uric acid stone formers; (b) alterations in eating habits resulting from socio-economic changes produced a decrease in the number of uric-acid stones, and consequent increase in CaOx and/or Ca-phosphate stones.

The dramatic decrease in the number of infection stones is, very probably, also related to better medical care of the population in general and, especially, of children and young and pregnant women[10]. Thus, the number of urinary infections which went undiagnosed and untreated in the past has diminished. However, it should be noted that the number of infection stones would be greater if we had included in this study stones with only a minimal quantity of struvite in the outer layers.

We can conclude that the pattern of the analysis of stones found in southern Portugal over the last eight years approximates that found in industrialised countries. However, if the reasons we have shown for this seem logical, we must not forget that other, more complex reasons, may be hidden by the multiple factors which influence the variation and composition of urinary stones. Furthermore, in dealing with symptomatic stone diseases, the sample is biased and, thus, takes no account of asymptomatic patients.

REFERENCES

1. JM Reis-Santos, "Composition and frequency of stone minerals in the South of Portugal", *in:* "Urinary Stone", R Ryall, JG Brockis, V Marshall, and B Finlayson (eds), Churchill Livingstone, Melbourne, (1984) pp 295-299.
2. Bl Otnes and D Montgomery, Method and reliability of crystallographic stone analysis. *Investigative Urol* 17:314 (1980).
3. EL Prien, Crystallographic analysis of urinary calculi. A 23 year survey study, *J Urol* 89:917 (1963).
4. DJ Sutor and SE Wooley, Composition of urinary calculi by x-ray diffraction, Collected data from various localities. Parts IX-XI, *Brit J Urol* 43:268 (1971).
5. DJ Sutor and SE Wooley, Composition of urinary calculi by x-ray diffraction. Collected data from various localities. Parts XII-XIV, Northern Ireland, South Africa and Kuwait, *Brit J Urol* 44:287 (1972).
6. A Nolin, B Lindell, P-O Granberg and N Lindvall, Urolithiasis - a study of its frequency, *Scand J Urol Nephrol* 10: 150, (1976).
7. WG Robertson, M Peacock, DH Baker et al, Studies on the prevalence and epidemiology of urinary stone disease in men in Leeds. *Brit J Urol.* 55: 595 (1983).
8. JF Larsen and J Phillips, Studies in the incidence of urolithiasis. *Urol Int* 13: 53 (1962).
9. R Scott, Prevalence of calcified upper urinary tract stone disease in a random population survey, *Brit J Urol* 59:117 (1987).
10. Al Shabad, On the increase of the percentage of females among patients with infection and calculus diseases and its possible causes, *Int Urol Nephrol* 7:179 (1975).

THE INFLUENCE OF DIET ON URINARY RISK FACTORS FOR STONES IN IDIOPATHIC CALCIUM STONE FORMERS AND HEALTHY SUBJECTS

Y. Berland[1], F. Leonetti[1], X. Thirion[2], P. Viot[2], J.P. Giordanella[2], B. Dussol[1] and R. Sambuc[2]

[1]Hôpital Ste-Marguerite and
[2]Département d'informations Médicales
Secteur Sud
Marseille, France

INTRODUCTION

The dietary intake and urinary solute excretion were evaluated in 49 idiopathic stone formers (SF) and 210 healthy subjects living in the south of France from June, 1990 to September, 1991. The SF consisted of 35 males and 14 females, age 45.6±14 (±SD) years. The healthy subjects consisted of 96 males and 114 females, age 42.4±10 years. The ratio of male/female was greater in the SF. Nutrients and calories were calculated by means of food-composition tables. Twenty-four h urine samples were collected from all subjects and analyzed for urinary excretion of sodium, calcium, phosphorus, uric acid, citrate, magnesium, and oxalate.

RESULTS

Dietary intake was similar in the two populations when corrected for sex. Hypercalciuria (urinary calcium >0.1 mmol/kg/24 h) was found in 45% of SF and 24% of healthy subjects. Urinary excretions of calcium, oxalate and uric acid were greater in SF than in healthy subjects ($p< 0.05$). Protein intake, evaluated by urine urea nitrogen, was also greater in SF than in healthy subjects ($p< 0.05$). A significant correlation was found between urinary calcium and urinary nitrogen, both in SF ($p< 0.01$) and in healthy subjects ($p< 0.001$). In contrast, a significant correlation between urinary calcium and sodium was found only in healthy subjects ($p< 0.001$).

DISCUSSION

The lack of a correlation between urinary calcium and sodium in SF supports the hypothesis that urinary calcium in idiopathic SF originates, in part, from sources other than diet or defective renal tubule reabsorption of calcium, such as may occur from increased bone resorption.

THE DRINKING WATER LINK BETWEEN ENDEMIC RENAL STONES, FLUOROSIS AND GOITRE: AN EPIDEMIOLOGICAL STUDY

M. Teotia, S.P.S. Teotia and R. Nath

PG Department of Human Metabolism and Endocrinology
LLRM Medical College
Meerut, India

INTRODUCTION AND METHODS

Endemic renal stone disease, fluorosis and goitre constitute major environmental health problems in India. An epidemiological study was undertaken to study the etiological relationship of these endemic diseases to the chemical composition of drinking water and the nutritional status of the community.

The total population studied and examined in detail included 25,800 subjects (males: n=13,002; females: n=12,798). Of these individuals, 9,200 were endemic for stone disease, 8,500 for fluorosis and 8,100 for goitre. A total of 4,367 samples of drinking water consumed by the population were analysed for Ca, Mg, fluoride, iodine, hardness, acidity, and alkalinity. The nutrient intakes from drinking water and food were estimated separately in each individual studied.

RESULTS

In the community endemic for stone disease, the prevalence of goitre was sporadic and fluorosis was practically non-existent. In the community endemic for fluorosis, goitre and stone disease were non-existent. In the areas endemic for goitre, fluorosis was non-existent and stone disease was sporadic. A positive correlation existed between the occurrence of stone disease, water hardness and its Ca content.

The mean contribution of water in daily Ca intake ranged from 200-400 mg/24 h. Multiple stepwise regression with hardness as the dependent variable and Ca, Mg, iodine, fluoride, and alkalinity as independent variables showed Ca positively correlating with hardness and iodine.

DISCUSSION

This study has confirmed the role of nutrition and the chemical composition of drinking water as important environmental factors in the etiology and epidemiology of renal stone disease, goitre, and fluorosis.

This is a major breakthrough in the environmental inter-relationship of these endemic disorders, and has important implications for national programmes aimed at their control and eradication.

OXALATE, GLYCOLATE AND GLYOXYLATE EXCRETION IN SAUDI STONE FORMERS AND NORMAL CONTROLS

H. Hughes, I. Al-Duraibi, D.M. Feuchuk and W.G. Robertson

King Faisal Specialist Hospital and Research Centre
Department of Biological and Medical Research
Riyadh, Saudi Arabia

INTRODUCTION AND RESULTS

The cause of the mild hyperoxaluria associated with urolithiasis in the Arabian Peninsula was investigated by measuring glycolate, glyoxylate and oxalate in urine samples from Saudi stone formers and normals. Elevation in the excretion of all compounds would indicate a metabolic abnormality as the major cause of the mild hyperoxaluria whereas elevation of urinary oxalate alone would implicate a dietary factor. Oxalate and glyoxylate were each estimated by reverse phase high performance liquid chromatography (HPLC) (Waters Chromatography Division of Millipore). Glycolate was estimated using HP anion-exchange chromatography (Dionex Ltd). Oxalate was determined as 2,3-dihydroxyquinoxaline derived by condensation with 1,2-diaminobenzene; glyoxylate was estimated as 6,7-methylenedioxy-2-hydroxyquinoxaline obtained by condensation with 1,2-diamino-4,5-methylenedioxybenzene. Glycolate was estimated conductimetrically by HP anion exchange chromatography. The results are shown in the table.

Table. Urinary oxalate, glycolate and glyoxylate excretion/24 h.

	Saudi Stone formers mean ± SD (n)	Saudi Non-stone formers mean ± SD (n)	Expatriates mean ± SD (n)
Oxalate (mmol/day)	0.69±0.19 (70)	0.53±0.15 (36)	0.36±0.10 (17)
Glycolate (mmol/day)	0.77±0.39 (25)	0.60±0.37 (12)	0.61±0.24 (20)
Glyoxylate (µmol/day)	17.70±18.9 (18)	9.60±6.0 (18)	nd

Specificity was confirmed by a wavelength-ratio technique for oxalate and glyoxylate and by an enzymatic technique for glycolate. The results indicated that the cause of the mild hyperoxaluria in Saudis is not a metabolic abnormality and is most probably related to a high dietary intake of oxalate.

CALCIUM OXALATE STONE DISEASE IN CHILDREN

M.F. Netzer, K. Spelsberg and T.J. Davies

Urology Clinic
University of Saarland
6650 Homburg/ Saar
Germany

INTRODUCTION AND RESULTS

Calcium oxalate nephrolithiasis is a rare disease in children and only a few of these cases have a known etiology. After initial treatment most of these young patients require long term follow-up to remain stone-free. The examination of renal metabolism is necessary with the first stone episode to detect the presence of enzymatic disorders contributing to stone disease. Most children with stone disease do not have obvious clinical features that explain their stone formation. In this study we excluded children with primary oxalosis, renal tubular acidosis or anatomical abnormalities to determine a possible metabolic reason for their stone disease.

We collected 24 h urine and serum specimens from 34 children with calcium oxalate nephrolithiasis and 15 stone-free children. The measurements for serum and urine levels of the common electrolytes and metabolites were performed. Additional parameters determined were oxalic acid, cystine and citrate in urine. Parathyroid hormone screening was also done.

Parameter	Non-Stone Former	Stone Former
24 h urine(/g creatinine)		
Calcium (mmol)	3.65± 0.90	5.96±3.06*
Oxalate (mmol)	0.23± 0.11	0.28±0.16
Citrate (mg)	772.00±167	812.00±267
Phosphate (mmol!	29.30±14.1	35.50±11.9

* significant differences. There were no significant differences in the serum parameters.

The only significant difference between healthy children and stone formers was a slight elevation in calcium excretion. These elevated values were not high enough to be classified as hypercalciuria compared to published data. We conclude that in the absence of disorders based on standards of urinary excretion of metabolites, other factors such as urinary inhibitors may be critical to stone formation in the pediatric population.

HYPOKALIURIC - HYPOCITRATURIC NEPHROLITHIASIS: A UNIQUE CLINICAL ENTITY OF NORTHEAST THAILAND

K. Tungsanga, P. Tosukhowong, P. Sriboonlue, V. Sridama, V. Prasongwatana and V. Sitprija

Departments of Medicine and Biochemistry
Chulalongkorn U Hospital
Bangkok and
Department of Biochemistry
Faculty of Medicine
Khon Kaen U
Khon Kaen, Thailand

INTRODUCTION AND RESULTS

The pathogenesis of nephrolithiasis in Northeast Thailand (NET) remains unknown. We prospectively studied 15 male healthy controls (G1) who lived in Khon Kaen City and 40 male renal stone formers (G2) who lived in rural villages around Khon Kaen City, NET. Their ages were 33±2 and 35±1 years respectively. No significant differences for serum creatinine, Na, Cl, HCO_3, Ca, P, Mg and uric acid between G1 and G2 were observed. The serum potassium in G1 was significantly lower than G2 (4.1±0.1 vs. 3.7±0.1 mmol/L, $p < 0.001$). The 24 h urine volume, pH, Ca, Mg, urate and oxalate did not differ significantly between G1 and G2. However, daily urinary sodium (153±13 vs. 86±8 mmol), urinary potassium (30±3 vs. 19±1 mmol) and urinary citrate (969±13 vs. 272±47 μmol) excretions were significantly greater in G1 than G2 ($p < 0.001$). Hypokaliuric hypocitratua (HKHC) was found in more than 80% of G2 cases. Urinary potassium correlated significantly with urinary citrate (r=0.552, p=0.0002). After oral supplementation with either potassium citrate or potassium chloride (n=27), the serum and urinary potassium increased remarkably ($p < 0.001$). An abnormal response to oral calcium loading test (CLT) compatible with renal and absorptive hypercalciuria (AH) was found in none and 3 cases of G1, and in 2 and 16 cases of G2, respectively.

Serum parathyroid hormone and urinary cAMP before and after CLT were normal in both groups. During ammonium chloride loading test (n=33), the urinary ammonium excretion on the 3rd day of acid load increased normally in all but one case. Direct food analysis of G1 (n=10) and G2 (n=10) revealed K intake of 32±5 and 25±4 mmol/day, respectively ($p < 0.01$). Sweat loss of K during regular daily activities in G1 (n=10) and G2 (n=10) were 3±1 and 12±1 mmol/day, respectively ($p < 0.01$).

CONCLUSION

We conclude that HKHC in association with AH is common among nephrolithiasis patients from NET. The data suggest that it s due to K depletion secondary to low intake and excessive sweat loss rather than to a renal tubular acidification defect or abnormal parathyroid function.

THE DISSOLUTION OF INFECTIVE STONE SALTS BY CITRATE SOLUTIONS USING A REPRODUCIBLE MODEL OF CATHETER ENCRUSTATION

W. Schmitz[1], G. Marklein[2] and A. Hesse[1]

[1]Division of Experimental Urology
Department of Urology
[2]Department of Medical Microbiology
University of Bonn, Germany

INTRODUCTION AND RESULTS

Struvite (MAP) and calcium phosphate (CaP) account for approximately 5% of all urinary stones. They are often encrusted on catheters, in association with nosocomial urinary tract infections, and often cause reinfection and functional disorders of the catheter.

Infection stone material was produced for 30 hours in a standardized '*in vitro* bladder model', using a sterile artificial urine supplemented with a nutrient solution and *Proteus mirabilis*, and employing physiological conditions (temperature, urinary flow rate, residual volume, flow direction). Urines were analysed for bacterial concentration and the pH value recorded. All experimental surfaces involved in the model were analysed by Infrared-Spectroscopy (IR) and quantitative chemical analysis (CH). The solutions containing citrate (Suby-G Solution and Solution-R) were tested for their ability to disintegrate encrusted material.

The concentration of bacteria and the pH increased significantly during the experiment (from 10^7 to 10^9 cfu/mL and from pH 6.3 to 9.0, respectively) and caused supersaturation with respect to infection stone material (MAP and CaP). Analysis of the encrustation showed that MAP comprised between 46.3 and 80.0% and CaP, between 11.8 and 50.2%, depending upon the nature of the experimental surface material used. Approximately 700mg of precipitated salt was found on each catheter and CH-analysis confirmed the percentages of MAP and CaP found from IR analysis. The solutions tested dissolved 70% (Suby-G) and 85% respectively (Solution-R) of the MAP. After irrigation, the hydrodynamic properties of the catheters were comparable to unused ones (98%), whereas catheters which had not been treated with citrate lost 45% of their drainage capacity.

CONCLUSIONS

The model presented is suitable for standardizing the formation of encrustation material, since it simulates most conditions occurring *in vivo*. It is therefore a useful tool to examine various factors affecting the precipitation of infective crystalline material on to urinary catheters. Furthermore, the two solutions tested may be useful in preventing such encrustation since they both efficiently dissolved struvite and calcium phosphate.

LABORATORY ANALYSIS AND CLINICAL RESULTS OF THE TREATMENT OF STRUVITE STONES: A TEN-YEAR STUDY

K. Jarrar[1], R.H. Boedecker[2] and W. Achilles[3]

[1]Department of Urology
[2]Institute of Medical Information
University of Giessen
[3]Department of Urology
University of Marburg, Germany

INTRODUCTION AND RESULTS

This study presents the results of a 10-year metaphylaxis of 19 former struvite stone formers, each having had two to three stone operations. In these patients, urine was acidified with L-methionine (Acimethin") using a dose of 1.5 to 3.0 gm per day. Every three months, 11 laboratory parameters relevant to stone formation were checked in 24 h urine collections. Six parameters were determined in serum, and urine samples were tested for infection.

Recurrent stone formation occurred in only two of the 19 patients. Another patient had three occasional infections. Statistical analysis of analytical data, which was supported by computer graphs, provided the following results for urinary parameters. They were described by geometric means and 95% tolerance intervals using ANOVA (analysis of variance). During therapy over 10 years, mean pH value decreased from 7.5 to 5.5 ($p< 0.001$). Significant increases were found in the excretion of citrate ($p< 0.004$), magnesium ($p< 0.001$), potassium ($p< 0.04$) and uric acid ($p< 0.04$). A tendency of increase was found for Ca, which, however, could not be confirmed to be statistically significant ($p< 0.08$).

In serum, changes of parameters could only be registered for Ca and phosphate. The mean value of Ca dropped during the first two to four years and then rose towards its final value ($p< 0.01$). However, at all times, total serum concentration stayed within its normal limits (2.3 -2.5 mmol/L). Serum phosphate dropped significantly from 1.1 to 0.8 mmol/L ($p< 0.001$). No relevant changes could be observed for serum creatinine, sodium, potassium and uric acid. In assessing the efficacy of L-methionine therapy, the drop of urinary pH to acid values is the most relevant factor.

DISCUSSION

The following essential points should be regarded in the metaphylaxis of infection stone formation: (1) complete removal of the stone; (2) removal of possible obstructions in the urinary tract; (3) specific antibacterial medication to inhibit urease; (4) acidification of urine and (5) inhibition of the resorption of phosphate.

COMPARATIVE METHODS FOR URINE CULTURE COLLECTION: STERILIZED GLASS TUBE VERSUS DISPOSABLE PLASTIC CUP

C.G. Novoa[1], S.T. Coelho[1], I.P. Heilberg[1], N.F. Yanashiro[2], A.B. Pereira[1] and N. Schor[1]

[1]Nephrology Division
[2] Central Laboratory Escola Paulista de Medicina
São Paulo, Brazil

INTRODUCTION AND RESULTS

Infection of the urinary tract[1-4] is common in our area and is usually more frequent in women. Urine collection for urine cultures in sterilized glass tubes involves difficulties in obtaining samples from women as well as having high costs due to the reutilization process.

In order to study the viability of collecting urine samples in disposable cups, paired collections of urine were taken from 128 adults in disposable cups and in test glass tubes under standardized aseptic conditions. In the first group 59 samples were sent to the laboratory for culture in cups sterilized with ethylene oxide and glass tubes sterilized in autoclave, and in the second group 69 paired samples were sent for culture in non sterile cups and sterile glass tubes under similar conditions.

In the first group, 38 samples had negative cultures, 15 were positive and five were contaminated in both tubes and cups. Only one sample was contaminated in the cup and not in the tube. Bacterial growth occurred to a similar extent in both groups in 67% of the samples, 20% grew more in the cup group and 13% grew more in the glass tube group. The specificity and sensitivity of the method was 100%. No significant difference was observed between cup and tube. In the second group, 22 samples were positive, 38 negative and seven contaminated in cup and tube. Only 2 samples were negative in cup and positive in tube. In 59% of cases the number of isolated bacterial colonies was similar in tube and cup; in 37% it was greater in tube and 5% greater in cup. Specificity was 95% and sensitivity 100%. No significant difference was observed between cup and tube in this group. Additionally, no difference was found between sterilized vs non-sterilized cups.

Similar results were observed for the urine cultures obtained in the tubes and in the cups. The data suggest that the cup is preferable for collecting urine samples because it is easy to use and has a low cost and thus, can be employed on a large scale.

REFERENCES

1. MG Bergeron and Y Marois, Benefit from high intrarenal levels of gentamicin in treatment of E.coli pyelonephrits. *Kidney Int.* 30:481, (1984).
2. PPS Earp, Infecsao urinaria, *JBM*, 57:36, (1989).
3. CM Kunin, Detection, Prevention and management of urinary tract, 4tr ed, Philadelphia, Lea & Feibiger, (1987).
4. M Turck, Urinary tract infections. *Hospital Practice*, 1:49 (1980).

RESULTS OF FAMILY STUDIES IN CYSTINE CALCULUS FOLLOW-UP

P. Brundig, H-J. Schneider, I. Steinhauser, U. Grimm and H. Christinck

Sonnebergerstrasse 2
D-2800 Bremen, Germany

INTRODUCTION

Family studies should be an indispensible part of the clinical investigation of patients with cystine stones.

METHODS

In a cystine calculus follow-up, an attempt was made to define pheno- and geno-types in 26 cystine-calculus patients and their family members by analysing amino acids in the urine.

RESULTS

Only in eight families could the familial genetic history be partly or completely elucidated. The results of quantitive determinations of cystine excretion determination by traditional analysis were in good agreement with those from amino acid determinations using chromatography.

CONCLUSION

This shows that useful information regarding familial cystinuria can be obtained using practical relevant approaches.

URINARY TRACT INFECTION IN IDIOPATHIC CALCIUM NEPHROLITHIASIS

G. Vagelli, G. Calabrese, V. Ferraris, A. Mazzotta, G. Pratesi, G. Buffa and M. Gonella

Service of Nephrology and
Division of Urology
Hospital of Casale
Monferrato, Italy

INTRODUCTION AND METHODS

Urinary tract infection (UTI) is well known to be a primary factor in the pathogenesis of struvite stones[1], yet little data exist on the effect of UTI on idiopathic calcium nephrolithiasis (ICN)[2]. The aim of this study was to evaluate the prevalence of UTI among patients with ICN and to determine the effect of UTI on the urinary risk factors in ICN.

In the last 5 years, 169 patients, 123 males (73%) and 46 females (27%), underwent metabolic screening for ICN in our Unit. Specimens for urine culture (midstream urine) were collected at the time of first examination and then at six-month intervals. The following clinical and metabolic parameters were evaluated: age, stone/year, urinary creatinine, calcium, calcium/creatinine, phosphate, uric acid, magnesium, oxalate, sodium, and potassium. The parameters were compared in patients (according to sex) with UTI (at least one episode) and without UTI using Student's *t* test.

RESULTS AND CONCLUSIONS

Of the 169 patients, 31 (18%) experienced at least one episode of UTI (males 11%; females 37%). The micro-organisms found in the cultures were: *E. Coli* 30%, *Staphylococci* 16%, *Proteus* 16%, *Klebsiella* 12%, *Enterococci* 12%, *Pseudomonas* 9% and others 5%. There were no differences in the clinical and metabolic parameters between patients with or without UTI.

These data confirm that ICN predisposes patients to UTI. Females with ICN are more prone than males to experience UTI. The clinical and metabolic parameters in ICN do not seem to be influenced by UTI.

REFERENCES

1. SP Lerner, MJ Gleeson, DP Griffith, Infection stones, *J Urol.* 141: 753 (1989).
2. K Holmgren, B Fellstrom, BG Danielson, S Ljunghall, F Niklasson, Stone analysis and urinary tract infection in renal stone patients, *in* "Urolithiasis and Related Clinical Research", PO Schwille, LH Smith, WG Robertson, W Vahlensieck, (eds), Plenum Press, New York pp. 399-402 (1985).

RISK OF UROLITHIASIS IN GULF-RETURNED KERALITES

S.V. Roshni, R.K. Vathsala, H.K. Moorthy, N.E. Thomas, C. Aravindakshan and Y.M. Fazil Marickar

Department of Surgery
Medical College Hospital
Trivandrum, India

INTRODUCTION AND METHODS

In India, the incidence of stone disease is believed to be less in the south. In the southern state Kerala, however, the incidence has been rising, especially in those individuals returning from the Gulf. The aim of this paper was to determine possible biochemical changes in these patients (Gulf SF) compared to the native patients.

Seventy-five Keralites who had been living in the Gulf countries for more than one year and who were attending the stone clinic were included in the study. An equal number of native Keralites who attended the stone clinic during the same period were taken as controls. On the first visit, 24 h urine Ca, oxalate (Ox), uric acid (UA), and citrate, and serum Ca and UA were estimated. Statistical analysis was performed using the Student's *t* test for unpaired data. *In vitro* CaOx crystal-growth studies were also performed, using the urine samples from the two groups of patients.

RESULTS AND DISCUSSION

Forty-nine per cent of the Gulf SF had some biochemical abnormality in urine or serum, alone or in combination, compared to only 39% of the native patients. The mean urinary Ox values in the Gulf SF and native patients were 80 and 69 mg/day, respectively ($p < 0.01$). The urinary UA levels (708 and 650 mg/day, respectively) and Ca levels (205 and 221 mg/day, respectively) were not different. The serum Ca and UA levels also did not differ markedly in the two groups. *In vitro* studies showed a significant lack of inhibition of CaOx crystal in urine samples from Gulf SF compared to urine samples from native patients.

It is clear, therefore, that the various risk factors of urinary stone disease were higher in Gulf SF than in the native population. This may be due to dietary habits in the Middle East countries which include a protein-rich diet. This results in an acid load which lowers urinary pH, citrate and inhibitors of stone formation, and raises urinary Ca, Ox and UA levels. Furthermore, high temperatures in the Gulf decrease urine volume. Increased exposure to sunlight may be another risk factor by increasing levels of vitamin D with subsequent enhanced Ca absorption and urinary Ca excretion and, thus, a further risk for Ca stones. The higher *per capita* income of Gulf SF may also contribute to the factors increasing risk of urinary stone disease.

THE BRAZILIAN MULTICENTRIC STUDY OF NEPHROLITHIASIS (MULTILIT)

M.S. Ancao, C.G. Novoa, S.T.S.N. Coelho,
S.M. Laranja, D. Sigulem, I.P. Heilberg and N. Schor

Nephrology Division
Escola Paulista de Medicina
São Paulo, Brazil

INTRODUCTION AND METHODS

The prevalence of nephrolithiasis in Brazil is about 5%[1] with a recurrency rate of 80%-90%[2]. Metabolic evaluation and clinical treatment can reduce this recurrency rate and morbidity. In Brazil, there are no statistical population data regarding nephrolithiasis. In order to evaluate some aspects of this disease, a national cooperative study was performed.

Nineteen clinical-nephrological centers participated in this study. The data entry sheets with 282 fields were designed to allow the participation of centers with different research facilities. Epi-Info v.5 for IBM-compatible microcomputers was chosen to create a database and to perform the statistics. The collected information was divided in 5 levels: (a) characteristics of the study population, clinical manifestations and dietary intake data, (b) urinalysis and image-diagnostic procedures, (c) calcium, uric acid, and creatinine concentrations in 24 h urine samples, (d) oral calcium load-test and (e) specific laboratory tests including net excretion of titrable acid, citrate, oxalate, phosphate, magnesium, urinary cyclic AMP, and serum parathyroid hormone.

RESULTS AND CONCLUSION

Although all centers are able to participate with the initial metabolic diagnosis (categories a-c), and most (63%) can perform specific laboratory studies, until now only 32% have effectively participated. In 249 cases evaluated, it was found that: there were no sex differences; 80% of patients were white; 60% had passed 1-10 calculi; 83% worked in normal environmental temperatures; and 86% were used to mild or regular physical activity. Thus, in Brazil, a national co-operative nephrolithiasis study is feasible and can yield preventive aspects of stone disease.

REFERENCES

1. MI Resnick, CYC Pak, *in* "Urolithiasis: A medical and surgical reference", WB Saunders Company, Philadelphia (1990).
2. FL Coe, J Keck, and ER Norton, The natural history of calcium urolithiasis. *JAMA* 238:1519 (1977).

CHARACTERIZATION OF A REFERRED POPULATION TO A NORTHERN-CALIFORNIA STONE CLINIC

M.E. Moran, H.E. Williams, P.A. Davis,
D.W. Bowyer, T.R. Wandzilak

University of California
Davis, Davis CA

INTRODUCTION AND METHODS

This study was undertaken to characterize a cohort of patients from northern California referred to our clinic for recurrent renal-stone disease. Patient data were obtained by history, analysis of serum, a 24 h urine collection, and from a 4 h urine collection following a 1 g oral Ca-load test. A total of 95 patients (males: n=38, mean age, 43 years; females: n=57, mean age, 48 years) were investigated.

RESULTS

Our study showed: (a) The frequency of stone types did not appear to differ significantly from that previously reported in the literature, except that the percentage of struvite stones declined; (b) There were no statistically significant differences in stone composition with respect to age or sex; (c) The frequency of hypercalciuria in the patient cohort differed markedly when this was diagnosed on the basis of a 24 h urine Ca (>300 mg) (17%), or when made on the basis of a post Ca-load urinary Ca/creatinine ratio (> 0.2) (44%); (d) Using the Ca/Creatinine ratio as a criterion for hypercalciuria, 20% of patients had no identifiable metabolic abnormality, 44% of patients had a single metabolic abnormality (that is, hypercalciuria, hyperuricosuria, hypocitraturia, or hyperoxaluria), and 36% had mixed metabolic abnormalities. Using 24 h urinary Ca as a criterion, 31% of patients had no identifiable metabolic abnormality, 51% of patients had a single metabolic abnormality, and 19% had mixed metabolic abnormalities; (e) When patient diagnosis was analyzed according to body mass index (BMI, wt/ht^2), there were statistically significant differences in the distribution of diagnoses between the normal weight, overweight and obese patients.

CONCLUSION

The results of this study suggest that the patient population in northern California differs from the cohorts which have been previously described.

RECURRENCE OF UPPER-URINARY-TRACT STONES IN NORTHEAST THAILAND

S. Borwornpadungkitti[1], P. Sriboonlue[2] and K. Tungsanga[3]

[1]Department of Surgery
Khon Kaen General Hospital
[2]Department of Biochemistry
Faculty of Medicine, University of Khon Kaen
Khon Kaen and
[3]Department of Medicine
Chulalongkorn University Hospital
Bangkok, Thailand

INTRODUCTION AND METHODS

Upper urinary tract stone disease (UUTS) is common in rural communities of northeast Thailand. In order to define rate of recurrence of UUTS, we conducted a prospective study of 127 consecutive patients having surgery for UUTS at Khon Kaen General Hospital. Fourteen cases were withdrawn because of inadequate or no follow-up data or because of nephrectomy after uncontrolled post-operative bleeding (2 patients). There were 115 cases eligible for study consisting of 83 males and 32 females age 35±10 (±SD) years.

RESULTS AND CONCLUSION

Pre-operative evaluation of these patients revealed abnormal intravenous pyelograms in 60% of cases (which consisted of impaired contrast excretion or visualization, hydronephrosis or small-sized kidney), significant urinary tract infection in 26%, renal failure in 15%, hypokalemia in 12%, and bilateral renal stones in 35%. Retained stones were found in 56% of patients post-operatively. Patients were followed at 6, 12, 18, and 24 months thereafter. Eight deaths occurred post-operatively or during follow-up because of uncontrolled urinary sepsis, cerebral embolism, uremia, and probable sudden unexplained-death syndrome. Overall, the mortality rate was 7%. Stone recurrence rates were 23% and 35% at 12 and 24 months, respectively. At 24 months, seven patients who were initially azotemic showed improved renal function, whereas five deteriorated. Recurrence of stones at 24 months was associated with the presence of initial urinary tract infection ($p< 0.05$) or initial azotemia ($p< 0.02$), and with presence of pyuria (>10 cells/HPF, $p< 0.001$) or proteinuria (+3 or +4, $p< 0.001$) at six months.

We conclude that UUTS patients in northeast Thailand have rates of stone recurrence of 23% and 35% at 12 and 24 months. The risk of recurrence was greater in those patients who had an initial urinary infection or renal failure, and pyuria or proteinuria at six months.

ETIOGENIC FACTORS FOR CALCIUM OXALATE LITHIASIS IN SOUTHWESTERN MACEDONIA

P. Ilievski, S. Ilievska, R. Nakovski, V. Jankovski and P. Krstev

Department of Urology
Medical Center
Bitola, Macedonia

INTRODUCTION AND RESULTS

Macedonia is located in the central region of the Balkan Peninsula and geographically belongs to an area that has a high incidence and prevalence of urolithiasis. Our study included analysis of 100 stones removed by surgery, and biochemical investigation of 60 stone formers (36 males and 24 females) and 20 volunteers (10 males and 10 females) from this region. Blood samples were taken for analysis of Na, K, Ca, Mg, Cl, phosphate and creatinine to exclude metabolic and endocrine causes of stone diseases, and a routine urinalysis was performed to exclude urinary tract infection. Twenty-four h urine samples were then collected and analysed.

Stone analysis in our area indicated that calcium oxalate (CaOx) was the main constituent in 51% of patients in the form of monohydrate and dihydrate, and 44% of the CaOx stones included the presence of amonium hydrogen urate, hydroxy carbonate, and Ca phosphate.

In 24 h urine samples in the stone formers, K and phosphate were normal in both sexes. However, increases in urinary Na (78% of males and 75% of females) Cl (64% of males and 50% of females) were found. In regard to Ca, 17% of males and 29% of females were hypercalciuric. A decreased urinary Mg excretion was found in 50% of males and 46% of females. Hyperoxaluria was found in 60% of all subjects. In normal subjects, an increased urinary excretion was found in 100% of males and 80% of females, and an increased urinary Cl in 80% of males and 70% of females. In addition in this population, 30% of males and 40% of females were hypomagnesuric.

DISCUSSION

CaOx stone disease in southwestern Macedonia is associated with increased urinary excretion of oxalate, Na, and Cl, and a low incidence of hypercalciuria. These findings are most likely secondary to ingestion of food and drinks rich in oxalate and Na, both easily absorbed from the intestine where there are deficiencies of Ca and Mg.

SECTION VIII: ESWL AND LITHOTRIPSY

ULTRASOUND VELOCITY - A MEASURE OF STONE STRENGTH?

N.P. Cohen[1], H.N. Whitfield[1], J.C. Shelton[2] and G.P. Evans[2]

[1]Department of Urology
St. Bartholomew's Hospital
West Smithfield
London, UK
[2]Interdisciplinary Research Centre
Queen Mary and Westfield College
University of London
London, UK

INTRODUCTION

Urinary stones exhibit a variable response to extracorporeal shockwave lithotripsy (ESWL) and endoscopic lithotripsy. Although fragmentation is affected by the patient's size, the type of shockwave generating system, mode of localisation etc, it is also known that the physico-chemical characteristics of calculi affect treatment outcome. However it is still unclear why some stones break more readily than others. Although chemical composition has been proposed as an important factor, a literature review revealed conflicting results.

In 1977, during the development of micro-explosion lithotripsy in Japan, Watanabe[1] reported a number of mechanical tests performed on 66 urinary calculi. Almost no correlation was found between stone composition (by infra-red spectroscopy) and the tensile strengths of calculi. However subsequent clinical papers have highlighted the role of composition in determining ease of fragmentation and retreatment rate. Calcium oxalate, particularly the monohydrate form[2], and brushite stones[3] are now well known to be harder to break with ESWL than stones of other compositions. They also have a greater retreatment rate[2,3].

MATERIALS AND METHODS

Hardness

Intact urinary stones are hard to find in the United Kingdom and our initial study concentrated on a method suitable for measuring the physical properties of urinary stone fragments. The technique used was to measure the microhardness of larger fragments collected after ESWL or percutaneous lithotripsy. The hardness of a material can be defined as its resistance to penetration by an indenter. Several different hardness tests exist and these vary in terms of the load and type of indenter used. For small indents produced by light loads the term microhardness is used. The most accurate of these methods is the Vickers hardness test, which uses a diamond pyramid indenter[4]. It can be shown theoretically that the Vickers measurement obtained is related to the yield stress of the

Urolithiasis 2, Edited by R. Ryall *et al.*,
Plenum Press, New York, 1994

material, ie, the stress at which the material begins to deform in an irrecoverable fashion. Hardness testing is a useful tool since it can be used to test small pieces of material, such as urinary stone fragments, which are of insufficient size for conventional mechanical testing. Microhardness testing of dry urinary calculi with the Knoop indenter has suggested that composition affects the degree of hardness[5]. However this was not found when using the Rockwell indenter[1].

In this study the Vickers hardness test was used on intact ureteric stones passed spontaneously, or stone fragments which were collected from patients undergoing ESWL or percutaneous nephrolithotomy (PCNL). Specimens were rehydrated, stored wet, and embedded in a fast setting polyester resin. Each resin block was cut with a diamond edged saw to obtain a flat stone surface and this was polished on silicon carbide paper and on a napped cloth impregnated with 6 μm diamond paste.

Hardness testing was performed on 52 specimens on a Shimadzu microhardness tester. A Vickers diamond pyramid indenter was applied for 10 seconds with a 75 g load (50g for struvite). The diagonals of the indent so formed were then measured at 400 times magnification. The Vickers hardness number was calculated using the equation

$$\mathrm{VHN} = \frac{1.8544 \times P}{d^2}$$

where P is the applied load in grams and d is the mean of the 2 measured diagonals in microns. Each specimen was indented at least 10 times.

RESULTS

The results confirmed that stone hardness varies with stone composition (Fig 1). Struvite was the softest naturally occurring stone. It proved difficult to polish because of its high porosity and loose structure. Calcium oxalate and brushite stones were the hardest. A commercially available stone phantom was tested and proved to be a poor model stone with hardness values considerably less than struvite. Some hardness values overlapped, probably as a result of the majority of stones being impure. Stone microstructure was visible under the light microscope and in some examples Vickers hardness numbers varied from core to shell structures, or from granular to smoother, homogeneous areas.

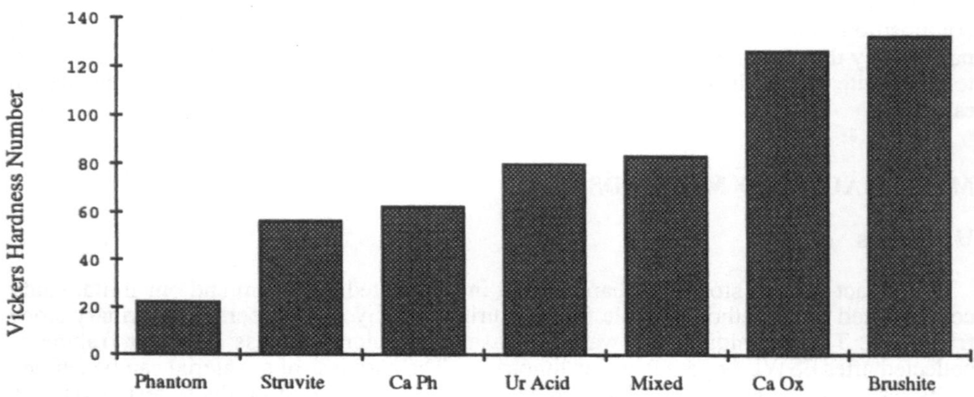

Figure 1. Stone hardness variation with chemical composition.

PREDICTION OF STONE COMPOSITION

The implication of these findings is that treatment of a particular stone can be made more efficient if its composition can be determined before ESWL. Various methods of discriminating stones have been examined. Dretler described differences on fine detailed *in vitro* radiographs but found that radiodensity did not correlate with ease of fragmentation in the lithotripter focus[2]. *In vitro* CT scanning can differentiate some stones, but not stones of mixed composition[6]. We have found that the composition of a stone can be made on a single pre-treatment urine specimen using scanning electron microscopy and X-ray energy dispersive spectroscopy (SEM and XES)[7]. This method has been helpful in predicting response to piezolithotripsy[8].

Ultrasound Velocity

An alternative method of stone characterisation is that determined by the speed of sound as it passes through a stone. Ultrasound, unlike SEM and XES, is widely available in hospitals. Furthermore it provides an *in vivo,* direct measure of the strength of the particular stone to be treated. This occurs because ultrasound passes through a material by exciting the very same bonds which determine the strength of that material. The usefulness of ultrasound velocity measurements lies in the fact that velocity is related in a predictable manner to the elasticity and density of the material through which it passes;

$$E = c^2 \times D$$

where E = Young's modulus of elasticity, C = ultrasound velocity and D = density. The Young's modulus of elasticity is a measure of how difficult it is to deform a material. It describes how stiff or flexible a material is. It can be shown theoretically that strength is related to the Young's modulus divided by ten[4]. Therefore, by measuring the ultrasound velocity and the density of a stone it is possible to obtain a value of its strength.

MATERIALS AND METHODS

Intact urinary stones were obtained from abroad and rehydrated in water under vacuum for 48 hours. A diamond tipped saw was used to prepare two parallel surfaces, and the stones were placed in a water bath between 1 MHz ultrasonic transducers, of 11 mm diameter. These were connected to a generator producing a short ultrasonic pulse of one nanosecond, a digital receiver, and PC IBM compatible computer. From the pulse transit time the corresponding velocity was calculated as described by Langton[9]. Density of the stones was measured using Archimedes' principle. Stones were then placed in tissue paper and centrifuged for one hour at 2000 revs per minute, to determine percentage weight loss as an indicator of porosity. From the equation above, a theoretical value for the Young's modulus of elasticity was determined.

RESULTS

Ultrasound velocity was found to increase for the series struvite to brushite (Fig 2).

Density increased in a similar fashion, although organic calculi generally had lower densities (Figure 3).

The Young's modulus values increased from struvite to brushite (Fig 4).

In contrast, porosity decreased from struvite to brushite, with the exception of the organic calculi which contained little water (Fig 5).

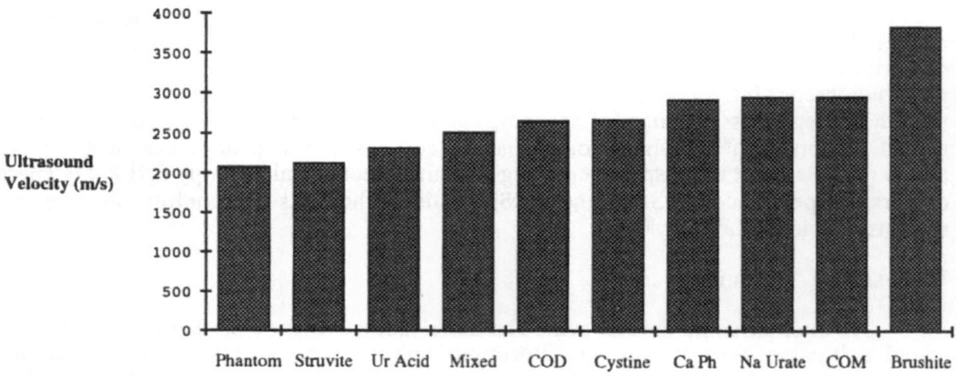

Figure 2. Ultrasound velocity for stones of different composition.

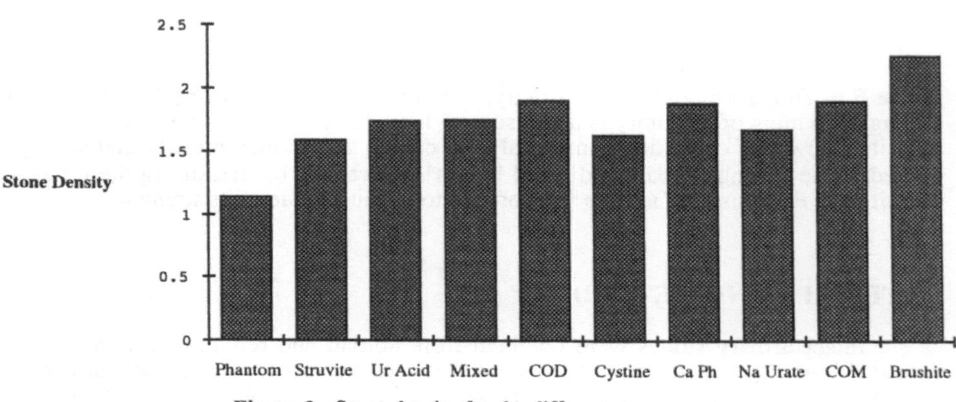

Figure 3. Stone density for the different stone groups.

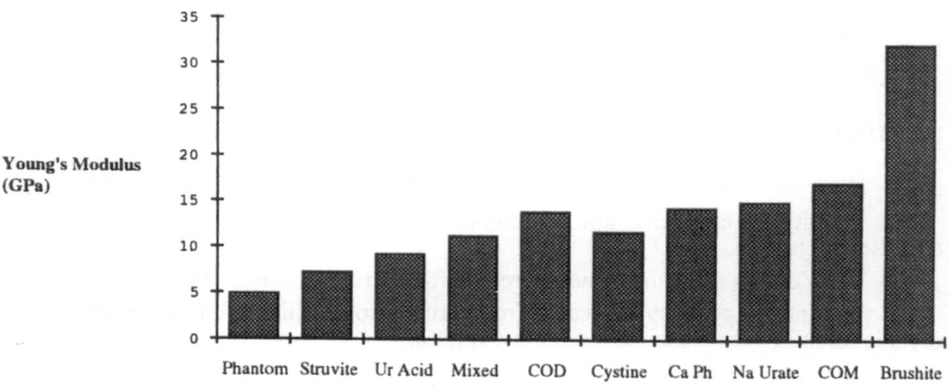

Figure 4. Young's modulus variation with stone composition.

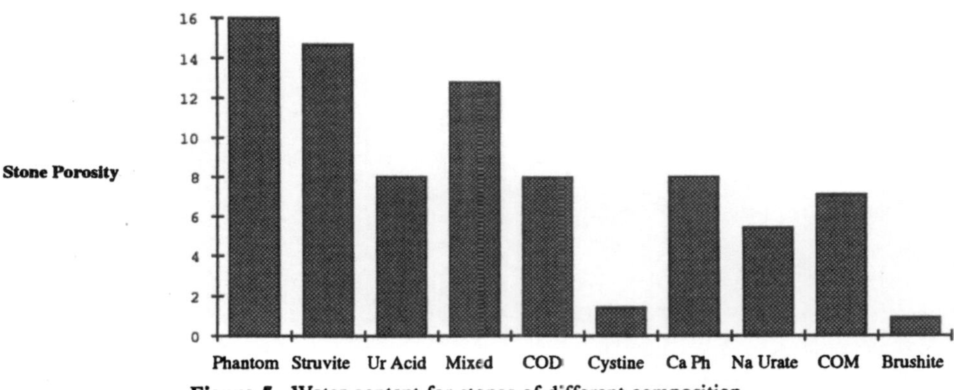

Figure 5. Water content for stones of different composition.

CONCLUSIONS

In conclusion we have found that microhardness, as an indicator of stone strength, is related to the chemical composition of the stone. Ultrasound velocity and the Young's modulus of elasticity of urinary calculi are also related to composition, exhibiting similar trends to microhardness. Therefore, stone composition is a crucial factor in predicting the outcome of lithotripsy treatments. The possibility of using ultrasound velocity measurements *in vivo* to predict the strength of calculi is now being investigated.

REFERENCES

1. S Murata, H Watanabe, T Takahashi, K Watanabe and S Oinuma, Construction and strength of Urinary Calculi, *Jpn J Urol* 68:249 (1977).
2. SP Dretler, Stone fragility - a new therapeutic distinction, *J Urol* 139:1124 (1988).
3. LW Klee, CG Britto and JE Lingeman., The implications of brushite calculi, *J Urol* 145: 715 (1991).
4. A Kelly, in: "Strong Solids," Clarendon Press, Oxford (1973).
5. LG Johrde and FH Cocks, Microhardness studies of renal calculi, *Mater lett* 3:111 (1985).
6. HD Mitcheson, RG Zamenhoff, MS Bankoff and EL Prien, Determination of the chemical composition of urinary calculi by computerised tomography, *J Urol* 130:814 (1983).
7. WG Bowsher, P Crocker, JWA Ramsay and HN Whitfield, Single urine sample diagnosis. A new concept in stone diagnosis, *Br J Urol* 65:236 (1990).
8. NP Cohen, H Pakhouse, ML Scott, WG Bowsher, P Crocker and HN Whitfield. Prediction of response to lithotripsy - the use of scanning electron microscopy and X-ray energy dispersive spectroscopy, *Br J Urol* 70:469 (1992).
9. CM Langton, AV Ali, CM Riggs, GP Evans and W Bonfield, A contact method for the assessment of ultrasonic velocity and broadband attenuation in cortical and cancellous bone, *Clin Phys Physiol Meas* 11:243 (1990).

EFFECT OF MEDICAL MANAGEMENT AND RESIDUAL FRAGMENTS ON RECURRENT STONE FORMATION FOLLOWING EXTRA-CORPOREAL SHOCK WAVE LITHOTRIPSY

J.K. Fine[1], C.Y.C. Pak[2] and G.M. Preminger[1,2]

[1]Division of Urology, Department of Surgery and
[2]The Center for Mineral Metabolism and Clinical Research
The University of Texas
Southwestern Medical Center
Dallas, Texas

INTRODUCTION

Before the advent of extracorporeal shock wave lithotripsy (SWL) and percutaneous nephrostolithotomy, patients underwent open surgical procedures for the removal of their stone burden. A successful surgical procedure was defined as complete removal of all stones without leaving any residual stone fragments. Patients with residual calculi were considered as treatment failures.

Yet, with the introduction of less invasive procedures for stone removal, success now appears to be determined by fragmentation rates and the size of remaining stone fragments. After the bulk of the bone burden is removed, little emphasis is placed on the fate of remaining stone material. Small stone fragments (< 5 mm in diameter) are considered by some as being "clinically insignificant". However, several studies have noted a dramatic increase in stone formation with residual calculi following SWL. If residual fragments can act as a nidus for further stone formation, why should the criteria of successful "stone-free" surgery be changed?

After SWL or any other modality of stone removal, the underlying physiologic or metabolic defect for recurrent stone formation persists. Growth of residual stones and formation of new calculi may be inevitable, due to supersaturation of stone-forming substances and/or lack of stone-inhibitors. However, appropriate medical therapy initiated after lithotripsy, even in the presence of residual calculi, might control active stone disease by reducing saturation or enhancing inhibitor activity.

Herein, we report our experience of long-term follow-up in 80 patients post-SWL; 69% remained on specific drug therapy, whereas 31% of patients were maintained on conservative measures alone. Thus, it was possible to examine the effectiveness of selective medical therapy on active stone formation during long-term follow-up. Specific attention was directed toward the significance of residual stone fragments and their effect on stone growth and recurrent stone formation.

Urolithiasis 2, Edited by R. Ryall *et al.*,
Plenum Press, New York, 1994

METHODS

Clinical data

We identified 131 patients in our Mineral Metabolism Clinic who had undergone SWL at various institutions throughout Texas and had been referred for metabolic evaluation and medical management of their stone disease. Among these patients, 92 were available for a personal interview and follow-up abdominal radiographs (9-79 months after lithotripsy). Inability to locate previous abdominal radiographs excluded 12 patients from the study. Therefore a total of 80 patients (53 male, 27 female) comprised the study population.

The average age of the patients was 45.7 years (18.8 to 73.9 years: median 45.3 years). The cause for their stone formation was determined by our outpatient protocol. Patients were then offered conservative as well as specific measures. Conservative measures, offered to all patients, comprised a high fluid intake to achieve a minimum urine output of 2 L/day and avoidance of dietary indiscretion. Specific medical therapy included thiazide diuretics or sodium cellulose phosphate for the management of hypercalciuria; allopurinol for the management of hyperuricosuria; potassium citrate for the management of hypocitraturia, renal tubular acidosis, chronic diarrheal syndromes, and gouty diathesis; and a-mercaptopropionylglycine for cystinuria. Some patients received more than one drug. During long-term follow-up post-SWL, 69% of the patients (N=55) continued on specific medical treatment (Med Rx), plus conservative measures. The remaining 25 patients chose to stop specific medical treatment within four months of SWL, preferring to be maintained on conservative measures alone (No Med Rx). All patients claimed that they were adhering to conservative measures.

Study protocol

Patients in this study were re-evaluated an average of 43.2 months (9-70 months; median 40.5 months) after SWL. All patients underwent a personal interview by one of the investigators to assess duration and compliance with recommended conservative and specific measures. Patients were also questioned on the activity of their stone disease (passage of stones or surgical procedures for removal of calculi) 36 months prior to SWL and up to the time of the first follow-up abdominal radiograph after lithotripsy. All information was confirmed by careful review of the clinical chart as well as the medical records of the referring physician. Results of the stone analysis were obtained upon review of medical records.

Pre-lithotripsy abdominal X-rays (within 1 month of SWL), as well as post-lithotripsy abdominal radiographs (immediate post-SWL, 1 to 3 months later, and long-term follow-up from 9-79 months) were reviewed personally by one of the investigators without knowing patient group allocation; results were corroborated later by another investigator. Pre-lithotripsy films were compared for the change in size of pre-existing stones or appearance or disappearance of stones. Immediate post-SWL films were examined for the presence of residual stone fragments. Long term follow-up X-rays were evaluated for the growth or loss of residual stone fragments and appearance of new stones. The size of residual fragments or new stones was determined by adding the length and width of the fragments/stones in millimeters (mm) and dividing by two. Overall stone burden was calculated by adding size of all individual residual fragments/stones. Change in stone burden was evaluated on post-lithotripsy abdominal radiographs by a scoring system. In every patient studied, the lithotripsy treatment was directed at all stones present within the renal collecting system. Thus, stone mass remaining after SWL represented residual fragments, not pre-existing stones.

Assessment of stone formation

During the follow-up period, new stone formation was represented by spontaneous stone passage in the absence of residual stone fragments, surgical removal of newly formed stones, stone passage without change in the number of residual fragments, or appearance of new stones or increase in size of stones/fragments noted on the abdominal radiograph. In each patient, individual stone formation rate (SFR) was calculated before and following lithotripsy by counting total numbers of stones spontaneously passed, new stones formed and existing stones showing growth, and dividing the stone number by duration in years. Median and mean SFRs were obtained for each group. For different groups of patients, group SFR was also determined. Reduction in group SFR was calculated as the percent change in group SFR from pre-lithotripsy to post-lithotripsy. Remission rates represented the % of patients in each group without active stone formation during the post-SWL period.

Cost of care for recurrent stones

The economics of recurrent stone formation was estimated from the cost of medical care for stone episodes resolved by stone passage, as well as the cost of stone removal procedures. The 'average' cost for stone passage was considered to be $2000 per event (cost of hospitalisation) while the cost of repeat SWL or percutaneous stone removal was taken at $12,000 per event. This simplistic analysis assumed that all patients had the same follow-up care (office visits and laboratory tests). Moreover, the cost of drugs was considered to be negligible compared to the above costs.

RESULTS

Patient group characteristics

After review of immediate post-lithotripsy abdominal radiographs, patients were divided into two groups, comprising 31 stone-free patients and 49 with residual fragments. These two groups were further subdivided into patients who continued with (Med Rx) and without (No Med Rx) specific medical therapy. The four groups of patients were similar with respect to median age, duration of follow-up post-SWL, and stone burden pre-SWL. Common derangements in all four subgroups were hypercalciuria and hypocitraturia, similar to the finding in our general stone forming population. The majority of patients (88%) had more than one 'metabolic' abnormality. Stone composition for the four groups was similar.

Effect of medical treatment

Specific medical therapy significantly reduced individual SFR in both stone-free and residual fragments groups. In the stone-free group, the median individual SFR significantly declined from 0.67 stones/patient/year before SWL to zero stones/patient/year while on medical therapy post-SWL ($p < 0.001$). Similarly, in the residual fragment group, the median individual SFR decreased from 1.17 stones/patient/year to zero stones in the medical treatment group ($p < 0.001$). In contrast, among patients receiving no medical treatment, the median individual SFR was not significantly different between pre-SWL and post-SWL. In the stone-free group, median SFR was lower following conservative treatment (No Med Rx) post-SWL (0.40 vs 0.83 stones/patient/year), but the change was not significant ($p = 0.06$). Similarly, in the residual fragment group, the difference in median SFR from 1.33 stones/patient/year pre-SWL to 0.77 stones/patient/year post-SWL with conservative treatment was not significant ($p = 0.07$).

The reduction in group stone formation rate from pre-SWL to post-SWL was greater with medical treatment (91 and 81%) than without specific medical therapy (35 and 17%). During the post-SWL period, a larger percentage of patients remained stone-free during medical treatment than without medical treatment. Thus, in the stone-free group, the remission rate of 89.5% during medical treatment was significantly higher than the 50% found without treatment ($p < 0.05$). Similarly, in the residual fragment group, the medically-treated patients had a significantly greater remission rate than the untreated patients (63.9 vs 23.1%, $p < 0.05$).

Among those with residual stone fragments, 27.8% of patients showed an increase in stone burden during medical treatment post-SWL, compared to 61.6% without medical treatment. Moreover, while 13.9% of the treated patients demonstrated a reduction in stone burden, none of the untreated patients showed decreased stone mass. The change in stone burden between the two groups was marginally significant. Among patients with 'clinically insignificant' fragments which remained after SWL, 16.0% of the medically treated patients demonstrated fragment growth, compared to 54.5%of the untreated patients ($p < 0.05$).

Effect of residual stones

In the medically-treated patients, mean individual SFR post-SWL was significantly higher in those with residual fragments than in stone-free patients (0.47 vs 0.09 stones/patient/year, $p<0.05$). Moreover, the remission rate of treated patients was significantly lower in those with residual fragments than in patients who were stone-free (63.9 vs 89.5% $p< 0.05$).

In contrast, among untreated patients (No Med Rx), there were no significant differences in the mean SFR post-SWL or in the remission rate between the residual fragment and the stone-free groups.

Economic impact

The cost of recurrent stones in the medically-treated stone-free group was $69/patient/year. However, in stone-free patients who did not remain on medical therapy, their cost of recurrent stones was $879/patient/year. For patients with residual stones, the cost of recurrent stone formation was $738/patient/year in those who received specific medical treatment, and $1959/patient/year in those continuing conservative measures alone.

DISCUSSION

This study was designed to ascertain the role of specific medical therapy and the effect of residual stone fragments on stone activity after SWL. Our results demonstrated that specific medical treatment significantly alleviated stone activity following SWL in patients who were stone-free and in those who had residual stone fragments. Thus, the median SFR significantly declined post-SWL from pre-SWL values in patients on medical treatment, whereas it did not change significantly in untreated patients. The remission rate was significantly higher in medically-treated than in untreated patients. Among patients with residual fragments, a minority of patients receiving medical treatment showed stone growth, whereas a majority of untreated patients had growth of their residual stones.

The continued stone formation following SWL in patients receiving no specific medical treatment is not unexpected. SWL should have no effect on underlying metabolic abnormalities which will predispose the patient to recurrent stone formation. Therefore, even if stone-free status is achieved following SWL, patients may still be at increased risk for active stone formation if the underlying metabolic or physiologic defect is not corrected. Indeed once a patient has achieved stone-free status following shock wave lithotripsy, recurrent stone formation is expected in 7-14% of patients per year, which is similar to the stone recurrence rates in previously reported epidemiologic studies of urinary stone formers.

Our data showing reduced stone activity with the initiation of appropriate medical therapy post-SWL are consistent with published reports of the medical management of stone disease. A 'metabolic' etiology for recurrent stone formation may be detected in the vast majority of patients with recurrent nephrolithiasis. Moreover, using a variety of specific medical treatments, underlying metabolic derangements may be corrected and recurrent stone formation may be controlled in most patients. Such drugs apparently exert the same physiological and clinical effects in patients who underwent SWL as in those who did not.

A noteworthy observation was the beneficial effect of medical treatment on stone activity post-SWL among patients with residual stone mass. This finding is consistent with our prior reports that appropriate medical treatment could reduce the recurrent stone formation rate and cause stone dissolution in patients with pre-existing stones. However, our study suggests that incomplete fragmentation and clearance of stones by SWL attenuates the beneficial response to medical treatment during the post-SWL period. Thus, among patients receiving such treatment post-SWL, the mean individual SFR was significantly higher and the remission rate lower in patients with residual stone mass than in those who were stone-free. In untreated patients, however, individual SFR and remission rate were not significantly different between those with residual stone mass and stone-free counterparts.

The above finding is compatible with extensive evidence in the literature suggesting that residual stone mass after any stone removal procedure may act as a nidus for existing stone growth and/or new stone formation. After anatrophic nephrolithotomy, a 53%

incidence of stone growth and/or recurrence in those patients who had incomplete stone removal following the initial procedure has been reported. Similar data have been presented for patients with residual fragments following percutaneous removal of staghorn calculi. In those patients who were rendered stone-free following percutaneous nephrolithotripsy, only 95 developed recurrent calculi. Yet, in patients who had residual stones following percutaneous procedures, 63% had either new stone formation or continued stone growth.

Similar findings regarding the fate of residual stone fragments have been reported by others. In one large study with 2 year follow-up, there was an 8-10% incidence of new stone formation in patients rendered stone-free following SWL. However, the annual incidence of stone growth increased to 20-22% in those patients who had residual fragments following lithotripsy. The incidence of stone growth and/or recurrent stone formation increased to 48-65% in patients with fragments composed of either struvite or uric acid. In two additional studies, the prevalence of new stone formation or residual stone growth following SWL ranged from 17-44%.

Of particular interest is the fate of so-called 'clinically insignificant residual stone fragments' defined as stone fragments less than 5 mm in diameter following SWL. Thus, among those with 'insignificant' fragments, the majority of patients on no medical treatment had growth of fragments following SWL, although a minority of patients showed growth while on medical therapy. Thus, such previously considered clinically insignificant stone fragments may pose a significant risk for recurrent stone formation.

During the first few years following the introduction of SWL, stone-free rates of greater than 90% were reported. However, as the indications for SWL have expanded with inclusion of larger calculi, a higher percentage of retained stone fragments have been reported. For stones greater than 1 cm in diameter, the stone-free rates have been reported to be about 75%. There has been a tendency to disparage the importance of residual stone mass, particularly small fragments less than 5 mm in diameter. This study indicates that more emphasis should be placed on rendering patients completely stone-free by SWL.

To the best of our knowledge, our current report is the first study to date which specifically investigates the long-term fate of residual stone fragments or the effects of specific medical therapy on stone growth/formation following lithotripsy. A major limitation of this study however, is that it represents a non-randomised, retrospective analysis. The four subgroups could have been more comparable with respect to patient number, demographics and residual stone burden. Patients on no specific medical treatment could have been less compliant and may have received less frequent follow-up care.

In conclusion, appropriate medical treatment applied after SWL can inhibit new stone formation or growth, whether or not residual stone fragments are present. However, the presence of residual stone fragments can attenuate the response to medical therapy. Thus, metabolic evaluation and specific medical therapy are warranted following SWL. Moreover, a concerted effort should be made to achieve a stone-free state following shock wave lithotripsy. Further validation of the importance of selective medical treatment and the complications associated with residual stone fragments warrants a prospective randomized trial.

SELECTED REFERENCES

- DM Newman, JW Scott and JE Lingeman, Two-year follow-up of patients treated with extracorporeal shock wave lithotripsy, *J Endourol* 2:163-171, (1988).
- RD Brown, CYC Pak and GM Preminger, Effect of lithotripsy on stone forming risk factors, *J Urol* 141:206A, (1989).
- GA Nijeholt, HK Tan and SE Papapoulos, Metabolic evaluation in stone patients in relation to extracorporeal shock wave lithotripsy treatment, *J Urol* 146:1478-1481, (1991).
- CYC Pak, F Britton, R Peterson, et al, Ambulatory evaluation of nephrolithiasis. Classification, clinical presentation and diagnostic criteria, *Am J Med* 69:19-30, (1980).
- GM Preminger, R Peterson, PC Peters, CYC Pak, The current role of medical treatment of nephrolithiasis: the impact of improved techniques of stone removal, *J Urol* 134:6-10, (1985).
- AM Fuchs, BA Wolfson, GJ Fuchs, Staghorn stone treatment with extracorporeal shock wave lithotripsy monotherapy: long-term results, *J Endourol* 5:45-48, (1991).

PROTECTIVE EFFECTS OF VERAPAMIL ON SHOCK WAVE INDUCED TUBULAR DAMAGE

W.L. Strohmaier, J. Koch, D.M. Wilbert and K-H. Bichler

Department of Urology
Eberhard-Karls-University
Tübingen, Germany

INTRODUCTION

Extracorporeal shock wave lithotripsy (ESWL) is a non-invasive, almost pain-free standard treatment modality for urolithiasis. However, ESWL is not completely free from side effects. Apart from local hematomas and a potential risk for hypertension, alterations of kidney function have been reported, including a transient decrease in glomerular filtration rate both in adults and children[1,2]. In addition, proximal and distal tubular impairment have been described, manifested as increased excretion of small molecular weight proteins and a decline in Tamm-Horsfall-glycoprotein, respectively[3,4]. Microscopic investigations on urine voided after ESWL frequently show vacuolized tubular cells[5]. Using an *in vitro* model with Madin Darby Canine Kidney (MDCK) cells we demonstrated protective effects of fosfomycin, a nephroprotective antibiotic, on shock wave induced tubular damage[6]. The present study was initiated to investigate the potential protective effect of verapamil, an organ protective calcium antagonist, against shock wave-induced impairment of renal tubular function *in vitro* and *in vivo*.

MATERIALS AND METHODS

In Vitro Studies

To study the effect of verapamil on shock wave induced tubular damage the MDCK-cell model was used as described previously[6]. Containers (1.1 mL) with cell suspensions (77 $x10^6$ cells, n = 24; verapamil (5 x 10^{-7} mol/L nutrient medium, n = 24 controls) were exposed to shock waves (0 - 256 impulses, Dornier lithotriptor HM 4, 18 kV). The concentrations of lactate dehydrogenase (LDH) and glutamic oxide trasaminase (GOT) in the nutrient medium were measured before and immediately after ESWL.

In Vivo Studies

Patients with renal pelvis or calyceal stones (< 2 cm in diameter) presenting for shock wave lithotripsy fulfilling the following criteria were included: creatinine < 1.5 mg/dL, no auxiliary measures pre or post lithotripsy, no application of contrast media during lithotripsy, no anesthesia, lithotripsy (i.e. absent postglomerular bleeding). These patients were randomly assigned to the verapamil (n = 12) or control group (n = 12). The duration of the study was three days. Verapamil (80 mg t.i.d. orally) was administered for two days starting the night before ESWL. Control patients received no medication.

To assess renal tubular function, 5 urine specimens (Ul-U5) were collected 24 and 12 h before (Ul,2) and immediately, 12 and 24 h after ESWL (U3-5). To evaluate the proximal tubular function a-l-microglobulin (alM) and N-acetyl-ß-D-glucosaminidase (NAG) were measured; for assessment of distal tubular function Tamm-Horsfall-glycoprotein (THG) was determined. To exclude significant postglomerular bleeding, apolipoprotein A I (Apo AI) was measured. Patients with positive Apo AI were excluded from the study, since postglomerular bleeding influences urinary protein patterns.

alM was measured using a synchrone enzyme linked immunosorbent assay (Elias Medizintechnik, Freiburg/Breisgau, Germany), NAG by a colorimetric assay according to Pott[7]. For the determination of THP we used an electroimmunodiffusion method[8]. Apo AI was measured by turbidimetry (TurbitimerR, Behringwerke AG, Marburg/Lahn, Germany). Creatinine was determined by a creatinine analyzer (Beckman Instruments, USA). For shock wave lithotripsy we used the Lithotriptor MFL 5000 (Dornier, München, Germany). Impulse rates and generator voltages were recorded.

For statistical analysis the Wilcoxon-Mann-Whitney test was used. Calculations were performed on a personal computer using the SAS-program.

RESULTS

In Vitro Studies

There was a significant dose-dependent increase of LDH and GOT concentrations in the nutrient medium after shock wave exposure. This increase was significantly less pronounced in the verapamil group (Fig. 1).

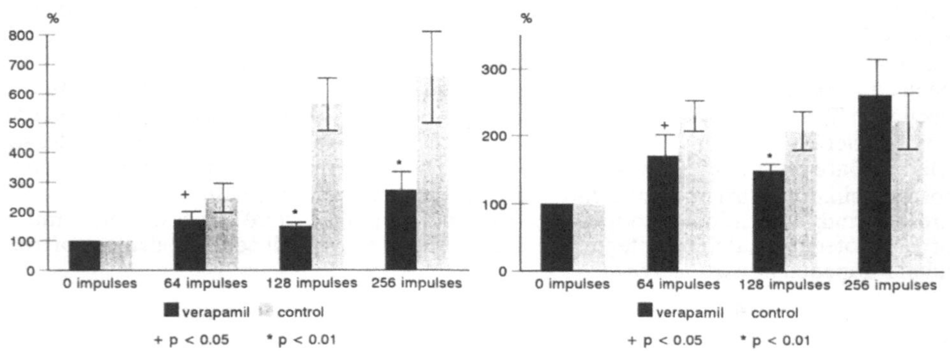

Figure 1. Changes in LDH and GOT after shock wave exposure in the verapamil and the control group.

In Vivo Studies

The verapamil and control groups did not differ with respect to the conditions of ESWL treatment (impulse rate and generator voltage). Baseline levels of alM, NAG and THG were not statistically different between the verapamil and control groups. Immediately after shock wave lithotripsy there was a significant rise of urinary alM and NAG in the controls. Verapamil treated patients showed only slightly, but not significantly elevated levels. The mean maximal rise in alM and NAG excretion after ESWL was significantly higher in the verapamil than in the control group (Fig. 2).

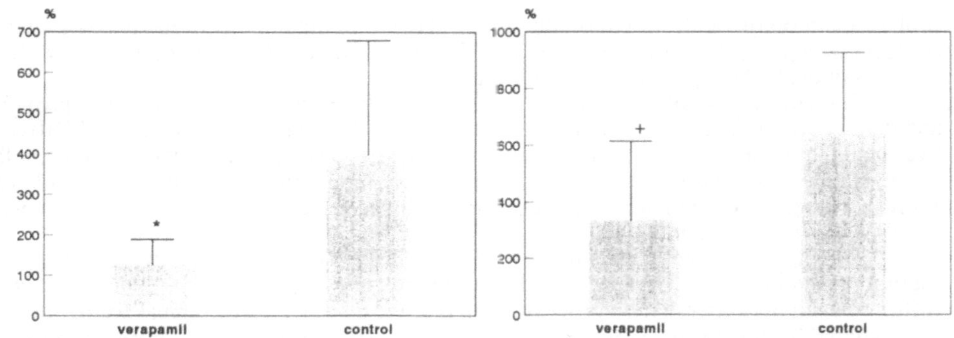

Figure 2. Maximal rise (mean ± SD in percent) of alM (left) and NAG (right) excretion after shock wave lithotripsy. Pretreatment levels = 100 %. Significant difference: * $p < 0.01$; + $p < 0.05$.

THG levels fell significantly after lithotripsy in the control group. Controls revealed only a slight, but not significant decline. Pretreatment levels were reached one day after ESWL in the verapamil group, whereas THG excretion in the control cohort did not recover completely during the observation period. The mean maximal decline of THG excretion after ESWL was significantly higher in the controls (Fig. 3).

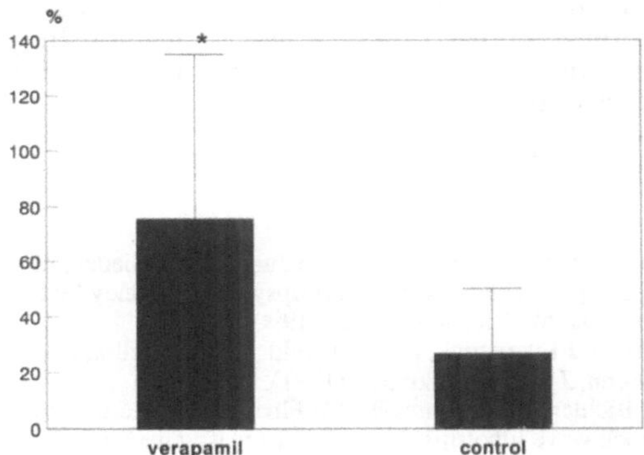

Figure 3. Maximal fall (mean ± SD%) of Tamm-Horsfall-glycoprotein excretion after shock wave lithotripsy. Pretreatment levels = 100 %. Significant difference: * $p < 0.01$.

DISCUSSION

After shock wave lithotripsy there was a significant rise in the excretion of alM and NAG. In evaluating proteinuria and enzymuria after shock wave exposure, possible contamination from postglomerular bleeding has to be considered. In the present study, only patients with mild hematuria and absent Apo AI excretion in the urine were included. Nevertheless, there was a significant rise of urinary proteins and enzymes after shock wave lithotripsy. This fact clearly demonstrates that proteinuria and enzymuria are certainly not exclusively caused by postglomerular blood loss. High energy shock waves obviously induce functional alterations of the renal tubules. Since alM and NAG are predominantly markers of the proximal tubules, their increase is representative of shock wave-induced impairment of proximal tubular protein resorption and impairment of the cell membrane integrity, respectively.

In accordance with previous results[3] the excretion of Tamm-Horsfall-glycoprotein fell after ESWL. This glycoprotein is synthesized in the distal tubule and the thick ascending limb of the loop of Henle and normally does not occur in the blood. An increased excretion of THG is representative of impaired distal tubular function[8]. Thus, high energy shock waves obviously damage the distal tubules as well.

The application of verapamil significantly reduced the signs of impaired proximal and distal tubular function. What mechanisms can explain this protective effect? Different treatment parameters, such as impulse rate and generator voltage were not considered in this study. Neither proteinuria nor enzymuria, markers of tubule function, were different before ESWL. Therefore we conclude that verapamil reduced the damaging effect of high energy shock waves on renal tubular cells.

Direct and indirect mechanisms may play a role. Verapamil may directly protect the membrane against radical-induced lipid peroxidation and inhibit a pathologically increased influx of calcium ions[13]. These processes (membrane disorders, free radical formation a.o.) are involved in shock wave induced cellular damage[9-11]. A similar direct protective effect of fosfomycin on shock wave induced tubular damage has been described previously[6].

Indirectly, verapamil may preserve tubular function by acting on the impaired renal hemodynamics after ESWL. Calcium antagonists increase the renal blood flow in states of pre-existing enhanced vascular resistance[14]. Since renal blood flow is decreased after shock wave lithotripsy[12], verapamil may restore the impaired renal perfusion by vasodilation. This effect could prevent tubular ischemia and subsequent functional impairment.

Our study demonstrates that verapamil protects against shock wave-induced renal tubular damage. Since tubular impairment after ESWL in the majority of patients is only a transient phenomenon without clinical signs, a routine application of verapamil does not seem justified. Sakamoto et al[4], however, reported that occurrence of more severe renal lesions after ESWL is more likely in the presence of pre-existing renal diseases, urinary tract infections, previous lithotripsies and in female patients. In the presence of these risk factors the administration of verapamil may be useful. This possibility should be investigated in prospective randomized studies.

REFERENCES

1. G Brien, P Kirschner, K Sydow, H Schroder, H Jung, T Schmiedel and K Sandring, Investigations on the effects of extracorporeal shock wave lithotripsy on the kidney by Lithostar, in: "Proc 1st Eur Symp Urolith" Excerpta Medica, Amsterdam (1990).
2. MT Corbally, J Ryan, J Fitzpatrick, RJ Fitzgerald, Renal function following extracorporeal lithotripsy in children, *J Ped Surg* 26: 539 (1991).
3. DM Wilbert, KH Bichler, WL Strohmaier, SH Fluchter, Glomerular and tubular damage after extracorporeal shock wave lithotripsy assessed by measurement of urinary proteins, *J Urol* 139: 326A (1988).
4. W Sakamoto, T Kishimoto, T Nakatani, et al, Examination of aggravating factors of urinary excretion of N-acetyl-ß-D-glucosaminidase after extracorporeal shock wave lithotripsy, *Nephron* 58: 205 (1991).
5. O Ryoji, F Ito, T Hayashi, Appearance of vacuolized cells in urinary sediment after extracorporeal shock wave lithotripsy, *J Urol* 142: 301A (1990).
6. WL Strohmaier, M Pedro, DM Wilbert, KH Bichler, Reduction of shock wave induced tubular alteration by fosfomycin, *J Endourol* 5: 57 (1991).
7. G Pott, H Zumkley, HW Intorp, Enzymaktivitatsveranderungen der bHexosaminidase im Urin bei Nierenkranken, Klin Wschr 54: 1001 (1976).
8. KH Bichler, H Haupt, G Uhlemann, HG Schwick, Human uromucoid, I. Quantitative immunoassay, *Urol Res* 1: 50 (1973).
9. M Delius, G Enders, Z Xuan, HG Liebich, W Brendel, Biological effects of shock waves: kidney damage by shock waves in dogs - dose dependence, *Ultrasound in Med & Biol* 14: 117 (1988).
10. JE Fegan, DA Husmann, GM Preminger, Preservation of renal function following extracorporeal shock wave lithotripsy, *J Endourol* 5, suppl 1: S46 (1991).
11. ME Moran, K Hynynen, MR Bottaccini, GW Drach, Cavitation and thermal phenomena associated with high energy shock waves - *in vitro* and *in vivo* measurements, *J Urol* 142: 388A (1990).
12. SJ Karlsen, B Smevik, J Stenstrom, KJ Berg, Acute physiological changes following exposure to extracorporeal shock waves, *J Urol* 143: 1280 (1990).
13. IT Mak, WB Weglicki, Comparative antioxidant activities of propanolol, nifedipine, verapamil, and diltiazem against sarcolemmal membrane lipid peroxidation, *Circ Res* 66: 1449 (1990).
14. RD Loutzenhiser, M Epstein, Renal hemodynamic effects of calcium antagonists, *J Cardiovasc Pharmacol* 12: 48, (1988).

RENAL MORPHOLOGICAL DAMAGE AND ITS ACCUMULATION BY REPEATED EXTRACORPOREAL SHOCK WAVES

H. Koga, T. Yamashita and S. Noda

Department of Urology
Kurume University School of Medicine
Kurume, Japan

INTRODUCTION

Extracorporeal shock wave lithotripsy (ESWL) has become popular for the treatment of upper urinary tract stones and repeated treatments for large stones are commonly used to resolve the problem of residual stone fragments. However renal damage caused by multisequential ESWL treatments and high frequency of shock waves is not clear. We have applied animal models to study by histopathological examination of renal morphology, both the damaging effects of changing shock frequency and of repeated shock treatments.

MATERIALS AND METHODS

Fifteen male mongrel dogs (14~16kg) were placed in a lateral position on an EDAP LT-01 lithotripter under general anesthesia with pentobarbital sodium of 25mg/Kg body weight. During these studies treatment doses were: generation power of 100%, frequency of shocks at 2.5, 5 and 20 shocks/second. For clinical treatment of renal stones the safe limit dose is regarded as 2.5 or 5 shocks/second for 1 hour focused on the renal pelvis using ultrasound scans.

The animals were divided into six groups (Table 1). Bilateral nephrectomies were performed immediately, 3, 7, and 60 days after the application of shock waves, and samples from the treatment area were then processed routinely for light and electron microscopy (TEM) for the observation of morphological changes.

RESULTS

Gross pathologies of early-phase-kidney were qualitatively the same in all six groups, including some radially oriented cortical hemorrhages and parenchymal damage with hemorrhagic foci in cortex and medulla. As the frequency rate was increased, the extent and the severity of the lesions tended to be heavier. In the late phase (sacrificed at 60 days) the kidneys were grossly normal except for slightly depressed lesions, which were hard, white, retracted and extended into the deep cortex on sections. These changes also tended to be more intense as the frequency rate and numbers of session were increased.

Group I, sacrificed immediately: Diffuse interstitial hemorrhage and tubular and venous dilatation with preservation of the normal kidney architecture were shown by light

Urolithiasis 2, Edited by R. Ryall *et al.*,
Plenum Press, New York, 1994

Table 1. Strategy for the application of shock waves

Group	Shocks/sec	Application Time (min)	No. of Sessions	Total Storage[2]
I	2.5	60	1	31
II	5	60	1	62
III	5	60	4[1]	250
IV	20	60	1	250
V	20	60	3[1]	750
VI	20	15	1	62

[1]Sessions at intervals of 7 days
[2]Total storage = total number of shocks applied x generation power (%) x application time (min) x a constant (0.002083).

microscopy. Electron micrography demonstrated interstitial edema and extravasation of red blood cells, and obstruction of peritubular capillaries with platelet plug formation and red blood cells.

Group I, sacrificed at 60 days: Diffuse interstitial fibrosis with chronic inflammatory cells and abundant deposit of hemosiderin were observed by light microscopy. Lobulated glomeruli and atrophic and hyalinized tubules were also seen in areas of minor damage. Under electron microscopy, the lumen of the interlobular artery was seen to be obstructed by swelling and stratification of endothelial cells. The interlobular artery was also surrounded by excessive fibrosis. Glomerulus was seen to be collapsed. Glomerular capillaries showed degenerated atrophy of endothelial cells and swelling of epithelial cells. There were focal foot process fusion and partial glomerulosclerosis.

Group II, sacrificed at 7 days: Light micrography demonstrated interstitial hemorrhage and dilated lumens of tubules with amorphous eosinophilic materials. Electron micrography of the distal tubule demonstrated balloonings of microvilli and increased cytoplasmic dense bodies. These were regarded as ischemic changes of the cell organelle. In one collecting tubule, ghost cells consisting of RBC plasma membranes and a cast of shedded cell organelle were observed. Stagnation of urinary flow in tubules was apparent.

Group IV, sacrificed immediately: In the tubular cell marked vacuolization and altered mitochondria with swelling, loss of matrix, breaks of cristae and marginal membrane, were observed by electron microscopy, suggesting deficiency of intracellular respiration.

Group IV, sacrificed at 3 days: Electron microscopy of the glomerulus demonstrated completely obstructed capillaries with loss of foot processes - so-called glomerulosclerosis.

Group IV, sacrificed at 60 days: By light microscopy the majority of glomeruli and tubules disappeared and excessive fibrosis was present throughout the scarred region. By electron microscopy most of the tubular epithelia vanished and only the basement membranes remained. Macrophage was seen in the tubular lumen.

Group III (equal to Group IV in total storage), sacrificed at 7 days: Light microscopy showed that except for an increase of amorphous eosinophilic material in tubules, almost normal kidney architecture was preserved, as in Groups I & II.

Group V, sacrificed at 7 days after the last session: Light microscopy showed intensive interstitial fibrosis, hyalinization of tubules and glomeruli, and fibrinoid degeneration of blood vessels. Electron micrograph demonstrated extensive fibrosis and degenerated tubule with marked vacuolization, fatty degeneration, secondary lysosomes and tortuous basement membrane.

Group VI (equal to Group II in total storage by shortening treating time), sacrificed at 7 days after the application of shock waves. Compared with Group II, more severe interstitial bleedings were obvious, suggesting that the application of a high frequency rate caused more severe damage to the kidney.

DISCUSSION

Macroscopic hematuria occurs in the majority of patients after ESWL. Likewise, in this study interstitial hemorrhage and casts of RBC in tubules were seen in all exposed kidneys. Our findings in the early phase were the rupture of interstitial capillaries and subsequent hypoxia of the surrounding renal tissue. As a result of hypoxia, morphological damage progresses according to the Figure. Karlsen et al[1] have stated that bleeding is related to damage of arteries and capillaries as well as veins, but our early-phase findings for damage of blood vessels are limited to capillaries. Platelet plug formation for the recovery of ruptured capillaries increases peripheral vascular resistance and causes obstructive change to the centric small arteries, such as interlobular arteries at late phase.

With a single session of 2.5 shocks/sec for 1 h, which is the minimal dose for a treatment of renal stone, morphological damage occurred. The damage tended to be more intense as the frequency rate was increased. Chaussy et al[2] did not find renal damage in dogs exposed to shock waves, but we have provided evidence of damage in association with macroscopic hematuria.

This study provides evidence that accumulation of damage occurs by repeated sessions. At low frequency, accumulation of damage is none or slight, but repeated sessions of high frequency intensifies damage, suggesting that accumulation of damage occurs. At equivalent ESWL energy, morphological damage is more severe with high frequency than with low frequency.

The frequency of shocks is thought to be a major factor in renal morphological damage, and so divided ESWL treatments at low frequency rate are considered advisable for clinical applications.

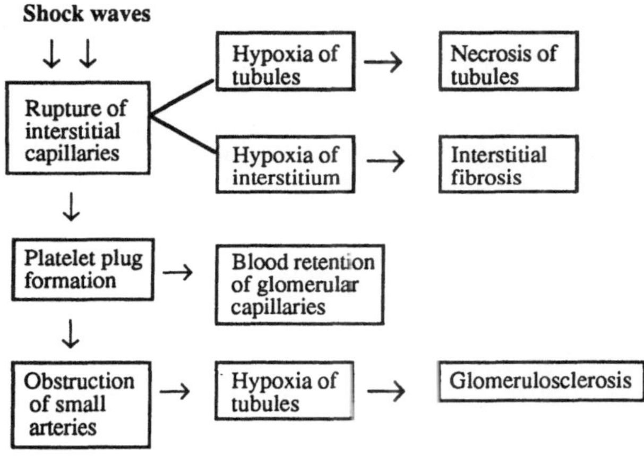

REFERENCES

1. SJ Karlsen, B Smevik, and T Hovig. Acute morphological changes in canine kidneys after exposure to extracorporeal shock waves, *Urol Res*, 19:105 (1991).
2. C Chaussy and E Schmiedt, Extracorporeal shock wave lithotripsy (ESWL) for kidney stones. An alternative to surgery?, *Urol Radiol*, 6:80 (1984).

URETERAL STONES: THE RESULTS OF PRIMARY *IN SITU* ESWL AND OUTPATIENTS PROCEDURE

H.P. Bastian, H.G. Therhag and W. Bonn

Department of Urology
St. Josef Hospital
Troisdorf, Germany

INTRODUCTION

The first clinical application of shock wave technology for the eradication of urinary stones radically altered the urological management of stone disease forever. Since patients prefer not to stay overnight in hospital, lithotripsy has moved to the outpatient setting; furthermore, the management is safe, cost-effective, efficacious and desirable in the majority of cases.

MATERIAL AND METHODS

From September 1989 to January 1992, 800 patients with 1021 ureteral stones were treated by *in situ* ESWL.The ESWL-Institute using a Lithostar Plus unit is a privately owned and operated center at the St. Josef Hospital. Patients treated on a same-day basis were allowed nothing by mouth after midnight. Next day, management in outpatient consisted of continuing to push fluids, straining of urine and a visit with the treating urologist. At that visit a plain abdominal film and renal ultrasonogram were obtained to evaluate the stone fragmentation and hydronephrosis. The average number of shock waves was 4200 (range 1500 to 6000). The size of the stones ranged from 5 to 25 mm.

RESULTS

The stone location in 2292 cases is presented in Fig.1.

The percentage of patients treated on an outpatient basis has gradually increased from 22.2% to 75.3% in the last 650 treatments (Fig.2).

Auxiliary procedures before ESWL were performed in 26.5% of the cases for a large stone mass or sepsis (Table 1).

Ancillary procedures post-ESWL were required in 6.9% of all outpatient treatments (Table 2).

Successful ureteral stone fragmention was achieved in 92% with a 13.6% retreatment rate (Table 3).

The retreatment rate was higher than 19.8% for calculi in the middle ureter. Rehospitalization was required for 4.8% of patients during the first week after treatment.

Urolithiasis 2, Edited by R. Ryall *et al.*,
Plenum Press, New York, 1994

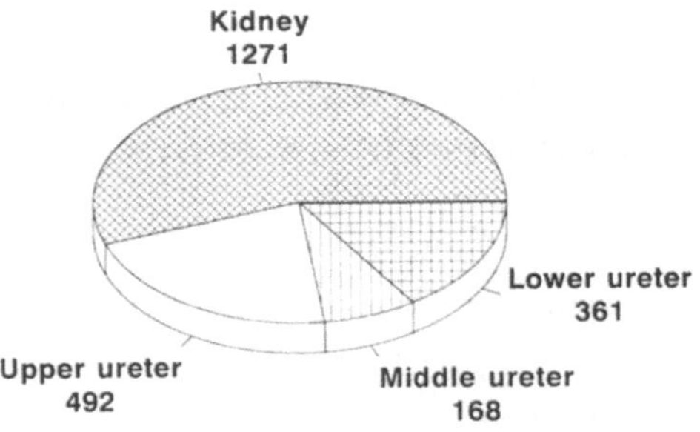

Figure 1. The anatomical location of of the stones in the 2292 cases.

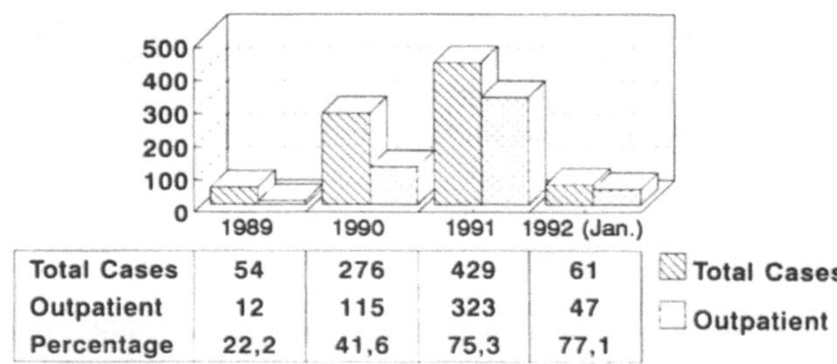

Figure 2. The percentage of patients treated on an outpatient basis from 1989 to January 1992.

Table 1. Auxiliary procedures before ESWL. Large stone mass or sepsis.

Measures	No.	%
None	751	73.5
Nephrostomy	87	8.5
Double-Y-Stent	183	18.0

Table 2. Ancillary procedures after ESWL.

Measures	No.	%
None	951	93.1
Nephrostomy	18	1.8
Double-Y-Stent	39	3.8
Ureteroscopy	11	1.1
Open surgery	2	0.2

Table 3. Results of treatment. 3-month follow-up.

	Upper ureter		Middle ureter		Lower ureter	
	No.	%	No.	%	No.	%
Free of stones	465	94.2	151	88.7	325	88.9
Residual stones	27	5.8	17	11.3	36	11.1

CONCLUSIONS

- Outpatient ESWL is safe and effective.
- Overall results are excellent.
- Treatment is cost-effective.
- Patients prefer outpatient treatment.
- *In situ* therapy is the method of choice for ureteral stones.
- Patients with multiple medical problems or who are recovering from sepsis should be treated with caution.

TREATMENT PHILOSOPHY AND RE-TREATMENT RATES FOLLOWING PIEZOELECTRIC LITHOTRIPSY

J. Fegan, L.A. Camp, W.T. Wilson, G.L. Miller and G.M. Preminger

Division of Urology
Department of Surgery and
Department of Radiology
The University of Texas Southwestern Medical Center
Dallas, Texas

INTRODUCTION

Second generation lithotripters have diminished or eliminated anesthesia requirements during treatment, thereby improving the comfort, convenience, and safety of therapy for urinary stone disease. Yet, as the need for anesthesia has been reduced, the efficiency of these machines has likewise diminished. Since, compared to first generation machines, second generation lithotripters often require a greater number of repeat treatment procedures and auxiliary procedures to render a patient stone-free, the potential exists to amplify renal injury from excessive re-treatment.

Treatment philosophies may affect reported re-treatment rates of second generation lithotripters, as much as the technology itself. Because second- and third-generation lithotripters have essentially made lithotripsy an office-based procedure, and because patients may sometimes travel hundreds of miles to stone centers, some institutions are re-treating patients at short, even daily intervals.

Some stones, which appear on plain X-rays of kidney, ureters and bladder (KUB) to need re-treatment at 24 h, may in fact be adequately fragmented when the patient is re-studied at a later date. A repeat treatment at 24 h would therefore expose such patients to an unnecessary procedure and an excessive number of shock waves. To test this hypothesis, we studied patients at 24 h and at two weeks post-lithotripsy to determine what percentage would have been unnecessarily re-treated at 24 h.

METHODS

We performed a prospective analysis of 100 consecutive patients who underwent extracorporeal piezoelectric lithotripsy (EPL) on the Wolf Piezolith 2300. Patients received 4,000 shocks at the full power setting (1,100 bar). All patients had a KUB 24 h following EPL.

Stones treated were within the intrarenal collecting system and ranged from 8 to 16 mm in diameter, with an average size of 11 mm. KUB radiographs obtained in our radiology department by standard methods were read in a blinded fashion, labelled only as pre-op films and post-lithotripsy studies. The dates of the follow-up KUBs were not available to the investigators during reading of the films.

To be included in the study, patients had to have had radiopaque stone(s) visible on KUB prior to treatment. In addition, the KUB taken 24 h post-EPL must have shown the stone(s) to be unchanged, or residual fragments measuring >4 mm in diameter. In other words, only those patients who appeared to need re-treatment based on the 24 h KUB were included in the study.

Exclusion criteria included radiolucent stones, re-treatment prior to two weeks (ie, due to renal colic), or 24 h films which demonstrated no remaining stone or showed fragments ≤4 mm (which might pass spontaneously). Films which were equivocal at 24 h (ie stone mass present but "smudged") were also excluded from analysis.

RESULTS

Of the original 100 patients, 42 met the inclusion criteria. Four additional patients were excluded because of the need for re-treatment prior to two weeks. All stones were located in the renal collecting system. Eighteen of the 42 patients (43%) who appeared to need re-treatment based on the 24 h film, were stone-free on the subsequent (two week) film without being re-treated. Stone-free status was documented by a follow-up ultrasound examination plus an intravenous pyelogram. Tomography was not performed routinely in our patient population.

DISCUSSION

Owing to their significantly reduced shock wave energy, second generation lithotripters, especially piezoelectric devices, require reduced or no anesthesia during the treatment of symptomatic renal calculi [1,2]. The major limitation of anesthesia-free or reduced anesthesia lithotripsy, however, is usually an increase in the re-treatment rate, since higher numbers of shock waves are needed for adequate fragmentation of calculi[3]. For example, for stones < 1.5 cm in diameter, the original Dornier HM-3 has a re-treatment rate of only 16%[4]. However, by widening the aperture of the focusing mechanism to reduce anesthesia requirements (modified HM-3 and HM-4), the re-treatment rate rises to 22-37%. The re-treatment rates with piezoelectric lithotripters are still higher, having been reported to range from 25-45%.

Additional advantages of second- and third- generation lithotripters include increased ease of treatment as well as the potential for reduced renal trauma, owing to the lower power of these shock wave devices. Because only sedation (and in some cases no anesthesia/analgesia at all) is required for treatment, lithotripsy is an outpatient or even an office procedure[5]. Moreover, the reduced shock wave energy significantly reduces the extent and severity of renal parenchymal damage compared to the higher-powered first-generation electrohydraulic lithotripters[6-7]. Yet, the increased safety and convenience of reduced-anesthesia piezoelectric lithotripters has led to treatment philosophies whereby patients are re-treated at short time intervals, even daily, to render them stone-free. This practice may lead to artificially high re-treatment rates (51% in one study) utilising a piezoelectric device. Moreover, animal studies in rabbit and non-human primate models have documented that an excessive number of shock waves administered during one treatment or short re-treatment intervals may lead to significantly more renal damage than 'spreading' the same number of shock waves over multiple treatments. These findings were similar for animals treated with either first-generation electrohydraulic or second-generation piezoelectric lithotripsy machines. Moreover, since all lithotripters appear to cause measurable renal injury (usually transient, but sometime permanent), the optimal treatment philosophy would be to render patients stone-free with as few treatments as possible with minimal disruption to their lives.

Re-treating patients at short (ie, 24-48 h) intervals does not seem to allow for adequate clearance of stone fragments and may potentially increase the amount of renal damage induced by the lithotripsy treatments. Moreover by re-treating our patients based on a KUB radiograph obtained at 24 h, we would have exposed 43% of the patients in our study to an unnecessary procedure with its attendant risks, increased costs and potential significant inconvenience to both patient and physician.

We therefore recommend that physicians performing extracorporeal shock wave lithotripsy consider re-treating patients only after assessing them for stone fragmentation/passage, at least one or two weeks following the initial lithotripsy procedure. Although the optimal time interval between lithotripsy treatments has yet to be determined, 1 to 2 weeks seems to be a reasonable period to allow fragments to clear, based on our experience and that of other investigators. Moreover, by leaving an interval of 1 to 2 weeks between treatments, one may significantly reduce the incidence of renal damage from high-energy shock waves. Finally, we would suggest that when assessing re-treatment rates reported by various investigators utilising different forms of shock wave energy, one also consider the treatment philosophies employed by the investigators, since more aggressive treatment patterns (ie, re-treatment intervals of only one to two days) may significantly overestimate the actual need for re-treatment with a particular lithotripsy device.

REFERENCES

1. M Marberger, C Turk and I Steinkogler, Painless piezoelectric extracorporeal lithotripsy, *J Urol* 139;695-699, (1988).
2. SC Kim, YT Moon and KD Kim, Extracorporeal shock wave lithotripsy monotherapy: experience with piezoelectric second generation lithotripter in 642 patients, *J Urol* 142:674-678, (1989).
3. J Rassweiler, A Westhauser, P Bub and F Eisenberger, Second-generation lithotripters: A comparative study, *J Endourol* 2:193-204, (1988).
4. JE Lingeman, D Newman , JH Mertz, PG Mosbaugh , RE Steele, RJ Kahnoski, TA Coury and JR Woods, Extracorporeal shock wave lithotripsy: the Methodist Hospital of Indiana experience, *J Urol* 135:1134-1137, (1986).
5. GM Preminger, Sonographic piezoelectric lithotripsy: More bang for your buck, *J Endourol* 3:321-327, (1989).
6. WT Wilson, GL Miller, LL McDougall, WA Erdman, JS Morris and GM Preminger, Renal magnetic resonance appearance after piezoelectric and electrohydraulic lithotripsy, *J Endourol* 4:407-413, (1990).
7. JS Morris, DA Husmann, WT Wilson, J Denstedt, PF Fulgham, RV Clayman and GM Preminger, A comparison of renal damage induced by varying modes of shock wave generation, *J Urol* 145:864-867, (1991).

GD-DTPA-ENHANCED DYNAMIC MRI FOR EVALUATION OF RENAL DAMAGE AFTER ESWL TREATMENT

T. Umekawa, T. Yamate, N. Amasaki, K. Kohri, M. Iguchi and T. Kurita

Department of Urology
Kinki University School of Medicine
Osaka-Sayama, Osaka, 589, Japan

INTRODUCTION

Extracorporeal shock wave lithotripsy (ESWL) is a non invasive therapy used worldwide to disintegrate urinary tract stones. It is the first-choice therapy for upper urinary tract stones. However, the influences of shock waves on the human body have not been studied sufficiently. In the present study, acute renal damage resulting from ESWL was evaluated with magnetic resonance imaging (MRI), including Gd-DTPA-enhanced dynamic MRI[1,2].

SUBJECTS AND METHODS

The subjects included 37 patients (12 females, 25 males; average age 49 years) selected at random from 112 renal stone patients who received an initial ESWL therapy at Kinki University School of Medicine. The patients who received nephrostomy before ESWL and those with stone street formation requiring additional postoperative measures were excluded.

The lithotriptor used was a Siemens Lithostar and the MRI system was a Shimadzu SMT-50; the static magnetic intensity was 0.5 tesla. MRI was examined under T1-weighted scan SE(500/30) and T2-weighted scan SE(2,000/12) before and 24 h after ESWL. Gd-DTPA-enhanced dynamic MRI was also examined under the same conditions. Gd-DTPA was injected intravenously and dynamic MRI was examined continuously at 30-second intervals by small tip angle gradient echo (STAGE) methods (100/23, flip angle 20 degrees, GdDTPA 0.1mmol/kg body weight: T2-d-MRI and 55/16, flip angle 60 degrees, Gd-DTPA 0.05mmol/kg body weight: T1-d-MRI). Changes in signal intensity over time were observed. These changes were plotted as time intensity curves.

Gd-DTPA is a recently developed contrast medium for MRI. Its particularly strong activity in shortening T1 enhances signal intensity in T1-weighted scans. However, if Gd-DTPA concentration becomes too high, its strong activity shortens T2, resulting in the susceptibility effect, that is, reduction in signal intensity. Hence, in imaging of the kidney with the conventional dosage (0.1mmol/kg) of Gd-DTPA, T2 is is shortened more than T1 because Gd-DTPA concentration is too high, and the signal intensity is not proportional to GdDTPA concentration. Thus, the enhancement of the signal intensity is difficult to judge. So we also attempted to obtain T1 weighted scans by decreasing the Gd-DTPA dosage to 0.05mmol/kg, about half of the conventional dosage, to avoid the susceptibility effect.

Urolithiasis 2, Edited by R. Ryall *et al.*,
Plenum Press, New York, 1994

RESULTS

Table 1 shows MRI results for the 37 patients. On the whole, one or more findings were observed in 25/37 (67%) patients and no changes were observed in the other 12 (33%). The most frequently observed finding was a perirenal fluid collection in 14/37 (38%) patients.

T2-d-MRI was done in 12 renal stone patients before and after ESWL. A vascular phase was observed 30 min after Gd-DTPA injection. The low intensity band was found 1 min after injection of Gd-DTPA and moved gradually from the renal cortex to the medulla. After 3 min, a reduction of renal pelvis intensity due to concentration of Gd-DTPA was observed. The low intensity band disappeared after ESWL in all 12 patients, and conventional MRI showed clear CMJ in patients.

T1-d-MRI was done in 7 renal stone patients. Images of the so-called renal cortex phase before and 20 seconds after administration of Gd-DTPA, collecting tubule images and excretion-phase images at 80 and 260 seconds after administration are shown.

Five of 7 cases showed no change following ESWL either with p-MRI or d-MRI (patients 1-5). In one of the remaining 2 cases, p-MRI showed the renal CMJ was retained but detected no other changes. With T1-d-MRI, however, elevation of the signal intensity in the treated kidney was earlier than in the non-treated kidney. The elevation signal lasted longer and excretion of Gd-DTPA was delayed. In the other case, which developed perirenal hematoma, elevation of the overall signal intensity after GdDTPA administration was lower in the treated kidney than in the non-treated kidney (patient 7).

Table 1. Acute renal effects of ESWL as shown by magnetic resonance imaging (n=37).

	No. Pts.	(%)
Perirenal fluid collection	14/37	(38)
Loss of corticomedullary junction	8/23	(35)
Increased signal intensity of muscle and other adjacent tissue	11/37	(30)
Renal enlargement	5/37	(14)
Subcapsular hematoma	1/37	(3)
Perinephric stranding	1/37	(3)
Hemorrhage into a pre-existing renal cyst	0/4	(0)

DISCUSSION

ESWL is the first-choice therapy for upper urinary tract stones, but has been reported to cause renal damage. Many studies using isotopes, computed tomography and blood chemical parameters have evaluated ESWL, and concluded that it is not a completely non-invasive therapy. We studied renal damage due to ESWL by MRI, including Gd-DTPA enhanced dynamic MRI for evaluating acute renal damage.The results suggest that d-MRI is highly sensitive and useful for studying intervals between therapeutic sessions and effects connected with long-term therapy.

REFERENCES

1. R Newman, R Hackett, D Senior, K Brock, J Feldman, J Sosnowski and B Finlayson, Pathologic effects of ESWL on canine renal tissue, *Urology* 29:194 (1987).
2. T Umekawa, K Kohri, N Amasaki, Y Ishikawa, M Takada, M Iguchi, and T Kurita, Renal damages after ESWL evaluated by Gd-DTPA-enhanced dynamic MRI, *Urol Int* 48:415 (1992).

LONG-TERM RESULTS OF EXTRACORPOREAL SHOCK WAVE LITHOTRIPSY, PERCUTANEOUS NEPHROLITHOTOMY AND OPEN SURGERY FOR UPPER URINARY TRACT STONE

A. Trinchieri[1], A. Mandressi[2], G. Zanetti[1], E. Montanari[1], G. Dormia[3], P. Luongo[3] and F. Rovera[1]

[1]Istituto di Urologia
Università di Milano
Milan, Italy
[2]UO di Urologia
Ospedale di Busto Arsizio, Italy
[3]Divisione di Urologia
Ospedale S Carlo Borromeo, Italy

INTRODUCTION

By the early 1980s percutaneous nephrolithotomy (PCNL) and extracorporeal shock wave lithotripsy (ESWL) had gained favour over open surgery. In particular, ESWL has become the operation of choice in the management of renal stones. The indications for ESWL have been extended to include the treatment of almost all renal stones[1]. It has also been claimed that ESWL in combination with PCNL should be the standard treatment of large volume staghorn stones[2]. The short term results of new techniques compare favourably with the operative series[3-6]. In contrast, there have been few investigations to determine the frequency of long term recurrences after ESWL and/or PCNL[7-9]. The incidence of late recurrence of stone following extracorporeal, endourological and open treatment is described and discussed.

MATERIALS AND METHODS

We carried out follow-up studies on patients treated for renal stones from 1980 to 1986. Between 1980 and 1983 renal stones were removed only through surgical lithotomy; since 1984 PCNL has been introduced, and at the beginning of 1985 ESWL became the primary method of treating stones. We reviewed 57 patients treated with ESWL monotherapy, 45 patients treated with PCNL (or by PCNL combined with ESWL for complete staghorn stones) and 59 patients treated by open surgery who all had at least three years of follow up. Stone morphology, burden and location of the stone were categorized according to Rocco's classification[10] in Table 1. In particular the location of the stone was defined as pyelic (C2), single or multiple calyceal associated or not with pyelic (C3), partial (C4) and complete staghorn (C5). Patients had follow-up studies 1 week, 3 months, 1 year, 2 years and 3 years postoperatively. At follow up patients underwent physical examination, serum creatinine, urinalysis, urine culture, plain abdominal film and renal ultrasound.

Table 1. Stones according to Rocco's classification

	Surgery	PCNL	ESWL
C2	18 (31%)	20 (44%)	15 (26%)
C3	22 (37%)	10 (22%)	35 (62%)
C4	12 (20%)	8 (18%)	7 (12%)
C5	7 (12%)	7 (16%) *	

* PCNL + ESWL

RESULTS

The residual stone rates after treatment and the 3 year recurrence rates of treatment with open surgery, PCNL and ESWL were examined in relation to location and size of the calculus (Table 2). The recurrence rate was lower in the patients who were free of residual fragments after ESWL or PCNL than in those who were not (Table 3).

Table 2. Residual stone rates after treatment and long-term recurrence

	Residual stone rates		
	Surgery	PCNL	ESWL
C2	0/18	6/20 (30%)	2/15 (13%)
C3	4/22 (18%)	4/10 (40%)	17/35 (49%)
C4	1/12 (8%)	5/8 (62%)	6/7 (87%)
C5	0/7	4/7 (57%)	-
	Long-term recurrence rates		
C2	6/18 (33%)	6/20 (30%)	3/15 (20%)
C3	9/22 (40%)	4/10 (40%)	15/35 (43%)
C4	5/12 (41%)	5/8 (63%)	3/7 (43%)
C5	3/7 (42%)	4/7 (57%)	-

DISCUSSION

We recommend ESWL as the primary procedure for almost all renal stones. The 3 year stone recurrence rates of 20 to 43% compare favourably with those of open surgery. PCNL has also proved to be effective for the removal of renal stones, but the morbidity of PCNL is greater than that of ESWL. PCNL should be a preferred treatment only for symptomatic patients with urinary obstruction and infection.

Staghorn calculi could be treated with PCNL followed by ESWL, but all stone fragments should be removed to ensure low recurrence rates. Complete clearance of large staghorn stones by combination therapy often requires so many adjunctive procedures that the cost-benefit ratio turns in favour of open surgery.

Therefore in our clinical experience, patients with large volume staghorn calculi involving all of the calyces still warrant open surgery, following which, if need be, ESWL can be utilized to fragment residual and inaccessible stones.

Table 3. Fate of residual fragments and recurrence rate of kidneys completely cleared after treatment (3 year follow up)

	Fate of residual fragments		
	Surgery	PCNL	ESWL
Free	-	4/19 (21%)	5/25 (20%)
Unchanged	-	6/19 (32%)	9/25 (36%)
Growth	5/5 (100%)	9/19 (47%)	11/25 (44%)
	Kidneys completely cleared after treatment		
Free	36/54 (67%)	20/26 (77%)	20/31 (65%)
Recurrence	18/54 (33%)	6/26 (23%)	11/32 (35%)

REFERENCES

1. CG Chaussy and GJ Fuchs, Current state and future developments of non-invasive treatment of human urinary stones with extracorporeal shock wave lithotripsy, *J Urol* 141:782 (1989).
2. H Schulze, L Elertle, J Graff et al, Combined treatment of branched calculi by percutaneous nephrolithotomy and extracorporeal shock wave lithotripsy, *J Urol* 135:1138 (1986).
3. JE Lingeman, D Newman, JHO Mertz et al, Extracorporeal shock wave lithotripsy: the Methodist Hospital of Indiana experience, *J Urol* 135:1134 (1986).
4. G Das, J Dick, MJ Bailey et al, 1500 cases of renal and ureteric calculi created in an integrated stone centre, *Br J Urol* 62:301 (1988).
5. DJ Jones, GL Russell, MJ Kellett et al, The changing practice of percutaneous stone surgery. Review of 1000 cases 1981-1988,*Br J Urol* 66:1 (1990).
6. JJ Smith III, JG Hollowell, RA Roth, Multimodality treatment of complex renal calculi, *J Urol* 143:891 (1990).
7. JW Segura, The role of percutaneous surgery in renal and ureteral stone removal, *J Urol* 141:780 (1988).
8. J Graff, W Diedrichs, H Schulze, Long-term follow up in 1,003 extracorporeal shock wave lithotripsy patients, *J Urol* 140:479 (1988).
9. M Marberger, W Stackl, W Hruby et al, Late sequelae of ultrasonic lithotripsy of renal calculi, *J Urol* 133:170 (1985).
10. F Rocco, A Mandressi, P Larcher, Surgical classification of renal calculi, *Eur Urol* 10:121 (1984).

GREY SCALE IMAGE DISCRIMINATION: AN ACCURATE METHOD OF STONE TARGETING?

N.P. Cohen, H.N. Whitfield, J.W.A. Ramsay and W.G. Bowsher

The Lithotripter Unit
St Bartholomew's Hospital
West Smithfield
London, UK

INTRODUCTION

Accurate stone targeting is desirable during lithotripsy because of concern over renal injury. A method has been developed to control shockwave release during real time imaging with ultrasound, so that firing occurs only when the stone is in the F2 focus.

PRINCIPLE

The human eye can only appreciate 16 shades of grey from white to black. However, the standard image generated on the ultrasound monitor of the lithotripter is in 64 shades. Consequently, more information is displayed than can be readily appreciated. Computerised analysis of renal ultrasound pictures has shown that hidden detail may be highlighted by grey scale analysis of the area of interest and that stone definition may be improved[1].

A grey scale image discriminator, the "Bart's Box", has been produced by Miller Systems, U.K. This modifies the conventional image (in 64 shades) generated by the standard lithotripter imaging system. The operator can select a particular grey level by manual controls on the Box, akin to highlighting a particular contour on a map. Each pixel from the original image is analysed and, depending on its relationship to the grey level selected, is then depicted in either black or white. A modified image of the stone within the kidney can be formed in just black and white. This modified image is displayed on a separate monitor. Manual controls allow this second monitor to display either the fully modified image, or a variable mix of the standard and modified images.

A memory facility has been incorporated within the "Bart's Box". Before ESWL treatment is commenced it is possible to "memorise" the image of the renal stone in the lithotripter focus. Subsequently during ESWL treatment the real time dynamic image is continuously compared with this stored image. Shockwave release is performed through a mechanical switch produced by Dornier Medizintechnik, Germany. This switch allows shockwaves to be fired only when the incoming image and the stored image are sufficiently similar, i.e. when the stone is within the F2 focus.

PATIENTS AND METHODS

To date, 43 patients have completed follow up after being treated with ESWL on a Dornier MPL 9000 lithotripter; 22 with the Bart's Box (group A) and 21 without (group B). All had single, uncomplicated radio-opaque stones in the renal collecting system which were localised by ultrasound. The number of shockwaves applied was determined by stone size; 1000 for stones <5mm, 1400 for stones 6-10mm, 1800 for stones 11-15mm, and 2000 for stones >15mm. Where possible, treatments were given at 18 kV output, beginning at 14 kV and increasing by one kV every 50 shockwaves. Patients were reviewed at 3 months and a plain KUB radiograph taken.

RESULTS

One stone in each group failed to fragment (>95% fragmentation rate). At 3 months 18/22 (83%) of group A were stone free on plain radiography and 17/21 (81%) of group B. No complications were seen in those treated in either group. No significant difference in success rates was found between the two groups.

DISCUSSION

Many factors are known to affect the outcome of stone treatment with ESWL - the stone's size, composition, and location, the patient's size, the power setting of the lithotripter generator, etc. This study has attempted to show a theoretical improvement which would be expected if more accurate stone targeting could be achieved. Unfortunately this series is too small at present to show a difference as measured by stone clearance alone. Therefore more patients are being recruited into this study. In addition, these patients are being studied to examine the degree of renal injury induced by ESWL with the MPL 9000 by measuring changes in enzymuria, proteinuria and renal clearance studies.

We hope to show that this image discriminator is beneficial in ultrasound guided treatment of renal stones. We believe this technique can reduce renal injury by allowing fewer shockwaves to be delivered to the surrounding renal tissue. With more shockwaves directly hitting the stone, fragmentation may become more efficient. In theory, this may also reduce the risk of complications from large stone fragments.

REFERENCE

1. WG Bowsher, JWA Ramsay and HN Whitfield, Ultrasound image analysis during ESWL. VIIth World Congress on Endourology and ESWL (1989).

SHOCK WAVE LITHOTRIPSY IN THE TREATMENT OF URINARY STONES IN RENOURETERAL ANOMALIES

E. Montanari, G. Zanetti, A. Trinchieri, A. Guarneri, M. Seveso and E. Austoni

Istituto di Urologia
Università di Milano
Milan, Italy

INTRODUCTION

Renoureteral anomalies can be seen as a pathogenetic factor of urolithiasis or simply as innocent bystanders revealed by symptomatic urolithiasis[1]. When ESWL was first attempted, these renoureteral anomalies were considered as absolute contraindications for this kind of treatment[2]. Nowadays, however, admission criteria for ESWL have been extended[3]. In the light of our own experience and of data from the literature we undertook the present review to examine whether these contraindications were still valid, with particular reference to calculosis associated with renoureteral duplication, horseshoe kidney and pelvic renal ectopy; we also wanted to assess the eventual therapeutic alternatives.

PATIENTS AND METHODS

We examined for the presence of renal stones associated with renoureteral anomalies, 2880 patients treated by ESWL from 1985 to December 1991 by different Dornier lithotripters: Original HM3 (1034 patients), Modified HM3 (1530 patients) and MPL 9000 (316 patients). We found renoureteral anomalies in 56 patients (1.9%) and we observed renoureteral calculosis associated mainly with complete and partial renoureteral duplication with and without ureterocele, with horseshoe kidney and with ectopic pelvic kidney (Table 1).

Table 1. Renoureteral Anomalies and Urinary Stones: 56 Patients (1.9%)

Ureteral duplication	**32 Patients (1%)**				
Complete	11	Complete + Ureterocele	2	Partial	19
Renal stone	6			renal stone	14
Lumbar stone	3			lumbar stone	3
Pelvic stone	2			pelvic stone	2
Horseshoe Kidney	**20 Patients (0.7%)**				
Ectopic Pelvic Kidney	**4Patients (0.1%)**				

Urolithiasis 2, Edited by R. Ryall *et al.*,
Plenum Press, New York, 1994

We performed pretreatment ureteral catheterization or stenting in 13 patients (40%) with renoureteral duplication, in 19 patients (95%) with horseshoe kidney, and in all 4 patients with ectopic pelvic kidney. Post treatment manoeuvres were performed only in 5 patients with renoureteral duplication (1 basket, 1 nephrostomy and 3 ureteral endoscopic meatotomy). From 1985 to 1987 13 patients with renal stones in horseshoe kidney have been treated in the supine position with the Dornier OHM3 and we did not localize the stone in 2 patients. Since 1988 we have treated 6 patients with HM3 in the prone position and have always localized the stone; 1 patient was treated with MPL 9000 in the supine position. All patients with renal stones in an ectopic pelvic kidney have been treated in the prone position, 3 of them by Dornier MPL 9000 and the other by Dornier HM3. Follow-up has been performed at discharge, 30 days, and 90 days.

RESULTS

Results are shown in Table 2.

Table 2.

Ureteral duplication	Discharge	30 days	90 days
Renal stones			
Stone free	5 (25%)	14 (70%)	
Passable fragments*	15 (75%)	6 (30%)	
Ureteral stones			
Stone free	6 (60%)	10 (100%)	
Passable fragments*	4 (40%)	-	-
Horseshoe Kidney Series 1985-87			
Stone free	2	4 (36%)	5 (45.5%)
Passable fragments*	8	6 (55%)	5 (45.5%)
Passable fragments	1	1 (9%)	1 (9%)
No treatment	2	-	-
Horseshoe Kidney Series 1988-91			
Stone free	1	3 (43%)	3 (43%)
Passable fragments*	6	4 (57%)	4 (57%)
Ectopic Pelvic Kidney			
Stone free	0	2	
Passable fragments*	4	2	

*Passable fragments = Diameter< 5 mm

DISCUSSION

In our experience, renoureteral duplication - whether complete or incomplete - unlike the horseshoe kidney does not constitute *per se* a pathology that predisposes to urolithiasis. In fact, the prevalence of renoureteral duplication in the general population is 0.8% and in our series 1%, while those for horseshoe kidney are 0.25% and 0.7%, respectively. It is well known, anyway, that urolithiasis risk is 8 to 10 times greater in horseshoe kidney because of infection, dilation or metabolic disorders associated with the anomaly[1,4-6]. In our opinion ESWL is the preferred method in the treatment of calculosis with complete or incomplete renoureteral duplicity: in renal stone treatment we frequently performed pretreatment catheterization to prevent post-treatment complications. In ureteral stone treatment for the lumbar tract, pretreatment catheterization allows the performance of "push and smash" or localization of the stone. For the pelvic and intramural tract, we prefer to employ MPL 9000[7]. In the ESWL of renal calculosis in horseshoe kidney, placing the patient in a prone position rather than supine always allows for correct aim and fragmentation[8]. Fragment clearance is insufficient, however, and only 42-45% patients are stone-free at 90 days follow-up. The percutaneous approach to horseshoe kidney guarantees a 77.7% rate of success (stone-free), which rises to 88.8% when associated with post-PCNL shock wave lithotripsy[9]. ESWL allows the resolution of 50% of cases of calculosis of the ectopic pelvic kidney if auxiliary manoeuvres are used prior to treatment.

REFERENCES

1. AD Perlmutter, AB Retik, SB Bauer, Anomalies of the upper urinary tract, in: "Campbell's Urology" WB Saunders, Philadelphia (1985).
2. CG Chaussy, E Schiedt, D Jocham, W Brendel, B Forsmann, V Walther, First clinical experience with extracorporeally induced destruction of kidney stones by shock waves, *J Urol* 127:417 (1982).
3. CG Chaussy, GJ Fuchs, Current state and future developments of non invasive treatment of human urinary stones with extracorporeal shock wave lithotripsy, *J Urol* 141:782 (1989).
4. JW Segura, PP Kelalis, EC Burke, Horseshoe kidney in children, *J Urol* 108:333 (1972).
5. A Mottola, C Selli, M Carini, A Natali, Lithiasis in horseshoe kidney. Acta Urol Belg 52:355 (1984).
6. WP Evans, MJ Resnick, Horseshoe kidney and urolithiasis, *J Urol* 125:620 (1981).
7. GG Tally, Experience with the Dornier HM4 and MPL9000 Lithotripters in urinary stone treatment, *J Urol* 144:622 (1990).
8. AD Jenkins, JY Gillenwater, ESWL, in the prone position: treatment of stones in the distal ureter or anomalous kidney, *J Urol* 139:91 1 (1988).
9. DJ Jones, JEA Wickham, MJ Kellet, Percutaneous nephrolitotomy for calculi in horseshoe kidneys, *J Urol* 145:481 (1991).

LITHOCLAST - A NEW ENDOLITHOTRIPTOR

K. Kleinschmidt, K. Miller, H.W. Gottfried and R. Hautmann

University of Ulm
Department of Urology
7900 Ulm, Germany

INTRODUCTION

With the advent of laser lithotripsy and miniscopes for the treatment of ureteral calculi, ultrasonic lithotripsy has lost its function as a universal lithotripsy system, as its probes are not compatible with the miniscopes. Still, laser lithotropsy has its disadvantages: purchase costs are high and the low fragmentation rates make it unsuitable for the primary treatment of bladder calculi and staghorn stones in the kidney. Thus, there is a continuing need for a universal lithotripsy system providing both miniaturized probes and high stone ablation rates.

MATERIAL AND METHODS

Over a period of 21 months we have used the lithoclast prototype EMS Switzerland for endolithotripsy in the ureter, kidney and bladder. The Lithoclast is a pneumatically operating device, providing semi-rigid probes in a size from 1.8 to 8 French. For the ureter the 9 - 11.5 French semi-rigid Wolf ureteroscope was utilized for most of the cases. For percutaneous surgery, standard nephroscope equipment was used. For bladder stone lithotripsy, a specially designed cystoscope with an angled eyepiece and a straight working channel is available.

RESULTS

A total of 60 patients was treated (see Table 1).

Forty one patients suffered from ureteral calculi; 24 were female, 17 male. In 26 cases stones were located in the distal ureter, in 8 cases in the middle, and in 7 cases in the

Table 1. Results by Lithoclast Lithotripsy

Calculi Position	complete	Fragmentation incomplete	none
Ureteral (n = 41)	35	3	3
Renal (n = 4)	4	-	-
Bladder (n= 15)	15	-	-

Urolithiasis 2, Edited by R. Ryall *et al.*,
Plenum Press, New York, 1994

proximal ureter. In all cases, the ureter was accessed by retrograde ureteroscopy; no dilatation of the orifice was necessary in any case. Twenty nine patients were treated under general or regional anesthesia and 4 patients under local anesthesia. Complete fragmentation was achieved in 35/41 patients and in 6 calculi no fragmentation, or incomplete fragmentation occurred, owing to stone dislocation. We observed 2 perforations of the ureter, which were managed by insertion of a double-J-stent; no further therapy was required for these complications.

Fifteen bladder stones were treated in 14 male patients and 1 female patient. Lithoclast-lithotripsy resulted in complete fragmentation of all calculi. The fragments were removed with an Ellic evacuator. As complications, 2 bladder mucosa lesions occurred, requiring no therapy.

Four renal staghorn calculi were treated by standard PCNL. The operations were done under general anesthesia. Again, Lithoclast lithotripsy resulted in complete fragmentation in all cases. No complications occurred. No technical failure was observed during the entire treatment period.

DISCUSSION

Technically, the Lithoclast works like a mini pneumatic hammer. No special energy source is required. The unit must be supplied with dry, clean air at a pressure between 3.5 and 5 bars, thus providing a working energy of about 85 mJ. Compressed air as an energy source is inexpensive and readily available throughout modern hospitals. The Lithoclast probe is compatible with most of the semi-rigid miniscopes currently available. However, it is not suitable for flexible ureteroscopes. The efficiency of the lithotripsy is excellent and the results are equal or even superior to laser lithotripsy[1-3]. This is true even for very hard stones, which may resist laser lithotripsy[3]. Our complication rate has been low and comparable to other reports[4].

Being equally effective for ureteral, bladder and renal calculi, the Lithoclast confirms its role as a universal lithotripsy system, thus providing a viable low cost alternative to the laser and to other lithotripsy systems.

REFERENCES

1. SP Dretler, Laser photofragmentation of ureteral calculi: analysis of 75 cases, *J Endourol* 1:9 (1987).
2. F. Eisenberger, K Miller, J Rassweiler, Stone therapy in urology, George Thieme Verlag Stuttgart (1991).
3. K Esuvaranathan, EC Tan, PK Tan, KH Tung, Does transurethral laser ureterolithotripsy justify its cost? *J Urol* 148:1091-1094 (1992).
4. JD Denstedt, PM Eberwein, RR Singh, The Swiss lithoclast: a new device for intracorporeal lithotripsy, *J Urol* 148: 1088-1090 (1992).

HYPERTENSION AFTER EXTRACORPOREAL SHOCK WAVE LITHOTRIPSY: LONG TERM FOLLOW UP

G. Zanetti, A. Trinchieri, E. Montanari, A. Guarneri,
E. Austoni and E. Pisani

Istituto di Urologia
Università degli Studi di Milano
Milan, Italy

INTRODUCTION

Hypertension has been reported to occur following extracorporeal shock wave lithotripsy (ESWL). Retrospective studies reported an incidence of new onset hypertension between 2.4 and 8.1%) and a significant rise in mean diastolic blood pressure was observed by some authors[1-5]. In order to assess the clinical impact of the potential risk of hypertension after ESWL, the present prospective study was performed.

MATERIALS AND METHODS

Sixty-eight patients (35 males, 33 females) mean age 45 years (range 15-67) who had been treated with ESWL for renal stones by the modified Dornier HM3 lithotripter were checked in a 24- to 36-month prospective study. None had been receiving pretreatment antihypertensive therapy. They were studied as follows: medical history, physical examination, ECG, renal ultrasound scan, and laboratory tests (urinary ratio of N-acetylglucosaminidase to creatinine, urinalysis). Blood pressure measurements were taken 3 times during each visit by a mercury random-zero manometer. These evaluations were performed 15 days prior to ESWL and 1, 6 and 24 to 36 months (mean 29 months) after the treatment. Following the Joint National Committee on Detection, Evaluation and Treatment of High Blood Pressure, the patient was defined as hypertensive if the diastolic blood pressure exceeded 95 mm Hg and the systolic pressure 160 mm Hg, and as borderline if the diastolic blood pressure was between 90 and 95 mm Hg and the systolic pressure between 140 and 160 mm Hg. Out of our group, 51 patients were normotensive before treatment, 9 borderline, and 8 hypertensive. These three groups of patients underwent, respectively, a mean of 2852±2209 shock waves (mean kV 20±1.2), 2872±1890 shock waves (mean kV 20.4±1.7) and 2980±2307 shock waves (mean kV 21.3±1.9). No more than 3,000 shock waves were applied in any session. For comparative statistical analysis, the t-test was used.

RESULTS

No significant increase in the mean diastolic blood pressure was observed after ESWL, but a slight increase in urinary NAG/creatinine ratio suggested mild renal tubular damage (Table 1). No correlation was found between the incidence of hypertension and the number of shock waves applied, the kilovoltage, urinary NAG variations, or initial blood pressure (Table 2).

Urolithiasis 2, Edited by R. Ryall *et al.*,
Plenum Press, New York, 1994

Table 1. Mean blood pressure (mmHg) at various follow up times after ESWL and urinary NAG/creatinine before and after ESWL.

	Basal	1 month	6 months	24-36 months
Diastolic	85±11	81±7	83±6	84±9
Systolic	128±17	126±12	130±12	132±15
	Basal	2nd day		
NAG/Cr (mU/mg)	5.7±5	7.9±8 *		

* p <0.05

Table 2. Treatment features, initial blood pressure and laboratory findings according to blood pressure at most recent follow up.

	Normotensive (n = 53)	Hypertensive (n = 6)	Borderline (n = 6	New Hypertensive (n = 3)
Shock waves (n)	2886±2226	2500±1152	3058±1053	2223±378
Kilovoltage (kV)	20±1	19±1.5	20±0	20±1
Diastolic BP(mmHg)	81±7	95±11	89±5	83±2
Systolic BP (mmHg)	126±10	159±8	149±7	128±7
NAG/Cr (mU/mg)				
Pretreat	6.1±5	5.0±5.3	4.4±4.9	5.0±2
Post-treat	9.5±9.1	6.1±7.9	7.0±0.5	6.0±2.7

Neither subcapsular nor perirenal hematomas nor morphological changes linked to ESWL treatment were found at the short- or long-term follow up. Among the originally normotensive group, three new cases of hypertension (6%) were observed during follow up, two at 1 month and one at 2 years after ESWL. Only one of these patients required pharmacological therapy. The hypertension prevalence by age in our group of patients was similar to that predicted from epidemiological data adjusted for age and sex. The number of the normotensive group developing a diastolic pressure of 100 mm Hg or above during 24 to 36 months of follow up was not significantly greater than expected. Specifically, according to the data of Miall and Chinn[6] two cases of new onset hypertension would be expected (4.3%), whereas three (5.8%) were observed. This difference is not statistically significant (chi square = 0.12, $p < 0.05$) (Table 3).

Table 3. The percentage of normotensive patients expected to have attained a diastolic blood pressure of 100 mm Hg or more, 24-36 months after ESWL, as calculated from the data of Miall and Chinn by sex, age and initial diastolic pressure, compared to the actual percent observed.

	Hypertensive	Normotensive
No. expected	2 (4.3%)	49
No. observed	3 (5.8%)	48

DISCUSSION

Our prospective study on a small number of patients did not demonstrate any significant mean diastolic blood pressure rise at long-term follow up. However,we believe that long-term surveillance is essential after ESWL.

A larger prospective trial is required to evaluate the existence and clinical impact of small pressure rises in treated patients.

Furthermore, identification of new onset hypertensive patients could offer the opportunity to study the risk factors related to post-ESWL blood pressure rises.

REFERENCES

1. JE Lingeman, J Woods et al, The role of lithotripsy and its side effects, *J Urol* 141:793 (1989)
2. B Liedl, D Jocham et al, Long term results in ESWL-treated urinary stone patients, *Urol Res* 16:256 (1988).
3. CM Williams, JV Kaude et al, Extracorporeal shock wave lithotripsy: long term complications, *Am J Roentg* 150:311 (1988).
4. BSI Montgomery, RS Cole et al, Does extracorporeal shockwave lithotripsy cause hypertension ? *Br J Urol* 64:567 (1989).
5. JE Lingeman, JR Woods and PD Toth, Blood pressure changes following extracorporeal shock wave lithotripsy and other forms of treatment for nephrolithiasis, *JAMA* 263:1789 (1990).
6. WE Miall and S Chinn, Screening for hypertension: epidemiological observations, *Br Med J* 3:595 (1974).

ESWL MONOTHERAPY FOR STAGHORN RENAL CALCULI

K. Yamamoto, H. Iimori, W. Sakamoto, T. Nakatani,
T. Sugimoto, S. Wada, M. Harima, Y. Katoh, T. Itoh,
S. Kasai and T. Kishimoto

Department of Urology
Osaka City
University Medical School
Osaka, Japan

INTRODUCTION

We started ESWL in July 1985 at Osaka City University Hospital. Based on our initial experience in treating patients with large renal stones, we chose ESWL monotherapy as a basic treatment for staghorn stones. When possible, we placed a double J stent before both initial ESWL and intentional repeated ESWL therapy were performed. From 1985 through 1991 we treated 104 Patients (107 stones) with a Dornier HM3 at Osaka City University Hospital and 13 patients (13 stones) with a Siemens Lithostar at Osaka Teisin Hospital. We retrospectively studied the therapeutic results of ESWL monotherapy in treating staghorn stones.

Patient profile

The 104 patients treated with the HM3 included 51 men and 53 women with a mean age of 50.4 years. There were 6 patients with a solitary kidney.

Table 1. Patient Profile.

	Men	Women	Total
Complete staghorn stone	31 cases (47.3 y)	34 cases (48.4 y)	65 cases (47.9 y)
Partial staghorn stone	28 cases (53.0 y)	27 cases (52.7 y)	55 cases (52.8 y)
Total	59 cases (50.0 y)	61 cases (50.3 y)	120 cases (50.2 y)

RESULTS

Of the 120 stones (117 patients) 34% of the kidneys were stone free and 27.5% had only small fragments less than 4mm in diameter, giving a total success rate of 61.7%. 5998 shock waves were required with the HM3 and 16300 for the Lithostar. Transurethral ureterolithotripsy was required in 24 renal units and percutaneous nephrostomy in 4. Percutaneous nephrolithotomy was required in 5 patients, and 32.5% of the 120 stones required this treatment. A case of nonvisualized kidney caused by xanthogranulomatous pyelonephritis before ESWL needed nephrectomy.

Details were analyzed in 107 staghorn stones treated with the HM3 since they comprised the majority of patients in this study, and these are shown in Table 2. The clinical results of complete staghorn stones (55 stones) were compared with partial staghorn stones (52 stones). The success rate was 54.5% and 71.1% respectively. The success rate was also analysed in relation to stone composition but no correlation was found.

Table 2. Success rate and clinical parameters in relation to stone size (long axis) in the patients treated with the HM3.

	<40mm	<50mm	<60mm	<70mm	<80mm	>/=80mm
Stone free	35.3%	32.0%	61.1%	25.0%	--	27.3%
residual stone <4mm diam.	35.3%	28.0%	27.8%	33.3%	57.1%	-
No. patients	17	25	18	12	7	11
No. shock waves	5344	6300	5131	6014	7350	8841

CONCLUSION

1. ESWL monotherapy is an effective initial treatment for staghorn stones.
2. Stone composition has a minor effect on success rate but the size and the volume of the stone may affect the rate of success.
3. Careful observation is necessary, especially in the case of massive staghorn stones.

TWO YEAR EXPERIENCE WITH THE LITHOCLAST FOR ENDOLITHOTRIPSY

M.Weber, M. Wöhr, R.D. Huber and D. Frohneberg

Department of Urology
Karlsruhe City Hospital
Karlsruhe, FRG

INTRODUCTION AND METHODS

The endolithotripsy of ureteral stones that have not responded to extracorporeal *in situ* treatment has been greatly facilitated by the development of semi-rigid fiberoptic "miniscopes". With these ureteroscopes of 6 and 7 F tip diameter and a working channel of 3 to 4 F, ultrasonic lithotripsy is no longer possible because of the size of the sonotrode. Alternative methods which have been used to date are Laser and Electrohydraulic Lithotripsy. For the last two years a newly developed endolithotripsy device has been investigated clinically. The Lithoclast (EMS, Le Sentier, Switzerland) is a pneumatically operated device, providing semi-rigid probes between 1.8 and 8 F for the treatment of ureteral as well as bladder stones. Because of the semi-rigid probes the device requires endoscopes with straight working channels. A 7 F semi-rigid ureteroscope and a 21 F lens cystoscope, both with an offset eyepiece, have been designed as special Lithoclast endoscopes (Richard Wolf, Knittlingen, FRG).

Between February 1990 and January 1992 a total of 52 patients were treated with the Lithoclast. Nineteen were female and 33 were male. Thirty eight patients had a ureteral calculus, 10 a bladder stone and in 4 patients preliminary experience was gained during percutaneous nephrolithotomy

RESULTS AND CONCLUSION

The 38 ureteral stones 34 (89.5%) were fragmented sufficiently. The remaining stones were dislocated and flushed back into the renal pelvis, inadvertently, before or after incomplete fragmentation. The 10 bladder stones were all fragmented sufficiently to be flushed out or extracted and in the 4 renal stones sufficient fragmentation was also achieved. No perforations due to the lithotripsy procedure itself were observed during any of the treatments.

In conclusion, the Lithoclast has proved to be a safe and efficient treatment modality for urolithiasis, with low maintenance and low risk. Its use during percutaneous treatments, however, does not give any advantage over ultrasonic lithotripsy since it is less effective, lacking the permanent suction and the much finer fragmentation of the sonotrode. For bladder stones, and even more so for ureteral calculi, it seems to have become the modality of choice, being much less traumatic than electrohydraulic lithotripsy and 10 times cheaper than a Laser Lithotripter.

ELECTROHYDRAULIC vs. LASER LITHOTRIPSY: IS NEW NECESSARILY BETTER?

R. Berlin, B.A. Feagins and G.M. Preminger

The University of Texas Southwestern Medical Center
Dallas, Texas

INTRODUCTION

Innovations in electrohydraulic lithotripsy (EHL) have allowed intracorporeal stone fragmentation with normal saline (NS) instead of 1/6 NS, as well as the introduction of small caliber EHL probes. These smaller probes are of similar size to the 600 micron (1.8 F) laser fibers. While NS irrigation and the smaller caliber probes are currently in use, basic *in vitro* studies have not yet been performed which demonstrate the efficiency of these innovations.

METHODS

Plaster of Paris stone phantoms (mixing ratio of 2:1) were numbered, desiccated (for 4 h at 80°C) and weighed. The stone phantoms were placed precisely 1 mm from the tip of the EHL probe and submerged in either NS or a solution of 1/6 NS. Stone phantoms received 20 shocks at a power setting of 75 volts using a Northgate SD200 EHL generator. Following fragmentation, the stone phantoms were redried and again weighed. Identical conditions were utilized to test phantom fragmentation induced by the 1.8 F pulsed dye laser fiber @ 120 mJ power.

RESULTS

Stone destruction with the EHL probes was significantly greater in the NS irrigant, compared to 1/6 NS ($p < 0.05$). The 1.4 F probe produced significantly more volume loss of 180 mm^3 compared to the 3 and 5 F probes (145 and 137 mm^3, $p < 0.05$). Moreover, the 1.4 F EHL probe produced significantly more volume loss than the 1.8 F laser fiber (180 vs. 112 mm^3, $p < 0.05$).

DISCUSSION

Our findings reaffirm the current practice of using NS as an irrigant medium for performing intracorporeal stone fragmentation with EHL. Moreover, the efficacy of the smaller 1.4 F EHL probes appears equivalent to that of the larger EHL probes and laser fibers. Combined with the advantage of allowing increased irrigant flow and significantly reduced expense compared to laser lithotripsy, the 1.9 F EHL probes would appear to be preferable for use during flexible or rigid uterorenoscopy.

IN SITU TREATMENT OF URETERAL CALCULI WITH EXTRACORPOREAL SHOCKWAVE LITHOTRIPSY - STILL FEASIBLE?

M. Wöhr, M. Weber, R.D. Huber and D. Frohneberg

Department of Urology
Karlsruhe City Hospital
Karlsruhe, FRG

INTRODUCTION AND METHODS

In most European centers, *in situ* treatment of ureteral calculi with extracorporeal shockwave lithotripsy (ESWL) is still the method of choice. Major advantages in minimally invasive endolithotripsy with very fine, semi-rigid fiberscopes of 6 and 7 F tip diameter have threatened this approach. In a retrospective study the ESWL treatment data were analysed in order to evaluate the feasibility of extracorporeal *in situ* treatment.

In 1991 a total number of 334 patients were treated with ESWL for ureteral calculi with the Siemens Lithostar. Two hundred and ninety four patients with an average age of 49.5 (20-89) yrs could be evaluated. One hundred and one stones were located in the proximal, 47 in the mid and 146 in the distal ureter. Stone localisation was achieved with fluoroscopy, the median number of shockwaves was 2,820, the median maximal energy 17.2 kV. Fifty three percent of patients went without any analgesia, 46% needed an i.v. analgesic and 0.6% were treated under general anesthesia.

RESULTS AND CONCLUSION

The 294 evaluable patients needed 490 ESWL treatments, which equals a rate of 1.66 treatments per patient, the overall re-treatment rate thus being 49.4%. A retrograde auxillary treatment was necessary in 49 patients (16%) consisting of 44 ureteroscopies and 7 "push & smash" treatments. eleven patients (3.7%) required a temporary nephrostostomy after ESWL-treatment. No patient required an open ureterolithotomy. No primary or secondary complications necessitating operative or intensive care treatment as a result of ESWL-treatment to the ureter were encountered in this series.

In conclusion, extracorporeal *in situ* treatment of ureteral calculi was, in this relatively large series, found to be a non-invasive treatment modality without any complications. It caused only mild or no real discomfort for the patient. An overall re-treatment rate of about 50%, however, is a major drawback compared to endolithotripsy techniques with a primary disintegration rate of about 90%. Despite this considerable difference in efficacy we still believe that, because of its entirely non-invasive nature even for repeated treatments, extracorporeal *in situ* treatment of ureteral calculi should remain the primary choice.

Although ureteroscopy may be 'minimal' it will always remain 'invasive'.

THE EFFECT OF HIGH ENERGY SHOCK WAVES ON OXALATE UPTAKE IN CULTURED RENAL CELLS

D.W. Bowyer, M.E. Moran, T.R. Wandzilak, P.A. Davis
and H.E. Williams

University of California
Davis
Davis, California

INTRODUCTION

Cellular damage, as evidenced by decreased cell viability, in response to high energy shock waves (HESW) has been demonstrated *in vitro*. However, effects of HESW on specific metabolic cell functions have not been described. We studied the effect of HESW on active cellular uptake of oxalate.

METHODS

We have previously described oxalate uptake in the LLCPK kidney cell line. LLCPK cells were plated at a density of 2.5 $\times 10^5$ cells/mL on 35x10mm tissue culture dishes in alpha MEM with 10% fetal calf serum and antibiotics. Uptake studies were performed on days 1,3 and 5 using a defined mannitol/Ca-EGTA buffer system. The cells were incubated with ^{14}C labeled oxalate to measure uptake. Trapping of oxalate was monitored using ^{3}H mannitol.

LLCPK cell suspensions (4×10^6 cells/2mL) in sterile polyethylene vials were positioned at the F2 focus of a Dornier HM3 lithotripter. Control vials were positioned in the same bath away from the F2. Treatment was 2000 shocks at 20KV. Viability of cells was assessed using trypan blue exclusion. Shock treated cell viability was 62±10% and untreated control viability was 88±5%. After treatment, shocked and control cells were plated at an equal viable cell number. Oxalate uptake after a 45 minute incubation period was determined.

RESULTS AND DISCUSSION

There was no significant difference in oxalate uptake between HESW treated and control cells at one, three or five days of culture. Confluency was reached after four days in both HESW treated and control cells.

These results demonstrate no detrimental effect of HESW on cell growth and no effect on the active metabolic uptake of oxalate in cultured renal cells.

RESIDUAL STONES FOLLOWING EXTRACORPOREAL SHOCK WAVE LITHOTRIPSY - THE CASE FOR CITRATE PROPHYLAXIS

K. Suzuki[1], R. Tsugawa[1] and R.L. Ryall[2]

[1]Department of Urology
Kanazawa Medical University
Ishikawa, Japan
[2]Department of Surgery
Flinders Medical Centre, South Australia

INTRODUCTION

The aim of this study was to evaluate the ability of sodium-potassium citrate (CG-120) to inhibit calcium oxalate crystal nucleation and growth on to stone fragments remaining after treatment.

METHODS

Fragments of calcium oxalate stones were obtained from 10 patients who had been treated with a Siemens LITHOSTAR. The continuous flow crystallization technique was adapted to induce calcium oxalate crystal nucleation and growth on to the surface of these fragments. Stone shards were left in a membrane filter cup in contact with the artificial urine for 3 h. The filter cup with fragments was removed, allowed to dry for 24 h and weighed. The inhibitory effect of sodium-potassium citrate was assessed by scanning electron microscopy and by determining the relative increase in crystalline mass at final concentrations of 0, 2, 4, 6, 8 and 10 mmol/L.

RESULTS AND DISCUSSION

In the absence of citrate, the median weight of the stone fragments increased by 4.3 mg. With increasing concentration of citrate the magnitude of the change in weight of the fragments decreased proportionately; at a concentration of 10 mmol/l the increase was 1.39 mg. Scanning electron micrographs of the stone fragments after 3 h incubation period showed that in all cases newly formed crystals were clearly visible upon the surface of the fragments. The number of newly formed crystal clumps decreased with increasing citrate concentration.

It was concluded that sodium-potassium citrate may provide an effective means of preventing the formation of new kidney stones resulting from the deposition of calcium oxalate on to residual stone fragments caused by shock wave lithotripsy and that the technique used in this study may be an efficient means of testing the efficacy of therapeutic agents to prevent stone recurrence in patients treated with extracorporeal shock wave lithotripsy.

SODIUM-POTASSIUM CITRATE IN THE TREATMENT OF RESIDUAL STONE FRAGMENTS RESULTING FROM EXTRA-CORPOREAL SHOCK WAVE LITHOTRIPSY

E. Cicerello, F. Merlo, A. Tessarolo and G. Anselmo

Kidney Stone Center
Division of Urology and Institute of Clinical Chemistry
Regional Hospital
Treviso, Italy

INTRODUCTION AND METHODS

Approximately 50% of renal stone patients subjected to extracorporeal shock wave lithotripsy (ESWL) show a tendency to recurrences because of residual fragments[1]. In contact with the supersaturated urine of recurrent calcium oxalate stone (CaOx) formers, fragments may growth or re-form into new stones, particularly as a possible consequence of their inadequate discharge following ESWL[2]. Alteration in body posture, the use of double pigtail ureteric stents, early retreatment ESWL, and percutaneous procedures have been suggested to facilitate their removal. However, a chemical approach for the prophylaxis of these recurrences has never been well evaluated. The aim of this study is to test the efficacy of sodium-potassium citrate for preventing the recurrences of CaOx stones after ESWL.

The trial included 20 patients with recurrent calcium oxalate stones who, 40 days after ESWL, showed residual fragments of the previously treated stone. In 11 patients the fragments were lodged in the lower calices, while 5 patients showed structural renal abnormalities. Sodium-potassium citrate was administered for 6 months (6-8 g daily in three divided doses). The patients were compared with a control group of 15 untreated patients with similar radiological and biochemical characteristics.

RESULTS AND DISCUSSION

At 6 months 68% of treated patients were free of stone fragments in comparison with 13% of the control group, and the Ca/Cit ratio was lower than that in the group of untreated patients ($t = 6.047$, $p > 0.001$).

Consistent with *in vitro* [3] evidence, our results confirm *in vivo* the effectiveness of sodium-potassium citrate in preventing the formation of new kidney stones by the deposition of calcium oxalate on to residual fragments after ESWL. The use of this treatment for preventing stone recurrence after ESWL is therefore recommended.

REFERENCES

1. AG Mulley, KJ Carlson and SP Dretler S, *AJR* 150:316 (1988).
2. B Pettersson, *Scand J Urol Nephrol* (Suppl.), 120:1 (1989).
3. K Suzuki, R Tsugawa, RL Ryall, *Br J Urol* 68:132 (1991).

EXTRACORPOREAL LITHOTRIPSY OF RENAL CALCULI WITH THE STORZ MODULITH™ SL20: A MULTICENTER STUDY

A. Finkbeiner[1], J.E. Fowler[2], R.J. Kahnoski[3], F.A. Klein[4] and M.I. Resnick[5]

[1]University of Arkansas
Little Rock, Arkansas
[2]University of Mississippi
Jackson, Mississippi
[3]West Michiagan Stone Center
Butterworth Hospital
Grand Rapids, Michigan
[4]University of Tennessee
Knoxville, Tennessee
[5]Department of Urology
Case Western Reserve University
Cleveland, Ohio

INTRODUCTION

During a 14 month period, a total of over 200 patients with renal calculi were treated at centers of the first 4 authors. Inclusion criteria were: 1) stone(s) in kidney or ureter suitable for lithotripsy or an alternative procedure; 2) 21-85 years of age; 3) informed consent; 4) anatomy permitting adequate focusing of stone; and 5) clear localization of the stone. Exclusion criteria included: 1) pregnancy; 2) presence of vascular calcification; 3) presence of a coagulopathy; 4) presence of a cardiac pacemaker; 5) obstruction distal to the stone; and 6) use of anti-inflammatory agents within 2 weeks of proposed treatment.

METHODS AND RESULTS

The Modulith™ SL20, a third generation electromagnetic lithotripter utilizes a unique cylindrical wave generator source and a parabolic reflector focusing system. The resulting large aperature angle reduces tissue stress and pain response without compromising fragmentation efficiency. The Modulith™ SL20 incorporates real-time coaxial ultrasound imaging and an integrated isocentric C-arm X-ray system with pulse progressive fluoroscopy for positioning and monitoring of fragmentation.

Treatment parameters and results, including success and stone free rates within a 90 day follow-up period, were analysed. Renal function was stable at the completion of the follow-up period. Success rate correlated with stone size. Only 7% of stones in the 6-9 mm category required more than 2 treatments, but 36% of staghorn stones required more than 2 treatments. Fragments >4mm were not retained in patients with 4-5 mm stones but were present in 52% of patients with staghorn stones following treatment. Follow-up of these patients continues.

THE MANAGEMENT OF URETERAL STONES USING EXTRACORPOREAL SHOCK WAVE LITHOTRIPSY AND URETEROSCOPY

S. Hayashida, T. Nasu, Y. Shinohara, H. Mitui, K. Imoto and K. Kitazima[1]

Tokuyama Central Hospital
Tokuyama and
[1]Tokushukai Hospital
Kagoshima, Japan

INTRODUCTION AND METHODS

Although extracorporeal shock wave lithotripsy (ESWL) and ureteroscopic stone manipulation (URS) have become routine procedures in the treatment of urolithiasis, the treatment of ureteral stones is still controversial. In an attempt to improve our results, we investigated the results of 1350 cases treated by ESWL and URS in our hospital.

A total of 1350 ureteral stones were treated by different procedures. 750 cases were treated by monotherapy ESWL (Group 1); 280 ureteral stones were pushed back to the renal pelvis and ESWL was performed (Group 2), and the other 320 cases were treated by URS (Group 3). EDAP LT. 0-1 and KARL STORZ rigid ureteroscopes were used in these series.

RESULTS AND CONCLUSION

In all the patients, stones were almost completely destroyed and removed from the ureter. However, several complications were observe:

Complication	Group 1	Group 2	Group 3
Residual microfragments in the kidney	0.8%	6.4%	4.1%
Severe ureteral stricture	0.0%	0.7%	1.6%
Prolapse of the stone fragments from the ureter	0.0%	0.0%	0.9%
Fever >38°C	2.0%	7.6%	15.0%
Admission more than 3 days	1.9%	93.1%	100.0%

We conclude that monotherapy ESWL is the best procedure in the treatment of ureteral stones and that the procedure should be carefully repeated if the stones are not fragmented during the first treatment. However, full endourological backup should be available to deal with stones that still do not respond to several sessions of ESWL.

EXTRACORPOREAL SHOCKWAVE LITHOTRIPSY FOR STAGHORN CALCULI

M. Moriyama, H. Miyamoto and S. Fukushima

Division of Urology
Yokohama Municipal Citizen's Hospital
Yokohama, Japan

INTRODUCTION AND METHODS

Eighteen patients with complete staghorn calculi underwent *in situ* ESWL with a Siemens Lithostar during a 42 month period at Yokohama Municipal Citizens' Hospital. These patients were submitted to an average of 6.4 treatment sessions with 26629 shock waves.

RESULTS

After ESWL, 10 patients had residual stones bigger than 10mm, 3 patients had residual stones more than 4mm but less than 10mm, 4 patients had residual particles less than 4mm and 1 patient had no residual stone.

No severe side effect was seen. Moderate fever was seen in 11 cases. Renal dysfunction was not seen.

CONCLUSION

We conclude that ESWL is a safe and useful method for treating staghorn calculi.

EXPERIENTIAL IMPROVEMENTS IN EXTRACORPOREAL LITHOTRIPSY ON MPL 9000, AN ULTRASOUND-MONITORED LITHOTRIPTOR

J. Talati and L.A. Khan

Department of Surgery (Urology)
The Aga Khan University Medical Center
Karachi, Pakistan

INTRODUCTION AND RESULTS

Between Jan 1990 and Oct 1991, 132 ureteric stones were treated on an ultrasound monitored lithotriptor (MPL 9000). Follow-up is available in 113 (89.4) patients. Eighty eight percent were stone-free without, and 96.6% with, post ESWL ancillary treatment. Improvement in results of extracorporeal lithotripsy (1991 vs 1990) are shown in the table.

Table. Comparison of ureteric stone management 1990 and 1991

	1990	1991
Stone free status		
After mono ESWL, at 90 days	84.4%	91.6%
Upper ureter	91.6%	95.6%
Lower ureter	72.7%	88.0%
Stone free rates at 30 days		
Upper ureter	65.8%	66.7%
Lower ureter	47.7%	55.5%
In situ treatment	68.3%	82.6%
DJ stent	33.0%	21.7%
Ratio of DJ Stent to ureteric obstruction	0.5%	0.4%
Post ESWL procedures	7.9%	5.8%
Patients needing more than 1 treatment	45.0%	14.5%

CONCLUSION

Analysis of results in 1990 suggests ureterolithotomy as the treatment of choice for larger than 1 cm ureteric stones with dilated upper urinary tract. Experience allows a surgeon to select treatment modality more effectively. This results in higher stone free rates in a shorter period of time with reduction repeat treatments and fewer pre- and post- ESWL interventions.

THE ENLARGEMENT OF RESIDUAL FRAGMENTS AFTER EXTRACORPOREAL SHOCK WAVE LITHOTRIPSY OF INFECTION STONES: UROLOGICAL STRATEGIES AND PATHOGENETIC HYPOTHESIS

L. Maccatrozzo, E. Cicerello, F. Merlo and G. Anselmo

Kidney Stone Center
Division of Urology
Regional Hospital
Treviso, Italy

INTRODUCTION AND RESULTS

The enlargement of residual infection stone fragments after extracorporeal shock wave lithotripsy (ESWL) led us to set up more articulate therapies and to re-evaluate the pathogenesis of infection stones on the basis of specific literature[1].

We reviewed 247 patients, representing 261 kidneys, with infection stones subjected to ESWL from September 1989 to September 1991. After a control period of 40 days post ESWL, 8 kidneys showed an increase in diameter of the residual fragments of the previously treated stone and a urinary tract infection. All patients were begun on specific parenteral antibiotic therapy and a nephrostomy was placed, followed by chemolysis with Solution G. After this treatment 2 patients were stone-free, 1 patient showed persistence of residual fragments and was retreated by ESWL; 4 patients were also subjected to percutaneous procedures, facilitated by nephrostomy tract. Besides chemolysis, 1 patient also required ureteroscopy because of a stone impacted in the prevesical ureter.

After discharge from hospital, all patients showed radiographic disappearance of calculi and sterile urine. An anatomic alteration retarding the clearance of residual fragments may be considered the cause of this complication[2]. In fact 5 of 8 had been previously subjected to calculus surgery and 3 had mild stenosis of the ureteropelvic junction, probably due to the inflammatory edema caused by the stone. Because of the persistence of infected fragments, resistant bacteria may act as a "nidus" for persisting urinary infection, maintaining the chemical conditions of stone formation and growth[3].

On the basis of our data, we may conclude that: a) these urological strategies provide the most effective approach for such patients; b) an accurate follow up is necessary to evaluate stone clearance in patients with distorted renal collecting systems, or following previous open surgery.

REFERENCES

1. EM Beck and RA Riehle, *J Urol* 145:6 (1991).
2. B Petterson, *Scand J Urol Nephrol (Suppl)*, 120: 1 (1989).
3. EK Michaels, JE Fowler, M Mariano, *J Urol* 140:254 (1988).

RADIOLOGICAL CHARACTERISATION OF URINARY STONES

N.P. Cohen[1], E. Kumala[1], J.A. Horrocks[1], G.J. Royle[2], R.D. Speller[2] and H.N. Whitfield[1]

[1]St Bartholomew's Hospital and
[2]University College Hospital
London, England

INTRODUCTION

Clinical experience with ESWL has shown that the size and composition of urinary calculi affect success. Unfortunately radiodensity on clinical radiographs is a poor guide to composition. We have used 2 X-ray techniques to attempt to categorise urinary calculi *in vitro*.

MATERIALS AND METHODS

23 stones were studied using a film/screen dual energy method with radiographs taken at two different X-ray energies. The images were analysed to try to find a correlation between the photoelectric attenuation coefficient and composition. A second technique used a low angle X-ray scattering method. This is based upon the phenomenon that a characteristic interference pattern is produced by different materials when placed within a polyenergetic X-ray beam. By accurately measuring the photon count at 6°, interference spectra were obtained for each stone. Composition was determined by standard infra-red analysis.

RESULTS AND DISCUSSION

Although the scattering patterns showed a number of characteristic peaks, attempts to correlate these spectra with the chemical compositions failed. This may have arisen from inaccuracy with IRS (performed on small, dry specimens), whereas the interference patterns were obtained from the whole stone. Similarly the dual energy X-rays did not clearly discriminate the major stone groups as classified by IRS. Further work will involve the comparison of the low angle scatter patterns with those produced by X-ray diffraction analysis of the urinary calculi.

URINARY RED BLOOD CELL MICROSCOPIC MORPHOLOGY AFTER ESWL: A PARAMETER OF GLOMERULAR DAMAGE

G. Zanetti, A. Trinchieri, E. Montanari, A. Guarneri, G.B. Fogazzi[1], P. Passerini[1], A. Currò and E. Austoni

Istituto di Urologia
Università degli Studi di Milano
[1]Divisione di Nefrologia
Ospedale Maggiore di Milano
Milan, Italy

INTRODUCTION

It has been shown that shock waves can cause damage to the vascular structure of the kidneys and tubules and also to the renal glomeruli[1]. We undertook a study of the microscopic morphology of urinary erythrocytes in a group of patients who underwent extracorporeal shock wave lithotripsy (ESWL) ,to assess possible glomerular damage.

METHODS

A group of 30 patients with renal calculi were considered for this survey, as all had undergone ESWL with the Dornier MPL 9000. The erythrocyte morphology and urinary examinations (N-acetyl-glucosaminidase, albumin/creatinine) were made on the day preceding treatment, and on days 1,3,7,14,21 and 28 following ESWL.

RESULTS AND DISCUSSION

Of 26 patients who did not demonstrate haematuria or who had haematuria with isomorphic erythrocytes before treatment, morphological assessment made immediately after ESWL showed isomorphic microhaematuria in 8, dismorphic erythrocytes with a percentage of dismorphism ranging between 10 and 50% in 10 and dismorphism of 10% or less, in 8.

In our survey a limited number of cases with 10-50% dismorphous erythrocytes indicates a trend suggestive of glomerular lesions as a result of ESWL. This view is supported by the observation in these cases of an increase, albeit of limited importance, of the albumin/creatinine levels in the first post ESWL days.

REFERENCE

1. P Jaeger, F Redha, G Uhlschmid and D Hauri, Morphological changes in canine kidneys following extracorporeal shock wave treatment, *Urol Res* 16:161 (1988).

ESWL AND URINARY INFECTION IN PRETREATMENT OF NON INFECTED PATIENTS

G. Zanetti, E. Montanari, A. Guarneri, A. Trinchieri, M. Seveso and E. Austoni

Istituto di Urologia
Università degli Studi di Milano
Milan, Italy

INTRODUCTION

Infective complications following extracorporeal shock wave lithotripsy (ESWL) can be prevented using appropriate antibiotic therapy. However, in many cases antibiotic prophylaxy seem to be excessively precautionary[1].

METHODS

We selected a group of 150 patients for prospective assessment of symptomatic and asymptomatic infections of the urinary tract (UTIs) following ESWL treatment for renal calculosis. Patients included in this group had a negative urine culture before ESWL and were free from recent UTIs. We used Dornier HM3 and MPL 9000 lithotriptors, delivering an average of 1800 shock waves, and did not administer any prophylactic antibiotics. An X-ray of the abdomen and a urine culture were both obtained on the day preceding the intervention and on the 30th day following.

RESULTS AND DISCUSSION

Third day urine cultures tested positive in 11 of the 150 patients examined, although only 5 of those cases were overtly febrile. The patients with positive urine culture underwent antibiotic therapy according to a sensitivity test for isolated pathogens. Fever cleared up following a short course of antibiotic therapy in the 5 symptomatic patients.

Our therapeutic standpoint for patients who have small stones, whose pre-ESWL urine cultures are negative, and who have no history of previous urinary tract infection or infected stones, is to avoid antibiotic prophylaxy prior to ESWL and to favour antibiotic therapy and monitoring of the post-ESWL urine culture to take care of eventual infective complications.

REFERENCE

1. B Pettersson and H-G Tiselius, Are prophylactic antibiotics necessary during extracorporeal shock wave lithotripsy?, *Br J Urol* 63:449 (1989).

IN SITU EXTRACORPOREAL SHOCK WAVE LITHOTRIPSY FOR URETERAL STONES

A. Al-Dayel, S. Egail, M. Ezzibdeh and I. Al-Oraifi

King Fahd Military Medical Complex
Dhahran, Saudi Arabia

INTRODUCTION

The use of extracorporeal shockwave lithotripsy has changed dramatically the treatment of ureteral stones requiring intervention. We reviewed our experience with *in situ* extracorporeal shock wave lithotripsy on 126 patients with 128 ureteral stones using Siemens lithostar.

METHODS

The treatment plan for ureteral stones included two principles: (i) a maximum of 3 sessions would be given to each stone, and (ii) any stone with no radiological evidence of fragmentation after session 3 would be removed surgically. Patients presenting with obstruction received urgent treatment. In patients with complicated obstruction a percutaneous nephrostomy tube was placed and these were then treated electively. Oral analgesia was used in all patients. Upper ureteric calculi were treated supine, while mid and lower ureteric calculi were treated prone. A maximum of 6000 shocks were given at any time. There were 33 upper, 33 mid and 62 lower ureteric stones. The size of the stones ranged from 4 - 30 mm. A total of 30 patients had a percutaneous nephrostomy tube inserted under local anaesthesia prior to ESWL treatment.

RESULTS

Of the stones 122 (95.3%) were treated successfully (70 required one session, 42 required 2, and 10 required 3 sessions). Treatment failed in 6 stones; 4 required ureteroscopy and 2, a ureterolithotomy. The number of treatment sessions increased in accordance with the degree of the ureteric obstruction (1.11 session for non-obstructed stones, 1.58 for partially obstructed and 1.66 sessions for obstructed stones), but no significant difference was noted with regard to the anatomical level of the stone. The overall complication rate was 10.8% and all were minor, requiring no intervention.

CONCLUSION

We conclude that *in situ* extracorporeal shock wave lithotripsy is an effective method of treating ureteral stones at all levels and at least 3 sessions should be given before surgical removal is considered.

URETERIC STONES MANAGED BY ESWL AND URETEROSCOPY (URS) UNDER GENERAL ANAESTHESIA OR LOCAL ANAESTHESIA

J.D. Desai, P.C. Patel, S.M. Shah and J. Amlanci

Lithotripsy and Endo-Urology Center
MP Shah Cancer Hospital
Ahmedabad, India

INTRODUCTION AND METHODS

Ureteric stones form 25% of our stone practice, and over the last three years we have treated 410 ureteric stone patients. Two hundred and seventy patients were treated by ESWL monotherapy, and the remaining 140, with lower ureteric stones, were treated by URS under general anaesthesia or IV sedation.

DISCUSSION

From our results we have concluded that the best treatment for upper and mid-ureteric stones is ESWL monotherapy using the Siemen's Lithostar, with which we have achieved an 88% success rate. For lower ureteric stones, ESWL monotherapy is less successful than URS manipulation (80% versus 98% for URS). *In situ* ESWL is preferred to push and bang ESWL.

URS in males should be performed under general anaesthesia. URS can be safely performed under IV sedation in female patients. We do not have any other series of URS under IV sedation for comparison.

THE ELECTROMECHANICAL IMPACTOR (EMI): A DEVICE FOR INTRACORPOREAL LITHOTRIPSY

S.P. Dretler and D.I. Rosen

Massachusetts General Hospital
Boston, MA USA

INTRODUCTION

The EMI was designed to be a safe, effective and inexpensive device for the fragmentation of ureteral calculi. It uses a 3.0F electrohydraulic (EHL) electrode encased in a .003 inch stainless steel casing with a spring and a conical end cap. It is flexible and the outer diameter is 5F. Water is pumped into the interior chamber and the EHL is discharged. The spark vaporizes water and creates a pressure wave which compresses the spring, forcing the pointed tip forward at 2000 cm/second over a distance of 2.5 mm. This discharge occurs at a rate of 6 Hz. By manipulating the mass of the tip (titanium) and the velocity of propulsion, the impact kinetic energy is controlled (KE = 1/2 MV2). The approximate force of tip impact measured by piezo-electric sensor is 1000 bar. This pressure is sufficient to fragment mulberry calcium oxalate monohydrate calculi at a fragmentation efficiency of .63 and .89 mg/pulse.

One thousand shocks are able to be discharged parallel to the wall of the pig ureter without muscle wall injury. Five pulses of the bare EHL fiber cause wall penetration. Clinically, fragile calcium oxalate dihydrate, rough cystine, struvite and uric acid stones fragment with less than 300 impulses under direct ureteroscopic vision. Mulberry calcium oxalate monohydrate and brushite may be fragmented using a balloon or basket to prevent cephalad migration. The life of each device is 800 pulses. Clinical trials at the Massachusetts General Hospital have been in progress. The fragmentation rate for ureteral calculi is 92% without evidence of ureteral injury. In the future, the device will be enlarged to 8 - 9F for fragmentation of bladder calculi and reduced in size to 3.5F for use in flexible endoscopes.

LASERLITHOTRIPSY

R. Muschter, A. Hofstetter, N. Schmeller, B. Liedl,
M. Kriegmair and W. Falkenstein

Department of Urology
Munich University
Munich, Germany

INTRODUCTION

The video shows the development of laserlithotripsy and our experiences and results both of *in vitro* and *in vivo* studies and clinical routine.

In vitro investigations reveal the different working principles of the wavelength used. *In vivo* investigations with regard to the risk of side effects show that plasma formation is inevitable when the laser radiation is applied directly to tissue. This results in lesions and hematoma and may even force an interruption of the treatment. So it became apparent that reliable protection against misapplication is desirable. Based on an optical feedback mechanism, a cut-off system for the laser beam is presented that is automatically activated prior to plasma formation in case of fiber contact with tissue.

For clinical routine, the indication is posed for all ureteral stones which cannot be passed spontaneously and cannot be treated with ESWL, but also - with increasing tendency - as primary therapy. For ureteroscopic access to the stone, arbitrarily thin rigid or flexible ureteroscopes can be used. Alternatively, a blind application under X-ray control only is possible.

RESULTS

The clinical results of more than 200 patients show a success rate of more than 90%, in accordance with literature references. Today's commonly used systems are dye and Alexandrite lasers.

CONCLUSION

Laser lithotripsy is, in conjunction with the stone-tissue recognition system, a reliable and safe method of stone treatment.

RENAL INSUFFICIENCY AFTER ESWL

G. Haupt[1], R. Schlick[2], H.J. Knopf[1], E. Allofs[2] and T. Senge[1]

[1]Departments of Urology
Ruhr-University
Bochum and
[2]Medizinische Hochschule
Hannover, Germany

INTRODUCTION AND SUBJECTS

ESWL is an effective and safe treatment for urolithiasis. Although side effects are rare they may sometimes be severe. We report two cases of renal failure after ESWL.

Patient 1: a 62 year old patient presented with symptomatic lower caliceal stones in both kidneys. General and urological history was uncomplicated. Routine evaluation showed a compensated renal insufficiency with serum creatinine values of 1.6 mg/dL (normal range below 1.2 mg/dL). Urine analyses were positive for protein (0.3 mg/dL). During a 10 day period ESWL was performed twice on the left and once on the right side using a Dornier lithotripter MFL 5000. A total dosage of 3000 SW was administered on the right, and 5000 SW at 18 and 20 kV on the left side. The day after the last treatment the creatinine rose to 2.8 mg/dL and continued to rise. One week after the last treatment hemodialysis had to be started. Kidney puncture was performed, and histological examination showed glomerulonephritis.

Patient 2: A 70 years old patient was admitted as an emergency with Kussmaul ventilation, bilateral hydronephrosis and caliceal stones, a leucocytosis with 40,000 wbc, and a creatinine of 800 µmol/L (normal range below 93 µmol/L). After a percutaneous nephrostomy on the left side and hemodialysis on two occasions the condition of the patient improved. Also, interstitial cystitis was diagnosed on histological grounds and a urinary diversion was planned. Before stone treatment was undertaken, a DJ-stent was placed in the right kidney. The creatinine was 300 µmol/L at that time. Within 8 weeks both kidneys were treated twice with ESWL using a HM 3 and a Modulith SL 20. The patient suffered sepsis, possibly due to infected intra- and subcapsular hematomas. The creatinine continued to rise and five months after his initial admission the patient was started on chronic dialysis. Stones were still present.

DISCUSSION

Renal insufficiency in these patients is probably not due to the ESWL treatment but the treatment may have accelerated the clinical course of their disease. Although ESWL is a very safe method its indication for use has to be well defined. In particular patients with pre-existing renal damage should undergo further evaluation before ESWL and need to be carefully monitored in between treatment episodes.

SECTION IX: INVESTIGATION, MEDICAL AND SURGICAL MANAGEMENT

CALCIUM SUPPLEMENTATION FOR ENTERIC HYPER-OXALURIA SECONDARY TO INTESTINAL BYPASS

V.R. Walker and R.A.L. Sutton

Department of Medicine
University of British Columbia
Vancouver, Canada

INTRODUCTION

Intestinal bypass (IBP) surgery for morbid obesity results in a urinary profile which is qualitatively similar to that found in patients with extensive small bowel disease. With ileal disease, steatorrhea and hyperoxaluria are frequent. Urinary Ca, Mg, and citrate are low but, despite the hypocalciuria, CaOx stone disease is common. Bambach et al[1] found a stone incidence of 6.7% in a population of 60 patients with ileal disease and small bowel resection in Leeds while Gelzayd et al[2] in 1968, in a much larger population (1468 patients), reported the incidence of nephrolithiasis as a complication of inflammatory bowel disease to be 6%[2].

In a population of patients with steatorrhea (from small bowel dysfunction) and intact colons, Dobbins and Binder found that 24 h urinary oxalate excretion was 83.7±12.3 mg (±SEM) (0.930 mmol)[3]. This hyperoxaluria is attributed to increased absorption of dietary oxalate in the colon[5] since hyperoxaluria is not found in patients with small bowel dysfunction and ileostomies[4]. The reasons for enhanced absorption of oxalate include preferential Ca binding by malabsorbed fatty acids, leaving oxalate free to be absorbed[5], and increased delivery of bile salts from the ileum which may enhance colonic permeability to oxalate[4], as well as possible deficiency of intestinal oxalate-metabolising bacteria, *Oxalobacter formigenes*[6].

We have studied a group of 8 IBP patients with recurrent CaOx stone disease. The hyperoxaluria in these patients is more marked than is usually reported in enteric hyperoxaluria owing to other causes, perhaps because of the smaller residue of functional small bowel. Stauffer et al showed that urinary oxalate excretion ws positively correlated with the length of resected ileum[7]. This has been observed by others as well[1,8]. The purpose of our study was not only to examine oxalate excretion in this IBP population, but also to determine to what degree it could be modified with increasing doses of Ca supplementation. Supplementation of dietary Ca provides additional Ca to bind with available dietary oxalate, rendering oxalate unavailable for intestinal absorption.

PATIENTS AND METHODS

Eight patients were studied (5F/3M) with a mean age of 44±8 (±SD) years, all of whom had undergone intestinal bypass surgery (14:4 bypass) for morbid obesity between 1974 and 1984. The protocol for study included collection of fasting blood and urine samples and 24 h non-fasting urine samples while subjects were consuming their normal

Urolithiasis 2, Edited by R. Ryall *et al.*,
Plenum Press, New York, 1994

diets and pursuing their normal activities. If patients had not previously been advised regarding consumption of a low oxalate diet, they were given dietary advice and asked to adhere to this for a week before collecting a further 24 h urine sample. One patient who could not comply with dietary oxalate restriction was given an oxalate-free formula diet (Ensure) for four days before making a further 24 h urine collection. All subjects were then given Ca supplementation at various doses ranging from 0.75 to 4.5 g of Ca/day. They were advised to take the supplement in divided doses with all meals for 5-6 days, collecting the 24 h urine samples of the last two days of supplementation. The Ca supplements given were Calcium Citrate (Vitaline Corporation, Ashland, OR 97520), Calcium Sandoz Forte (Sandoz Canada Inc, Dorval, Que, H9R4P5, and TUMS (Ca carbonate) (Smith Kline Beecham, Weston, Ont, Canada).

As the dose of supplementary Ca was increased in each patient, the urine was monitored for Ca (along with other urinary analytes) in order to detect hypercalciuria. Plasma oxalate was also monitored following supplementation at the higher levels in some patients. Urine samples were collected in acid containers (HCl) and analysed using standard procedures. Plasma and urinary oxalate and glycolate were measured by recently-modified high-pressure liquid chromatographic methods[9].

RESULTS

All patients had normal electrolytes, blood urea nitrogen, Ca and albumin levels in fasting baseline blood samples. As seen in Table 1, modest increases in plasma creatinine were seen in 4/8 patients, decreases in plasma Mg were seen in 4/8 patients, plasma oxalate was above the normal range in 6/8 patients, and plasma glycolate was normal.

Table 1. Fasting plasma oxalate, glycolate and Mg in relation to plasma creatinine in IBP patients.

Patient (Age, sex)	Creatinine	Oxalate	Glycolate	Mg
		µmol/L		mmol/L
1 (47, F)	132*	7.0*	3.9	0.7*
2 (45, F)	110*	6.4*	3.9	0.7*
3 (31, F)	101*	2.5	6.5	0.7*
4 (42, F)	62	4.2*		0.9
5 (41, F)	72	6.8*	6.0	0.8
6 (47, M)	85	2.2	3.0	0.6*
7 (40, M)	98	3.4*	4.0	0.8
8 (62, M)	125*	9.1*	5.3	0.8
Mean±SEM	98±8	5.3±0.9	4.7±0.4	0.75±0.03
Normal: Female	40-90	1.0-2.6	1.4-7.4	0.8-1.2
Male	90-110	0.8-3.2	1.9-7.5	0.8-1.2

*Outside the normal range

Table 2 shows baseline 24 h urinary excretion in the IBP patients. These patients had urinary Ca and citrate levels below the normal range and urinary Mg was low normal. Urinary oxalate was high, averaging almost three times the limit of normal. Urinary glycolate, however was normal. Baseline fractional excretion (FE) of oxalate following overnight fast was significantly higher than normal ($p < 0.01$) at 143±10% (±(SEM) (normal range, 7-169%) but FE of glycolate was normal (59±12%, normal range, 20-79%). The FE of oxalate positively correlated with plasma creatinine, $r=0.82$ ($p < 0.01$) and plasma oxalate correlated positively with urinary oxalate/creatinine, $r=0.95$ ($p < 0.001$).

Table 2. 24 h Baseline urinary excretion in IBP patients.

	mmol/day	Ion/Creatinine
Ca	2.38±0.28*	0.183±0.028
Oxalate	1.471±0.275*	0.100±0.011
Mg	2.10±0.30	
Citrate	0.95±0.25*	
Glycolate	0.524±0.067	0.035±0.002
Risk Index	1261.0±120*	

Values shown are Means±SEM, n=8. *Mean values outside normal range. Note: ion/creatinine ratio, mmol/mmol.

In the two patients who had not previously been following an oxalate-restricted diet, one reduced his 24 h urinary oxalate excretion following dietary oxalate restriction from 2.342 mmol to 1.080 mmol, a 54% reduction. The other was not able to comply with dietary restriction but when placed on oxalate-free formula diet alone, he reduced his 24 h urinary oxalate from 2.870 mmol to 0.508 mmol, an 82% fall.

In response to Ca supplementation of 3 g Ca/day, urinary oxalate was reduced by half ($p< 0.01$) and urinary Ca was increased ($p< 0.05$). However, this increase in urinary Ca was minimal and 24 h urinary Ca remained in the low-normal range. Significant increases in urinary Mg and citrate were not seen (paired t test). However, the Tiselius Risk Index[10] was reduced ($p< 0.01$) (Table 3).

Table 3. Changes in urinary excretion in IBP patient following 3 g/day Ca supplementation.

	Baseline (mmol/day)	3 g Ca/day
Ca	2.22±0.30	3.27±0.58*
Oxalate	1.525±0.333	0.730±0.169**
Mg	2.16±0.36	2.97±0.67
Citrate	0.83±0.27	2.08±0.60
Glycolate	0.546±0.108	0.654±0.163
Risk Index	117.01±105	711.0±151**

Values shows are means±SEM, n=6. *p<0.05; ** p< 0.01 (paired t test)

Figure 1 shows changes in urinary oxalate (expressed as a ratio with urinary creatinine) in each of the patients as Ca supplementation was increased from 0.75-4.5 g /day. In the three female patients who took the highest dose of supplement (4.5 g/day; total Ca intake, 5-6 g/day), urinary oxalate fell 76-89% from baseline oxalate values of 0.980-1.223 mmol/day to 0.123-0.263 mmol/day (normal females, 0.269±00.081 (±SD). Figure 2 shows the associated changes in urinary Ca at this time. None of the patients increased their urinary Ca excretion beyond the normal range for their sex, even at the highest doses of supplement. At the 4.5 g dose, urinary Ca ranged between 2.4 and 4.2 mmol/day in the three patients.

Some of the patients changed the form of their Ca supplement for reasons of personal preference or cost. In these patients, the response did not differ at a given dose with regard to oxalate and Ca excretion. However, in response to the Ca Sandoz preparation, urinary citrate was increased to a greater degree than that found for TUMS.

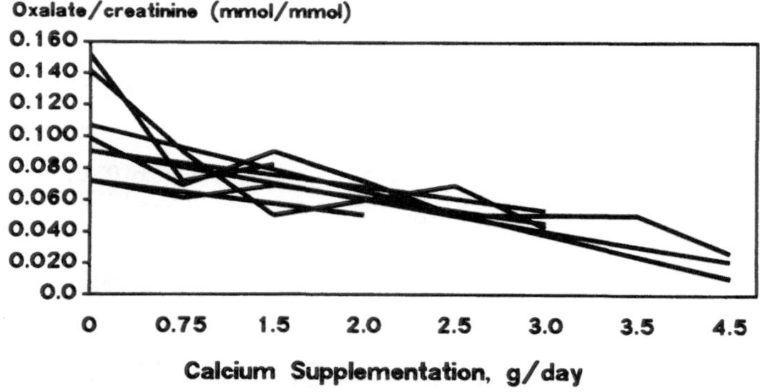

Figure 1. Effect of Ca supplements on urinary oxalate/creatinine ratios in IBP patients. Normal ratios are <0.040.

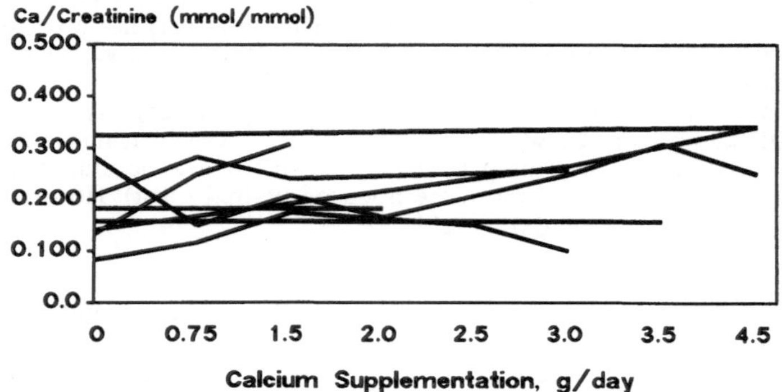

Figure 2. Effect of Ca supplements on urinary calcium/creatinine ratios in IBP patients. Normal ratios are <0.620.

DISCUSSION

Patients who elect to undergo intestinal bypass (IBP) surgery for morbid obesity do so accepting both acute and chronic complications. One of the most frequent complications in the IBP patient is hyperoxaluria and kidney stone disease, with CaOx being the major stone type even though urinary Ca excretion is below or low normal. As observed by others[5], removal of oxalate from the diet as we observed in our patient on an oxalate-free formula diet restored urinary oxalate to normal. This further demonstrates that the hyperoxaluria in these patients is secondary to increased absorption of dietary oxalate[4,5,7]. Plasma and urinary glycolate levels were not different in the IBP patients from normal values. Nor was the fractional excretion of glycolate different from normal. This also supports the contention that the hyperoxaluria associated with small bowel dysfunction is not secondary to altered glycolate production.

Management of the hyperoxaluria by dietary oxalate, in addition to fat restriction[11], is difficult since the IBP patient population are, in general, not compliant with regard to dietary restrictions. The patients in our study have been taking Ca supplements for up to two years. Urinary oxalate was markedly reduced to levels within the normal range at the highest dose given (4.5 g Ca supplement/day) and approaching the normal range at somewhat lower doses (3-3.5 g Ca supplement/day). At the same time, despite the very high levels of Ca given, urinary Ca remained low or well within the normal range. Normal dietary Ca intake in all of the patients ranged from 1-1.5 g/day so at the highest doses of supplement given, the total Ca intake would have amounted to 5-6 g of Ca/day.

Calcium supplementation as a means of reducing urinary oxalate in ileal disease was suggested by Earnest et al in 1975[12]. They found that administration of 3.6 g of Ca supplement 8 hyperoxaluric patients lowered urinary oxalate 56%, a value similar to that of ours at this dose range, and increased urinary Ca 75%, an increase greater than ours (~47%). However, their patients included individuals with small bowel disease and bypass and differences would be expected. Barilla et al found that in short term studies in two patients with ileal disease, Ca supplementation (36 g Ca/day) while lowering urinary oxalate 41% and 33%, resulted in elevations in urine Ca of 190% and 193% and were sceptical about such therapy having a role in treatment of Ca urolithiasis in enteric hyperoxaluria[13]. In studies in 8 jejuno-ileal bypass patients, Stauffer found that urinary oxalate was reduced to 81 mg/day when total Ca intake (supplement plus diet) was 3000 mg[11]. In all of these studies, the method of Ca supplement administration was not described. We gave supplementation in divided doses with all meals so that Ca would be readily available for binding with dietary oxalate.

We have measured intestinal transit time in some of our patients using ^{111}-Indium chloride as a marker[14] and found that it can be less than 5 hours. It appears that this rapid transit time and the limited surface area of small bowel remaining following bypass surgery may result in an upper limit to the amount of Ca that can be absorbed so that once a maximum absorption rate has been reached, the remaining Ca would remain in the intestine to bind with intestinal oxalate, rendering it unavailable for absorption.

REFERENCES

1. CP Bambach, WG Robertson, M Peacock and GH Hill, Effect of intestinal surgery on the risk of urinary stone formation, *Gut* 22:257 (1981).
2. EA Gelzayd, RI Breuer and JB Kirsner, Nephrolithiasis in inflammatory bowel disease, *Am J Digestive Disease* 13:1027 (1968).
3. JW Dobbins and HJ Binder, Importance of the colon in enteric hyperoxaluria, *NEJM* 296:298 (1977).
4. VS Chadwick, K Modha and RH Dowling, Mechanism for hyperoxaluria in patients with ileal dysfunction, *NEJM* 289:172 (1973).
5. DL Earnest, G Johnson, HE Williams and WH Admirand, Hyperoxaluria in patients with ileal resection: an abnormality in dietary oxalate absorption, *Gastroenterology* 66:1114 (1974).
6. KJ Allison, HM Cook, DB Milne, S Gallagher and RV Clayman, Oxalate degradation by gastrointestinal bacteria from humans, *J Nutrition* 116:455 (1986).
7. JQ Stauffer, MH Humphreys and GJ Weir, Acquired hyperoxaluria with regional enteritis after ileal resection: role of dietary oxalate, *Ann Intern Med* 79:383 (1973).
8. H Andersson, I Bosaeus, S Fasth, R Hellberg and L Hultn, Cholelithiasis and urolithiasis in Crohn's disease, *Scand J Gastroenterology* 22:253 (1987).
9. L Hage, VR Walker and RAL Sutton, Plasma and urinary oxalate and glycolate in healthy subjects, *Clin Chem* 39:134 (1993).
10. HG Tiselius, An improved method for the routine biochemical evaluation of patients with recurrent calcium oxalate stone disease, *Clin Chim Acta* 122:409 (1982).
11. JQ Stauffer, Hyperoxaluria and intestinal disease. The role of steatorrhea and dietary calcium in regulating intestinal oxalate absorption, *Am J Digestive Dis* 22:921 (1977).
12. DL Earnest, HE Williams and WH Admirand, A physicochemical basis for treatment of enteric hyperoxaluria, *Trans Assoc Am Physicians* 88:224 (1975).
13. DE Barilla, C Notz, D Kennedy, and CYC Pak, Renal oxalate excretion following oral oxalate loads in patients with ileal disease and with renal and absorptive hypercalciurias, *Am J Med* 64:579 (1978).
14. B Lentle, VR Walker, D Lyster and RAL Sutton, Gastrointestinal transit time made easy, *Clin Nuclear Med* 16:718 (1991).

PRIMARY HYPEROXALURIA: LONG-TERM OUTCOME IN PATIENTS TREATED WITH ORTHOPHOSPHATE AND PYRIDOXINE

D.S. Milliner and L.H. Smith

Division of Nephrology
Mayo Clinic
Rochester, Minnesota, USA

INTRODUCTION

Primary hyperoxaluria is a rare, autosomal recessive disorder. Although recognized earlier, it was not until the early 1950s that sufficient numbers of cases were described to permit its identification as a distinct clinical entity. During the decades since, two subtypes have been recognized. Type I primary hyperoxaluria is secondary to a deficiency of alanine: glyoxylate aminotransferase (AGT). There may be partial or complete absence of enzyme, deficiency of enzymatically active AGT protein, or mistargeting of the enzyme from peroxisomes to mitochondria where it is ineffective[1]. Type II primary hyperoxaluria is due to a deficiency of D-glyceric dehydrogenase. To date, only 16 patients with type II primary hyperoxaluria have been described in the literature. In both type I and type II, the result of the hepatic enzyme deficiency is a marked increase in oxalate production.

The landmark paper of Hockaday et al in 1964[2] described the clinical course of 64 patients with primary hyperoxaluria. Hockaday's observations regarding the usual clinical course have been substantiated in the years since. Increased oxalate production initially leads to hyperoxaluria with supersaturation of the urine for calcium oxalate. Calcium oxalate stones form in the kidneys and urinary tract. With time, calcium oxalate deposition in renal parenchyma, complicated by stone-related infection, obstruction, or loss of renal mass resulting from scarring or surgical excision, typically leads to renal insufficiency. Once renal function is compromised, the concentration of oxalate in plasma increases, and there is widespread deposition of oxalate throughout body tissues (oxalosis), eventually culminating in death. In Hockaday's original series, 53% of the patients had died at a mean age of 13.4±9.7 years. Despite the more recent widespread availability of renal replacement therapies in the form of maintenance dialysis and transplantation, patients with primary hyperoxaluria have fared poorly[3,4]. Dialysis is inefficient at oxalate removal[5], and patients on long-term maintenance dialysis programs develop progressive oxalosis. Furthermore, transplanted kidneys are frequently destroyed by calcium oxalate deposits, and three year allograft survival is only 23% in patients with primary hyperoxaluria[6].

In 1963, EW Frederick and colleagues[7] suggested phosphate for treatment of patients with primary hyperoxaluria. Orthophosphate has been shown to decrease the amount of urine calcium and increase urine pH, phosphate, citrate, and CO_2, thus leading to decreased supersaturation for calcium oxalate. In addition, the increased pH, increased citrate, and increased pyrophosphate resulted in an increased concentration of calcium phosphate and calcium oxalate inhibitors in the urine[8]. In patients with primary hyperoxaluria, crystalluria improved markedly following orthophosphate treatment[8]. Pyridoxine is a cofactor in the glyoxylate to glycine pathway. When given in pharmacologic doses, pyridoxine appears to augment the AGT enzyme pathway, thus decreasing oxalate production in a subset of patients with primary hyperoxaluria[9]. A program of combined orthophosphate and pyridoxine for long-term treatment of primary hyperoxaluria was suggested in 1967[10]. Promising results in five patients were presented at the First International Congress of Urolithiasis in 1968 by LH Smith and colleagues[11]. Four of their original patients are included in this report.

RESULTS

We here report the long-term outcome of 32 patients with primary hyperoxaluria, with particular attention to results of 309 patient years of treatment with orthophosphate and pyridoxine. The mean age at onset of symptoms was 8.6 years (range < 1-23), mean age at diagnosis was 13.3 years (range < 1-39), and age at last evaluation 23.8 years (range 2.8-58). Of the 32 patients, six had onset of symptoms at less than one year of age and 19 had onset of symptoms prior to 10 years of age. Eleven of the patients have type I primary hyperoxaluria. Six were confirmed by liver biopsy with enzyme studies of the patient or an affected first degree relative, 3 by increased urine glycolate, and 2 by complete pyridoxine responsiveness. Eight patients have type II primary hyperoxaluria. Two of these were confirmed by liver biopsy of the patient or an affected first degree relative, and 6 were identified by an increase in urine glycerate with normal urine glycolate. The remaining 13 patients did not have liver biopsy or urine glycolate or glycerate determinations. Twenty four h urine oxalate excretion in those with confirmed type I pH was 2.2±0.93 mmol/1.73 m^2/ 24 hours (mean±SD), in those with type II, 1.63±0.42 mmol/1.73 m^2/24 h, and in those of indeterminate type, 2.06±0.88.

Twenty-seven of the patients were treated with orthophosphate and pyridoxine before the onset of renal failure, and one received orthophosphate and pyridoxine for 14 years following successful renal transplantation. Treatment was initiated at an orthophosphate dose of 40.3±13.8 mg/kg/day and pyridoxine 5.8±9.2 mg/kg/day. Of the group as a whole, the change in urine oxalate after pyridoxine was +0.7±0.5 %. In the two patients who were pyridoxine responsive, urine oxalate returned to near normal (0.52 and 0.57 mmol/1.73 m^2/24 hours). Patients were treated for a mean of 10.0±8.9 years (median 9.0, range 1-25). At last follow-up, the dose of orthophosphate was 33.4±11.9 and of pyridoxine 2.7±1.3 mg/kg/day. The medications were well tolerated, and the only apparent adverse effects were transient loose stools in several patients which did not require discontinuation of therapy, and an increase in serum phosphorus in one patient with renal insufficiency. No hyperphosphatemia was seen prior to the onset of renal insufficiency in any patient.

The regimen of orthophosphate and pyridoxine has been successful in controlling stone forming activity. At last follow-up, all patients but one were free of active stone formation. Twenty-two of the 27 patients treated before the onset of renal insufficiency have maintained excellent renal function. Mean creatinine clearance at last follow-up evaluation was 94.7±24.2 mL/min/1.73 m^2 (median 96, range 57-147). Three of these patients have a solitary kidney. Six women have had 12 uncomplicated, term pregnancies.

Concomitant conditions have included vesicoureteral reflux with recurrent infections (2, with parenchymal renal scars in 1), nephrotic syndrome (1), acromegaly (1), ulcerative colitis (1), breast carcinoma (2), endometrial and ovarian carcinoma (1), and hypothyroidism (1).

Among treated patients, 5 have progressed to end stage renal disease (Table I). Three have had successful transplantation (two renal, one combined kidney/liver), and a fourth is awaiting transplantation. Patient 5 required chemotherapy for breast carcinoma. Orthophosphate and pyridoxine were discontinued due to chemotherapy-induced nausea and vomiting. Renal failure developed rapidly and she subsequently died of systemic oxalosis.

Five patients were diagnosed as having primary hyperoxaluria at or near end stage renal disease and were not treated initially because of potential adverse effects of phosphates with reduced renal function (Table 1). Patient 6 died at age eight before dialysis or transplantation were available. Patient 7 had successful renal transplantation and has had excellent allograft function subsequently for 14 years. The three remaining patients died following unsuccessful transplantation attempts.

Table I. Outcome of ESRD in patients with primary hyperoxaluria.

Pt.	Age Onset	AgeRx	AgeESRD	Transplant Type	Success	Vital Status/Age	Comments
1	2 9/12	3	27	K	Yes	Living/28	Medication non-compliance
2	1	6	18	None		Living/18	--
3	8 6/12	11	16	K	Yes	Living/22	ATN after nephrolithotomy
4	11	22	31	K/L	Yes	Living/41	Single kidney
5	21	24	34	None		Dead/34	Breast CA. Oxalosis
6	5	no Rx	8	None		Dead/8	Oxalosis
7	21	no Rx	22	K	Yes	Living/36	--
8	18	no Rx	23	K/L	No	Dead/30	Oxalosis
9	10	no Rx	27	L	No	Dead/37	Oxalosis
10	7	no Rx	27	K	No	Dead/37	Oxalosis

Patients 1, 3, and 7 have received orthophosphate and pyridoxine since transplantation. Patient 4 has normal urine oxalate following successful combined kidney and liver transplantation. All four have excellent allograft function with a mean iothalamate clearance at last follow-up of 64.7±19.6 mL/min/1.73 m^2.

Table 2 compares actuarial renal and patient survival among treated patients in our series with actuarial patient survival of the original series of Hockaday[2]. Given the change in end stage renal disease management and currently available modalities, renal survival in the 1990s should probably be compared with patient survival of 1964.

We conclude that long-term outcome in patients with primary hyperoxaluria who are treated with orthophosphate and pyridoxine is much improved when compared with historical controls.

Table II. Actuarial patient and renal survival of 27 patients with primary hyperoxaluria treated with orthophosphate and pyridoxine and in 64 untreated patients[2].

	Orthophosphate/Pyridoxine		No Rx
Age	Pt. Survival	Renal Survival	Patient Survival
10 yrs	1.00	1.00	.80
20 yrs	1.00	0.87	.43
30 yrs	1.00	0.78	.29
40yrs	0.86	0.66	.13

REFERENCES

1. CJ Danpure, and PR Jennings, Further studies on the activity and subcellular distribution of alanine: glyoxylate aminotransferase in the livers of patients with primary hyperoxaluria type 1, *Clin Sci* 75:315-322 (1988).

2. TDR Hockaday, JE Clayton, EW Frederick, LH Smith Jr, Primary hyperoxaluria, *Med* 43:315-345 (1964).
3. J Brodehl, K Latta, Primary hyperoxaluria Type I, *Eur J Pediatrics* 149:518-522 (1990).
4. I Helin, Primary hyperoxaluria - An analysis of 17 Scandinavian patients, *Scand J Urol Nephrol*, 14:61-64 (1980).
5. RWE Watts, N Veall, P Purkiss, Oxalate dynamics and removal rates during haemodialysis of peritoneal dialysis in patients with primary hyperoxaluria and severe renal failure, *Clin Sci* 66:591-597 (1984).
6. M Broyer, FP Brunner, H Brynger, SR Dykes, JHH Ehrich, Kidney transplantation in primary oxalosis: Data from the EDTA Registry, *Nephrol Dial Transplant* 5:332-336 (1990).
7. EW Fredrick, MT Rabkin, RH Richie, LH Smith, Studies in Primary Hyperoxaluria I. *In vivo* Demonstration of a defect in glyoxalate metabolism, *N Engl J Med* 269 (16):821-829 (1963).
8. LH Smith, Applications of physical, chemical, and metabolic factors to the management of urolithiasis. *in*: Urolithiasis Research, H Fleisch, WG Robertson, LH Smith, and W Vahlensieck, eds Plenum Press, New York 199-211 (1976).
9. DA Gibbs, and RWE Watts, The action of pyridoxine in primary hyperoxaluria, *Clin Sci* 38:277-286, (1970).
10. LH Smith Jr, and HE Williams, Treatment of primary hyperoxaluria, *Mod Treat* 4:522-530, (1967).
11. LH Smith, JD Jones, and FR Keating, Jr, Primary hyperoxaluria, *in*: Proceedings of the Renal Stone Research Symposium, A Hodgkinson and BEC Nordin, eds J and A Churchill, Ltd, London 297-307 (1969).

THE CLINICAL SPECTRUM OF AMMONIUM URATE RENAL CALCULI

W.H. Dick[1], L.H. Smith[2], D.M. Wilson[2], G.M. Preminger[3] and J.E. Lingeman[1]

Departments of Nephrology and Urology
[1]Methodist Hospital of Indiana
Indianapolis, Indiana
[2]Mayo Clinic
Rochester, Minnesota and
[3]University of Texas Southwestern Medical Center
Dallas, Texas

INTRODUCTION

Ammonium acid urate may occur with urealytic urinary-tract infections where it is most often found with magnesium ammonium phosphate[1]. Ammonium urate may be present in sterile urine with endemic bladder stones which were common in certain areas of nineteenth-century Europe and are occasionally found today in India, Turkey, or southeast Asia[2-6]. In addition, this calculus was reported by us to exist in sterile urine of patients with a certain form of laxative abuse[7]. A new cause for this form of urinary calculus has been discovered in three patients with ileal resection and ileostomy/or massive colonic resection. Their clinical profiles, which are strikingly similar to the patients with laxative abuse, are presented in the following report.

All patients had over 50% ammonium urate in the calculus (only one was under 90%). The protocol consisted of history and physical examination, IVP or CT scan, non-contrast nephrotomograms, fasting urinalysis and chemistry profile, urine culture, urine for phenolphthalein, and 24 h urine studies, when the patient was consuming their usual diet and fluids. Urine analyses included volume, calcium, phosphorus, uric acid, oxalate, citrate, magnesium, sodium and potassium.

RESULTS

Serum chemistry studies

Serum creatinine, calcium, phosphorus, uric acid, albumin, sodium, and chloride were normal. However, serum bicarbonate was decreased in the intestinal-surgery group with a mean value of 21.5 mmol/L and potassium was at the low end of the normal range. Prior studies in the laxative-abuse patients (including two new ones) showed similar results, except that the bicarbonate level was 24 mmol/L, decreased but statistically normal.

Table 1. Twenty-four h urine studies

Patient number	Vol mL	Ca	P	Ur	Ox	Cit	Mg	Na	K
		mg					mmol		
Laxative patients									
1	430	43	650	400	31	25	6	2	18
2	1500	82	435	285	7	210	35	26	17
3	900	182	737	252	13	21	40	8	7
4	550	213	869	253	28	91	65	13	52
5	1100	318	1078	418	28	66	140	16	13
6	940	91	964	386	31	40	85	139	36
7	1230	100	635	423	17	354	81	47	18
8	740	64	494	598	14	217	27	0	7
9	730	23	383	395	33	12	71	6	21
10	590	117	1127	543	18	11	54	46	-
11	700	120	881	531	34	89	102	32	50
Mean	855	123	750	408	23	103	64	30	24
Normal Mean	1400	160	880	450	28	620	100	120	60
Amonium Urate GI Patients									
1	447	35	608	528	10	8	46	1	7
2	505	41	515	248	21	6	22	13	-
3	1140	89	638	422	30	13	22	54	27
Mean	697	55	587	399	20	9	30	23	17
Ca Ox GI Patients									
1	2530	106	612	531	132	46	65	124	
2	1650	45	1221	577	94	20	43	114	
3	1350	195	964	317	68	51	61	214	
4	2480	45	694	471	108	80	123	208	
5	950	53	817	323	66	22	22	161	
Mean	1792	89	862	444	94	44	63	164	

Urine studies

The laxative-abuse group and the intestinal-surgery patients were statistically similar. When compared to normals, the laxative-abuse patients had abnormal urine volumes and excretions of sodium, citrate, magnesium and potassium. The intestinal-surgery patients had abnormal calcium, phosphorus, citrate, sodium, and magnesium excretions. Urine volumes and potassium excretions were highly abnormal (Table 1). Urine cultures were negative in all patients except in one laxative-abuse patient who had 1,000 colonies of *Proteus mirabilis*. Urine phenolphthalein was negative in the three intestinal-surgery patients, and positive in all nine of the laxative-abuse patients in whom it was measured.

Case Histories of Intestinal Surgery Patients

Each patient had at least two ileal resections, two had ileostomies, and the third had most of the colon removed. All had passed stones consisting of over 90% ammonium urate in sterile urine. All had decreased serum bicarbonate and abnormal 24 h urine volumes and excretions of sodium, potassium, citrate, and magnesium. Therapy with loperazole and citrate medication has stopped formation of new calculi.

DISCUSSION

American clinicians rarely see ammonium-urate calculi. Among 10,000 stones analyzed in 1962 in the United States, only 8 contained 50% or more ammonium urate[8]. When considered at all, ammonium-urate calculi are thought of as co-existing with struvite under conditions of urealytic bacterial infection. If enough urease-producing bacteria are inhabiting the urinary tract, a large amount of ammonium is produced from hydrolysis of urea. The resultant highly-alkaline urine can precipitate phosphorus, leading to a magnesium-ammonium-phosphate calculus. During periods of slightly lower urine pH, ammonium can combine with urate if enough of this ion is present[1]. Thus, struvite calculi can contain various mixtures of ammonium urate, usually less than 50%.

A second form of this calculus has not been reported in this country. Prior to the twentieth century, endemic stones consisting mainly of ammonium urate were found in certain areas of England and Europe. These endemic stones still exist in India, Turkey, Thailand, Taiwan, and Indonesia. When children are fed a diet of breast milk and cereal, there is a potential risk of stones consisting mostly of ammonium urate. The low animal protein/high vegetable diet promotes an alkaline urine. Urine uric acid was found to be high in these children[9], as it is in most children because of rapid tissue turnover[10]. Because the phosphate intake is very low in these children, ammonia serves as the main acid buffer in the urine and ammonium is excreted in higher amounts. Thus, a risk of ammonium-urate stone production arises.

Another type of ammonium-urate stone occurs in phenolphthalein laxative abuse. Laxative abuse may result in metabolic acidosis[11], in which an increase in ammonium may occur[12]. Intracellular acidosis caused by potassium depletion may also increase ammonium excretion[13]. In our patients, significant reductions in urine volume, citrate, sodium, magnesium, and potassium were seen. High ammonium excretion resulted from low intracellular potassium, as judged by the low urine potassium and the low-normal serum potassium. Urate combined with ammonium because of the paucity of urine sodium and potassium. Urine analytes in five of the laxative-abuse patients returned to normal when the laxatives were discontinued, which supports this theory.

We now describe a new group of patients with ammonium-urate stones following ileal resection and ileostomy/or massive colon resection. Numerous studies have documented the risk of urinary calculi following gastrointestinal disease[14-17]. The incidence of calculi formation in the general population is increased two- to four-fold. Of particular liability is the association of ileal bypass, ileal resection or ileostomy, and the rate of stone disease[18]. Both ileal resection and ileostomy are associated with decreased urine volume and citrate. Depending on the length of small bowel removed, ileal resection is normally associated with decreased urine calcium and increased urine oxalate excretion. Urine studies in ileostomy patients generally demonstrate a low urine volume with an extremely acid urine[19]. Thus, in these patients, even normal uric-acid excretion rates are a risk for uric-acid calculi because at pH 4.5, only 100 mg/L of uric acid in urine can be held in solution. When the group of three ammonium-urate patients with ileal resection and ileostomy, or three-fourth colon resection, is compared to a group of five patients from the Kidney Stone Clinic at Methodist Hospital with ileal resection and one-third colon resection (Table 1), it can be seen that significant differences are found in urine volume and the excretion of oxalate and sodium.

The 24 h urine excretion in our gastrointestinal patients was virtually identical to that in the laxative-abuse subjects. Once again, as a consequence of mild metabolic acidosis, excess ammonia appears in the tubule lumen and becomes ammonium. Because urinary sodium and potassium ions are low in the laxative-abuse patients, urate must combine with ammonium. Previous research has shown that the ideal pH for formation of ammonium-urate calculi is 6.2-6.3[1]. At higher pH, larger concentrations of urate must be present and, at lower pH, mixtures of uric acid and ammonium urate may be seen, a rare clinical event. Bowyer et al[20], in pH studies of ammonium and urate molecules in urine demonstrated that little ammonium urate exists in solution under pH 5.7.

Thus, the theory of ammonium-urate stone formation for the group of intestinal-surgery patients is the same as that for the laxative-abuse subjects. The risk factors include extreme volume depletion with markedly abnormal urine volume and sodium excretion, mild metabolic acidosis with low urine citrate and high ammonium excretion, and potassium depletion secondary to intestinal loss resulting in low urine-potassium levels. These conditions, maintained at an ideal pH for a sustained length of time, may allow ammonium urate stones to form. Therapy with loperazole and citrate medication has controlled new stone growth.

REFERENCES

1. M Teotia and DJ Sutor, Crystallisation of ammonium acid urate and other uric acid derivatives from the urine, *Brit J Urol* 43: 381 (1971).
2. R Van Reen, Geographic and nutritional aspects of endemic stones, *in:* "Urinary Calculus", JG Brockis and B Finlayson, eds, PSG Publishing Co, Littleton (1981).
3. AL Aurora, V Ramalingaswami and PD Gaitonde, Bladder stone disease in children in Delhi area, *J Urol* 91: 347 (1964).
4. DA Anderson, The nutritional significance of primary bladder stones, *Brit J Urol* 34: 160 (1962).
5. A Valyasevi, SB Halstead and S Dhanamitta, Studies of bladder stone disease in Thailand. VI. Urinary studies in children, 2-10 years old, resident in a hypo-and hyperendemic area, *Amer J Clin Nutrition* 20: 1362 (1967).
6. K Thalut, A Rizal, JG Brockis, RC Bowyer, TA Taylor and ZS Wisniewski, The endemic bladder stones of Indonesia - epidemiology and clinical features, *Brit J Urol* 48:617 (1976).
7. WH Dick, JE Lingeman, GM Preminger, LH Smith, DM Wilson and WL Shirrell, Laxative abuse as a cause for ammonium urate renal calculi, *J Urol* 143: 244 (1990).
8. LC Herring, Observations on the analysis of ten thousand urinary calculi, *J Urol* 88: 545 (1962).
9. WG Robertson, Urolithiasis: epidemiology and pathogenesis, *in:* "Tropical Urology and Renal Disease", I Hussain, (ed), Churchill Livingstone, Edinburgh, (1984).
10. WG Robertson, A Hodgkinson and DH Marshall, Seasonal variations in the composition of urine from normal subjects: a longitudinal study, *Clin Chim Acta* 80: 347 (1977).
11. JR Oster, BJ Materson and AI Rogers, Laxative abuse syndrome, *Amer J Gastroenterology* 74: 451 (1980).
12. RF Pitts, Control of renal production of ammonia, *Kidney Int* 1: 297 (1972).
13. RL Tannen and J McGill, Influence of potassium on renal ammonia production, *Amer J Physiol* 231: 1178 (1976).
14. JJ Deren, JG Porush, MF Levitt, and MT Khilnani, Nephrolithiasis as a complication of ulcerative colitis and regional enteritis, *Ann Intern Med* 56: 843 (1962).
15. EA Gelzayd, RJ Breuer and JB Kirsner, Nephrolithiasis in inflammatory bowel disease, *Am J Digestive Disease* 13: 1027 (1968).
16. LH Smith, H Fromm and AF Hofmann, Acquired hyperoxaluria, nephrolithiasis, and intestinal disease. Description of a syndrome, *NEJM* 286: 1371 (1972).
17. DM Wilson, Gastrointestinal disorders and urolithiasis, in: "Stone Disease: Diagnosis and Management" SN Rous, (ed), Grune & Stratton, Inc, Orlando (1987).
18. CP Bambach, WG Robertson, M Peacock and GL Hill, Effect of intestinal surgery on the risk of urinary stone formation, *Gut* 22: 257 (1981).
19. HJ Kennedy, EWL Fletcher and SC Truelove, Urinary stones in subjects with a permanent ileostomy, *Br J Surg* 69: 661 (1982).
20. RC Bowyer, RK McCullough, JG Brockis, and GD Ryan, Factors affecting the solubility of ammonium urate, *Clin Chim Acta* 95: 17 (1979).

A NEW THERAPEUTIC AGENT FOR CYSTINURIA

T. Koide, M. Utsunomiya, S. Yamaguchi and T. Yoshioka

Department of Urology
Osaka University School of Medicine
Osaka 55, Japan

INTRODUCTION

We have already demonstrated the effectiveness of tiopronin (α−mercaptopropionylglycine, Thiola®,) for cystine stone dissolution and for preventing stone formation in cystinuric patients[1]. Thiol compounds, such as D-penicillamine, tiopronin and captopril convert urinary cystine by a mixed disulfide exchange reaction, into a cysteine complex which is more soluble than cystine. Therefore, thiol compounds can dissolve cystine stones and effectively prevent cystine stone formation. Bucillamine (2-mercapto-2-methylpropanoyl-L-cysteine), which is a new therapeutic agent for rheumatoid arthritis, having been used for the last 5 years in Japan[2], is a dithiol compound, while tiopronin and D-penicillamine are monothiol compounds (Fig. 1). Approximately 40% of bucillamine given *per os* is excreted into urine as various metabolites. Three major metabolites have been identified. About 10% of bucillamine appeared in urine as a non-metabolized free form and about 15% as a metabolized monothiol compound[3]. A small amount appeared as a metabolized non-thiol compound. Theoretically, bucillamine is a more effective medicine for cystinuria than tiopronin, because of its number of thiol groups. We studied the effect of bucillamine on urinary cystine both *in vitro* and *in vivo* clinical trials and compared the result with that of tiopronin in this study.

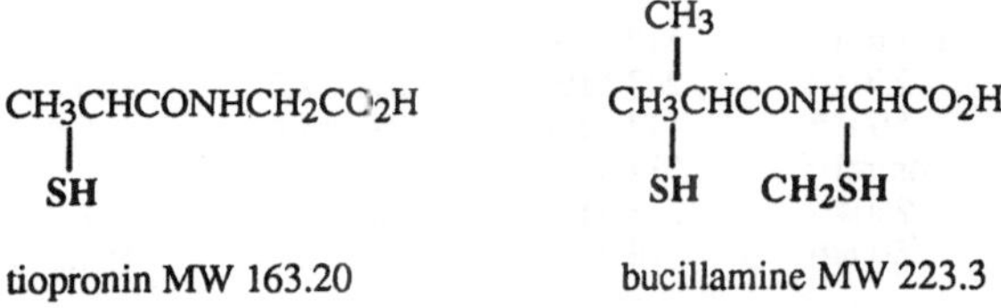

Fig.1. Chemical structure of tiopronin and bucillamine

MATERIALS AND METHODS

In vitro study: The effect of bucillamine was tested in whole urine. L-cystine was added to the urine at a concentration of 500μg/mL and either 250μg/mL and 500μg/mL of tiopronin or bucillamine was added to the urine. The urine was incubated for up to 4 hours at 37°C. Cystine and cysteine concentrations were determined before and 4 hours after the incubation.

In vivo study: Three cystinuric patients agreed to participate the study with their informed consent. They were a 13-year-old female, a 20-year-old female and a 38-year-old male. Since the influence of tiopronin or bucillamine could be washed out within a day, urine collection was done according to the following schedule. The subjects were requested to collect 24 h urine under the continuous tiopronin medication, then to stop taking any medicine. A 24 h urine sample was collected on the second day after quitting medication. Then, 600mg of bucillamine was given in the next morning and every 2 hr urine was collected for 8 hours after this single administration. Bucillamine was given orally at the same doses as tiopronin for 2 days from the following day. Twenty four hr urine was collected on the second day of bucillamine administration. Daily doses of tiopronin and bucillamine were 800mg in the 13-year-old patient and 1,600mg in the other 2 patients. Both tiopronin and bucillamine were given in four divided doses. In this manner, we obtained control 24 h urine, 24 h urine under tiopronin medication, 24 h urine under bucillamine medication and each of the four 2 h urine samples after single administration of bucillamine. Effectiveness of bucillamine was compared with tiopronin by analyzing the 24 h urine samples under the three different conditions. The duration of the effect of bucillamine was studied by measuring cystine concentration in the each 2 h urine sample. Urine was collected in plastic bottles which contained 100mL of 2N HCl to avoid reversed mixed disulfide exchange reaction during the urine collection.

The concentrations of cystine and cysteine were determined by high pressure liquid chromatography (HPLC) with an o-phthaldehyde (OPA) post-labeled method.

RESULTS

In vitro study: The concentration of cystine was markedly reduced by both tiopronin and bucillamine. The reduction of cystine was thought to be caused by formation of cysteine-tiopronin or cysteine-bucillamine (or - monothiol bucillamine metabolite) complex through a mixed disulfide exchange reaction. In the *in vitro* study, the relative activity of bucillamine was 5 to 12% stronger than that of tiopronine and calculated relative molecular activity of bucillamine was approximately 40 to 50% stronger than that of tiopronin (Table 1). In other words, bucillamine dissolved urinary cystine much more effectively than tiopronin at the same molecular weight and a little more effectively than tiopronin at the same mg concentration.

Table 1. Cystine reducing activity of tiopronin and bucillamine *in vitro*.

	amount of thiol (μg/mL)	concentration of L-cystine (μg/mL)	dissolved L-cystine (%)	relative activity	relative molecular weight
	control	512.7±9.4			
tiopronin	250*	243.5±9 6	52.5±2.7	100.0	100.0
(MW 163.2)	500**	117.8±20.0	77.1±3.6	146.9	73.5
	control	517.5±7.2			
bucillamine	250#	213.9±16.3	58.7±2.6	111.8	152.7
(MW 223.3)	500##	98.3±15.4	81.1±2.7	154.5	105.5

* tiopronin 1.53μg/mL, **tiopronin 3.06μg/mL, # bucillamine 1.12μg/mL, ##bucillamine 2.24μg/mL

In vivo study: About 40 per cent of bucillamine is excreted into urine during 4 to 6 hours after the single administration of 400mg[3]. In the *in vivo* study also, urinary cystine reduction by bucillamine was observed clearly up to 6 hours after bucillamine administration. The reduction of 24 h urinary cystine level was much higher than that by tiopronin, despite the fact that the same amount of bucillamine was administered (Fig.2). As shown in Fig.3, two of the three patients showed a marked reduction of urinary cystine level by bucillamine in comparison with tiopronin, although the reduction in patient No 3 was only slight. The reducing activity of bucillamine in patients was more marked than the result of *in vitro* study.

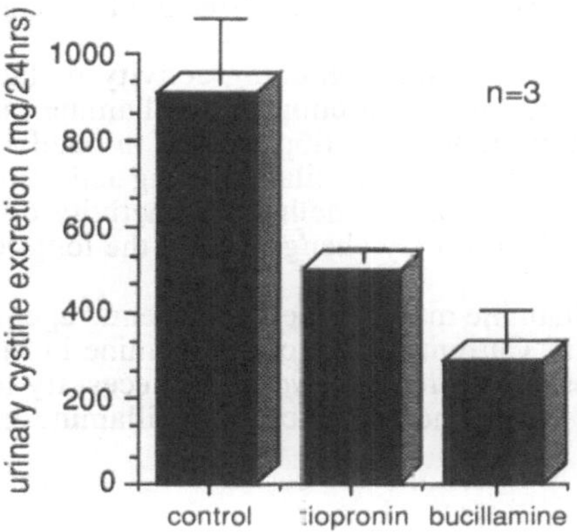

Fig 2. Comparison of the effect of tiopronin and bucillamine on 24 hr urinary cystine.

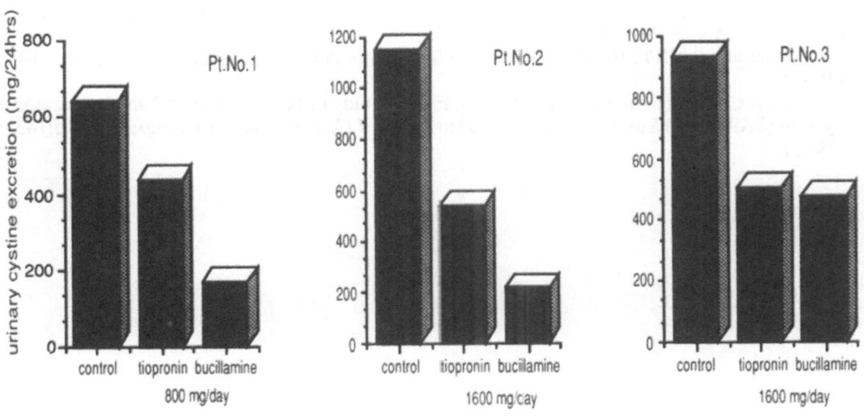

Fig 3. Effect of tiopronin and bucillamine on 24 hr urinary cystine in individual patients.

DISCUSSION

Monothiol compounds, such as D-penicillamine and tiopronin convert cystine into soluble form by the mixed disulfide exchange reaction, and have been used for cystinuric patients for long time. A new dithiol compound, bucillamine, should be more effective in reducing the urinary excretion of cystine by this reaction. It has already been confirmed that about 40% of bucillamine administered orally is excreted into urine, with about 10% in the non-metabolized free form and about 15% as metabolized monothiol compound[3]. In other words, bucillamine can change urinary circumstances more effectively than tiopronin, wherein urinary cystine tends to convert into a cysteine-thiol complex. In an *in vitro* study using natural urine, bucillamine converted cystine into soluble form effectively in comparison with tiopronin. When the same amount of bucillamine was added to the urine, it showed a slightly stronger solubilizing power than tiopronin, while the calculated dissolution activity per molecular weight was much higher with bucillamine than with tiopronin. In the *in vivo* study bucillamine showed an excellent result in cystinuric patients, particularly in two. The response was greater than expected from the result of the *in vitro* study. The result indicated that bucillamine can dissolve cystine approximately twice as effective as tiopronin at the same mg amount. The reason for this difference between the *in vitro* and *in vivo* studies might be the result of metabolism of bucillamine to a monothiol compound and the fact that total free thiol radical in urine is greater than in the case of tiopronin.

In the third patient, the cystine reducing activity of bucillamine was similar to tiopronin. This might be because absorption of bucillamine may vary in individuals or because methylation of thiol radicals in tiopronin or in bucillamine may be different *in vivo* in this particular patient following similar reducing activity of urinary cystine by both tiopronin and bucillamine. As for the methylation activity of thiol radicals, it may be different among individuals or it may change during the long-term treatment. This point needs to be clarified.

In conclusion, bucillamine may be a new therapeutic agent for cystinuria in place of the monothiol compounds. Current dosage of bucillamine in rheumatoid arthritis is only 300 mg/day. Much larger doses probably would be necessary to dissolve cystine stones. The long term efficacy, safety and tolerance of bucillamine in patients with cystinuria needs to be evaluated.

REFERENCES

1. T Koide, J Kinoshita, M Takemoto, S Yachiku and T Sonoda, Conservative treatment of cystine calculi: Effect of oral alpha-mercaptopropionylglycine on cystine stone dissolution and on prevention of stone recurrence, *J Urol* 128:513-516, (1982).
2. S Kashiwazaki and SY, Bucillamine: A new immunomodulator, *Int J Immunotherapy*v 3: 1-6, (1987).
3. S Sugawara, M Ishigami, and T Kageyama, Phase I study of N-(2-mercapto-2-methylpropionyl)-L-cystine (SA96), (I) Single administration study, *Jpn J Clin Pharmacol Therapeutics* 16:143-152, (1985).

THE TREATMENT OF IDIOPATHIC RECURRENT UROLITHIASIS WITH FISH OIL AND EVENING PRIMROSE OIL - A DOUBLE-BLIND STUDY

A.C. Buck, W.S. Smellie, A. Jenkins, R. Meddings, A. James and D. Horrobin

Department of Urology
Glasgow Royal Infirmary
Glasgow, UK

INTRODUCTION

Epidemiological studies suggest that Eskimos have a low incidence of atherosclerosis and other degenerative diseases, including stone disease[1-4]. This immunity which the Inuit shares with the Japanese of Okinawa island has been attributed to their fish-rich diet with its high concentration of eicosapentaenoic acid (EPA)[5]. EPA is the precursor poly-unsaturated fatty acid for the Ω-3 series of prostaglandins. High concentrations in the diet will act as a replacement for and competitive inhibitor of arachidonic acid, the precursor of the Ω-6 series of pro-aggregatory and pro-inflammatory dienoic metabolites. In a preliminary experimental and open clinical study, fish oil (EPA) inhibited experimental nephrocalcinosis and reduced urinary calcium and oxalate excretion in hypercalciuric recurrent stone formers. Recent studies have shown that the beneficial effects of Ω-3 fatty acids (EPA) can be potentiated in the presence of evening primrose oil containing γ-linolenic acid (GLA), a precursor poly-unsaturated fatty acid of the monoenoic metabolites (ie, PGE1). GLA has also been reported to inhibit formation of 2-series prostaglandins from arachidonic acid. To study the influence of fish oil and evening primrose oil on urinary solute excretion a double-blind, placebo-controlled study was performed in 40 recurrent idiopathic stone formers.

PATIENTS AND METHODS

A study was performed in 40 idiopathic recurrent stone formers - 34 males and 6 females whose ages ranged from 24-72 years (mean 50.1±SD 15.6 yrs). Two weeks prior to the commencement of the study the patients were stabilised on an 800 mg calcium diet which was maintained for the 12 week duration of the study. The patients were randomly allocated to receive indistinguishable capsules of either fish oil (6 gm/day = eicosapentaenoic acid - 1.08 g; docosahexaenoic acid - 360 mg), evening primrose oil (6 gm/day = γ-linoleic acid - 480 mg), a combination of evening primrose oil and fish oil (6 gm/day = γ-linoleic acid - 408 mg, eicosapentaenoic acid - 204 mg, docosahexaenoic acid - 132 mg), or sunflower oil (6gm/day).

The following investigations were performed, before treatment, at 6 weeks and at 12 weeks: fasting serum calcium, phosphate, urate, urea and electrolytes, creatinine, insulin, PTH, cAMP, 1,25(OH)$_2$D$_3$; 24 urinary volume, calcium, oxalate, phosphate, citrate, urate, magnesium, sodium, potassium, creatinine clearance and Ca/Cr ratio was measured on 2 consecutive occasions. Following an overnight fast intestinal calcium absorption was determined by measuring the fractional rate of ^{45}Ca absorption using a single isotope technique and expressed as a fraction of the isotope dose/hour using the formula: Radiocalcium absorption rate (fraction of dose/h) = 1.17 f 1 2.54 f^2, where f = X administered dose/L plasma at 1 h x body weight (kg) x 0.0015[7].

The data for solute excretion before and after 12 weeks treatment were analysed for statistical significance between the groups by a two way analysis of variance.

RESULTS

Of the 40 patients, 35 were fully evaluable at 12 weeks; 5 patients were excluded for protocol violations. All the patients were normocalcaemic and had normal levels of serum PTH (< 5 pmol/L). The fasting serum insulin was consistently raised in 14/35 patients (>13mU/L). The serum $1,25(OH)_2D_3$ levels were within the normal range. There was no correlation between serum $1,25(OH)_2D_3$ levels and urine calcium excretion or ^{45}Ca intestinal absorption. There were no changes in any of the serum biochemical parameters with any of the treatment regimens. The pre-treatment fractional ^{45}Ca rate of absorption was raised in 16 patients [46%] (> 0.3/hour) .

All 4 groups were comparable with regard to solute excretion before treatment (P=0.8).

There was a significant decrease in urinary calcium excretion at 12 weeks in the patients treated with fish oil from a pre-treatment mean value of 9.5±1.04 mmol/24 h to 6.2±1.0 mmol/24 h, (p< 0.0041), with a concomitant decrease in Ca/Cr ratio from 0.61±0.04 to 0.53±0.03. A significant decrease in urinary calcium excretion also occurred in the patients treated with a combination of evening primrose oil and fish oil (mean 8.64±0.54 mmol/24 h to 5.7±0.51 mmol/24 h). The Ca/Cr ratio fell from 0.68±0.06 to 0.62±0.08. There were no significant changes in urinary calcium excretion or the Ca/Cr ratio in patients treated with evening primrose oil or in the control group treated with sunflower oil (Fig. 1).

At 12 weeks calcium excretion in the fish oil treated group and the combination of fish oil and evening primrose oil treated group were significantly different from the calcium excretion in the other two groups (p< 0.04). However, two way analysis of

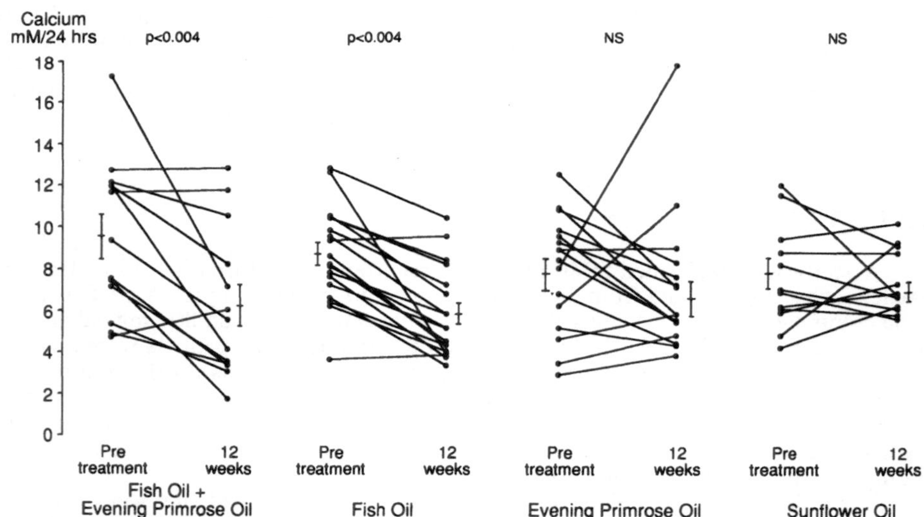

Figure 1. the 24 h urinary calcium excretion before and after 12 weeks of treatment in the 4 groups of patients. A significant reduction in 24 h urinary excretion occurred in the groups treated with fish oil or a combination of fish oil and evening primrose oil.

variance indicated that fish oil alone was responsible for the effect in urinary calcium excretion ($p< 0.004$). The interaction with evening primrose oil was not significant ($P=0.62$).

There were no changes in the urinary excretion of phosphate, urate, oxalate, citrate, magnesium or urine pH in any of the groups.

A significant increase in intestinal ^{45}Ca was observed in all the groups at 12 weeks (figure 2) ($p< 0.001$).

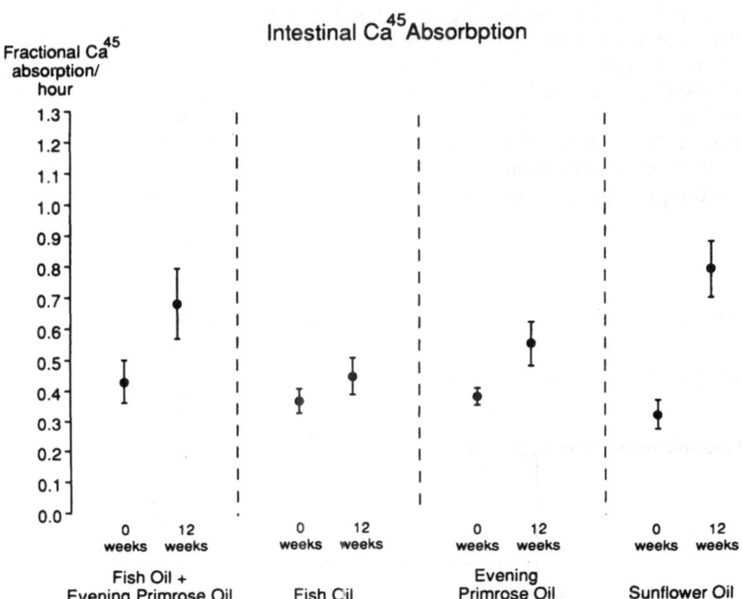

Figure 2. The fractional rate of ^{45}Ca abscrption was seen to increase significantly ($p< 0.001$) in all groups with no statistical differences between the groups.

DISCUSSION

There are three major families of polyunsaturated fatty acids; oleic, linoleic and α-linolenic. Linoleic and α-linolenic acid are essential fatty acids and are the precursors of the Ω-6 and the Ω-3 series of prostaglandin metabolites respectively. An inappropriate intake resulting in a lower polyunsaturated to saturated fatty acid ratio, and/or defective metabolism of essential fatty acids (EFAs) may be responsible for many of the diseases that are common in Western societies. The main source of polyunsaturated fatty acids (PUFA) in Western diets is the Ω-6 fatty acid, linoleic acid, which can be converted by a series of alternating desaturation and elongation reactions to arachidonic acid. Arachidonic acid itself is present in animal fat and vegetable oils, whereas the Ω-3 derived essential fatty acids, which include eicosapentaenoic acid (EPA) and docosahexaenoic acid (DHA), are found mainly in cold water fish and fish oils, such as salmon, mackerel, herring and tuna. The PUFAs are incorporated into the phospholipid bio-layer of every cellular membrane; plasma, nuclear, mitochondrial and lysosomal, as the esters of phosphotidyl-choline, phosphotidyl-serine, phosphotidyl-ethanolamine and phosphotidyl-inositol. The composition of the membrane is dependent, therefore, on the availability of precursor fatty acid and on the release and re-acylation mechanisms controlling the metabolic cascade. EFAs confer on cell membranes properties of fluidity, flexibility, and permeability, and modulate the behaviour of ion channels and membrane bound receptors and enzymatic interactions.

Humans cannot rapidly synthesize EPA from precursor α-linolenic acid, but can assimilate pre-formed EPA present in marine oils, into their lipid stores to bring about a change in the eicosanoid pathway from the Ω-6 to the Ω-3 series of metabolites[8,9]. A diet supplemented with fish or fish oil, rich in EPA, results in an Eskimo-like pattern of plasma lipids[10-12]. In both experimental animals and humans, EPA is rapidly and readily incorporated into cell membrane phospholipids at the expense of arachidonic acid when fish oil enriched diets are administered, whilst a high affinity of EPA for cyclo-oxygenase results in competitive inhibition of this enzyme in the synthesis of metabolites derived from arachidonic acid[13,14]. The inhibitory effect of EPA on the oxygenated products of arachidonate metabolism was shown to be equal to that of indomethacin[12]. The metabolites of EPA, i.e. PGE3, PGI3, TXA3 and LTC5 etc., express biological activity that is different from and opposite to that of their structurally analagous counterparts derived from arachidonic acid, (PGE2, PGF2α, PGI2, TXA2 etc.). There is now substantial evidence that GLA (γ- linolenic acid), is derived from evening primrose oil, a precursor for the monoenoic series (PGE1), which have properties similar to the eicosanoids derived from IPA (figure 3). Thus, a combination of GLA and EPA will enhance levels of the desirable monoenoic (PGE1) and trienoic metabolites (PGE3) at the expense of the pro-aggregatory and pro-inflammatory dienoic (PGE2) series[15,16].

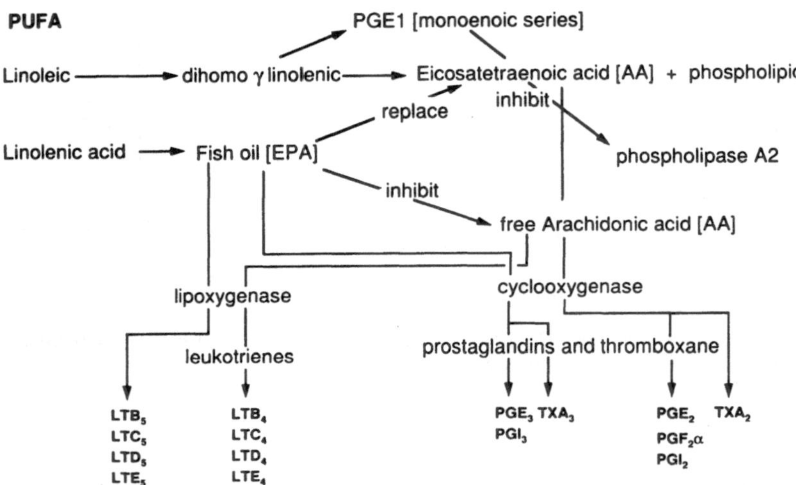

Figure 3. Schematic diagram showing the pathways for PUFA metabolism. Linoleic acid is the EFA source of arachidonic acid, the precursor fatty acid of the dienoic series of metabolites. Eicosapentaenoic acid, (EPA), derived from marine oil, competitively inhibits arachidonic acid resulting in the preferential synthesis of the trienoic series. The actions of EPA are enhanced in the presence of the monoenoic series (ie,PGEl) derived from dihomo-γ-linolenic acid (DGLA). Evening primrose oil is a rich source of DGLA.

Prostaglandin synthesis in the renal cortex and medulla is principally via the cyclo-oxygenase pathway to the Ω-6 series. Our previous studies have demonstrated that prostaglandins (PGE2) influence the renal handling of calcium and that non-steroidal anti-inflammatory agents reduce urinary calcium excretion in experimental animals and in hypercalciuric recurrent stone formers[17,18]. Several studies have shown raised urinary PGE2 levels associated with hypercalciuria and that both urinary PGE2 and urinary calcium levels were reduced with NSAIDs[19-22]. Incorporation of EPA (marine oil) and GLA (evening primrose oil) in the diet as the major poly-unsaturated fatty acid substitute for arachidonate will result in a switch in the synthesis of prostanoid metabolites of the dienoic series (i.e. PGE2) to those of the monoenoic (PGEl) and trienoic (PGE3) series.

With regard to renal calcium excretion this transformation should effectively alter the metabolic response to stimuli which result in the PGE2-related hypercalciuria associated with idiopathic urolithiasis. In a previous study[6] we showed that fish oil in a dose of 10 g/day over an 8 week period significantly reduced urinary calcium and oxalate excretion in a group of hypercalciuric recurrent stone formers, while in a streptozotocin diabetic rat model GLA (evening primrose oil) significantly reduced urine calcium excretion (unpublished observations). In this double-blind placebo controlled study in stone formers on a fixed calcium intake, fish oil alone or in combination with evening primrose oil significantly reduced urinary calcium excretion. Whilst the statistical analysis using a two-way analysis of variance indicated that the main effect was due to the fish oil, it has to be remembered that the amount of marine EPA in the combination capsule was only 20X of the amount of EPA in the fish oil capsule alone. Thus, the hypocalciuric effect in the patients on the combination of EPA and GLA has to be ascribed to the potentiating effect of evening primrose oil on fish oil.

A possible mechanism for the hypocalciuric effect of fish oil could be its influence on intestinal calcium absorption. To study the effect of fish oil and evening primrose oil on gut calcium absorption, intestinal calcium absorption was monitored by means of an established radioisotope absorption technique[7]. An interesting observation was the significant increase in the rate of intestinal ^{45}Ca absorption at 12 weeks in all 4 groups of patients. However, despite the increase in calcium absorption in the fish oil and combination group, urinary calcium excretion in these patients decreased, which would indicate that the hypocalciuric effect of EPA and GLA is an expression of renal calcium handling independent of gut calcium absorption. Further studies are in progress to investigate the role of EPA and GLA in bone metabolism together with long-term studies on the effect of prolonged dietary supplementation on stone recurrence rates.

REFERENCES

1. M Modlin, Urinary sodium and renal stone, "Renal Stone Research Symposium", A Hodgkinson & BEC Nordin, eds, J & A Churchill Ltd, London, pp 209-220 (1968).
2. JB Arthaud, Cause of death in 339 Alaskan natives as determined by autopsy, *Arch Pathol Lab Med* 90: 433 (1970).
3. N Kromann, A Green, Epidemiological studies in the Upernavik district, Greenland, *Acta Med Scand* 208: 401 (1980).
4. HM Sinclair, Essential fatty acids and chronic degenerative diseases, *in* "'Nutrition and Killer Diseases" J Rose, ed, NJ Noyes, pp 69-83 (1983).
5. Y Kagawa, M Nishizawa, M Suzuki, et al, Eicosapolyenoic acids of serum lipids of Japanese islanders with low incidence of cardiovascular disease, *J Nutr Sci Vitaminol* 28: 441 (1982).
6. AC Buck, RL Davies, T Harrison, The protective role of eicosapentaenoic acid (EPA) in the pathogenesis of nephrocalcinosis, *J Urol* 146:188 (1991).
7. RM Francis, M Peacock, SA Barkworth, DH Marshall, A comparison of the effect of sorbitol and glucose on calcium absorption in postmenopausal women, *The American J of Clinical Nutr* 43: 72 (1986).
8. JT Bernert, H Sprecher, Studies to determine the role rates of chain elongation and desaturation play in regulating the unsaturated fatty acid composition of rat liver lipids, *Biochim Biophys Acta* 398: 354 (1975).
9. HO Bang, J Dyerberg, JN Hjorne, The composition of food consumed by Greenland Eskimos, *Acta Med Scand* 200: 69 (1976).
10. J Dyerberg, HO Bang, E Stoffersen, S Moncada, JR Vane, Eicosapentaenoic acid and prevention of thrombosis and atherosclerosis, *Lancet* ii: 117 (1978).
11. M Thongren, A Gustafson, Effects of 11 week increase in dietary eicosapentaenoic acid on bleeding time, lipids and platelet aggregation, *Lancet* ii: 1190 (1981).
12. HM Sinclair, The relative importance of essential fatty acids of the linoleic and linolenic families: studies with an Eskimo diet, *Prog Lipid Res* 20: 897 (1982).
13. P Hoffman, H-J Mest, What about the effects of dietary lipids on endogenous prostanoid synthesis? A state of the art review, *Biomed Biochim Acta* 46: 639 (1987).
14. DH Hwang, M Boudreau, P Chanmugam, Dietary linolenic acid and longer-chain n-3 fatty acids: comparison of effects on arachidonic acid metabolism in rats, *J Nutr* 118: 427 (1988).
15. BA Nassar, MS Manku, YS Huang, DK Jenkins, D Horrobin, The influence of dietary marine oil (polepa) and evening primrose oil (Efamol) on prostaglandin production by the rat mesenteric vasculature, *Prostaglandins Leukotrienes and Medicine* 26: 253 (1987).

16. DF Horrobin, Interactions between n-3 and n-6 essential fatty acids (EFAs) in the regulation of cardiovascular disorders and inflammation, *Prostaglandins Leukotrienes and Essential Fatty Acids* 44: 127 (1991).
17. AC Buck, WF Sampson, CJ Lote, NJ Blacklock, The influence of renal prostaglandins on glomerular filtration rate (GFR) and calcium excretion in urolithiasis, *Br J Urol* 53: 485 (1981).
18. AC Buck, CJ Lote, WF Sampson, The influence of renal prostaglandins on urinary calcium excretion in idiopathic urolithiasis, *J Urol* 129: 421 (1983).
19. M Houser, B Zimmerman, M Davidson, et al, Idiopathic hypercalciuria associated with hyperreninaemia and high urinary prostaglandin E, *Kidney Int* 26: 176 (1984).
20. H Hirayama, K Ikegami, T Shimomura, T Yamamoto, The possible role of prostaglandin E2 in the urinary stone formation. Proc. XX Congres *Int Soc Urol* pp 231-233 (1985).
21. L Calo, S Cantaro, F Marchini, et al, Is hydrochlorthiazide-induced hypocalciuria due to inhibition of prostaglandin E2 synthesis?, *Clin Sci* 78: 321 (1990).
22. C Henriquez-La Roche, B Rodriguez-Iturbe, G Parra, Increased urinary excretion of prostaglandin E2 in patients with idiopathic hypercalciuria is a primary phenomenon, *Clin Sci* 83: 75 (1992).

SIMULTANEOUS TREATMENT OF CALCIUM OXALATE AND URIC ACID STONE DISEASE IN SAUDI ARABIA

W.G. Robertson, H. Hughes, I. Husain[1], S. Al-Faqih[1]
A. Arafat[1], A. Chakrabarti[1], A. Shamsuddin[1] and
L.S. Tipton

King Faisal Specialist Hospital & Research Centre
[1]King Khalid University Hospital
Riyadh, Saudi Arabia

INTRODUCTION

Urolithiasis is a major public health problem in most of the oil-rich Gulf countries. In Saudi Arabia, for example, it is estimated that over 20% of men will have at least one episode of stones before the age of 60. The majority of stones are idiopathic in origin and consist of calcium oxalate (CaOx) and/or uric acid; calcium phosphate (CaP) stones, on the other hand, are relatively rare.

There are several epidemiological reasons why the prevalence of the disorder is so high in this part of the world. Firstly, populations of the region are generally affluent and have a high consumption of animal protein and purine. This, in turn, leads to urine which tends to be acidic and which has a high content of uric acid. Hypocitraturia is a common finding as a consequence of the high acidity of urine.

Secondly, the intake of oxalate, which is derived mainly from local salad vegetables[1], is about three times the corresponding figure in most Western countries. The marked effect of this excessive intake on urinary oxalate is compounded by the fact that dietary calcium is only about 60% of that in the West. This enhances the intestinal absorption of oxalate and increases urinary oxalate to the extent that 60 to 70% of the population have mild hyperoxaluria (> 0.5 mmol/day).

As a corollary to this, however, hypercalciuria is an uncommon finding, even in stone formers, unlike the situation in most Western countries. The reasons for this are three-fold: (i) dietary calcium is low[1] and (ii) the circulating levels of vitamin D are also low[2]. This leads to calcium being poorly absorbed[3]. Indeed, urinary calcium is so low that it partly protects the population from forming more stones than they already do. (iii) The third reason for the already high prevalence of stone disease is the hot, dry climate. In patients who have not previously attended a urology clinic and been instructed to drink more fluid, the tendency to pass a low urine volume is quite marked. This is accentuated by the local popular misconception that the high incidence of stones is "something to do with the water".

In summary, then, the main urinary risk factors which appear to account for the high prevalence of calcium oxalate and/or uric acid stones in the population of Saudi Arabia are extensive mild hyperoxaluria, hypocitraturia, hyperuricosuria, an acid urine and a tendency to a low urine volume. Hypercalciuria does not appear to be a major risk factor in the population.

Urolithiasis 2, Edited by R. Ryall *et al.*,
Plenum Press, New York, 1994

Treatment Programmes

The very different urine biochemistry reported in stone formers from the affluent countries of the Middle East compared with that in Western stone formers raises an interesting challenge in terms of devising a form of medical treatment suitable for the prevention of the recurrence of both CaOx and uric acid stone formation, since most of the existing procedures which are currently employed in the West would be unlikely to be beneficial in this environment. Thus, altering the diet to reduce the consumption of animal protein, purine and oxalate would probably fail, since dietary habits and social customs are fairly conservative in the region; a low calcium diet would only exacerbate the situation by further increasing urinary oxalate; thiazide diuretics would probably not have much beneficial effect since urinary calcium excretions are already low; phosphate supplements would be unlikely to be efficacious since urinary calcium excretions are already low and, in any case, if they did bind calcium in the intestine they might cause more oxalate to become available for absorption in the colon; magnesium supplements might be beneficial in those patients with low urinary magnesium excretions but many of the stone formers have been found to have normal or even high magnesium excretions; allopurinol might reduce the degree of hyperuricosuria and so decrease the risk of uric acid stones but its value in the treatment of CaOx is highly debatable. The only form of preventative therapy which would be likely to be of some value would be alkaline potassium citrate supplements. These would alkalinise the urine and stimulate citrate production in the renal tubules. This measure would solubilise uric acid (in spite of the hyperuricosuria) and would eliminate the hypocitraturia by normalizing urinary citrate.

None of the above procedures, however, including the use of potassium citrate supplements, would correct the mild hyperoxaluria which has been identified in a high proportion (80 to 90%) of stone formers and is probably the main cause of their CaOx stone disease. The challenge, therefore, was to develop a form of treatment which would normalize urinary oxalate in addition to correcting the hypocitraturia, solubilizing uric acid and preventing a significant increase in the risk of forming other types of stone.

Two procedures were devised which take into account the specific dietary, urinary and metabolic circumstances existing in the Middle East. In particular, advantage was taken of the observations that dietary calcium and its intestinal absorption tend to be low in Saudis. Under these circumstances, it was suggested that calcium supplementation might be used to block the intestinal absorption of oxalate by precipitating and/or complexing the anion in the gut without the risk of causing significant hypercalciuria.

Super Citracal Study

In the first treatment, calcium citrate [$Ca_3(C_6H_5O_7)_2.4H_2O$] supplements were administered in the form of a soluble effervescent tablet [Super Citracal (Mission Pharmacal, San Antonio, Texas)]. This was prescribed at two dose levels - the first supplying an additional 480 mg (12 mmol) of calcium and 1512 mg (8mmol) of citrate per day to the diet and the second supplying double these quantities.

A total of 25 male Saudi idiopathic stone formers (aged 20 to 50 years) were recruited into the study. They had had at least one renal stone within the previous 12 months and the stone had been identified as consisting of CaOx and/or uric acid. All had good renal function (serum creatinine < 150 µmol/L) and no identifiable metabolic disorder (such as primary hyperparathyroidism, renal tubular acidosis or hereditary hyperoxaluria). None had a history of a urinary tract infection or was receiving any other medication at the time of study. All gave written consent to be included in the trial and agreed not to be informed whether they had been entered into the treatment group or into the placebo group.

The patients were allocated at random to the two groups on a 4:1 rotation basis. They were then studied for two weeks on their free, home diet. A 24 h urine was collected at the end of each basal week. During the following two weeks the treatment group, consisting of 20 patients, was given a total of 12 mmol/day additional calcium in the form of two tablets/day of effervescent Super Citracal, each dissolved in a glass of water (250 mL), the first taken with the midday meal and the second with the evening meal. Placebo tablets made up of lactose and matched for size and weight with the Super Citracal tablets were given to the control group consisting of five patients. The two groups of patients were studied for two weeks on their respective regimens and a 24 h urine collected at the end of each week.

Twelve of the patients who were members of the treatment group were studied for a further two weeks at a dosage of Super Citracal which was twice that used during the first part of the study. This provided a total supplementation of 960 mg (24 mmol) calcium/day and 3024 mg (16 mmol) citrate/day. Twenty four hour urines were collected at the end of each week.

Calcium Carbonate/Polycitra-K Study

In the second treatment, a combination of one sachet of 1250mg calcium carbonate (providing 12.5 mmol of calcium) and one sachet of polycitra-K crystals [a mixture of $K_3C_6H_5O_7.H_2O$(10.2 mmol) and $C_6H_8O_7.H_2O$(4.8 mmol)] was used. The two sachets were dissolved in a glass of water (2580 mL) and stirred before drinking. The pH of the mixture was 7.05. This combined medication was given twice daily, the first dose with the midday meal and the second with the evening meal. Altogether the treatment provided an additional 1000 mg/day (25 mmol) of calcium and 60 mmol/day of potassium.

A total of 20 male Saudi idiopathic stone formers, fitting the same criteria as the patients used in the Super Citracal study, were recruited into the investigation. No placebo group was included on this occasion. The subjects were studied for two weeks on their free, home diet using a similar protocol to that employed in the Super Citracal study. A 24 h urine was collected from each patient at the end of each basal week. During the subsequent two weeks all 20 patients were given daily calcium and alkali supplements as described above and a 24 h urine was again collected from each subject at the end of each week.

Analysis of Urine

All urine samples were analysed for volume, pH, calcium, magnesium, sodium, potassium, phosphate, oxalate, citrate, uric acid and creatinine. The relative supersaturation of urine with respect to calcium oxalate (log RS CaOx), calcium phosphate (log RS CaP) and uric acid (log RS UA) was calculated for each patient during the basal and treatment periods. The maximum volume of CaOx crystals precipitable from each urine was calculated from a knowledge of the log RS CaOx value and the oxalate/calcium ratio of the urine[4].

The mean values for the excretion data on the two basal days, the two placebo days, the two low-dose Super Citracal days, the two high-dose Super Citracal days and the two calcium carbonate/Polycitra-K days were calculated separately for each patient and the group data computed and compared using the Wilcoxon non-parametric matched-pairs test. To simplify the presentation in this paper, only the data for the 12 patients who completed all three phases of the Super Citracal study are reported. The data from the placebo group are omitted since the mean values for all urinary constituents during the placebo period did not differ from those of the same patients during the basal period.

RESULTS

Table 1 summarises the urinary changes from the basal state produced by the low dose and high dose of Super Citracal and by the administration of calcium carbonate /Polycitra-K.

Low-Dose Super Citracal

There were very few changes in urinary biochemistry following the ingestion of the low dose of Super Citracal. There was a tendency to a reduced excretion of oxalate and a slight increase in urinary volume which together led to small decreases in log RS CaOx and in the oxalate/calcium ratio. Together these small reductions produced a significant fall in CaOx crystalluria ($p< 0.05$).

Table 2 summarises the overall effects of these changes on various indices of the risk of stone formation.

Table 1. A summary of the changes in urine composition produced by Super Citracal and by calcium carbonate/Polycitra-K.

Treatment	Δ pH	Δ Ca	Δ Ox	Δ Cit	Δ Mg	Δ P
Super Citracal (low dose)	ns	↑ns	↓ ns	ns	ns	↓*
Super Citracal (high dose)	ns	↑*	↓ ***	ns	↑*	↓*
Ca+Polycitra-K	↑***	↓ns	↓**	↑***	ns	↓***

ns = not significant; * $p< 0.05$; ** $p< 0.01$; *** $p< 0.001$

Table 2. A summary of the changes in the indices of stone risk produced by Super Citracal and by calcium carbonate/Polycitra-K.

Treatment	logRS UA	logRS CaP	logRS CaOx	Ox/Ca	CaOx crystals
Super Citracal (low dose)	ns	ns	↓ns	↓ns	↓*
Super Citracal (high dose)	ns	ns	↓ns	↓***	↓***
Ca+Polycitra-K	↓***	↑***	↓***	↓ns	↓***

ns = not significant; * $p< 0.05$; ** $p< 0.01$; *** $p< 0.001$

High-Dose Super Citracal

In the studies using the higher dose of Super Citracal, many of the trends observed at the lower dose became more marked and, in most instances, significant. The most notable were the reductions in urinary oxalate ($p< 0.001$) and in the oxalate/calcium ratio ($p< 0.001$). Indeed, all the urinary oxalate excretions except one fell into the normal range for urinary oxalate in the West.

There was an increase of about 40% in urinary calcium ($p< 0.05$) corresponding to an intestinal absorption of about 7% of the additional calcium load. This is a much smaller absorbed fraction than would have been found in a similar study on stone formers in the West.

There was a small but significant increase in the urinary excretion of magnesium ($p< 0.05$), presumably as a result of competition between magnesium ions and the increased throughput of calcium ions for reabsorption sites in the renal tubule. Phosphate excretion, on the other hand, decreased ($p< 0.05$) probably because of precipitation in the intestine of a proportion of dietary phosphate in the form of calcium phosphate, thereby making it unavailable for absorption.

No significant changes were found in urinary pH or in the excretion of citrate, indicating that calcium citrate does not supply a sufficient load of alkali to the kidney to increase the pH of the tubular fluid or to stimulate the renal production of citrate.

In spite of the marked decrease in urinary oxalate, log RS CaOx fell slightly but not significantly. This was due to the fact that the increase in urinary calcium took place over the lower end of the urinary calcium range where the relationship between supersaturation and urinary calcium concentration is at its steepest[4]. Thus the effect of the decrease in oxalate on supersaturation was almost offset by the increase in calcium. However, because of the highly significant decrease ($p < 0.001$) in the oxalate/calcium ratio (from 0.15 to 0.06), there was an almost 50% reduction in the volume of CaOx crystals produced ($p < 0.001$) even although supersaturation remained almost unchanged[4]. There was no beneficial effect of the high dose of Super Citracal on uric acid supersaturation, however, because of the failure of the drug to alkalinise urine.

Calcium Carbonate/Polycitra-K

In the study using calcium carbonate plus Polycitra-K there was a marked increase ($p < 0.001$) in urinary pH from a mean value of 5.87 on the basal diet to 6.71 during the treatment period. The supplements also caused an increase in the excretion of potassium ($p < 0.001$) and doubled that of citrate ($p < 0.001$) into the high-normal Western range. There was a reduction in urinary oxalate ($p < 0.01$), as anticipated, but this was less marked than that observed using the equivalent load of calcium in the form of Super Citracal. One unexpected finding was a small (although not significant) fall in urinary calcium and a highly significant reduction ($p < 0.001$) in urinary phosphate of approximately 9 mmol/day.

This latter observation probably yields a clue as to (i) why urinary calcium not only did not increase but actually fell slightly and (ii) why urinary oxalate did not decrease as much as it did during treatment with the high dose of Super Citracal which added the same amount of calcium to the diet. If it is assumed that the reduction in urinary phosphate was caused by the precipitation of calcium phosphate in the gut induced by the excessively high concentrations of calcium and alkali, then approximately 13.5 mmol of the additional calcium (54%) would become unavailable either for absorption or, more importantly, for precipitation with oxalate. In practice, therefore, the patients are "receiving" less than half of the anticipated amount of calcium and so it is not surprising, perhaps, that urinary oxalate did not fall by as much as it did on the high dose of Super Citracal where little or no alkalinization of the gut contents took place. Urinary magnesium remained unchanged in this study probably because urinary calcium did not increase as in the Super Citracal study.

The net effects of these alterations in urinary biochemistry were to reduce markedly both log RS UA ($p < 0.001$) and log RS CaOx ($p < 0.001$). There was also a small (but not significant) fall in the oxalate/calcium ratio of urine and a highly significant reduction in the volume of CaOx crystalluria ($p < 0.001$). The only adverse effect of the treatment was the increase in log RS CaP as a result of the marked rise in urinary pH although the effect was partly modulated by the accompanying reduction in urinary phosphate and the increase in citrate.

In summary, Super Citracal on its own does not provide sufficient alkali to increase urinary pH in order to solubilize uric acid and to correct the hypocitraturia, although it causes a highly satisfactory reduction in urinary oxalate and in calcium oxalate crystalluria. The combination of calcium carbonate and Polycitra-K used in this study, on the other hand, was probably too alkaline since much of the added calcium was lost in the gut through precipitation with phosphate and urine became sufficiently alkaline to cause concern that there might be an increased risk of CaP stone-formation. If the two treatments could be combined by using the low dose of Super Citracal plus half the quantity of calcium carbonate and the full amount of Polycitra K as used in the current study, then all the criteria would probably be satisfied. If this still produced a urine which was too alkaline, it might be necessary to eliminate the calcium carbonate altogether and to use the high dose of Super Citracal in conjunction with Polycitra-K. This is currently under investigation.

ACKNOWLEDGEMENTS

The authors wish to thank Sheikh Rafiqu El-Hariri for his generous financial support during the study.

REFERENCES

1. WG Robertson, M Nisa, et al, The importance of diet in the aetiology of primary calcium and uric acid stone formation: the Arabian experience, *in:* VR Walker, RAL Sutton, ECB Cameron, CYC Pak, WG Robertson, eds, "Urolithiasis, Plenum Press, New York, p735 (1989).
2. NJY Woodhouse and WL Norton, Low vitamin D levels in Saudi Arabians. *King Faisal Specialist Hosp J* 2:127,1982.
3. VR Walker, N Bissada et al, Urinary calcium excretion in Saudi Arabia. *in:* VR Walker, RAL Sutton, ECB Cameron, CYC Pak, WG Robertson, eds,"Urolithiasis", Plenum Press, New York, p717 (1989).
4. WG Robertson and H Hughes, Importance of mild hyperoxaluria in the pathogenesis of urolithiasis - new evidence in the light of the Arabian experience, *Scan Microsc* (in press).

URINARY TRACT STONE AFTER URETEROSIGMOIDOSTOMY

A. Trinchieri, M. Cogni, E. Patelli, A. Maggioni, F. Rovera, R. Nespoli, G. Zanetti and E. Austoni

Istituto di Urologia
Università degli Studi di Milano
Milan, Italy

INTRODUCTION

Patients with urinary diversion have an increased risk of forming renal calculi. The incidence of renal stones after ureterosigmoidostomy (USS) increases with the length of follow up and varies in different series from 8 to 18%[1]. Renal calculi after urinary diversion are attributed to stasis and chronic infection, but the presence of metabolic complications may increase the risk of forming renal calculi.

The present study was designed to study variations of the so-called "urinary risk factors" that occur following urinary diversion through intestinal segments.

MATERIALS AND METHODS

The series included 93 patients (68 M, 25 F; age range 32-84 years) who had undergone USS or ileal conduit, at times ranging from 2 to 8 years following their operation. Group I consisted of 72 patients who underwent USS. Goodwin's technique was used in all cases. Group II comprised 21 patients who were operated with ileal conduit. Carcinoma of the bladder was the most common indication for operation. Only 6 patients were operated on for benign disease. Acid-base balance was assessed by arterial blood gas analysis. Specimens of venous blood taken after fasting were obtained from each patient. Plasma potassium, sodium, calcium, phosphate, urate and creatinine were assayed. Patients with ileal conduit collected 24 h urine samples and patients with USS collected mid-morning urine-feces samples. Urine-feces samples were centrifuged to separate solid feces and the clear supernatant was removed for analysis. Sodium, potassium, calcium, magnesium, phosphate, oxalate, urate, citrate, glycosaminoglycans, creatinine and pH were measured in the urine and in the supernatant of urine-feces samples. Urinary values were calculated as mg per mg of creatinine. In addition serum samples were obtained and 24 h urine were collected from normal control subjects (14 M, 19 F; age range 51-73 years).

RESULTS

Five patients had urinary tract calculi after USS (7%). None of the patients with ileal conduit had stones. Serum and urinary values are summarized in Table 1. Patients with USS had a higher amount of urinary calcium, as well as sodium and potassium, than healthy controls. Urine glycosaminoglycans were highly elevated in both groups.

Table 1. Blood and urinary values in patients after ureterosigmoidostomy (USS), in patients with ileal conduit (IC) and in healthy controls (N).

	USS	IC	N
Blood			
pH	7.3±0.37	-	-
K (mEq/L)	4.4±0.5	4.5±0.5	-
Na (mEq/L)	147.0±2	14.05±2	-
Ca (mg/dL)	9.28±0.52	9.28±0.46	-
PO_4(mg/dL)	2.6±0.4	2.5±0.4	-
Mg (mg/dL	1.76±0.26	1.83±0.23	-
Urine			
K/Cr (mEq/mg)	0.09±0.07**	0.04±0.01	0.05±0.02
Na/Cr (mEq/mg)	0.45±0.54**	0.19±0.10	0.14±0.06
Ca/Cr (mg/mg)	0.25±0.29*	0.07±0.03	0.15±0.09
Ua/Cr (mg/mg)	0.48±0.57	0.48±0.74	0.40±0.16
Cit/Cr (mg/mg)	0.38±0.68	0.12±0.13	0.44±0.26
Mg/Cr (mg/mg)	0.07±0.11	0.03±0.02	0.08±0.03
GAGs/Cr (mg/mg)	0.43±0.55***	0.12±0.07**	0.02±0.01
USS or IC versus N	p <0.005 *, p <0.001 **, p <0.0001 ***		

DISCUSSION

Acidosis

In patients subjected to USS, metabolic acidosis occurs in up to 75% of cases[2,3]. Three mechanisms may account for the metabolic acidosis that develops with urinary diversion through the intestine:

a) absorption of ammonium ions present in the urine or generated from urea by urease containing organisms;
b) secretion of bicarbonate in exchange for urinary chloride;
c) renal acidifying defect due to obstruction, pyelonephritis or both.

Koch et al demonstrated that the acid load is primarily the result of ammonium reabsorption[4]. On the contrary, bicarbonate secretion probably does not cause clinically significant additional acidosis.

Calcium metabolism

Metabolic acidosis reduces renal tubular reabsorption, stimulates net bone resorption and, ultimately, increases urinary calcium excretion. Clearance and micropuncture studies demonstrate the presence of a component of tubular calcium reabsorption situated beyond the proximale tubule, which is inhibited by chronic acidosis. On the other hand, use of an alkalinizing agent such as sodium bicarbonate may also result in the increase of calcium excretion. In fact, our patients with USS ingested 1-7.5 g of sodium bicarbonate daily in order to control acidosis. The sodium load may cause hypercalciuria by decreasing calcium tubular reabsorption. On this basis the administration of potassium alkali should be considered to prevent renal stones when dealing with acidosis after USS. In contrast to the action of sodium alkali, potassium alkali at the same equimolar dosage reduces urinary calcium excretion.

Inhibitors of crystallization

Metabolic acidosis has been demonstrated to exert an effect on urinary excretion of magnesium and citrate, but we found no significant difference in magnesium and citrate excretion between patients with USS and controls. The significance of this finding is not clear and remains to be determined by future investigations. Our study showed a higher urinary glycosaminoglycan excretion in patients with urinary diversion through the intestine. The explanation for this is the secretion of glycosaminoglycans from the intestinal mucosa.

Ileal conduit

Metabolic acidosis is a much less frequent complication of ileal conduit. In fact, the metabolic effects of urinary diversion through the intestine depend on the surface area and the time of the exposure to urine.

CONCLUSIONS

Data from the present investigation reveal an increased metabolic risk of forming stones following USS, owing to high urinary excretion of calcium. It would seem that metabolic acidosis is the underlying metabolic abnormality which increases calcium excretion. Oral sodium bicarbonate is effective in controlling acidosis, but the sodium load may cause hypercalciuria by decreasing calcium tubular reabsorption. The administration of potassium alkali should be considered to prevent renal stones when dealing with acidosis after USS.

REFERENCES

1. WE Goodwin and PT Scardino, Ureterosigmoidostomy, *J Urol* 118:169 (1977).
2. H Zincke, JW Segura. Ureterosignoidostomy: critical review of 173 cases, *J Urol* 113: 324 (1975).
3. T Kalbe, AR Tricker, P Friedl et al, Ureterosigmoidostomy: Long-term results, risk of carcinoma and etiological factors for carcinogenesis, *J Urol* 144:1110 (1990).
4. MO Kock, E Gurevitch, DE Hill, WS McDougal, Urinary solute transport by intestinal segments: a comparative study of ileum and colon in rats, *J Urol* 143:1275 (1990).

RENAL TARGETING OF PHOSPHOCITRATE VIA A GAMMA-GLUTAMYL PRODRUG

J. Meehan and J. Sallis

Dept. Biochemistry
University of Tasmania
GPO Box 252C Hobart
Tasmania, Australia

INTRODUCTION

Recurrent stone formation is often a consequence of inadequate clearance of residual fragments following shock-wave or surgical procedures. A need exists, therefore, to prophylactically restrict re-growth. In earlier studies we demonstrated that combined phosphocitrate (PC) and antibiotic treatment can minimize such a process for infection stones in rats[1]. Conventional intraperitoneal (ip) treatment with PC, however, is marred by considerable losses incurred through normal biodistribution processes. These combine to reduce efficacy, and necessitate an increase in dosage.

The use of γ-glutamyl prodrugs to confer kidney specificity on agents containing amino is well documented[2]. Although such agents are preferentially metabolized by γ-glutamyl transpeptidase (GGT) activities present in the proximal tubules, the nature of the prodrug moiety is crucial. Simple γ-glutamyl prodrugs are metabolized to the parent compound in the lumen, whilst N-acetyl-γ-glutamyl derivatives are hydrolyzed intracellularly[2]. Clearly, prodrugs of this type would have interesting implications for a compound such as PC. Outlined here is the synthesis and preliminary biological assessment of the N-acetyl-γ-glutamyl prodrug of PC (NAG-PC).

EXPERIMENTAL METHODS

Synthesis of N-Acetyl-γ-Glutamyl PC Prodrug

As summarized in Figure 1, PC was esterified then selectively de-protected to give the mono-α-methyl ester, and hydrazinolysis was used to introduce nitrogen into the compound. This amino terminal was subsequently linked to N-hydroxy succinimide (NOHS)-activated N-Acetyl-α-benzyl-glutamate. The coupled product was hydrogenolytically de-protected at 20 psi over Pd/C and ultimately crystallized from ethanol to give NAG-PC. Structural confirmation by NMR was performed in D_2O at 300.13 MHz.

Enzymatic Hydrolysis *In Vitro*

Enzymatic transfer of the glutamyl residue to hydroxylamine by GGT and acylase was determined spectrophotometrically as the ferric/hydroxamate chromophore[2,3]. Appropriate blanks and controls were included where acylase, or both acylase and GGT were omitted from the reaction mixture.

Urolithiasis 2, Edited by R. Ryall *et al.*,
Plenum Press, New York, 1994

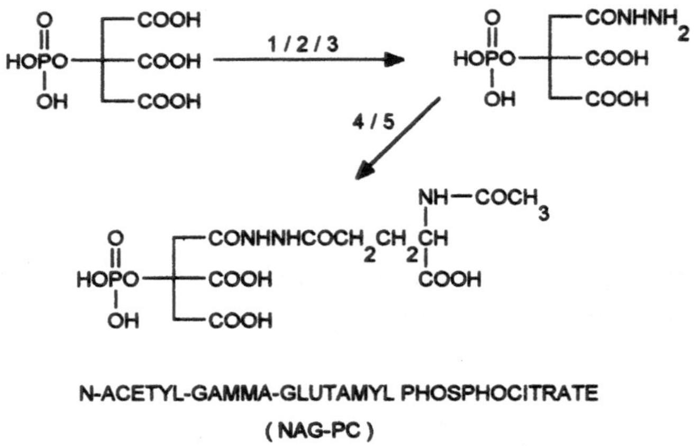

N-ACETYL-GAMMA-GLUTAMYL PHOSPHOCITRATE

(NAG-PC)

Figure 1. Synthetic scheme for the production of NAG-PC. 1. = $MeOH/H^+$; 2. = NaOH; 3. = $NH_2NH_2.H_2O$; 4.= NOHS-N-acetyl-α-benzyl-glutamate; 5.= H_2/Pd-C. ^{1}H-NMR results (δppm): 2.02 (s, 3H); 2.48-2.81 (m, 4H); 3.15-3.21 (d, 2H, J_{ab} = 16.3 Hz); 3.33-3.38 (d, 2H); 3.70 (m, 1H).

Inhibition of Nephrocalcinosis

Nephrocalcinosis was developed by administering calcium gluconate to rats, a procedure known to induce severe calcification of the kidneys[4]. On termination of the experiment kidneys were weighed and freeze-clamped, and the tissues digested. Calcium was measured by atomic absorption spectroscopy following appropriate dilution with a lanthanum solution

RESULTS

The synthetic strategy outlined in Figure 1 produced NAG-PC in 35% yield, and to our knowledge this is the first report of the use of a γ-glutamyl derivative for kidney targeting of an agent such as PC. ^{1}H-NMR analysis confirmed the structure, the presence of PC being validated by the spin-spin coupling constant between protons in the doublet centred at 3.15-3.38 ppm of 16.3 Hz.

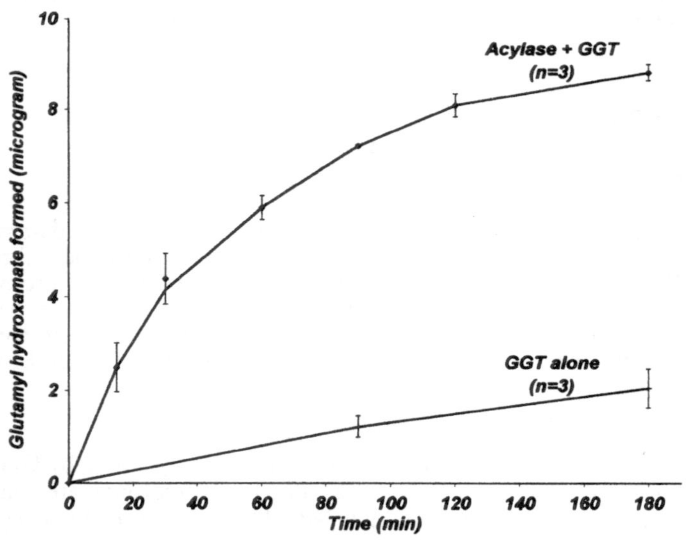

Figure 2. *In vitro* enzymatic hydrolysis of NAG-PC (GGT = γ-glutamyl transpeptidase).

Results for the *in vitro* hydrolysis of the derivative are shown in Figure 2. The prodrug moiety was demonstrated to be susceptible to enzymatic cleavage in the presence of both GGT and acylase. The combined enzymes were far more hydrolytically efficient than GGT alone.

Restriction of nephrocalcinosis is illustrated in Figure 3. When given at a dose corresponding to 50 mg PC/kg body weight/day, NAG-PC was almost twice as effective as PC ($p < 0.0005$), inhibiting Ca accumulation by 30% compared to 16% for sodium PC.

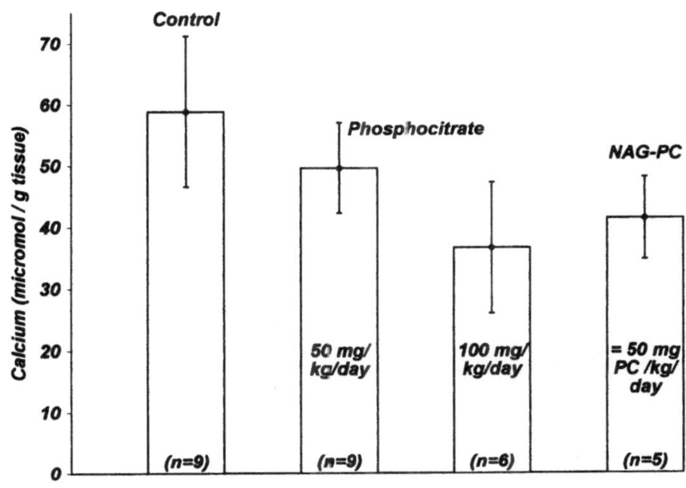

Figure 3. Comparative inhibition of nephrocalcinosis by PC and NAG-PC (n = rats / group).

CONCLUSIONS

This research demonstrates that a γ-glutamyl derivative of PC can be readily synthesized, and highlights the benefits of conferring site specificity on PC. Further tissue distribution studies are now required to fully delineate the overall tissue distribution and metabolism of the compound. Research into other acylated and non-acylated γ-glutamyl derivatives of PC is continuing as these may bestow further promising characteristics onto the drug.

ACKNOWLEDGEMENTS

This research was supported in part by a NHMRC project grant.

REFERENCES

1. JD Sallis, R Thomsom, B Rees and R Shankar, Reduction of infection stones in rats by combined antibiotic and phosphocitrate therapy, *J Urol.*140:1063 (1988).
2. JC Drieman, HHW Thijssen and HJ Struyker-Boudier, Renal selective N-acetyl-γ-glutamyl prodrugs.II, *J Pharm ExpTher*, 252:1255 (1990).
3. SDJ Magnan, FN Shirota and HT Nagasawa, Drug latentiation by γ-glutamyl transpeptidase, *J Med Chem* 25:1018(1982).
4. MR Brown and JD Sallis, N-sulpho-2-amino tricarballylate, a new analogue of phosphocitrate: metabolic studies and inhibitory effect on renal calcification, in: *Urolithiasis and Related Clinical Research*, pp. 891, PO Schwille et al, eds, Plenum Press, NY (1985).

THERAPEUTIC EFFECT OF *TRIBULUS TERRESTRIS* (CHOTA GOKHRU, AN AYURVEDIC DRUG) IN THE MANAGEMENT OF EXPERIMENTAL HYPEROXALURIA

D. Sangeeta[1], H. Sidhu[1], S.K.Thind and R. Nath

Departments of Biochemistry
Postgraduate Institute of Medical Education and
[1]Research and Panjab University
Chandigarh, India

INTRODUCTION

Ayurveda is the oldest existing medical system of traditional medicine, having its heritage in ancient India. It is recognized by the World Health Organisation and is still practised. Hyperoxaluria is one of the primary risk factors of urolithiasis. In various systems of traditional medicine, a number of indigenous plants are being used in India and other countries. One such plant is *Tribulus terrestris* (family zygophyllaceae) common in arid zones of India. Although its extract is known to have diuretic and antilithic properties[1] its mode of action is not fully understood. Hence, the present study was undertaken to evaluate the mode of action of *T. terrestris* in hyperoxaluric rats.

MATERIALS AND METHODS

Male wistar rats weighing 200-250g after acclimatization, were divided into four groups of 6 animals each and given the dietary regimen for 30 days as stated:

Group I Regular pellet diet
Group II Pellet diet and aqueous extract of *T. terrestris* (5g/kg BW) by gastric intubation (g.i)
Group III Pellet diet and sodium glycolate (100mg/100g BW)
Group IV Pellet diet, sodium glycolate and *T. terrestris* extract (5g/ kg B.W) by g.i

Twenty-four hour urine samples of each group were collected on days 0 and 30 and analysed for creatinine, oxalate, calcium, phosphorus, uric acid and citrate as routinely done in our laboratory[2]. After 30 days, animals were sacrificed under anaesthesia. Liver and kidney tissues were utilized for estimating the enzymes of oxalate biosynthesis viz. glycolate oxidase (GAO), glycolate dehydrogenase (GAD) and lactate dehydrogenase (LDH) by the method of Murthy et al[3].

For the preparation of aqueous extract of *T terrestris,* one part of the powdered drug and 16 parts of water were boiled down to 1/4th of the original volume and evaporated to dryness. The dried residue was dissolved in a solution of gum acacia (1.5% in DDW) and Tween-20 (12 drops).

RESULTS AND DISCUSSION

The effect of sodium glycolate feeding alone and in combination with the *T terrestris* extract on urinary excretion of oxalate and oxalate synthesising enzymes of liver and kidney are shown in the table.

Table Effect of *T Terrestris* administration on urinary oxalate and oxalate biosynthesizing enzymes of liver and kidney.

	Group I	Group II	Group III	Group IV
Urinary oxalate (mg/24h)	0.302 ±0.028	0.265 ±0.019	0.869 b*** ±0.087	0.377 c** ±0.028
Liver GAO (U^1/mg Pr)	3.46 ±0.33	3.45 ±0.23	6.46 b*** ±0.53	4.70 c*** ±0.49
Liver GAD (U^2/mg Pr)	0.270 ±0.005	0.250 ±0.013	0.420 b*** ±0.020	0.140 c*** ±0.008
Liver LDH (U^3/mg Pr)	1.33 ±0.050	1.44 ±0.070	1.73 b* ±0.120	1.70 ±0.060
Kidney LDH (U^3/mg Pr)	2.72 ±0.060	2.54 a* ±0.050	2.35 b** ±0.090	2.79 c*** ±0.090

*$p < 0.05$; **$p < 0.01$; ***$p < 0.001$ [a]I vs II; [b]I vs III; [c]III vs IV
$U^{1,2}$=one nanomole of glyoxylate/oxalate produced per minute at 37°C respectively
U^3= 0.1 OD/min at 340nm at 37°C

In our experimental model sodium glycolate feeding induced significant ($p < 0.001$) hyperoxaluria compared to controls. When the drug was administered along with sodium glycolate, urinary oxalate levels did not alter significantly in comparison to group I. Other urinary parameters were comparable in all the groups. The mechanism of action of this drug in preventing hyperoxaluria could well be elucidated by assaying enzymes of oxalate biosynthesis viz GAO, GAD and LDH in liver and kidney as done earlier[3]. Sodium glycolate feeding significantly increased liver GAO, GAD and LDH levels compared to controls, thereby indicating the major contribution of liver in converting the precursor glycolate into oxalate. The increased body pool of oxalate is removed by kidney leading to hyperoxaluria. In contrast to liver enzymes, the major oxalate synthesizing enzyme of kidney i.e. LDH was significantly lowered in sodium glycolate fed animals, which could be due to inhibition of this enzyme by high levels of renal oxalate resulting from increased liver enzyme activities. Inhibition of oxidation reaction of LDH by oxalate has been demonstrated by Banner and Rosalki[4]. In the animals fed the drug along with sodium glycolate, the liver GAO and GAD were significantly decreased compared to group III animals, thereby explaining the decreased endogenous production of oxalate from its precursors. The lowering of oxalate body pool may also be promoting the normalization of kidney LDH. Thus the beneficial effect of the drug is apparently mediated through its effect on enzymes of oxalate biosynthesis.

REFERENCES

1. G Santha Kumari and GYN Iyer, Preliminary studies on the diuretic effects of *Hygrophila spinosa* and *Tribulus terrestris*, *Ind J Med Res* 55:714 (1967).
2. H Sidhu, In: Biochemical mechanism of hyperoxaluria and role of urinary inhibitors of calcium oxalate crystallization: "Ph.D. thesis, PGIMER, Chandigarh, India" (1985).
3. MSR Murthy, HS Talwar, SK Thind, and R Nath, Effect of sodium glycolate and sodium pyruvate on oxalate synthesizing enzymes in rat liver and kidney, *Ann Nutr Metab* 27:355 (1983).
4. MR Banner and SB Rosalki, Glyoxylate as a substrate for lactate dehydrogenase, *Nature* 213:726 (1967).

METABOLIC ABNORMALITIES IN PATIENTS WITH RENAL STAGHORN STONES

J.R. Zanchetta, C.E. Bogado and E. del Valle

Instituto de Investigaciones Metabolicas
Buenos Aires
Argentina

INTRODUCTION

Renal calculi can be considered a cause or an effect of renal infection; the latter role produces the so-called staghorn type stone, with characteristic chemical composition (struvite and other related phosphates). Its cause is the enzymatic alkalinization of urine by urease-producing bacteria[2]. The infection-mediated stone grows very fast (sometimes called "stone cancer"). In the case of renal infection being caused by renal calculi the stones are of mixed composition and associated with metabolic disturbances.

As these two patterns need different treatments, it is necessary to perform proper metabolic evaluation to decide the proper approach to therapy. We describe here our experience with the metabolic evaluation of patients with staghorn renal stones.

MATERIALS AND METHODS

We studied 2532 idiopathic stone forming patients with our ambulatory protocol[1]. All patients were kept for 7 days on a diet which contained 1200 mg of calcium, 200 mg of magnesium, 800 mg of phosphate and 100 meq of sodium per day. The diet supplied 1500 Kcal per day with a content of 50% in carbohydrates, 20% in proteins, and 30% in lipids. The meals were selected to provide required quantities of phosphate, magnesium and sodium but only 200 mg of calcium. The rest of the calcium was taken orally in 2 phases of 500 mg with breakfast and supper.

At the sixth and seventh days of the diet two 24 h samples which were kept in plastic containers without preservatives at 4°C during the collection period, were obtained from each patient. On the morning of the 8th day, after a 12 h fast, they were instructed to drink 300 mL of distilled water; a blood sample was obtained 1 h later, and a new urine sample 2 h later. Calcium, magnesium, sodium, phosphate and creatinine were measured in all blood and urine samples. Blood ionized calcium was recorded as well.

Calcium and magnesium excretions were calculated from 24 h values and expressed as mg/100 mL glomerular filtrate (GF); sodium excretion was expressed as mmol/100 mL GF. Results from 2 h samples were used to calculate phosphate renal threshold and phosphate tubular resorption.

Urolithiasis 2, Edited by R. Ryall *et al.*,
Plenum Press, New York, 1994

RESULTS

Staghorn renal calculi (determined by X-ray examinations) were present in 182 (7.2%) patients. The following data correspond to patients of this group. Stone location was bilateral in 34 patients and unilateral in 133, whereas 15 patients had had a previous contralateral total nephrectomy. Urinary infection was demonstrated in 122 (67%) patients; in 60 subjects it was absent at the time of the study and could never be demonstrated. In 40 patients we were not able to find evidence of any metabolic disturbance, and thus, urinary infection was considered the sole cause of stone formation. On the other hand, metabolic abnormalities were detected in 142 (78%) cases.

Specific metabolic disturbances are shown in Table 1. The decreasing frequency order found was: Hypercalciuria (72 patients), uric acid metabolism disturbances (hyperuricosuria and/or hyperuricemia, 30), both hypercalciuria and uric acid metabolism disturbances (18 subjects). Primary hyperparathyroidism was diagnosed in 10 subjects, hypocitraturia in 8, and low magnesium excretion in 4 cases.

Table 1. Metabolic abnormalities in patients with staghorn calculi.

	n	%
Hypercalciuria	72	39.6
Uric acid metabolism	30	16.5
Hypercalciuria + uric acid metabolism disturbances	18	9.9
Primary Hyperparathyroidism	10	5.5
Hypocitraturia	8	4.4
Hypomagnesuria	4	2.1

Specific treatments were given to those patients with metabolic disturbances, according to the following criteria: thiazides were given to hypercalciuric patients (12.5 to 25 mg/day), allopurinol to those with uric acid metabolism disturbances (300 mg/day), 2 g/day of magnesium to those who had low magnesium excretion, and 40 to 80 meq/day potassium citrate to hypocitraturic subjects. Hyperoxaluria was treated by a diet with a low content of this ion, and cystinuria was treated by a combination of hyperdiuresis, potassium citrate, low sodium diet and penicillamine.

No recurrence or growth of the pre-existing stone (X-ray examination) could be demonstrated during 53±21 months of follow-up, taking into account that in 20% of the cases surgical removal of the stone was needed.

CONCLUSIONS

Our results show that, although staghorn calculi are produced by urinary tract infection, specific metabolic abnormalities can be found in a substantial number of these patients. Therefore, metabolic evaluations are desirable even for infection stones.

REFERENCES

1. JR Zanchetta and CE Bogado, Excrecion urinaria de sodio y magnesio en la hipercalciuria idiopatica, *Medicina* 51(4), 296 (1991).
2. DP Griffith, and CA Osborne, Infection (urease) stones, *Min Elect Metab*, 13, 278 (1987).

EFFECT OF ACE INHIBITOR ON CYSTINE EXCRETION IN CYSTINURIA

N. Amasaki, K. Kohri, M. Iguchi, K. Kataoka, T. Umekawa, T. Yamate and T. Kurita

Department of Urology
Kinki University School of Medicine
Osaka 589
Japan

INTRODUCTION

In patients with cystinuria, recurrence of calculi is prevented by increasing the urine volume, alkalinizing the urine and administering *a*- mercapto-propionylglycine (MPG)[1]. Long-term high-dose administration of MPG has been known to cause side effects such as hepatopathy, dermatopathy and nephropathy. On the other hand, captopril, an angiotensin-converting enzyme inhibitor (ACE inhibitor) used as a hypertensive drug, exhibits sulfhydryl-binding (SH-binding) in the same way as MPG, and produces a disulfate compound by reacting with cystine. Solubility of this disulfate is about 4 times as high as that of MPG[2]. In this study, ACE inhibitor was administered to patients with cystinuria and its efficacy evaluated.

SUBJECTS AND METHODS

We performed 3 experiments to study the effect of ACE inhibitor on cystinuria. First, 10 patients (6 men and 4 women, mean age 27.3 yr), who underwent follow-up observation, were treated with ACE inhibitor. All patients had not been receiving standard therapy for cystinuria, such as forced fluid and urinary alkalinization. ACE inhibitor was administered to these cases in doses of 50-150 mg (2-6 tablets) per day for 2 months and the urine cystine concentration was measured before and after administration.

Second, 5 different concentrations (0, 500, 1000, 2000 and 3000 µmol/L) of the ACE inhibitor and MPG were added to 1000 µmol/L of cystine at 20°C *in vitro* at pH 7.0, and the cystine and disulfate produced were measured by high-performance liquid chromatography (HPLC). Third, concentrations of urinary cystine were measured hourly after administration of 50 mg ACE inhibitor in two patients with cystinuria.

RESULTS

First, a significant decrease in cystine concentration in the urine was observed in the period after ACE inhibitor administration (Fig. 1).

No side effect was observed except for one case. A decrease of 20 mmHg was observed in both systolic and diastolic pressure.

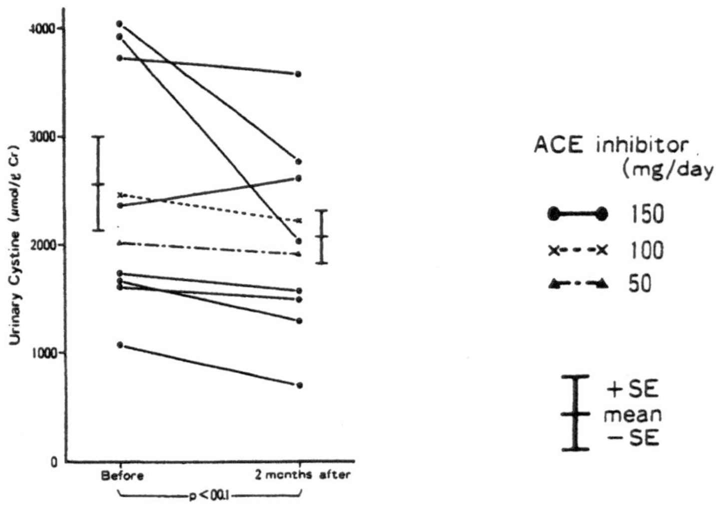

Figure 1. Effect of ACE inhibitor over two months.

Second, Figure 2 shows the result of HPLC. Peak 3 at the left of the bottom of the figure shows mixed disulfate, peak 2 cysteine, which is an oxide compound of cystine, and peak 1 cystine. From the top of the figure down, the amount of ACE inhibitor was increased. Peaks of disulfate and cysteine were increased in proportion to the addition of the amount of ACE inhibitor. The concentration was obtained from each peak respectively. It can be seen that the concentration of cystine decreased proportionally to the concentration of the ACE inhibitor.

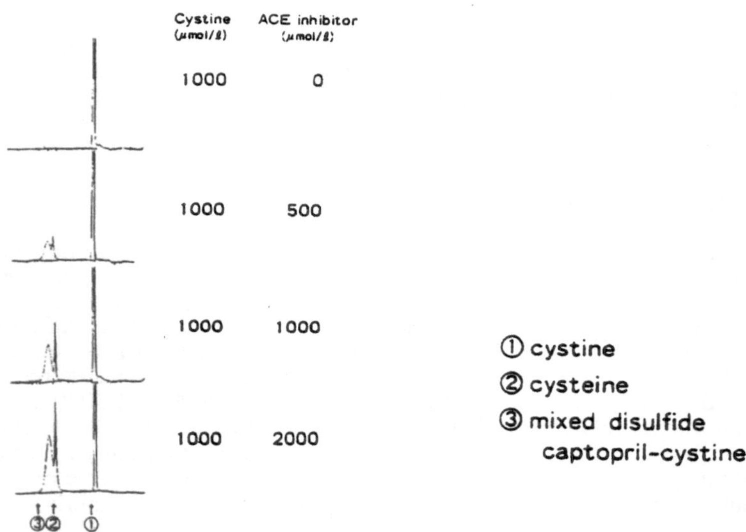

Figure 2. The result of HPLC (0, 500, 1000, 2000 and 3000 μmol/L of the ACE inhibitor and MPG were added to 1000 μmol/L of cystine at 20°C *in vitro* at pH 7.0 and the cystine and disulfate produced were measured by HPLC).

The reaction of cystine with MPG was also investigated. The decrease of concentration of cystine was proportional to the dosage of MPG; however, it was smaller than that in the case of ACE inhibitor addition.

Third, in order to find out how these changes are reflected in the volume of cystine

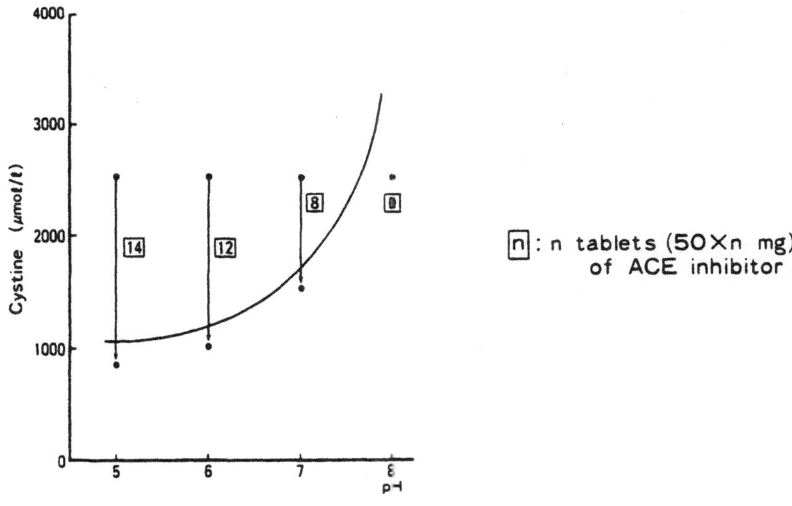

Figure 3. Solubility curve and dose of ACE inhibitor.

excreted in patients with cystinuria, sequential changes in the urine concentration of cystine after administration of the 2 tablets (50 mg) of ACE inhibitor were investigated. After taking the ACE inhibitor, the urine concentration of cystine fell immediately. The urinary cystine was gradually decreased for 5 hours and then returned to the pretreatment level 8 hours later.

DISCUSSION

We evaluated the effect of ACE inhibitor on cystine excretion in patients with cystinuria. The amount of urinary cystine decreased after administration, suggesting that this medication is useful for the prevention of cystine calculi formation. From the result of these 3 experiments and the solubility curve of cystine, it is thought that the effective dosage for decreasing urinary cystine excretion is 8 tablets (200 mg) at pH 7, 12 tablets (300 mg) at pH 6, and 14 tablets (350 mg) at pH 5. These doses were set on the assumption that cystine would be dissolved completely. When ACE inhibitor is to be administered over a long period or at a high dose in the future, observation and investigation will be required to monitor side effects.

Figure 3 also emphasizes the importance of increasing both the volume and alkalinity of urine, which have been adopted as measures to prevent cystine calculi. From second experiment, it is noteworthy that when MPG was administered in the same dose as that of the ACE inhibitor, the urinary volume was relatively higher than that excreted with the ACE inhibitor alone. For this reason, combined use of these 2 drugs may help to reduce the severity of side effect.

REFERENCES

1. K Kinosita, Therapy with MPG on cystinuria, *Jpn Med,* 30: 2324 (1972).
2. RC Weast, MJ Astle, CRC 'Handbook of Chemistry and Physics', CRC Press Inc, section C, 726 (1982).

TREATMENT OF CYSTINE NEPHROLITHIASIS WITH ALPHA-MERCAPTO-PROPIONYLGLYCINE

A. Trinchieri, P. Luongo[1], F. Rovera, R. Nespoli, G. Zanetti and E. Austoni

Istituto di Urologia
[1]Università degli Studi di Milano
Milan, Italy

INTRODUCTION

The overall frequency of cystine stone formers is 1-6% of stone forming patients[1,2]. Cystine stones form by the precipitation of cystine when present in excess in the urine. Cystinuria is an inherited disorder of amino acid transport affecting the epithelial cells of the renal tubules and gastrointestinal tract. The estimated incidence in the general population varies from 1:200 to 1:1370, but patients in whom cystine stones form comprise only 3 to 59 % of all cystinuric patients[3-6]. Extracorporeal shock wave lithotripsy, electrohydraulic and ultrasonic endoscopic lithotripsy and pulsed dye lasers have limited efficacy in pulverizing cystine calculi owing to its uniform crystal structure. On the contrary, pharmacological treatment of cystine calculi is possible due to new, more effective drugs with fewer side effects. We used α-mercaptopropionylglycine (α-MPG) for cystinuric patients and we have obtained favorable results, as reported in this paper.

MATERIALS AND METHODS

Cystinuria was assessed in 2000 consecutive patients with renal stones by using the cyanide-nitroprusside test (Brand test). Pathological cystinuria was present in 31 patients (1.5%). Patients with positive cyanide-nitroprusside test were further studied for identification of urine amino acids by quantitative ion-exchange chromatography. In addition 24 h urines were collected from patients with cystine stones and normal controls. Sodium, potassium, calcium, magnesium, phosphate, oxalate, urate, citrate, glycosaminoglycans, creatinine and pH were measured. Finally we studied 85 members from 24 families of patients with cystine stones. Twenty four family members excreted excessive amounts of cystine, but only 5 of them (21%) had cystine calculi. Twenty two patients were treated with 1-1.5 g α-MPG daily.

RESULTS

Urinary volume and oxalate were significantly higher in cystine stone formers than in controls, whereas urinary magnesium was significantly lower (Table 1). Alpha-mercaptopropionylglycine reduced stone formation from 0.93 to 0.46 stones/patient/year. Only six patients had side effects of sufficient severity to require withdrawal (proteinuria, nephrosis, fever and chills, hematological abnormalities).

Table 1. Urinary values in 31 patients with cystine stones and in 146 healthy controls.

	Patients with cystine stones	Controls
Urinary volume (mL/day)	2131±1320 **	1072±340
Urinary pH	5.90±0.44	5.90±0.36
Urinary potassium (mEq/day)	55±34	48±15
Urinary sodium (mEq/day)	203±113	170±59
Urinary calcium (mg/day)	167±91	178±86
Urinary phosphate (mg/day)	671±240	679±224
Urinary uric acid (mg/day)	454±137	506±187
Urinary citrate (mg/day)	457±225	465±173
Urinary GAGs (mg/day)	33±25	18±7
Urinary oxalate (mg/day)	20.7±11.9 **	7.1±5.7
Urinary creatinine (mg/day)	1284±419	1280±438
Urinary magnesium (mg/day)	56±18 *	93±42

p< 0 05 *   p< 0.01 * *

DISCUSSION

Firstly it must be stressed that in all patients with nephrolithiasis, a screening test for cystinuria should be performed. In fact, cystinuric patients often have calculi composed of substances other than cystine. For this reason cystinuria is often diagnosed many years after the onset of the initial symptoms. Secondly we were unable to show any difference in urinary risk factors for calcium stones between cystine stone formers and controls. Urinary volume was higher in patients than in controls because they had been advised to increase their fluid intake prior to our evaluation. On the other hand the higher daily urinary excretion of oxalate in cystine stone formers may also be explained by the positive correlation between urinary oxalate excretion and urinary volume. Although in cystine stone formers urinary magnesium excretion was less than normal, our findings seem not to be in accordance with the suggestion that calcium oxalate crystallization may play a role in cystine stone formation. In contrast, it seems more likely that the mineral content of cystine stones may be an accidental contaminant. Finally, the treatment of cystine stones is designed to increase cystine solubility by alkalinizing the urine and reducing the concentration of cystine in urine by increasing urine volume and by use of disulfide exchange such as D-penicillamine and α-MPG.

In particular, α-MPG appears to be as effective as D-penicillamine in lowering the urinary cystine excretion to safe levels, and has fewer side effects.

The percentage of patients discontinuing α-MPG because of adverse side effects was 27%. Further studies are required to define the role of captopril in maintenance treatment of cystinuria.

REFERENCES

1. RJ Caldwell, JI Townsend, MJV Smith, Genetics of cystinuria in an inbred population, *J Urol* 119:531 (1978).
2. RS Malek, PP Kelalis, Pediatric nephrolithiasis, *J Urol* 113:545 (1975).
3. H Ito, M Murakami, T Miyauchi, I Mori, K Yamaguchi, T Usui, J Shimazaki, The incidence of cystinuria in Japan, *J Urol* 129:1012 (1983).
4. H Bostrom, K Tottie, Cystinuria in Sweden school children, *Acta Paed* 48:345 (1959).
5. JC Crawhall, EP Saunders, CJ Thompson, Heterozygotes for cystinuria, *Ann Hum Genet* 29:257 (1966).
6. HB Lewis, The occurrence of cystinuria in healthy young men and women, *Ann Intern Med* 6:183 (1932).
7. JA Sloand, JL Jr Izzo, Captopril reduces urinary cystine excretion in cystinuria, *Arch Intern Med* 147:1409 (1987).

INFLUENCE OF TOTAL PARENTERAL NUTRITION ON URINARY CALCIUM OXALATE SATURATION AND THE DEVELOPMENT OF NEPHROCALCINOSIS IN PRETERM INFANTS

B. Hoppe[1], A. Hesse[2], T. Neuhaus[1], I. Forster[1], S. Fanconi[1],
N. Blau[1] and E. Leumann[1]

[1]University Children's Hospital
Zurich, Switzerland and
[2]Experimental Urology
Department of Urology
University of Bonn
Bonn, Germany

INTRODUCTION

Nephrocalcinosis is a well-known complication in pre-term infants[1,2]. Hypercalciuria induced by furosemide administration was at first considered to be the main cause of nephrocalcinosis[1,2]. Dexamethasone therapy and long-term oxygen requirements have also been considered causative[3,4]. Total parenteral nutrition (TPN) was presumed to increase the risk of renal calcification because of increasing urinary oxalate excretion[5].

PATIENTS AND METHODS

Urinary lithogenic and inhibitory substances were studied prospectively in 27 preterm infants: 16 had TPN and 11 received breastmilk (BMN) with an additional glucose-electrolyte infusion. Twenty-four h urine samples were collected on day 2 (Period A), day 3 (Period B), and once between days 4 and 10 (Period C). Urinary oxalate was measured by high-performance liquid chromatography[6], and urinary calcium oxalate (CaOx) saturation was calculated using the computer program, Equil 2, according to Finlayson[7]. Renal ultrasonography was performed every second week until discharge.

RESULTS

No significant differences between TPN and BMN were observed except for the mean birth weight which was significantly lower in the TPN infants (TPN: 1,526g; BMN: 2,085g ($p< 0.05$). Because birth weights differed, urinary excretion rates were expressed per kg body weight.

The urinary calcium/creatinine ratio increased in TPN (from A: 0.91; C: 1.68 mol/mol, $p< 0.05$) and was significantly ($p< 0.001$) higher at period C than in BMN (A: 0.52; C: 0.36, Fig. 1). The total urinary calcium excretion (mmol/kg/24 h) also increased in TPN (from A: 0.057 to C: 0.117) and decreased in BMN (from A: 0.045 to C: 0.020).

The urinary oxalate/creatinine ratio was persistently higher in TPN (A:203; C:202 mol/mol) than in BMN (A: 98; C: 137, Fig. 1). Urinary oxalate excretion (µmol/kg/24 h) increased slightly in both groups, being persistently higher in TPN (A: 10.0; C: 11.4) than in BMN (A: 7.5; C: 8.4).

The mean urinary citrate excretion (mmol/kg/24 h) increased in both groups to a comparable level at period C (TPN: 0.035, BMN: 0.031). Urinary citrate/creatinine ratios remained constant in TPN (0.44 mol/mol), whereas they increased significantly in BMN (from A: 0.26 to C: 0.49, $p < 0.05$). The calcium/citrate ratios rose considerably in TPN, but decreased in BMN to a significantly ($p < 0.005$) lower level than in TPN (Fig. 1).

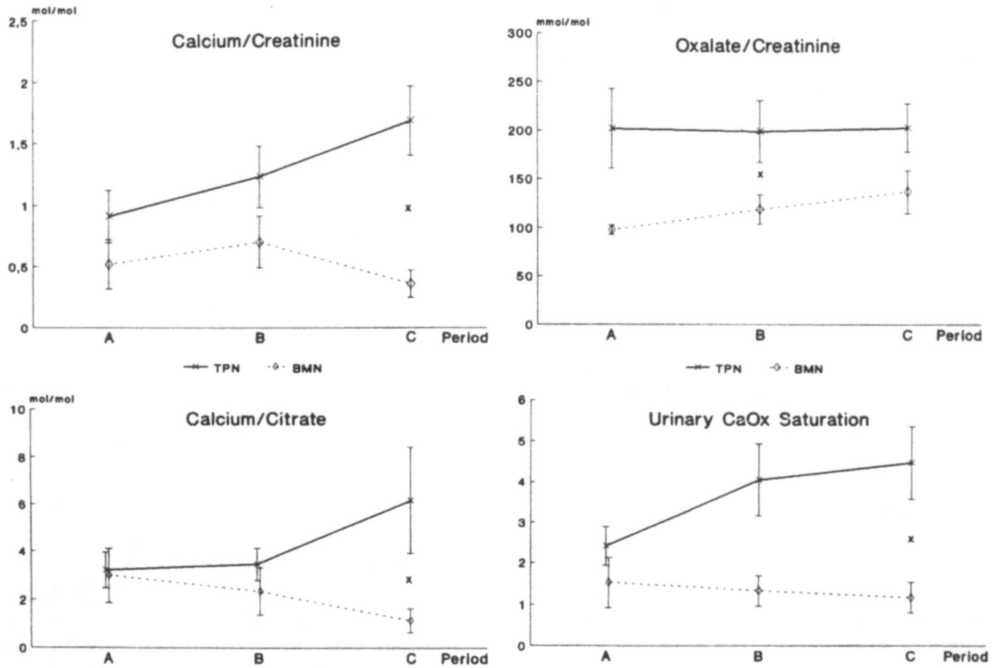

Figure 1. The urinary calcium/creatinine, oxalate/creatinine, and calcium/citrate ratios (mol/mol), and CaOx saturation (±SEM) in 24 h urines from pre-term infants receiving either TPN or BMN at Periods A, B and C. x = significant difference between TPN and BMN (see text).

The urinary CaOx saturation was similar in both groups at period A, and then increased in TPN (A: 2.4; C: 4.5, $p < 0.05$), whereas it decreased in BMN (A: 2.1; C: 1.5) to a significantly ($p < 0.01$) lower value than in TPN (Fig. 1). Slight medullary nephrocalcinosis developed in two infants with TPN and persisted until discharge. The mean daily calcium intake was similar in both groups of infants, whereas protein, sodium, phosphorus and ascorbic acid intakes were significantly ($p < 0.05$) higher in TPN.

CONCLUSION

We observed a considerably higher risk of CaOx precipitation and of nephrocalcinosis in pre-term infants on TPN than in those on BMN. The higher CaOx saturation was due to increased calcium excretion, higher urinary calcium/citrate ratios and persistently higher oxalate/creatinine ratios. Both groups studied had similar (low) doses of furosemide, but the infants on TPN had a considerably higher daily intake of protein, sodium, phosphorus and ascorbic acid. Additional studies in pre-term infants are needed to evaluate the influence of nutritional factors on urinary lithogenic and inhibitory parameters, and on the risk of nephrocalcinosis.

REFERENCES

1. KG Hufnagle, SN Khan, D Penn, A Cacciarelli, P Williams, Renal calcifications: a complication of long term furosemide therapy in pre-term infants, *Pediatrics* 70:360, (1982).
2. F Ezzedeen, RD Adelmann, CE Ahlfors, Renal calcification in pre-term infants: pathophysiology and long term sequelae, *J Pediatr* 113:532 (1988).
3. MD Kamitsuka, D Peloquin, Renal calcification after dexamethasone in infants with bronchopulmonary dysplasia, *Lancet* 337:626 (1991).
4. A Short, RWI Cooke, The incidence of renal calcification in pre-term infants, *Arch Diseases Children* 66: 412 (1991).
5. TJ Campfield, GL Braden, Urinary oxalate excretion by very low birth weight infants receiving parenteral nutrition, *Pediatrics* 77:860, (1989).
6. A Classen, A Hesse, Measurement of urinary oxalate: an enzymatic and ion chromatographic method compared, *J Clin Chem Clin Biochem* 25: 95 (1987).
7. PG Werness, CM Brown , LH Smith, B Finlayson, Equil 2: A basic computer program for the calculation of urinary saturation, *J Urol* 134:1242 (1985).

VALUE OF BONE MINERAL ANALYSIS IN PATIENTS WITH UROLITHIASIS BY SINGLE PHOTON ABSORPTIOMETRY, DUAL PHOTON ABSORPTIOMETRY, AND DUAL ENERGY X-RAY ABSORPTIOMETRY - COMPARISON OF PRIMARY HYPERPARATHYROIDISM AND IDIOPATHIC UROLITHIASIS, AND ON THE POSSIBILITY OF DIFFERENTIATING IDIOPATHIC HYPERCALCIURIA

M. Takeda, Y. Katayama, T. Tsutsui, T. Komeyama, H. Go, S. Wakatsuki and S. Sato

Department of Urology
Niigata University School of Medicine
Asahimachi 1
Niigata 951, Japan

INTRODUCTION

In the investigation of urolithiasis, it is very important to examine bone which contains most of the calcium (Ca) in the body. Bone mineral density (BMD) has hitherto been measured with a microdensitometry (MD) technique[1] or a single photon absorptiometry (SPA) technique in patients with primary hyperparathyroidism (PHP)[2-4] and idiopathic urolithiasis[5-7]. These reports show that 2 types of idiopathic hypercalciuria (IC) - renal hypercalciuria (RH) and absorptive hypercalciuria (AH) - can be differentiated with the MD or SPA technique. The differential diagnosis of IC, which is a very important cause of urolithiasis, is now performed by the Ca restriction and load test (so-called Pak test)[8], but the test is time-consuming and demanding. So, assessment of BMD as a substitute for the Pak test might be preferable. However, bone fractures first occur in vertebral bones, and later in peripheral bones in metabolic bone disease, including primary hyperparathyroidism,[9,10]. Because the dual photon absorptiometry (DPA) technique,which uses 2 different energy types of γ-ray, and dual energy X-ray absorptiometry (DEXA), which uses two types of X-ray but does not use radioisotope, can examine the BMD of deep bone[11], these are more useful to examine BMD in IU patients than SPA and MD. In this report, BMD of patients with PHP and IU were assessed by SPA, DPA, and DEXA techniques simultaneously and the data obtained with the 3 techniques compared.

SUBJECTS

Controls consisted of 48 men and 58 women (20-29, 8M, 6F; 20-39, 12M, 4F; 40-49, 12M, 15F; 50-59, 7M, 10F; 60-69, 6M, 16F; 70-79, 3M, 7F). BMD was measured with SPA, DPA, and DEXA techniques in the control group. Forty-two patients (25F, 17M, ranging from 21 to 75 years old, mean 46.7 years old) had IU and 10 patients (7F

Urolithiasis 2, Edited by R. Ryall *et al.*,
Plenum Press, New York, 1994

and 3M, ranging from 28 to 73 years old, mean 49.5 years old) had PHP. In the PHP group, 9 patients had the clinical type of urolithiasis and the other the clinical type of both bone disease and urolithiasis. Fifteen of the IU group patients were classified as having IC according to the following criteria. On a normal diet, hypercalcuria in men is > 250 mg/day, in women, is > 200 mg/day, and in both sexes combined, is > 4 mg/kg/day.

METHODS

SPA was performed at the distal 1/3 of the left radial bone with the Norland Cameron 178. DPA was performed at the 3rd lumbar vertebral body with the Dualomex HC-1. DEXA was performed with Hologic QDR-1000 whole body X-ray bone densitometer. Bone mineral density (g/cm^2, BMD) was measured as a parameter.

Serum Ca and 24 h excretion of urinary Ca were measured with the o-cresyl phthalein complexon (OCPC) method. Serum inorganic phosphate (IP) and 24 h excretion of urinary IP were measured with an enzymatic method. Serum hypersensitive PTH (PTH-HS) was measured with an RIA kit (Yamasa Co, Ltd). Statistical analysis was performed using Wilcoxon rank-sum test. The Pak test[8] was performed in 10 of the 15 IU patients to differentiate AH from RH.

RESULTS

As assessed by SPA, DPA and DEXA, the BMD of women decreased markedly after the age of 50.

In patients with urOlithiasis, by SPA, DPA and DEXA, the relative BMD of the IU group was significantly higher than that of the PHP group, while that of patients with normocalciuria was higher than that of the PHP group. By DPA and DEXA, the relative BMD of the IC group was significantly higher than that of the PHP group. By SPA, there was no difference in relative BMD between the IC patients and patients with normal calciuria, while the relative BMD of the PHP group tended to be lower than in the IC group.

Five patients with IC were classified into the RH group and the AH group after the Pak test. Table 1 shows the relative BMD in the two types of IC. Although relative BMD as assessed by SPA did not differ between the two types, a significant difference was found between the two types when assessed by both DPA and DEXA.

Table 1. Relative BMD and types of hypercalciuria.

		Relative BMD	
SPA	Renal hypercalciuria	0.97+0.05	} N.S.
	Absorptive hypercalciuria	0.98±0.17	
DPA	Renal hypercalciuria	0.77±0.05	} p< 0.01
	Absorptive hypercalciuria	1.02±0.12	
DEXA	Renal hypercalciuria	0.81±0.08	} p<0.01
	Absorptive hypercalciuria	1.05±0.09	

There was a positive correlation between urinary excretion of Ca and the relative BMD by both DPA and DEXA, but not BMD by SPA. There was no significant correlation between urinary excretion of IP and the relative BMD by SPA, DPA and DEXA. No correlation was found between serum PTH-HS and relative BMD by SPA, DPA and DEXA.

DISCUSSION

As inorganic bone contains 90% of the Ca and 85% of the IP in the body, the mineral content of bone must change in any disease that alters the balance between Ca and IP. BMD has been studied in PHP, with many reports that BMD is lower than normal, even in patients without any abnormal radiological findings[1-4]. In the present study as well, BMD of the PHP group was lower than that of normal subjects and the IU group, both by SPA and DPA. In addition to the patients with PHP, bone densitometry was utilized to evaluate pathogenesis of urolithiasis, two major causes of which are hypercalciuria (AH) and renal hypercalciuria (RH)[8]. Parathyroid function is normal or suppressed in the former, and patients with this condition would not be expected to have a low BMD. In contrast, the latter, which is associated with increased parathyroid function, would be expected to show a low BMD. Although a significant inverse correlation should have been found between relative BMD and serum PTH-HS, our study did not show this.

In this study, the above speculation is confirmed by DPA, but not by SPA. This disagreement is attributed to two factors. First, pathological changes occur early in central bones rather than peripheral bones[9]. Second, DPA may more sensitively detect changes in bone and also detect them earlier than SPA. Because restriction of Ca intake causes reduction of bone mineral content and bone fracture in the RH group, differential diagnosis of AH and RH must be performed early.

Although DPA is superior to SPA, it has some disadvantages, such as radiological exposure and decreased reproducibility due to natural decay of the radioisotopes. From this standpoint, dual energy X-ray absorptiometry (DEXA), which uses X-ray in place of radioisotope, is superior to DPA and will be widely used in future.

Finally, some patients in the normal calciuria group showed decreased BMD, both by SPA and DPA. The cause of the decreased BMD is unclear, but we should examine its cause, and determine its relation to urolithiasis.

ACKNOWLEDGEMENTS

We thank the staff of the Section of Nuclear Medicine, the Department of Urology, Niigata University School of Medicine.

REFERENCES

1. W Sakamoto, H Kawashima, T Nishijima, T Riishimoto and M Maekawa, A study of bone mineral content in patients with calcium urolithiais by microdensitometry, *Jap J Urol* 79: 495 (1988).
2. CYC Pak, A Stewart, R Raplan, H Bone, C Notz and R Browne, Photon absorptiometric analysis of bone density in primary hyperparathyroidism, *Lancet* 7923: 7 (1975).
3. M Fuss, C Gillet, J Simon, JC Vandewalle, A Schoutens and P Bergmann, Bone mineral content in idiopathic renal stone disease and in primary hyperparathyroidism, *Eur Urol* 9: 32 (1983).
4. SM Gupta, JL Belsky, RP Spencer, J Frias, P Rotch, T Halpin and NE Herrera, Parathyroid adenomas and hyperplasia dual nuclide scintigraphy and bone densitometry studies, *Clin Nucl Med* 10: 243 (1985).
5. EM Alhava, M Juuti and P Rarialainen, Bone mineral density in patients with urolithiasis, *Scand J Urol Nephrol* 10: 154 (1976).
6. S Lawoyin, S Sismilich, R Browne and CYC Pak, Bone mineral content in patients with calcium urolithiasis, *Metabolism* 28: 1250 (1979).
7. M Fuss, T Pepersack, JV Geel, J Corvilian, JC Vandewalle, P Bergmann, and J Simon, Involvement of low-calcium diet in the reduced bone mineral content of idiopathic renal stone formers, *Calc Tissue Int* 46: 9 (1990).
8. CYC Pak, Physiological basis for absorptive and renal hypercalciuria, *Am J Physiol* 237: F415 (1979).
9. B Krolner, SP Nielson and B Lund, Measurement of bone mineral content (BMC) of the lumbar spine. 2. Correlation between forearm BMC and lumbar spine BMC, *Scand J Clin Lab Invest* 40: 665 (1980).
10. RB Mazess WW Peppler, RW Chesney TA Lange, U Lindgrin and E Smith Jr, Does bone measurement on the radius indicate skeletal status?, *J Nucl Med* 25: 281 (1984).
11. WL Dunn, HW Wahner and BL Riggs, Measurement of bone mineral content in human vertebrae and hip by dual photon absorptiometry, *Radiology* 136: 485 (1980).

EXAGGERATED CALCIURIC RESPONSE TO AN ACUTE ACID LOAD IN PATIENTS WITH CALCIUM RENAL LITHIASIS

M. Normand[1], P. Houillier[2], A. Peuchant[1], J.L. Cayotte[1] and M. Paillard[2]

[1]Service de Nephrologie
Clinique Saint-Martin
Pessac
[2]Departement de Physiologie
INSERM U 356,
Hôpital Broussais
Paris, France

INTRODUCTION

The incidence of calcium renal lithiasis is increasing in industrialized countries[1], and this increase appears to be linked to modifications in food intake. Indeed, there is a strong relationship between the consumption of animal proteins and the risk of developing calcium stone[2,3]. The putative lithogenic factors are a rise in calciuria, oxaluria, uricosuria, and a fall in urinary pH and citraturia[4]. The increase in acid load, a consequence of the metabolism of animal proteins, has been considered as the cause of enhanced calciurias. In fact, many workers have studied, both in calcium stone formers and in controls, the calciuric response to a protein-rich diet or an oral chronic acid load[2,4-7]. By contrast, only few studies have dealt with the calciuretic response induced by an acute acid load. The goal of this study was therefore to address this latter point.

PATIENTS AND METHODS

Nine controls (C) (4M, 5F, mean age 36±4 yrs; range 27-46, mean body weight 64±10 kg) and 13 recurrent calcium stone formers (SF) (9M, 4F, mean age 43±12 yrs; range 30-65, mean body weight 67±17 kg) were studied. All patients had normal renal function. None had diabetes, hypertension, or renal malformations, or had taken any medication for the 15 days prior to the experiment. All patients had recurrent renal stone disease (2 stones: 4 patients; at least 3 stones: 9 patients). Composition of stones, analysed by infrared spectrophotometry, was: whewellite (2), weddellite (3), carbapatite (8).

The patients were on free diets before the experiment. On the day preceding the acid load, they collected their 24 h urine. At 8 am on the morning of the experiment, after an overnight fast, they voided and drank 400 mL of slightly mineralized water: Four 2 h urine specimens were then collected for determination of urinary flow rate, pH, bicarbonate, titratable acid (TA), and ammonium (NH_4), calcium, phosphate, citrate, and creatinine concentrations. Venous blood was drawn without a tourniquet through an indwelling catheter at the mid point of each period for the measurement of pH, pCO_2, and sodium, potassium, chloride, and creatinine concentrations. The acid load (NH_4Cl: 2 mmol/ kg B.W.) was given orally at 10 am together with a light breakfast.

Urolithiasis 2, Edited by R. Ryall *et al.*,
Plenum Press, New York, 1994

Both pH and pCO_2 were measured with an automated gas analyzer, titratable acid and ammonium concentrations by titration methods, calcium and phosphate concentrations by colorimetry, citrate by enzymatic reaction, creatinine concentration by Jaffe's method, and sodium and potassium concentrations by specific electrodes. Bicarbonate concentration was calculated according to the equation of Henderson-Hasselbach. Net acid excretion rate was calculated as follows: TA + NH_4 excretion rates - bicarbonate excretion rate.

RESULTS

Fasting calciuria (expressed as Ca/Cr) was nearly identical in the two groups (0.146±0.070; SF, vs 0.144±0.080 mg/mg; C, NS). However, after the acid load, the urinary calcium excretion was greater in the SF group than in the controls (0.378±0.113 vs 0.253±0.108 mg/mg during the second period post acid loading, $p < 0.01$ (Fig. 1). Similarly, the maximum rise in urinary calcium excretion was significantly higher in the SF group than in the controls (0.234±0.099 vs. 0.130±0.046 mg/mg, $p < 0.01$).

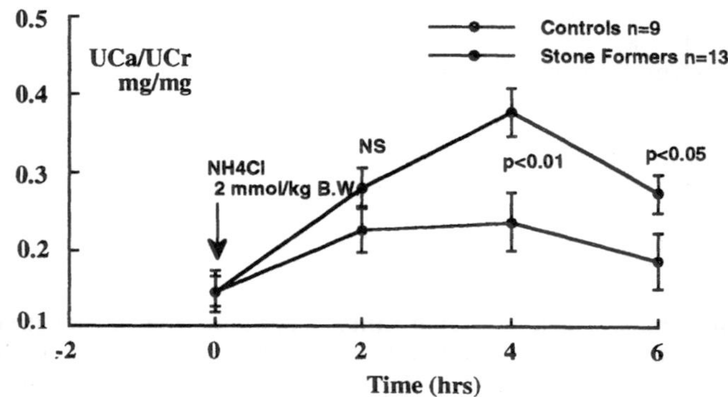

Figure 1. Evaluation of urinary calcium excretion after an acute acid load.

The fasting plasma bicarbonate concentrations were 28.2±2.2 mM in the SF group and 26.6±2.6 mM in the controls (NS). The maximal decrease in plasma bicarbonate concentration after acid loading was 4.0±1.8 mM in the SF group and 4.2±1.7 in the controls (NS). Similarly, the increase in TA excretion (16.9±3.5 vs 26.4±4.0 mol/min), NH_4 excretion (26.9±4.7 vs 23.8±4.6 µmol/min) and NAE (41.7±8.1 vs 50.3±6.4 µmol/min) were identical in the controls and stone formers, respectively.

Lastly, fasting urinary citrate excretion (expressed as citrate/creatinine) was similar in the two groups (0.33±0.17 vs 0.42±0.10 mg/mg, NS). The maximum decrease in urinary citrate excretion was identical (0.11±0.06 vs 0.10±0.08 mg/mg, NS).

DISCUSSION

Our results demonstrate that:

- an acute acid load induces a prompt increase in urinary calcium excretion
- this increase is higher in patients with recurrent calcium lithiasis than in controls
- there is no defect in urinary acidification mechanisms in stone formers.

These results extend to an acute situation those previously demonstrated in the chronic situation[5,7].

Three mechanisms may be involved in the acute increase in calciuria:

(1) a rise in ultrafiltrable plasma calcium concentration induced by an enhancement of bone calcium release[6].

(2) an increase in glomerular filtration rate leading to a rise in filtered load of calcium and

(3) a decrease in tubular calcium reabsorption[6,8].

The present study, however, does not allow us to distinguish between these three mechanisms. Further investigations are therefore necessary to examine the role of each mechanism and the possible involvement of parathyroid hormone.

REFERENCES

1. DA Andersen, Environnmental factors in the aetiology of urolithiasis, *in*: "Urinary Calculi", Cifuentes Delatte, Rapado, and Hodgkinson, eds, Karger, Basel (1973).
2. S Goldfarb, The role of diet in the pathogenesis and prophylaxis of calcium nephrolithiasis, *Kidney Int* 34: 544 (1988).
3. WG Robertson and M Peacock, The pattern of urinary stone disease in Leeds and the United Kingdom in relation to animal protein intake during the period 1960-1980, *Urol Int* 39: 394 (1982).
4. Y Katoh, Influence of dietary animal protein on renal stone disease, *Nippon Hinyokika Gakkai Zasshi* 6: 823, (1989).
5. J Lemann, Jr, RW Gray, WJ Maierhofer, and HS Cheung, The importance of renal net acid excretion as a determinant of fasting urinary calcium excretion, *Kidney Int* 29: 743 (1986).
6. J Lemann, Jr, RW Litzow, and EJ Lennon, Studies on the mechanism by which metabolic acidosis augments urinary calcium excretion in man, *J Clin Invest* 46: 1318 (1967).
7. J Lemann, Jr, Urinary calcium excretion and net acid excretion: effects of dietary protein, carbohydrate and calories, *in*: "Urolithiasis and Related Clinical Research", PO Schwille, LO Smith, WG Robertson, and W Wallensieck, eds, Plenum Press, New York, (1985).
8. RAL Sutton, NLM Wong, and JH Dirks, Effects of metabolic acidosis and alkalosis on sodium and calcium transport in dog kidney, *Kidney Int* 15: 520 (1979).

DO IDIOPATHIC STONE FORMERS HAVE LOW BONE MASS?

J. Casez[1], C. Hug[1], B. Hess[1], D. Ackermann[2] and Ph. Jaeger[1]

[1]Policlinic of Medicine and
[2]Clinic of Urology
University Hospital
3010 Berne
Switzerland

INTRODUCTION

Several investigators[1-7], using a variety of techniques such as ^{241}Americium γ-ray attenuation, ^{125}I-photonabsorptiometry neutron activation analysis or quantitative computerized tomodensitometry have suggested that calcium renal stone formers have a lower bone mass than normal subjects. The present study was carried out to evaluate in an unselected population of idiopathic calcium stone formers the magnitude of the decrease in bone mass, at different sites of the skeleton, using dual-energy X-ray absorptiometry (DXA).

PATIENTS AND METHODS

Ninety-nine male idiopathic recurrent calcium stone formers were studied as outpatients on a free-choice diet. Daily calcium intake was derived from a questionnaire and mean urinary excretion rate of calcium based on two 24 h urine collections. Forty-seven patients were hypercalciuric (HCSF) (UCaV>0.1 mmol/kg/24 h) and 52 normocalciuric (NCSF). Bone mineral density (BMD) was measured at lumbar spine, upper femur and distal tibia. A skeletal score was calculated from these BMD measurements. Results were compared to those obtained in 234 healthy male controls.

RESULTS

BMD at any site was similar in HCSF, NCSF and controls. The skeletal score, however, was significantly ($p < 0.04$) lower in stone formers than in controls. Six of the 10 patients with the lowest skeletal score gave a history of low calcium diet for 2 to 8 years. After exclusion of these 6 patients, the difference in skeletal score between calcium stone formers and controls was no longer significant.

CONCLUSION

Low bone mass in idiopathic calcium stone formers could well be a myth. Therefore at the present stage of our knowledge, measurement of bone density in patients with

Urolithiasis 2, Edited by R. Ryall *et al.*,
Plenum Press, New York, 1994

urolithiasis should probably be restricted to those stone formers who have been on a low calcium diet for a prolonged period of time, although the deleterious action of that diet on bone remains to be better defined.

ACKNOWLEDGEMENTS

This study was supported in part by the Swiss National Funds for Scientific Research, Grant Number 32-33543.92.

REFERENCES

1. EM Alhava, M Juuti and P Karljalainen, Bone mineral density in patients with urolithiasis, *Scand J Urol Nephrol*10: 154-156 (1976).
2. S Lawoyin, S Sismilich, R Browne and CYC Pak, Bone mineral content in patients with calcium urolithiasis, *Metabolism* 28: 1250-1254 (1979).
3. B Lindergard, S Colleen, W Mansson, C Rademark and B Rogland, Calcium loading test and bone disease in patients with urolithiasis, *Proc EDTA* 20: 460-465 (1983).
4. J Barkin, DR Wilson, MA Manuel, A Bayley, T Murray and J Harrison, Bone mineral content in idiopathic calcium nephrolithiasis, *Mineral Electrolyte Metab* 11: 19-24 (1985).
5. P Bataille, JM Achard, A Fournier, et al, Diet, vitamin D and vertebral mineral density in hypercalciuric calcium stone formers, *Kidney Int* 39: 1193-1205 (1991).
6. M Fuss, C Gillet, J Simon, J-C Vandewalle, A Schoutens and P Bergmann, Bone mineral content in idiopathic renal stone disease and in primary hyperparathyroidism, *Eur Urol* 9: 32-34 (1983).
7. R Pacifici, M Rothstein, L Rifas, K-HW Lau, DJ Baylink, LV Avioli and K Hruska, Increased monocyte interleukin-1 activity and decreased vertebral bone density in patients with fasting idiopathic hypercalciuria, *J Clin Endocrinol Metab* 71: 138-145 (1990).

RENAL HANDLING OF CITRATE IN PATIENTS WITH PRIMARY HYPERPARATHYROIDISM AND CALCIUM NEPHROLITHIASIS

M. Marangella, C. Vitale, D. Cosseddu, M. Bruno, A. Tricerri,
M. Manganaro and F. Linari

Renal Stone Laboratory
Mauriziano Umberto I Hospital
Turin, Italy

INTRODUCTION

The risk of calcium stone formation in the course of primary hyperparathyroidism (PHPT) is known to arise from increases in excretion of promoters such as calcium[1], phosphate[2] and perhaps oxalate. In contrast, the role of inhibitors has so far received less attention and very little is known about citrate, which is a main inhibitor of calcium salt crystallization in urine. Therefore, we have studied patterns of citrate excretion and renal handling in patients with PHPT and calcium nephrolithiasis and their changes after successful parathyroidectomy (PTX).

PATIENTS AND METHODS

Forty-five outpatients (12 males, 33 females, aged 50.4±10.6 years) were studied. All were recurrent stone formers (9.3±8.2 stones/patient) and most of them had undergone surgery for urolithiasis-related complications (1.4±1.3 operations/patient). PHPT was confirmed by successful PTX in all. No patient had diseases known to influence citrate excretion, such as intestinal malabsorption, potassium deficiency, renal tubular acidosis, or urinary tract infection or was taking diuretics. Ninety healthy subjects (44 males, 46 females, aged 39.0±14.7 years) served as controls.

Patients were studied while on their home diets, before and at least three months after successful surgery, according to the following schedule: fasting blood was assayed for intact PTH, alkaline phosphatase (APh), calcium, phosphate and creatinine. Main chemistries were determined on 24 h urine as described elsewhere[3]. Fasting urines were assayed for calcium, creatinine and hydroxyproline. Total nitrogen excretion (TNE) and net acid excretion (NAE) were calculated as previously reported[3]. Citrate was measured in serum and urine by an enzymatic method, using citrate lyase[3]. Renal function was assessed as creatinine clearance. Calcium citrate soluble complexes as well as urine state of saturation with calcium oxalate (β_{CaOx}) and brushite (β_{bsh}) were assessed by our computer method[3]. Results are means±SD. Paired or unpaired Student's t tests were used to assess significance of differences.

RESULTS

PTX restored to normal both PTH and other relevant blood and urine biochemistries, indicating that all the patients considered were cured (Table 1).

Table 2 lists data of renal function and diet-related urine constituents. Patients' GFR was slightly lower than that of controls and did not change after PTX. Whole and animal protein, and sodium and potassium intakes were similar in the three study groups, as can be taken by urine measurements.

Table 1. Main biochemistries and hormones, Means±(SD).

	Before PTX	After PTX	Controls
sCalcium (mmol/L	2.8 (0.2)*	2.4 (0.1)	2.3 (0.2)
uCalcium (mmol/24h)	9.7 (4.2)*	3.9 (2.3)	4.1 (1.8)
sPhosphate (mmol/L)	0.8 (0.2)*	1.0 (0.1)	1.0 (0.2)
$TmPO_4/GFR$ (mg/100 mL)	1.9 (0.5)*	2.9 (0.5)	3.1 (0.7)
ALP (U/L) 215.0 (89)#	147.0 (57.0)	126.0 (65)	
PTH (pg/mL) 158.0 (93)*	49.0 (23.0)	33.0 (11)	
OHPro/Cr (mg/mg)	0.047 (.04)*	0.022 (.01)	0.021 (.01)

*p <0.001 and #p <0.05 vs both after PTX and controls. s = serum, u = urine

Table 2. GFR and diet-related urine constituents, Means±(SD).

	Before PTX	After PTX	Controls
GFR (mL/min) 88 (26)	85 (23)	113 (19)	
TNE (mmol/24h)	684 (236)	704 (204)	758 (277)
TAE (mEq/24h)	49 (24)	45 (19)	52 (23)
uSulfate (mmol/24h)	18 (7)	16 (5)	20 (7)
Uric Acid (mmol/24h)	2.9 (1.0)	2.8 (0.8)	2.5 (0.8)
uSodium (mmol/24h)	175 (61)	170 (66)	173 (62)
uPotassium (mmol24h)	52 (16)	48 (16)	59 (18)

Table 3. Renal handling of citrate in PHPT, Means±(SD).

	Before PTX	After PTX	Controls
sCit (mmol/L) 117 (35)	126 (47)	97 (28)	
uCit (mmol/24h)	3.2 (1.4)*	2.2 (1.1)*	2.9 (1.3)
Cit Cl (mLmin) 21 (11)	13 (8)*	19 (9)	
Cit Cl/Cr Cl (%)	23 (13)*	18 (19)	17 (10)

*p <0.05 vs stone formers. s = serum, u = urine.

The main features of renal handling of citrate are listed in Table 3. Serum citrate was significantly higher in patients than in controls, irrespective of PTH levels. Citrate excretion and clearance decreased by about 30% after PTX. Fractional excretion of citrate (CitCl/CrCl) was higher in PHPT patients before PTX and returned to normal after PTX. There was a positive correlation between PTX induced changes of calcium and citrate 24 h excretions: $\delta Cit(\%) = 10.83 - 0.64\ \delta Ca(\%)$; $r = 0.48$, $p < 0.01$. The fraction of citrate complexed to calcium in urine was 58±10% of total citrate pre-PTX (42±11% in controls, $p < 0.001$) and decreased to control values, i.e. 39±13% after PTX. Therefore, the PTX induced decreases in β_{CaOx} and β_{bsh} (Table 4) were independent of citrate changes.

Table 4. PTX induced changes in urine saturation with calcium salts, Means±(SD).

	Before PTX	After PTX	Controls
β_{CaOx}	9.6 (6)*	4.5 (3.2)	5.2 (2.7)
β_{bsh}	3.4 (2.1)	1.2 (1.0)	2.0 (1.1)

*p <0.01 vs both after PTX and controls

DISCUSSION

Only a few previous studies have addressed the issue of citrate excretion in PHPT: urine citrate was reported in a former paper by Hodgkinson[4] to be increased, whereas Nicar et al[5] found this to be unchanged compared to controls. Ljunghall et al[6] reported high baseline citrate and a decrease upon PTX. These latter results are in full agreement with ours: citrate excretion, as mmol/100 mL GFR, was high pre PTX and returned to normal after PTX. It has been shown that acid-base and potassium balances are major determinants of citrate excretion and that systemic alkalosis increases it. PHPT is associated with high bone resorption and this may in turn induce alkalosis. It could be speculated that this is the mechanism for hypercitraturia. However, net acid (and potassium) excretions, both pre and post PTX, were similar to controls' means. Instead, from our results, it is tempting to speculate that the observed changes in citrate excretion and renal handling depended on changes in tubular handling of calcium: greater tubular delivery of calcium would result in increased amounts of poorly reabsorbable calcium citrate complexation, thereby increasing citraturia. Therefore, derangements in citrate excretion in this subset do not appear to be directly conducive to calcium nephrolithiasis.

REFERENCES

1. AE Broadus, Primary hyperparathyroidism, *J Urol* 141 723 (1989).
2. CYC Pak, MJ Nicar, R Peterson, A Zerwekh and W Snyder, A lack of unique pathophysiologic background for nephrolithiasis of primary hyperparathyroidism, *J Clin Endocrinol Metab* 53 536 (1981).
3. M Marangella, C Vitale, D Cosseddu, C Martini, M Petrarulo, F Linari, Renal handling of citrate in chronic renal insufficiency, *Nephron* 57:439 (1991).
4. A Hodgkinson, The relation between citric acid and calcium metabolism with particular reference to PHPT and idiopathic hypercalciuria, *Clin Sci* 24:167 (1963).
5. MJ Nicar, C Skurla, K Sakhaee, CYC Pak, Low urinary citrate excretion in nephrolithiasis, *Urology* 21:8 (1983).
6. S Ljunghall, BG Danielson, G Johansson and L Wibell, Renal stone formation in PHPT. Role of tubular disfunction, *in* "Urolithiasis", Plenum Press, New York (1981).

EXOGENOUS CALCITONIN (CT) - A PROBE FOR THE STATE OF MINERAL METABOLISM IN IDIOPATHIC RECURRENT CALCIUM UROLITHIASIS (RCU)

I. Berger and P.O. Schwille

Mineral Metabolism and Endocrine Research Laboratory
Departments of Surgery and Urology
University of Erlangen
Germany

INTRODUCTION

Calcitonin (CT), one of the three major calciotropic hormones, binds to renal cortical membranes[1], and the kidneys account for about two-thirds of its metabolism[2]. Interestingly, the possible contribution of endogenous CT to the pathophysiology of RCU, a major health problem worldwide, is unsettled. Subgroups of RCU patients may exhibit postprandial or fasting hypercalciuria, together with high urinary sodium[3]. Calciuria, magnesiuria, and natriuresis increase in response to exogenous calcitonin[4]; this would fit with the concept that CT, when present in excess in RCU, at least partially accounts for the hypercalciuria and natriuresis[5]. On the other hand, reduced calcium eflux from bone and subsequent hypocalcemia are well documented CT effects. These contrast with the reported increased bone resorption in unclassified (according to calciuria) RCU patients[6], which makes it unlikely that there is an excess of CT in this disorder. Previously, we failed to demonstrate inappropriate hypercalcitoninemia following an oral calcium challenge in normo- and hypercalciuric RCU[7] patients and others have made similar observations when using the same approach[8]. From all this, we hypothesized that it is not the endogenous CT secretion but rather the CT responsiveness of target organs that could play some critical role in determining urine composition. For instance, exposure of end organs to a defined high CT load may allow more insight into the uptake and degradation of CT, as reflected by the subsequently developed CT blood levels and the associated order of magnitude of response parameter(s). This work reports on nephrogenous cyclic AMP (NcAMP) and the mineral metabolic urinary environment observed after CT administration in RCU patients and volunteers.

METHODS

Two groups of males (Controls, n = 7; RCU, n = 7), matched for age (Controls: 31±2; RCU: 31±2 years) and body mass index (Controls 24.3±1.2; RCU 25.8±1.2 kg/m^2) were recruited. On separate days, after an overnight fast, individuals were mildly hydrated with water; after a baseline untimed period a 3 h creatinine clearance was followed. Venous blood was taken at 0, 90, 180 min for measurement of analytes (plasma cAMP, creatinine, etc); the mean value was used for calculations (creatinine clearance; NcAMP). At 0 min, either 2 mL saline, or 100 IU salmon CT in saline was injected intramuscularly. Reliable laboratory procedures were employed throughout, including determination of plasma salmon CT by specific radioimmunoassay and calculations according to established formulas.

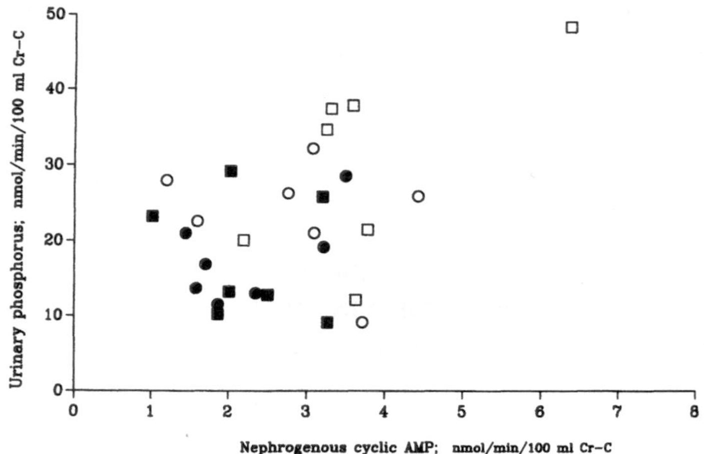

Figure 1. Relationship between NcAMP and urinary phosphorus excretion, both normalized for creatinine clearance (Cr-C). Circles: Controls; squares: RCU. Filled symbols: saline; open symbols: calcitonin.

RESULTS

Following CT administration, plasma salmon CT was elevated above the pre-injection level (at 0 min) in each individual, the group mean in RCU being lower than in controls (data not shown). The associated state of calcium homeostasis in both controls and RCU is characterized by a fall in serum integrated total calcium[9], a rise in serum integrated intact PTH[9], and urinary cAMP and NcAMP (this study), that was expected (regulatory hyperparathyroidism); however, those changes, and additional ones (Table 1) were more marked in RCU than in controls.

Table 1

Variables	CONTROLS; n = 7 Saline	Calcitonin	Mean change %	RCU; n = 7 Saline	Calcitonin	Mean change %
Nephrogenous cyclic AMP[1]; $nmol.min^{-1}.100\ mL^{-1}$ Cr-C[2]	2.2 (0.3)	2.7 (0.5)	+23	2.70 (0.2)	3.8 (0.5)[b]	+41
Excretion and pH[1]						
Volume; mL	277 (67)	595 (136)[b]	+115	271 (85)	511 (135)[b]	+89
pH;	6.45 (0.26)	7.18 (0.07)[b]	+11	7.33 (0.21)[a]	7.61 (0.07)	+4
Sodium; mmol/g[3]	120 (27)	259 (26)[b]	+116	138 (34)	288 (41)	+109
Potassium; mmol/g	95 (17)	92 (4)	- 3	68 (14)	94 (19)	+38
Calcium; mg/g	87 (12)	163 (14)[b]	+87	100 (20)	150 (30)[b]	+50
Magnesium; mg/g	66 (8)	87 (10)	+32	52 (17)	36(5)[a]	-31
Phosphorus; mg/g	449 (62)	549 (66)	+22	350 (77)	715 (102)[b]	+104
Oxalate; mg/g	19 (3)	31 (9)	+63	26 (7)	34 (13)	+31
citrate; mg/g	378 (47)	443 (42)	+17	362 (47)	612 (90)[b]	+69
Relative supersaturation products[1,4]						
Calcium oxalate	0.68 (0.11)	0.62 (0.12)	-9	0.89 (0.14)	0.66 (0.12)[b]	-26
Brushite	0.28 (0.16)	0.59 (0.18)	+111	0.83 (0.14)[a]	0.84 (0.09)	+1
Octacalcium phosphate	0. 10 (0.31)	0.84 (0.21)	+740	1.34 (0.29)[a]	1.39 (0.16)	+4

[1]Mean values (SEM); [2]creatinine clearance; [3]urine creatinine; [4]according to the nomogram method
[a]p <0.05 vs CONTROLS (unpaired data); [b]p <0.05 vs saline (paired data)

PTH and NcAMP were positively associated (n = 28, r = 0.39, p <0.05[9]), but the positive association between NcAMP and urinary phosphorus was stronger (n = 28, r = 0.53, p <0.01; figure 1); note that the open square symbols were shifted to the right. Underlying the latter are the elevated urinary NcAMP and phosphorus (Table 1), which indicate that the renal tubular pathway linking brush border membrane adenylate cyclase - intracellular cyclic AMP production - phosphodiesterase - phosphaturia, was in fact more activated in RCU than in controls. Other more impressive discrepancies among urinary response parameters to CT include citrate (4-fold higher mean increase in RCU than in controls), magnesium (mean fall in RCU, but mean rise in controls), pH and octacalcium phosphate supersaturation (no change in RCU, but dramatic rise in controls).

COMMENTS AND CONCLUSIONS

The interpretation of acute changes in urinary composition of RCU versus controls needs to take into consideration that, due to CT excess in blood, PTH begins to rise. Since this latter was greater in RCU[9], several of the changes in their urine composition could be simply ascribed to documented renal tubular actions of PTH, and not necessarily to those of CT. These may include the smaller rise of calciuria (as a result of the smaller filtered load of calcium and enhanced tubular reabsorption of calcium), the considerably higher citraturia (from PTH-induced mobilization of bone citrate), and, within limitations, the greater phosphaturia (see above). On the other hand, the time course of excess of blood CT and PTH was such that only early within the 180 min observation period was there a peak in CT, and only toward its end a peak in PTH[9]. Therefore, it appears likely that the majority of the present data are largely a reflection of some concerted tubular action of the two hormones.

Magnesium homeostasis is to a great extent maintained by kidneys, more specifically in the ascending limb of Henle's loop. In the rat, CT exerts a direct effect on adenylate cyclase at this anatomic location, resulting in increased magnesium reabsorption[10], as does hypocalcemia[10]. In our study in man, the magnesiuric response in controls may reflect dominance of PTH-mediated tubular calcium reabsorption over magnesium reabsorption, or some species difference. By contrast, CT-induced hypomagnesiuria in RCU may be related to their more alkaline urinary pH, owing to decreased proximal tubular bicarbonate reabsorption and hence increased bicarbonate delivery to distal sites. In view of the numerous interacting factors it is difficult to discern effects that are specific for CT. Therefore, studies on end organ responsiveness toward calciotropic hormones would benefit from a superior technique, eg a calcium clamp, analogous to the glucose clamp procedure. In conclusion, exogenous CT proved successful as a tool in elucidating differences in the renal response of RCU patients, compared with controls. The results suggest that unequal CT uptake by peripheral organs or alternatively, altered degradation and metabolic clearance of CT, are at least in part responsible for the observed differences.

ACKNOWLEDGEMENT

The work was supported by Deutsche Forschungsgemeinsehaft, Bonn, Germany.

REFERENCES

1. PK Seitz, ML Thomas, CW Cooper, Binding of calcitonin and calcitonin gene-related peptide to calvarial cells and renal cortical membranes, *J Bone Min Res* 1: 51 (1986).
2. RE Simmons, JT Hjelle, C Mahoney, LJ Deftos, W Lisker, P Kato, R Rabkin, Renal metabolism of calcitonin, *Am J Physiol* 254: F593 (1988).
3. PO Schwille, G Rümenapf, B Schreiber, Urinary sodium in renal calcium lithiasis - An approach to its relationship with urinary calcium, and the possible role of atrial natriuretic peptide (hANP), *Fortschr Urol Nephrol* 26: 167 (1987).

4. M Walser, Divalent cations: Physicochemical state in glomerular filtrate and urine and renal excretion, *in* "Handbook of Physiology, section 8 Renal Physiology" ed American Physiological Society, Washington DC (1973).
5. RS Kestenbaum, Y Winikoff, PL Lismer, The role of calcitonin in calcium stone formation, *Nephron* 38: 154 (1984).
6. PO Schwille, G Rümenapf, Idiopathic recurrent calcium urolithiasis - Clinical problems and suggested approaches in an ambulatory stone clinic, *in* "Renal Tract Stone - Metabolic Basis and Clinical Practice" ed Churchill Livingstone, Edinburgh, London, Melbourne, New York (1990).
7. PO Schwille, G Rümenapf, J Schmidtler, R Kohler, Fasting and post-calcium load serum calcium, parathyroid hormone, calcitonin, in male idiopathic calcium urolithiasis - Evidence for a basic disturbance in calcium metabolism, *Exp Clin Endocrinol* 90: 71 (1987).
8. M Fuss, T Peppersack, J Corvilain, P Bergman, D Willens, J Simon, J J Body, Stimulation of calcitonin secretion by an oral calcium load test in normal subjects and in idiopathic renal stone formers, *Bone Min* 6: 102 (1991).
9. PO Schwille, I Berger, Acute effects of exogenous calcitonin on calcium homeostasis and bone metabolism in idiopathic calcium urolithiasis, In preparation.
10. MP Ryan, The renal handling of magnesium, *in* "Metal Ions in Biological Systems" ed Marcel Dekker, New York, Basel (1990).

OSTEOCALCIN RESPONSE TO LOW CALCIUM DIET: A HELPFUL TOOL FOR THE DIFFERENTIAL THERAPY OF HYPERCALCIURIA

W.L. Strohmaier, H. Konakci and K-H. Bichler

Department of Urology
Eberhard-Karls-University
Tübingen, Germany

INTRODUCTION

The treatment of idiopathic hypercalciuria is controversial. Low-calcium diets can induce osteopenia[1], and thiazide diuretics have side effects such as disturbances of uric acid and glucose metabolism, hypokalemia, and hypotension. Calcium-load tests and the measurement of serum osteocalcin levels on a free diet are not reliable enough to indicate appropriate management with calcium-restricted diet or thiazide therapy, respectively[2]. This study was initiated to investigate whether the osteocalcin response to a calcium-restricted diet, reflecting bone turnover, is a potential tool in the metabolic investigation of hypercalciuria. Osteocalcin is a vitamin K-dependent, non-collagenous protein synthesized by osteoblasts[3]. Its synthesis is probably regulated by parathyroid hormone and 1,25-dihydroxyvitamin D. Serum levels ($T_{1/2}$, 4-5 min) are determined by the activity of osteoblasts and mineralization. The highest concentrations are observed in states with increased osteoblastic activity and decreased mineralization[3].

PATIENTS AND METHODS

Fifty-two calcium stone formers (44 males, 8 females) ranging in age from 25-74 years were investigated. All patients underwent a calcium-load test using the protocol of Pak et al[4] whereby dietary calcium was restricted (400 mg Ca/day) for 7 days. Osteocalcin levels were determined on a free diet (OC1) and after 7 days on a low- calcium diet (OC2). Blood samples were taken in the morning. For osteocalcin determinations, a radio-immunoassay (INCSTAR Corporation, Stillwater, Minnesota, USA) was used. The osteocalcin response to a low-calcium diet was expressed as OC2-OC1 x 100%. Bone mineral density of lumbar vertebrae 2-4 was measured by dual photon absorptiometry. Bone mineral density was expressed as percent of the normal mean for age and sex. Statistical analysis was performed using the Student's *t* test for unpaired data.

RESULTS

Table 1 shows the results of the calcium-load test. The osteocalcin response to a calcium-restricted diet was positive and significantly higher ($p< 0.05$) in patients with renal hypercalciuria and normocalciuria than with absorptive hypercalciuria type I and II who showed a negative response or none at all (Fig. 1). This reflects increased bone turnover on a low calcium diet in the former groups. There were, however, two patients with absorptive hypercalciuria with a highly positive OC-response (+182 and +71%). Long-term follow-up (up to 2 years) in these individuals showed a steady decrease in bone mineral density while they were consuming a low-calcium diet.

Bone mineral density was significantly lower ($p < 0.05$) in patients with renal hypercalciuria than the other groups (81 vs 91- 98 %).

Table 1. Age, sex, and classification of hypercalciuria in 52 calcium stone formers.

Diagnosis		n	sex		age ($\bar{x}$, range)
diet dependent	normocalciuria	10	f m	3 7	50,6 , 36-61
	abs. hypercalciuria II	13	f m	2 11	49,2 , 27-74
diet independent	renal hypercalciuria	22	f m	3 19	38,9 , 25-54
	abs. hypercalciuria I	7	f m	- 7	44,0 , 34-62

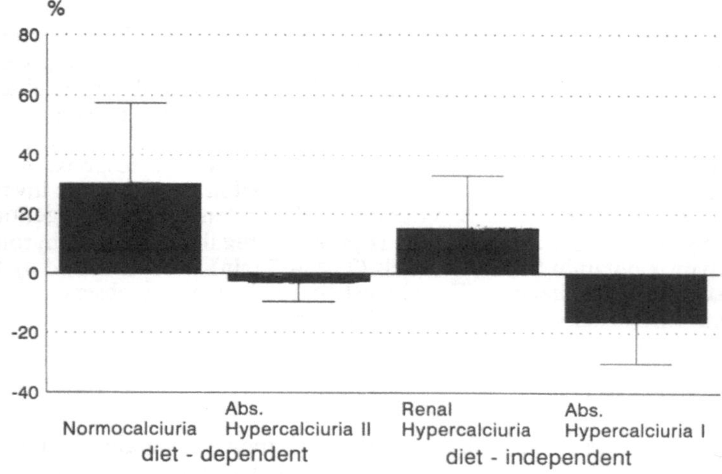

Figure 1. Osteocalcin response to a low-calcium diet (means ±SD) in 52 calcium stone formers.

DISCUSSION

Contrary to serum osteocalcin on a free diet[2], the osteocalcin response to a calcium-restricted diet differentiates the subgroups of hypercalciura. Generally, patients with renal hypercalciuria and normocalciuria show positive osteocalcin responses. Since an increase in serum osteocalcin reflects increased bone turnover[3], the positive osteocalcin response indicates a negative calcium balance in these patients, and a risk for developing osteopenia when treated with a low-calcium diet. Patients with absorptive hypercalciuria generally showed either no response or a negative osteocalcin response to a low-calcium diet, indicating that there was no negative calcium balance. There were in this group, however, two exceptions who demonstrated a highly-positive osteocalcin response. Long-term follow-up on a low-calcium diet showed a steady decrease in bone mineral density in these patients. This fact clearly demonstrates that the osteocalcin response is more reliable as a means of determining the appropriate management for idiopathic hypercalciuria than is the calcium-load test. Furthermore, it seems unjustified to recommend a calcium-restricted diet to all calcium stone formers.

For practical purposes, we suggest that hypercalciuric patients with positive osteocalcin responses be treated with thiazide diuretics, regardless of the calciuric response to a low-calcium diet. Absorptive hypercalciurics with a negative osteocalcin response should be treated with a low-calcium diet. We are currently pursuing this concept of differential therapy of hypercalciuria.

REFERENCES

1. M Fuss, T Pepersack, P Bergamn, T Hurard, J Simon and J Corvilain, Low calcium diet in idiopathic urolithiasis: a risk factor for osteopenia as great as in primary hyperparathyroidism, *Br J Urol* 65: 560 (1990).
2. WL Strohmaier, M Schreiber, K-H Bichler, Osteocalcin levels in calcium urolithiasis, *in:* "Proc 1st European Symp Urolithiasis," Excerpta Medica, Amsterdam (1990).
3. RA Venbrocks, "Osteocalcin", Enke, Stuttgart (1989).
4. CYC Pak, R Kaplan, H Bone, J Townsend and O Waters, A simple test for the diagnosis of absorptive, resorptive and renal hypercalciuras, *N Engl J Med* 292: 497 (1975).

INFLUENCE OF MORPHOLOGIC AND URODYNAMIC FACTORS ON CALCIUM-CONTAINING STONE FORMATION

T. Yamate, K. Kohri, M. Iguchi, Y. Ishikawa and T. Kurita

Department of Urology
Kinki University School of Medicine
Osaka-Sayama
Osaka, 589, Japan

INTRODUCTION

Studies on the pathogenesis of urolithiasis have mainly been performed on such aspects as metabolism of calcium[1] and other urinary constituents[2,3], and crystallization processes. Although recurrent stone formers are the most appropriate subjects in such studies, stone formation in these patients cannot always be explained by metabolic disorders alone. The stone incidence in the lower calyx of the kidney is significantly higher than in other regions of the kidney[4] and stone formation usually recurs on the same side. It is also known that long-term recumbency is one of the causes of stone formation[5]. Based on these findings, it is speculated that morphologic and urodynamic disorders of the urinary tract may have some bearing on stone formation and passage. In the present study, we examined the relationship between morphologic and urodynamic disorders of the urinary tract and the incidence of stone formation.

MATERIALS AND METHODS

In the morphological study, the subjects consisted of 50 males who had experienced spontaneous passage of unilateral calcium-containing stones. Intravenous pyelography (IVP) showed a normal bilateral pelvi-caliceal system (PCS) in all patients. The IVP's of the stone and non-stone forming side were compared with attention to a number of factors described in Table 1.

In the urodynamic study, an 8-Fr microtransducer-tipped catheter was inserted into the ureter and renal pelvis of renal stone formers. The pressures in the renal pelvis, the pelvi-ureteral junction (PUJ), and the upper third of the ureter were measured.

RESULTS

The results are shown in Table 2. Comparison of the PCS revealed a significant difference in the following parameters on the stone-producing side compared to the non-stone forming side of patients. On the stone side, higher values were found for Np ($p< 0.01$), Nm ($p< 0.05$), and Nb ($p< 0.01$). Also, L_1 and L_2 were longer ($p< 0.05$ and $p< 0.01$, respectively). Furthermore, larger values were found on the stone side with regard to Ac, Ar and At ($p< 0.01$).

Table 1. Definition of parameter

Np	Number of minor calyces.
Nm	Number of major calyces.
Nb	Number of branches starting from the major calyx up to the minor calyx. All branches are counted and their total number is registered.
L_0	Circle radius in the renal pelvis being the largest possible circle constructible without intersecting the contours of the renal pelvis.
L_1	Upper calyx radius. From the center of the circle in the renal pelvis, L_1 is the radius to the most cranially-situated papilla.
L_2.	Lower calyx radius. From the center of the circle in the renal pelvis, L_2 is the radius to the most caudally-situated papilla.
Ac	Total calyx area. Area of the renal pelvis limited by the narrowest sites of the major calyx.
Ar	Renal pelvis area. Area of the renal pelvis limited by the narrowest sites of the calyx necks and L_2 toward the ureter.
At	Total area which includes Ac+Ar.

Table 2. A comparison of IVP measurements taken on the stone-producing side of the patient compared to the non-stone forming side.

Parameter Normal side	Stone side	Non-stone forming side
Np	10.4±2.6	9.2±2.3**
Nm	2.4±0.5	2.2±0.4#
Nb	16.6±4.7	14.5±4.1**
L_0, mm	6.9±2.2	6.5±2.1
L_1, mm	37.8±8.2	36.5±7.1*
L_2, mm	31.0±4.9	28.3±4.5**
Ac, mm^2	372.0±136.6	323.2±133.4**
Ar, mm^2	412.1±185.2	362.6±163.5**
At, mm^2	782.1±263.9	683.9±211.8**

Values shown are Means±SD.
*$p<0.05$, ** $p<0.01$, #$p<0.02$

A

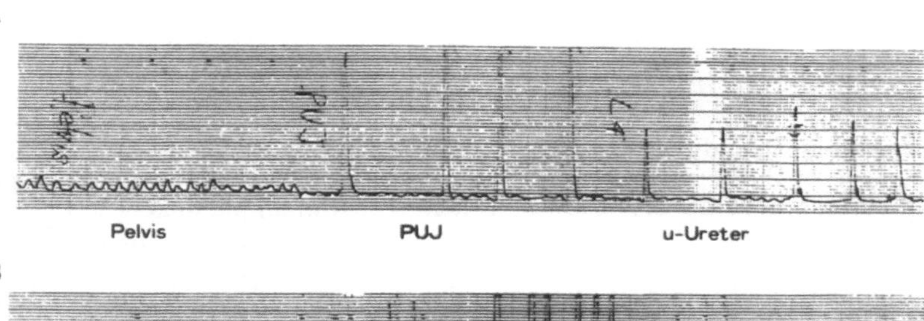

B

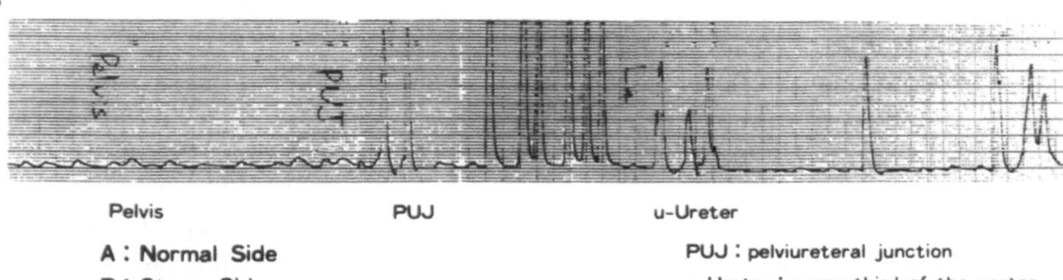

A : Normal Side
B : Stone Side

PUJ : pelviureteral junction
u-Ureter : upper third of the ureter

Figure 1. Urodynamic evaluation in upper urinary tract showing the normal side of a patient (A) compared to the stone forming side (B).

Figure 1 A shows that in the urodynamic study, the pressure was not different in the non-stone forming side in this case study, and the rhythm of the wave was regular. In contrast, in Fig. 1B, it is seen that in the stone forming side of this patient, the rhythm of wave was irregular, and biphasic spikes were sometimes observed. The average pressures were not significantly different. However, the range of pressures was considerable.

DISCUSSION

It is speculated that this difference in pressure causes stone formation in the stone forming side, and the inflammatory change caused by the presence of a kidney stone affects the dynamics of the urinary tract. This study was a new attempt to clarify the pathogenesis of stone formation.

REFERENCES

1. CYC Pak, *in:* "Hypercalciurias, Calcium, Calcium Urolithiasis", Plenum Press, New York (1978) p.37.
2. SG Welshman and MG McGeown, Urinary citrate excretion in stone formers and normal controls, *Br J Urol* 48: 7 (1976).
3. FL Coe, Hyperuricosuric calcium oxalate nephrolithiasis, *Kidney Int* 13: 418 (1978).
4. Y Ishikawa, Y Katoh, S Mitsubayashi, T Matsuura, K Kohri, T Akiyama and T Kurita, Clinical results of percutaneous nephro-ureterolithotripsy trans-urethral-ureterolithotripsy, Nishinippon, *J Urol.* 52: 9 (1990).
5. J Hamburger, G Richt and H deMontera, *in:* "Nephrology", Saunders, Philadelphia 1: 426 (1968).

PROTECTIVE EFFECTS OF CALCIUM ANTAGONISTS ON VITAMIN-INDUCED NEPHROLITHIASIS

W.L. Strohmaier[1], R.D. Seeger[1], H. Oßwald[2] and K-H. Bichler[1]

[1]Department of Urology
[2]Institute of Pharmacology
Eberhard-Karls-University
Tübingen, Germany

INTRODUCTION

The active vitamin D, 1,25-dihydroxycholecalciferol (DHCC), plays an important role in the pathogenesis of calcium urolithiasis. Increased 1,25-DHCC levels are found in 30 - 80 % of hypercalciuric patients[1-3]. The underlying mechanisms, however, are not understood completely. In an experimental study we investigated whether cellular processes are important for the pathogenesis of vitamin D-induced nephrolithiasis and whether cytoprotective drugs as calcium antagonists can influence these processes.

MATERIALS AND METHODS

Male Wistar rats (body weight 250 - 300 g) were randomly assigned to one of the following groups: (1) 1,25-DHCC (n = 8), (2) 1,25-DHCC + calcium antagonist Goe 6070 (Goedecke AG, Berlin, Germany) (n = 8), (3) control (n = 8). 1,25-DHCC (Roche, Basle, Switzerland) was administered for 6 days (120 pmol/24 h s.c.), Goe 6070 for 6 days (1 mg/kg/body weight/24 h) by gavage. Goe 6070 is a new 1,4-dihydronaphthyridin currently under development[4]. Clearance studies (inulin, creatinine, calcium, phosphate) were performed on day 6 and kidneys were taken for histological examination (Kossa staining) with determination of the calcification index (CI)[5] and measurement of calcium tissue concentration. Statistical analysis was performed using the Wilcoxon Rank Sum test.

RESULTS

1,25-DHCC induced substantial concrement formation (CI 2.4±0.8), which was significantly ($p<0.01$) less pronounced in the 1,25-DHCC + Goe 6070 group (CI: 1.4±0.8. Controls: CI 0.4±0.3). Concrement formation started within the tubular cells in the region of the cortico-medullary junction (fig. 1).

1,25-DHCC rats showed significantly higher calcium tissue concentrations in the kidney than 1,25-DHCC + Goe 6070 animals (fig. 2). The glomerular filtration rate (GFR) was reduced by 1,25-DHCC. This could be almost completely inhibited by concomitant application of Goe 6070 (fig. 2).

1,25-DHCC animals showed increased fractional excretions and increased serum levels of calcium and phosphate. These parameters were not changed by Goe 6070.

Urolithiasis 2, Edited by R. Ryall *et al.*,
Plenum Press, New York, 1994

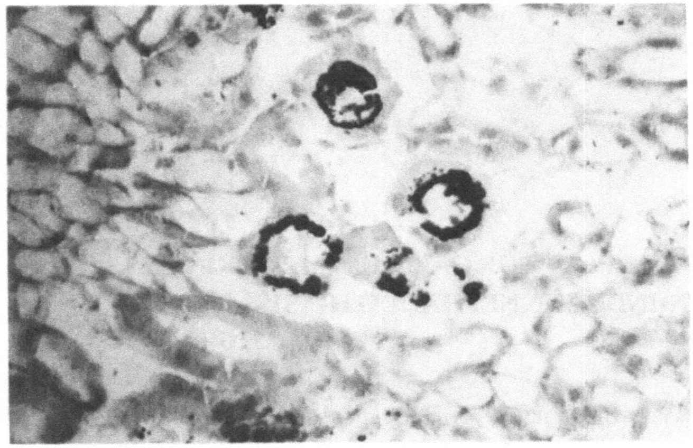

Fig. 1. Intracellular and membrane standing calcifications in renal tubules of the corticomedullary junction area (Kossa staining, magnification 160 x).

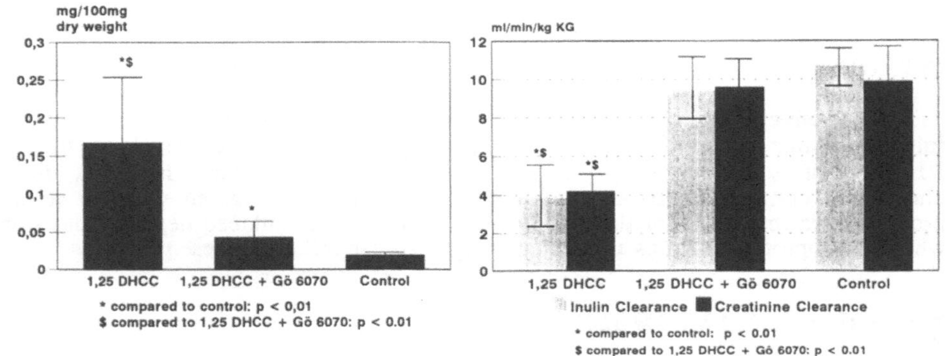

Fig. 2. Calcium concentrations in the kidneys (left) and glomerular filtration rate (right) of 1,25DHCC, 1,25-DHCC plus Goe 6070 and control rats (means±SD). Significant differences: vs. control * $p< 0.01$. 1,25-DHCC vs. 1,25-DHCC plus Goe 6070 \$ $p< 0.01$.

DISCUSSION

Our results demonstrate that 1,25-DHCC induced marked concrement formation in the corticomedullary junction area which started within the tubular cells. The GFR was reduced drastically. Since no signs of tubular obstructions could be observed, the reduced GFR may be attributable to a 1,25-DHCC-induced increase in the contractility of mesangial cells[6].

These effects of 1,25-DHCC on concrement formation and GFR were ameliorated by concomitant application of the calcium antagonist Goe 6070. The underlying mechanism of action may be an interference with an increased cellular influx of calcium caused by 1,25-DHCC. Since biochemical parameters such as serum levels and fractional excretions of calcium and phosphate were not influenced by the calcium antagonists, cellular processes seem to be more important for vitamin D-induced nephrolithiasis than biochemical alterations.

ACKNOWLEDGEMENTS

We thank Dr G Satzinger, Goedecke AG, Berlin, Germany for the generous gift of Goe 6070 and Dr H Weisser, Roche, Basle, Switzerland for providing us with 1,25-DHCC.

REFERENCES

1. K-H Bichler, WL Strohmaier, M Schreiber, S Korn and I Gaiser, Hyperkalziurie und Vitamin D-Stoffwechsel. *Fortschr Urol Nephrol* 22 342 (1984).
2. KL Insogna, AE Broadus, BE Dreyer, AS Ellison and JM Gertner, Elevated production rate of 1,25-dihydroxyvitamin D in patients with absorptive hypercalciuria, *J Clin Endocrinol Metab* 61: 490 (1985).
3. FB Stapleton, CB Langman, J Bittle and LA Miller, Increased serum concentrations of 1,25(OH)2 vitamin D in children with fasting hypercalciuria, *J Pediatr* 110: 234 (1987).
4. H Oßwald, W Steinbrecher, G Weinheimer, B Wagner, J Kleinschroth, J Hartenstein and G Satzinger, Antihypertensive activity of a novel calcium antagonist, Goe 5584-A, in rat models of hypertension, *J Cardiovasc Pharmacol* 12II, suppl.6: 154 (1988).
5. K-H Bichler, WL Strohmaier in: "Nephrocalcinosis, Calcium Antagonists and Kidney" Springer, Berlin (1988).
6. T Weinreich, J Merke, M Schonermark, H Reichel, M Diebold, GM Haensch and E Ritz, Actions of 1,25-dihydroxyvitamin D3 on human mesangial cells, *Am J Kidney* Dis 18: 359 (1991).

RESPONSE TO AN ORAL CALCIUM LOAD IN STONE FORMERS, THEIR SPOUSES AND FIRST DEGREE BLOOD RELATIVES

P. Kaul[1], S. Vaidyanathan[2], S.K. Thind[1] and R. Nath[1]

[1]Departments of Biochemistry
[2]Department of Urology
Postgraduate Institute of Medical Education and Research
Chandigarh, India

INTRODUCTION

Hypercalciuria has long been considered a common abnormality in many calcium stone formers. A familial pattern of renal stone disease[1,2] with a polygenic form of inheritance has been reported[3]. However, the question as to whether the genetic pattern or the environment (dietary habits, etc) play the major role in calcium oxalate urolithiasis is still controversial[2,4]. In the present study, the response to an oral calcium load in stone formers, their spouses and first degree blood relatives was studied to explore the importance of genetic and environmental factors.

PATIENTS AND METHODS

Thirty idiopathic calcium stone formers, their 16 first degree blood relatives and 20 spouses of the stone formers (the latter two groups had no present or past history of stone disease) were studied. The stone former, his/her first degree blood relatives and spouses were all living in the same house, sharing the same type of food and pursuing a similar type of occupation. All the subjects had normal renal function as evaluated by blood urea and serum creatinine levels. Blood sugar, serum calcium and phosphorus were within normal range and urine culture was sterile. None of the participants was taking any drug for any other concomitant disease.

Three days prior to the calcium load test, subjects were asked to eat a standard diet consisting of 2400 Kcal/day, protein 55g/d, fat 15g/d and calcium 400 mg/d. On the fourth day subjects were asked to present after an overnight fast for the calcium load test. Twenty four h urine samples were collected for baseline urinary calcium levels.

Calcium Load Test

2g calcium (as dicalcium phosphate) was given orally (salt mixed with 50 mL of distilled water) to all the subjects. They were allowed to take standard breakfast 2 h after the load. Urine was collected for next 8 hours and a blood sample was drawn 4 h after the load. Blood and urine samples were analysed for calcium, phosphorus, urea, creatinine and uric acid by routine procedures.

Urolithiasis 2, Edited by R. Ryall *et al.*,
Plenum Press, New York, 1994

RESULTS AND DISCUSSION

The mean age, sex ratio and mean rise in calcium excretion (over the basal calcium excretion) for the three groups of subjects are provided in Table 1. There was no significant difference in the age of participants among the 3 groups studied. The mean rise in calcium excretion (over the basal excretion) was in the order stone formers > first degree relatives > spouses.

Table 1. Age, sex ratio and the mean rise in calcium excretion (over the basal calcium excretion) in stone formers, their first degree blood relatives and spouses.

Parameters	Stone formers (n=30)	First degree blood relatives (n=16)	Spouses (n=20)
Mean age	38.0±13	41.0±16	38.0±11
Male:Female	18:12	13:3	10:10
Mean rise in calcium excretion (mg/g creatinine)	20.07±11***#	#13.31±9**	5.8±5

Results are given as Mean ±S.D. *** p<0.001, ** p<0.01 compared to spouses.
# p<0.05 compared to first degree blood relatives.

Eight hour urinary calcium excretion after the calcium load ranged from 68-253mg in stone formers, 40-161mg in first degree blood relatives and 40-86 mg in spouses (Fig.1). Taking 100mg/8h as the upper limit for normal urinary calcium excretion, 66.6% of the stone formers, 25% of the first degree blood relatives and none of the spouses were found to be hypercalciuric after calcium load.

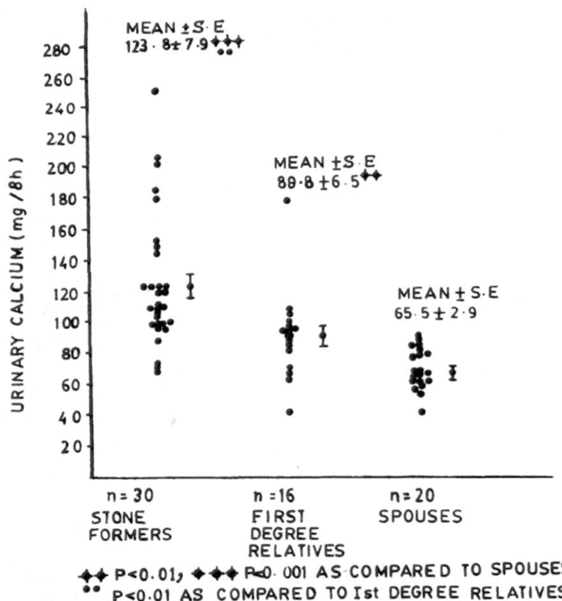

Figure 1. Effect of oral administration of 2g calcium on urinary calcium excretion in stone formers, their first degree relatives and spouses.

Eight hour urinary calcium excretion was significantly higher in stone formers ($p < 0.001$) and first degree relatives ($p < 0.01$) compared to spouses. Stone formers excreted higher ($p < 0.01$) urinary calcium compared to their first degree relatives as well (Fig. 1).

Serum calcium, phosphorus, uric acid, urea and creatinine were all within the normal range before and after the load (data not shown).

Greater urinary calcium excretion (after the calcium load) in stone formers and their first degree blood relatives, and the incidence of hypercalciuria observed in the 3 groups indicate hyperabsorption of calcium in stone formers and their first degree blood relatives compared to spouses of stone patients. An increased absorption of calcium in stone formers compared to normal controls (after the calcium load) has also been noticed by other workers[5]; however, they did not compare the effects on the blood relatives or spouses. The present study suggests that predisposition towards renal stone disease is greater in those who share the genetic pattern of stone formers than those who share the environmental conditions only. Hence the first degree blood relatives of stone formers should be regarded as being at greater risk, and timely clinical and metabolic evaluation would be justified for early detection and prevention.

REFERENCES

1. B Wikström, U Backman, BG Danielson et al, *in* "Urolithiasis and Related Clinical Research", PO Schwille et al, ed. Plenum Press, New York (1985).
2. RK Marya, RC Dadoo and NK Sharma, *Urol Int* 36: 245 (1981).
3. M Resnick, DB Pridgen and HO Goodman, *N Engl J Med* 278: 1313 (1968).
4. RW White, RD Cohen, FP Vince et al, *in* : "Renal stone symposium", A Hodgkinson and BEC Nordin, eds, Churchill Living Stone, London (1969).
5. A Ahmed, AK Pendse, PN Sharma et al, *Ind J Clin Biochem* 6: 113 (1991).

ATTENUATION OF HYPOCALCIURIC RESPONSE TO LONG-TERM THIAZIDE AND AMILORIDE IN CALCIUM UROLITHIASIS WITH RENAL HYPERCALCIURIA

N. Colleoni, G. Arrigo, E. Gandini, A. Luciano and G. D'Amico

Division of Nephrology
Ospedale S Carlo
Via Pio II 3
20153 Milano, Italy

INTRODUCTION

The effect of long-term low dose hydrochlorothiazide (tz) and amiloride (am) therapy on recurrence rate and metabolic factors was evaluated in 32 well defined patients with hypercalciuria not dependent on diet.

PATIENTS AND METHODS

Patients were studied during a control phase (c) and after 1, 4, 7 and 10 years of tz 50 (25) mg/day and am 5 (2.5) mg/day. Urinary calcium (UCa), sodium (UNa), chloride (UCl), citrate (Ucit), oxalate (UOx), creatinine (UCr); serum creatinine (SCr), potassium (SK) and bicarbonate (SHC) were measured. At the beginning, the recurrence rate, mean (±SD) was 7 stones/pt±1.5[1] stone/pt/year.

Statistical Analysis

Data were evaluated with analysis of variance for repeated measures and with correlation coefficient tests.

RESULTS

Results are shown in Tables 1 and 2. Compliance with therapy seems to have been good, as shown by mild alkalosis and hypokalemia, corrected by K supplement-ation since year 7. Three patients discontinued the drugs for reversible side effects.

UCa decreased significantly after 1, 4 and 7 years of treatment, while at year 10 we observed an attenuation of the hypocalciuric effect, with normal fasting calciuria: UCa/UCr = 0.10(0.03) mg/100 mL glomerular filtrate. We did not find correlations between UCa and UNa nor between UCa and UCl; there was an obvious correlation between UNa and UCl throughout. SCr increased significantly, although it remained in the normal range. UOx decreased and Ucit increased during the treatment.

The long-term remission rate was 75% (1.2 stones/pt vs 5.2 predicted). Four of the 15 patients continued to have stones during long-term treatment, although with reduced frequency; they had hypercalciuria at year 10.

Table 1

Years	n	UCa (mg/kg bw) Mean (SD)	<3	3-4 % of pts	>4	UNa (mEq/day)	UCl (mEq/day)	UCit (mg/day)	UOx (mg/day)
C	32	5.1 (1.7)	0	25	75	205 (72)	190 (74)		
1	32	3.4 (1.7) *	50	16	34	203 (88)	214 (71)		
4	28	3.4 (1.6) *	50	13	37	231 (117)	232 (99)		
7	21	4.0 (1.8) *	37	11	52	205 (78)	227 (83)	382 (60)	27 (9)
10	15	4.2 (1.7)	27	20	53	203 (92)	218 (103)	419 (51)	23 (9)

p< 0.05 Dunnett t test

Table 2.

Years	n	SCr (mg/dL)	SK (mEq/L	SHC (mEq/L)
C	32	1.03 (0.2)	3.9 (0.5)	26.3 (3.0)
1	32	1.10 (0.1)	3.6 (0.4) *	28.6 (2.1) *
4	28	1.16 (0.2) *	3.6 (0.3) *	28.4 (3.1) *
7	21	1.15 (0.1) *	3.6 (0.3) *	28.4 (1.9) *
10	15	1.16 (0.2) *	3.8 (0.4) *	28.5 (2.9) *

* p< 0.05 Dunnett t test

DISCUSSION

The results suggest that some patients with renal hypercalciuria lose the hypocalciuric effect of hydrochlorothiazide and amiloride during long-term treatment. As fasting calciuria at year 10 was normal, we suppose that the attenuation of the hypocalciuric effect was due to intestinal calcium hyperabsorption, as previously described in patients with absorptive hypercalciuria[1]. Otherwise amiloride did not prevent thiazide induced hypokalemia and alkalosis as described by others[2]. During long-term low-dose thiazide and amiloride therapy the kidney stone recurrence rate decreased significantly. It is possible that the patients who continued to form stones, developed absorptive hypercalciuria.

REFERENCES

1. GM Preminger and CYC Pak, Eventual attenuation of hypocalciuric response to hydrochlorothiazide in absorptive hypercalciuria, *J Urol* 137:1104 (1987).
2. U Alon, LS Costanzo, JCM Chan, Additive hypocalciuric effects of amiloride and hydrochlorothiazide in patients treated with calcitriol, *Mineral Electrolyte Metab* 10:379 (1984).

BONE EFFECTS OF CALCIUM RESTRICTION DIET IN CHILDREN WITH ABSORPTIVE HYPERCALCIURIA

H.C. Perrone[1], J. Toporovski[1], A.C. Bianco[3], Marilia Marone[3], C. Langman[4] and N. Schor[2]

[1]Nephrology Division
Department of Pediatrics of Santa Casa
[2]Department of Medicine
Escola Paulista de Medicina
[3]Unidade de Densitometria Ossea
São Paulo, Brazil
[4]Mineral Laboratory-Children's Memorial Hospital
Chicago IL, USA

INTRODUCTION

A major risk factor for nephrolithiasis is hypercalciuria, present in 50-70 % of hypercalciuric children[1]. Hypercalcuria may occur as a results of either increased intestinal absorption of dietary calcium (absorptive hypercalciuria-AH)[2] or decreased renal tubular resorption of filtered calcium (renal hypercalciuria-RH)[3,4]. The treatment of AH is by reducing dietary calcium intake. However, low calcium intake is potentially associated with decreased bone mineral mass of growing children and its effects have not been prospectively studied in children with hypercalciuria.

The purpose of the present investigation was to evaluate the bone mineral density (BMD) of children with idiopathic hypercalciuria (IH) subjected to chronic calcium restriction intake. Plasma concentrations of 1,25-dihydroxyvitamin D, 25-hydroxyvitamin D, osteocalcin and PTH were also measured.

MATERIALS AND METHODS

We prospectively studied 20 children with hypercalciuria, aged from 4 to 13 years. All patients were from the outpatient clinic and were eating either a calcium restricted diet containing 400-500 mg Ca/day or rice bran (10-15 g/day) along with dairy products. In addition, a moderate sodium intake was recommended. Bone mineral density (BMD) of lumbar vertebrae (L2-L4) was measured by dual photon absorptiometry (DPX-L, Lunar Corp. Madison-WI) every six months for a period of 42 months. Plasma osteocalcin was measured by specific radioimmunoassay (Incstar Corp, Stillwater-MN); PTH was by a two-site immunoradiometric assay for the biologically intact 184 PTH molecule (Nichols Institute, S. Juan Capistrano-CA); calcidiol and calcitriol were measured by competitive protein binding assay, as previously described[5].

RESULTS

Table 1 shows that, as expected, urinary calcium was significantly increased in uncontrolled children. However, there was no significant difference between hypercalciuric and normocalciuric children in the plasma concentrations of osteocalcin, PTH, 1,25 -dihydroxyvitamin D and 25-hydroxyvitamin D.

Bone mineral density of lumbar vertebrae was negatively correlated to urinary calcium excretion as well as to osteocalcin levels.

Table 1. 24 h Urinary calcium excretion in uncontrolled (UC) and controlled (C) children, plasma osteocalcin (Gla-p), PTH, 1,25$(OH)_2$-vitamin D and 25(OH)-vitamin D.

	UCa mg/kg/day	Gla-p ng/mL	PTH pg/mL	1,25D3 pkg/mL	25D2 ng/mL
C	2.7±0.5[a]	7.0±1.6	11.4±2.9	45.7±5.	26.1±1.8
UC	4.9±0.8*	6.9±1.1	10.4±2.0	46.1±4.6	25.9±2.4

a= mean±SE, *p <0.05 vs. C

DISCUSSION

We have found that children with IH due to AH present with normal plasma concentrations of PTH, 25(OH)-vitamin D, 1,25$(OH)_2$- vitamin D and osteocalcin. However, children spine BMD negatively correlates with urinary calcium, and thus, IH might be considered an important risk factor for bone resorption and/or deficient development.

During therapy, despite the restricted calcium diet (400-500 mg/day), PTH and 1,25$(OH)_2$-vitamin D serum levels were normal, suggesting that during the course of the low calcium diet a new steady-state was achieved and a new hormonal/calcium balance was obtained. Future studies are necessary to establish whether a strict calcium metabolism control may affect bone mineralization.

REFERENCES

1. FB Stapleton, Idiopathic hypercalciuria in children, *Seminars Nephrol* 3:116 (1983).
2. CYC Pak, R Kaplain, H Bone, J Townsend, O Waters, A simple test for the dagnosis of absorptive, resorptive and renal hypercalcurias, *New Engl J Med* 292, 497-500 (1975).
3. HC Perrone, H Ajzen, J Toporovski, N Schor, Metabolic disturbance as a cause of hematuria in children. *Kidney Int* 39:707 (1991).
4. LC Hymes, BL Warshaw, Idiopathic hypercalciuria: renal and absorptive subtypes in children, *Am J Dis Child* 138:176 (1989).
5. TA Reinhardt, RL Horst, JW Orf, BW Hollis, A microassay for 1,25 dihydroxyvitamin D not requiring high performance liquid chromatography: application to clinical studies, *J Clin Endocrinol Metab* 58:191 (1984).

ASSESSMENT OF NUTRIENT INTAKE BY CALCIUM STONE FORMER PATIENTS

L.A. Martini, I.P. Heilberg, L. Cuppari, S.A. Draibe and N. Schor

Nephrology Division
Escola Paulista de Medicina
São Paulo, Brazil

INTRODUCTION

Idiopathic stone disease of the urinary tract is commonly associated with an increased urinary excretion of calcium, oxalate and uric acid, among other promoters of crystallization. Dietary factors are known to modify the urinary electrolyte profile. For instance, a higher consumption of animal protein can increase urinary calcium, uric acid and oxalate, and decrease urinary citrate[1,2]. Carbohydrate induces hypercalciuria and promotes intestinal calcium absorption. Therefore, it is reasonable to suggest that dietary management is one of the most important goals for the prophylaxis of kidney stone disease. The present study was undertaken to evaluate dietary intakes of calcium, carbohydrate, and protein of animal and vegetable origin in calcium stone formers.

MATERIALS AND METHODS

Patients consisted of 95 calcium stone formers (SF), 57 of whom were absorptive hypercalciurics (AH) and 38 were fasting hypercalciurics (RH). Thirty-four normal subjects served as controls. Four-day dietary histories were obtained from all subjects, which included three weekdays (72 h) and Sunday (24 h). Subjects were instructed to write down their total daily food intake in normal household measures. Nutrient intakes were calculated using a computer program developed in our Servce. An estimation of salt intake was based on 24 h urinary sodium excretion. Body weight and height were measured, and body mass index was determined.

RESULTS

Table 1 describes the groups of SF and normal subjects with regard to their nutrient intake based on four-day dietary histories.

DISCUSSION

The present series of patients consumed 436-574 mg of calcium/day, reaching 55-70% of suggested Recommended Dietary Allowance (RDA) values for calcium intake. We observed that SF consumed higher amounts of animal protein than normal subjects whether during the week or on Sunday. However, the mean total protein to body weight

Urolithiasis 2, Edited by R. Ryall *et al.*,
Plenum Press, New York, 1994

ratio of these patients (1.12 g/kg/24 h) was within the RDA for adults. The mean salt intake was higher than the RDA in all groups, reaching 10 to 14 g/24 h. The two subgroups of SF were similar with regard to nutrient intake. Overconsumption of carbohydrate or energy (KCAL) either during the week or week-end was not found.

Therefore, protein as well as salt intake may be more important factors than Ca intake in stone formation. The study showed how specific a dietary habit can be for a given patient or population, requiring different and more rational dietary recommendations.

Table 1. Nutrient intake in SF and AH and RH subgroups compared to that in normal subjects.

		Normals	SF	AH	RH
		(34)	(95)	(38)	(57)
Age, years		39±2	44±1'	44±2	44±2
Weight, kg		66±2	69±1	71±2	68±2
Body Mass Index		26±1	27±0	28±1	26±1
24 h Dietary Intake					
Calcium, mg	W:	569±36	542±29	554±51	534±34
	WE:	436±39	565±30	574±52	556±37
CHO, g	W:	213±15	236±8	232±13	239±11
	WE:	218±15	244±9	246±14	242±12
KCAL	W:	1714±115	1850±64	1812±102	1875±82
	WE:	1702±114	1971±71*+	1913±112	2010±92
Total prot.	W:	66±5	70±2	66±4	72±3
g	WE:	63±5	77±3*+	67±4	83±4*+
Prot./Kg	W:	1.01±0.07	1.02±0.04	0.94±0.05	1.08±0.06
g/kg	WE:	0.95±0.06	1.12 ±0.05*+	1.00±0.06	1.21±0.06*+
HBV, g	W:	42±4	44±1	42±3	45±2
	WE:	41±3	55±3*	48±3	60±4*+
LBV, g	W:	23±2	26±1	24±1	27±2
	WE:	21±2	23±2*	22±2	24±2*
NaCl, g	W:	20±3	13±4	12±3	14±5

Values are Means±SEM, n in parenthesis, CHO=carbohydrate; KCAL=energy; HBV=high-biological-value protein; LBV=low-biological-value protein. NaCl=salt intake. 'Statistically different from normal subjects, $p<0.05$; *weekend vs week within group, $p<0.05$; +weekend vs weekend, SF vs normal subjects, $p<0.05$.

REFERENCES

1. PN Rao, J Prevendiville, PA Buxton, DG Moss and NJ Blacklock, Dietary management of urinary risk factors in renal stone formers, *Br J Urol* 54:578 (1982).
2. S Goldfarb, The role of diet in the pathogenesis and therapy of nephrolithiasis, *in*: "Endocrinology and Metabolism Clinics of North America", Lynwood H Smith, ed, Philadelphia (1990).

INCIDENCE OF POUCH STONES AND RISK FACTORS FOR UROLITHIASIS IN PATIENTS RECEIVING CONTINENT DIVERSION OR NEOBLADDER USING INTESTINE - COMPARISION WITH ILEAL CONDUIT

M. Takeda[1] ,Y. Katayama[1] , T. Tsutsui[1],
T. Komeyama[1], T. Kawasaki[1], T. Mizusawa[1],
H. Takahashi[1], S. Sato[1] and S. Nakamura[2]

[1]Department of Urology
Niigata University School of Medicine
[2]Section of Urology
Niigata Citizen Hospital
Japan

INTRODUCTION

Recently, the use of continent urinary reservoir (pouch) as a urinary drainage operative method has been spreading, leading to improved quality of life in patients undergoing total cystectomy[1]. Various methods of pouch operation have been devised[2-4]. The Rock pouch[2] is a representative method, in which a high incidence of pouch stone formation has been reported[5,6]. Although the use of metal staples and Marlex mesh is believed to promote stone formation, the structure of the pouch and deposits of intestinal mucus have also been implicated[7,8]. Apart from these factors, no other risk factors for urolithiasis in patients with pouch stones have been identified. In the present study, risk factors and inhibiting factors for urolithiasis were retrospectively examined in patients undergoing the pouch operation or augumentation cystoplasty using the intestine, to explore the causes and mechanisms of pouch stone formation.

PATIENTS AND METHODS

Twenty-two patients undergoing total cystectomy (12 pouch - P group, 10 conduit - C group) at Niigata University Hospital were examined. Patients of the R group ranged in age from 13 to 60 (mean 54.5) years, with 9 men and 3 women. The period after operation ranged from 2 to 38 months (mean 10.8). In 6 patients non-absorbable artificial material (metal staple: TA-55,US Surgical Corporation, USA, and Dacron fabric: USCI, Bard Japan, Japan) were used for intestinal invagination, and in 6 patients non-absorbable artificial materials were not used. Patients of the C group ranged from 30 to 76 (mean 64.5) years. There was no difference in age or sex between the P and C groups. Urodynamic evaluation showed that maximum pouch capacity ranged from 220 mL to 700 mL (mean 440.8 mL), and only 8 patients showed abnormal waves over 20 cmH_2O. No patient had any previous history of urolithiasis. Every patient showed mild pyuria, but no bacterial growth in the urine.

Twenty-four h urine samples were collected under aseptic conditions more than 2 months after operation or after postoperative adjuvant chemotherapy in cases with malignancy. The mean urinary excretion of oxalate[10], citrate[11], calcium[12], phosphate[13], uric acid[14] and magnesium[15] were measured quantitatively in 2 or 3 24 h urines. The control group consisted of 40 to 100 people randomly selected without checking history of urolithiasis. Data were analyzed with the chi-square test. If the value of an inhibiting factor for urolithiasis was under the lower limit of normal, it was classified as a risk factor. Urinary pH and urine culture were examined more than three months after the pouch operation, before stone formation. Stone formation was checked using routine X-ray.

RESULTS

Ten patients (83.3%) in the P group and 6 (60%) in the C group had at least 1 risk factor, while 5 (41.7%) in the P group and 4 (40%) in the C group had 2 or more. There was no significant difference in the incidence of risk factors between the two groups. Upper urinary tract infection did not develop in any patient. Pouch stones developed in 4 patients (33.3%) in the P group and none in the C group. All stone formers had 2 or more risk factors. In the P group, the incidence of stones in patients with 2 or more risk factors was significantly higher than in patients with 1 or none ($p < 0.025$, chi-square test). The use of absorbable materials did not affect the incidence of stone formation. Stones were removed after destruction using an ultrasonic lithotriptor, and were analyzed by infra red spectroscopy. Every stone contained Ca-phosphate, and other components were Mg-ammonium phosphate, Ca-bicarbonate, and Ca-oxalate. There were no significant correlations between urinary pH and the presence of urinary tract infection, and the incidence of pouch stone formation. No patient showed clinical manifestations compatible with enteric hyperoxaluria[16] or urinary tract infection, except for mild pyuria.

REFERENCES

1. B Goldwasser and GD Webster, Continent urinary diversion, *J Urol* 134: 227 (1985).
2. NG Kock, AE Nilson, LO Nilson, LJ Norlen and BM Philipson, Urinary diversion via a continent ileal reservoir: Clinical results in 12 patients, *J Urol* 128: 469 (1982).
3. RG Rowland, ME Mitchell, R Bihrle, RJ Rahnoski and JE Piser, Indiana continent urinary reservoir, *J Urol* 137: 1136 (1987).
4. PA Thuroff, H Riedmiller, GH Jacobi, GH and R Hohenfellner, 100 cases of Mainz pouch: continuing experience and evolution,*J Urol* 140:283 (1988).
5. DG Skinner, G Lieskovsky and SD Boyd, Continuing experience with the continent ileal Reservoir (Kock pouch) as an alternative to cutaneous urinary diversion: an update after 250 cases, *J Urol* 137: 1140 (1987).
6. D Ginsberg, JL Huffman, G Lieskovsky, S Boyd and DG Skinner, Urinary tract stones: A complication of the Rock pouch continent diversion, *J Urol* 145: 956 (1991).
7. G Lieskovsky, SDBoyd and DG Skinner, Management of late complications of the Rock pouch form of urinary diversion, *J Urol* 137: 1146 (1987).
8. JW Thuroff, P Alken, U Engelmann, GH Jacobi, GH and R Hohenfelner, The Mainz-pouch (mixed augmentation ileum 'n zecum) for bladder augmentation and continent diversion, *World J Urol* 3:179 (1985).
9. S Shishido, M Nagato, M Hayashi, H Imamura, R Rudo, and I Chino, 3 cases of stone formation in the pouch of Kock continent ileal reservoir,*Jap J Clin Urol* 45: 514 (1991).
10. A Ichiyama, E Nakai, E and T Funai, Spectrophotometric determination of oxalate in urine and plasma with oxalate oxidase, *J Biochem* 98:1375 (1985).
11. TT Nielsen, A method for enzymatic determination of citrate in serum and urine, *Scand J Clin Lab Invest* 36:513 (1976).
12. J Stern and WHP Lewis, The colorimetric estimation of calcium in serum with o-cresolphthalein complexone, *Clin Chim Acta* 2: 576 (1957).
13. Y Machida and T Nakanishi, Utilization of bacterial xanthine oxidase for inorganic phosphorous determination, *Agric Biol Chem* 46: 807 (1982).
14. T Senoo, R Rawashima, N Nakariki, K Eguchi, R Sato and T Saito, Study of uric acid determination. - comparison between BCMA and MEHA methods, *Jap J Med Tech* 33: 1561 (1984).
15. C Bohuon, Microdosage du magnesium dans divers milieux biologiques, *Clin Chim Acta* 7: 811 (1962).
16. FL Coe and JH Parks, *in* "Nephrolithiasis: Pathogenesis and Treatment", 2nd ed, Year Book Medical Publishers, Inc, Chicago, (1988).

LONG-TERM RESULTS OF SODIUM-POTASSIUM CITRATE THERAPY (OXALYT-C^R) FOR PREVENTING CALCIUM OXALATE STONES

W. Hauser, W. Aulitzky and J. Frick

Department of Urology
General Hospital
Salzburg, Austria

INTRODUCTION

The purpose of our study was to investigate the effect of a long term treatment in the metaphylaxis of calcium oxalate (CaOx) stone formers with sodium-potassium-citrate (Oxalyt C^R).

METHODS

Forty CaOx stone formers were treated with sodium-potassium citrate and 40 CaOx stone formers had no metaphylaxis except fluid intake. Oxalyt-C^R was given each day according to the optimal pH value (6.8 - 7.4) over 1 year and longer. At this urine pH citrate excretion increased, and soluble calcium-citrate complexes were formed. Clinical, biochemical, radiographic and sonographic examinations were performed every three months during therapy. Usually the patients were stone-free at the initiation of therapy, although some cases had small stones or fragments after ESWL

RESULTS

Our results showed that the citrate excretion increased and the calcium-citrate ratio decreased significantly with alkali-citrate therapy. The stone frequency decreased from 67% to 5.6%. There was no change in the group without Oxalyt-C^R in calcium and citrate excretion or stone frequency. As we finished the metaphylaxis the stone relapse rate increased and the calcium-citrate ratio and the citrate excretion reached pre-treatment values.

DISCUSSION

Patients who respond favourably to alkali citrate treatment should have permanent alkali citrate metaphylaxis, maybe lifelong, to reduce the stone frequency and regulate the citrate and calcium excretion.

The pattern of alkali citrate administration which will be most efficacious remains to be defined.

DISSOLUTION OF CALCIUM CALCULI WITH POTASSIUM CITRATE

S. Womack, J. Fine, G. Miller, N. Mandel, C. Pak and G. Preminger

The University of Texas Southwestern Medical Center
Dallas, Texas

INTRODUCTION

We have noted radiological evidence of diminution and even complete disappearance of pre-existing calcium stones in a number of patients being treated with potassium citrate (K-Cit) for the prevention of new stone formation. Two prospective studies were initiated to more closely investigate this clinical phenomenon.

METHODS

In an "open" study, 46 patients with documented renal stones containing calcium were continued/started on K-Cit, 40 - 80 meq/day, in divided doses. The second study included 15 patients who were randomly placed on either K-Cit or placebo (KCl), 80 meq/day, in a double-blinded fashion. All 61 patients were followed for an average of 32 months, with serial 24 h urine collections and abdominal X-rays obtained at 8 month intervals. All X-rays were analyzed for changes in stone size by 3 independent (blinded) investigators: positive response = ↓ stone size; no response = no change in stone size; negative response = ↑ stone size.

Group	N	+ Response	No Response	- Response
Open	46	21 (45.7%)	17 (26.2%)	8 (17.4%)
KCit	8	4 (50.0%)	2 (25.0%)	2 (25.0%)
Placebo	7	1 (14.7%)	3 (42.9%)	3 (42.9%)

RESULTS AND DISCUSSION

Statistical analysis of urine chemistry data revealed a significantly higher mean urinary citrate excretion in the + responders compared to the - responders ($p < 0.05$). Analysis of stones passed before and after K-Cit reveal crystallographic alterations.

The positive correlation between the elevation of urinary citrate, secondary to oral K-Cit therapy and the reduction in size and/or complete disappearance of renal calcium stones has a valid physicochemical basis, as citrate is known to complex calcium.

It appears that K-Cit therapy is effective in preventing new stone formation and in some patients may promote stone dissolution.

ALKALI CITRATE METAPHYLAXIS FOR RECURRENT CALCIUM OXALATE STONE FORMERS - A PROSPECTIVE RANDOMIZED STUDY

J. Hofbauer, K. Höbarth, M. Eisenmenger and M. Marberger

Department of Urology
University of Vienna Medical School
Vienna, Austria

INTRODUCTION AND METHODS

A prospective randomized study was performed for the first time to assess the metaphylactic efficacy of alkali citrate therapy in calcium oxalate stone formers. Patients with regular stone formation - ie, at least one stone annually over the previous three years - were recruited and randomized into two groups of five each. They were given either general metaphylactic instructions (abundant liquid intake and dietary restrictions) plus alkali citrate (group I) or metaphylactic instructions alone (group II).

Three single doses of alkali citrate were adapted individually to obtain a urinary pH of a mean of 7 to 7.2. The study lasted for 36 months. Follow-ups were performed every three months.

Of group I, the data of 16 patients were suitable for evaluation; alkali citrate therapy was discontinued in four patients due to gastrointestinal side-effects, and the remaining five patients did not comply. In group II, the criteria were met by 22 patients.

RESULTS

The mean rate of stone formation at study-entry was 21.2±22.8 in group I and 17.5±18.7 in group II at a phase of regular stone formation of 8.2±5.2 and 9.7±4.5 years, respectively. In group I, the rate of stone formation was reduced from 2.1 to 0.7 stones per patient per year. However, the same was true for group II, the corresponding figures being 1.8 and 0.7. Despite a statistically significant higher urinary citrate excretion in group I than in group II, no such difference was seen between the two groups regarding recurrent stone formation (Student's t-test).

DISCUSSION

The fact that stone formation decreased in both groups (group I, 29%, group II, 25%) suggests that in a number of patients the phase of regular stone formation terminated spontaneously during the study, rather than being caused by alkali citrate therapy. An objective benefit of alkali citrate could not be established.

INNATE HUMORAL IMMUNITY IN THE FIGHT AGAINST POST-OPERATIVE SEPSIS: IS IT ENOUGH ?

D. Neilson, P.N. Rao, B. Oppenheim and G.R. Barclay

Department of Urology
University Hospital of South Manchester
Manchester, UK

INTRODUCTION

Systemic sepsis occurs in about 30% of cases of percutaneous nephrolithotomy (PCNL) despite antibiotic prophylaxis. It is usually due to the release of endotoxin, or lipopolysaccharide (LPS), from the cell wall of gram-negative bacteria. The core region of LPS is relatively constant in structure between bacterial species and patients with high levels of antibodies to this region would therefore be expected to be more resistant to developing post-op sepsis.

METHODS

30 patients undergoing PCNL had pre-op measurement of serum IgG IgM anti-LPS-core antibodies and this was expressed as a percentage of laboratory standard serum. They were divided into those with low humoral immunity (mean IgG & IgM <100%, N=11) and high humoral immunity (mean IgG & IgM >100%, N=19). Serial blood samples were taken for endotoxin assay and patients were monitored for signs of sepsis (T >38°C & P >90).

RESULTS AND DISCUSSION

More patients in the high immunity group developed signs of sepsis (7/19 versus 1/11, p <0.05) suggesting there is no protective effect. However, their higher starting antibody level probably reflects stimulation of the immune system by the presence of LPS, as 9/19 in the high immunity group had a raised pre-op LPS level (>13pcg/mL) compared to 0/11 in the low group p <0.05). Similarly, there were more patients who had a raised LPS level as a result of the operation in the high immunity group (13/19 versus 3/11, p <0.05) and also the levels of LPS attained were significantly higher. This suggests that they had to face a larger LPS stimulus as a result of the operatioN.

Thus the body appears to respond to LPS stimulation by increased antibody production but not to a level sufficient to prevent sepsis from increased LPS released at the time of surgery. The augmentation of humoral immunity may be of potential benefit in these cases.

ANTIBIOTIC PROPHYLAXIS FOR ENDOUROLOGICAL SURGERY - A NEW TWIST

D. Neilson, B.A. Crawley, B.A. Oppenheim and P.N. Rao

Departments of Urology and Bacteriology
University Hospital of Southern Manchester
Manchester, UK

INTRODUCTION

Endotoxaemia, a potential cause of clinical septicaemia, occurs in up to 40% of cases of percutaneous nephrolithotomy (PCNL). The role of antibiotic prophylaxis in preventing this endotoxaemia is not clear. Whilst antibiotics may decrease the bacterial load they may also increase the the endotoxin release from killed bacteria.

METHODS

33 patients undergoing PCNL were divided into 3 groups, 27 of these were randomised to receive gentamicin either 24 hours pre-op (Group l; N = 13) or with induction of anaesthesia (Group 2, N = 14); the other 6 patients (Group 3) had previously been started on antibiotics for clinical reasons. Serial blood samples were cultured and assayed for endotoxin. Pre-operative urine cultures were performed. Patients were monitored for clinical signs of sepsis post-operatively.

RESULTS AND DISCUSSION

The results indicate that antibiotic administration by itself causes endotoxin release. Whilst antibiotic prophylaxis starting 24 hours pre-op reduces the incidence of post-op rise in endotoxin and of positive blood cultures, it appears to make no difference to the numbers who become clinically septic! The presence of sterile urine pre-operatively similarly does not necessarily protect against post-op endotoxaemia or clinical sepsis.

BLOOD PRESSURE IN IDIOPATHIC CALCIUM NEPHROLITHIASIS

G. Vagelli, G. Calabrese, A. Mazzotta, G. Pratesi and M. Gonella

Service of Nephrology
General Hospital
Casale Monferrato, Italy

INTRODUCTION

Idiopathic calcium nephrolithiasis (ICN) and hypertension (HY) are both common in affluent countries. Hypercalciuria is reported to be associated with HY[1] and hypertensive subjects tend to develop renal stones more often than normals[3]. Our aim was to evaluate possible differences in blood pressure (BP) between ICN patients and normals and to find possible relationships between BP and metabolic parameters in ICN patients and normals.

METHODS AND RESULTS

Twenty-eight consecutive ICN patients (M 18, F 10) and 17 age and sex matched normals underwent BP recording under conditions of rest and tranquillity (10-15 min); mean arterial pressure (MAP) was calculated. Serum calcium, phosphate and uric acid, urinary calcium, phosphate, uric acid, oxalate, magnesium, creatinine, sodium and potassium were evaluated. Possible relationships between MAP and metabolic parameters were found by calculating the correlation coefficient (r).

MAP was significantly higher in ICN patients than in normals (100±10 vs 94±10, $p <0.05$). In ICN patients MAP was directly correlated with urinary sodium ($p <0.01$), phosphate and potassium ($p < 0.05$ in both cases). In normals MAP was directly correlated with urinary calcium ($p <0.01$), uric acid and creatinine ($p <0.05$ in both cases).

The finding of a higher BP in ICN patients than in normals may have some relationship with the tendency of HY patients to develop renal stones. The positive relationship between MAP and urinary calcium in normals, already shown in HY patients[1], is lost in ICN patients, supporting the notion of an impaired calcium metabolism in such patients. The positive relationship between MAP and urinary sodium in ICN patients suggests a higher sodium intake, which can be a risk factor in ICN[3].

REFERENCES

1. P Strazzullo, V Nunziata, M Cirillo, Abnormalities of calcium metabolism in essential hypertension, *Clin Sci* 65: 137 (1983).
2. G Tibblin. High blood pressure in men aged 50, A population study of men born in 1913, *Acta Med Scand*, suppl. 470, 1 (1967).
3. E Duranti, M Sasdelli, Effects of dietary sodium on lithogenic risk factors, *in* "Urolithiasis", VR Walker, RAL Sutton, ECB Cameron, CYC Pak, WG Robertson, eds , Plenum Press, 747 (1989).

BENEFICIAL EFFECT OF MAGNESIUM OXIDE ADMINISTRATION ON CALCIUM OXALATE UROLITHIASIS

S.R. Khan[1], C.J. Su[2], P.N. Shevock[1] and R.L. Hackett[1]

[1]Department of Pathology
College of Medicine
University of Florida, Gainesville, Florida, USA
[2]Tri-Service General Hospital
Taipei, Republic of China

INTRODUCTION

Magnesium (Mg) is a well known inhibitor of calcium oxalate (CaOx) crystallization *in vitro*, but its role in urinary stone formation remains controversial. Experimental Mg depletion has been demonstrated to accelerate and exaggerate the process of CaOx crystal deposition in the renal tubules of rats with experimentally induced hyperoxaluria. However fewer data are available concerning inhibitory effects of Mg in rats that are hyperoxaluric, but not hypomagnesiuric. We studied urinary CaOx relative supersaturation (RSS) and deposition of CaOx crystals in kidneys of rats receiving 1.0% ethylene glycol (EG) in their drinking water and rat chow supplemented with magnesium oxide (200mg/100g). Administration of MgO at this dose for 2 weeks had inconsistent and insignificant effects on urinary oxalate, CaOx RSS and crystal deposition in the rat kidneys.

EXPERIMENTAL PROTOCOL AND RESULTS

The above study was expanded, and rats receiving 1.0% EG were dispensed an increased dosage of MgO (500mg/100g chow) for 4 weeks. Increasing the oral magnesium dose did not result in a significant increase in urinary excretion of magnesium, but by day 14, there was a significant decline in urinary excretion of oxalate. There was a modest increase in urinary citrate and pH. All rats receiving EG demonstrated crystalluria. From the group receiving EG only, 3 of 4 rats sacrificed on day 15, and 2 of 4 rats sacrificed on day 29, had crystal deposits in their kidneys. None of the 8 rats who received both EG and MgO had crystals deposited in their kidneys.

CONCLUSION

The study demonstrates that administration of Mg can be beneficial against CaOx nephrocalcinosis and that magnesium acts by reducing the urinary excretion of oxalate.

ACKNOWLEDGEMENTS

This work was supported by NIH grants #RO1DK41434 and #PO1DK20586

SODIUM POTASSIUM CITRATE IN THE MEDICAL MANAGEMENT OF UROLITHIASIS: A UNIVERSAL METAPHYLACTIC AGENT?

T. Esen, M. Akinci and S. Tellaloglu

Department of Urology
Medical Faculty of Istanbul
Turkey

INTRODUCTION

Sodium potassium citrate is a proven metaphylactic agent in urolithiasis[1]. To establish its role as a universal agent we evaluated 250 consecutive stone patients.

MATERIALS AND METHODS

190 adults and children with urolithiasis underwent metabolic evaluation and specific metaphylaxis using Na-K-Citrate (Uralyt-U) in more than 75% of cases. The efficacy of the program was determined by Ca/Citrate ratio, Ca/Mg ratio and stone recurrence rate.

RESULTS

Hypocitraturia was the most common causal factor (45%) and accompanied other causal factors in 43-75% of the cases. Seven patients developed new stone activity within an average follow-up of 26 months. Recurrence rate decreased from 0.5 to 0.025, and Ca/citrate and Ca/Mg improved significantly (0.7 vs 0.5 and 2.19 vs 1.47 respectively).

DISCUSSION

Na-K-Citrate produces a good metaphylactic response in more than 75% of the stone population, including idiopathic urolithiasis. Although not tested in this study, it is possible that Na-K-citrate could influence the fate of residual fragments following ESWL in a beneficial fashion.

REFERENCE

1. M Butz, Rational prevention of calcium urolithiasis, *Urol Int* 41:387 (1986).

THE DAY-TO-DAY MANAGEMENT OF STONE PROPHYLAXIS USING THE URIMHO SELF-TEST DEVICE

G.A. Rose, R.R. Nauth-Misir, B.J. Landers, J.S. Virdi and J.E.A. Wickham

Institute of Urology
London

INTRODUCTION AND RESULTS

Although stone formers are advised to maintain a dilute urine at all times, it is difficult to know if their urine is sufficiently dilute. Urimho (Gyrus Medical Ltd) is a hand-held portable device enabling patients to self-monitor urinary conductivity.

The patient places the tip of the instrument into the urinary stream, removes it, and presses a button whereupon it displays a red, yellow or green level of conductivity. Yellow represents a conductivity of 14-17mS equivalent to an osmolality of 390-500 mosmol/Kg. Red indicates higher values, and green lower values. Since it is thought desirable for stone formers to maintain urine s.g. below 1.015, patients are advised to keep urine in the green zones.

Twenty two known recurrent stone formers, aged 20-72 years, had been advised (with one exception) to consume a high fluid intake and were prescribed the Urimho device. One patient was receiving bendrofluazide throughout, but otherwise no other medication was given. Each patient was requested to test himself using the device on each voiding and to record the results in the compliance record chart provided with the kit.

Week 1	Normal pattern of fluid consumption.
Weeks 2 & 3	Drink 200-250mL of water/hour by day and 600mL before retiring.
Week 4	Increase water intake by; - 200mL per hour if in red 2 zone, 100mL per hour if in red 1 zone, slightly if in yellow zone.

Readings in weeks 1 and 4 were then compared. Results from 972 and 998 voidings were recorded in weeks 1 and 4 respectively. Red tests fell from 452 (46.5% in week 1 to 261 (26.8%) in week 4, while green tests rose from 520 (53.5%) to 737 (73.9%). The most significant reductions in urine concentration occurred in the 6-9 am, 9-12 noon and 3-6 pm periods of week 4 ($p< 0.001$, < 0.0001 and < 0.0001 respectively). Eighty six percent of patients showed a fall in the number of red readings and 68% a rise in the number of green readings. Two patients showed poor compliance. They had had recurrent stones for 40 and 50 years respectively, and were over 70 years of age. Another was unable to drink a cup of water every hour. Excluding these patients, results were considerably improved. All but one patient reported no problems in use of the Urimho. The exception found some flickering between readings, which was, however, cured by replacing the battery.

The device is helpful in stone prophylaxis by enabling patients to comply with the advice to increase water intake, by keeping below the therapeutic threshold of urine concentration.

LONG TERM EFFECT OF CONSERVATIVE MANAGEMENT

Y.M. Fazil Marickar, S. Jayadevan, H.K. Moorthy,
S.V. Roshni and N.E. Thomas

Department of Surgery
Medical College Hospital
Trivandrum, India

INTRODUCTION

This study was undertaken to find out whether appropriate chemotherapy will correct biochemical abnormalities, reduce or alleviate crystalluria and help partial dissolution and passage of stones 6-11 mm in size. This was a prospective study involving a randomised trial for 19 years.

METHODS

Two hundred randomly selected proven stone patients, who did not require urgent surgery, PCNL, ESWL or ureterorenoscopy were included in the study. Their biochemical parameters, namely urinary calcium, uric acid, oxalate, creatinine and magnesium and serum uric acid were assessed. Appropriate chemotherapy was instituted: pyridoxine (40-120 mg/day) for hyperoxaluria, allopurinol (100-300 mg/ day) for hyperuricosuria/ hyperuricaemia, hexasodium hexapotassium citrate (7.5 g/day) for hypocitraturia/ hyperuricosuria, sodium bicarbonate (3-7.5 g/day) for renal tubular acidosis and a combination of drugs for mixed abnormalities. The patients were followed up for a period ranging from 3 to 18 years by clinical assessment, crystal studies, urine and blood biochemistry, radiography and ultrasonography. They acted as their own controls and the progress was compared to their response prior to the starting of chemotherapy.

RESULTS AND CONCLUSION

Of the 43 patients with caliceal/infundibular stones, 37 passed the whole stone or fragments. Three patients showed no response to conservative treatment and three had only partial response. Of the 107 patients with ureteric stones, 23 had stones in the upper one third of the ureter, 32 in the middle one third and 52 in the lower one third. Chemotherapy was successful in eliminating the stones in 87%. Fifty patients were recurrent stone formers, 96% of whom stopped forming stones. All the crystalluric patients had control of crystalluria after the treatment. Drug trial was abandoned in 3% of patients and appropriate intervention was carried out. Treatment failed in 10% of patients.

It is concluded that scientifically based conservative management will benefit the majority of stone patients. Regular follow up with ultrasound studies can identify back pressure so as to terminate medical treatment. A systematic protocol of therapy will prevent a large number of invasive interventions in urinary stone disease.

ROLE OF MUSAPITH IN UROLITHIASIS

R.K. Vathsala, H. K. Moorthy, C. Aravindakshan,
P. Koshy[1], P.L. Vijayammal[2] and Y.M. Fazil Marickar

Department of Surgery
Medical College Hospital
[1]RRL, Trivandrum
[2]Dept. of Biochemistry
University of Kerala
Trivandrum, India

INTRODUCTION AND METHODS

The aim of this paper was to determine the efficacy of musapith (plantain stem), a commonly used indigeneous plant extract, in the therapy of urolithiasis by native physicians in India.

Fifty urinary stone patients consumed 200 mL of different varieties of musapith juice daily for one month. Their early morning and random urine samples were collected before and after consumption of musapith, and studied microscopically for crystalluria and analysed for calcium oxalate and uric acid. The difference between the pre and post consumption urine values was statistically analysed using the Wilcoxon signed rank test. *In vitro* calcium oxalate crystal growth studies were also undertaken and the crystals grown were examined under the scanning electron microscope.

RESULTS AND DISCUSSION

The extent of crystalluria after consumption of musapith juice rose significantly ($p < 0.01$) and new crystal formation was evident in 20% of patients. Calcium oxalate monohydrate (COM) crystals increased more than calcium oxalate dihydrate (COD) crystals. The number of COM and COD crystals significantly increased after consumption of musapith juice ($p < 0.007$, 0.001). The mean levels of urinary oxalate and uric acid before and after consumption of musapith juice were 42.7 and 60 mg/day and 484.20 and 773.80 mg/day respectively ($p < 0.01$ and < 0.03). *In vitro* studies showed the crystal number and size to be significantly higher with urine samples of patients who consumed musapith and SEM studies showed more aggregation of crystals was also greater.

Musapith contains significant amount of oxalate. Other parts of the plantain tree also to contain high quantities of oxalate (420 and 480 mg% according to ICMR standards). Since high oxalate diets increase diuresis it is possible that plantain stem simultaneously causes diuresis and supersaturation of urine with respect to calcium oxalate, thereby causing marked calcium oxalate crystalluria. Musapith juice is not to be recommended as an anticalculogenic agent.

DICLOFENAC SODIUM IN ACUTE URETERIC COLIC

H.K. Moorthy, N.E. Thomas, S.V. Roshni,
S. Jayadevan and Y.M. Fazil Marickar

Department of Surgery
Medical College Hospital
Trivandrum, India

INTRODUCTION AND METHODS

Acute ureteric colic is thought to be produced by a calculus producing back pressure in kidneys and by increased production of prostaglandins. There is no ideal drug for acute ureteric colic as responses to various analgesic/ antispasmodic drugs vary between patients. This study was designed to assess the role of parenteral diclofenac sodium in the treatment of this painful condition.

Included in the study were 240 patients with acute ureteric colic. They were given intramuscular injections of diclofenac sodium (75 mg), hyoscine (20 mg), analginpitofenone combination - *Baralgan* (0.02 mg), pethidine (100 mg), avapyrozone (24 mg), pentazocine (30 mg), paracetamol (300 mg) or atropine (0.6 mg) alone or in combination after random selection of patients. The time of onset, duration and degree of relief were observed for the next 24 hrs. The efficiency of each drug was critically evaluated using analysis of variance for onset of relief, Chi-square test for comparing degree of relief of all drugs together and Wilcoxon rank sum test for comparing individual values.

RESULTS AND DISCUSSION

The time of onset and degree of relief produced by the different drugs were significantly different ($p < 0.05$). Diclofenac produced total relief of pain in 79% and moderate relief of pain in 21% of patients with a mean onset of action at 18 minutes, lasting for 10-12 hrs. Hyoscine produced relief in 50% of the patients with an average onset of action at 85 minutes and lasting for 2-4 hrs. *Baralgan* produced similar relief. Narcotic analgesics like pethidine produced moderate relief in 12 minutes lasting for 1 hr in 66% of patients. Avapyrazone, paracetamol and pentazocine and atropine also did not significantly reduce the colic. Atropine alone had no effect but when given along with other analgesic combinations produced late onset of relief lasting for 12-24 hrs. When the values of patients receiving diclofenac sodium were compared individually with the values of other drugs, significant statistical differences were observed. The mean onset of action of diclofenac sodium in dull pain was significantly faster (mean - 10 min) than in colicky pain (mean 24 min).

Thus parenteral diclofenac sodium was found to be the best drug for the treatment of acute ureteric colic with the earliest onset of action, maximum degree of relief, fewer side effects and longest action.

ORAL CITRATE THERAPY IN UROLITHIASIS

N.E. Thomas, S.V. Roshni, H.K. Moorthy,
S. Sindhu, R.K. Vathsala and Y.M. Fazil Marickar

Department of Surgery
Medical College Hospital
Trivandrum, India

INTRODUCTION AND METHODS

Hypocitraturia is a proven risk factor in recurrent urolithiasis. Citrate in urine chelates calcium ions and reduces the amount of calcium available for oxalate stone formation, thereby lessening the chances of stone formation. This study was conducted to determine the efficacy of different oral citrate drugs in altering urinary risk factors. The aim of the study was to identify a satisfactory citrate preparation for stone patients to increase the urinary citrate level.

Sixty patients with proven urinary stone disease and hypocitraturia (<150 mg% which is our laboratory standard) were included in the study. Fifty patients were given hexa sodium hexa potassium hydrogen citrate *(Uralyt - U)* in a dose of 2.5 g to 7.5 g per day adjusted to maintain urine pH between 6.2 to 6.7. Ten patients were given potassium citrate *(Cital - K)* in a dose of 3 - 4 g per day so as to maintain the urine pH above 6.5. The biochemical effect of the drugs was assessed by estimating the changes in the levels of 24 h urine calcium, oxalate, uric acid and citrate, and serum calcium and uric acid and alteration in the crystalluric status before and after administration of the therapeutic agents. Differences between the two groups were analysed using the paired Student's t test.

RESULTS

Of the patients, 95% who had taken either of the oral citrates showed an increase in the citrate levels. The mean urine citrate level before *Uralyt - U* was 112 mg/day and after treatment it was 256 mg/day, while the citrate level before *Cital - K* was 98 mg/day increasing to 211.4 mg/day. The mean urine uric acid level in patients before starting *Uralyt - U* was 798 mg/day which was reduced to 510 mg/day after treatment. The pre treatment uric acid level in *Cital - K* group was 465.9 mg/day rising to 591.2 mg/day after the drug ($p < 0.001$).

DISCUSSION

Oral citrate drugs are primarily thought to act by increasing the urinary pH. This increase in pH facilitates excretion of more citrates. We conclude that *Uralyt - U* has better efficacy and therapeutic value compared to *Cital - K* but is costly and not available in India. *Cital* K carries the risk of increasing the urine uric acid.

VEERATHARADI IN UROLITHIASIS

C. Aravindakshan, N.J. Bai[1], C.P.R. Nair[2], H.K. Moorthy and Y.M. Fazil Marickar

Departments of Surgery and
[1]Biochemistry
Medical College
[2]Dhanwanthari Matom
Trivandrum, India

INTRODUCTION AND METHODS

Indigenous Indian medicine, namely Ayurveda, uses medicines called *Kashayam* prepared from herbal plants found in almost all parts of India for treatment of human urinary stone disease. This study was undertaken to find out the efficacy of such a drug, *Veeratharadi Kashayam,* prepared according to the procedure given in Ashtangahridaya, the authoritative text of Ayurveda. The ingredients included 21 Ayurvedic plants found in the countryside of South India. The drug is also available commercially.

The inhibitory effect of*Veeratharadi Kashayam* in the growth of calcium oxalate crystals using the modifications of conventional silica gel medium was studied. Two concentrations of the *Kashayam* (1 mL and 5 mL) were tested and deionised water was added to the control. Thickness of crystal column and the number and the size of the crystals were studied serially for one month in the two groups and data were compared using Student 't' test.

RESULTS AND DISCUSSION

Crystal column thickness was greatest in the control and less in the test group. Both test groups showed inhibition of calcium oxalate crystal size compared to the control. The differences between these values were statistically significant at $p < 0.05$ and $p < 0.001$ respectively. When the *Kashayam* concentration was increased there was a further decrease in column thickness. The control group had a mean crystal size of 201.60 µm and the test group containing 5 mL kashayam showed a mean size of 122.67 µm ($p < 0.001$). The test group containing 1 mL of *Kashayam* produced crystals with a mean size of 148.67 µm (n.s.).

From the *in vitro* studies it is evident that *Veeratharadi Kashayam* inhibits calcium oxalate crystal growth and the inhibition is greater when the drug concentration is high. The *in vitro* crystal growth studies have proved beneficial in assessing inhibitory properties of different drugs used in different systems of medicine. Since the drug is a mixture of different plant extracts, further studies are needed to identify which of these ingredients are inhibiting calcium oxalate crystal growth.

THERAPEUTIC EFFICACY OF COMBINED SUPPLEMENTATION OF MAGNESIUM AND PYRIDOXINE IN CALCIUM OXALATE STONE FORMERS

V. Rattan[1], S.K. Thind[1], S. Vaidyanathan[2] and R. Nath[1]

[1]Department of Biochemistry and
[2]Department of Urology
Postgraduate Institute of Medical Education and Research
Chandigarh, India

INTRODUCTION AND METHODS

Hypomagnesuria is a common finding in stone formers[1]. An inhibitory effect of magnesium on the formation of calcium containing urolithiasis has been emphasized[2]. Pyridoxine supplementation decreases the urinary excretion of oxalate in idiopathic calcium oxalate stone formers[3].

Magnesium oxide (300 mg/day) and pyridoxine-HCl (10mg/day) were orally administered to 16 calcium oxalate stone patients. Various parameters of blood (Na, K, Mg, urea, creatinine, Ca, P, uric acid, AST, ALT and ALP) and urine (creatinine, Na, K, Mg, uric acid, Ca, P, oxalate and citrate) were measured at 0, 30, 60, 90 and 120 days after treatment.

RESULTS AND DISCUSSION

Serum magnesium increased ($p < 0.01$) from 0.82±0.11 to 0.92±0.07 mmol/L at 30 days and remained constant thereafter. All other parameters of blood remained constant. Combined treatment led to a significant increase in the excretion of Mg ($p < 0.01$ for 30 days; $p < 0.001$ for 60, 90 and 120 days) and citrate ($p < 0.05$ for 30 days, $p < 0.001$ for 60, 90 and 120 days) compared to the pretreatment values. A significant decrease in the urinary excretion of oxalate ($p < 0.001$ for 60, 90 and 120 days) was also observed. All other parameters of urine remained unaltered. Thus the combined supplementation may be more beneficial to stone patients than pyridoxine administration alone.

REFERENCES

1. D Wangoo, V Rattan, SK Thind and R Nath, *Magnesium Res.* 4:211 (1991).
2. K Kohri, J Garside and NJ Blacklock, *Brit J Urol* 61:107 (1988).
3. MSR Murthy, S Farooqui, HS Talwar, SK Thind and R Nath, *Int J Clin Pharmac Ther Toxicol* 20:434 (1982).

THERAPEUTIC ROLE OF DICLOFENAC-SODIUM IN MANAGEMENT OF HYPOCITRATURIA AND HYPOPYROPHOSPHATURIA IN IDIOPATHIC STONE FORMERS

S. Sharma[1], S. Vaidyanathan[2], S.K. Thind[1] and R. Nath[1]

[1]Department of Biochemistry and
[2]Department of Urology
Postgraduate Institute of Medical Education and Research
Chandigarh, India

INTRODUCTION AND METHODS

Diclofenac-sodium, a non-steroidal anti-inflammatory drug presents a new strategy in the treatment of urolithiasis by virtue of its effects on calcium excretion[1]. However, its effect on inhibitory potential of urine in stone formers has not been studied. In the present study we have clinically evaluated the effect of diclofenac sodium upon 2 urinary inhibitors of calcium oxalate crystallization.

Twenty four h urine was collected for baseline estimation of calcium, citrate and pyrophosphate. Diclofenac-sodium was then administered in a dose of 50 mg tds for seven days.

RESULTS AND DISCUSSION

No adverse effects were observed from this therapy. Twenty four h urinary citrate excretion increased from 1.25±0.05 mmol to 1.40±0.05 mmol ($p< 0.05$) post-treatment. Similarly pyrophosphate excretion increased from 54.32±21.4 µmol to 78.31±28.03 µmol subsequent to diclofenac-sodium therapy ($p< 0.01$). The mean excretion of calcium remained unchanged in 17 of 20 patients who manifested normocalciuria before treatment. However, in 3 patients exhibiting hypercalciuria (> 300mg/24h), urinary excretion of calcium decreased after the treatment. These results demonstrate that diclofenac-sodium therapy in the management of renal stone disease may be beneficial.

REFERENCE

1. C Collette, L Aquiore, L Mommier, and A Mimrau, *Renal Physiol.* 6:68 (1982).

PROPHYLACTIC TREATMENT OF RECURRENT CALCIUM STONES WITH POTASSIUM SODIUM CITRATE

A. Trinchieri, F. Rovera, R. Nespoli, G. Zanetti and E. Austoni

Istituto di Urologia
Università degli Studi di Milano
Milan, Italy

INTRODUCTION AND METHODS

Recently, several workers using treatment with alkaline citrate, either as potassium citrate or as potassium sodium citrate, have reported the control of stone formation in the majority of patients[1].

We selected 38 patients with idiopathic calcium stones for treatment with potassium sodium citrate (5-10 g/daily). Before treatment, venous blood and 24 h urine samples were obtained for determination of pH, potassium, sodium, calcium, phosphate, urate, oxalate (urine), citrate (urine) and creatinine. The investigations at follow up visits included urological examination, measurement of blood and urine levels of electrolytes, urine culture and X-ray examination.

RESULTS AND DISCUSSION

After two years of follow up 19 patients dropped out (50%). A total of 19 patients were treated for 24 months or more. Urinary pH, potassium, sodium and citrate significantly increased. A recurrence rate of 47% at 2 year follow up was observed.

The lack of stone prophylaxis, despite the increase of citrate output, was probably due to the inability of potassium sodium citrate to reduce calcium excretion, which can be explained by a reduced tubular reabsorption of calcium owing to the sodium load. The reduced clinical efficacy might also be explained by the fact that low amounts of potassium sodium citrate were administered to patients who experienced gastrointestinal side effects at full dosage.

REFERENCE

1. CYC Pak, C Fuller, K Sakhaee, GM Preminger and F Britton, Long-term treatment of calcium nephrolithiasis with potassium citrate, *J Urol* 134:11 (1985).

DIETARY MODERATION AS A MEANS OF IMPROVING URINARY RISK FACTORS OF RECURRENT CALCIUM OXALATE STONE FORMATION

R.W. Norman and J.N. Hughes

Dalhousie University
Halifax
Nova Scotia, Canada

INTRODUCTION

The impact of diet and fluid intake on recurrent calcium oxalate stone disease has long been recognized but is not often considered as the primary intervention. This study was designed to evaluate the effectiveness of nutritional intervention measured in terms of dietary compliance, and a reduction in urinary risk factors over time.

METHODS

Fifty-nine hypercalciuric stone formers with recurrent calcium oxalate stone disease were instructed on diets designed to limit excess intakes of calcium (400-600 mg/d), oxalate (40-60 mg/d) and animal protein (30-50 g/d) and to increase fluid intake to 3L per day with half of that intake as water. Four day food records and 2 corresponding 24 h urine collections were obtained prior to the first clinic visit and 3 months after dietary instruction.

RESULTS

Decreases were confirmed in the dietary calcium from 847±46 (Mean±SEM) to 672±41 mg/d ($p < 0.0005$), oxalate from 118±31 to 70±33 mg/d (n.s.) and animal protein from 52±2.4 to 45±2.8 g/d ($p < 0.02$); there was an increase in fluid intake from 1531±100 to 2064±25 ml/d ($p < 0.0001$).

This dietary moderation resulted in decreases in the 24 h urinary excretions of calcium from 9.7±0.3 to 7.7±0.4 mmol/d ($p < 0.0001$), oxalate from 0.51±0.02 to 0.41±0.02 mmol/d, ($p < 0.003$) and uric acid from 6.0±0.2 to 5.8±0.3 mmol/d (n.s.). Urine volumes increased from 1483±53 to 1865±120 mL/d ($p < 0.002$).

CONCLUSION

In conclusion, moderate dietary intervention was effective in improving major risk factor abnormalities and should reduce the risk of stone formation.

INFLUENCE OF L-METHIONINE ON THE RISK OF STRUVITE AND BRUSHITE STONE FORMATION

A. Hesse[1], F. Struwe[1], D. Bach[2], K. Klocke[1] and H. Heynck[1]

[1]Division of Experimental Urology
Department of Urology
University of Bonn
[2]St. Agnes Hospital
Bocholt, Germany

INTRODUCTION AND METHODS

Physiological urine acidification with L-methionine (L-M) is a successful accompanying therapy in urinary tract infection and phosphate stone formation. In cases of acute infection, residual stones or methaphylaxis it is of great importance to maintain sufficient acidification throughout 24 h. The aim of the present study was to investigate pharmacokinetics and pharmacodynamics of L-M and the beneficial effect of urine acidification on the risk of struvite and brushite stone formation.

A daily progress chart of serum and urine composition was drawn up for 12 healthy male individuals under standardized conditions with and without (control day) one oral administration of 1500 mg (10.15 mmol) L-M. L-M and sulfate in serum and urine were determined by HPLC. Relative supersaturation (RS) of struvite, brushite and calcium oxalate was estimated by the EQUIL program.

RESULTS AND DISCUSSION

Serum L-M increased 1.5 h after ingestion and dropped continuously to reach the initial concentration after 24 h. Serum sulfate concentration also increased after 1.5 h and remained above the control value for a period of 22.5 h. At the same time the urinary pH-value decreased and remained at a significantly lower level for 8h, while sulfate excretion increased markedly 2 h after L-M intake and remained significantly above the control day value for 22 h. The 24 h urinary excretion of sulfate increased by 10.2 mmol, while the increase of L-M amounted to 8.2 µmol (= 0.8% of L-M administration). RS values of struvite were low during 24 h compared to the initial values and decreased in 24 h urine by 30%, and 25% respectively. RS of calcium oxalate did not alter.

Urine acidification with L-M is physiological and efficient. L-M is absorbed rapidly and metabolized after an oral dose. As sulfate excretion in 24 h urine correlates with L-M intake, complete metabolism of L-M can be assumed. To maintain acidity, L-M should be administered 4-6 hourly during the day and 8-10 hourly at night. In addition to acidification, further measures (eg, decrease in urinary Ca, PO4) are necessary for therapy of brushite stones.

HEMATURIA DUE TO HYPERURICOSURIA (HU): 30 MONTHS FOLLOW-UP

H.C. Perrone[1], J. Toporovski[1] and N. Schor[2]

[1]Nephrology Division from Pediatric Department
Santa Casa (SP) and
[2]Department of Medicina Escola Paulista de Medicina
São Paulo, Brazil

INTRODUCTION AND RESULTS

HU occurs in up to 32% of patients with CaOx stones, but has not been fully investigated as a possible cause of hematuria in pediatric patients[1,2]. We prospectively studied 30 children with HU without idiopathic hypercalciuria (IH) at the beginning of the evaluation. The follow-up was from 6 to 30 months. Twenty four h urine collections were obtained at the beginning of the study and repeated each 6 months. Ultrasound evaluation was performed annually. Treatment consisted of dietetic measures (increased water intake and low purine) and/or allopurinol (50-100 mg/day) and/or potassium citrate (0.5 mEq/kg/day), depending on the urinary uric acid levels and persistence of hematuria.

Months (n)	0 (30)	6 (25)	12 (20)	18 (16)	24 (12)	30 (7)
Clinical presentation (%)						
Macrohematuria	86	9	11	9	0	0
Microhematuria	14	17	6	9	0	0
HU	100	68	45	25	8	0
IH	0	17	22	18	17	0
Nephrolithiasis	0	13	6	0	0	0

In addition to IH, HU seems to be a cause of hematuria in children[2]. Stones occurred after 6 and 12 months (13% and 6% respectively) in the untreated children, indicating that treatment decreases the risk for urolithiasis in children with HU. Moreover at least 2 24 h urinary Ca and uric acid measurements should be performed in every patient with hematuria in order to prevent unnecessary and invasive diagnostic investigation. Results also show that treatment prevents stone events and decreases hematuria recurrence.

REFERENCES

1. MS Bayle, CR Mancheno, Hyperuricosuria and microhematuria in childhood, *Am J Dis Child* 143: 878 (1989).
2. HC Perrone, J Toporovski, H Ajzen, N Schor, Metabolic disturbance as a cause of hematuria in childhood, *Kidney Int* 39:707 (1991).

UNILATERAL HYDRONEPHROSIS: DMSA AND DTPA SCAN IN RENAL STONE FORMERS

L. Gandolpho, I.P. Heilberg, M.C. Monteiro and N. Schor

Nephrology Division
Escola Paulista de Medicina
São Paulo, Brazil

INTRODUCTION AND RESULTS

Hydronephrosis, a consequence of renal stone disease, may lead to obstruction. Eventual loss of renal function may be related to the duration of hydronephrosis, associated urinary tract infection (UTI), previous surgery and/or calculi size. The present study was undertaken to evaluate renal function[1] and morphology by scintigraphic scan.

Eleven patients (8 males, 3 females) with unilateral hydronephrosis detected by intravenous urogram and/or renal ultrasound were submitted to both Dynamic Renal Scintigraphy ^{99m}Tc-DTPA (diethylene triamine pentacetic acid) and Static Renal Scintigraphy ^{99m}Tc-DMSA (dimethyl succinic acid).

The average duration of hydronephrosis was between 2 - 48 months and calculi size between 1 to 2.5 cm.

DTPA: 5 patients showed functional deficit (3 with cortical damage and 4 with organic obstructive pattern), 6 were normal (4 with normal cortical function and 4 with functional obstructive pattern).

DMSA: 9 patients showed renal scar (4 with unilateral functional loss and 5 without functional loss) and 2 were normal.

CONCLUSIONS

Preliminary data suggest that there was no relation between duration of unilateral hydronephrosis, calculus size, association with recurrent UTI and previous surgery with the functional alteration pattern detected by either DMSA or DTPA. The presence of cortical damage and organic obstruction may be indicative of poor prognosis for renal function. These radioisotope studies provide a better indication of urgency and most appropriate surgical procedure for each case[2].

REFERENCES

1. AC Buck, MA MacLeod and NJ Blacklock. The advantages of ^{99M}Tc DTPA (Sn) in dynamic renal scintigraphy and measurement of renal function, *Brit J Urol* 52:174 (1980).
2. HN Whitfield, KE Britton, I Kelsey et al, The obstructed kidney: correlation between renal function and urodynamic assesment, *Brit J Urol* 49:615 (1977).

THE VALUE OF REPEATED ANALYSES OF 24 HOUR URINE IN RECURRENT CALCIUM UROLITHIASIS

K. Höbarth, J. Hofbauer, N. Szabo and M. Marberger

Department of Urology
University of Vienna Medical School
Vienna, Austria

INTRODUCTION AND RESULTS

Samples (n=441) of 24 h urine of 49 recurrent calcium stone formers (with a mean of 6.3 episodes of stone formation or passing at the time of first examination) with a mean follow-up period of 80.4 months were retrospectively analyed. Patients with hyperparathyroidism, renal tubular acidosis or hyperuricaemia were not included in the study. Evaluating the first two samples (with or without calcium load), which were taken for classifying metabolic disorders, type I or II absorptive hypercalciuria (HCU) or hyperuricosuria (HUCU) was detected in 29 patients; they received a specific drug therapy (thiazide or allopurinol) for a mean period of 45.7 months. Twenty patients without initial metabolic disorders were given general metaphylactic recommendations (high fluid intake, no excessive calcium, oxalate, protein and purine intake).

During follow-up when various 24 h urine collections were measured (in an outpatient setting while on the usual fluid and dietary intake), HCU or HUCU was found in 57% of the therapy group and in 40% of all patients with initially normal urine findings. Seventy-three % of all patients were metabolically active (mean 7.7 recurrences). In the therapy group, 62% formed stones during therapy (mean 9.5 recurrences). After completion of therapy, a further 65% developed recurrences. In the general metaphylaxis group, the same percentage (65%) of patients was found to be metabolically active. Of all patients, 27% were metabolically inactive, 50% exhibiting HCU or HUCU in the various 24 h follow-up urine collections.

DISCUSSION

The diagnostic and therapeutic value of repeated 24 h urine analyses in the long-term follow-up of recurrent calcium stone formers is limited. Initial serum and urine examinations are indicated to exclude hyperparathyroidism or renal tubular acidosis. Patients with absorptive HCU or hyperuricosuric calcium urolithiasis apparently do not benefit from specific drug therapy, and analysis of various 24 h urine collections fail to provide additional information about the risk of stone recurrences, since in 75% of all cases both normal and abnormal 24 h urine findings were seen in the same patient at different times.

THE COURSE OF UROLITHIASIS OVER FIFTEEN YEARS IN CALCIUM OXALATE STONE FORMERS UNDER SPECIAL, GENERAL OR NO METAPHYLAXIS

J. Hofbauer, K. Höbarth and M. Marberger

Department of Urology
University of Vienna Medical School
Vienna, Austria

INTRODUCTION AND METHODS

The course of urolithiasis was retrospectively analysed in 66 recurrent calcium oxalate stone formers with a mean rate of stone formation of 16.5±14 stones over a period of 15 years. Due to the fact that all patients were recurrent stone formers, general metaphylaxis was supplemented from the outset by special metaphylaxis, which depended on the respective metabolic dysfunction, ie, thiazide in cases of hypercalciuria and allopurinol in hyperuricosuria. Special metaphylaxis lasted over a mean period of 5.5±2.1 years. Subsequently, 32 patients received general metaphylaxis alone and were followed up at regular intervals over a mean of 5.7±2.6 years. Of this group, 22 were classified as metabolically active stone formers. Thirty four patients were lost to follow-up due to non-compliance; however, 20 could be re-examined after a mean of 6.3±2.2 years. During this time, they had not observed metaphylaxis.

RESULTS

The rate of recurrent stone formation among the 11 metabolically active patients amounted to 75% both under special and general metaphylaxis. By contrast, the same rate was 54% among the metabolically inactive stone formers under special and general metaphylaxis, 27% during the subsequent period of general metaphylaxis and regular follow-ups, and 21% in the group of patients who observed neither follow-up nor metaphylaxis. The phase of regular stone formation amounted to 12.5±5.8 years.

DISCUSSION

The results of this retrospective analysis revealed a spontaneous decrease in recurrent stone formation over the course of disease. It appears that during the phase of regular stone formation, which we postulate is an autonomous, so far unexplained process, response can be expected neither to special nor to general metaphylaxis. The decrease in recurrences can be explained by spontaneous termination of the phase of regular stone formation in a number of patients.

IS THERE A ROLE FOR THE ORAL CALCIUM TOLERANCE TEST IN A ROUTINE STONE CLINIC?

R.C. Bowyer[1], P.J. England[2] and T. A. Taylor[2]

Departments of [1]Biochemistry and
[2]Urology
Royal Perth Hospital
Perth, Western Australia

INTRODUCTION

The calcium tolerance test described by Pak and coworkers[1] appeared to us to offer a relatively simple means of classifying recurrent calcium stone formers according to their type of hypercalciuria, thus leading to the earlier introduction of the most appropriate treatment modality.

RESULTS

Forty-four recurrent calcium stone formers agreed to undergo testing, along with 39 age- and sex-matched controls. The concept of a "net response" to the load was introduced because 3 h excretion values were higher in some patients than the 2 h values. The net response[1] in patients' calcium varied from -12.4 to 72.4 µmol/L glomerular filtrate with a reference range of 0.4 to 43.2. There was a significant difference ($p < 0.05$) between the mean value of patients and that of controls. The fasting (baseline) excretion of calcium in patients varied from 2.1 to 65.9 µmol/L glomerular filtrate with a reference range of 3.0 to 32.5.

The classification of patients into "renal leakers" and "hyperabsorbers" follows from the pattern of calcium parameters in individuals. No patient exhibited the renal leaker pattern with high fasting excretion and high net response. Only three out of 44 patients had results consistent with the hyperabsorber classification with normal fasting excretion and elevated net response. Both of these indices were within the reference range in 31 out of 44 cases.

CONCLUSION

It is concluded that the oral calcium tolerance is not useful in classifying recurrent calcium stone formers.

REFERENCE

1. CYC Pak, H Bone, J Townsend and O Waters, A simple test for the diagnosis of absorptive, resorptive and renal hypercalciurias, *N Engl J Med* 292: 497 (1975).

AMBULATORY METABOLIC EVALUATION OF 1515 STONE FORMING PATIENTS IN ARGENTINA

E. del Valle, C.E. Bogado, F.R. Spivacow and J.R. Zanchetta

Instituto de Investigaciones Metabolicas
Buenos Aires, Argentina

INTRODUCTION

Stone forming patients were placed on a controlled diet with 1200 mg Ca, 800 mg P, 200 mg Mg and 100 mmol Na per day during 7 days. Two 24 h urine samples were collected on the sixth and seventh days. On the eighth day, following a 12 h fast, a blood sample was taken, 300 mL of distilled H_2O were drunk and a 2 h urine sample was obtained. In all blood and urine samples, calcium, sodium, magnesium, phosphate, creatinine and urate were quantified. In 24 h samples oxalate and citrate were also determined. Hypercalciuria was defined by daily calcium excretion above 300 mg in men and 220 mg in women. Patients with a fasting calcium to creatinine ratio or calcium per 100 mL GFR value above 0.11 were considered to have renal hypercalciuria. Hyperuricosuria was diagnosed as a 24 h urinary uric acid above 750 mg in women and 800 mg in men, hyperuricemia as a serum uric acid higher than 7 mg% and hypomagnesuria as a daily magnesium excretion below 70 mg. A 24 h citrate excretion less than 350 mg was considered hypocitraturia and an oxalate excretion over 50 mg, hyperoxaluria.

RESULTS

Main findings expressed as percent of patients were: hypercalciuria 32% (renal 13.6%, absorptive 18.4%); uric acid metabolism disturbances, 16.5%; both hypercalciuria and uric acid disturbances, 9.5%; low magnesium excretion, 14.7%; hypocitraturia alone or associated with other abnormalities, 16.8% (as a sole finding 7.3%); primary hyperparathyroidism, 2.5%; hyperoxaluria, 9.3%; cystinuria, 0.5%. We could not find evidence of metabolic disturbance in 6.2% of the patients.

CONCLUSION

This protocol can be performed in an ambulatory setting and allows a diagnosis in 93.8% of the patients.

THE WEEKEND EFFECT

A.L. Rodgers, L.J. Barbour, B.M. Pougnet, C. Lombard and R.L. Ryall[1]

Chemistry Department
University of Cape Town
South Africa and
[1]Department of Surgery
Flinders Medical Centre
Bedford Park, South Australia

INTRODUCTION AND METHODS

Since the presence of any element in a urinary calculus suggests that it may have fulfilled some pathogenic role, the present study was undertaken to establish whether differences exist between stone formers (SF) and normals (N) with respect to their urinary excretion of such elements.

The concentrations of 10 different elements were measured in the early morning urines of 19 male SF and 20 male N. The procedure was repeated on three consecutive days (Monday, Tuesday, Wednesday). Concentrations were determined by inactively coupled plasma atomic emission spectroscopy and atomic absorption spectroscopy.

RESULTS

The results showed that the concentrations of several elements in the urine of N were significantly higher than those in SF, but only on Mondays. This observation was termed the "weekend-effect".

DISCUSSION

It is suggested that N may indulge in dietary excesses over the weekend while SF's, as a result of dietary instruction, may regulate their diets throughout the week.

ACKNOWLEDGEMENTS

We wish to thank the South African FRD, the MRC and the University of Cape Town.

ULTRA SOUND STUDY VERSUS IV UROGRAM IN URETERIC COLIC

H.K. Moorthy, N.E. Thomas, S.V. Roshni and Y.M. Fazil Marickar

Department of Surgery
Medical College Hospital
Trivandrum, Indida

INTRODUCTION AND METHODS

Several unwarranted investigations are usually performed in patients with acute ureteric colic owing to suspected stones in the urinary tract. The present study was undertaken to identify the efficacy of ultrasonographic scanning (USS) as an alternative to emergency intravenous urogram (IVU) in such patients. 150 patients attending the stone clinic with acute ureteric colic formed the test group, while 50 other patients with acute ureteric colic formed the control group. In the test group, all patients had a plain X-ray of the kidney, ureters and bladder (KUB) and USS done. In the control group, an IVU was performed. The advantages of using two approaches in deciding treatment were evaluated.

RESULTS AND DISCUSSION

KUB taken in the emergency room showed radio opaque stones in 32% of patients in both groups. This included 13% of stones in the pelvi-caliceal system, 7% in the upper one third of ureter and the remaining 12% in the lower third of ureter. In the KUB, a back pressure effect, as evidenced by the presence of significant renomegaly, was seen in 37% of the patients in both groups. USS done in the test group showed pelvicalyceal dilatation of mild to moderate degrees in 78% of the patients, but could identify the sono-dense shadows only in the renal and upper ureteric zones. It was difficult to visualize stones less than 0.5 cm size. An IVU performed at the time of acute ureteric colic failed to show the nephrogram due to temporary renal shut down in the majority of cases. In the rest of the cases, IVU showed pelvi-caliceal system dilatation with or without dilatation of the upper ureters, similar to the observations made by the USS. In functioning kidneys, corticomedullary distinction was possible both by USS and IVU.

Thus, USS coupled with a proper KUB was found to be an alternative for IVU in patients with acute ureteric colic. The advantages of USS over IVU are (1) its better cost-utility ratio (2) reduced radiation hazard and (3) its usefulness in follow up of patients on conservative management by repeating the scan at frequent intervals. The problem of non-functioning kidneys due to acute renal shut down was not encountered on using USS. The advantages claimed for IVU are that small calculi producing obstruction can be identified by the block in the ureter without any back pressure in the kidneys, and its ability to identify anatomical abnormalities of the urinary system. But this is needed only prior to intervention in patients with large stones.

LOCAL CHEMOLYSIS OF OBSTRUCTIVE URIC ACID STONE BY 0.1M THAM AND 0.02% CHLORHEXIDINE

Y.H. Lee, L.S. Chang, M.T. Chen and J.K. Huang

Division of Urology
Department of Surgery
Veterans General Hospital
Kaohsiung and Taipei
Taiwan, ROC

INTRODUCTION

Uric acid stones can be dissolved by an alkaline solution, usually 0.1M sodium bicarbonate or 0.1M tris (hydroxymethyl) aminomethane (THAM). However, the side effects of urinary tract infection and pyrexia are infrequently encountered. We tried a new strategy combining 0.1M THAM and 0.02% chlorhexidine for chemolytic purpose to minimize complication rate.

METHODS

From January 1990 to December 1991, 10 patients (5 males and 5 females) with obstructive uric acid stones were enrolled in this study. The mean age was 62 years (range 50 - 75). A total of 23 sessions of local chemolysis were done; 13 sessions were adjuvant chemolysis after extracorporeal shock wave lithotripsy and 10 sessions were primary treatment. All chemolyses were performed by 0.1M THAM and 0.02% chlorhexidine via a pigtail catheter through a percutaneous nephrostomy. The infusion rate was kept around 50 mL/hr by an infusion pump and the intrapelvic pressure was maintained less than 25 cm water pressure.

RESULTS

All uric acid stones were dissolved effectively by 0.1M THAM and 0.02% chlorhexidine. The complication rate was relatively low. Pyrexia developed only twice during the 23 sessions of local chemolysis (8.7%). Intravenous pyelography and comprehensive renal function test (^{131}I-orthoiodohippurate test) all revealed significant improvement at 3 months follow up.

CONCLUSION

A chemolytic solution of 0.1M THAM and 0.02% chlorhexidine is an effective and safe agent to dissolve obstructive uric acid stones and could be used routinely for local chemolysis.

COMPLEX THERAPY OF CYSTINE UROLITHIASIS IN 1992

D. Frang, M. Berényi and A. Hamvas

Semmelweis University of Medicine
Budapest, Hungary

INTRODUCTION AND RESULTS

In spite of potent medicaments and sophisticated instruments, and methods such as extracorporeal shock wave lithotripsy (ESWL), percutaneous nephrolithotomy (PCNL) and ureterorenoscopy (URS), the treatment of extraordinarily hard cystine stones has remained one of the greatest problems of urolithiasis.

Within the past 16 years 61 patients with cystine stones visited our clinic. Meanwhile 5 patients moved to other hospitals and 13 vanished from our registration. We are now reporting on 43 patients. Of these, 36 were treated by open surgery (mainly before 1987) involving 52 operations, including 9 nephrectomies (mostly before 1985), 17 PCNL (after 1987) and 4 ESWL (after 1988). Five patients died. Of the 43 patients, 78 surgical interventions were done for stones (1-7 operations per person, average=1.8 operations/patient). Three persons had cystine bladder calculi which were transurethrally removed. According to our data, cystine stone formation, even now, should be considered a malignant disease.

To date our patients can be divided into the following groups according to the severity of disease:

1. Those not requiring oral treatment because they had been stone-free for years.
2. Those who as a result of oral treatment are free of stones or very rarely pass a small stone.
3. Those who, in spite of oral treatment, formed stones (<1,5 cm in diameter) which were removed by ESWL or PCNL monotherapy, or by their combination.
4. Untreated patients, usually with bilateral large renal stones, who were relieved of stones by a combination of ESWL, PCNL, URS or open surgery.

IN VITRO MAGNETIC RESONANCE SCANNING OF URINARY CALCULI

N.P. Cohen, H.N. Whitfield, G. Dolke, Y.Y. Ng and P. Armstrong

Departments of Urology and Radiology
St Bartholomew's Hospital
London, ECIA 7BE

INTRODUCTION

Clinical experience with ESWL has shown that the size and composition of urinary calculi affect success. Knowing the composition of a stone before treatment is therefore desirable. We have examined the role of a clinical Nuclear Magnetic Resonance (NMR) body scanner to discriminate stone groups.

MATERIALS AND METHODS

23 intact urinary stones were rehydrated in water under gentle vacuum for 48 hours. They were placed in a phantom consisting of water-filled plastic specimen jars. Imaging was conducted with a 0.08 Tesla system by Mallard and Davis. T1 weighted sequences were performed with a repetition time of 400 msec, echo 40 msec, and T2 weighted sequences with a repetition time of 1500 msec and 40/80 msec echo. The images were analysed and graded according to grey scale; (0 = white, 3 = black). A similar scale was used to estimate water content. Composition was determined by standard infrared spectroscopy.

RESULTS

Only stones greater than 8 mm could be graded accurately. Grading did differ for the two largest stone groups studied. 5 oxalate stones were light grey on T1- and black on T2-weighted imaging. All 5 struvite stones large enough to grade appeared dark grey on T1- and black on T2-weighted images. Other stones (sodium urate, mixed) had a similar appearance to struvite. The presence of water in some stones could be seen to fill spaces between laminations and in large porous spaces.

CONCLUSION

Our results suggest that *in vitro* NMR scanning is able to differentiate urinary calculi made of oxalate from stones of other chemical compositions. This is of clinical interest as oxalate stones are often difficult to fragment with either ESWL or endoscopic lithotripsy.

NEW MEDICAL AND UROLOGICAL MANAGEMENT OF MEDULLARY SPONGE KIDNEY DISEASE

E. Cicerello, F. Merlo, P. Checchin, L. Maccatrozzo and G. Anselmo

Kidney Stone Center
Division of Urology
Regional Hospital
Treviso, Italy

INTRODUCTION AND METHODS

Medullary Sponge Kidney Disease (MSKD) is a congenital anomaly of the renal medulla characterized by multiple small cysts often containing stones[1]. In the past, urological therapy was performed only in cases of complicating calculosis, and medical prophylaxis was considered only in the presence of metabolic anomalies or urinary infections. The present study was carried out to evaluate the effectiveness of ESWL with regard to stone clearance, and the results of medical therapy in the prevention of calculosis complicating this disease.

Eleven MSKD patients with symptomatic renal stones (in 9 patients bilateral nephrolithiasis) underwent metabolic screening. Two patients had hypercalciuria, 1 patient hyperuricosuria, 3 patients urinary infection and 1 patient type I renal tubular acidosis evaluated by the ammonium chloride loading test. Seven patients (11 treated kidneys) with stones of more of 3 mm were subjected to ESWL and obtained satisfactory stone disintegration. All patients were begun on sodium-potassium citrate (2-3 g 3 times day) and subjected to radiological and biochemical examinations every 6 months. A chemotherapeutic evening prophylaxis with cinoxacine (300 mg/day) was added to those patients who had a urinary infection.

RESULTS AND CONCLUSIONS

At 24 months follow-up 8 patients (14 kidneys) were totally stone free; 3 patients (6 kidneys) showed a reduction of nephrocalcinosis. The stone-free rate in the ESWL treated group was 71% (5 of 7 patients).

In MSKD patients our study shows that: a) ESWL should be considered an effective treatment for stones of more than 3 mm; b) sodium potassium citrate is effective for the dissolution of small stones and for preventing recurrences in the absence of metabolic anomalies.

REFERENCE

1. LW Welling and JJ Grantham, in: "The Kidney", edited by Brenner & Rector, Vol II: 1676 (1991).

MANAGEMENT OF ENCRUSTATION OF INDWELLING URETERAL STENTS

A. Al-Dayel, M. Ezzibdeh, S. Egail and I. Al-Oraifi

King Fahd Military Medical Complex
Dhahran, Saudi Arabia

INTRODUCTION

Indwelling internal stents often become encrusted with insoluble urinary salts. The crystal deposits lead to blockage of the internal lumen and enlargement of the external diameter. The drainage function of the stent is greatly impaired and injury to ureteral mucosa can occur when the stent is removed.

METHODS AND RESULTS

Seven patients with Double-J ureteral stent developed extensive encrustation, five at the lower end and two at both ends. All patients complained of recurrent attacks of haematuria, although none had a urinary tract infection, and five patients had their stents for more than twelve weeks. Encrustation was detected on plain X-ray films in 4 patients, and by renal tomogram intravenous urogram in 3 patients. The upper J encrustation disintegrated with extracorporeal shock wave lithotripsy. The lower end encrustation was treated with cystoscopic electrohydraulic lithotripsy. The main constituents of the collected encrusted material were similar to the major components of the stones, two urate, one cystine and four calcium oxalate.

DISCUSSION

We believe that extracorporeal shock wave lithotripsy is an effective method for the treatment of encrustation of the upper J of internal ureteral stents and recommend regular follow up and early removal of indwelling ureteral stents.

STAGHORN CALCULI (SC): LONG TERM FOLLOW-UP

O.V.B. Andrade, S.T.S. Coelho, V.T.C. Teixeira and N. Schor

Nephrology Division
Department of Medicina
Escola Paulista de Medicina
São Paulo, Brazil

INTRODUCTION AND RESULTS

Staghorn calculi are associated with a high incidence of morbidity. Staghorn stones account for between 5 and 37% of all renal tract calculi[1]. The relationship with urinary tract infection (UTI) is well known[2,3].

We evaluated retrospectively 32 outpatients (PTS) with SC from 1982 to 1991 with a mean follow-up of 14±4 months (mean±SE). There were 15 male and 17 female patients, (2-86 years old) with a mean age of 40±6 and 44±3, retrospectively. We analysed data regarding history, previous surgical manipulations, screening for metabolic disturbance, laboratory data and other associated diseases.

The mean duration of symptoms was 11±2 years. A family history was noted in 44%, and 47% had passed at least one stone. UTI was observed in 60% and, in 32% of the PTS it appeared to be the sole aetiological factor for stone formation. However, in 36% of the PTS other metabolic factors were probably the cause of the renal stones rather than a UTI. Only 9% of patients failed to have any identifiable metabolic disturbance. A high urinary pH (pH 6.0) was observed in 55% and a reduced creatinine clearance was noted in 25%. *Escherichia coli* was the principal agent causing UTI. 50% underwent surgery and in 53% systemic hypertension was noted.

DISCUSSION

Our data indicated that there was a long time interval between the onset of symptoms and the diagnosis of SC, and a high family incidence of stone disease and hypertension. In only 32% of cases did a UTI appear to be the principal aetiological factor and *Eschericha coli* was the most frequently isolated organism. Thus, in PTS with SC it is necessary to look for a metabolic disturbance in order to establish a rational treatment program that can potentially modify the natural evolution of this severe disease.

REFERENCES

1. MI Resnick, Evaluation and management of infection stones, *Urol Clin North Am* 8:265 (1981).
2. MJ Gleeson, DP Griffith, Infection Stones, *In:* "Urolithiasis. A Medical and Surgical Reference", ed MI Resnick and CYC Pak, WB Saunders Co, 113 (1990).
3. DP Griffith, Struvite Stones, *Kidney Int* 13:372 (1978).

URINARY EVALUATION IN BRAZILIAN RENAL STONE FORMER PATIENTS

N.A. Cunha and N. Schor

Nephrology Division
Escola Paulista de Medicina
São Paulo, Brazil

INTRODUCTION AND RESULTS

Many factors participate in stone formation, and metabolic alterations are the most frequent causes in nephrolithiasis[1]. Promoters and inhibitors[2] must be determined in order to select suitable treatment[3]. The aim of this report is to investigate some constituents implicated in urolithiasis in Brazilian patients compared to normal subjects. We studied 24 h urine samples from normal subjects (98 males, 78 females) and renal stone formers (SF, 98 males, 78 females). Urinary creatinine was determined using the method of Jaffe, and Ca and Mg were measured by atomic absorption. An automatic enzymatic method was used for citrate and uric acid (UA) measurements.

Both urinary Ca and UA were higher in SF than in normal subjects as seen in the table, and the inhibitors, urinary citrate and Mg were lower in males. Citrate excretion but not Mg was also lower in female SF. These data indicate that an imbalance exists between promoters and inhibitors in SF, and a higher mean urine volume in SF was not able to prevent stone formation.

	Stone formers		Normal subjects	
	Female	Male	Female	Male
Volume, mL	1713±90*	1886±88#	1059±44	1260±65*
Creatinine, mg	1171±27	1677±46*	1118±25	1564±51*
Ca, mg	217±11*	230±11#	115±6	154±8*
Uric acid, mg	551±20*	689±22*#	495±15	587±23*
Citrate, mg	295±14*	250±14*#	560±29	326±16*
Mg, mg	66±3	60±3#	60±2	86±3*

Values are Means ±SEM, n in parentheses. Significantly different from females*, from males#, $p < 0.05$.

REFERENCES

1. DR dos Santos, ME Pinheiro and N Schor, Diagnostico metabolico da litiase renal, *Rev Ass Ned Brasil* 33 161 (1987).
2. F Grases, JJ Gil and A Conte, Urolithiasis inhibitors and calculus nucleation, *Urol Res* 17 163 (1989).
3. K Sakhaee, N Nicar, R Hill and CYC Pak, Contrasting effect of potassium citrate and sodium citrate therapies on urinary chemistries and crystallization of stone forming salts, *Kidney Int* 24:348 (1983).

AUTHOR INDEX

SUBJECT INDEX